Col·lecció Homenatges

Homage to Ramon Margalef

or

Why there is such pleasure in studying nature

Oecologia aquatica, 10

UNIVERSITAT DE BARCELONA

PUBLICACIONS

BIBLIOTECA DE LA UNIVERSITAT DE BARCELONA. Dades catalogràfiques.

Homage to Ramon Margalef or Why there is such pleasure
in studying nature. - (Homenatges/Universitat de Barcelona;
8) (Oecologia aquatica; 10)

Referències bibliogràfiques
ISBN: 84-475-0019-5

I. Ros, Joandomènec, ed. II. Prat, Narcis, ed.
III. Margalef i López, Ramon V. Col·lecció
1. Ecologia-Reculls d'escrits

Entitat Editora
UNIVERSITAT DE BARCELONA

1a. Edició: BARCELONA Juliol 1992

Disseny de la col·lecció: Cesca Simón

Composició de text: Ormograf S.A.

Impressió: Gramagraf S.C.C.L.

Dipòsit Legal: B: 21. 543-92

ISBN: 84-475-0019-5

Tiratge: 1.600 exemplars

Direcció i Administració de la publicació
PUBLICACIONS DE LA UNIVERSITAT DE BARCELONA
Gran Via, 585
08007 Barcelona.

HOMAGE TO RAMON MARGALEF

OR, WHY THERE IS SUCH PLEASURE IN STUDYING NATURE

Frontispiece:

Ramon Margalef as seen by Mercè D'Ocón. From a 1983 poster announcing a conference by Professor Margalef at the Joanot Martorell High School, Esplugues de Llobregat (Barcelona). By kind permission of the artist.

HOMAGE TO RAMON MARGALEF

OR, WHY THERE IS SUCH PLEASURE IN STUDYING NATURE

Oecologia aquatica, 10

Edited by Joandomènec Ros and Narcís Prat

Department of Ecology, University of Barcelona

Barcelona, 1991

Oecologia aquatica

is published at irregular intervals by the Department of Ecology of the University of Barcelona

Oecologia aquatica

can be obtained by means of exchange with similar scientific journals or by subscription. The price of an ordinary volume is 2,000 pts. (Spain) or US$ 25 (abroad), postage included. The price of this volume is 5,000 pts. (Spain) or US$ 62 (abroad).

All correspondence should be addressed to:

The Secretary,
Oecologia aquatica
Department of Ecology
University of Barcelona
Av. Diagonal, 645
08028 Barcelona - SPAIN

CONTENTS

FOREWORD

When a scientific journal like *Oecologia aquatica* reaches its tenth issue, it is perhaps not an occasion for extraordinary celebration. However, if it turns out that this issue coincides with a series of unusual circumstances (for example, it is the journal of a university department that is published at irregular intervals, it is produced by good will rather than by plentiful means, it dates from 1973, it is a vehicle of international diffusion of part of the scientific research produced in the Department of Ecology at the University of Barcelona and it is common currency to exchange with some 300 periodicals in the field of ecology, limnology and oceanography throughout the world), then the perspective changes somewhat.

Moreover, if the editors hasten to confess that this modest milestone of issue nº 10 was really taken as an excuse to pay tribute to Professor Ramon Margalef, who was the founder, the first director and the driving force behind the journal, we can be forgiven for wanting to celebrate.

A tribute like this one is not linked to any particular commemoration, although we could easily find more than one date worth remembering: Professor Margalef became professor of Ecology, the first in Spain, in 1967 (nearly 25 years ago) and the Department of Ecology at the University of Barcelona opened its doors for the first time in the academic year 1969-70 (more than twenty years ago). If we look at those of Margalef's papers which have become landmarks in the development of ecology, they include "La teoría de la información en ecología", which appeared in 1957, and "On certain unifying principles in ecology", published in 1963. As for his books, *Perspectives in ecological theory*, which made his way of understanding the biosphere internationally known, and which was to have a powerful influence on biologists the world over, was published in 1968. His monumental *Ecología* appeared in 1974, *La Biosfera, entre la termodinámica y el juego* in 1980 and the encyclopaedic *Limnología* in 1983. And now (1991) Margalef has just published another book, *Teoría de los sistemas ecológicos*, and he has two more books on ecology in the making.

Professor Margalef was 70 in 1989; under the Spanish University system (now being reformed) he had retired two years earlier. He was named Emeritus Professor immediately, and he has continued working ever since (studying, among other things, phytoplankton in the Catalano-Balearic hydrographic front, writing scientific books and articles, giving classes and lectures). This present tribute, a collection of articles on ecology, limnology and oceanography, is aimed more at celebrating this continuing activity than remembering time past as is so often the case.

Gathering a series of articles signed by old and new pupils, collaborators and friends of Professor Ramon Margalef, into one special issue of *Oecologia aquatica* seemed to us to be the best way of repaying but a small part of our debt to the man who has taught us for many years and in many fields of the science of ecology. Indeed, from the lecture halls of the University of Barcelona and the plethora of courses he has given in universities and research establishments throughout the world; from his leadership in oceanographic and limnological campaigns and research projects at the Department of Ecology of Barcelona and at the former Institut d'Investigacions Pesqueres; from his scientific books and articles and from his popular publications and exhibits on the sciences of which he is a master; from the forums and seminars on science and thought and from the everyday contact with him in his department, those of us who have been privileged to learn from him, to work at his side and to discuss aspects of the workings of nature, have always received fertile ideas, a macroscopic vision of nature, invaluable information on very diverse subjects, words of encouragement and at all times a climactic model to which to tend towards, although we know that, as is the case with the ecological climax, we will never reach it.

In mid 1989 the editors of this volume invited past and present colleagues in the department and pupils of Margalef at the University, at the Institut de Ciències del Mar and at other research centres the world over, to contribute to a special issue of the departmental scientific journal. The list of authors and subjects grew during the rest of that year and most contributions had been received by the beginning of 1991.

The editorial work required greater time and dedication than the editors were able to offer in their present academic circumstances, and the issue has come out somewhat later than intended; apologies both to Professor Margalef and to the authors.

The present list of authors is as it is, and not otherwise, because the edition was closed after a certain number of contributions were received. We could not produce a larger volume, nor did it seem proper to limit the number of pages available to the authors when they were asked that their articles should aim to report the state of the art in various fields, or provide general overviews of their own fields. We are aware that many colleagues who wished to were unable to contribute on this occasion. Let them prepare for the next!

We would like to add three clarifications, perhaps unnecessarily. Firstly, on the format of this volume; it is a special issue of *Oecologia aquatica*, as we said before, and at the same time, an independent book in the series *Homenatges* of the University of Barcelona. Secondly, on the use of English as the only language; not only is it the language we prefer for our journal, but we also intend this book to have the broadest possible circulation. We hope that the articles included will reach a wide audience, not so much to guarantee the dissemination of our research (but that too), but rather because the book is dedicated to Margalef. Even though all the authors habitually use Catalan or Spanish as their language, Professor Margalef long ago taught us the importance of the universality of our scientific message.

The third point has to do with the title of the book; we want it to reflect the naturalist of the old school in Ramon Margalef, and at the same time encapsulate his admiration of another naturalist who has recently left us, G. E. Hutchinson. Indeed, as ecologist, limnologist and complete naturalist, Hutchinson was the author of an article, among many others, that was to be fundamental in ecological theory: "Homage to Santa Rosalia, or why are there so many kinds of animals?" (*Amer. Natur.*, 1959). In turn, Margalef dedicated to Hutchinson an article on diversity, later to become a landmark in ecological theory, and which paraphrased the title of that classical article: "Homage to Evelyn Hutchinson, or why there is an upper limit to diversity" (1972; see complete reference in Appendix I). We have tried to add to the saga of "Homage to...", and we would be pleased if, in the future, this volume pointed the way to new challenges in the fields of ecology, limnology and/or oceanography.

Many people have made this issue possible. We wish to thank all of them. In the first place, the 52 authors, who have responded as best they could to the editors' recommendations and/or demands, and withstood the editors' urgency at the beginning, and have then had to forgive us for the delay in publication at the end. Then Pilar Comín, who as factotum typist has fought with texts, formulae, tables and bibliography and has introduced hundreds of last-minute changes. Anna Maria Domingo has drawn or redrawn several of the figures. Marina Mir forged a link for us with Mercè D'Ocón, who was kind enough to supply us with the frontispiece. Maria Mir de Margalef provided us with photographs of her husband, from which we have chosen the one at the beginning of the appendix.

This special volume would not have been possible without the enthusiastic collaboration of the Rector of the University of Barcelona, Josep Maria Bricall, who, when the proposed tribute was little more than a vague idea, offered us the institutional support of our university with these words: "If it is for Margalef - whatever you need". That "whatever you need" meant unlimited access to the Language Advisory Service of the University of Barcelona (whose translators and correctors, and especially Robin Rycroft, have overseen the quality and uniformity of the English version), and the Publications Service at the University of Barcelona, who have put the professional skills of their team to work on this book. The Comisión Interministerial de Ciencia y Tecnología has also contributed to the publication costs.

The Rector's comment has also been our watchword. In the past two years we have put this book together with enthusiasm, and we have tried to publish it in the greatest possible secrecy, so that it would be a surprise to the man it honours (which is quite a feat, since Margalef is always in the department and wants to know what everybody is doing; the authors who sought his opinion on the originals they present here have resorted to the most outlandish excuses not to reveal the journal they were destined for; the editorial tasks had to be done at home rather than in the department). Knowing Margalef as we do, we feared that, once he found out what we were preparing for him, he would have persuaded us not to "waste our time".

We are sure that we have not wasted our time. Both from the variety of articles that are collected here and from their standard, and especially because of the love that all of us, authors and editors, have put into this publication, we hope that this tribute, which is simple and in the form that Margalef most appreciates - scientific creation - pleases him, and will give him some hours of intellectual enjoyment (as the subtitle of the book suggests). We all eagerly await his seminal comments, his proper suggestions, even his ironic criticisms. Perhaps they can generate another special issue of *Oecologia aquatica.*

Joandomènec Ros and Narcís Prat
Barcelona, December 1991

Homage to Ramon Margalef; or, Why there is such pleasure in studiyng nature (J. D. Ros & N. Prat, eds.). 1991. *Oecologia aquatica*, 10: 1-12

PRESENT TRENDS IN RIVER STUDIES

NARCÍS PRAT

Departament d'Ecologia. Facultat de Biologia. Universitat de Barcelona. Av. Diagonal, 645. 08028. Barcelona. Spain.

Received: February 1990.

SUMMARY

A review of recent studies on river ecology is presented. Emphasis is laid on the importance of basin processes which are key factors in defining the structure and function of river communities. At present rivers are viewed as parts of ecosystems where disturbance and spatial heterogeneity play a central role. Biotic interactions (predation, competition) do not seem to be important at the ecosystem level because of the frequency and intensity of disturbances in the rivers. In some of them, however (the lower part of the largest rivers), or during short periods of time, herbivores and predators (especially fish) may have great involvement in the production and structure of the community.

On the other hand, much recent work has been done in the temperate zone, and some of the more recent ideas may well have to be re-examined in the future, when more comprehensive studies of small streams in tropical areas become available. The lack of knowledge about Mediterranean rivers is also emphasized.

Many of the recent ideas in river ecology have emerged in discussions at meetings. Those of the North American Benthological Society and the Regulated Rivers Symposia deserve special mention as forums for the exchange of fruitful ideas. Finally, tribute is paid to Prof. Margalef for his pioneer studies on river ecology and his always stimulating and iconoclastic ideas on the general rules of science.

KEY WORDS: River ecology, basin, disturbance, spatial heterogeneity, biotic interactions, regulated rivers, Mediterranean rivers.

INTRODUCTION

The last twenty years have seen a spectacular development in the ecology of rivers. Constant repetition of the same type of descriptive studies has given way to synthetic studies which are based on previous hypotheses. One of the results of this is an increase in the number of books about river ecology over the last few years. A large part of ecology theory that had only just been introduced in the study of rivers has been applied to them on a large scale, especially by means of *in situ* experiments and studies carried out in the laboratory. Biotic interactions, hardly considered as determining factors in population dynamics until the mid-seventies, have gathered importance as one of the aspects that may sometimes be crucial to the organization of communities of organisims living in rivers. Nowadays the ecology of rivers is one of the most thrilling areas of ecology, and the number of studies published increases year by year. For this reason, it is both easy and difficult to synthesize them.

MARGALEF (1960) was one of the first to focus upon a few valid ideas which would place river ecology within the conceptual framework of general ecology. Even today, his paper is still compulsory reading for those who wish to view rivers within a wider context. These original ideas, corrected in the light of new discoveries, were later developed in his book *Limnología* (MARGALEF, 1983). In this book and in many other papers, we can appreciate Professor Margalef's wide knowledge of the subject, set within the more general framework of limnology and ecology, as he himself has emphasized recently (MARGALEF, 1990). It is indeed in this latter publication that Margalef complains of the scant attention that general ecology has paid to river ecology.

A LITTLE HISTORY

There were very few books available on river ecology until the beginning of the seventies. Those by HYNES (1960, 1970) deserve mention, as his research team was one of those that worked most intensely on the subject. Articles of synthesis from those years dealt especially with the subject of zonation (HUET, 1959; ILLIES & BOTOSENEANU, 1963) and almost completely ignored the dependence of a river on its basin. One of the first books to bring together the ideas of many authors is a good example of this kind of omission and of the state of the subject at that time (WHITTON, 1975). MARGALEF (1960) had already argued the importance of considering the river within the wider framework of its basin and studying the longitudinal organisation of the river within the context of ecological succession, a subject that was later broached by FISHER (1983). Without doubt the holistic approach to the study of hydrographic catchment areas (LIKENS & BORMANN, 1985) was responsible for the main subject of study in the seventies being the dependence of the river on its basin (FISHER & LIKENS, 1973; PETERSEN & CUMMINS, 1974). Hynes' Baldi lecture at the SIL Conference in Canada (HYNES, 1975) had a very thought-provoking title in this respect. Some years later, the subject was approached again, but now with a more precise development, at another SIL conference (again as the central communication). It was possible to see what had been produced by a decade of research focused on the idea of the basin as an ecological unit, especially in headwater rivers (LIKENS, 1984). At this point we should pay a special tribute to the limnology conferences of the SIL (*Societatis Internationalis Limnologicae*), which has been a meeting place for ideas and experiences in limnology every three years since 1922. The *Verhandlungen* (Proceedings) that are published after every conference, although often criticised for their bulk and sometimes for their lack of "quality", are a reflection of the development of world limnology over the preceding three years, and have thus a testimonial value which transcends their scientific value.

Studies by various North American groups at the end of the seventies have produced, as their most outstanding consequence, the emergence of a hypothesis which was applauded without reserve by some and criticized by others (especially the Europeans): the hypothesis of the river as a continuum (VANOTTE *et al.*, 1981). This had the virtue of recognizing the work of some of the pioneers (among them Margalef) and of provoking a few years of intense study of rivers, together with various controversies on the subject (WINTERBOURN *et al.*, 1981; BARMUTA & LAKE, 1982; STATZNER & HIGHLER, 1985; RYDER & SCOTT, 1988). This frenzy is reflected in numerous publications and more than a dozen books, some of which are monographs which integrate data to one specific river, or to a determinate geographical area (BARNES & MINSHALL, 1983; FONTAINE & BARTELL, 1983; WHITTON, 1984; WELCOME, 1985; DAVIES & WALKER, 1987). New ideas and methodologies have emerged from this, which situate the study of rivers in its optimum moment at present. Studies published in the *Journal of the North American Benthological Society* (JNABS) bear witness to this moment, from among which we draw special attention to a recent monograph (STANFORD & COVICH, 1988), or the conferences on regulated rivers and studies published in the journal that specializes in this subject, *Regulated rivers*.

River ecology is, however, still the ecology of rivers from temperate zones, and for that reason it is only part of our understanding of a much wider topic. The present state of knowledge of the great African and Latin American rivers is still poor, but some of these rivers have merited their own more or less recent monographs, like the Amazon (SIOLI, 1984). As for the Mediterranean rivers, we could say that we have hardly past the descriptive stage, with the exception of a few studies which have recently been done on Californian rivers (RESH, in press).

THE HYDROGRAPHIC BASIN AS AN ECOLOGICAL UNIT

Nowadays nobody would dispute the importance of the processes which occur in the basin in determining how the energy flow is structured in a river. For certain basins, the integration of data from many years allows us a comprehensive view of the processes in the basin, and of how these connect with those of the river, which are of extraordinary interest (e.g. SWANK & CROSSLEY, 1987). This is particularly clear in the case of headwater rivers when the river flows through forest regions. The Hubbard Brook experiments have provided a solid set of data which show the dependence of energy flux in these rivers on the input of alochthonous material, and how plant leaves cannot be processed by animals without first being colonised by bacteria and fungi (KAUSHIK & HYNES, 1971; CUMMINS *et al.*, 1973; PETERSEN & CUMMINS, 1974; ROUNICK & WINTERBOURN, 1983). The nutrient flow in the river is closely related to this process (MEYER *et al.*, 1981). The movement of the nutrients in the river is affected by their transport downstream and by the retention they may undergo if they are used by organisms, or in the sediments. This has led to the interesting concept of nutrient spiralling (NEWBOLD *et al.*, 1982), which although highly attractive from the theoretical point of view is hardly applicable to natural rivers.

This point of view has recently come in for criticism, as all the experiments are normally carried out on dry leaves, whereas in areas where willows and elders are abundant the input of leaves is mainly in the form of fresh material. This does not decompose so rapidly, and may have a different nutritional value (GESSNER & SCHWOERBEL, 1989). In the same way the study of the decomposition of tropical forest leaves that are poor in nutrients also shows the importance of other factors. Thus, the leaves of certain trees that are rich in tannins take much longer to decompose than those that live in nutritive soils and are poor in these compounds. Experimental transfer of the leaves of tropical rivers to temperate rivers demonstrates the importance of the characteristic flora of each river in the decomposition, as the leaves that have been transferred always decompose with greater difficulty (STOUT, 1989). These experiments open very interesting new perspectives in a subject which could have been considered exhausted.

However, even in highly shaded rivers, primary production can be important and may explain the permanence of certain species in the river (MEYER & LIKENS, 1987). On the other hand, when there is no forest cover, or when light can penetrate at certain times of the year, primary production can also be important. The energy flux may then become dominated by the incorporation of carbon by the plants and not by alochthonous inputs or by transport from the upper parts of the river (MINSHALL, 1978). The dependence of primary production on the river gradient and the importance of this in the seston transported by the current were thoroughly studied by NAIMAN (1983 a, b).

Regularities in the organisation of the energy flux along the length of the river and the change from coarse materials in the upper reaches to fine materials in the lower parts form one of the central axes of the hypothesis of the river as a continuum. The comparison between different rivers (in near natural conditions or without important disturbances by man) has led to one of the arguments that most support this hypothesis (MINSHALL *et al.*, 1983, 1985), in spite of the criticisms it received.

THE RIVER: AN ECOSYSTEM WITH DISTURBANCE AND HETEROGENEITY

It may sound like a cliché to say that a river is a system organized along a horizontal axis. This is doubtless the reason that sets it apart from lakes, which have always been aquatic ecosystems *par excellence*; UNESCO even defines limnology as the study of lakes in its fields of knowledge.

The factor that generates the river identity is the existence of a current. The possibility of carrying materials and organisms downstream (and thus, the possibility that communities might disappear) has been counteracted by a multitude of adaptations on the part of organisms, from morphological changes to behavioural attitudes which have been studied convincingly in laboratory rivers and using microphotography (STATZNER & HOM, 1982; STATZNER *et al.*, 1989; EYMAN, 1988).

For the same reason, the substrate recolonisation in a river after disturbance is relatively fast, and in little more than two weeks an abundant, well-structured population can be found. This is very clearly evidenced in experiments of artificial colonisation of rivers (LAMBERTI & RESH, 1985; PRAT *et al.*, 1986). A general model of colonisation and extinction has been put forward for the colonisation of rivers (MINSHALL & PETRERSEN, 1985). In the first few days of colonisation, the model seems to fit well with the process, but afterwards the situation can evolve in many different ways, in response to external changes, or the importance of biotic interactions, especially as a response to herbivore or predator pressures.

Changes that exceed the limit of resistance of any organism to the water flow (floods), or those related to the disappearance of water (drought) are extremes in the face of which we must wonder how communities respond. Thus, when faced by great floods, many organisms possess different adaptations (like taking refuge in the interstitial zone), the importance of which has recently been pointed out (HYNES, 1983; WILLIAMS, 1981; DANIELOPOL, 1988; STANFORD & WARD, 1988). After a strong disturbance (excessive flow, drought), the river regenerates in a short time a community similar to the previous one (WILLIAMS & HYNES, 1976, 1977). Thus, thanks to their adaptability to extreme changes, rivers as ecosystems recover easily, unless these changes (disturbances) are either very strong or too frequent. The hypothesis of the mean disturbance (CONNELL, 1978) has been suggested as one of the keys to understanding the changes in specific richness and diversity in rivers (WARD & STANFORD, 1983a). Some studies have provided convincing data on the importance of disturbance in community structure in different types of rivers (FISHER & GRIMM, 1988; WALLACE *et al.*, 1986; ROBINSON & MINSHALL, 1986), although in other cases the effect of disturbances does not appear to be so determinant (HAWKINS, 1988). A recent study by GRIMM & FISHER (1989) clearly shows the concepts of resistance and resilience may be applied to rivers in the study of a long series of disturbance events, thus placing rivers inside the framework of general ecological theory.

The greatest complexity of a river cannot be understood, however, unless its enormous spatial variability is considered, both in the longitudinal course and also within a specific zone. The difficulties involved in the estimation of the population density of the species inhabiting rivers are well known. Even with standard errors of 40%, it is necessary to take more than 20 samples to estimate the density of a given species (RESH, 1979). The use of transects clearly reveals the importance of this spatial heterogeneity and the problem of estimating populations (MILLET *et al.*, 1987; MUÑOZ *et al.*, 1985), for which solutions have been attempted with stratified sampling or grids of random samples (IVERSEN *et al.*, 1985). The microdistributions produced in algal populations related to the microtopography of the substrates are also interesting (KREJCI & LONG, 1986). The action of the animals, like the presence of tubes of chironomids (WARD, 1985), and the importance of the size of individuals (OSBORNE & HERRICKS, 1987) has been also emphasized as being responsible for spatial heterogeneity.

Therefore, the most promising models in river communities studies are the combination of changes due to disturbances with structuration in space. The model of TOWNSEND (1989) is a good starting point in this respect, and may well replace the classical models of zonation. This model contemplates the entire range of systems, from the simplest system describable in terms of the Lotka-Volterra equations, to the most complex system, which would be one with great spatial heterogeneity and intermediate disturbances. The hypothesis of mean disturbance is again emphasised in this model.

The most generalised conclusion for the majority of rivers is that the frequency the intensities of disturbances produced at random falls within a range that does not permit its internalisation by the organisms. Gradual assimilation of these disturbances by life is not possible (in the sense of MARGALEF, 1986; Fig. 3). Thus the organisation of the system in rivers has little to do, in general, with biotic interactions, but rather with the type, frequency and intensity of random disturbances.

THE ROLE OF BIOTIC INTERACTIONS

Some of the most thought-provoking ideas in river ecology in the last few years have come from laboratory experiments. In many cases, an attempt has been made to show the importance of biological interactions (competition, predation, herbivory) in the structure and complexity of communities; thus in the experiments the density of predators or of possible competing species was manipulated. Most of the experiments carried out to date (both in the laboratory and *in situ*) show the real importance of herbivors for the algae or macrophytes on which they feed and also the key effect of the presence of predators. On the other hand it appears that competition does not has a particularly outstanding role in rivers of the temperate zone (ALLEN, 1984). The possible importance of parasitism in the regulation of some insect populations has recently been suggested (CUMMINS & WILZBACH, 1988). Although it is probably of limited importance, the presence of symbionts or of mutualism has also been pointed out in certain cases, like that observed between *Nostoc* and a species of chironomid growing in their interior (*Nostococladius*). The *Nostoc* colonies with chironomids have a different morphology and grow much more than those that do not have them (PRINGLE, 1985).

In the case of the interaction between herbivores and the communities on which they feed, changes have been described both in the specific composition (which results in the dominance of small diatoms if the herbivore density is high), and in the production and biomass of plant communities (e.g. HILL & KNIGHT, 1987; COLLETTI *et al.*, 1987; LAMBERTI *et al.*, 1987; LOWE & HUNTER, 1988; OGILVIE, 1988; STEINMAN *et al.*, 1987a, 1987b; RICHARDS *et al.*, 1988). A general model of the effect of herbivores on production of algal communities was suggested by LAMBERTI & MOORE (1984). This model assumes a decrease in biomass with an increase in pressure from herbivores and greater productivity at intermediate densities. The model seems to fit well to many of the experimental results reported to date. The possible influence of differences in periphyton composition on the animal communities that feed on them has also been pointed out (DUDLEY, 1988).

The effect of predators in artificial rivers and its importance for the macroinvertebrate communities has been emphasised on many occasions in river ecology (PECKARSKY, 1984; MERRIT & WOTTON, 1988; RUTHERFORD & MACKAY, 1986). In some cases data are available on their effect on the microdistribution and behaviour of the prey (MICHAEL & CULVER, 1987). As has been mentioned above, it has been more difficult to find evidence that competition is a key factor in rivers (ALLEN, 1984), although some examples have been presented from laboratory studies (LAMBERTI *et al.*, 1987). Only in cases of competition for space, at moments when population density is high, it can be assumed that competition may be an important factor. One of the most frequently cited cases is the intraspecific competition among trichoptera to spin their webs (*Hydropsyche*), or interspecific competition between simuliidae and hydropsychidae (HERSHEY & HILTER, 1988).

Nevertheless, as RESH (in press) has shown, densities that appear to be critical for herbivores, predators or competition to act, deduced from laboratory studies, are rarely reached in rivers. Thus in rivers of the temperate zone that are subject to considerable thermal and hydrological fluctuations, biotic interactions appear to be relatively unimportant (except the herbivores in some cases or the predators, especially fish) in modelling the community, whereas external factors (which we have referred to as disturbances) and heterogeneity in space appear to be the key to the organisation of communities in rivers (LAKE, 1985; LAKE *et al.*, 1988; GRIMM & FISHER, 1989).

Studies on the importance of behaviour in the interactions between organisms has not received much attention in rivers. Some data on the interactions between *Baetis* and the chironomids that form their tubes in the zones in which these are grazing (recorded on film), show that this aspect deserves much greater attention (HERSHEY, 1987).

REGULATED RIVERS

An important factor in the development of river ecology in recent years has been the boost given to the general understanding of it by limnologists who have studied the effects of river regulation by dams. The model of the river as a continuum has always assumed a river with gradual transformations along its whole length; most of the rivers in the world, however have been dammed (WARD & STANFORD, 1979), hence the presence of reservoirs has a clear incidence on the characteristics of the river downstream.

Depending on the type of dam and the depth at which the water flows into the river, the effects on the communities are predictable according to a relatively simple scheme (WARD & STANFORD, 1984). The river can be viewed not as a continuum, but rather as a series of discontinuities and WARD & STANFORD (1938b) and STANFORD *et al.*, (1989) have proposed a measure of this discontinuity. A similar idea has been applied in some Catalan rivers and yields positive results in simple parameters such as conductivity (SABATER & ARMENGOL, 1989). In fact the changes produced by the dams could be classified as disturbances, which would allow us to place regulated rivers within the framework of the more general ecological context.

The various conferences on regulated rivers have been a good forum for the debate of these and other aspects of rivers, and have contributed in a significant way to the recent advances in river ecology (WARD & STANFORD, 1979; LILLEHAMMER & SALVEIT, 1984; CRAIG & KEMPER, 1987; PETTS, 1989). There is also a monographic book on the subject (PETTS, 1984).

LONGITUDINAL ORGANIZATION

Longitudinal organization has always been one of the classic subjects within the study of rivers. Words like rhython and potamon became common currency in the sixties. The study by ILLIES & BOTOSENEANU (1963) cited above is the best example of this research line. Here we should also remember some of the pioneering work done by MARGALEF (1958) in this field.

Longitudinal zonation, which assumes the change from some communities to others along the length of the river, was heavily criticised in some cases (HAWKES, 1975) and the idea of the river as a continuum seemed to relegate the application of zonation to those cases where serious alterations in the characteristics of the environment were produced, such as pollution. More recently, however, the subject has drawn new interest, especially because of a better application to rivers of the concept of hydraulics.

In effect, changes in gradient may occur in certain rivers in relatively short stretches within their length. This may lead to important differences in shear stress at the bottom of the river, and consequently limit the presence of many species belonging to higher or lower stretches of the river. The sites at which the gradients of the rivers change would be where species from both zones would be found, and thus these will be zones of higher specific richness. STATZNER (1981) and STATZNER & HIGLER (1986) are the most ardent defenders of this idea, which allows a somewhat different interpretation of the change in species richness along the river, which derived from the idea of the river as a continuum (VANOTTE *et al.*, 1981).

The study of the lower parts of rivers has brought to light the importance of alluvial plains in determining the characteristics of the river, and their interest as filters between zones adjacent to the river and the river itself (CHAUVET & DECAMPS, 1989). These studies confirm the importance of ecological processes in flood plains and in the lower stretches, as they are known in large rivers (SIOLI, 1984), and open up possibilities for the study of processes of succession, as proposed by MARGALEF (1990).

APPLICATION OF ECOLOGICAL THEORY TO MEDITERRANEAN RIVERS

Most of the studies carried out to date have been in the temperate zone. The rivers that we could include within the regions of

Mediterranean climate have not received equal treatment until recently (GIUDICELLI *et al.*, 1985), if the studies carried out in California and in Australia are excepted; even in the latter case only very recent attention has been paid to them (BUNN & BOUGHTON, in press). The rivers situated around the Mediterranean Sea, with a typically torrential hydraulic regime, have been predominantly descriptive. They have not been placed explicitly within the comprehensive scheme of river ecology (PRAT *et al.*, 1986). Furthermore, most of these rivers have been greatly transformed by the action of man, and so there are very few left to compare as a model. Moreover, only long-term research can provide us with complete ideas on their behaviour, and it is rarely possible to go on with the same type of research for 20 years or more. The only data of this kind appear to be those of RESH *et al.* (in press), for Californian rivers that may dry up in some years, or be subjected to sharp increases in their volume.

Severe disturbances that are repeated annually (floods and droughts) with varying intensity are key factors in Mediterranean rivers. However, the intensity of the floods and the severity of the droughts vary from one year to the next and from one river to another. This is linked to a thermal regime that permits the development of the majority of the species in the river with relatively short cycles if we compare them with the same species living at higher latitudes.

As a result of this, and of the ease of dispersion of the species, Mediterranean rivers undergo spectacular changes in their community similar to those of rivers from desert zones such as those studied by FISHER *et al.* (1983). But in some years, or in some rivers, there is no period of drought and the bed of the river does not dry up. Also, in many years the floods are not great enough to eliminate all the community from the river.

As a consequence, the Mediterranean rivers offer a much greater diversity and a much higher complexity than rivers of temperate zones, although the species with a fast life cycle (particularly Diptera) appear to dominate over those with a more long and complex life cycle (for example, Trichoptera and Ephemeroptera).

LOOKING TO THE FUTURE

In the next few years river ecology should devote more attention to large and small rivers in the tropical zones, both on the plains and at higher altitudes, in order to determine whether it is really in terms of disturbances and spatial heterogeneity that the structure and evolution of communities are to be explained. Particularly the alluvial plains of the great rivers should attract our attention, as on the one hand they are threatened with extinction, and on the other they promise great scientific interest, especially in ecotones. A broad perspective on the progress and questions for the future can be found in the special volume of the *JNABS* mentioned above (STANFORD & COVICH, 1988) and in a recent book (NAIMAN & DÉCAMPS, 1990).

Another area which promises to be of interest is the in-depth study of the importance of biochemical factors and related subjects at the molecular level, including population genetics by means of the use of enzymes (ARSUFFI & SUBERKROPP, 1988; BARLOCHER & PORTER, 1986; FINDLAY *et al.*, 1986).

Studies in which the sampling unit is the stone are also of interest, and this may be related to the colonised area, an aspect with a certain tradition in river ecology (STOUT & VANDERMEER, 1975), which has recently been developed in a more complex way (WRONA *et al.*, 1986).

Without doubt laboratory studies of rivers will continue to be important, especially as far as they are able to reflect and lay down the rules of the possible signification of certain processes and interactions (e.g. CIBOROWSKI & CRAIG, 1989; REITER, 1989).

But we must not forget those overall views that remind us that the tree should not be allowed to interrupt our perception of the wood. The ideas developed by Margalef, his view on this or any other subject, always as critical as they are passionate, are a stimulus and a perennial reminder of the insatisfaction that is so characteristic of scientists. This paper is heartily devoted to him.

REFERENCES

BOOKS ON RIVER ECOLOGY

BARNES, J.R. & MINSHALL, G.W. (eds.). 1983. *Stream ecology, application and testing of general ecological theory*. Plenum. New York.

BREHM, J. & MEIJERING, M.P.D. 1982. *Fliesswasserkunde Einführung in die Limnologie der Quellen, Bäche und Flusse*. Quelle & Meyer. Heidelberg.

CRAIG, J. & KEMPER. J.B. 1987. *Regulated streams. Advances in ecology*. Plenum. New York.

DAVIES, B.R. & WALKER, K.F. (eds.). 1987. *The ecology of river systems*. Monogr. Biologicae. Junk. The Hague.

FONTAINE, T.D. & BARTELL, S.M. (eds.). 1983. *Dynamics in lotic systems*. Ann Arbor Science. Ann Arbor.

HART, C.W. & FULLER, S.L.H. (eds.). 1974. *Pollution ecology of freshwater invertebrates*. Academic Press. New York.

HELLAWELL, J.M. 1978. *Biological surveillance of rivers*. Water Research Centre. Stevenage.

HELLAWELL, J.M. 1986. *Biological indicators of freshwater pollution and environmental management*. Elsevier. London.

HYNES, H.B.N. 1960. *The biology of polluted waters*. Liverpool University Press. Liverpool.

HYNES, H.B.N. 1970. *The ecology of running waters*. Liverpool University Press & Univ. Toronto Press. Liverpool.

LILLEHAMMER, A. & SALVEIT, S.J. 1984. *Regulated rivers*. Oslo Univ. Press. Oslo.

LOCK, M.A. & WILLIAMS, D.D. (eds.). 1981. *Perspectives in running water ecology*. Plenum Press. New York.

MAITLAND, P.S. 1978. *Biology of freshwaters*. Blackie. Glasgow.

MASON, C.F. 1981. *Biology of freshwater pollution*. Longman. London.

MARGALEF, R. 1983. *Limnología*. Omega. Barcelona.

Mc CAFFERTY, W.P. 1981. *Aquatic entomology*. Science book int.

MERRIT, R.W. & CUMMINS, K.W. 1984. *An introduction to the aquatic insects of North America*. (2nd. ed.). Kendall. Hunt.

NAIMAN, R.J. & DÉCAMPS, H. (Eds.). 1990. *The ecology and management of aquatic-terrestrial ecotones*. UNESCO & Parthenon. Paris.

OGLESBY, R.T., CARLSON, C.A. & Mc CANN, J.A. 1972. *River ecology and man*. Academic Press. New York.

PETTS, G.E. 1984. *Impounded rivers*. Wiley. Chichester.

PETTS, G.E. (ed.). *Regulated rivers: Research and manegement*. 4. Wiley. London.

RESH, V.H. & ROSENBERG, D.M. (eds.). 1984. *The ecology of aquatic insects*. Praeger. New York.

SIOLI, H. (ed.). 1984. *The Amazon. Limnology and landscape ecology of a mighty tropical river and its basin*. Junk. The Hague.

STELLA, E. 1984. *Fondamenti di Limnologia*. Edizioni dell'Ateneo. Roma.

SWANK, W.T. & CROSSLEY, D.A. (Eds.) 1987. *Forest hydrology and ecology at Coweeta*. Springer. New York.

WARD, J.W. & STANFORD, J.A. (eds.). 1979. *The ecology of regulated streams*. Plenum Press. New York.

WARREN, C.E. 1971. *Biology and water pollution control*. Saunders. Philadelphia.

WELCOMME, R.L. 1985. *River fisheries*. FAO Fisheries Technical Paper.

WHITTON, B.A. (ed.). 1975. *River ecology*. Univ. California Press. Palo Alto.

WHITTON, B.A. (Ed.) 1984. *The ecology of European rivers*. Blackwell. London.

CITED PAPERS

ALLAN, J.D. 1984. Hypothesis testing in ecological studies of aquatic insects. In: *The ecology of aquatic insects*. (Resh, V.H. & Rosenberg, D.M., eds.): 484-507. Praeger. New York.

ARSUFFI, T.L. & SUBERKROPP, K. 1988. Effects of fungal mycelia and enzymatically degraded leaves on feeding and performance of caddisfly (Trichoptera) larvae. *J. N. Am. Benthol. Soc.*, 73(3): 205-211.

BARLOCHER, F. & PORTER, C.W. 1986. Digestive enzymes and feeding strategies of three stream invertebrates. *J. N. Am. Benthol. Soc.*, 5(1): 58-66.

BARMUTA, L.A. & LAKE, P.S. 1982. On the value of the river continuum concept. *New Zealand J. Mar. Freshw. Res.*, 16: 227-231.

BUNN, S.E. & BOUGHTON, W.C. In press. Ecological significance of stream flow patterns of some Australian rivers. *Verh. internat. verein. Limnol.*, 24.

CHAUVET, E. & DESCAMPS, H. 1989. Lateral interactions in a fluvial landscape: The river Garonne, France. *J. N. Am. Benthol. Soc.*, 8 (1): 9-17.

CIBOROWSKI, J.J.H. & CRAIG, D.A. 1989. Factors

influencing dispersion of larval black-flies (Diptera: Simuliidae): Effect of current velocity and food concentration. *Can J. Fish. Aquat. Sci.*, 46: 1329-1341.

COLLETTI, P.J., BLINN, D.W., PICKART, A. & WAGNER, V.T. 1987. Influence of different densities of the mayfly grazer *Heptagenia criddlei* on lotic diatom communities. *J. N. Am. Benthol. Soc.*, 6(4): 270-280.

CONNELL, J.H. 1978. Diversity in tropical rain forests and coral reefs. *Science*, 199: 1302-1310.

CUMMINS, K.W., PETERSEN, R.C., HOWARD, F.D., WUYCHECK, J.C. & HOLT, V.C. 1973. The utilization of leaf litter by stream detritivores. *Ecology*, 54(2): 336-345.

CUMMINS, K.W. & WILZBACH, M.A. 1988. Do pathogens regulate stream invertebrate populations? *Verh. internat. verein. limnol.*, 23: 1232-1234.

DANIELOPOL, D.L. 1988. Groundwater fauna associated with riverine aquifers. *J. N. Am. Benthol. Soc.*, 8(1): 18-35.

DUDLEY, T.L. 1988. The roles of plant complexity and epiphyton in colonization of macrophytes by stream insects. *Verh internat. verein. limnol.*, 23: 1153-1158.

EYMANN, M. 1988. Drag on single larvae of the black fly *Simulium vittatum* (Diptera: Simuliidae) in a thin, growing boundary layer. *J. N. Am. Benthol. Soc.*, 7(2): 109-116.

FINDLAY, S., MEYER, J.L. & SMITH, P.J. 1986. Incorporation of microbial biomass by *Peltoperla* sp. (Plecoptera) and *Tipula* sp. (Diptera). *J. N. Am. Benthol. Soc.*, 5(4): 306-310.

FISHER, S.G. 1983. Succession in streams. In: *Stream ecology: application and testing of general ecological theory* (J.R. Barnes & G.W. Minshall, eds.): 7-27. Plenum Press. New York.

FISHER, S.G. & LIKENS, G.E. 1973. Energy flow in Bear Brook, New Hampshire: An integrative approach to stream ecosystem metabolism. *Ecol. Monographs.*, 43: 421-439.

FISHER, S.G., GRAY, L.J., GRIMM, N.B. & BUSCH, D.E. 1982. Temporal succession in a desert stream ecosystem following flash flooding. *Ecol. Monographs*, 52: 93-110.

FISHER, S.G. & GRIMM, N.B. 1988. Disturbance as a determinant of structure in a Sonoran Desert stream ecosystem. *Verh. internat. verein. limnol.*, 23: 1183-1189.

GESSNER, M.O. & SCHWOERBEL, J. 1989. Leaching kinetics of fresh leaf-processing in streams. *Arch. Hydrobiol.*, 115(1): 81-90.

GIUDICELLI, J., DAKKI, M. & DIA, A. 1985. Caractéristiques abiotiques et hydrobiologiques des eaux courantes méditerranéennes. *Verh. internat. verein. limnol.*, 22: 2094-2101.

GRIMM, N.B. & FISHER, S.G. 1989. Stability of periphyton and macroinvertebrates to disturbances by flash floods in a desert stream. *J. N. Am. Benthol. Soc*, 8(4): 293-307.

HAWKES, H.A. 1975. River zonation and classification. In: *River ecology* (B.A. Whittton, ed.): 312-374. Blackwell. London

HAWKINS, C.P. 1988. Effects of watershed vegetation and disturbance on invertebrate community structure in western Cascade streams; implications for stream ecosystem theory. *Verh.. internat. verein. limnol.*, 23: 1167-1173.

HERSEY, A.F. 1987. Tubes and foraging behaviour in larval Chironomidae: implications for predator avoidance. *Oecologia (Berlin)*, 73: 236-241.

HERSEY, A.E. & HILTNER, A.L. 1988. Effect of a caddisfly on black fly density: interspecific interactions limit black flies in an artic river. *J. N. Am. Benthol. Soc.*, 7(3): 188-196.

HILL, W.R. & KNIGHT, A.W. 1987. Experimental analysis of the grazing interaction between a mayfly and stream algae. *Ecology*, 68: 1955-1965.

HUET, M. 1959. Profiles and biology of western European streams as related to fish management. *Trans. Am. Fish. Soc.*, 88: 153-163.

HYNES, H.B.N. 1975. The stream and its valley. *Verh. internat. verein. limnol.*, 19: 1-15.

HYNES, H.B.N. 1983. Groundwater and stream ecology. *Hydrobiologia*, 100: 93-99.

ILLIES, J. & BOTOSENEANU, J. 1983. Problèmes et métodes de la classification et de la zonation écologiques des eaux courantes, considerées surtout du point de vue faunistique. *Mitt. int. verein. theor. angew. limnol.*, 12: 1-57.

IVERSEN, T.M., THORUP, J. HANSEN, T., LODAL, J. & OLSON, J. 1985. Quantitative estimates and community structure of invertebrates in a macrophyte rich stream. *Arch. Hydrobiol.*, 120: 291-301.

KAUSHIK, N.K. & HYNES, H.B.N. 1971. The fate of dead leaves that fall into streams. *Arch. Hydrobiol.*, 68: 465-515.

KREJCI, M.E. & LOWE, R.L. 1986. Importance of sand grain mineralogy and topography in determining micro-spatial distribution of epipsammic diatoms. *J. N. Am. Benthol. Soc.*, 5(3): 211-220.

LAKE. P.S. 1985. Water quality management and the ecological nature of rivers. *Aust. Soc. Limnol. Bull.*, 10: 41-50.

LAKE, P.S., MORTON, D.W. & MILNE, B.J. 1988. Interspecific associations in the faunal assemblages of stones of an upland Australian stream. *Verh. internat. verein. limnol.*, 23: 1384-1387.

LAMBERTI, G.A. & MOORE, J.W. 1984. Aquatic insects as primary consumers. In: *The ecology of aquatic insects* (V.H. Resh & D.M. Rosenberg, eds.): 164-195. Praeger. New York.

LAMBERTI, G.A. & RESH, V.H. 1985. Comparability of introduced tiles and natural substrates for sampling lotic bacteria, algae and macroinvertebrates. *Freshwat. Biol., 15: 21-30.*

LAMBERTI, G.A., FEMINELLA, J.W. & RESH, V.H. 1987. Herbivory and intraespecific competition in a stream caddisfly population. *Oecologia*, 73: 75-81.

LAMBERTI, G.A.., ASHKENAS, L.R., GREGORY, S.V. & STEINMAN, A.D. 1987. Effects of three herbivores on periphyton communities in laboratory streams. *J. N. Am. Benthol. Soc.*, 6(2): 92-104.

LIKENS, G.E. 1984. Beyond the shoreline: a watershed-ecosystem approach. *Verh. internat. verein. limnol.*, 22: 1-22.

LIKENS, G.E. & BORMANN, F.H. 1985. An ecosystem approach. In: *An ecosystem approach to aquatic ecology* (G.E. Likens, ed.): 1-8. Springer. New York.

LOWE, R.L. & HUNTER, R.D. 1988. Effect of grazing by *Physa integra* on periphyton community structure. *J. N. Am. Benthol. Soc.*, 7(1); 29-36.

MARGALEF, R. 1958. "Trophic" typology versus biotic typology as exemplified in the regional limnology of Northern Spain. *Verh. internat. verein. limnol.*, 13: 339-349.

MARGALEF, R. 1960. Ideas for a synthetic approach to the ecology of running waters. *Internat. Rev. Ges. Hydrobiol.*, 45: 133-153.

MARGALEF, R. 1986. Reset successions and suspect chaos in models of marine populations. In: *Int. Symp. Long term changes Mar. Fish. Pop.*: 321-343. Vigo.

MARGALEF, R. 1990. Limnology: reconsidering ways and goals. In: *Scientific perspectives in theoretical and applied limnology* (R. de Bernardi, G. Giussani & L. Barbanti, eds.). *Mem Ist. Ital. Idrobiol.*, 47: 57-56.

MAYER, M.S. & LIKENS, G.E. 1987. The importance of algae in a shaded headwater stream as food for an abundant caddisfly (Trichoptera). *J. N. Am. Benthol. Soc.*, 6(4): 262-269.

MERRIT, R.W. & WOOTTON, R.S. 1988. The life story and behaviour of *Limnophora riparia* (Diptera: Muscidae) a predator of larval black flies. *J. N. Am. Benthol. Soc.*, 7: 1- 12.

MEYER, J.L., LIKENS, G.E. & SLOANE, J. 1981. Phosphorus, nitrogen and organic carbon flux in a headwater stream. *Arch. Hydrobiol.*, 91: 28-44.

MICHAEL, D.I. & CULVER, D.A. 1987. Influence of plecopteran and megalopteran predators on *Hydropsiche* (Trichoptera: Hydropsychidae) microdistribution and behaviour. *J. N. Am. Benthol. Soc.*, 6(1): 46-55.

MILLET, X., MUÑOZ, I. & PRAT, N. 1987. The use of transect sampling in estimating chironomid abundance and distribution. *Ent. scand. Suppl.*, 29: 323-330.

MINSHALL, G.W. 1978. Autotrophy in stream ecosystems. *Bioscience*, 28: 767-771.

MINSHALL, G.W. & PETERSEN, C.R. 1985. Towards a theory of macroinvertebrate structure in stream ecosystems. *Arch. Hydrobiol.*, 104: 49-76.

MINSHALL, G.W. PETERSEN, C.R. CUMMINS, K.W., BOTT, T..L., SEDELL, J.R., CUSHING, C.E. & VANNOTE, R.L. 1983. Interbiome comparison of stream ecosystem dynamics. *Ecol. Monographs.*, 53: 1-25.

MINSHALL, G.W.., CUMMINS, K.W., PETERSEN, R.C., CUSHING, C.E., BRUNS, D.A., SEDELL, J.R. & VANNOTE, R.L. 1985. Developments in stream ecosystem theory. *Can. J. Fish. aquat. Sci.*, 42: 1045-1055.

MUÑOZ, I. PRAT, N., MILLET, X. & MARTINEZ-ANSEMIL, E. 1986. Heterogeneidad espacial en la distribución de los macroinvertebrados a lo largo de un transecto en el rio Llobregat (NE España). *Limnetica*, 2: 135-145.

NAIMAN, R.J. 1983. A geomorphic approach for examining the role of periphyton in large watersheds. In: *Periphyton in freshwater ecosystems*, (R. Wetzel, ed.): 191-198. Junk. The Hague.

NAIMAN, R.J. 1983. The influence of stream size on the food quality of seston. *Can. J. Zool.*, 61: 1995-2010.

NEWBOLD, J.D., O'NEILL, R.V., ELWOOD, J.W. & VAN WINKLE, W. 1982. Nutrient spiralling in streams: implications for nutrient limitation and invertebrate activity. *Amer. Natur.*, 120: 628-652.

OGILVIE, G. 1988. The effects of periphyton manipulation on the microdistribution of grazing invertebrates. *Verh. internat. verein. limnol.*, 23: 1101-106.

OSBORNE, L.L. & HERRICKS, E.E. 1987. Microhabitat characteristics of *Hydropsiche* (Trichoptera: Hydropsichidae) and the importance of body size. *J. N. Am. Benthol. Soc.*, 6(2): 115-124.

PATRICK, R. 1970. Benthic stream communities. *Am. Scient.*, 58: 546-549.

PECKARSKY, B.L. 1984. Predator-key interaction among aquatic insects. In: *The ecology of aquatic insects* (V.H. Resh & D.M. Rosenberg, eds.): 196-254. Praeger. New York.

PETERSEN, R.C. & CUMMINS, K.W. 1974. Leaf processing in a woodland stream. *Freshwat. Biol.*, 4: 343-368.

PRAT, N., GONZÁLEZ, G., MILLET, X. & PUIG, M.A. 1985. El Foix, entre l'eixutesa i la contaminació. *Estudis i monografies*, 11: 1-89. Publicacions Diputació de Barcelona.

PRAT, N.., GONZÁLEZ, G., MILLET, X. & PUIG, M.A. 1986. Community structure and colonization process in a Mediterranean drainage basin. *Proc. 3rd. European Ent. Congress.*: 121-124. Amsterdam.

PRINGLE, C.M. 1985. Effects of chironomid (Insecta: Diptera) tube-building activities on stream diatom communities. *J. Phycol*, 21: 185-194.

REITER, M.A. 1989. Development of benthic algal assemblages subjected to differing near-substrate hydrodinamical regimes. *Can. J. Fish. Aquatic. Sci.*, 46: 1375-1382.

RESH, V.H. 1979. Sampling variability and life story features: basic considerations in the design of aquatic insect studies. *J. Fish. Res. Bd. Ca.*, 36: 290-311.

RESH, V.H. JACKSON, J.K. & McELRAVY, E.P. in press. Disturbance, annual variability and lotic benthos: examples from a Californian stream influenced by a Mediterranean climate, *Jubilee Symposium Ins. Ital. idrobiol.. Pallanza.*

RICHARDS, C. & MINSHALL, W.G. 1988. The influence of periphyton abundance on *Baetis bicaudatus* distribution and colonization in a small stream. *J. N. Am. Benthol. Soc.*, 7(2): 77-86.

ROBINSON, C.T. & MINSHALL, W.G. 1986. Effects of disturbance frequency on stream benthic community structure in relation to canopy cover and season. *J.N. Am. Benthol. Soc.*, 5(3): 237-248.

ROOKEJ, B. 1986. Macroinvertebrates associated with macrophytes and plastic imitations in the Erasmosa river, Ontario, Canada. *Arch Hydrobiol*, 106: 307-325.

ROUNICK, J.S. & WINTERBOURN, M.J. 1983.Leaf processing in two beech forest streams:Effects of physical and biotic factors on litter breakdown.

Arch. Hydrobiol., 106: 448-474.

RUTHERFORD, J.E. & MACKAY, R.J. 1986. Pattern of pupal mortality in field populations of *Hydropsiche* and *Cheumatopsyche* (Tricoptera: Hydropsychidae). *Freshwat. Biol.*, 16: 337-350.

RYDER, G.I. & SCOTT, D. 1988. The applicability of the River Continuum Concept to New Zealand streams. *Verh. internat. verein. limnol.*, 23: 1441-1445.

SABATER, F. & ARMENGOL, J. 1989. Measuring discontinuities in the river Ter. *Regulated rivers*, 3(1-4): 133-142.

STANFORD, J.A., HAVER, F.R. & WARD, J.V. 1988. Serial discontinuity in a large river system. *Verh. internat. verein. limnol.*, 23: 1114-1118.

STANFORD, J.A. & WARD, J.V. 1988. The hyporheic habitat of river ecosystems. *Nature*, 335: 64-66.

STANFORD, J.A. & COVICH, A.L. 1988. Community structure and function in temperate and tropical streams. *J. N. Am. Benthol. Soc.*, 7(4); 10-259.

STATZNER, B. 1981. The relation between "hydraulic stress" and microdistribution of benthic macroinvertebrates in a lowland running water system, the Schierenseebrooks (North Germany). *Arch Hydrobiol.*, 91: 192-218.

STATZNER, B. & HOLM, T.B. 1982. Morphological adaptations of benthic invertebrates to stream flow. An old question studied by means of a new technique (Laser Doppler Anemometry). *Oecologia (Berlin)*, 53: 290-292.

STATZNER, B. & HIGLER, B. 1985. Questions and comments on the river continuum concept. *Can. J. Fish. aquat. Sci.*, 42: 1038-1044.

STATZNER, B. & HIGLER, B. 1986. Stream hydraulics as a major determinant of the benthic invertebrate zonation patterns. *Freshwat. Biol.*, 16: 127-139.

STATZNER, B., BERHNART, H.H. & BECKER, C. 1989. Hydraulic stream ecology. Fundamentals, applications and future perspectives. In: *Limnology in FRG* (W. Lampert & K.O. Rothhaupt, eds.): 148-151. C. Druck. Kiel.

STEINMAN, A.D., McINTIRE, C.D., GREGORY, S.V., LAMBERTI, G.A. & ASHKENAS, L.R. 1987. Effects of herbivore type and density on taxonomic structure and physiognomy of algal assemblages in laboratory streams. *J. N. Am. Benthol. Soc.*, 6(4): 175- 188.

STEINMAN, A.D., McINTIRE, C.D. & LOWRY, R.R. 1987. Effects of herbivory type and density on chemical composition of algal assemblages in laboratory streams. *J. N. Am. Benthol. Soc.*, 6(3): 189-197.

STOUT, R.J. 1989. Effects of condensed tannins on leaf processing in mid-latitude and tropical streams: A theoretical approach. *Can. J. Fish. Aquat. Sci.*, 46: 1097-1106.

STOUT, J. & VANDERMEER, A. 1975. Comparison of species richness for stream-inhabiting insects in tropical and mid-latitude streams. *Amer. Natur.*, 109: 263-280.

TOWNSED, C.R. 1989. The patch dynamics concept of stream community ecology. *J. N. Am. Benthol. Soc.*, 8(1): 36-50.

VANNOTE, R.L., MINSHALL, G.W., CUMMINS, K.W., SEDELL, J.R. & CUSHING, C.E. 1980. The river continuum concept. *Can. J. Fish. Aquat. Sci.*, 37:130-137.

WALLACE, J.B., VOGEL, D. & CUFFNEY, T.F. 1986. Recovery of a headwater stream from an insecticide-induced communtity disturbance. *J. N. Am. Benthol. Soc.*, 5(2): 115-126.

WARD, A.K. 1985. Factors affecting distribution of *Nostoc* in Cascada mountain streams of western Oregon, USA. *Verh. internat. verein. limnol.*, 22: 2799-2804.

WARD, J..W. & STANFORD, J.A. 1983a. The intermediate disturbance hypothesis: an explanation for biotic diversity pattern in lotic ecosystems. In: *Dynamics of lotic ecosystems* (T.D. Fontaine & S.M. Bartell, eds.): 347-356. Ann Arbor Scien. Publ. Ann Arbor, Michigan.

WARD, J.W. & STANFORD, J.A. 1983b. The serial discontinuity concept in lotic ecosystems. In: *Dynamics of lotic ecosystems* (T.D. Fontaine & S.M. Bartell, eds.): 29-42. Ann Arbor Scien. Publ. Ann Arbor. Michigan.

WARD, J.W. & STANFORD, J.A.. 1984. The regulated stream as a testing ground for ecological theory. In: *Regulated Rivers* (A. Lillehammer & S.J. Salveit, eds.): 23-38. Oslo Univ.Press. Oslo.

WILLIAMS, D.D. 1981. Migrations and distributions of stream benthos. In: *Perspectives in Running Water Ecology* (M.A. Lock & D.D. Williams, eds.): 155-207. Plenum. New York.

WILLIAMS, D.D. & HYNES, H.B.N. 1976. The recolonization mechanisms of stream benthos. *Oikos*, 27: 265- 272.

WILLIAMS, D.D. & HYNES, H.B.N. 1977. The ecology of temporary streams. II. General remarks on temporary streams. *Int. Rev. ges. Hydrobiol.*, 62: 53-61.

WILZBACH, M.A., CUMMINS, K.W. & KNAPP, R.A. 1988. Toward a functional classification of stream invertebrate drift. *Verh. internat. verein. limnol.*, 23: 1244-1254,

WINTERBOURN, M.J., ROVNIK, J.S. & COWIE, B. 1981. Are New Zealand stream ecosystems really different? *New Zealand J. Mar. Freshw. Res.*, 15: 321-328.

WRONA, F.J., CALOW, P., FORD, I., BAIRD, D.J. & MALTBY, L. 1986. Estimating the abundance of stone dwelling organisms: a new method. *Can. J. Fish. Aquat. Sci.*, 403(10): 2025-2035.

Homage to Ramon Margalef; or, Why there is such pleasure in studying nature (J. D. Ros & N. Prat, eds.). 1991. *Oecologia aquatica*, 10: 13-20.

NEW APPROACHES TO MEDITERRANEAN FLUVIAL COMMUNITIES

M. ÀNGELS PUIG[1], MARINA ABOAL[2] & ADOLFO DE SOSTOA[3]

[1] Centro de Investigaciones del Agua (CSIC). 28500 Arganda del Rey. Madrid. Spain
[2] Departamento de Biología Vegetal. Universidad de Murcia. Campus de Espinardo. 30100 Murcia. Spain.
[3] Departament de Biologia Animal. Facultat de Biologia. Universitat de Barcelona. Av. Diagonal, 645. 08028 Barcelona. Spain.

Received: March 1991.

SUMMARY

The organization of fluvial benthic communities in the Iberian Mediterranean system shows a series of factors which differ in part from present models suggested in the current literature. The effects of disturbances can be observed in models including the gradient of organization (WARD & STANFORD, 1983; VANNOTE *et al.*, 1980; SABATER *et al,*. in press) in the major and permanent northwest basins, but not in temporary systems.

Temporality does usually not represent a handicap as far as the organization of the algae communities and the macroinvertebrates communities is concerned,: equivalent diversities have been observed in temporary and permanent systems. But a latitudinal gradient is observed among the temporary systems, with an increment to the south of the diversity. Mediterranean fluvial systems have integrated floods as periodical perturbations in the dynamics of benthic communities, which have a high resistance to flood-related disturbance and a high subsequent resilience.

KEY WORDS: Mediterranean rivers, fluvial communities, algae, invertebrates.

INTRODUCTION

The study of fluvial systems in the Iberian peninsula goes back to the work of MARGALEF (1960). Following his path, one approach to the ecology of Catalan fluvial basins has been developed at the Department of Ecology at the University of Barcelona. This approach has involved different perspectives, such as physico-chemical dynamics (PRAT *et al.*, 1982, 1984; SABATER, 1987; PUIG *et al.*, 1987; SABATER *et al.*, 1989, 1991) and the composition and dynamics of benthic fluvial communities (PEÑUELAS & SABATER, 1987; SABATER & SABATER, 1988; PRAT *et al.*, 1984; PUIG, 1984; GONZÁLEZ, 1990). Nowadays, a broader knowledge of the Iberian fluvial systems has been possible thanks to this group and its influence throughout other universities. as well as its collaboration with several universities, and research centers.

This report attempts to elucidate some aspects associated with the structure and organization of the benthic communities of fluvial streams based on previous taxonomic knowledge, especially of algae and macroinvertebrates. A discussion of the longitudinal organization of the fluvial benthos and the use of diversity as a possible parameter for testing the theoretical proposal of SABATER *et al.* (1989, 1991), in relation to disturbance distances (WARD & STANFORD, 1983), is presented. In this model alternative statistical techniques have been used. The role of fluctuations in the complexity of communities has also been considered. Differences between the degrees of temporality, whether the river dries each year or over a longer timespan, have been elucidated as basic with regard to the effects of floods. For the first case, possibly catastrophic flood-related phenomena can be absorbed into the annual cycle and assimilated by the communities dynamics; thus catastrophe or disruption are avoided.

STUDY AREA

The basins of the rivers Ter, Tordera, Matarraña (Ebro basin), Vinalopó, Segura and Yeguas (Guadalquivir basin) were studied. In some cases the work was conducted in the principal course of the basin and in others in smaller sub-basins. All of them are included in figure 1; minimal data of their structural references are shown in Table I.

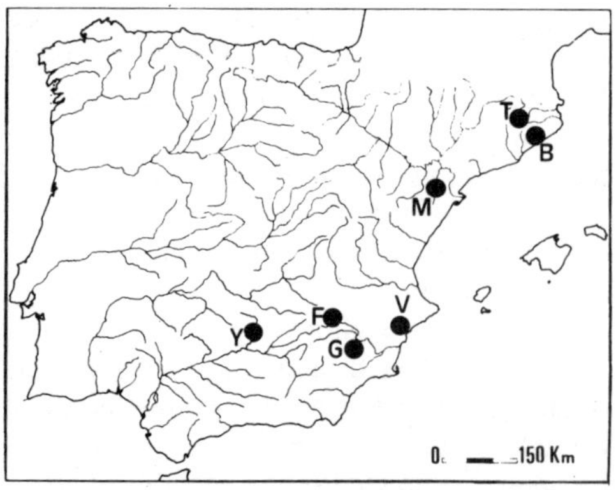

FIGURE 1. Location of study areas in Eastern Spain. Letters refer to rivers and are identified in Table I.

TABLE I. Location and temporal characteristics of the sampling stations (P = permanent stretch; T = temporary stretch; T/P = temporary river with summer pools).

Loc.	River	Basin	Alt. m(a.s.l.)	Dist. to the spring(km)	Type
T7	Ter	Ter	1400	7.0	P
T10	Ter	Ter	915	20.2	P
T12	Ter	Ter	720	38.7	P
T15	Ter	Ter	680	44.9	P
T18	Ter	Ter	540	62.1	P
T21	Ter	Ter	490	71.1	P
T22	Ter	Ter	460	88.1	P
T30	Ter	Ter	160	135.35	P
T38	Ter	Ter	98	148.15	P
T45	Ter	Ter	60	158.15	P
T46	Ter	Ter	30	173.35	P
T47	Ter	Ter	20	186.35	P
T3	Rigart	Ter	1100	14	T
T5	Freser	Ter	740	24.7	T
T9	Tort	Ter	1000	11.3	P
T27	Major	Ter	470	14.0	P
B1	Buscons	Tordera	420	1..5	T
B2	Buscons	Tordera	240	7.0	T/P
B3	Buscons	Tordera	100	11.0	T/P
M1	Matarraña	Ebro	650	8.0	P
M2	Matarraña	Ebro	500	18.0	P
M3	Matarraña	Ebro	170	76.0	T/P
M4	Matarraña	Ebro	95	98.0	T/P
V	Vinalopó	Vinalopó	820	7.0	T/P
F1	Mundo	Segura	1100	20.0	P
F2	Mundo	Segura	1000	40.0	P
F3	Mundo	Segura	997	44.0	P
F4	Mundo	Segura	850	54.0	P
G	Argos	Segura	1050	3.2	T
Y1	Cereceda	Yeguas	700	8.0	P
Y2	Pradillo	Yeguas	660	8.4	P
Y3	Yeguas	Guadalquivir	0	12.5	P
Y4	Términos	Yeguas	570	7.9	T
Y5	Yeguas	Guadalquivir	550	18.0	T
Y6	Yeguas	Guadalquivir	300	51.5	T/P
Y7	Fresnedoso	Yeguas	220	19.0	T
Y8	Yeguas	Guadalquivir	200	4.7	T/P

MATERIALS AND METHODS

Diversity was estimated by Shannon's index (STATZNER, 1981), but working at three different levels: diversity estimated from stationary sampling, accumulated diversity in the treatment of part of the data of the fluvial axes, and annual diversity, studied by monthly sampling.

For the statistical analysis processing, principal component analysis (PCA; BMDP, 1984) was used.

RESULTS AND DISCUSSION

LONGITUDINAL ORGANIZATION OF THE COMMUNITIES

Comparing the values of diversity in the stationary data for the different basins, from the heads of the rivers up to the mouths (Table II) we do not find greater values of diversity in the middle watercourse than in the headwaters.

These results contradict the theories of the "River Continuum Concept" (VANNOTE *et al.*, 1980) and the "Serial Discontinuity Concept" (WARD & STANFORD, 1983). In both theories it is presupposed that from a certain value of initial diversity at the head of the river there is an increase up to a maximum inside the mean course of the fluvial axis. This area will, theoretically, show the maximum spatial heterogeneity. This heterogeneity is generated by the effect of natural disturbance in this zone, because of the fact that the inputs are in part absorbed due to the inertia attained by the watercourse.

The data obtained for the basin of the Ter, Matarraña, Yeguas (Table II) or Llobregat (PRAT *et al.*, 1983) are clear examples of the high diversity of the heads of the rivers and have practically the same values as the mean sections for the communities of macroinvertebrates.

There are two principal causes for this situation. The first is that the permanent rivers (Ter and Llobregat) lie in a displacement of the diversity curve toward the heads. These are systems with a maximal climatic and structural gradient for the first twenty kilometers while in the rest of the fluvial courses it is similar. In fact, they are rivers in which only one head zone is clearly appreciated and all the rest is an undefined structure, as happens in all Iberian Mediterranean rivers. The mouth *senso stricto* practically constitutes a small independent entity.

The second cause is the regime of temporality of some rivers, in which the

TABLE II. Diversity values in the study stretches. A, algae community; I, invertebrate community.

Locality	Div. (H')	Acc. div.	Comm.	Time
F-1	3.51	3.51	A	
F-2	3.75	4.37	A	
F-3	3.62	4.43	A	
F-4	3.68	4.50	A	
M-1	3.31			winter
M-2	3.25			spring
M-3	3.37			summer
M-4	3.28			fall
Y-1	4.18	4.18	I	
Y-3	4.37	5.25	I	
Y-5	2.68	5.43	I	
Y-6	4.25	5.56	I	
Y-8	3.87	5.75	I	
Y-2	3.62		I	
Y-4	3.87		I	
Y-7	4.12		I	
B-1	2.43	2.43	I	
B-2	3.12	3.37	I	
B-3	2.75	3.56	I	
V	2.93	2.93	I	Aug.
V	2.87	3.10	I	Sept.
V	4.0	4.06	I	Oct.
V	3.40	4.27	I	Dec.
V	3.37	4.40	I	Jan.
V	3.56	4.50	I	Feb.
V	3.60	4.56	I	Apr.
V	3.60	4.56	I	May
T-7	3.09	3.09	I	Oct.
T-10	3.035	3.76	I	Oct.
T-12	3.079	3.97	I	Oct.
T-15	2.175	3.94	I	Oct.
T-18	2.097	3.36	I	Oct.
T-21	2.36	3.49	I	Oct.
T-22	0	3.50	I	Oct.
T-30	3.082	3.73	I	Oct.
T-38	3.248	3.88	I	Oct.
T-45	2.478	3.87	I	Oct.
T-46	2.958	3.94	I	Oct.
T-47	2.380	4.01	I	Oct.
T-3	2.97		I	Oct.
T-3	2.87		I	Nov.
T-3	2.66		I	Jan.
T-3	2.45		I	Apr.
T-3	dry		I	July
T-5	3.5		I	Oct.
T-5	1.37		I	Nov.
T-5	1.93		I	Jan.
T-5	1.93		I	Apr.
T-5	dry		I	July
T-9	4.27		I	Oct
T-9	3.87		I	Nov.
T-9	3.81		I	Jan.
T-9	3.68		I	Apr.
T-9	3.12		I	July
T-27	3.81		I	Oct.
T-27	3.06		I	Nov.
T-27	1.75		I	Jan..
T-27	3.40		I	Apr.
T-27	3.47		I	July

first portion is permanent and the rest, up to the confluence or mouth, is temporal. In this case great structural heterogeneity is produced in the permanent portion with which the highest diversity is also associated.

For the algae communities, the results support the trend of showing the maximal diversity in the middle section of the fluvial system. The results obtained in the Matarraña river are the exception. Due to the calcareous characteristics of the system, with precipitation of travertine in the middle section of the watercourse, this river maintains the same tendency of similar diversity in the heads and successive sections. Here, it is necessary to consider the homogenization of the substratum that the precipitation of calcium carbonate produces; for the communities of epilithic algae, the variation of the stream speed affects neither the general composition nor their diversities (Fig. 2; ABOAL & PUIG, in press).

LONGITUDINAL ORGANIZATION AND DISCONTINUITY DISTANCES ASSOCIATED WITH THE MACROINVERTEBRATE COMMUNITY IN THE TER RIVER

If the accumulated diversity along the fluvial axis is used for a particular stage of the year, as far as changes in the communities of the fluvial zoobenthos are concerned, an increment from the heads up to the mouths of the rivers is observed, with moments of inversion due to different effects of human action, and also a clear absorption of the diversity increment in the final sections of the basin (Fig. 3).

Comparing the curve obtained with the ones proposed for the Ter by SABATER *et al.* (1991), similar trends of variation are appreciated from the chemical distances. We can define the intersite distances for the diversity (H') along the x_i to x_j stretch by $dH'_{ij} = (H_i - H_j) / (Hmax - Hmin)$. This distance was used to reveal discontinuities along the Ter river, when this synthetic parameter is related to the geographic distance (X), also calculated using the preceding equation (Fig. 3). But in this case it is impossible to summarize it by an empirical equation.

If a methodology like that used by these authors is applied, due to the scores of stations corresponding to the axis of PCA defining the longitudinal gradient of the fluvial course for the basin (Axis 1, PC I), a similar tendency of the curves can be seen. Hence it is possible to propose an alternative model for the changes in the communities, similar to that proposed by SABATER *et al.* (1991) for the physico-chemical behaviour of the Ter river basin (Fig. 4). Variation in the catchment area along the river can be described by the equation:

$$PC\ I' = 9{,}7\ (X)^{0.20} \quad r^2 = 0{,}76 \quad p > 0{,}001$$

The tendency of the curves of accumulated diversity for other basins does not always show the initial absorption effect produced in the lower section of the Ter. This may be due to the shorter length, smaller drainage courses, or the fact that they are temporal basins with a clearly different dynamic. In this last case, because

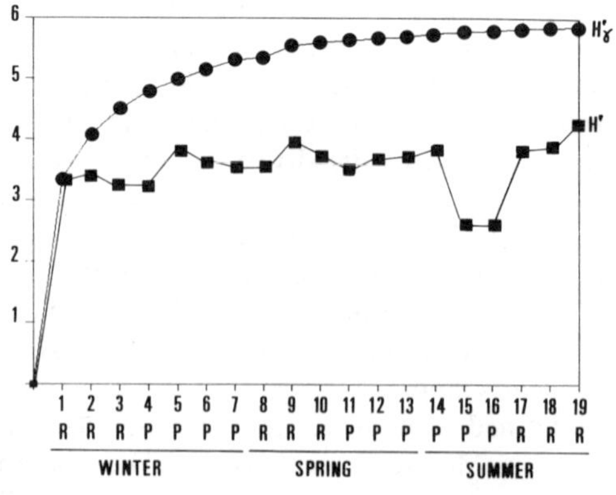

FIGURE 2. Patterns of diversity (H') and accumulated diversity (H'γ) for algae communities at the riffle (R) and pool (P) areas of Matarraña river (M2, travertine precipitation stretch).

they are normally constituted in the Mediterranean slope for permanent karstic heads, temporal lower sections and very variable transitional zones, the communities found in the temporal zones are almost completely different from those of the permanent zone, and due to this high values of accumulated diversity are associated with them. This tendency is noted for the algae communities (Segura river, Matarraña river), and also for the macrobenthos communities (Table II).

DIVERSITY AND TEMPORALITY

We will initially consider the set of clear temporal fluvial systems, with a part of their course dried out every year for variable periods of time. These periods are normally from three to five months, according to the geographic area and the annual rainfall. These rivers are: the Matarraña, the Tordera, the Yeguas and different sub- basins of the Segura river.

In these temporal systems, the results obtained are different if the area completely dries out or if broad ponds of water remain through the summer period. In all cases, maximum diversity is observed during the winter period, with lower values in summer associated with the communities developed in the ponds (Table II).

There are no differences observed among the values obtained for the strict temporal areas and for those with ponds if the global diversity during the flooding period of each station is considered.

We will now consider the results for the permanent station, or the permanent sub-basins studied in relation to the temporal ones. For the algae communities, no differences in diversity are found between the permanent systems and the temporal systems with ponds. Considering diversity in a complete annual cycle, a tendency for lower values of diversity is observed in the strict temporal courses, but it is not the same in the temporal systems with ponds in relation to the permanent ones (Fig. 5).

The communities of macrobenthos in the permanent systems considered in this work also maintain a maximum of winter diversity or spring-winter diversity,

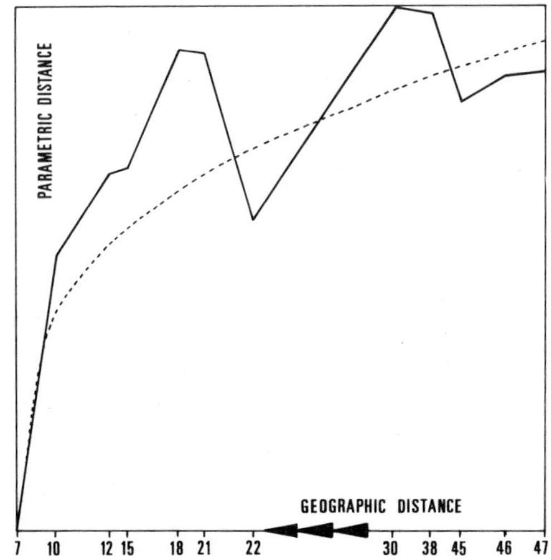

FIGURE 3. Plot of geographical against parametrical distances of accumulated diversity in the Ter river (see text for explanation).

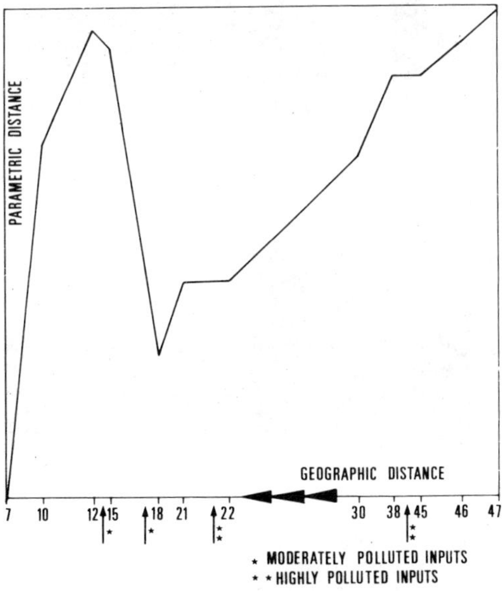

FIGURE 4. Plot of the PC I scores of PCA of each site against the geographical ones in the Ter river, using the benthic invertebrate community data.

depending on the individual case. The summer diversity values are considered for the total set of invertebrates and a wide range is appreciated, inside which the temporal and permanent rivers cannot be clearly separated. Besides, a latitudinal gradient is observed among the temporal systems and this gradient focusses the increment of the diversity of the communities in the temporal Mediterranean system to the south (Fig. 5).

RIVER FLOODS AND DIVERSITY

In the previous section the possible effect of the seasonal cyclic fluctuations on the organization of the communities was discussed. But this type of fluctuation may

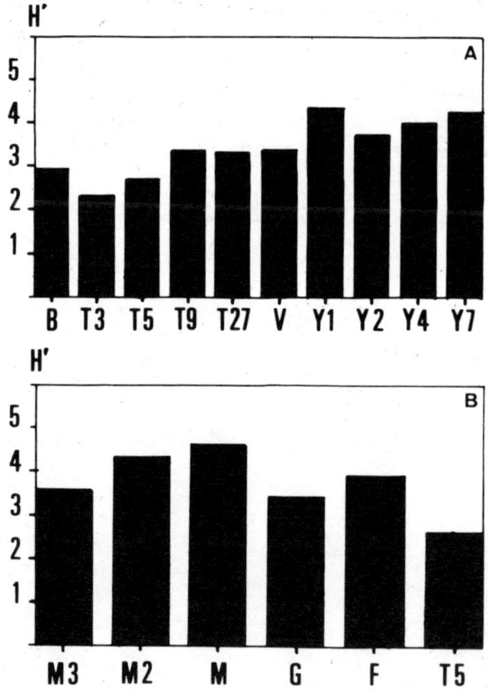

FIGURE 5. Different values of diversity during the annual cycle observed in part of the river basins. A, benthic invertebrate communities; B, algae communities. (B= Tordera river basin; T= Ter river basin; V= Vinalopó river; Y= Yeguas river basin; M= Matarraña river; G and F= Segura river basin).

not only occur during the dry period; for a great number of sections, sub-basins and Mediterranean basins the effects of the annual floods have been considered the principal kind of fluctuations, though of variable intensity over the years.

The basin of the Matarraña river can be considered a typical example of this kind of system. Autumn is the season presenting the lowest values of diversity in the zones permanently under flood effects without any subsequent decrease; indeed, a fast increase in diversity is observed. Detailed study of the benthic communities of this river for more than an annual cycle has allowed us to show that the cycle of the majority of the species is related to the floods. The floods start the initiation of almost all the winter species, while those present all through the year have certain mechanisms which allow them to compensate for the destructive or dispersive effect of this catastrophe (e.g. living in the precipitations of the travertines, burying in the bed river, etc.) (RECASENS & PUIG, 1987; PUIG *et al.*, 1990).

Data from this basin are totally different from those of permanent Mediterranean basins where floods of no cyclical periodicity occur. In these other systems (PUIG *et al.*, 1987) the effects of floods are more or less disruptive, in part demolishing the communities, and the periods of recuperation can be long (six to eight months, depending on the communities and the zones).

In fact, the fluvial Mediterranean systems are a clear example of the integration of periodical disturbances in the dynamics of the communities.

The temporal sub-basins are as adapted to the autumn maximal rainfalls as the permanent ones. For both systems a flood is only a maximum greater than the common one during an annual cycle. It must be pointed out that associated with the seasonal concentration slope of rainfalls, maximum in the south of the Mediterranean area and less notable in the northern basins, a reduction in the response time for the reorganization of communities after a flood is also registered. In these systems the stream communities appear to have a high resistance to flood-related disturbance and high resilience after disturbances, contrary to the data obtained in Australia by LAKE & BARMUTA (1986).

In fact, we prefer to follow MARGALEF (1960), "... to see the river systems as a mosaic of series that reflected or were constrained by local and regional terrestrial and aquatic ecosystems", but in the Iberian Mediterranean fluvial systems we do not see "an orderly progression from simple, immature head waters to complex, mature downstream reaches" (MARGALEF, 1963), "with a corresponding increase in species diversities as the system became more mature" (MARGALEF, 1968). The headwaters and the downstream reaches exhibited the same degree of immature or mature pattern.

ACKNOWLEDGMENTS

We would like to acknowledge the influence of Prof. R. Margalef in some of the ideas expressed in this article. This research was sponsored by the Comisión Asesora de Investigación Científica y Técnica under contract 478/81 and by the United States-Spain Joint Committee for Scientific and Technological Cooperation under contract CCA8411065.

REFERENCES

ABOAL, M. & PUIG, M.A. (in press). Dynamique de peuplements de diatomées épilitiques en rivières temporaires: le cas du Matarranya (NE de l'Espagne). *Archiv. für Hydrobiologie*

BMDP. 1984. Statistical Software, Inc. Program BMDP.4M. Factor Analysis (version 1987). University of California.

LAKE, P.S. & BARMUTA, L.A. 1986. Stream benthic communities: Persistent presumptions and current speculations. In: *Limnology in Australia* (De Decker, P. & Williams, W.D, eds.): 263-276. Junk. The Hague.

GONZÁLEZ, G. 1990. *Sistemática y ecología de los Simuliidae (Diptera) de los ríos de Catalunya y de otras cuencas hidrográficas españolas.* Ph.D. Thesis. University of Barcelona.

MARGALEF, R. 1960. Ideas for a synthetic approach to the ecology of running waters. *Intern. Rev. gesamt. Hydrobiologie*, 45:133-153.

MARGALEF, R. 1963. On certain unifying principles in ecology. *Amer. Nat.*, 97:357-374.

MARGALEF, R. 1968. *Perspectives in Ecological Theory.* University of Chicago Press. Chicago.

PEÑUELAS, J. & SABATER, F. 1987. Distribution of macrophytes in relation to environmental factors in the Ter River, N.E. Spain. *Intern. Rev. gesamt. Hydrobiologie,* 72: 41-58.

PRAT, N., PUIG, M.A., GONZÁLEZ, G. & TORT, M.J. 1982. *Predicció i control de la qualitat de les aigües dels rius Besós i Llobregat. I: Els factors físics i químics del medi.* Servei del Medi Ambient de la Diputació de Barcelona. Barcelona.

PRAT, N., PUIG, M.A., GONZÁLEZ, G. & MILLET, X. 1983. Chironomid longitudinal distribution and macroinvertebrate diversity along the Llobregat river (NE Spain). *Mem. Amer. Entomol. Soc.*, 34:267-278.

PRAT, N., PUIG, M.A., GONZÁLEZ, G., TORT, M.J. & ESTRADA, M. 1984. The Llobregat: A Mediterranean river fed by the Pyrenees. In: *The ecology of European rivers* (Whitton, B.A., ed.): 527-552. Blackwell. London.

PUIG, M.A. 1984. *Efemerópteros y Plecópteros de los ríos de Cataluña.* Ph.D. Thesis. University of Barcelona.

PUIG, M.A., ARMENGOL, J., GONZÁLEZ, G., PEÑUELAS, J., SABATER, S. & SABATER, F. 1987. Chemical and biological changes in the Ter River induced by a series of reservoirs. In: . *Advances in Regulated Stream Ecology* (Craig, J.F. & Kemper, J.B. eds.): 373-382. Plenum Press. New York.

PUIG, M.A., FERRERAS-ROMERO, M. & GARCÍA-ROJAS, A. 1990. Morphological variability of *Tyrrhenoleuctra minuta* (Klapalek, 1903) in South Spain. In: *Mayflies and Stoneflies* (Campbell, L.C., ed.): 357-360. Kluwer. The Hague.

RECASENS, L. & PUIG, M.A. 1987. Life cycles and growth patterns of Trichoptera in the Matarraña, a karstic river. In: *Proceedings of the 5th International Symposium on Trichoptera* (Bournaud, M. & Tachet, H., eds.): 247-251. Junk. Dordrecht.

SABATER, F. 1987. *Estudi integrat del riu Ter i la seva conca: Les característiques de l'aigua i els factors que les determinen.* Ph. D. Thesis. University of Barcelona.

SABATER, S. & SABATER, F. 1988. Diatom assemblages in the river Ter. *Archiv für Hydrobiologie,* 111:397-408.

SABATER, F., ARMENGOL, J. & SABATER, S. 1989. Measuring discontinuities in the Ter River. *Regulated Rivers: Research & Management,* 3:

133-142.

SABATER, F., ARMENGOL, J. & SABATER, S. 1991. Physicochemical disturbances associated with spatial and temporal variation in a Mediterranean river. *J. North Amer. Benth. Soc.* (in press).

STATZNER, B. 1981. Shannon-Weaver diversity of the macrobenthos in the Schierenseebrooks (North Germany) and problems of its use for the interpretation of the community structure. *Verh. intern. Verein. Limnol.*, 21: 782-786.

VANNOTE, R.L., MINSHALL, G.W., CUMMINS, K.W., SEDELL, J.R. & CUSHING, C.E. 1980. The river continuum concept. *Canadian J. Fish. Aq. Sci.*, 37: 130-137.

WARD, J.V. & STANFORD, J.A. 1983. The serial discontinuity concept of lotic ecosystems. In: *Dynamics of lotic ecosystems* (Fontaine, T.D. & Bartell, S.M., eds.): 29-42. Ann Arbor Science Publishers. Ann Arbor.

Homage to Ramon Margalef; or Why there is such pleasure in studying nature 21
(J. D. Ros & N. Prat, eds.). 1991. *Oecologia aquatica.* 10: 21-33

USING THE RESCALED RANGE ANALYSIS FOR THE STUDY OF HYDROLOGICAL RECORDS: THE RIVER TER AS AN EXAMPLE

JOAN ARMENGOL[1], SERGI SABATER[1], ANTONI VIDAL[2] & FRANCESC SABATER[1]

[1] Departament d'Ecologia. Facultat de Biologia. Universitat de Barcelona. Av. Diagonal, 645. 08028 Barcelona. Spain
[2] Junta d'Aigües. Abastament d'Aigua del Ter. Cardedeu. 08440 Barcelona. Spain

Received: November 1990

SUMMARY

Water discharges into the Ter river have a strong influence on the water chemistry and on the dynamics of benthic communities. For this reason, the temporal patterns followed by monthly runoff from 1954 to 1988 have been studied. From a statistical point of view, discharges show a very regular seasonal cycle, but in practice, climatic events introduce a source of variability on runoffs. To determine whether the series of 420 monthly discharges fluctuates with some regularity or at random, a rescaled range analysis was made. The persistence (H) of the series of discharges was measured by the equation $R / S = (T / 2)^H$. In fact, H is a measure of the existence of clear trends or periodicities in the records of persistent stocastic processes. An additional measure of the persistence was obtained by means of the relationship D = 2 - H, where D is the local fractal dimension. For comparative reasons, a random series of the same length, mean and standard deviation was generated and analysed in the same way as the Ter discharges were. The results were H = 0.68 and D = 1.38 and H = 0.44 and D = 1.52, for the Ter data and for the random series respectively. These results show that the simulated series is a case of ordinary Brownian motion while the Ter discharges series has an intermediate value of persistence. Because of the value of H in the Ter series, there are periods whose values were estimated by the autocorrelation function and the periodogram of the series. The results show the existence of short fluctuations of 3, 6 and 12 months determining the seasonality of the annual cycle, and large fluctuations with periods of 5.5, 8.6, 10.1 and 11.7 months which can be considered an expression of cycles of circa 7 and 11 years.

KEY WORDS: River discharge, fractal dimension, Hurst exponent, fractional Brownian motion, persistence.

. . . In the case of the landscape, rain falling at random and moving from each point of the surface along the steepest profile, through erosion and convergence of runoffs, generates a secondary "hydrologic" relief. Will the fractal quality be preserved ? (MARGALEF, 1988)

INTRODUCTION

In Mediterranen ecosystems, annual cycles are always described in terms of regular fluctuations associated with seasonal climatic changes. Nevertheless, this pattern is usually modified by irregular climatic fluctuations that cause sudden changes in their dynamics. Temporal ponds and rivers are among the most affected ecosystems because they are mainly controlled by the periodicity and intensity of the rains. The turnover time of the water acts as a key factor in most of the processes taking place (SCHINDLER, 1988; ARMENGOL, 1988).

In rivers, strong rains increase the intensity of the advective transport of solid

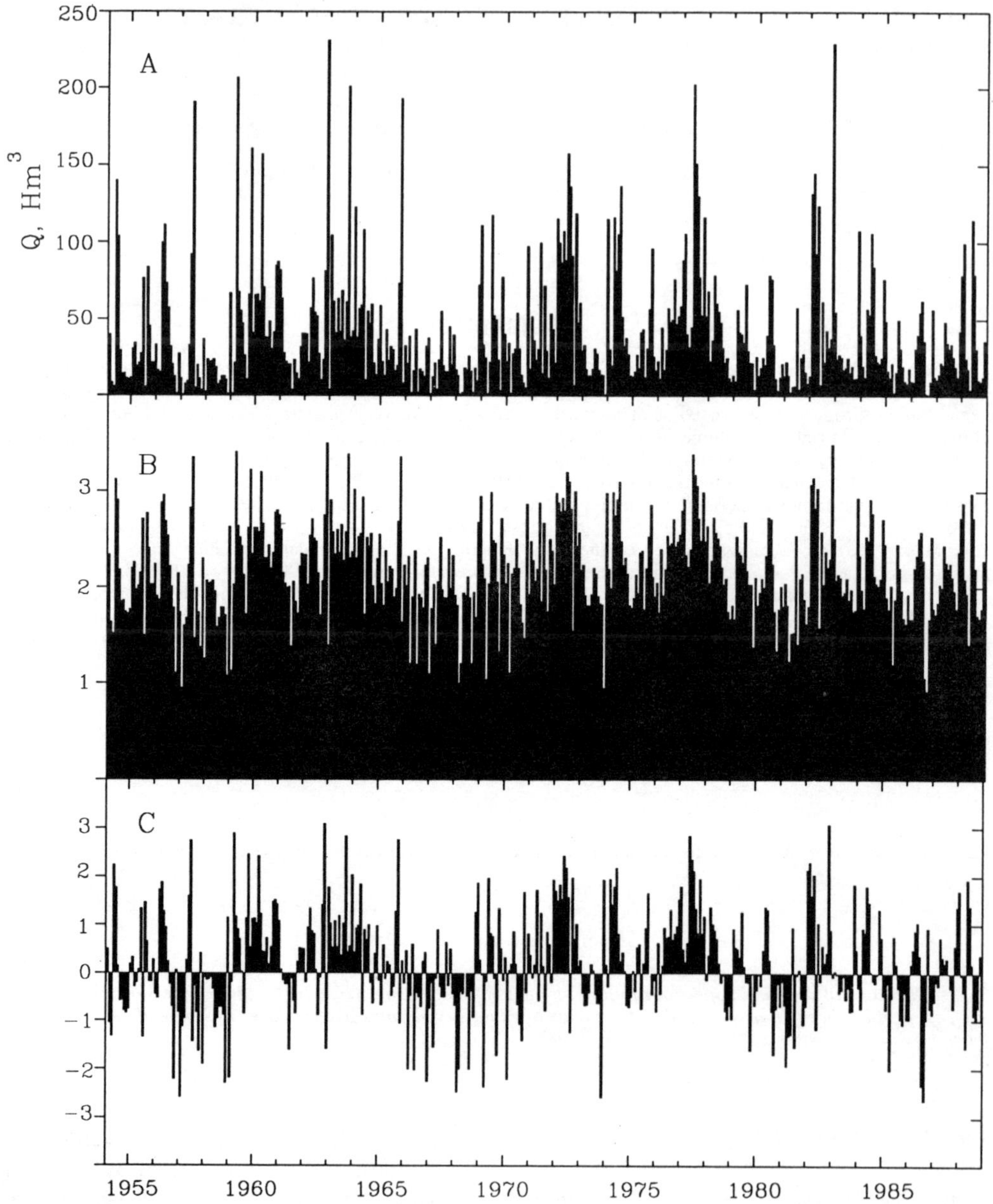

FIGURE 1. River Ter monthly discharges for the 1954 - 1988 period. A) original series of data in Hm^3 $month^{-1}$. B) normalized series by $Y = X^{0.22}$. C) the standardized sequence.

and dissolved substances. Because of this, the spates act as huge inputs of external energy that simplify the structure of the river. Most of the organisms are carried down and the chemical composition of water becomes more similar along the river. The annual cycle of the communities can be seen as a sequence of populations with short life span, well adapted to seasonal changes but able to reconstruct the community after each event (MARGALEF, 1980). If longer periods of time are considered, changes in the communities can be seen as a sequence of periods of organisation (secondary successions) followed by quick disruptive episodes (backward steps) (MARGALEF, 1977).

In the Ter river, water flow is the most important factor determining the annual variability in the environmental conditions (SABATER & ARMENGOL, 1986) and in the dynamics of phytobenthic (SABATER *et al.*, 1988) and macroinvertebrate (PUIG *et al.*, 1987) communities. These relationships are well established for short periods of time, such as the annual cycle, but for longer periods it is assumed that climatic fluctuations have a similar effect. If we accept these relationships for the Ter river, the long-term study of a record of hydrological events as large as possible can provide us with some insight into how it

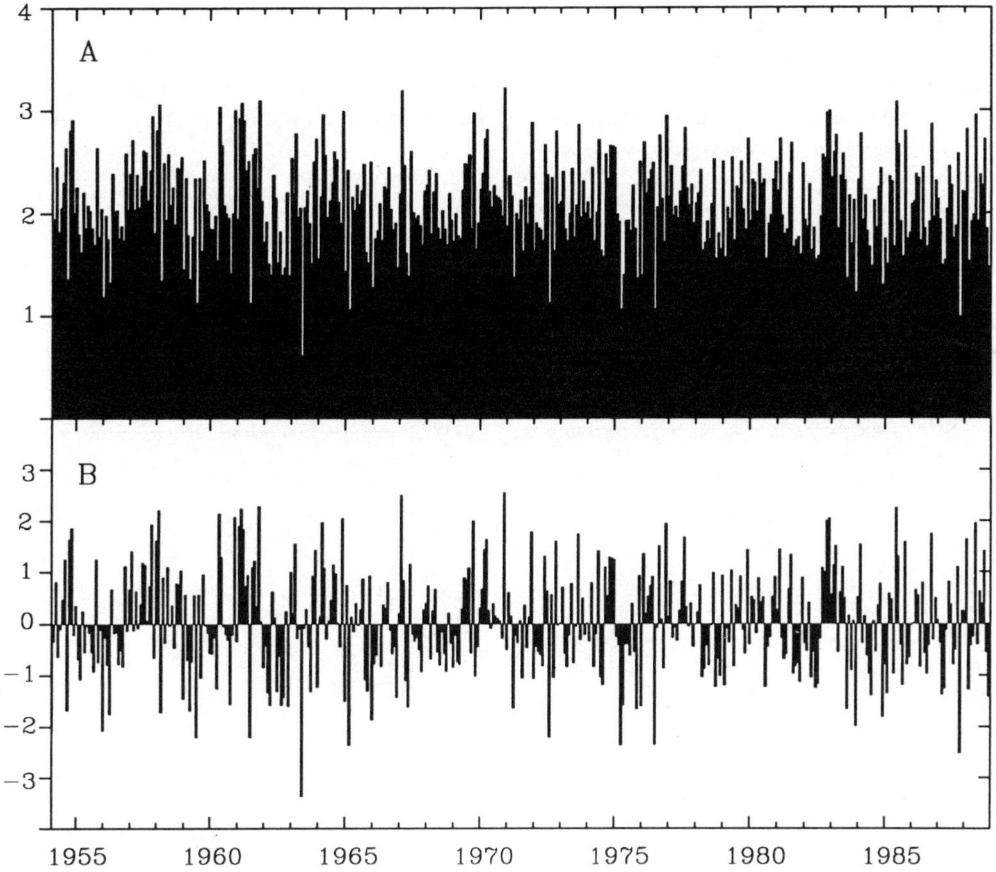

FIGURE 2. Simulated sequence of 420 values (35 years x 12 months) normally distributed with the same mean and standard deviation as the data of figure 1B . B) the standardized series.

functions from an ecological point of view. For longer periods it is well established that the water discharges affect the chemical pollution and the eutrophication levels in the reservoirs situated in the middle stretch of the river (VIDAL, 1977, 1987, 1989; ARMENGOL *et al.*, 1986).

Long-term records of hydrological events show erratic behaviour at different time scales (FEDER, 1988). These records can be studied by means of the rescaled range analysis (HURST *et al.*, 1965) and characterized by an exponent H, called the Hurst exponent, which is a measure of the persistence or "memory" of one series (MANDELBROT & VAN NESS, 1968). MANDELBROT (1985) has shown that this record is a self-affine fractal curve with a dimension D = 2 - H. For a practical perspective, the H exponent measures how far a temporal series of data departs from a random series.

We have used a record of 35 years of monthly average discharges to analyse the persistence in the data sequence. The aim of this study is to determine the temporal trend followed by the series and how this key factor carried weight on the river considered as an ecosystem. It is well known that wet and dry years appear in clusters, the "Joseph effect" (MANDELBROT & WALLIS, 1969); for this reason the existence of temporal "cycles" has been studied here by means of autocorrelation and spectral analysis.

MATERIAL AND METHODS

The record of hydrological data used in this study is composed by the monthly discharges at Roda de Ter station from 1954 to 1988 (Fig.1). The data were obtained from the Dirección General de Obras Hidráulicas gauging yearbooks (D.G.O.H., 1972, 1973, 1974 and 1975; D.G.O.H., 1976) for the period 1965-1977, and were provided by the Servei Hidrològic de la Generalitat de Catalunya for the periods 1954-1965 and 1978-1988.

Roda de Ter is situated in the middle stretch of the Ter river, 88.6 Km from the source (river length 200,3 Km), where it has a drainage area of 1,476.8 Km^2 (48.11 % of the total). At this point the Ter is a fourth order river. This station was selected because it is situated above the reservoirs, thus excluding their regulatory effect, and at the same time the drainage area is large enough to avoid the effect of sudden discharge due to a local rainfall.

The hydrological characteristics of the

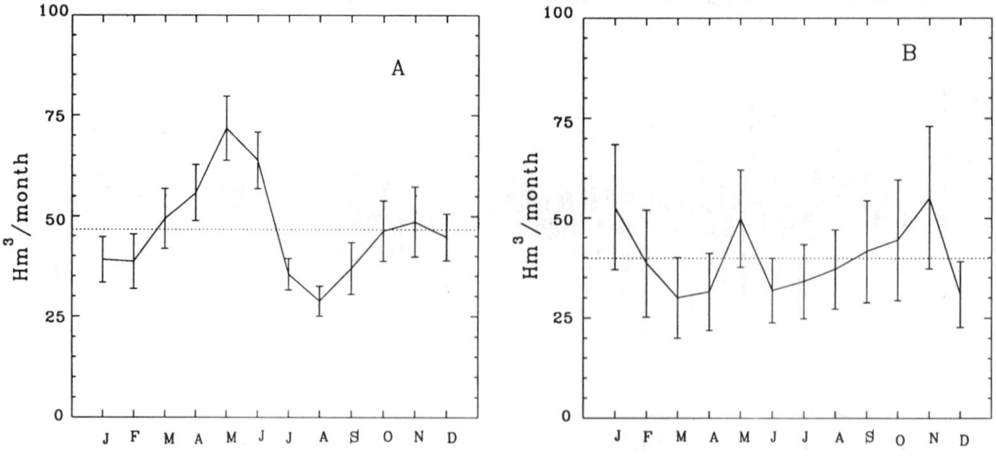

FIGURE 3. Average annual cycle showing the mean monthly discharges with 95 % confidence interval: A) for Ter discharges (Fig. 1B) and B) for random series (Fig. 2 A).

Ter are typical of the Mediterranean region, but the proximity of the Pyrenees also has an influence. The annual cycle is characterized by two periods of maximum water flow; the first, in May (28.6 ± 6.0 Hm^3 $month^{-1}$, $p < 0.05$ for all the data of this paper), is the highest due to the effect of the spring rains plus the thaw of the snow accumulated in the Pyrenees, while the second, in November, is less important (17.6 ± 5.7 Hm^3 $month^{-1}$) and exclusively due to the autumn rains.

Two minima are situated in between, the lower in August (11.4 ± 2.6 Hm^3 $month^{-1}$) and the other in February (14.2 ± 3.9 Hm^3 $month^{-1}$). The average monthly water flow for the period studied was 41.3 ± 3.74 Hm^3 with ranges between 231.4 Hm^3 and 0.70 Hm^3, while, if the data are grouped by years, a mean of 554.6 ± 35.7 Hm^3 is obtained with a maximum of 1180 Hm^3 in 1972 and a minimum of 258 Hm^3 in the year 1957.

The Ter hydrological record does not follow a Gaussian distribution (the kurtosis is 4.912 and the skewness is 2.001). For this reason, it is not possible to use many of the parametrical statistical methods. In order to transform the original data into a new series with a near normality distribution, a power transformation has been sought by means of the Box & Cox

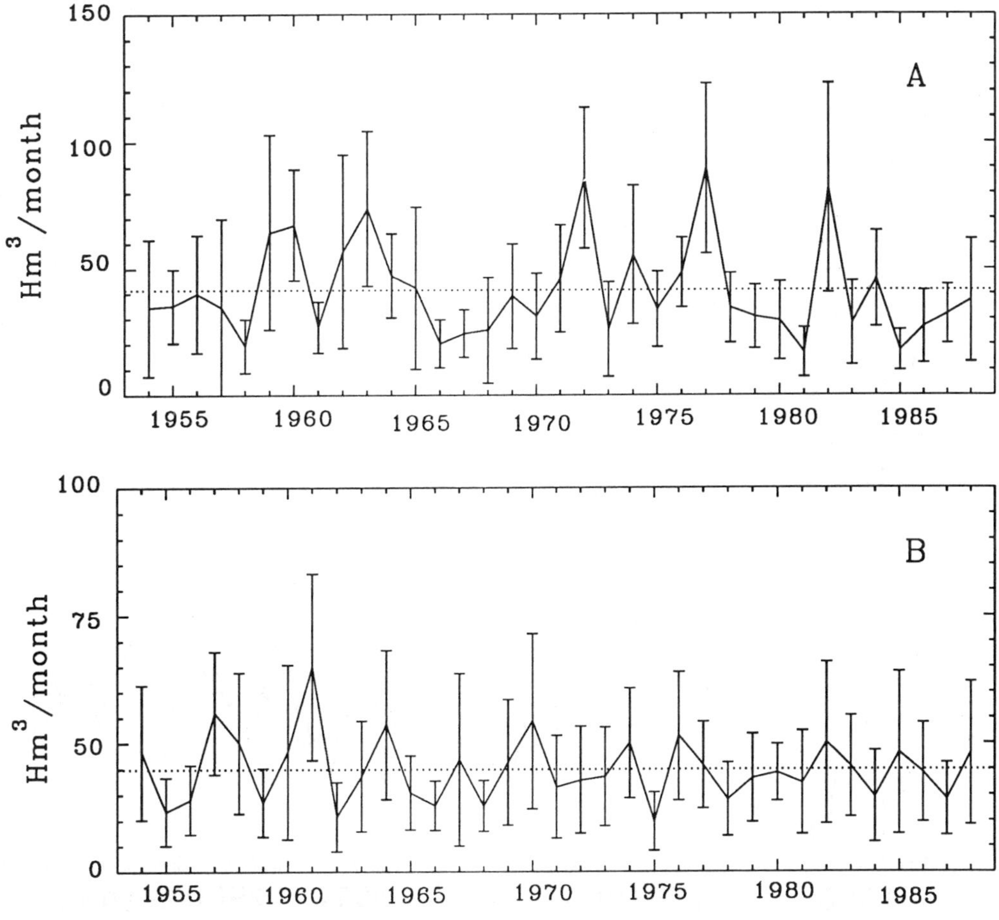

FIGURE 4. Mean annual water flow with 95 % confidence interval: A) for Ter series (Fig. 1B) and B) simulated sequence (Fig. 2A).

method (BOX & COX, 1964). The closest transformation to near normality of the data is $y = x^{0.22}$. The new series (Fig. 1) has a mean of 2.112 and an standard deviation of 0.449, while the kurtosis and the skewness measure 0.058 and 0.028, respectively. Finally, the series used in the analysis has been standardized (Fig. 1).

To illustrate how far the hydrological series departs from a random record, a new Gaussian series of the same length, mean and standard deviation has been constructed, but following a random walk process (Fig. 2).

For comparison of the two series the seasonal and annual distribution of the data are shown in figures 3 and 4.

RESULTS

Not all the rainfall runs immediately into the rivers; a certain amount of water remains stored in the watershed and contributes to the discharge in the driest months. In the same way, during dry periods, the waterflow is higher than expected due to rainfall. As a first approach, there is a buffer effect on discharge due to the storage capacity of the drainage basin. The rescaled range analysis measures the importance of this effect by means of the Hurst exponent.

If $x_1, x_2, \ldots x_T$ is a series of monthly discharges for a period of time T, it may be characterized by the mean ($\bar{x}$), the standard

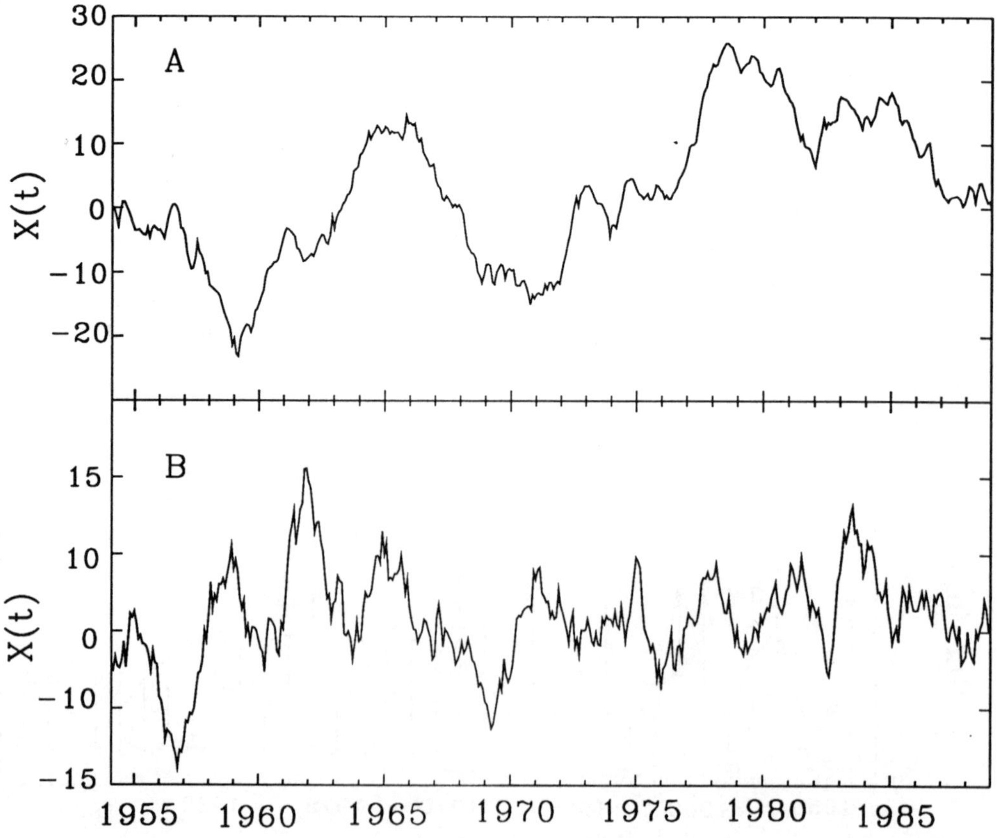

FIGURE 5. Accumulated departures from (zero) mean for: A) the Ter hydrological data, and B) the random series.

deviation (S) and the range (R), which is the difference between the maximum and the minimum departures of the water flux (x_i) from the mean ($\overline{x}$). Formally,

$$\overline{x} = \sum_{i=1}^{T} (x_i) / T \qquad (1)$$

$$S = [1/T (\sum_{i=1}^{T} (x_i-\overline{x})^2)]^{1/2} \qquad (2)$$

$$x(t) = \sum_{i=1}^{T} (x_i-\overline{x}) \qquad (3)$$

$$R(T) = \underset{i=1}{\overset{T}{Max}}\, x(i) - \underset{i=1}{\overset{T}{Min}}\, x(i) \qquad (4)$$

The rescaled range, R/S, is well described by the empirical relation (HURST *et al.*, 1965):

$$R / S = (a\,T)^H \qquad (5)$$

where H is the Hurst exponent.

In figure 5 the accumulated departures from the mean, x (t), are plotted against time for the standardized discharges of the Ter river (Fig. 5a) and for the random series (Fig. 5 b). The figures obtained are two cases of *fractional Brownian motion*, and are generalized random functions, x (t), each one characterized by the exponent H, which ranges between 0 and 1 (MANDELBROT & VAN NESS, 1968; MANDELBROT, 1987). Fractional Brownian motions are very useful to describe time series followed by various natural phenomena such as river discharge, rainfall, tree rings, distribution of varves in lake sediments or the frequency of sun spots (FEDER, 1988). This kind of motion can also be used to generate new surfaces like watersheds or continents, mountain profiles, the shape of lakes or the form, number and distribution of islands in archipelagos (MANDELBROT, 1975). VOSS (1985) has introduced an algorithm called *successive random addition* for generating this kind of curve for a given value of H, although there are many other ways of doing so (e.g. KIRKBY, 1987). From the practical point of view, the case H = 0.5 is ordinary Brownian motion, and when $0.5 < H \leq 1$ the curves show a *persistence* as high as the value of H. Finally, it has been demonstrated (MANDELBROT, 1985) that these curves are statistically self-affine scaled sets and their fractal dimensions are not uniquely defined. In other words, there is no global fractal dimension, but there are local ones and this is because the fractional Brownian motions are scaled in two variables; in this case river discharge and time. For each case the fractal dimension is:

$$D = 2 - H \qquad (6)$$

At this point it is interesting to describe how figures 5a and 5b were obtained. We began with two series of data and after their transformation into fractional Brownian motions they had the appearance of profiles or chains of mountains. In fact, these are generated irrespective of whether the data series are real or simulated. Can we consider that these profiles reflect something intrinsic or characteristic of the basin?

Obviously they cannot be considered as an idealized reconstruction because we have time series and the basin has a spatial plan. Nevertheless, our tentative answer is affirmative. Can the path of the runoff to the rivers be imagined from figures 5a and 5b? The answer is yes, we only have to imagine how the water runs through the artificial landscapes of figure 5. The random series (Fig. 5 b) shows a more abrupt relief, high peaks and narrow valleys, suggesting a quick response to rainfalls and low storage capacity. In other words, there is no persistence or memory. On the other hand, the real series (Fig. 5 a) has a more even profile, with large plateaus and wide valleys, where water can be retained or stored for more time, meaning persistence or memory.

In practice the Hurst exponent is calculated fitting the power equation $R/S=(aT)^H$ to a set of values of R/S and T obtained from different time lags. In figure 6 the results for the Ter hydrological data, H = 0.68, and random series, H = 0.44, are shown. As was expected, the random series has a value of H near 0.5 which corresponds to ordinary random motion. On the other hand, the Ter data show an intermediate value of H which is very similar to the value found by HURST (1965), H = 0.72 using 94 series of discharges ranging between 10 and 100 years of data. For comparison, the Loire, using the mean monthly flow between 1863 and 1966, has H = 0.69, the Rhine at Basle for the period 1808-1966 has H = 0.5, while for the Nile the value of 0.9 is the highest measured to date (MANDELBROT & WALLIS, 1969).

It is possible to achieve a complementary measure of the Hurst exponents for the hydrological and random series by means of the relationship D = 2 - H, in which D is the local fractal dimension of each record

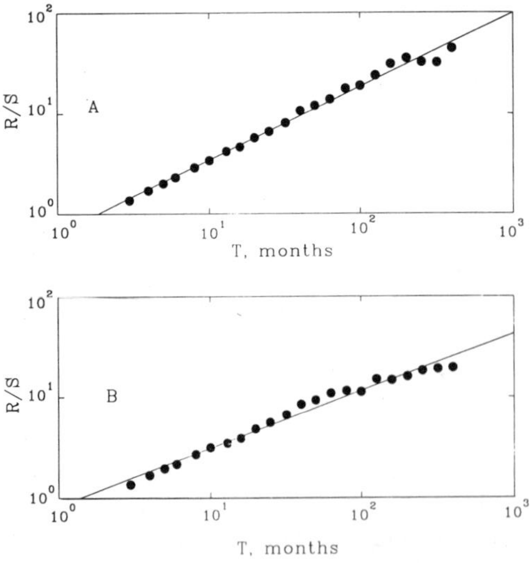

FIGURE 6. R / S as a function of lag T (filled circles) for the series of figure 1 B (A) and 2 A (B). The lines represent fits of the Hurst law $R / S = (aT)^H$ to the observed R / S. For the Ter series (A), H is 0.68, with a correlation coefficient of 0.955, while for the random series H is 0.44, with a value of r = 0.915.

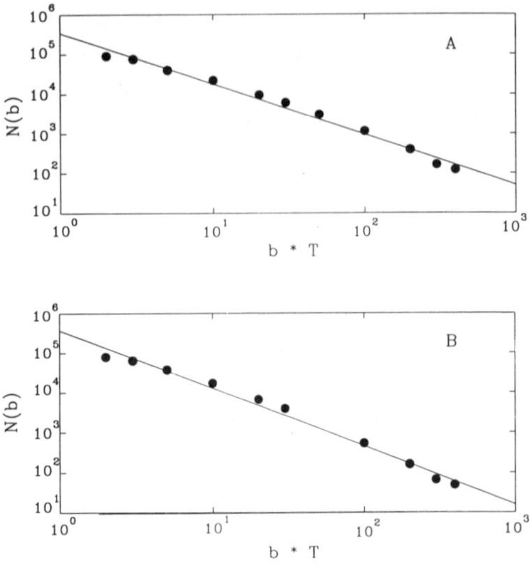

FIGURE 7. Number of "boxes" N (b) as a function of the time scale obtained for the accumulative normalized departures of the Ter series (Fig. 5A), and of the simulated sequence (Fig. 5B). In both cases a minimum "box" of T = 1 month and a = 0.001 Hm^3 have been used and its size was increased by 13 different multiplicative values of b. The curves are fitted to the equation $N(b) = b^{-D}$. In the Ter series (A) the box dimension is 1.38 with a correlation coefficient of 0.997. In contrast, in the random series (B) these values are 1.52 and 0.998 respectively.

of accumulated departures, x(t) (Fig. 5), considered as a Brownian function. As these functions are self-affine records scaled in two variables, x(t) and t, only local fractal dimensions depending on time lags can be calculated. The box counting method to determine the fractal dimension considers that if we cover the x (t) records with a grid of boxes of width bt in time, and of length ba in ordinates, *the box dimension* (D) is defined by the equation,

$$N(b; a', t) = b^{-D} \qquad (7)$$

where N (b; a', t) is the number of boxes we need for covering the record, a and t are the minimum values of the parameters x (t) and t respectively, while b is a multiplicative factor which determines the increasing size of the boxes. In practice, the box dimension (D) is obtained by fitting the equation (7) to a series of values of N(b) and b by linear regression.

Figures 7a and 7b show the results we obtained using the box-counting method from data in figure 5. Because time is an integer, the minimum box width selected was t = 1 month, while a value of 0.001 Hm^3 was chosen for height. The obtained results for these values of a and t are indicated in figures 7 a and 7 b. The fractal dimension for the Ter hydrological record is 1.38 and for the random series 1.52. Both results comply with the expected fractal values according to equation (6) showing slight differences due to the statistical approach.

Since both H and D change when another scale of boxes is taken, the evolution of the box dimension for other sizes of boxes is considered. The value of D remains almost constant when the height of the box is between 0.01 and 0.0001 Hm^3. However, D grows very quickly when the time scale is changed and values higher than two months are considered as the minimum width of the boxes. In figure 8 the values of D are always higher for random series than for the Ter discharge series, but both curves grow asymptotically towards a theoretical maximum of D = 2 (MANDELBROT, 1987). In conclusion, there is a loss of the sensitivity of method when the selected time lag is longer than the minimum time intervals of the data.

High values of persistence can be considered as a measure of the strength of the periodical cycles throughout time - the "Joseph effect"- but they do not give information about the frequency of such cycles. To determine these frequencies, we have calculated the autocorrelation function (ACF) and the periodogram of the original series of hydrological data. The results of the ACF are shown in figure 9. Because of the noise of the series, the ACF (Fig. 9 a) does not show any characteristic period. To avoid this effect, the series was filtered by moving average (MA) process using three different spans: 3 months (MA3), 6 months (MA6) and 1 year (MA12). In each case, a certain amount of information is lost but the most important cycles persist (Fig. 9 b-d). The results show the existence of high correlations for periods of 1, 2 and 3 years, then there is a gap until around 5.5, 8.6, 10.1 and 11,7 years. These longer term fluctuations can be considered as an expression of the cycles of *circa* 7 and 11 years that are characteristic in many climatic events in all the Mediterranean region (MARGALEF, 1977). Because of the high correlations obtained for short time periods, these cycles do not appear in the ACF. For this reason the periodogram and the spectral density of the Ter discharges, logarithmically transformed, have been calculated by Fourier analysis. In the spectral density a Tukey window (DENMAN, 1975) with a span of 12 was used. The results (Fig. 10) show little difference from the values obtained by ACF for long-term cycles, that is 5.8, 8.7 and 11.7 years. Nevertheless, this method gives a more precise measure of the importance of seasonal cycles. The cycles of 3 and 6 months are the most persistent,

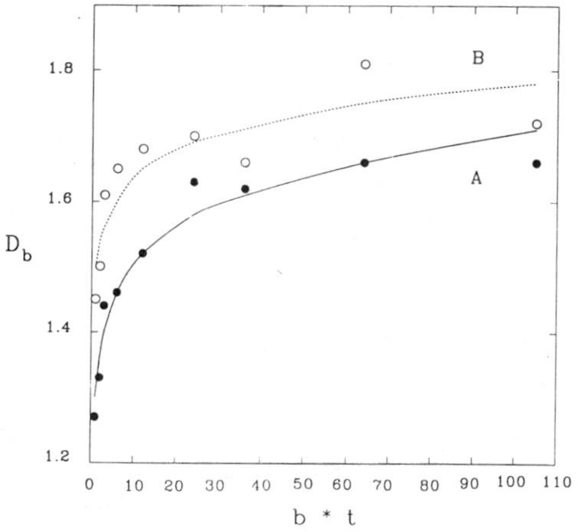

FIGURE 8. Evolution of the box-dimension as a function of the time scale of the "boxes": A) for the Ter data (filled circles) and B) for the random sequence (hollow circles). In both cases a minimum value of 0.001 for x (t) have been taken. The data are fitted to a logarithmic equation to show their trend.

but there are also other short cycles of around these periods which can be produced by small differences in the seasonality of each year.

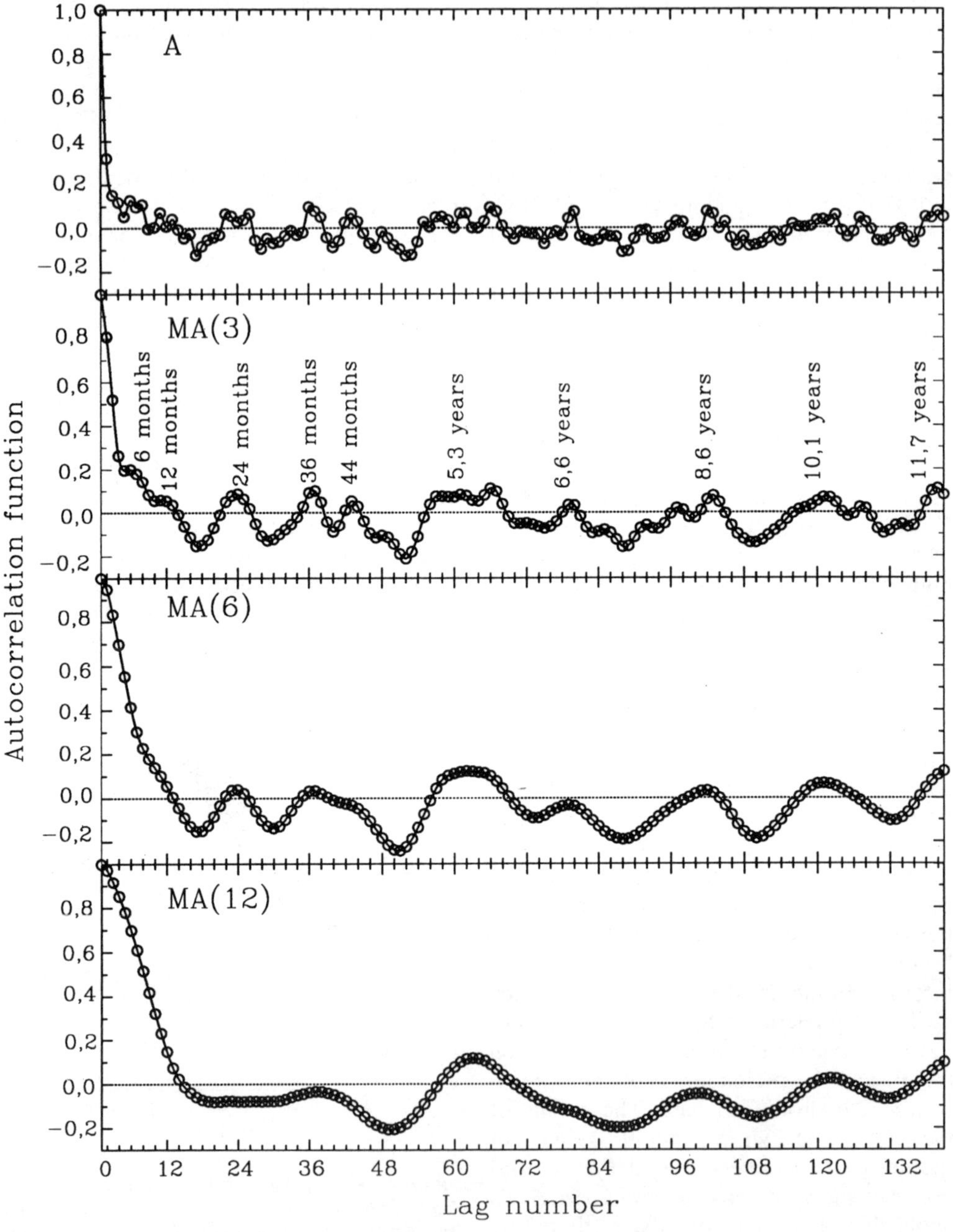

FIGURE 9. Autocorrelation functions of the River Ter monthly water flow (A) and the same series but filtered by a moving average process with a 3 month span MA (3), 6 month span MA (6) and a 1 year span MA (12). The periods which represent the most conspicuous peaks are indicated.

DISCUSSION

To date, the morphology of the catchment has been considered the main cause of the persistence in hydrological fluctuations. This is the simplest version of the storage model: the water runs off more slowly in the floodplains and even reliefs than in mountainous areas with high slopes. In a more realistic model the discharges are the sum of a series of independent processes in decreasing order of importance. For instance, the first factor may be the existence of natural stores of geological origin (karsts); the second, morphology (the slopes); the third, climatic factors (snow accumulation); the fourth, microclimatic factors (soil moisture), and so on, until all possible factors have been taken into account. The fractal nature of the basin may also contribute to the fractal behavior of the river discharge (FEDER, 1988). The result is that the discharge of a river depends not only on recent precipitation but also on earlier rainfalls and, at a theoretical limit, it will be independent of recent rainfall.

In lakes, the intensity of the fluctuations produced by climatic oscillations is regulated by their size (GOLDMAN *et al.*,

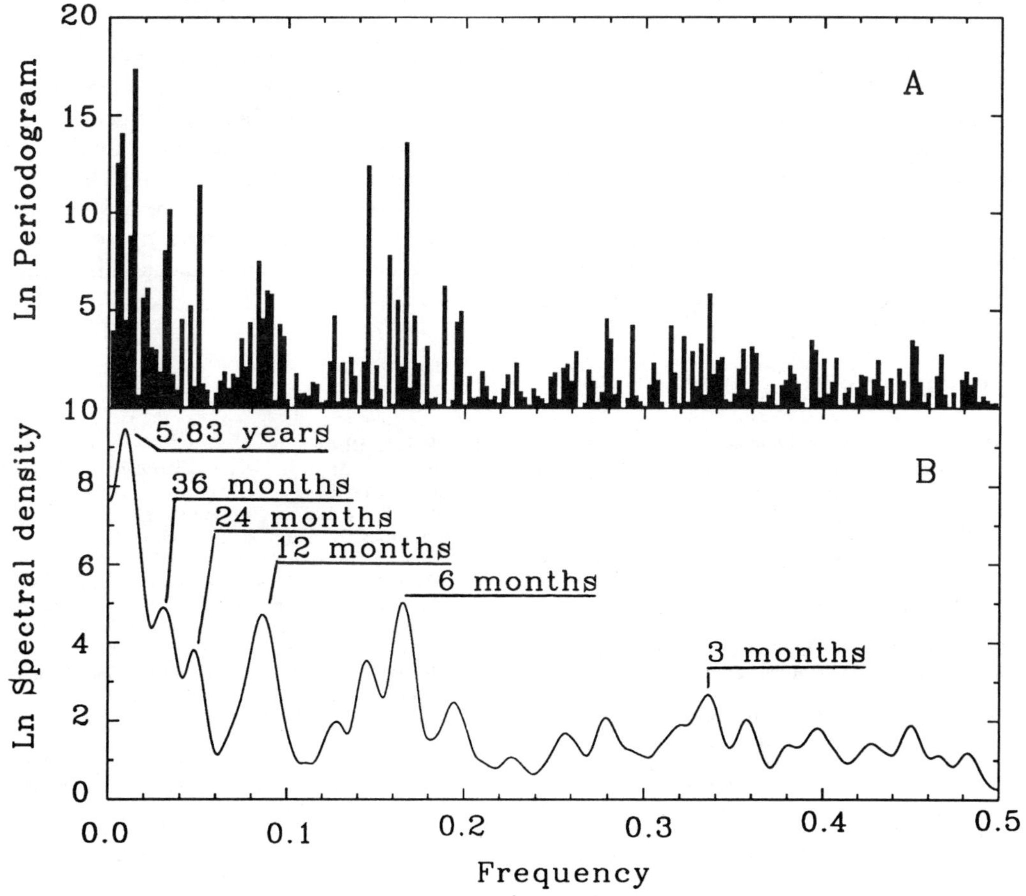

FIGURE 10. Periodogram (A) and spectral density (B) plots of the results obtained by a Fourier analysis from the original Ter data, logarithmically transformed. The periods which correspond to the largest cycles are indicated.

1988; POWELL, 1989). In a similar way, the size of the catchment is an important factor determining the persistence of the hydrological series due to the possibility of integrating events taking place in different subbasins.

It is interesting to consider the effect of various human activities on the basin to determine what effects may be expected on persistence. The channeling of rivers leads to destruction of floodplains; the pumping of groundwater destroys various natural stores, while the construction of reservoirs increase them. In a similar way, the areas without vegetation and with eroded soils do not have water retention capacity and the persistence is close to values of 0.5.

As for the communities, hydrological events are a key factor in determining their persistence in the river. According to MARGALEF (1980), the organisms make their own Fourier analysis to foresee the environmental fluctuations. In fact, first they make a factorial analysis to identify the key factor and the correlated ones. Accordingly, communities in a river are an example of physical - biological coupling. The species try to couple their biological cycles to temporal scales of variability of physical (= climatic) events. First of all, they need to know if prediction is possible. If it is, that means $0.5 < H < 1$ or $0 < H < 0.5$, and then natural selection can act. This coupling is as strong as the value of H.

In conclusion, it is easy to be a prophet in the Nile with $H = 0.9$, and it may be possible to have some chance in the Ter ($H = 0.73$), but it is better not to bet on rivers with a value of H close to 0.5.

ACKNOWLEDGEMENTS

We express our gratitude to Professor R. Margalef for giving us an idea of the value of the study of environmental fluctuations in understanding the structure and dynamics of communities. His ideas (MARGALEF, 1977, 1980, 1983, 1988) were a stimulus for the study of the Ter river hydrological series. It is a pleasure to dedicate this paper to him.

REFERENCES

ARMENGOL, J. 1988. *Manipulació física i orgànica de masses d'aigua continentals*. Tretzè Congrès de Metges i Biòlegs de Llengua Catalana. Llibre de Ponències: 194-207.

ARMENGOL, J., CRESPO, M., MORGUÍ, J.A. & VIDAL, A. 1986. Phosphorus budgets and forms of phosphorus in the Sau reservoir sediment: An interpretation of the limnological record. *Hydrobiologia*, 143: 331-336.

BOX, G.E.P. & COX, D.R. 1964. An analysis of transformations. *J. Royal Statistical Soc.* (B), 26: 211-252.

DENMAN, K.L. 1975. Spectral Analysis: A Summary of the Theory and Techniques. *Fish. Mar. Serv. Res. Dev. Tech. Rep.*, 539

D.G.O.H. 1972, 1973, 1974, 1975. *Aforos. 10 Cuenca del Pirineo Oriental. Anuarios* 66-67 a 69-70.

D.G.O.H. 1976. Aforos y aguas subterraneas. 10 Cuenca del Pirineo Oriental. Período 1970-71 a 1973-74.

FEDER, J. 1988. *Fractals*. Plenum Press. NewYork

GOLDMAN C.R., JASSBY, A. & POWELL, T.M. 1989. Interannual fluctuations in primary production: Meteorological forcing at two subalpine lakes. *Limnol. Oceanogr.*, 34:308-321.

HURST, H.E, BLACK, R. & SIMAIKA, Y.M. 1965. *Long-Term Storage: An experimental Study*. Constable. London.

KIRKBY, M.J. 1987. The Hurst effect and its implications for extrapolating process rates. *Earth Surface Processes and Landforms*, 12:57-67

MANDELBROT, B. B. 1975. Stochastic models of Earth relief, the shape and the fractal dimension of coastlines, and the number-area rule for islands. *Proc. Nat. Acad. Sci. USA.*, 72:3825-3828

MANDELBROT, B. B. 1985. Self-affine fractals and fractal dimension. *Phys. Scr.*, 32:257-260

MANDELBROT, B. B. 1987. *Los objetos fractales*. Tusquets. Barcelona

MANDELBROT, B. B. & VAN NESS, J. W. 1968. Fractional Brownian motions, fractional noises and applications. *SIAM Review*, 10:422-437

MANDELBROT, B. B. & WALLIS, J. R.1968. Noah, Joseph and operational hydrology. *Water Resources Research*, 4:909-918

MARGALEF, R. 1977. *Ecología*. Omega. Barcelona

MARGALEF, R. 1980. *La biosfera, entre la termodinámica y el juego*. Omega. Barcelona.

MARGALEF, R. 1983. *Limnología*. Omega. Barcelona

MARGALEF, R. 1988. *An Image of the Biosphere*. Unpublished manuscript.

POWELL, T.M. 1989. Physical and biological scales of variability in Lakes, Estuaries and the Coastal Ocean. In: *Perspectives in Ecological Theory*. (J. Roughgarden, M. May & S.A. Levin, eds.): 157-176 Princeton University Press. Princeton

PUIG, M. A; GONZÁLEZ, G. & RECASENS, L. 1987. Modelos de distribución de plecópteros, efemerópteros, tricópteros y simúlidos en el río Ter. *Limnetica*, 3: 125-132

SAUPE, D. 1988. Algorithms for random fractals. In: *The science of fractal images* (H.O. Peitgen & D. Saupe, eds.): 71-136. Springer. New York.

SCHINDLER, J. E. 1988. Freshwater Ecosystem: A Perspective. In: *Concepts of Ecosystem Ecology*. (L.R. Pomeroy & J.J. Alberts, eds.): 57-74. Springer. New York

SABATER, F. & ARMENGOL, J. 1986. Chemical characterization of the Ter river. *Limnetica*, 2:75-84

SABATER, S., SABATER, F. & ARMENGOL, J. 1988. Relationships between Diatom Assemblages and Physico- chemical Variables in the River Ter (N.E. Spain). *Int. Revue ges. Hydrobiol.*, 73:171-179

VIDAL, A. 1977. Eutrofización del embalse de Sau en el transcurso de sus primeros años (1963-1972). *Trab. II Asamblea Nac. Geodesia y Geofísica*: 1955-1997.

VIDAL, A. 1987. Impacto de distintas fuentes de fósforo en la eutrofización del embalse de Sau. *XVIII Jornadas Comité Español de la Detergencia, Tensoactivos y Afines*. Barcelona.

VIDAL, A. 1989. El cas de l'embassament de Sau. In: *Sistemes Naturals. Història Natural dels Països Catalans*. XIV. (J. Terradas, N. Prat, A. Escarré & R. Margalef, eds.): 434-441. Enciclopèdia Catalana. Barcelona.

VOSS, R. F. 1985. Random fractals forgeries. In: *Fundamental Algorithms in Computer Graphics*. (R.A. Earnshaw, ed.): 805-835. Springer. Berlin.

(J. D. Ros & N. Prat, eds.). 1991. *Oecologia aquatica*, 10: 35-43

THE HYPORHEIC ZONE CONSIDERED AS AN ECOTONE

FRANCESC SABATER[1] & PETER B. VILA[2]

[1] Departament d'Ecologia. Facultat de Biologia. Universitat de Barcelona. Av. Diagonal, 645. 08028 Barcelona. Spain.
[2] Institute of Ecology. University of Georgia. Athens. Georgia 30602. USA.

Received: August 1990

ABSTRACT

Physical, chemical, or biological exchanges that occur between the groundwater system and surface waters pass through the interstitial sediments underlying the river channel. This paper assesses the role these interstitial sediments can have in lotic ecosystem functioning. In addition to regulating energy and material exchange in lateral and upstream-downstream linkages, hyporheic sediments support an important biocenosis and can be an important site for evolution and colonization. The hyporheic zone is a transitional zone of boundary control between adjacent ecological systems and may be considered as an ecotone.

KEY WORDS: Hyporheos, rivers, meiofauna, ecotone.

INTRODUCTION

Interstitial alluvial sediments of lotic systems and sediments associated with riverine aquifers harbor numerous fauna (KARAMAN, 1935; HERTZOG, 1936; CHAPPUIS, 1946; MOTAS, 1958; HUSMANN, 1966). Previous studies have been descriptive and have emphasized the organisms' vertical and horizontal distribution. While the habitat has been considered as a biocenosis, little attention has been given to the role the biocenosis has in fluvial system dynamics.

This article considers the hyporheic habitat as an ecotone and emphasizes the importance of hyporheic sediments on mediating interactions between running waters and contiguous groundwaters. Similarly, WARD (1989) conceptualized the vertical dimension in lotic systems as exchanges of matter and energy between the channel and contiguous groundwater. Since interstitial alluvial sediments may extend laterally from the river channel for large distances (STANFORD & WARD 1988), they may have a large influence on neighboring regions and impact lotic function.

THE HYPORHEIC ZONE AS A SUPPORT FOR AN IMPORTANT BIOCENOSIS

Since the 1950's, European and North American researchers have reported an important biocenosis, called hyporheos, in the gravel beds of rivers and lake shores (ORGHIDAN,1955; WILLIAMS & HYNES, 1974). SCHWOERBEL (1961) defines the hyporheic zone as a practical term for a superficial groundwater habitat underlying a running water bed. Subsequent work has indicated a true hyporheal fauna confined below the water-substrate interface (HYNES, 1983). Besides an important biomass of bacteria and fungi, common fauna are protozoa, invertebrate worms, rotifera, crustacea and insect larvae, which may be distributed as deep as 70 cm vertically within the substrata (ANGELIER, 1953) and up to 60 cm laterally on the shore (HYNES, 1970).

There is a numerous and widespread fauna in the hyporheic zone. For example, O'DOHERTY (1988) and PERLMUTTER (1988) describe a community rich in species and with densities of up to 40,000 ind m^{-2} in the low order streams of the Southern Appalachian Mountains, U.S.A. A remarkable example is given by STANFORD & GAUFIN (1974). In their study the stonefly *Paraperla frontalis* is found only in subsurface samples and leaves the interstitial medium for emergence. A large variety of taxa (Sabater, unpublished data) were obtained at eighteen sites along the river Ter (Spain) from a depth of 20-50 cm with a Bou Rouch pump (BOU, 1974). Proportions of each taxon at each site are shown in figure 1.

In rivers with extensive hyporheos, standing crop biomass could easily exceed benthic epigean biomass. For example, STANFORD & GAUFIN (1974) found that the hyporheic zone extends laterally up to two km and to a depth of 10 m from the river. As expected when sampling a new habitat, new species not found in surface water are encountered in the hyporheic layers (DUMONT, 1983; SABATER 1987a, 1987b; SABATER & DE

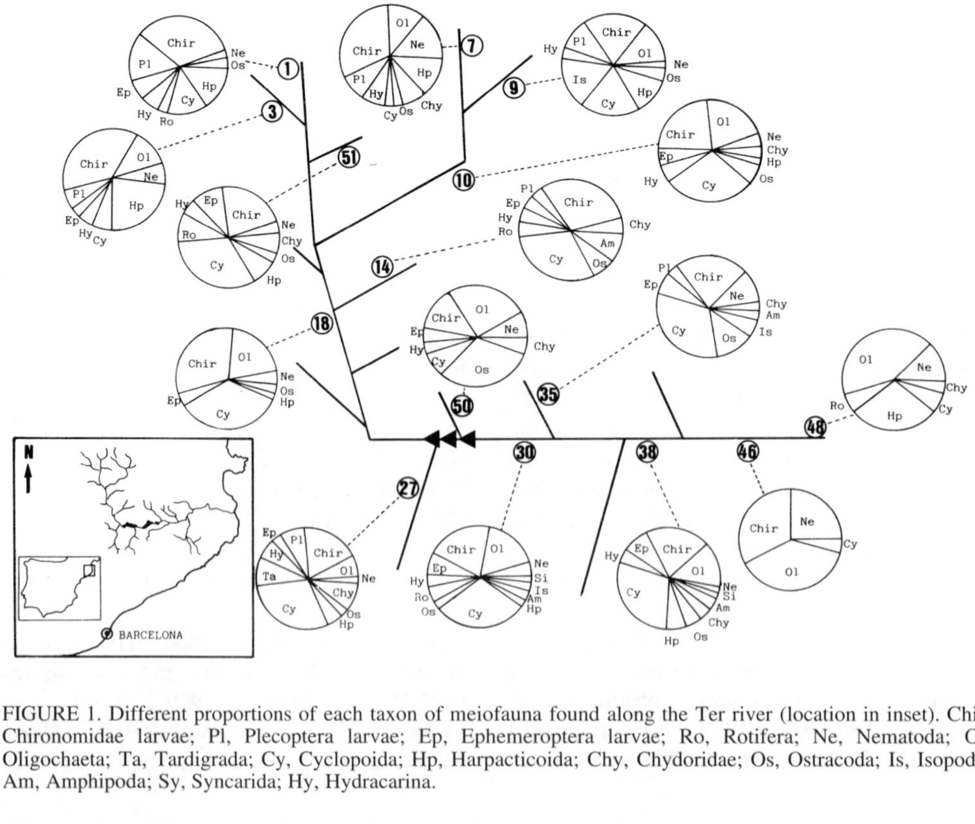

FIGURE 1. Different proportions of each taxon of meiofauna found along the Ter river (location in inset). Chir, Chironomidae larvae; Pl, Plecoptera larvae; Ep, Ephemeroptera larvae; Ro, Rotifera; Ne, Nematoda; Ol, Oligochaeta; Ta, Tardigrada; Cy, Cyclopoida; Hp, Harpacticoida; Chy, Chydoridae; Os, Ostracoda; Is, Isopoda; Am, Amphipoda; Sy, Syncarida; Hy, Hydracarina.

MANUEL, 1989).

A mixed community of stygobiont and epigean river species colonize the hyporheic zone temporarily or permanently (CHAPPUIS, 1950; WILLIAMS & HYNES, 1974; DANIELOPOL, 1976, 1984, 1989; DOLE, 1983; PENNAK & WARD, 1986). Hypogean (subterranean) taxa, called stygobionts, are abundant in the hyporheic zone. These organisms have developed specific morphological specializations and requirements (see MARGALEF, 1976; HOLSINGER, 1988). An important reference on stygofauna is BOTOSANEANU (1986).

Surface animals may enter the interstitial medium to find food or refuge from physical (SLACK, 1955) or biological adversities. STANFORD & WARD (1988) observed overlap between epigean and stygobiont taxa in several wells located in the hyporheic zone. CREUZE des CHATELLIERS & MARMONNIER (1990) found that physical perturbations caused high numbers of Cladocera and Ostracoda to enter interstitial layers. VILA (1989) considered that the large number of Cladocera species found in the surface sediments of several Indiana streams could, in part, be explained by retreat to deeper sediments after an increase in discharge.

THE ROLE OF THE HYPORHEIC HABITAT IN PROVIDING A PATHWAY FOR ACTIVE COLONIZATION

Surface flow and groundwater have a dynamic and seasonally variable influence on the hyporheic habitat due to hydrologic disturbances such as spates and droughts, that alter the physical structure of hyporheic sediments. A disturbance can affect energy flows into and out of the sediments. For example, high discharge could restructure sediments and either deplete or recharge sediments with detritus. Consequently, community structure and faunal distribution in the hyporheic zone are dynamic and vary between sites and seasonally within a site depending upon the disturbance regime (DOLE, 1983, 1985; WILLIAMS, 1984). Sampling regimes must be structured accordingly and such dynamics complicate extrapolations of the results of studies that only consider a single point in time or space.

Biota may navigate through the interstitial waters by following several factors or combination of factors; thermal or ion concentration gradients (STANFORD & WARD, 1988); food quality or quantity (e.g. detritus); or by following a preferred prey species. Organisms can also transform their environment, e.g. pelletization of sediment that affects water circulation (DANIELOPOL, 1984).

The hyporheic zone can be a mixing ground important in evolution, large scale migratory patterns and survival in a habitat without large predators. RUFFO (1961) considered that the interaction of groundwaters and water in the hyporheic zone could allow for the migration of organisms between different aquatic systems on an evolutionary time scale (Fig. 2). So, although not presently connected, modern faunal assemblages may have been shaped from interactions permitted by this "interstitial highway" (WARD, 1989). ROUCH & DANIELOPOL (1987) suggested that some surface species inhabiting interstitial waters are already preadapted morphologically and

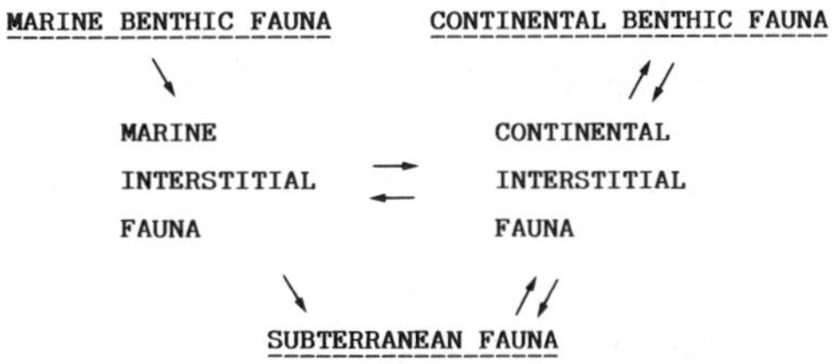

FIGURE 2. Scheme showing pathways on an evolutionary time scale for the migration of organisms between groundwaters, hyporheic zone, and surface waters.

physiologically to colonize groundwater habitats. For example, SKET (1985) pointed out that several crustacean species are able to tolerate low oxygen conditions. DANIELOPOL (1989) mentioned other examples.

The study of phylogenetically related species may give an insight into the origin and evolution of the stygobionts as well as the degree of adaptation to the subsurface environment. MARGALEF (1983) suggested that many hyporheal species found refuge from cold weather in meridional basins during the Pleistocene glaciations. These basins were later centers of dispersion and recolonization.

IMPORTANCE OF THE HYPORHEIC ZONE IN AFFECTING PROCESSES OCCURRING IN THE RIVER CHANNEL

The hyporheic zone can have a dynamic flow regime. At high discharge, water moves from the river channel into the hyporheic zone, whereas at low discharge groundwater is discharged from the hyporheic zone into the river. Groundwater outflow contributes to the basal flow in most river channels (HYNES, 1983). There is evidence that the thickness of the hyporheic zone is related to the amount and extent of discharge and recharge areas in the river channel (GODBOUT & HYNES, 1982). Larger areas of discharge and recharge occur in coarse sediments than in sandy sediments (MESTROV & LATTINGER-PENKO, 1981).

Changes in the geomorphology along the longitudinal river profile result in a pattern of alternate areas of upwelling and recharge. Further understanding of species distribution and processes such as nutrient dynamics will require more detailed knowledge of matter flux and hydrodynamics at each hydrologically distinct stream section. For example, preliminary mass transport calculations indicate that nutrient discharge from the hyporheic zone may be crucial for the enhancement of phytobenthos production (STANFORD & WARD, 1988). The nutrient flux in upwelling zones may potentially result in biological "hot spots" (increased production, high species number), paralleling those reported in the upwelling areas of marine systems. In streams, however, site specific variables such as strong lateral or upstream inputs may prove to be overriding factors.

The hyporheic zone may also be a nutrient sink. HILL (1981) reported a fifty percent reduction in nitrate attributed to bacterial denitrification in deep sediments. Hyporheic metabolism may contribute significantly to total metabolism and could impact surface recharge nutrient dynamics. However, few studies have considered its importance (GRIMM & FISHER, 1984).

Conversely, nutrient uptake in surface waters may impact recharge waters. For example, GRIMM *et al.* (1981) attributed nitrate uptake in desert stream reaches to algal activity. There has been little work on the role of potamoplankton on nutrient or carbon dynamics, especially in the lower reaches of large rivers. Transfer of floodplain bacterial inputs into ciliate biomass (CARLOUGH & MEYER, 1989) would increase carbon spiraling distance as animals drift downstream. As in lakes, there may also be tight coupling between the "leaking" of nutrients from zooplankton and phytoplankton excretion and uptake as the "slugs" move downstream. Recent channelization and reduced floodplain interaction due to human activities in lower reaches may have reduced the role of the hyporheic zone in mediating interchange between surface and groundwater and increased the importance of water column processes.

WILLIAMS & HYNES (1974) considered the hyporheic zone as a sink for fine particulate organic matter (FPOM; e.g. < 50 µm) from the channel. However, FPOM decomposition supports the ground water food chain and could constitute a nutrient subsidy to primary production in areas of the river channel where hyporheic waters are discharged (STANFORD & WARD, 1988). STANFORD & WARD (1988) found that the annual nitrogen load to channel water in the Flathead River is increased 75% at baseflow due to aquifer discharge. RUTHERFORD & HYNES (1987) suggested that dissolved organic carbon (DOC) fluctuations at baseflow can be explained by the stream beds capacity to remove surface water DOC. LUSH & HYNES (1978) and LOCK *et al.* (1984) observed rapid DOC uptake by substrata suggesting an active biological uptake in the hyporheic zone.

HUSMANN (1978) has stressed that well oxygenated hyporheic areas are biologically active zones that contribute to the self purification of organically polluted waters; similar processes occur in waste treatment plants. Such considerations cannot be ignored in studies related to the spiraling concept (NEWBOLD *et al.*, 1982). However, CROCKER & MEYER (1987) have stressed the important role of hyporheic sediments in generating DOC from *in situ* POM decomposition.

In general, nutrient concentrations are significantly higher in interstitial waters than in surface waters (HYNES, 1983; VALET *et al.*, 1990). Phosphate and nitrogen concentrations in the River Ter are higher in the hyporheic than in surface waters (Fig. 3). Given the importance of nutrients to productivity, a large hyporheic nutrient pool could result in an important contribution to the system metabolism. This suggests that physical and biological

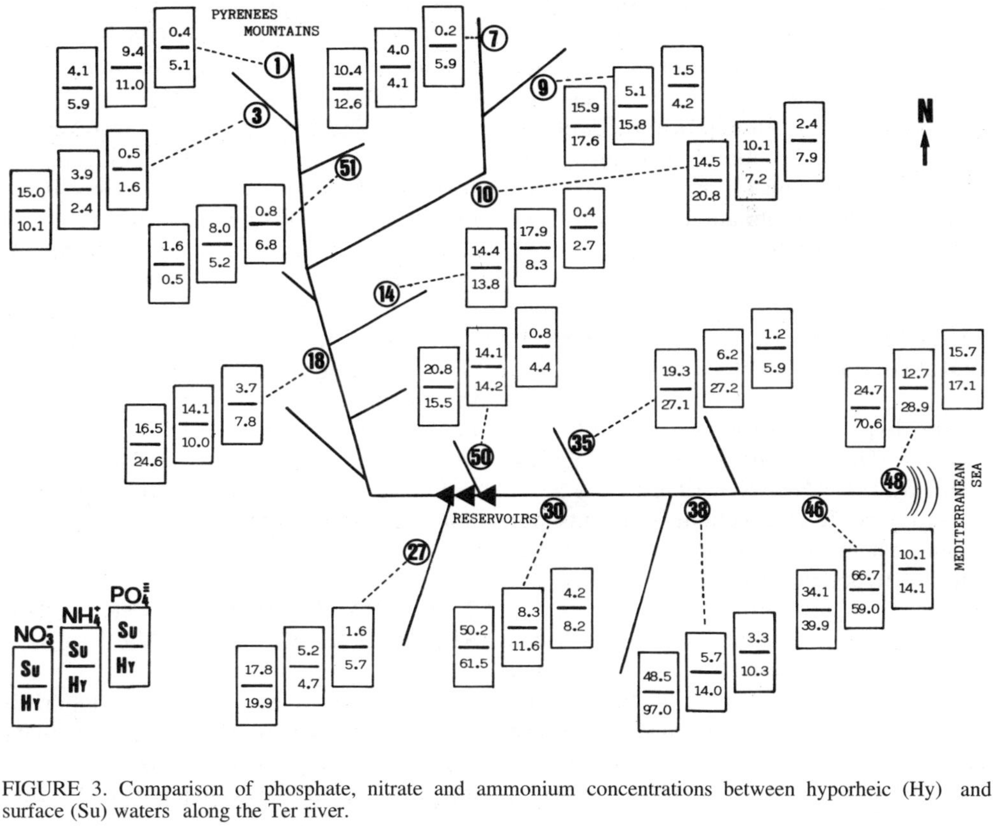

FIGURE 3. Comparison of phosphate, nitrate and ammonium concentrations between hyporheic (Hy) and surface (Su) waters along the Ter river.

processes occurring within the hyporheic zone should be included in a conceptualization of stream ecosystem function. Even preliminary research comparing nutrient levels in interstitial waters with levels entering the river channel would increase our understanding of nutrient dynamics in streams.

The great variety of organisms inhabiting the hyporheic zone utilize food carried by both active (i.e., mediated by organisms) and passive transport in waters moving vertically and laterally from the stream channel. The vertical distribution of these animals is determined by external energy, the velocity of flow, and the diffusion of gases or matter through the alluvial sediments along the river (MARGALEF, 1983). Distinct vertical and horizontal faunal distribution patterns may even be strongly correlated with environmental gradients (FENCHEL, 1978; WILLIAMS, 1989). Even though standing crop biomass in the hyporheic zone may or may not exceed benthic biomass in surface sediments, the impact of the interstitial organisms on ecosystem processes could be substantial. For example, meiofauna were estimated to account for approximately one half the carbon assimilation in Mirror Lake (STRAYER & LIKENS, 1986).

THE HYPORHEIC ZONE WITHIN A BOUNDARY PERSPECTIVE

The hyporheic zone has the following features (DANIELOPOL, 1982; HYNES, 1983; BOTOSANEANU, 1986):

Darkness. It is functionally a sink for organic matter. By affecting water turnover rate the hyporheos can impact energy and matter transport through the hyporheic zone. It is an open functional system since it mediates physical, chemical and biological exchange between the ground water system and the surrounding surface environment; it is trophically impoverished and energy is mainly, or exclusively, allocthonous; physical variations in factors such as temperature and water flow are dampened and lag behind those occurring in neighboring surface habitats; there are gradients or discontinuous patches in physical parameters such as granulometry, conductivity or water salinity.

The widespread distribution and mobility of the hyporheic fauna coupled with the spatial and temporal independence of major external inputs of physical and chemical parameters from the sediments suggests that a precise definition and demarcation of the borders of the hyporheic zone is difficult, if not impossible (WILLIAMS, 1984; STANFORD & WARD, 1988; DANIELOPOL, 1989; TRISKA *et al.*, 1989). It it therefore difficult to define ecologically a distinct hyporheic zone.

We propose that the hyporheic zone be viewed in a functional process framework as a significant component of the riverine landscape. Using ideas presented in NAIMAN *et al.* (1988), we believe that the hyporheic zone can be envisioned as a zone of transition between adjacent ecological systems. The hyporheic zone provides an example of boundary control on the function of the lotic ecosystem, since it can modify the direction, character and magnitude of materials and information exchanged between surface and subterranean zones along the river. Knowledge of the degree and strength of surface and groundwater interaction with the hyporheic zone would help us to understand processes such as nutrient loading and the regulation of matter and exchange in lateral and upstream-downstream linkages.

THE HYPORHEIC ZONE CONSIDERED AS AN ECOTONE

As a transitional zone, the hyporheic zone possesses specific physical, chemical and biological properties defined by space and time scales that can regulate the flow of energy and matter between different resource patches. The functional properties are as follows: 1) It can support high faunal densities and provide a critical habitat for some species. 2) It may be a refuge in which riverine zoobenthos escape environmental disturbances (e.g., spates, drought). 3) It contains a mixed assemblage of aquatic epigean fauna and other fauna from surrounding systems. 4) It contains a genetic pool for evolution. 5) It extends spatially allowing for species migration. 6) It is an important area for energy and material flux (e.g., nutrient cycling and decomposition). 7) It contains high concentrations of biologically important resources (e.g., nutrients, carbon). 8) It influences and is impacted by the hydrologic regime in the surrounding surface and groundwater areas. 9) It tends to dampen the amplitude of physical parameters. 10) It possesses the ability to resist change or a resilience to disturbance.

Finally, groundwater studies have two main aims of fundamental biological importance. First, interstitial biotopes provide evolutionary pathways linking groundwaters and adjacent surface freshwaters and marine environments (DELAMARE-EBOUTTEVILLE, 1957; PENNAK, 1968; HUSMANN, 1971). Attributes of the hyporheic zone are of considerable interest in investigating questions regarding the ecology, evolution and zoogeography of hyporheic fauna and questions concerning functional processes. Second, hypogean forms may be important as indicators of groundwater quality (DANIELOPOL, 1980).

ACKNOWLEDGEMENTS

We gratefully acknowledge the influence and ideas of Professor R. Margalef and would like to thank him for the encouragement and enthusiasm he communicated to us for the study of biological systems and in particular for the study of the interstitial sediments.

REFERENCES

ANGELIER, E. 1953. Recherches écologiques et biogéographiques sur la faune des sables submergés. *Arch. Zool.ogie Expér. Génér.*, 90: 37-162.

BOTOSANEANU, L. 1986. *Stygofauna Mundi.* Brill. Leiden.

BOU, C. 1974. Les méthodes de récolte dans les eaux souterraines interstitielles. *Annls. Speleol.*, 29: 611-619.

CHAPPUIS, P. A. 1946. Un nouveau biotope de la faune souterraine aquatique. *Bull. Sect. Sci. Acad. Roumanie*, 29: 1-8.

CARLOUGH, L. A. & MEYER, J. L. 1989. Protozoans in two southeastern blackwater rivers and their importance to trophic transfer. *Limnol. Oceanogr.*, 34: 163-177.

CHAPPUIS, P. A. 1950. La récolte de la faune souterraine. *Notes Biospéleologiques*, 5: 7-37.

CREUZE des CHATELLIERS, M. & MARMONIER, P. 1990. Macrodistribution of Ostracoda and Cladocera in a by-passed channel: Exchange between superficial and interstitial layers. *Stygologia*, 5:17-24.

CROCKER, M. T. & MEYER, J.L. 1987. Interstitial dissolved organic carbon in sediments of a Southern Appalachian headwater stream. *J. North Amer. Benthol. Soc.*, 6: 159-167.

DANIELOPOL, D. L. 1976. The distribution of the fauna in the interstitial habitats of riverine sediments of the Danube and the Piesting (Austria). *Internat. J. Speleology*, 8: 23-51.

DANIELOPOL, D. L. 1980. The role of the limnologist in groundwater studies. *Internat. Rev. ges. Hydrobiol.*, 69: 729-746.

DANIELOPOL, D. L. 1982. Phreatobiology reconsidered. *Polsk. Archi. Hydrobiol.,* 29: 375-386.

DANIELOPOL, D. L. 1984. Ecological investigations on the alluvial sediments of the Danube in the Vienna area. A phreatobiological project. *Verh. Internat. Verein. Theor. Angew. Limnologie*, 22: 1755-1761.

DANIELOPOL, D. L. 1989. Groundwater fauna associated with riverine aquifers. *J. North Amer. Benthol. Soc.*, 8: 18-35.

DELAMARE-DEBOUTTEVILLE, C. 1957. Lignées marines ayant pénétré dans les eaux souterraines continentales. Un problème de biogéographie actuelle. *C. R. Somm. Soc. Biogeogr.*, 296-297: 53-67.

DOLE, M. J. 1983. Le domaine aquatique souterrain de la plaine alluviale du Rhône à l'Est de Lyon. 1. Diversité hydrologique et biocénotique de trois stations réprésentatives de la dynamique fluviale.

Vie Milieu, 33: 219-229.

DUMONT, H. J. 1983. Discovery of groundwater inhabiting Chydoridae (Crustacea: Cladocera), with the description of two new species. *Hydrobiologia*, 106: 97-106.

FENCHEL, T. M. 1978. The ecology of micro and meiobenthos. *Ann. Rev. Ecol. Sys.*, 9: 99-121.

GODBOUT, L. & HYNES, H.B.N. 1982. The three dimensional distribution of the fauna in a single riffle in a stream in Ontario. *Hydrobiologia*, 97:87-96.

GRIMM, N. B. & FISHER, S.G. 1984. Exchange between interstitial and surface water: Implications for stream metabolism and nutrient cycling.. *Hydrobiologia,* 111: 219-228.

GRIMM, N. B., FISHER, S.G. & MINCKLEY, W.L.. 1981. Nitrogen and phosphorus dynamics in hot desert streams of Southwestern USA. *Hydrobiologia*, 83: 303-312.

HERTZOG, L. 1936. Crustacés des biotopes hypoges de la valle du Rhin d'Alsace. *Bull. Soc. Zool. France,* 61: 350-372.

HILL, A. R. 1981. Nitrate-nitrogen flux and utilization in a stream ecosystem during low summer flows. *Can. Geogr.*, 25: 225-239.

HOLSINGER, J. R. 1988. Troglobites: The evolution of caved-weling organisms. *American Scientist*, 76: 147-153

HUSMANN, S. 1966. Versuch einer ökologischen Gliederung des interstitiellen Grundwassers im Lebensbereiche eigener Praung. *Arch. Hydrobiologie*, 62: 231-268.

HUSMANN, S. 1971. Eine Rammpumpe zur Untersuchung des Subaquatischen Stygorheals. *Gewasser Abwasser*, 50/51: 115-119.

HUSMANN, S. 1978. Die Bedeutung der Grundwasserfauna für biologische Reinigungs-vorgange im Interstitial von Lockergestein. *G.W.F. Wass. Abwass.*, 119: 293-302.

HYNES, H.B.N. 1970. *The ecology of running waters*. Liverpool University Press. Liverpool.

HYNES, H.B.N. 1983. Groundwater and stream ecology. *Hydrobiologia*, 100: 93-99.

KARAMAN, S. 1935. Die Fauna unterirdischen Gewasser Iugoslawiens. *Ver. Internat. Verein. Theor. Angew. Limnologie*, 7: 46-73.

LOCK, M. A., WALLACE, R.R., COSTERTON, J.W., VENTULLO, R.M. & CHARLTON, S.E. 1984. River epilithon: toward a structural-functional model. *Oikos*, 42: 10-22.

LUSH, D. L. & HYNES, H.B.N. 1978. The uptake of dissolved organic matter by a small spring stream. *Hydrobiologia*, 60: 271-275.

MARGALEF, R. 1976. Paralelismo entre la vida de las cavernas y la de las grandes profundidades marinas. *Bol. Soc. Hist. Nat. Baleares*, 21: 10-20.

MARGALEF, R. 1983. *Limnología.* Omega. Barcelona.

MESTROV, M. & LATTINGER-PENKO, R. 1981. Investigation of the mutual influence between a polluted river and its hyporheic. *Int. J. Speleol.*, 11: 159-171.

MOTAS, C. 1958. Freatobiologia o nova ramura a limnologiei (Phreatobiology a new field of limnology). *Nature*, 10: 95-105.

NAIMAN, R. J., DECAMPS, H., PASTOR, J. & JOHNSTON, C.A. 1988. The potential importance of boundaries to fluvial ecosystems. *J. North Amer. Benthol. Soc.*, 7: 289-306.

NEWBOLD, J.D., MULHOLLAND, P.J., ELWOOD, J.W. & O'NEILL, R.V. 1982. Organic carbon spiralling in stream ecosystems. *Oikos*, 38: 266-272.

O'DOHERTY, E. 1988. *The ecology of meiofauna in an Appalachian headwater stream.* Ph.D. Thesis. University of Georgia.

ORGHIDAN, T. 1955. Un nou domenniu de viata acvatica subterrana: "biotopul hiporeic". *Bull. Sect. Biol. Agr., Sect. Geol. Geogr.*, 7: 657-676.

PENNAK, R. W. 1968. Historical origins and ramifications of interstitial investigations. *Trans. Amer. Microsc. Soc.*, 87: 214-218.

PENNAK, R. W. & WARD, J.V. 1986. Interstitial fauna communities of the hyporheic adjacent groundwater biotopes of a Colorado mountain stream. *Arch. Hydrobiologie*, 74: 356-396.

PERLMUTTER, D. 1988. *Meiofauna in stream leaf litter; patterns of occurrence and ecological significance.* Ph.D. Thesis. University of Georgia.

ROUCH, R. & DANIELOPOL, D.L. 1987. L'origine de la faune aquatique souterraine, entre le paradigme du refuge et le modèle de la colonisation active. *Stygologia*, 3: 345-372.

RUFFO, S. 1961. Problemi relativi allo studio della fauna interstiziale iporreica. *Bull. Zool.*, 28: 73-319.

RUTHERFORD, J. E. & HYNES, H.B.N. 1987. Dissolved organic carbon in streams and groundwater. *Hydrobiologia*, 154: 33-48.

SABATER, F. 1987a. On the interstitial cladocera of the River Ter (Catalonia, NE Spain), with a description of the male of *Alona phreatica. Hydrobiologia*, 144: 51-62.

SABATER, F. 1987b. Some interstitial species of the Crustacean communities of the Ter and Ebre rivermouths (NE Spain). *Misc. Zool.*, 10:113-119.

SABATER, F. & De MANUEL, J. 1988. Interstitial Isopoda of the River Ter (Catalonia, NE Spain) and description of two new species, *Microcharon margalefi* n.sp. and *Proasellus phreaticus* n.sp. *Stygologia*, 4: 267-278.

SCHWOERBEL, J. 1961. Über die Lebensbedin-gungen und die Besiedlung des hyporcheischen Lebensraumes. *Arch. Hydrobiologie*, 25: 182-214.

SKET, B. 1985. Why all cave animals do not look alike. A discussion on adaptive value of reduction processes. *Nat. Speological Soc. Bull.*, 47: 78-85.

SLACK, K.V. 1955. A study of the factors affecting stream productivity by the comparative method. *Investigations of Indiana Lakes and Streams*, 4: 3-47.

STANFORD, J. A. & GAUFIN, A.R. 1974. Hyporheic communities of two Montana rivers. *Science*, 185: 700-702.

STANFORD, J. A. & WARD, J.V. 1988. The hyporheic habitat of river ecosystems. *Nature*, 335: 64-66.

STRAYER, D. L. & LIKENS, G.E. 1986. An energy budget for the zoobenthos of Mirror Lake, New Hampshire. *Ecology*, 67: 303-313.

TRISKA, F.J. KENNEDY, R.J., AVANZINO, R.J., ZELLWEGER, G.W. & BENCALA, K.E. 1989.

Retention and transport of nutrients in a third-order stream in northwestern California: hyporheic processes. *Ecology*, 170: 1893-1905.

VALETT, H.M., FISHER, S.G. & STANLEY, E.H. 1990. Physical and chemical characteristics of the hyporheic zone of a Sonoran Desert stream. *J. N. Am. Benthol. Soc.*, 9: 201-215.

VILA, P. B. 1989. The occurrence and significance of Cladocera (Crustacea) in some streams of Central Indiana, USA. *Hydrobiologia*, 171: 201-214.

WARD, J. V. 1989. The four-dimensional nature of lotic ecosystems. *J. North Amer. Benthol. Soc.*, 8: 2-8.

WILLIAMS, D. D. 1984. The hyporheic zone as a habitat for aquatic insects and associated arthropods. In: *The ecology of aquatic insects* (J.J. Resh & J.J. Rosenberg, eds.): 430-454. Praeger. New York.

WILLIAMS, D.D. 1989. Towards a biological and chemical definition of the hyporheic zone in two Canadian rivers. *Freshwater Biology,* 22: 189-209.

WILLIAMS, D. D. & HYNES, H.B.N. 1974. The occurrence of benthos deep in the substratum of a stream. *Freshw. Biology*, 4: 233-256

Homage to Ramon Margalef; or, Why there is such pleasure in studying nature 45
(J. D. Ros & N. Prat, eds.). 1991. *Oecologia aquatica*, 10: 45-60

SIZE AS A FACTOR IN CENTRIC DIATOMS DISTRIBUTION: THE SPANISH RESERVOIRS AS AN EXAMPLE

SERGI SABATER

Departament d'Ecologia. Facultat de Biologia. Universitat de Barcelona. Av. Diagonal, 645. 08028 Barcelona. Spain.

Received: June 1990

SUMMARY

Centric diatoms are relatively abundant in the phytoplankton of Spanish reservoirs, and therefore they can be considered as indicative of the phytoplankton populations as a whole. Taxa distribution may often be related to some major environmental characteristics of the reservoirs. In the case of some species of the genus *Aulacoseira* (formerly belonging to *Melosira*), *A. distans*, *A. italica* and *A. subarctica* are confined to N-NW reservoirs, where there are waters of low mineral content, while *A. granulata* is more widespread, mainly preferring higher mineral content and/or nutrient richer waters. Among the smaller centrics, *Cyclotella comensis* and *C. radiosa* are the most frequent, but occur only in the hard-water reservoirs (East Spain, alkalinity > 1 meq l^{-1}). *C. radiosa* was located in northern reservoirs, while *C. comensis* preferred warmer waters. Other taxa had a more restricted distribution, but their presence was indicative of peculiar environmental conditions. For example, *Chaetoceros muelleri* is restricted to the phytoplankton of a meromictic, brackish water reservoir.

Some morphological observations of the cell valve (size, number of striae or areolae, number and disposition of fultoportulae) have been made with SEM on *C. comensis* and *C. radiosa* populations. The data come from a large number of reservoirs in the east of Spain, and were tentatively related to environmental variables of the reservoirs using statistical analyses. Cell diameter was revealed as the most directly related morphological character. Differences in size between populations of each species were consistent, as confirmed through an analysis of variance. Using the certainty that size has characteristic values in every population, its relationship with environmental variables has been explored using a stepwise regression analysis. Temperature is in fact a common factor affecting the size of organisms. Dependence on calcium content seems more unusual, and it may be considered as an accurate expression of the effect of ionic strength of the water on diatom cell size.

KEY WORDS: Phytoplankton, centric diatoms, reservoirs, cell size, environmental gradients.

INTRODUCTION

Certain notable biological manifestations may contribute to an understanding of the environmental factors which influence the composition and structure of phytoplankton assemblages. Among them, the distribution patterns of the assemblages themselves (or of some representative taxa) provide a general reflection of such relationship (HARRIS, 1986). Further, more detailed knowledge of certain environmental factors can be obtained by studying the responses of a single population. Morphological features of the population should be considered among others. The use of suitable techniques (e.g. SEM) can provide optimum sets of observations in some groups of organisms, such as centric diatoms. Although this perspective contemplates only a small part of the whole ecosystem, the particular response of these organisms should provide useful information on this subject.

The two assumptions outlined above undoubtedly offer a complementary vision of the influence of the environment on organisms. MARGALEF's outstanding paper (1969) surmised, from different taxa and populations of *Cyclotella* and *Melosira*, that variations in population size (taken as an expression of the morphological character of the population) were mainly related to certain environmental characteristics. He pointed to water salinity and water temperature as the factors most determinant in affecting size. The main objective of the present work is to test whether size (and certain related elements of the diatom valve) may be used as an appropriate indicator of environmental conditions in Spanish reservoirs. Phytoplankton assemblages in Spanish reservoirs have been grouped with respect to two major environmental factors (SABATER & NOLLA, in press). These were the ionic strength of the water and the trophic state of the reservoirs. In accordance with this, five groups of phytoplankton assemblages were outlined, ranging from slightly mineralized to hard water reservoirs, combined with the range from oligotrophic to eutrophic conditions. Centric diatoms are found in high numbers in each of these five groups, and their occurrence, relative abundance and morphology may be used as representative indices of the major environmental conditions defining the groups.

MATERIAL AND METHODS

The present study uses the observations performed on the phytoplankton samples collected at a depth of 5 m in a set of 104 reservoirs distributed all over Spain (SABATER & NOLLA, 1991). A HITACHI 2300 scanning electron microscope (SEM) was used to count the number of striae of small centrics, as well as in the observation of some potentially significant structures; fultoportulae (strutted processes), rimoportulae (labiate processes), areolae, granules and spines. The total number of striae per valve was counted instead of density of striae, in order to avoid statistical artifacts and consequent misinterpretation of results (THERIOT, 1988). Samples were not subjected to any previous oxidation treatment prior to observation with SEM.

Because not all the samples containing centric diatoms could be observed with SEM, results concerning distribution are sometimes necessarily incomplete. Diameter was measured with SEM when possible, but also with light microscopy (LM) in order to complete a statistically significant data set. For LM some drops of distilled-cleaned frustules were air-dried and mounted with Naphrax for observation at x1000. The number of specimens counted in each reservoir is given in Table II.

RESULTS

SOME TRENDS IN THE DISTRIBUTION OF CENTRIC DIATOMS IN PHYTOPLANKTON ASSEMBLAGES

The ease with which taxa of genus *Aulacoseira*, formerly a part of the genus *Melosira*, may be identified with the inverted microscope facilitates the accurate determination of their distribution in Spanish reservoirs. *A. distans*, *A. distans* var. *tenella*, *A. italica* and *A. subarctica* are confined to the central and western part of the northern half of the Iberian peninsula (Figs. 1; 1'-5). *A. granulata* is distributed in the rest of the peninsula, but is also found in certain reservoirs in the north, co-occurring with the former taxa. These are reservoirs of groups B and C (SABATER & NOLLA, 1991), that is to say of low ionic concentration but with a high proportion of eutrophic taxa in their phytoplankton assemblages. The dividing line between the *A. distans*-group (including *A. italica* and *A. subarctica*) and *A. granulata* corresponds approximately to the division between reservoirs with alkalinity lower or higher than 1 meq l^{-1}. A similar distribution for these taxa was already observed fifteen years ago (MARGALEF *et al.*, 1982).

Cyclotella comensis and *C. radiosa* (valid synonym for *C. comta* as shown by HÅKANSSON, 1988) are the most frequent taxa among the small centric diatoms of the phytoplankton in the reservoirs (Figs. 8-14; 15-19). Although their distribution (Fig. 2) may hide some gaps, it is noteworthy that they appeared only in reservoirs of groups D and E, which are of hard water. As may be observed from figure 2, both species are completely absent in the western reservoirs. As a usual trend, *Cyclotella comensis* is more frequent than *C. radiosa*. While *C. radiosa* has been found only in the north, *C. comensis* occupies all the area. However, *C. comensis* is not highly abundant when appearing together with *C. radiosa*. Difference in water temperature is the main factor influencing the relative distribution of *C. comensis* and *C. radiosa*. This is the only environmental variable to reach the level of significance required for an stepwise

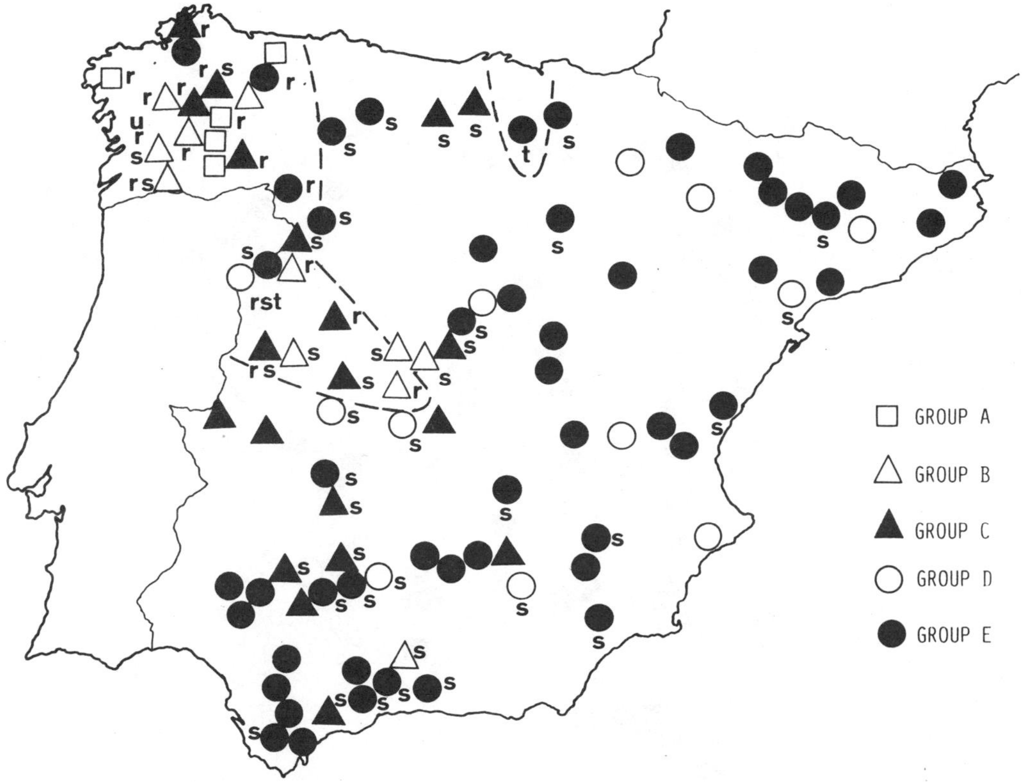

FIGURE 1. Distribution of *Aulacoseira* taxa in Spanish reservoirs: *A. distans* (Ehrenberg) Simonsen and its var. *tenella* (Nygaard) Ross (r); *A. granulata* (Ehrenberg) Simonsen (s); *A. italica* (Ehrenberg) Simonsen (t); *A. subarctica* (O. Müller) Haworth (syn. *A. italica* var. *subarctica* O. Müller) (u). Their presence is indicated with respect to the grouping of phytoplankton assemblages of SABATER & NOLLA (1991), that cluster the reservoirs from oligotrophic to eutrophic, either in those of slightly mineralized waters (A to C) or in those of hard waters (D and E). The broken line segregates the zone of occurrence of the *distans*-group from that of *A. granulata*.

regression analysis.

Cyclotella meneghiniana and *Stephanodiscus hantzschii* are found fairly frequently (Figs. 3; 20 and 23), but usually only in small numbers.

The distribution of other centric taxa of scarce appearance is shown in figure 3. *Chaetoceros muelleri* (Figs. 21-22) has been found only in a small saline reservoir, where water conductivity was about 5,700 µS cm^{-1} (ARMENGOL *et al.*, 1990). *Cyclotella stelligera* (figs. 27'-29) was limited to six reservoirs, always with a low number of individuals. *Cyclostephanos dubius* (Fig. 24), *Cyclotella atomus* (Figs. 6-7), *Stephanodiscus rotula* (Figs. 26-27) and *Thalassiosira pseudonana* (Figs. 29-32) may be considered as occasional in the phytoplankton of reservoirs, at least during the period of our study.

SIZE AND RELATED CHARACTERISTICS AS ECOLOGICAL INDICATORS

Some ecological questions arise from morphological considerations. Intriguing morphological variations are observable in nearly every one of the taxa quoted above. For instance, *Cyclotella stelligera* var. *pseudostelligera* has a remarkable variability concerning the degree of flatness and definition of its central area (see Figs. 27'-28). Such variability is not a special feature of the populations inhabiting the Spanish reservoirs. BELCHER *et al.* (1966) observed it in a population of an English

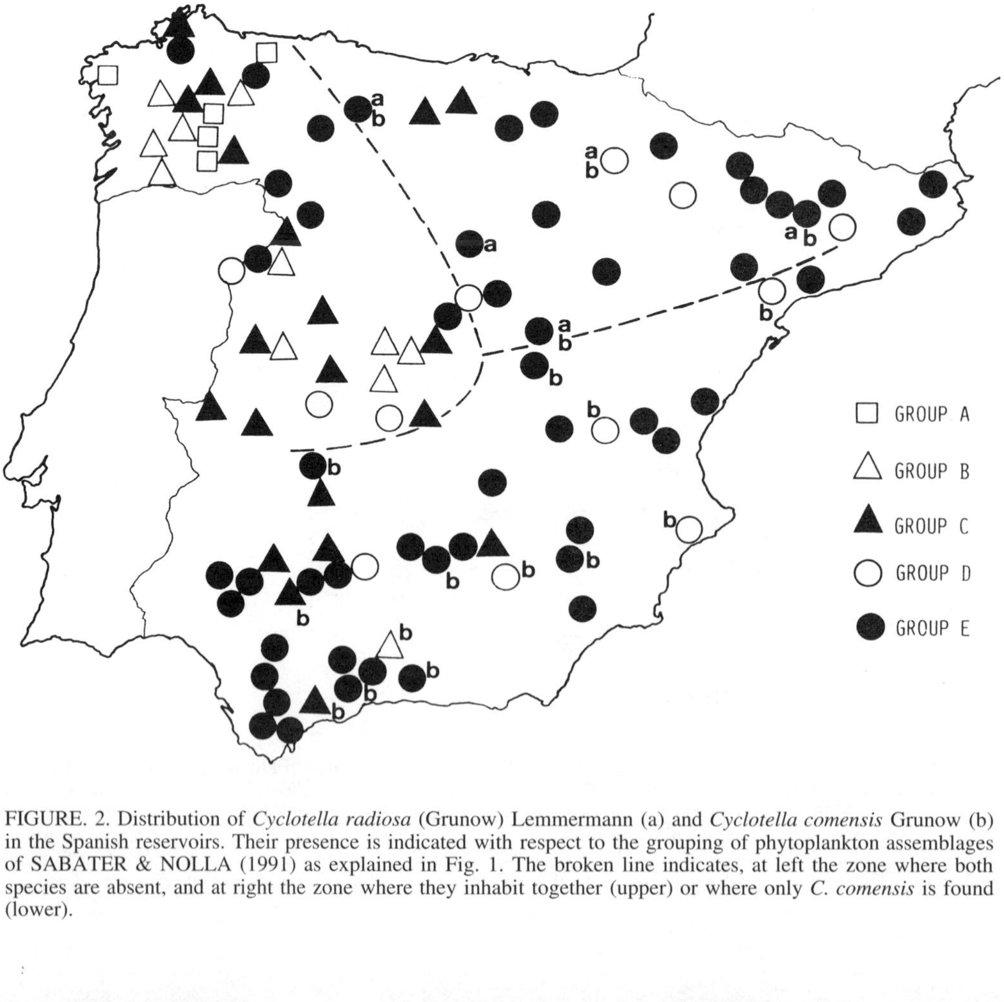

FIGURE. 2. Distribution of *Cyclotella radiosa* (Grunow) Lemmermann (a) and *Cyclotella comensis* Grunow (b) in the Spanish reservoirs. Their presence is indicated with respect to the grouping of phytoplankton assemblages of SABATER & NOLLA (1991) as explained in Fig. 1. The broken line indicates, at left the zone where both species are absent, and at right the zone where they inhabit together (upper) or where only *C. comensis* is found (lower).

pond, suggesting a relationship between sculpturing of the valves and availability of silica. HAWORTH & HURLEY (1986) have figured the heterovalvy of flat and ondulated valves in the same specimen, although they consider that prominency of one over the other may have some relation to phosphorus or silica during population growth. Moreover, complicating an accurate ecological interpretation, the segregation between forms of this taxon and others belonging to *Thalassiosira pseudonana* (Figs. 29-32) becomes sometimes quite difficult (e.g. in reservoir 91). Both taxa are inhabitants of eutrophic waters (KISS, 1984; SABATER & KLEE, 1990) and taxonomical boundaries between them are not sufficiently clear, at least when small specimens are considered (SABATER & KLEE, 1990).

The high occurrence of *Cyclotella comensis* and *C. radiosa* in the Spanish reservoirs provides enough material for the study of their morphology in connection with ecological implications. *Cyclotella comensis* shows a high degree of morphological variability in the populations studied (Figs. 8-14). The three tripunctate cavities on the valve face which are predicted by Grunow's description of the taxon (KLING & HÅKANSSON, 1988) are difficult to observe in the Spanish populations. In fact, form, arrangement and number of cavities are variable from

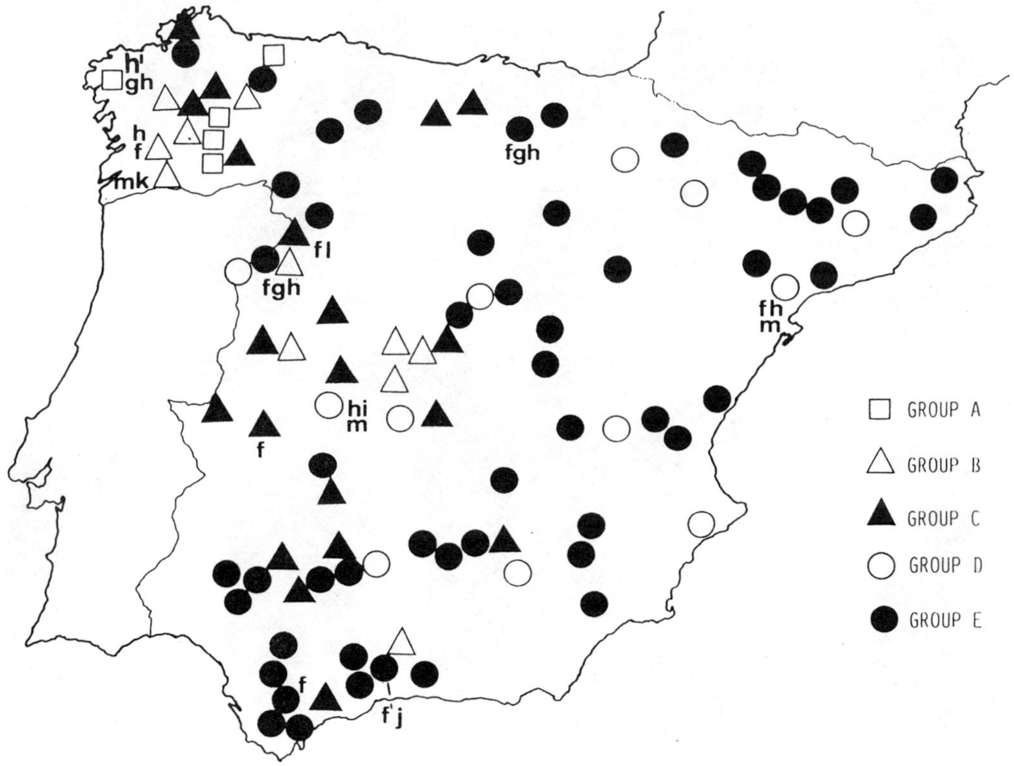

FIGURE 3. Occurrence of the less frequent centric diatoms in the Spanish reservoirs (grouping as explained in figure 1): *Cyclotella meneghiniana* Kützing (f); *Stephanodiscus hantzschii* Grunow (g); *Cyclotella stelligera* Cleve & Grunow (h) and its var. *pseudostelligera* (Hustedt) Haworth & Hurley (h'); *Cyclotella atomus* Hustedt (i); *Chaetoceros muelleri* Lemmermann (j); *Cyclostephanos dubius* (Fricke) Round (k); *Stephanodiscus rotula* (Kützing) Hendey (l); *Stephanodiscus minutulus* (Kützing) Cleve & Müller (m); *Thalassiosira pseudonana* Hasle & Heimdal (n).

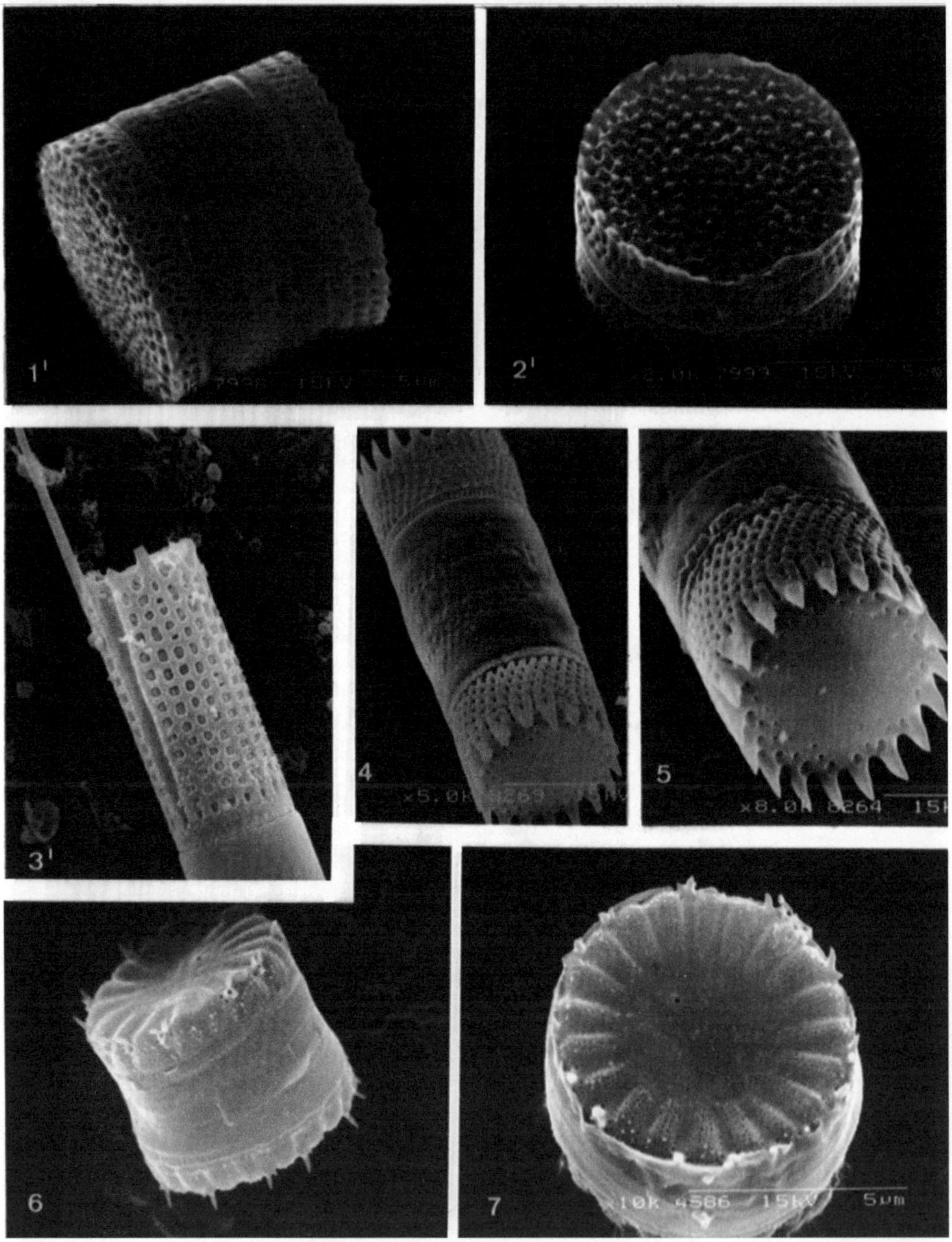

FIGURES 1'-2'. *Aulacoseira distans* var. *tenella*. Fig. 1', exterior view of a valve (x 3000, reservoir 16). Fig. 2', side view (x 4800, r99). Fig. 3'. *Aulacoseira granulata*. Exterior view of a limiting cell (x 2250, r35). Figs. 4-5. *Aulacoseira subarctica* (syn. *A. italica* var. *subarctica* O. Müller), two approximative views of a same filament. Fig. 5 shows the peripheric ring of areolae in the valve face (Fig. 4, x 3000; Fig. 5, x 4800; both from r97). Figs. 6-7. *Cyclotella atomus*. Fig. 6, side view (x 4800, r50). Fig. 7, valve view (x 6000, r50).

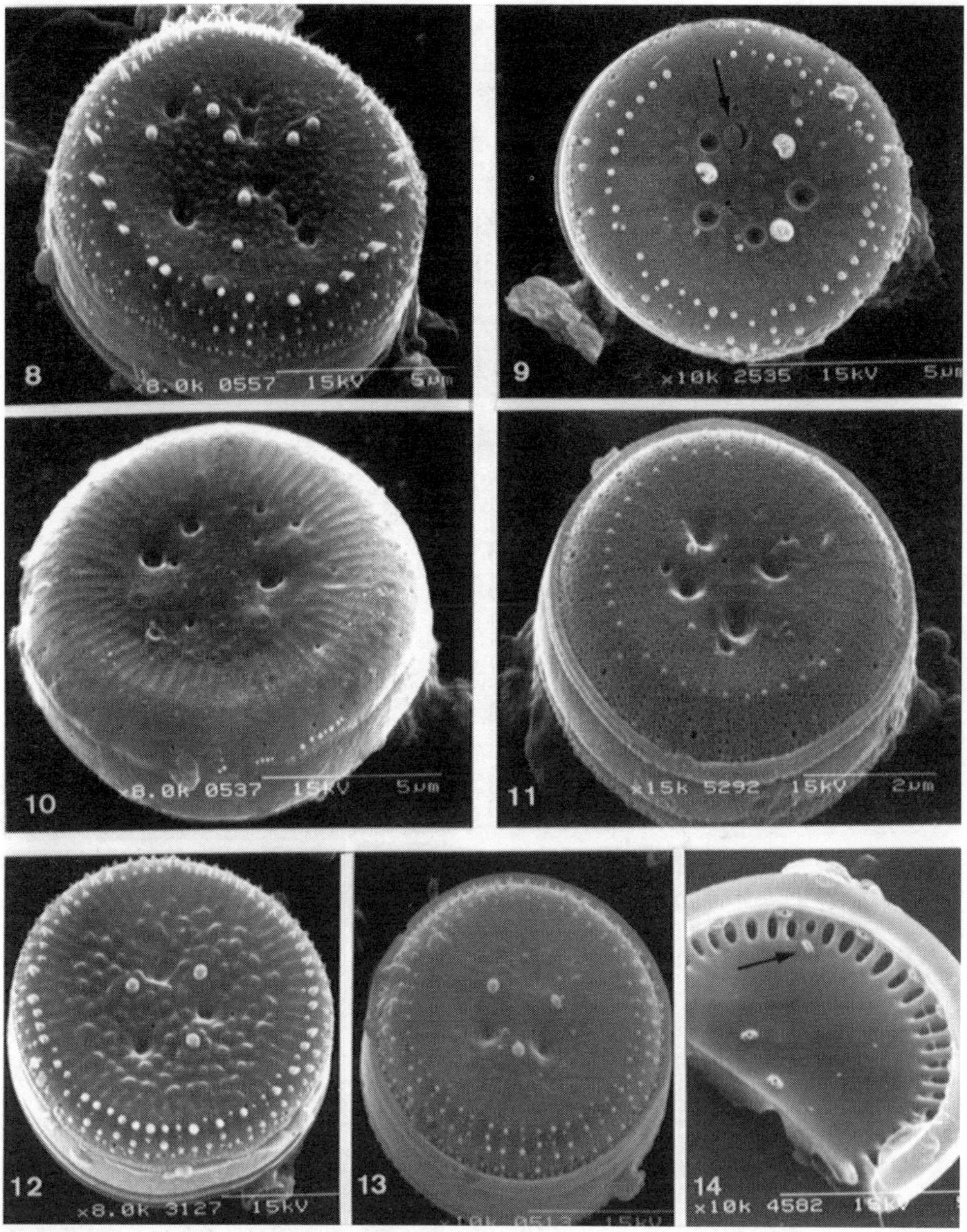

FIGURES 8-14. *Cyclotella comensis*. Figs. 8-13, exterior views of representative specimens covering the whole range of morphological variation. Figs 8 (x 4800, r73) and 9 (x 6000, r73), valves of verrucose surface, with granules on the face and spines on the rim of the valve; granules are variable with respect to their number and width. Note the crater-like cavities in fig. 9, one of them with a tapering flap of silica (arrowed). Fig. 10, an extreme example of specimen bearing a smooth valve face and without granules or spines (x 4880, r63). Fig. 11, valve bearing four cavities on its face (x 9000, r74). Fig. 12, valve showing a conspicuous verrucose face (x 4800, r94). Fig. 13, a specimen with a reduced central area (x 6000, r72). Fig. 14, interior view of a broken valve, showing two central fultoportulae, a rimoportula on a marginal position (arrowed), and some marginal fultoportulae, located every three to four costae (x 6000, r48).

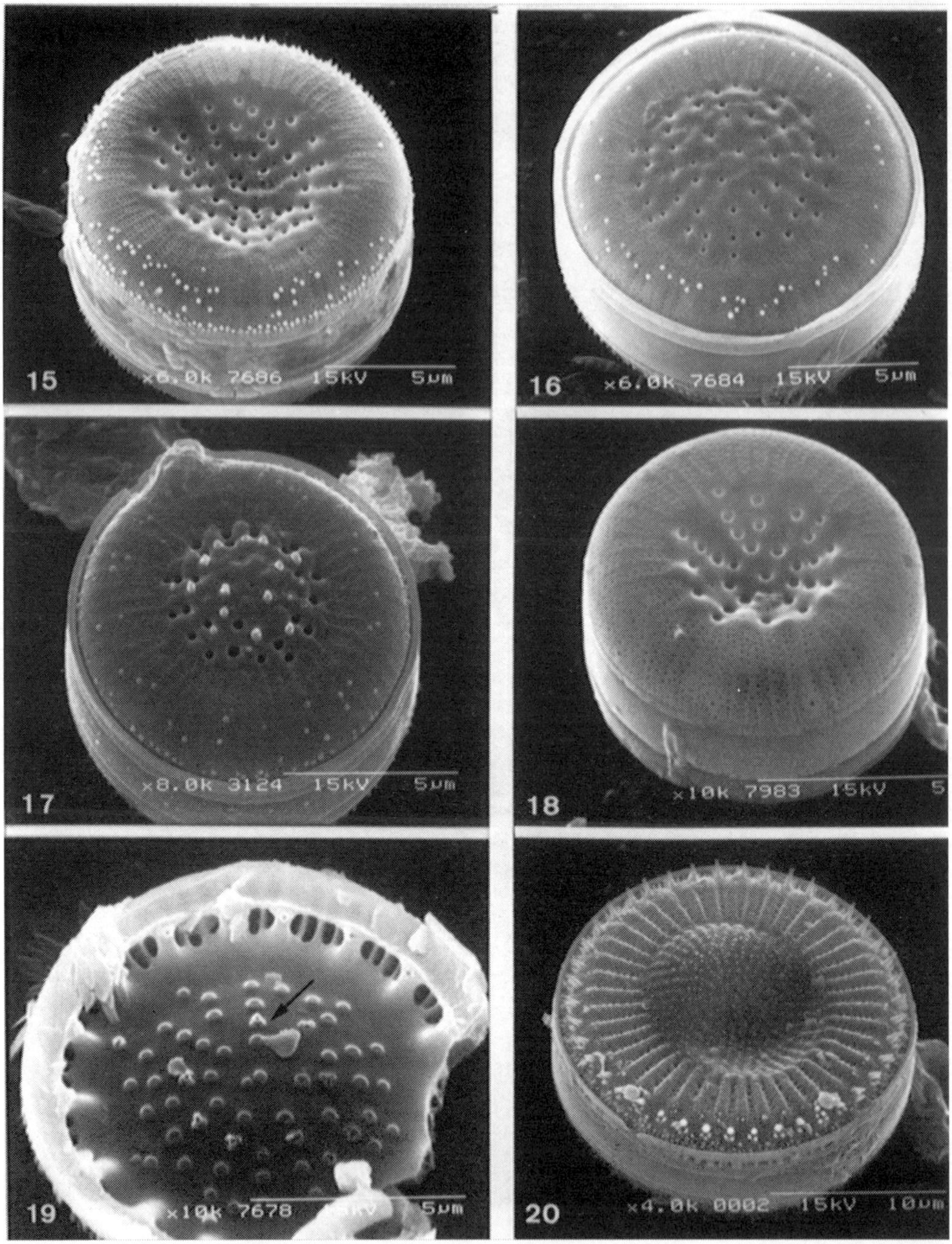

FIGURES 15-19. *Cyclotella radiosa*. Figs. 15 and 16, lacunate and scutate valves of specimens with a wide central area; note the two kinds of pores radially arranged; some larger (fultoportulae), other small (loculi; both at x 3600, r94). Figs. 17 and 18, scutate and lacunate valves with narrow central area and wider larger pores (x 4800, r8; x 6000, r3). Fig. 19, interior view of a valve showing in its central part the two kinds of pores; the loculi with their domed criba, and the central fultoportulae (arrowed), bearing each one three struts (x 6000, r94). Fig. 20. *Cyclotella meneghiniana*. Exterior view of a typically undulated valve (x 2400, r103).

FIGURES 21-22. *Chaetoceros muelleri*. Girdle view of valves showing the outer surface and the emergence of setae (Fig. 21, x 9000; Fig. 22, x 6000; both from r103). Fig. 23. Group of *Stephanodiscus hantzschii* (x 1500, r15). Fig. 24. *Cyclostephanos dubius*. Exterior view of a valve with scattered spines (x 6000, r20). Fig. 25. *Stephanodiscus minutulus*. Exterior view; the external opening of the central fultoportula is arrowed (x 9000, r50). Figs. 26-27. *Stephanodiscus rotula*. Exterior and interior views, showing the central fultoportulae distinctive of this species (x 3600, x 3000; both from r26).

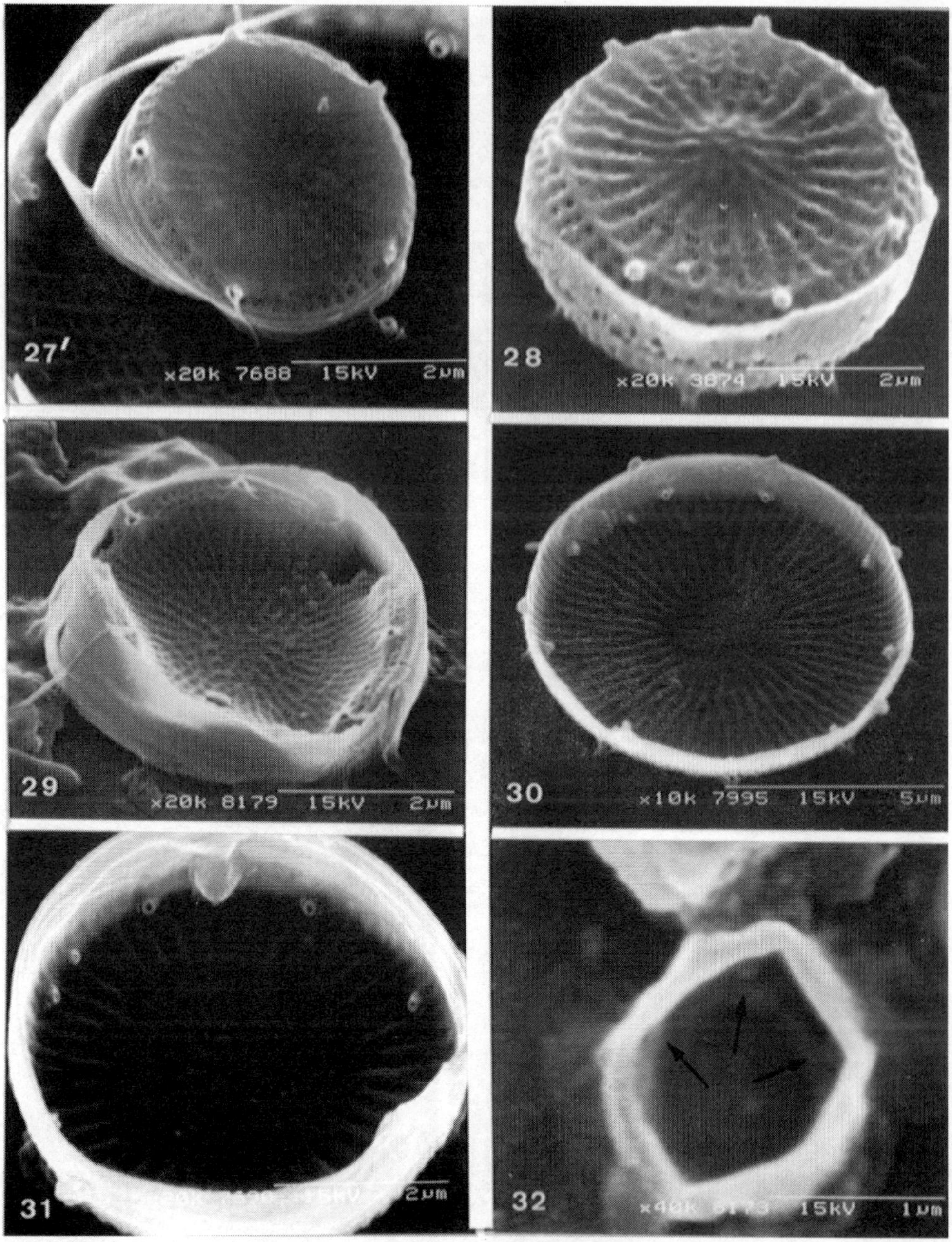

FIGURES 27'-28. *Cyclotella stelligera* var. *pseudostelligera.* Fig. 27'. External view of a valve of *C. stelligera* var. *pseudostelligera* (form *woltereck*i Hustedt) placed over a *Stephanodiscus*; note the smoothness of the surface (x 1200, r91). Fig. 28. Undulated form of *Cyclotella stelligera* var. *pseudostelligera.* Figs. 29-32. *Thalassiosira pseudonana.* Fig. 29. External view of a sculptured valve (x 1200, r15). Figs. 30-31. Interior views; note that the central fultoportulae have two struts (x 6000, r29; x 12000, r91). Fig. 32, very small forms (1.6 mm Ø), tentatively identified as specimens of *Thalassiosira pseudonana*, of a weakly silicified structure; arrowed there are three marginal fultoportulae.

TABLE I. Regression functions which express the dependence of some rotational elements of the diatom valve against diameter in *Cyclotella comensis* and *Cyclotella radiosa*.

a) *Cyclotella comensis*

COSTAE = 8.42 DIAMETER $^{0.84}$	$r^2 = 0.910$, $p < 0.001$, n = 18 (reservoir 74)
COSTAE = 8.23 DIAMETER $^{0.86}$	$r^2 = 0.874$, $p < 0.001$, n:= 12 (reservoir 72)
COSTAE = 5.76 DIAMETER $^{1.08}$	$r^2 = 0.976$, $p < 0.001$, n = 12 (reservoir 78)
COSTAE = 9.6 DIAMETER $^{0.77}$	$r^2 = 0.774$, $p < 0.001$, n = 23 (reservoir 73)

b) *Cyclotella radiosa*

COSTAE = 10.9 DIAMETER $^{0.64}$	$r^2 = 0.980$, $p < 0.001$, n = 5 (reservoir 8)
DIAMETER CENTRAL AREA = 0.33 DIAMETER$^{1.17}$	$r^2 = 0.976$, $p < 0.001$, n = 33 (reservoirs 3, 8, 31, 94)
NUMBER AREOLAE = 3.84 DIAMETER $^{0.95}$	$r^2 = 0.343$, $p < 0.001$, n = 32 (reservoirs 3, 8, 31, 94)

population to population, but a rather high variation appears in any given one. In some cases morphology of cavities is peculiar (Fig. 9), but this seems to be more closely related to the process of cellular division and cell wall morphogenesis than to environmental causes (SCHMID, 1985). Moreover, marginal fultoportulae are arranged irregularly, even in the same occurring valve, varying from two to five costae (Fig. 14). Variability of these characters is very high, and no stable quantitative pattern has been established.

Number and dimensions of granules - and of spines too - on the central and/or marginal part of the valve face in *C. comensis* are also variable. The occurrence of granules seems to follow a more-or-less stable pattern, because they appear in some populations, but not in others. A possible relationship between the appearance of granules and changes in salinity has been suggested by A.M. Schmid (personal communication). Studies of specific cultures with populations bearing granules or not may lead to an understanding of such regularity.

Variability of morphological characteristics was also noteworthy in populations of *Cyclotella radiosa* (Figs. 15-19). The width of the central area, the number and diameter of the areolae covering the valve face, and the presence or absence of granules follow divergent patterns from reservoir to reservoir. The marginal fultoportulae are usually arranged irregularly, from two to five costae, but then diverge from the classical description (HÅKANSSON, 1988).

The ecological interest of all these structures is sometimes difficult to establish, often because it is hard to count them in a statistically significant number. On the other hand, it is easier to measure the number of costae and the diameter, which facilitates the search for a pattern.

The dependence of some rotational elements of the valve with respect to the diameter was investigated. A linear and positive regression against diameter is expected, as has been summarized by THERIOT (1988). In the case of the total number of costae, the significance is not so high in *C. comensis* and in *C. radiosa*. The exponential functions obtained with the respective logarithmically transformed values had an r^2 of 0.769 in the former, but only of 0.25 in the latter (Figs. 33 and 34). Slopes of the regression lines ranged from 0.77 to 1.08 for *Cyclotella comensis* (Table I); with *C. radiosa* the relation was not significant in a high number of reservoirs (Table I).

This supports the statement of MARGALEF (1983) that the density of any given element is related to a dimension of the valve following an exponential function with a slope of 0.66. One possible consequence of such an empirical assumption is that morphological characteristics of the valve are not completely dependent on (nor completely independent of) cell size.

This might be true of the number of striae, although interpretation is rather complex. Ecologically, variation of number of striae has been related to changes in, for instance, water temperature (KACZMARSKA & FRYXELL, 1986) or water salinity (GEISSLER, 1982; MAKAROVA, 1982). But in these cases size, when considered (cf. MAKAROVA, 1982), also appears to be affected by the same environmental factor.

Additionally, in those characteristics for which slope will move significantly from the stated value, a different ecological interpretation may be made. For instance, in *C. radiosa* the regression of the central area diameter against size has a slope of 1.2 while, for the number of areolae, the slope is 0.95. This may suggest a close dependence of such associated structures on valve size, and it therefore has no real independent significance.

In order to examine any relationship between size and environment, the variations in diameter in *C. comensis* and *C. radiosa* populations were analyzed among certain reservoirs (Table II), with two objectives: first, to determine whether there were consistent differences in size among populations; second, if so, to relate them to certain environmental parameters.

Both objectives start from the discussion of cell size reduction. The idea that the cells undergo a progressive reduction of size during vegetative reproduction has been widespread since the classic work of PFITZER (1871). Although GEITLER (1932) justified that auxospore formation would be a way for recovering size, BETHGE (1925) showed that auxospore formation had no direct implication for the size spectrum of a given population. Anyway, if such a reduction exists, it does not happen as a regular process (PATRICK, 1966). In certain unialgal diatom cultures lasting several years no decrease in size of the species occurred (WIEDLING, 1948; MARGALEF, 1969). Therefore, the assumption of size reduction cannot be taken as a rule. MARGALEF (1969) hypothesized that the average diameter of a given population of *Cyclotella* or *Melosira* would be an expression of environmental properties, more than the result of the sexual cycle. Recently, size variation has been used as a palaeolimnological indicator of past conditions (STOERMER *et al.*, 1989), or as indicative of the trophic state

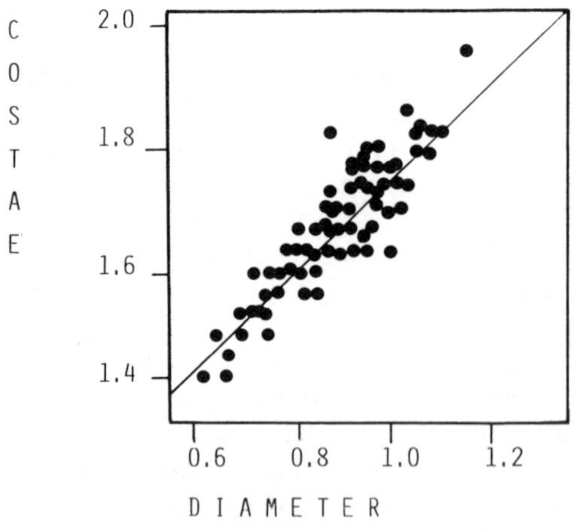

FIGURE 33. Regression function describing relationship of number of costae with respect to diameter (log transformed values) in *Cyclotella comensis* (Costae = 9.28 diameter$^{0.79}$; $r^2 = 0.769$, $p < 0.001$, $n = 97$).

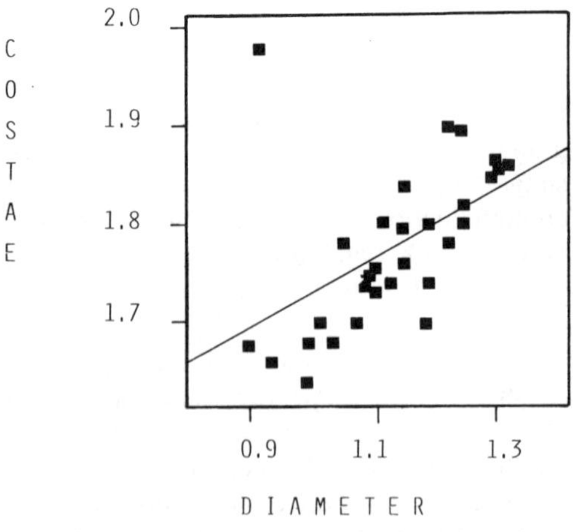

FIGURE 34. Regression function describing relationship of number of costae with respect to diameter (log transformed values) in *Cyclotella radiosa* (Costae = 22.5 diameter$^{0.38}$; $r^2 = 0.25$, $p < 0.004$, $n = 30$).

TABLE II. Means and coefficients of variation of some morphometric characters in *Cyclotella radiosa* and *Cyclotella comensis* populations.

a) *Cyclotella radiosa*

RESERVOIR	DIAMETER	NUMBER OF COSTAE	NUMBER OF AREOLAE	DIAMETER OF CENTRAL AREA	FREQUENCY
3	8.88 (0.25)	59.7(0.23)	38.7 (0.54)	5.5 (0.23)	73
94	14.44 (0.19)	63.4 (0.12)	54.1 (0.15)	8.3 (0.22)	24
8	12.39 (0.18)	59.4 (0.15)	43.0 (0.24)	6.5 (0.23)	59
31	11.27 (0.19)	44.5 (0.05)	27.3 (0.14)	4.5 (0.10)	54

b) *Cyclotella comensis*

RESERVOIR	DIAMETER	NUMBER OF COSTAE	FREQUENCY
3	6.8 (0.27)	56.3 (0.2)	10
73	9.4 (0.34)	46.8 (0.27)	98
74	7.47 (0.22)	47.1 (0.24)	68
72	8.9 (0.19)	53.8 (0.15)	62
78	8.1 (0.14)	53.3 (0.18)	62
94	10.3 (0.15)	52.5 (0.13)	23
8	9.8 (0.27)	39 (0.1)	12
63	8.49 (0.14)	53.0 (0.14)	11

in lakes (THERIOT *et al.*, 1988). However, the complexity of such a question is obvious (cf. LEWIS, 1984; ROUND *et al.*, 1990). As has been summarized by GEISSLER (1982) through studies on *Stephanodiscus hantzschii*, size may be related either to the process of division or to ecological conditions. The environment would act by selecting a given morphotype, or conversely, a phenogenic response of the population would follow the environmental change (MARGALEF, 1969). In any case, we may use size as an indicator of environmental conditions, but we should be aware that other implications may be involved. In nature, differences of time scale between the generation period of a given species and the changes of environmental conditions are important. Otherwise, genetic variation is high. Therefore, when working with natural populations, the difficulty of obtaining reliable results increases.

Size distribution of frequencies did not follow a continuous pattern in every local population, but showed discrete peaks. This is shown by three different populations of *Cyclotella radiosa* (Fig. 35), which discourages the idea of the existence of a continuous growth linked to an auxosporulation cycle for those populations. Consequently, differences in size among populations (Table II) were found to be consistent both in *Cyclotella comensis* and in *C. radiosa*, as is shown with an analysis of variance ($p < 0.00001$ in the two sets of data).

In addition, in three reservoirs where both *Cyclotella* coincided, a representative number of valve diameters were counted. In spite of the low number of replicates, a covariance in size of the two *Cyclotella* was also found to be significant by another

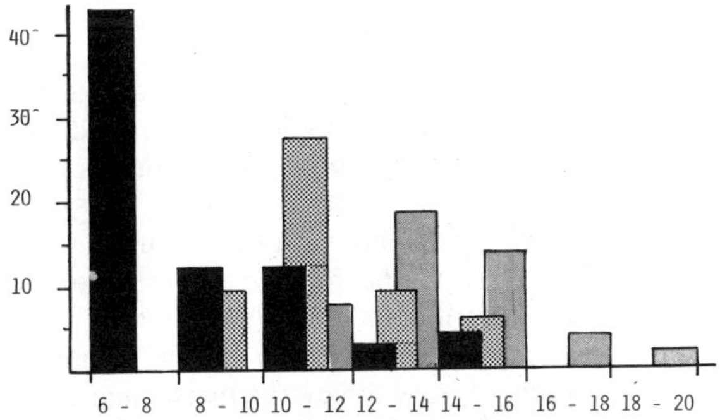

FIGURE 35. Distribution of *Cyclotella radiosa* valve diameter (x axis represents every 2 mm categories; y axis represents the number of valves examined) in three different populations: r3 (black), r8 (dotted), r94 (shaded).

anova ($p < 0.1$). This may be taken as a test of the value of the diameter as an indicator of the environmental variation affecting the two species populations. In the words of MARGALEF (1969), "correlation between changes of size in species that live in association should be one of the best criteria of significance when size is used as an ecological indicator".

Significance of the variance analysis was also high when other characters of the diatoms valve were considered: diameter of the central area ($p < 0.003$) and number of areolae ($p < 0.09$) in *C. radiosa*, number of areolae ($p < 0.003$) and total number of costae ($p < 0.009$) in *C. comensis*.

Using the fact that size has characteristic values in every population, its relationship with environmental conditions has been explored through stepwise regression analysis. In a first, exploratory analysis only water temperature, depth of the photosynthetic layer, conductivity, nitrogen as PON and planktonic chlorophyll-a were considered. These variables were selected because they are major descriptors of physico-chemical and biological properties of the reservoir. Transformed log values were used for all the variables involved in the analysis. The following expression is obtained when both *Cyclotella* populations are considered as a whole:

$$\log \text{diameter} = 2.31 - 0.53 \log \text{temperature} - 0.29 \log \text{conductivity} \qquad (1)$$

Hence, water temperature and water mineral content (as indicated by conductivity) may tentatively be pointed out as the major factors affecting size in the *Cyclotella* populations. Furthermore, in order to clarify the relative importance of some contributors of the total ion concentration with respect to the diatom variation of size a new stepwise analysis was carried out. Log-transformed values of total alkalinity, silicate (as total silicates and as reactive soluble silicate), calcium, carbonate, bicarbonate, and sulphate concentrations, were added to the previous variables. In this case, size of both *Cyclotella* populations was significantly dependent on temperature and calcium content in the reservoirs:

$$\log \text{diameter} = 2.43 - 0.78 \log \text{temperature} - 0.16 \log \text{calcium} \qquad (2)$$

Similar expressions are obtained when populations of the two species are considered in particular. While temperature has been found to affect size of organisms elsewhere (MARGALEF, 1974), calcium is highlighted among other components of water mineral content as influencing size. Although the abundance of this element in the eastern Spanish reservoirs is striking, the relative abundance of some other ions (e.g. sulphates) is also high there (ARMENGOL *et al.*, in press a). Neither silicates, which are important components of the diatom cell wall, are revealed as affecting size. The relationship expressed by the stepwise regression allows us to hypothesize some possible physiological implications of calcium concentration with respect to the valve size. However, only detailed observations with cultures would bring a more complete response to this assumption.

Apart from the influence I assume that some physico-chemical factors may exert on the size of a diatom population, some other factors can not be disregarded. MANN (1988) suggests that selection of a given population size in the pennate diatom *Nitzschia sigmoidea* would be related to the action of parasites. Predation, which would appear to affect larger cell classes more than small ones, may also potentially affect the size of a population. Therefore, it will be interesting to compare such outlined dependence on the populations of the reservoirs with other sets of data elsewhere where conditions or/and species are different.

ACKNOWLEDGEMENTS

This work has been sponsored by the CAICYT of the Ministry of Education and Science (project number PB85-0166). My gratitude to Rolf Klee (Munich), A.M. Schmid (Salzburg) and E.Y. Haworth (Windermere) for their comments on different aspects of centric diatoms. The Microscopical Service of the University of Barcelona provided their facilities and technical help. Dr. Ramon Margalef has contributed with his personal learning and papers to the ideas which I have tried to express in this study.

REFERENCES

ARMENGOL, J., RIERA, J.L., MORGUí, J.A. in press. Major ionic composition in the Spanish reservoirs. *Ver. Internat. Verein. Limnol.*, 24.

ARMENGOL, J., CATALAN, J., GABELLONE, N., JAUME, D., DE MANUEL, J., MARTÍ, E., MORGUÍ, J.A., NOLLA, J., PEÑUELAS, J., REAL, M., RIERA, J.L., SABATER, F., SABATER, S. & TOJA, J. 1990. A comparative limnological study of the Guadalhorce reservoirs system (Málaga, SE Spain). *Scientia Gerundensis.*, 16(2): 27-41

BELCHER, J.H., SWALE, E.M.F. & HERON, J. 1966. Ecological and morphological observations on a population of *Cyclotella pseudostelligera* Hustedt. *J. Ecol.*, 54: 335- 340.

BETHGE, H. 1925. *Melosira* und ihre Planktonbegleiter. *Pflanzenforschung*, 3: 1-80.

GEISSLER, V. 1982. Experimentelle Untersuchungen zur Variabilität der Schalenmerkmale bei einigen zentrischen Süsswasser Diatomeen. *Nova Hedwigia Beih.*, 73: 211-247.

GEITLER, L. 1932. Der Formwechsel der Pennaten Diatomeen (Kieselalgen). *Arch. Protistenk*, 78: 1-226.

HÅKANSSON, H. 1988. A study of species belonging to the *Cyclotella bodanica/comta* complex (Bacillariophyceae). In: *Proc. 9th. Internat. Diatom Symp.* (F. Round, ed.): 329-354. Biopress. Bristol. Koeltz. Koenigstein.

HARRIS, G.P. 1986. *Phytoplankton ecology*. Chapman & Hall. New York.

HAWORTH, E.Y. & HURLEY, M.A. 1986. Comparison of the stelligeroid taxa of the centric diatom genus *Cyclotella.* In: *Proc. 8th. Internat. Diatom Symp.* (D.G. Mann, ed.): 43-57. Koeltz. Koenigstein.

KACZMARSKA, I. & FRYXELL, G.A. 1986. The diatom genus *Nitzschia*; morphological variation of some small bicapitate species in the Gulf Stream Warm Core rings. In: *Proc. 8th. Internat. Diatom Symp.* (D.G. Mann. ed.): 237-252. Koeltz. Koenigstein.

KISS, K.T. 1984. Occurrence of *Thalassiosira pseudonana* Hasle et Heimdal (Bacillariophyceae) in some rivers in Hungary. *Acta Bot. Hungarica*, 30 (3-4): 277-287.

KLING, H. & HÅKANSSON, H. 1988. A light and electron microscope study of *Cyclotella* species (Bacillariophyceae) from Central and Northern Canadian lakes. *Diatom Res.*, 3(1): 55-82.

LEWIS, W.M. 1984. The diatom sex clock and its evolutionary significance. *Amer. Nat.*, 123: 73-80.

MAKAROVA, I.V. 1982. Principles of the systematics of *Thalassiosira* Cleve and the significance of its taxonomic characters. In: *Proc. 6th. Internat Diatom Symp.* (R. Ross, ed.): 1-12. Koeltz. Koenigstein.

MANN, D.G. 1988. Why didn't Lund see sex in *Asterionella*? A discussion of the diatom cycle in nature. In: *Algae and the aquatic environment* (F.E. Round, ed.): 384-412. Biopress. Bristol.

MARGALEF, R. 1969. Size of centric diatoms as an ecological indicator. *Mitt. Internat. Verein. Limnol.*, 17: 202-210.

MARGALEF, R. 1974. *Ecología*. Omega. Barcelona.

MARGALEF, R. 1983. *Limnología.* Omega. Barcelona.

MARGALEF, R., PLANAS, D., ARMENGOL, J., VIDAL, A., PRAT, N., GUISET, A., TOJA, J. & ESTRADA, M. 1976. *Limnología de los embalses españoles*. Dirección General de Obras Hidraúlicas. Ministerio de Obras Públicas. Madrid.

MARGALEF, R., MIR, M. & ESTRADA, M. 1982. Phytoplankton composition and distribution as an expression of properties of reservoirs. *Can. Wat. Res. J.*, 6: 26-50.

PATRICK, R. 1966. *The diatoms of the United States*. Acad. Natural Sciences of Philadelphia.

PFITZER, E. 1871. Untersuchungen über Bau und Entwicklung der Bacillariaceen (Diatomaceen). In: *Botanische Abhandlungen aus dem Gebiet der Morphologie un Physiologie*. Bonn

ROUND, F.E., CRAWFORD, R.M. & MANN, D.G. 1990. *The diatoms: Biology and morphology of the genera*. Cambridge University Press. Cambridge.

PLANAS, D. 1975. Distribution and productivity of the phytoplankton in Spanish reservoirs. *Ver. Internat. Verein. Limnol.*, 19: 1860-1870.

RIERA, J.L., MARTÍ, E. & MORGUÍ, J.A. in press. Changes in the trophic state of the Spanish reservoirs during the last fifteen years. *Ver. Internat. Verein. Limnol.*, 24.

SABATER, S. & KLEE, R. 1990. Observaciones sobre diatomeas centrales del fitoplancton del río Ebro, con especial interés en algunas pequeñas *Cyclotella. Diatom Res.*, 5 (1): 141-154

SABATER, S. & NOLLA, J. 1991. Distributional patterns of phytoplankton in Spanish reservoirs: first results and comparison after fifteen years. *Ver. Internat. Verein. Limnol.*, 24: 1371-1375

SCHMID, A.M. 1985. Valve morphogenesis in diatoms. In: *Schalen in Natur und Technik* (K. Bach & B. Burkhardt, eds.): 300-316. Mitt. Instituts für leichte Flächentragwerke, 28.

STOERMER, E.F., EMMERT, G. & SCHELSKE, C.L. 1989. Morphological variation of *Stephanodiscus niagarae* Ehr. (Bacillariophyta) in a Lake Ontario sediment core. *J. Paleolimnology*, 2: 227-236.

THERIOT, E. 1988. An empirically based model in rotational elements in centric diatoms with comments on ratios in phycology. *J. Phycol.*, 24: 400-407.

THERIOT, E., HÅKANSSON, H. & STOERMER, E.F. 1988. Morphometric analysis of *Stephanodiscus alpinus* (Bacillariophyceae) and its morphology as an indicator of lake trophic status. *Phycologia*, 27 (4): 485-493.

WIEDLING, S. 1948. Beitrage zur Kenntnis der vegetativen Vermehrung der Diatomeen. *Botanika Notiser*, 1948 (3): 322-354.

(J. D. Ros & N. Prat, eds.). 1991. *Oecologia aquatica*, 10: 61-76

CONTRIBUTION OF PHYTOPLANKTON AND PERIPHYTON TO THE PRODUCTION IN A RESERVOIR OF S.W. SPAIN

JULIA TOJA[1] & MARÍA ADELA CASCO[2]

[1] Departamento de Biología Vegetal y Ecología. Facultad de Biología. Universidad de Sevilla. Ap. 1095. 41080 Sevilla. Spain.
[2] Cátedra de Plantas Celulares. Facultad de Ciencias Naturales y Museo. Pº del Bosque, s/n. 1900 La Plata. Argentina

Received: February 1990

SUMMARY

The phytoplanktonic and periphytic communities of La Minilla reservoir were studied from September 1987 to August 1988. The aim of the present paper is to assess their contribution to total primary production.

The sampling station for phytoplankton was situated in the middle of the reservoir, 500 m from the dam, above the deepest point. The periphyton was studied in a closed bay using slates as experimental substrata. Two kinds of samplers were used with these experimental substrata: potential production samplers (PPS) that always had the same position with respect to the water top, and real production samplers (RPS) that always had the same position with respect to the bottom.

The dominant species in both communities were small and unicellular: *Cyclotella ocellata*, *Rhodomonas minuta* and *Chrysidalis peritaphrena* in phytoplankton, and *Achnanthes minutissima*, *A. linearis* and *Geitleribactron periphyticum* in periphyton.

The production of both communities was estimated by the amounts of chlorophyll *a*. In order to compare the contribution of each community to the total production, the volume of water in the reservoir was estimated and the surface able to be colonized was determined by echosounding.

The annual mean of potential production of the periphytic community was the 5% of the total production (maximum 16% in spring) and the annual mean of real production was 2.44% of the total production (maximum 6% in spring). The differences between potential and real periphytic production were due to the water level fluctuations in autumn and winter and because of grazing in spring and summer.

KEY WORDS: Phytoplankton, periphyton, reservoirs, production

INTRODUCTION

In spite of the numerous studies that have been carried out in the last 20 years in Spanish reservoirs, only occasionally has anything been done on benthic microalgae, such as the work of ALVAREZ COBELAS (1982) on the floristic composition of El Vellón reservoir. The high fluctuations of water level that occur in most of the Spanish reservoirs are considered to be unfavourable for the development of well-structured benthic communities and so their contribution to the total production is of little importance. As yet, though, there are not sufficient studies about this subject. The situation of the studies on benthos in reservoirs and their relation to plankton in the rest of the world also leaves much to be desired.

On the basis of the initial concept that periphyton and phytobenthos are poor communities in the Spanish reservoirs due to water level fluctuations, a study was initiated in a reservoir in the south of Spain (Fig. 1) with the aim, among others, of determining the contribution of periphyton and phytobenthos to the total primary production.

Considering the water level fluctuation and the input of nutrients as related factors and limiting to both benthic and planktonic production, we find a series of situations in which water fluctuation affects the benthic organisms negatively, by preventing the settlement of mature communities as they undergo changes in their environment. And it affects the plankton positively since floods mean an input of nutrients and elements that these organisms use more rapidly than the benthic ones. Plankton is then more ready to reproduce and it grows in a way that shades and limits the algae on the bottom surface.

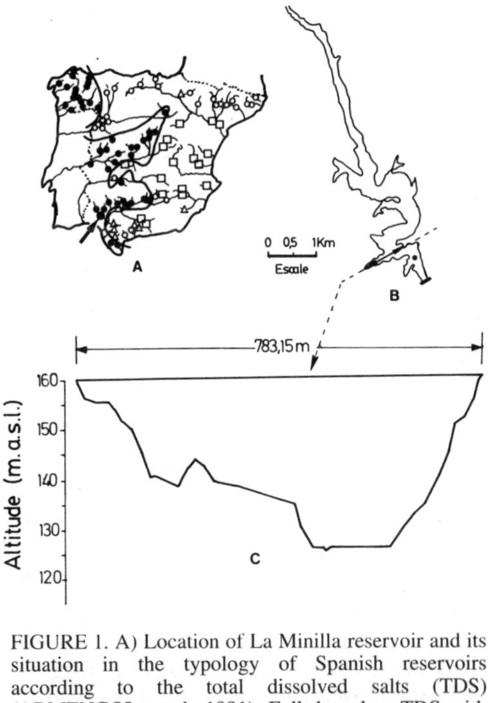

FIGURE 1. A) Location of La Minilla reservoir and its situation in the typology of Spanish reservoirs according to the total dissolved salts (TDS) (ARMENGOL *et al.*, 1991). Full dots, low TDS with high silicate concentrations; open dots, Bicarbonate waters with intermediate TDS; squares, Sulfate waters with high TDS, and triangles, very high TDS with dominance of chloride. B) Location in La Minilla reservoir of the phytoplankton sampling point (dot) and the transect (arrow) for the study of periphyton. C) Section of the transect measured by echosounding.

Our studies were centered on La Minilla reservoir (Fig. 1) since its general limnology is well known (TOJA, 1980a, 1980b, 1983, 1984, 1990; TOJA *et al.*, 1983; SANCHO & GRANADO, 1988; GABELLONE & GUISANDE, 1989). Several parameters have been characterized as high constraints for the development of periphyton: 1) the great water level fluctuation (12,4 m during the period of study) that prevents the settlement of macrophytes; 2) the scarcity of nutrients; 3) low water transparency due to inorganic seston; 4) the instability of the substratum and steep slopes.

The goal of this paper is to assess the contribution of periphyton and phytoplankton to the total primary production of the reservoir.

GEOGRAPHY AND MORPHOMETRY

La Minilla reservoir is located in the Rivera de Huelva river, a tributary of the Guadalquivir river, at 37°43'58''N and 6°9'56''W. It holds a maximum capacity of 60 Hm^3, has a maximum depth of 37 m and a mean depth of 20.83 m. The perimeter at full capacity is 14.5 km. This reservoir is used for the water supply of Seville.

Most of the drainage basin of La Minilla reservoir presents Silurian rocks with some granitic upgrowth and other plutonic rocks. The reservoir is settled on Silurian slates.

La Minilla is included in the Group I of Spanish reservoirs, that have water with low total dissolved salts (less than 250 mg l^{-1}) with a balanced relative composition $CO_3 > SO_4 = Cl$. It is characterized by high silicate concentrations and lower buffering capacity (MARGALEF *et al.*, 1976; ARMENGOL *et al.*, 1991).

MATERIALS AND METHODS

Samples were collected monthly from September 1987 to August 1988. A Van Dorn sampler of 6 l of capacity was employed to collect the samples for chemical and phytoplanktonic analysis. The phytoplanktonic study was carried out in the deepest area, at 500 m from the dam and at several depths from 0 m to 35 m. This was the only vertical profile used to characterize the reservoir and it was based on the work by TOJA (1990) who from an extensive sampling concluded that the most prominent differences took place seasonally and not spatially.

The benthic studies were developed at the bay crossed by the transect designed from an extensive benthic sampling (CASCO & TOJA, 1991; Fig. 1). Three samples were collected for the analysis of nutrients: one was obtained at the mean depth of the potential production sampler (PPS) (at 2 m depth) and the other two were obtained from the real production samplers (RPS) at 1,5 and 3 m depth.

Temperature was measured *in situ* with a mercury thermometer; conductivity with a portable YSI conductivimeter and light extinction was compared using a 30 cm diameter Secchi disc; the extinction light coefficient was calculated using k=1.7/Dm (MARGALEF, 1983) and k=2.5/Dm (MARGALEF *et al.*, 1976) equations. The samples for chemical analysis were stored in a refrigerator at 4°C . All analyses were made before 72 hours. Alkalinity was measured by titration using H_2SO_4 with mixed methyl red bromocresol green indicator (APHA, 1971). Soluble phosphate was determined according to the ascorbic acid method (MURPHY & RILEY, 1963). Nitrate was measured by reduction to nitrite through a Cd-Cu column (MORRIS & RILEY, 1963) and nitrite by the colorimetric method of SHINN (1941), all of them from STRICKLAND & PARSONS (1965). Ammonium was determined using Nessler reactive after precipitation with $ZnSO_4$ and OHNa (APHA, 1971).

In order to study periphyton, clear slates from the littoral of the reservoir were employed, since this type of substratum composes almost all the littoral of the reservoir at least up to the level of the photic zone. Below 10 m of depth the proportion of sediment increased. At the sampling area chosen, based on the extensive sampling, the following experimental surfaces were placed: potential production samplers (PPS), which comprised a float from which the substrata were suspended at 5 different depths: 0.5-1-1.5-2.5-3.5 m, and real production samplers (RPS) with substrata at two levels (R1: 1-1.5 m; R2: 2.5-3.5 m of depth) but in which the distance to the bottom was constant in spite of the water level changes (Fig. 2). Every substratum was collected and removed monthly, always placing the RPS at the depths already mentioned.

Enough substrata were placed at each depth to produce two replicates of the analysis, both for the identification and

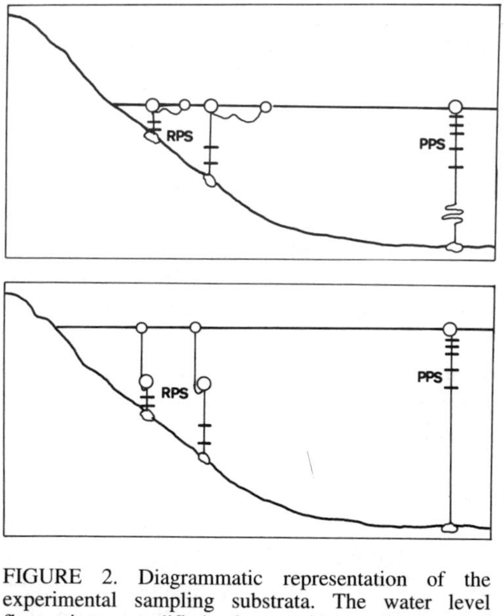

FIGURE 2. Diagrammatic representation of the experimental sampling substrata. The water level fluctuations modified the position of the real production samplers (RPS) with respect to the water surface and the position of potential production samplers (PPS) with respect to the bottom.

counting of species and for the photosynthetic pigment determination.

In order to foresee the situation in which the RPS were dried or below the photic zone, stones or sediment were collected using a dredge or a core sampler respectively on each sampling day at different depths (up to 30 m). The amount of photosynthetic pigments were determined in the same way as for the experimental surfaces.

Phytoplankton was fixed with Lugol and the number of cells were counted following UTERMÖHL's (1958) method. A volume of 25 or 50 ml of each sample was sedimented and the fields were counted at random under a 400x magnification. To determine the floristic composition of the periphyton, samples were fixed with a formalin solution of 4%. On each stone an area was scratched whose surface had been calculated by wrapping it in aluminium foil and its weight compared with the weight of a known area of the same kind of foil. The use of this method allowed the inclusion of every crack on the stone in the determination of the surface for colonization. The solution obtained was homogenized and then aliquots of 50fl were taken and fields were counted at random at 400x under the optical microscope. Colonies, coenobies and filaments in both plankton and benthos have been considered as individuals.

In the photosynthetic pigment determination, 1 litre was filtered for the phytoplankton at each depth through a Whatman GF/C filter. Two pieces of rock of a known area were scratched for the benthos determination. Methanol was used in both analyses as solvent. The methanolic extracts were preserved in a freezer for less than 10 days. Then they were filtered through Whatman GF/C filters and their absorbance was read on a Perkin Elmer 550 SE UV/VIS spectrophotometer. After acidification with HCl 0.4 N the absorbance at 750 and 665 nm was read again in each sample for the correction of the phaeopigments (LORENZEN, 1967; VARELA, 1981).

The surface and volume available for benthic and planktonic colonization respectively, were determined from the bathimetric data obtained by echosounding (50 KHz, 12° transducer, pulse length: high 1.1 msec, low 0.3 msec) along the transect chosen (Fig. 1; HAKANSON, 1981). Four sounding tracks were performed and the perimeter of the bottom was estimated as the mean value obtained from the 4 echoprofiles with respect to the maximum height of 160 m asl. A strip of one meter in width was used to calculate the surface and volume to be colonized according to the water level and the depth of light penetration.

The total amount of planktonic chlorophyll in this strip was calculated by multiplying the chlorophyll *a* concentration of each sampling depth by the water volume of the corresponding strip. The results of each strip were all added up to the limit of the euphotic zone. The benthos was obtained in a similar way, by calculating the total surface to be colonized at each sampling depth and by multiplying it by the corresponding chlorophyll concentration. If the photic zone were wider than the sampling depth, the chlorophyll obtained from the natural substratum was then employed, since, as previous studies had shown (Casco, unpublished data), substrata acquired their stability of colonization before one month of exposure.

RESULTS

Physico-chemical data obtained in the photic zone from the open water and from the bay where the samplers had been placed are shown on Table I.

The temperature of the open water photic zone ranged from 11.80 ^{o}C in February 1988 to 27.10 ^{o}C in July 1988. Thermal stratification started in March and

TABLE I. Data of some physico-chemical parameters obtained from the water taken in the sampling point of plankton (at a depth of 2 m), and from the potential samplers (PPS, at 2 m of depth) and from the real samplers (RPS1 at 1.5 m and RPS2 at 2.5 m of depth). T: temperature (°C). C: Conductivity (µS cm^{-1}). A: Alkalinity (meq l^{-1}). No: nitrate (µg-at N l^{-1}). Ni: Nitrite (µg-at N l^{-1}). A: Ammonia (µg-at N l^{-1}). P: Phosphate (µg-at P l^{-1}).

Date	T	C	A	No	Ni	A	P
Open Water							
22-09-87	23.5	276	1.87	0.97	0.12	63.26	0.46
28-10-87	19.5	260	1.88	0.39	0.36	1.07	0.50
27-11-87	13.9	256	1.83	9.32	0.13	24.33	0.07
08-01-88	12.0	160	0.93	28.03	0.64	5.10	1.33
08-02-88	11.8	173	1.15	28.63	0.24	6.20	1.80
16-03-88	13.9	182	1.26	22.29	0.10	1.30	0.00
16-04-88	17.8	215	1.57	8.25	0.38	5.26	0.00
20-05-88	21.6	218	1.65	5.25	0.38	6.76	0.00
24-06-88	24.3	216	1.68	1.50	0.04	5.06	0.00
27-07-88	27.1	218	1.83	4.58	0.06	3.35	0.20
27-08-88	26.2	250	1.96	0.95	0.09	4.82	0.00
PPS							
22-09-87	26.0		2.08	0.00	01.00	0.71	1.88
28-10.87	17.5		1.64	9.69	9.69	15.31	0.00
27-11-87	13.0	11.8		0.56	0.56		0.00
08-01-88	10.8	11.8	0.71	0.63	0.63	2.83	1.09
08-02-88	10.0	120	1.10	0.41	0.41	6.64	1.02
16-03-88	13.0	148	1.33	0.19	0.19	2.98	0.00
16-04-88	15.0	228	1.55	0.76	0.76	6.43	0.22
20-05-88	16.0	190	1.66	0.23	0.23	3.93	0.07
24-06-88	23.0	218	1.69	0.13	0.13	7.97	0.11
27-07-88	28.0	272	1.81	0.15	0.15	7.89	0.34
27.08-88	26.0	2770	1.95	0.27	0.27	0.67	3.68

Date	T	C	A	No	Ni	A	P
RPS1							
22-09-87	26.0		2.46	0.03	0.03	1.43	0.18
28-10-87	15.7		1.55	0.00	0.00	4.84	0.00
27-11-87	13.2	118	1.00	0.48	0.48		0.00
08-01-88	11.0	118	0.71	0.63	0.63	3.11	1.08
08-02-88	10.0	120	1.19	0.33	0.33	7.31	0.85
16-03-88	18.0	151	1.35	0.30	0.30	2.98	0.00
16-04-88	17.5	228	1.30	0.75	0.75		2.00
20-05-88	16.0	190	1.64	0.21	0.21	5.71	0.07
24-06-88	24.0	210	1.69	0.12	0.12	6.04	0.27
27-07-88	28.0	258	1.83	0.12	0.12	6.18	0.00
27-08-88	26.0	180	1.62	0.27	0.27	1.12	4.52
RPS2							
22-09-87	25.2		2.36	0.00	0.00	1.78	0.18
28-10-87	17.5		1.57	0.00	0.00	4.60	0.00
27-11-87	13.2	118	1.00	0.48	0.48		0.00
08-01-88	10.5	118	0.71	0.63	0.63	2.83	1.09
08-02-88	10.0	120	1.14	0.36	0.36	11.29	0.73
16-03-88	15.0	153	1.33	0.25	0.25	2.98	0.00
16-04-88	14.5	228	1.30	0.46	0.46	1.90	0.16
20-05-88	16.0	190	1.64	0.21	0.21	5.71	0.07
24-06-88	23-7	210	1.71	0.09	0.09	3.30	1.10
27-07-88	28.0	258	1.81	0.10	0.10	13.43	1.46
27-08-88	15.0	180	1.62	0.27	0.27	1.12	4.52

the thermocline was established at a depth of between 5 and 10 m in April. The maximum gradient was produced in July with a decrease of 1.96 °C per meter.

Conductivity ranged from 160 to 276 µS cm^{-1} in the open water. However, the bay showed a rather different pattern. Conductivity here is slightly below that of the open water during rainy periods due to the direct effect of run-off inputs and to a delay in the mixing with the water that arrives from the rest of the drainage basin. The values were the same only in July, then the conductivity in the bay started to increase at a proportional rate as a result of evaporation and resuspension of the sediment when the water level fell.

Nitrogen and phosphorus were similarly concentrated in the open water and the littoral and their variation with time followed a similar pattern. However, there were differences at certain moments which are mostly due to the removal of sediments from the shore caused by the water level fluctuations. These fluctuations can be observed in figure 3, where the water level variation of the reservoir is indicated for the study period, as well as the location of the RPS (from their emplacement to their removal) with respect to the water surface and the limit of the euphotic zone. The heavy rainfall during the autumn-winter period of 1987-88 determined that the reservoir was practically full during the rest of the study. The water level started to fall only during July and August, but it was always higher than in the summer of 1987 when the width of the fluctuation reached 17 m.

This heavy rainfall also provoked a high increase in suspended inorganic seston during December, January and February, limiting the width of the photic zone to a maximum of 2.5 m. During this period the turnover rate of water in the epilimnion was also very high: the water in the photic zone was renewed 4.77 times in January and 3.64 times in February. Thus, plankton was negatively affected in such a way that in January the amount of chlorophyll was very low (0.5 mg m^{-3}) and in February only 1.25 mg m^{-3} of chlorophyll were detected.

Table II shows the dominant species in both plankton and the littoral areas of the reservoir for every sampling date. It also indicates the number of individuals per millilitre in each area and the total number of ind. ml^{-1} in the community. The plankton community was represented by an integrated sample of the whole photic zone, whereas the benthos was represented by the average result from the samplers. The maximum density in the plankton was 4239 indiv ml^{-1} in October 1987 and the minimum was 123 ind. ml^{-1} in February 1988 (with a chlorophyll concentration of 7.42 and 1.25 mg m^{-3} respectively). There is a seasonal tendency to establish the maximum values on the surface as a result of a lower light intensity in winter and, particularly during this winter, when the turbidity constrained the development of organisms to the upper layers, *Cyclotella ocellata* was the dominant species during almost the whole sampling period, although its dominance was substituted by *Rhodomonas minuta* during the winter of 1987-88. The motility of this last species enables it to migrate to the upper layers where light intensity is less limiting. This situation has already been verified (TOJA, 1984). However, light intensity is not the only cause of this shift in the dominance of species: alkalinity seems to be of greater importance (Table I). MARGALEF *et al.* (1976) mentioned that *Cyclotella ocellata* is a better competitor in calcium-enriched waters.

Aphanocapsa elachista is present almost all the year and it is generally limited to the hypolimnion. In October it extended to the whole profile and co-dominated with

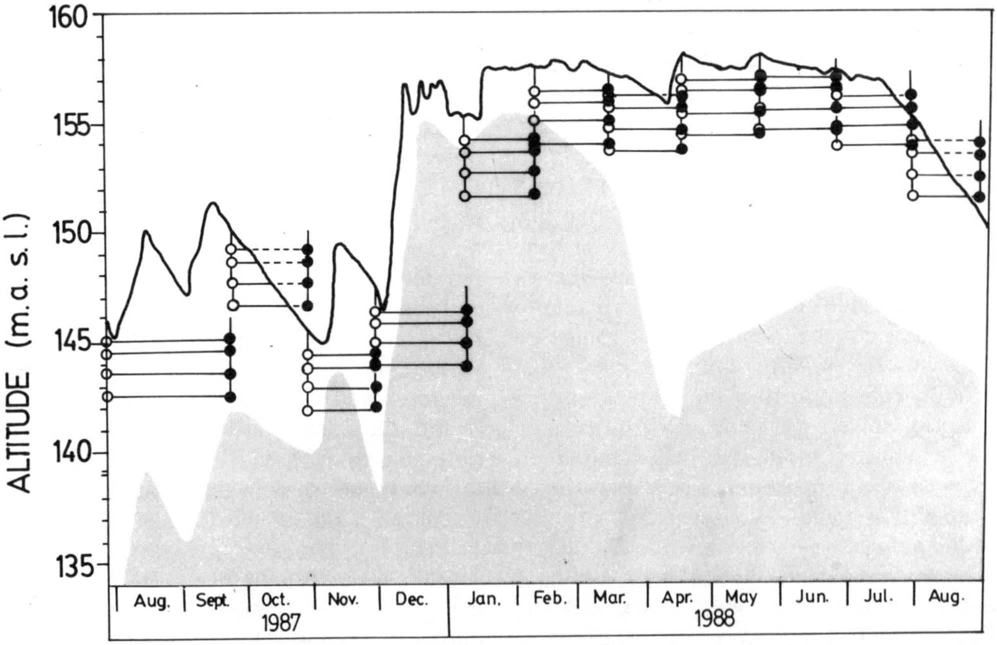

FIGURE 3. Variation of the water level during the study period. The location of the periphyton real production samplers (RPS) at the moment of their installation (open dots) and removal (full dots) have been marked with respect to their distance to the water surface. The depths without light have been shadowed.

TABLE II. Dominant species and total number of individuals per millilitre on each sampling date. For the phytoplankton community the mean value of ind ml^{-1} from the photic zone has been used. For potential and real production benthos samplers the mean value of ind cm^{-2} in all samplers has been used. The coenobia and colonies have been counted as one individual. Only live individuals have been considered.

SEPTEMBER 1987					
Cyclotella ocellata	1410	Lyngbya sp.	1061148	G. periphyticum	313544
Chrysidalis peritaphrena	1063	Lyngbya perelegans	244156	Achnanthes linearis	135112
Oocystis marssoni	294	Achnanthes linearis	186516	Achnanthes minutissima	6499
		G. periphyticum	138800	Lyngbya perelegans	5914
Total	3491	Total	1845294	Total	488576
OCTOBER 1987					
Lyngbia limnetica	1137	Achnanthes minutissima	172645	G. periphyticum	15658
Aphanocapsa elachista	991	G. periphyticum	142100	Achnanthes linearis	2315
Cyclotella ocellata	810	Achnanthes linearis	86165	Lyngbia perelegans	1795
Total	4239	Total	451680	Total	28174
NOVEMBER 1987					
Rhodomonas minuta	1097	A. minutissima	244742	A. minutissima	174497
Cyclotella ocellata	466	Achnanthes linearis	117480	Achnanthes linearis	127653
Lyngbia limnetica	190	Synedra acus	57943	G. peryphyticum	127628
Total	2422	Total	492595	Total	4300927
JANUARY 1988					
Rhodomonas minuta	374	A. minutissima	1994200	A. minutissima	249
Cyclotella ocellata	14	Navicula placentula	30027	A. linearis	61
		Achnanthes linearis	7629	Navicula placentula	37
Total	401	Total	2082615	Total	710
FEBRUARY 1988					
Cryptomonas ovata	28	A. minutissima	1319276	Navicula placentula	54
Rhodomonas minuta	22	Achnanthes plinensis	42082	A. minutissima	21
Trachelomonas intermedia	11	Gomphonema herculeana	41150	Achnanthes linearis	17
Total	123	Total	1465647	Total	143
MARCH 1988					
Chrisidalis peritaphrena	299	A. minutissima	4650476	A. minutissima	156799
Cyclotella ocellata	259	Chloroccocal sp. 1	127710	Cymbella laevis	74975
Navicula sp. 3	149	Achnanthes lanceolata	28725	Amphora veneta	61142
Total	1446	Total	4897814	Total	838927
APRIL 1988					
Cyclotella ocellata	792	A. minutissima	4650476	A. minutissima	6330098
Rhodomonas minuta	734	Achnanthes linearis	636944	Achnanthes linearis	366695
Oocystis parva	112				
Total	2858	Total	4186725	Total	1071124
MAY 1988					
Mycrocystis minutissima	486	A. minutissima	7668923	A. minutissima	7872246
Rhodomonas minuta	233	Achnanthes linearis	280588	Achnanthes linearis	94018
Cyclotella ocellata	180				
Total	1115	Total	7975112	Total	7975100
JUNE 1988					
Aphanocapsa elachista	725	A. minutissima	5250513	A. minutissima	1680203
Rhodomonas minuta	576	Achnanthes linearis	280588	Achnanthes linearis	1538127
Chrysidalis peritaphrena	313				
Cyclotella ocellata	180				
Total	2294	Total	7325441	Total	3256618
JULY 1988					
Cyclotella ocellata	723	A. minutissima	1421200	A. minutissima	1093640
Rhodomonas minuta	347	Achnanthes linearis	1476822	Achnanthes linearis	1025433
Chrysidalis peritaphrena	108				
M. tennuissima	99				
Total	2872	Total	2986417	Total	2243158
AUGUST 1988					
Chrysidalis peritaphrena	723	A. minutissima	1436783		
Cyclotella ocellata	276	Achnanthes linearis	303244	DRY	
M. tennuissima	245				
Total	1421	Total	2234315		

Lyngbya limnetica. The presence of this species decreased during the mixing period and was constrained to the hypolimnon until June when it co-dominated again, this time with *Rhodomonas minuta*.

With respect to the benthos, in general, it was noted that the number of species of diatoms without mucilaginous stalks predominated and that there was little representation of filamentous chlorophytes. In most cases, the resulting physiognomy was of a community developed in only one stratum.

Species of *Achnanthes* (especially *A. minutissima*) were present in great numbers during all the sampling period both in the potential and in the real samplers. Even after April 1988, *A. minutissima* and *A. linearis* composed almost the total community (Table II). Diversity reached a maximum in March but decreased after May. A conspicuous population of *Chlorohydra* sp. was established in the potential samplers during July (Table III). This allowed the chlorophyll level to be maintained though the number of individuals in the periphyton decreased. *Geitleribactron periphyticum* is another important species in this community, but it shows a greater relative abundance in the real samplers, where it grew in a layer.

The major qualitative difference between both experimental substrata was observed during April, and this was also the only time in which large green algae (*Pseudoulvella americana*, *Coleochaete scutata*, *C. soluta* and *Stigeoclonium* sp.) were present.

Generally in plankton, the maximum amount of chlorophyll during winter corresponded to the surface (Fig. 4) especially between January and April as a result of the lower light intensity and the turbidity of the water and of the output of water between 2 and 4 m in depth (154-156 m asl) in this particular winter. This brought about the exploitation of these levels (TOJA, 1982). When, in summer, the photic zone increased to an average of 10 m, the maximum amount of chlorophyll was obtained at a depth of between 5 and 10 m, which showed a certain relation to the depth of the thermocline.

The experimental surfaces were at a constant distance from the surface (Fig. 2). Thus, potential production of benthos was not affected by fluctuations in the water level, but rather by the turbidity and the availability of nutrients (Table I). In figure 4 chlorophyll *a* values are shown.

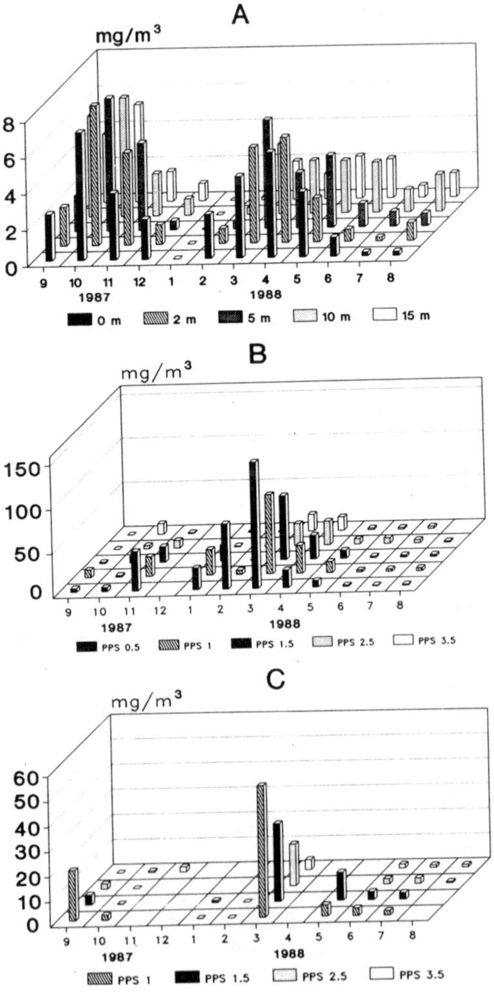

FIGURE 4. Variation of the chlorophyll concentrations in the plankton samples (A) and the potential production (B) and real production (C) benthos samplers.

TABLE III. Fauna associated with the potential and real production samplers of periphyton. 3: abundant; 2: frequent; 1: scarce.

	PPS 1987-88											RPS 1987-88										
	S	O	N	J	F	M	A	M	J	J	A	S	O	N	J	F	M	A	M	J	J	A
Rhizopoda																						
Diflugia sp.	.	1	.	.	.	.	.	.	.	1	.	.	.	.	.	.	.	.	.	.	1	.
Amoeba sp.	.	.	2	.	.	.	1	.	.	.	1	.	.	.	.	.	.	1	1	.	.	.
Pseudoheliozoa																						
Vampyrella sp.	.	.	.	.	.	2	.	.	.	.	.	.	.	.	.	.	3	.	.	.	.	.
Ciliophora																						
Vorticella sp.	3	2	.	1	1	.	.	.	.	1	.	3	2	2	1	.	1	.	.	.	.	.
Codonella sp.												.	1	.	.	.	.	.	.	.	.	.
Coelenterata																						
Chlorohydra sp.	.	.	.	.	.	1	.	.	3	1	.	.	.	.	.	.	.	.	.	.	.	.
Nematoda	.	.	.	.	.	.	.	.	.	.	.	.	.	.	.	.	.	.	.	.	2	2
Mollusca																						
Cl. Gasteropoda	.	.	.	.	.	.	1	.	.	.	.	.	.	.	.	.	.	.	.	.	.	.
	.	.	.	.	.	1	2	2	.	.	.	.	.	.	.	.	.	.	.	.	.	.
Ephemeroptera																						
Caenis luctuosa	2	2	2	.	.	.	.	.	.	.	.	.	.	.	.	.	.	.	.	.	.	.
Diptera																						
F. Chironomidae	1	1	.	.	.	.	1	1	1	.	.	.	.	.	.	.	.	.	.	.	3	2

Chlorophyll increased from August to November 1987 and reached a value of 50 mg m^{-2} at a depth of 0.5 m possibly because of an increase in nutrients due to the water input of the first rainfalls. Then, the massive rainwater discharge in winter produced a decrease, although much less marked than that of the plankton since the benthic community was not washed along by the water input. Nevertheless, the diminished photic zone limited the production to the upper layers, as can be proved by the great difference in chlorophyll concentration from the substratum at 0.5 m with respect to the rest of surfaces during February. The maximum concentrations were detected in March and April (up to 130 mg m^{-2} of chlorophyll) and were linked to the availability of nutrients and the increase of the photic zone. A much steeper decline than in the plankton can be observed after May. This may be produced by a lesser availability of nutrients or by a negative effect of light on the epiphytic community. Productivity was kept relatively low during the rest of the summer and values similar to the previous year were only attained in August.

The RPS, however, were subjected to the water level fluctuations in such a way that they were above the water surface for certain periods of time (October-November 1987; July-August 1988) or else they were so deep that they were found below the photic zone (December 1987, January-February 1988).

Although the real production samplers always showed lower values of chlorophyll (maximum of 70 mg m^{-2}), they evolved in a similar way to the potential production ones (Fig. 4). However it has to be noted that the surfaces at 1-1.5 and 2.5 m were lost in November 1987.

From March to July 1987 (Fig. 3) both PPS and RPS were subject to the same conditions of water level, light and

TABLE IV. Volume and surface that can be colonized by the phytoplankton and periphyton communities in a section of 1 m of width of the transect considered. Data of total amounts of chlorophyll *a* of phytoplankton and periphyton in the photic zone at each sampling date and both potential and real percentage of the total chlorophyll at each sampling date. The altitude of the water level (m asl) and the depth of the photic zone (z) are also given.

Date	Volume m^3	Area m^2	Plankton mg Cla m^{-3}	PPB mg Cla m^{-2}	PPB %	RPB mg Cla m^{-2}	RPB %	level m asl	Z m
22-09-88	4532	80	14401.44	282.02	1.92	363.84	2.46	152.00	8.12
27-10-87	3238	43	3237.95	90.72	0.37			146.50	5.56
21-11-87	8288	131	32768.39	541.79	1.62	189.78	0.57	149.20	10.83
08-01-88	1275	68	510.19	1746.40	77.39	65.58	11.38	156.35	1.81
08-02-88	1671	32	1986.12	400.24	16.77	33.17	1.64	158.38	2.30
16-03-88	3548	102	19541.81	3167.63	13.94	1259.01	6.05	158.60	5.09
16-04-88	7257	200	30267.04	1808.88	5.63	1296.72	4.10	158.80	14.08
20-05-88	8045	191	24112.99	407.47	1.66	293.21	0.98	158.90	12.20
24-06-88	6528	158	11942.42	219.17	2.38	203.92	1.67	158.16	9.75
29-07-88	6199	103	5216.11	215.51	3.96	173.71	3.22	155.50	9.71
29-08-88	4295	61	4945.25	107.13	2.12	51.13	1.02	150.86	7.31
Annual mean	4995	106	15448.43	823.54	5.06	387.60	2.44		

nutrients. Yet, the real production was lower than the potential production. This can only be explained by selective predation. The real production samplers always showed a greater colonization by animals (Table III), possibly due to a higher accessibility for the benthic fauna, than the potential production samplers, which were suspended at a medium depth.

Although real and potential benthic productivity per m^2 was higher than planktonic productivity per m^3, especially in spring, the volume available for planktonic colonization is much greater than the benthic surface available during the different seasons. Thus, the total amount of planktonic chlorophyll in absolute terms is much higher than the total benthic chlorophyll (Table IV).

The benthic percentage contribution to the total production is especially high in spring, reaching 16% for the PPS. However, the maximum value for the RPS is 6% (except during the unusual situation of winter time, when only the benthos is productive as the plankton is strongly affected).

The mean surface to be colonized by the benthos during the year was approximately 2% of the volume to be colonized by the plankton. On calculating the mean annual amount of chlorophyll both in the plankton and benthos from the RPS, it can be observed that the benthic chlorophyll represented 2.44% of the total chlorophyll, that is, the benthic production per m^2 is approximately equal to the planktonic one per m^3 of volume. These results are in accordance with those obtained by KAIRESALO (1980) in the oligotrophic lake Pääjärvi (Finland). KAIRESALO (1980) also found that the benthic share to the total production is 6%, 75% being from plankton and the rest from macrophytes. The latter component does not exist in La Minilla and it is rare in the rest of Spanish reservoirs.

DISCUSSION

Most of the work on algae living on the bottom surface, attached or not, has been developed in lakes and rivers. Descriptive studies have been carried out both at the community and species levels (MARGALEF, 1947a; DOUGLAS, 1958; CASTENHOLZ, 1960; TELL, 1972, 1979; ETTL, 1973; ROSENMARIN & GELIN, 1978; SULLIVAN, 1979; KOUWETS,

1980; HUDON & BOURGET, 1981; MUNTEANU & MALY, 1981; ROUNIK & GREGORY, 1981; BLUM, 1982; ANTONIETTI, 1983; KAIRESALO, 1984; MEULEMANS & ROSS, 1985; MEULEMANS, 1988). However, in spite of this wealth of descriptive data, there is little research work into the causes that determine the presence, attachment and growth of periphyton (HILLBRICHT-ILKOWSKA *et al.*, 1972; MARCUS, 1980; ROOS, 1983; AUSTIN & DENISEGER, 1985; KAIRESALO *et al.*, 1985; KAIRESALO & KOSKIMIES, 1987) as well as few *in situ* studies on primary production (WETZEL, 1963; HILLBRICHT-ILKOWSKA *et al.*, 1972; DAVIS BROWN, 1976; CATTANEO & KALFF, 1979; AIZAKI, 1979; APESTEGUIA & MARTA, 1979; ROGERS & BREEM, 1981; CARIGNAN & KALFF, 1982; FONTAINE & NIGH, 1983; KAIRESALO, 1983; DELBECQUE, 1983; BUTTON, 1986; ANTONIETTI *et al.*, 1987) or on the role of the periphyton in the functioning of the system analyzing the energetic metabolism of the littoral communities (WETZEL, 1960; CONFER, 1971; ALLEN, 1971; KAIRESALO, 1977; CATTANEO & KALFF, 1980; KAIRESALO, 1980).

It is widely known that research into the periphytic communities is very difficult. There is a great spatial heterogeneity in the microdistribution of organisms and in the different habitats from lotic and lentic environments. The methodology to deal with each piece of research is also highly diverse and has been the subject of numerous studies (MARGALEF, 1947b, 1949; KEVERN *et al.*, 1960; CASTENHOLZ, 1961; WETZEL, 1965; WETZEL & WESTLAKE, 1969; DUMONT, 1969; HICKMAN, 1969; BOTT & BROCK, 1970; HANSMANN *et al.*, 1971; JONES, 1974; GITTINS, 1976; VARELA, 1981; ELORANTA, 1983; DUFF *et al.*, 1984; ANTONIETTI & FERRINI, 1986; CASCO & D'AMELIO, 1987). Yet it is still necessary to formulate new methods and to adapt those already existing, mainly to explain their functioning.

The study of periphyton started long ago, but its advances have been slow and fragmented. Slow, because much of the work was devoted to interpreting the communities by describing and systematizing them, and fragmented because of the multiplicity of ways to deal with this work, chiefly due to the difficulty in applying proper methodologies. This brings about great trouble when it comes to comparing studies from different authors, as the aim of each of them needs to be explained.

When comparing the photosynthetic activities of different communities either absolute or relative production rates have been used (ELSTER, 1965). Relative production rates can be defined using various biomass parameters of which total cell volume and chlorophyll *a* concentration are the most common (MUNAWAR & MUNAWAR, 1975; KAIRESALO, 1980). In our case, the similar size of dominant species in both plankton and benthos permitted the use of chlorophyll *a* concentrations of each subsystem for this comparison.

Few data existing in the bibliography can be compared with the results from this work. All of them refer to periphyton in lakes but not in reservoirs. However, as has already been mentioned, some similarity with KAIRESALO's (1980) results exist both in the productivity per unit of volume and surface and in the percentage of benthic algae to the total production. MEULEMANS (1988) cited FLICK & KEINER's (1981) work about the contribution of epiphytes attached to *Phragmites communis* in Lake Maarsseveen, where it was estimated to represent 1% of total production. Chlorophyll values from this community range from 10 to 200 mg m^{-2} of substratum, which is in a similar order of magnitude to the values found in this study. MARGALEF (1983) cited Moss's results in some epilithic communities: 127-1200 mg m^{-2}; Szczepanska's results in the periphyton of lakes of Masuria: 300 mg m^{-2}; Moss and Feljöldy's chlorophyll values from several periphytic communities: 110-2350 mg m^{-2}, and Moss's results in communities from mud and sand substrata: 281-286 mg m^{-2}. The chlorophyll concentration of samples from soft sediment found in this study ranged from 0.53 to 85.13 mg m^{-2}, varying according to each season and the depth of the sample.

The chlorophyll values obtained in this work from the samplers are always bordering on the lower limits of the data previously mentioned in the literature, possibly because of the lack of nutrients in the PPS and the effect of the water level fluctuation in the RPS. More similar to our data are TUITE's (1981) results found in a chain of lakes in Kenya and Tanzania, where he found average values of 45 mg m^{-2} in the photic zone.

The results found in this study indicate that the total planktonic production is much higher than the real benthic production. This could be chiefly due to the great difference existing between the volume and the surface to be colonized (which is very small because of the steep slopes of the basin) since both productivities per surface and volume units are similar. Therefore it is expected that in reservoirs with other morphology, that is, with a smaller mean depth, the share of benthos can be much higher. The water level fluctuation in La Minilla also contributes to a considerable decrease in production.

At first, the depth of the photic zone was calculated using k=2.5/Dm as the light extinction coefficient, suggested by MARGALEF *et al.* (1976) as the most suitable way to calculate it for reservoirs. However, the chlorohyll data obtained here both from RPS and PPS indicated that, in fact, below this limit there was enough light to support the development of periphyton so we adopted the equation k=1.7/Dm (MARGALEF, 1983) to fix a limit that is now adjusted to the results obtained, and is also confirmed by the results of RULL *et al.* (1984).

If the water level did not fall (as in a lake of similar morphology) the contribution of benthos would duplicate PPS. However, it would always be small and similar to the data found in oligotrophic lakes (KAIRESALO, 1980; MARGALEF, 1983). Not all of the Spanish reservoirs suffer the wide level fluctuations like those located in the Mediterranean area and devoted to water supply and irrigation. Therefore, the share of the benthos may be somewhat higher than that reported here.

Logically, the higher relative importance of plankton makes the physico-chemical characteristics of the water vary according to the planktonic activity and not the benthic activity. But, from a production point of view, the benthos also plays its role, especially in the littoral area where this production can be considerable for the associated fauna. When considering only the littoral area (up to the limit of the photic zone), the relationship surface/ volume increases and, then, so does also the relative benthic contribution. The annual mean of real production is aproximately 22% of the total in this area of La Minilla.

The effect of fauna on the samplers had been clearly observed when it was proved that real production is approximately half the potential production in the seasons favourable to snails (spring) and nematodes (summer), the rest of the factors being equal for both samplers. Table III shows that the abundance of these organisms is always higher in the RPS than in the PPS, except in the case of *Caenis luctuosa* which was only recorded in winter at a depth of 0.5 m in the PPS. However, this can easily be explained by the increase of water level in the reservoir in this season. Then, the RPS remained too deep to be colonized.

The effect of nematodes was very clear since an inverse relationship between their number (greater in the deepest sustrata) and the number of algae cells existed. The higher relative abundance of *Gleitleribactron periphyticum* in the RPS was probably a result of lower grazing than that suffered by diatoms, which would allow greater proliferation.

It is notable that in July 1988 the number of cells in the PPS was much lower than that of other samples collected near them (Table II), though chlorophyll concentration was similar (Fig. 4). This may be explained by a large population of *Chlorohydra* sp. that proliferated during this month. Then, the chlorophyll concentration could be overestimated in relation to May, June and August when fewer *Chlorohydra* sp. were observed.

It can be concluded from these results that benthic production is impeded more by a rise in water level than by a fall, especially if it is a sudden change. When the water rises the recently flooded areas must be colonized again because either they do not present algae or the algae are in resistant forms. However, previous data from the same reservoir (CASCO & TOJA, 1991) and from direct samples from benthos collected during the study period show that below the photic zone (even down to a depth of 30 m) a considerable amount of living algae can be found. They are rapidly activated at the moment in which the water level falls and they begin to receive light. SUNDBACK & GRANELI (1988) demonstrated that algae can survive for several weeks in the dark and can be rapidly activated when they are exposed to light. MOSS (1977) reported that epipelic algae can tolerate darkness for 3 weeks if they can find aerobic conditions (which is the normal situation in the hypolimnion of La Minilla) and, though their survival is reduced in anaerobic conditions, they can maintain a photosynthetic potential for several days in the absolute absence of oxygen. With relation to the specific composition it is noteworthy that there was a higher number of *Lyngbya perelegans*, *Lyngbya* sp. and *Geitleribactron periphyticum* in deeper samplers than in surface ones.

Finally, though our studies were carried out in one place in the reservoir it is very likely that the benthic production calculated for this section was very similar to the one that would be obtained when considering the total reservoir. Previous studies indicate that there are no significant differences in quantity or composition of phytoplankton (TOJA, 1989) or periphyton (CASCO & TOJA, 1991) along the longitudinal axis of the reservoir. Also, several echosounding tracks performed in 1981 (GUILLÉN & TOJA, unpublished data) show that, generally, the different sections of the reservoir have a similar morphology to the transect designed for this study, except in the tail of the reservoir. which is in fact, the old river bed.

The results of this study show that even in a reservoir where there are such unfavourable conditions for the benthos as in La Minilla, the benthic contribution to the total production is, at least in spring, noticeable. Reservoirs with fewer fluctuations in water level or with a lower mean depth might present a much more considerable benthic production; thus we claim that this subsystem should be paid more attention than to date.

ACKNOWLEDGEMENTS

The Seville Water Supply Company (EMASESA) financed part of this work and supplied physico-chemical data of the water column. Dr. R. Margalef contributed with his suggestions to the design of this study. Dr. C. Granado collaborated in the determination of the reservoir bathimetry. L. Serrano helped with the English writing. The authors thank all of them. M. A. Casco carried out this study with the support of a grant from the CSIC-CONICET.

REFERENCES

APHA. 1971. *Standard Methods for the Examination of Water and Wastewater*. 30th ed. American Public Health Ass. New York.

AIZAKI, M. 1979. Growth rates of microorganisms in a periphyton community. *Jap. J. Limnol.*, 40: 10-19.

ÁLVAREZ COBELAS, M. 1982. *Una limnología de la cuenca del embalse de El Vellón (Madrid)*. Ph. D. Thesis. Universidad Complutense.

ALLEN, H.L. 1971. Primary productivity, chemo-organotrophy, and nutritional interactions of epiphytic algae and bacteria on macrophytes in the littoral of a lake. *Ecological Monographs*,: 97-127.

ANTONIETTI, R. 1983. A rapid extraction technique for photosynthetic pigments and ATP from periphyton communities. In: *Periphyton of freshwater ecosystems (Wetzel, R.G., ed.): 287-289. Junk. The Hague.*

ANTONIETTI, R. & FERRINI, C. 1986. An automatic method for the measurement of the effects of short stresses on the microbial ATP pool. *J. Microbiological Methods*, 6: 21-25.

ANTONIETTI, R., VERONESI, S., FERRINI, C., TAROZZI, L. & PISI, R. 1987. Liquid nitrogen pretreatment in ATP extraction from sediments. *Verh. Internat. Verein. Limnol.*, 23: 910-914.

APESTEGUIA, C. & MARTA, C. 1978. Producción de perifiton en ambientes acuáticos del río Paraná Medio. I. Técnicas de anlisis. *Ecología*, 3: 43-55.

ARMENGOL, J., RIERA J.L. & MORGUI, J.A. 1991. Major ionic composition in the Spanish reservoirs. *Verh. Internat. Verein Limnol.*, 24: 1363-1366

AUSTIN, A. & DENISEGER, J. 1985. Periphyton community changes along a heavy metals gradient in a long narrow lake. *Environ. Exp. Botany*, 25 (1): 41-52.

BLUM, J.L. 1982. Colonization and growth of attached algae at the Lake Michigan water line. *J. Great Lakes Res.*, 8 : 10-15.

BOTT, T.L. & BROCK, T.D. 1970. Growth and metabolism of periphytic bacteria: Methodology. *Limnol. Oceanogr.*, 15: 333-343.

BUTCHER, R.W. 1932a. Studies on the ecology of rivers, II. The micro-flora of rivers with special reference to the algae on the river bed. *Ann. Bot.*, 46: 813-861.

BUTCHER, R.W. 1932b. An apparatus for studying the growth of epiphytic algae with special reference to the River Tees. *Trans. North. Naturalist's Union*, 1: 1-15.

BUTTON, D.K. 1986. Affinity of organisms for substrate. *Limnol. Oceanogr.*, 31 (2): 453-456.

CARIGNAN, R. & KALFF, J. 1982. Phosphorus release by submerged macrophytes: significance to epiphyton and phytoplankton. *Limnol. Oceanogr.*, 27: 419-427.

CASCO, M.A. & D'AMELIO, D. 1987. Un muestreador permanente de perifiton para la obtención simultánea de muestras con diferente tiempo de exposición. *Actas IV Congreso Español Limnología: 405-414.*

CASCO, M. A. & TOJA, J. 1991. Benthic microalgae of La Minilla reservoir (SW Spain). *Verh. Internat. Verein. Limnol.*, 24: 1386-1389

CASTENHOLZ, R.W. 1960. Seasonal changes in the attached algae of freshwater and saline lakes in the lower Grand Coulee, Washington. *Limnol. Oceanogr.*, 5: 1-28.

CASTENHOLZ, R.W. 1961. An evaluation of submerged glass method of estimating production of attached algae. *Verh. Intern. Verein. Limnol.*, 14: 155-159.

CATTANEO, A. & KALFF, J. 1979. Primary production of algae growing on natural and artifical aquatic plants: a study of interactions between epiphytes and their substrate. *Limnol. Oceanogr.*, 24(6): 1031- 1037.

CATTANEO, A. & KALFF, J. 1980. The relative contribution of aquatic macrophytes and their epiphytes to the production of macrophyte beds. *Limnol. Oceanogr.*, 25 (2) : 280-289.

CONFER, J.L. 1971. Interrelations among plankton, attached algae, and the phosphorus cycle in artificial open systems. *Ecological Monographs*, 42(1): 1-23.

DAVIS BROWN, H. 1976. A comparison of the attached algal communities of a natural and an artificial substrate. *J. Phycology*, 12(3): 301-306.

DELBECQUE, E.J.P. 1983. A comparison of the periphyton of *Nuphar lutea* and *Nymphaea alba*. The distribution of diatoms on the undersides of floating leaves. In: *Periphyton of freshwater Ecosystems* (Wetzel, R.G., ed.): 41-47. Junk. The Hague.

DOUGLAS, B. 1958. The ecology of the attached diatoms and other algae in a small stony stream. *J. Ecol.*, 46: 295-322.

DUFF, J.H., STANLEY, K.C., TRISKA F.J. & AVANZINO, R.J. 1984. The use of photosynthesis-respiration chambers to measure nitrogen flux in epilithic algal communities. *Verh. Internat. Verein. Limnol.*, 22(3): 1436-1443.

DUMONT, H. 1969. A quantitative method for the study of periphyton. *Limnol. Oceanogr.*, 14(2): 303-307.

ELSTER, H.J. 1965. Absolute and relative assimilation rates in relation to phytoplankton populations. In: *Primary Productivity in Aquatic Ecosystems*: (Goldman, C.R., ed.): 77-104. Univ. of California Press. Berkeley.

ELORANTA, P.V. 1983. Periphyton pigment analyses using paper chromatography. In: *Periphyton of freshwater Ecosystems*. (Wetzel, R.G., ed.): 291-298. Junk. The Hague.

ETTL, H. 1973. Cytological investigation methods for the growth of *Cladophora glomerata*. *Pol. Arch. Hydrobiol.*, 20(1):181- 184.

FONTAINE III, T.D. & NIGH, D.G. 1983. Characteristics of epiphyte communities on natural and artificial submersed lotic plants: substrate effects. *Arch. Hydrobiol.*, 96 (3): 293-301.

GABELLONE. N. & GUISANDE, C. 1989. Relationship between grain size and fractions of inorganic phosphorus in a reservoir. *Aquatic Science*, 54(9): 306-316.

GITTINS, B.T. 1976. A littoral artificial substrate. *Br.*

Phycol. J., 11: 383-386.

HANSMANN, J., LANE, J. & HALL, J. 1971. A direct method of measuring benthic primary production in streams. *Limnol. Oceanogr.*, 16: 822-826.

HAKANSON, L. 1981. *A manual for lake morphometry*. Springer. Berlin.

HICKMAN, H. 1969. Methods for determining the primary productivity of epipelic and epipsammic algae associations. *Limnol. Oceanogr.*, 14 (6): 936-941.

HILLBRICHT-ILKOWSKA, A. 1972. Field experiment on the factors controlling primary production of the lake plankton and periphyton. *Ekol. Polska*, 20(25): 315-326.

HILLEBRAND, H. 1978. Factor analysis of temporal and spatial patterns of multicellular green freshwater algae. *Verh. Internat. Verein. Limnol.*, 20: 2341-2346.

HILLEBRAND, H., & VAN DIJK, C.P.J. 1979. Patterns of freshwater algae on artificial substrates at different sampling scales. *Hydrobiological Bull.*, 13(2/3): 118-119.

HUDON, C. & BOURGET, E. 1981. Initial colonization of artificial substrate: community development and structure studied by scanning electron microscopy. *Can. J. Fish. Aquat. Sci.*, 38: 1371-1384.

JONES, J.G. 1974. A method for observation and enumeration of epilithic algae directly on the surface of stones. *Oecologia,* 16: 1-8.

KAIRESALO, T. 1977. On the production ecology of epipelic algae and littoral plankton communities in Lake Pääjärvi, Southern Finland. *Ann. Bot. Fennici*, 14: 82-88.

KAIRESALO, T. 1980. Comparison of *in situ* photosynthetic activity of epiphyte, epipelic and planktonic algal communities in an oligotrophic lake, southern Finland. *J. Phycol.*, 16: 57-62.

KAIRESALO, T. 1983a. Photosynthesis and respiration within an *Equisetum fluviatile* L. stand in Lake Pääjärvi, southern Finland. *Arch. Hydrobiol.*, 96(3): 317-328.

KAIRESALO, T. 1983b. Dynamics of epiphytic communities on *Equisetum fluviatile* L.: Response to short-term variation in environmental conditions. In: *Periphyton of freshwater Ecosystems*. (Wetzel, R.G., ed.): 153-160. Junk. The Hague.

KAIRESALO, T. 1984. The seasonal succession of Epiphytic communities within an *Equisetum fluviatile* L. stand in Lake Pääjärvi, Southern Finland. *Int. Revue Ges. Hydrobiol.*, 69(4): 475-505.

KAIRESALO, I., KOSKIMIES, I., LETOVAARA & VÄHÄ-PIIKIÖ, I. 1985. Consequences of fertilization within a littoral *Equisetum fluviatile* L. stand in lake Pääjärvi, southern Finland. *Verh. int. Verein. Limnol.*, 22(5):2904-2908.

KAIRESALO, T. & KOSKIMIES, I. 1987. Grazing by oligochaetes and snails on epiphytes. *Fresh. Biol.*; 17:317-324.

KEVERN, N. R., WILHM, J.L. & VAN DYNE, G.M. 1960. Use of artificial substrate to estimate the productivity of periphyton. *Limnol. Oceanogr.*, 11:499-502.

KOUWETS, F. A. C. 1980. Floristic and ecological notes on some little-known unicellular and colony-forming algae from a Dutch moorland pool complex. *Cryptogamie:Algologie*, 1(4):293-300.

LORENZEN, C. J. 1967. Determination of chlorophyll and phaeopigments spectrometric equations. *Limnol. Oceanogr.*, 12:343:346.

LUCHINI, L. 1974 Diatomeas epífitas de algunas macrófitas del lago Situación (Prov. Chubut). *Physis* Secc. B, 33(86):127:139 .

MARCUS, D. M. 1980. Periphytic community response to chronic nutrient enrichment by a reservoir discharge. *Ecology*, 61(2):387-399.

MARGALEF, R. 1947a. Limnosociología. *Monogr. Ciencia Moderna*, 4. Inst. Esp. Edaf. Ecol. y Fisiol. Vegetal. Madrid.

MARGALEF, R. 1947b. Los métodos para la investigación de comunidades acuáticas adnadas y especialmente las formadas por organismos microscópicos. (perifiton, pecton). *Collectanea Botanica*, I,III(15): 247-259.

MARGALEF, R. 1949. A new limnological method for the investigation of thin layered epilithic communities. *Hydrobiologia*, 1:215-216.

MARGALEF, R. 1983. *Limnología*. Omega. Barcelona.

MARGALEF, R., PLANAS, D., ARMENGOL,J., VIDAL, A., PRAT, N., GUISET, A., TOJA, J. & ESTRADA, M. 1976. *Limnología de los embalses españoles*. D. G. Obras Hidraúlicas. MOPU. Madrid.

MEULEMANS, J.T. 1988. Seasonal changes in biomass and production of periphyton growing upon reed in Lake Maarsseveen I. *Arch. Hydrobiol.*, 112 (1): 21-42.

MEULEMANS, J.T. & ROOS, P.J. 1985. Structure and architecture of the periphytic community on dead reed stems in Lake Maarsseveen. *Arch. Hydrobiol.*, 102(4): 487-502.

MOSS, B. 1977. Adaptations of epipelic and epipsammic freshwater algae. *Oecologia (Berl.)*, 28: 103-108.

MOSS, B. 1980. *Ecology of freshwaters*. Blackwell. Oxford.

MUNAWAR, M. & MUNAWAR, I.F. 1975. The abundance and significance of phytophlagellates and nannoplankton in the St. Lawrence Great Lakes. I. Phytophlagellates. *Verh. Internat. Verein. Limnol*, 19: 705- 723.

MUNTEANU, N. & MALY, E.J. 1981. The effect of current on the distribution of diatoms settling on submerged glass slides. *Hydrobiologia*, 78: 273-282.

ROOS, P.J. 1983a. Dynamics of periphytic communities. In: *Periphyton of freshwater Ecosystems*. (Wetzel, R.G., ed.): 5-10. Junk. The Hague.

ROOS, P.J. 1983b. Seasonal changes in diatom-structure of periphyton from two localities in Lake Maarsseveen: community indices. In: *Periphyton of freshwater Ecosystems*. (Wetzel, R.G., ed.): 59-61. Junk. The Hague.

ROSEMARIN, A.S. & GELIN, G. 1978. Epilithic algal presence and pigment composition in naturally ocurring and artificial substrates in Lakes Trummen and Fiolen, Sweden. *Verh. Int. Verein. Limnol.*, 20: 808-813.

ROUNICK, J.S. & GREGORY, S.V. 1981. Temporal changes in periphyton standing crop during an unusual dry winter in streams of the Western Cascades, Oregon. *Hydrobiologia*, 83: 197-205.

SANCHO ROYO, F. & GRANADO LORENCIO, C. 1988. *La pesca en los embalses andaluces*. Inst. Desarrollo Regional. Univ. Sevilla.

SECKT, H. 1931. Fenómenos de epifitismo en algas de agua dulce. *Rev. Univ. Nac. Córdoba*, XVIII: 86-133.

STRICKLAND, J.D.H. & PARSONS, T.R. 1965. A practical handbook of sea water analysis. *Bull. Fish. Res. Board. Canada*, 167.

SULLIVAN, M.J. 1979. Taxonomic notes on epiphytic diatoms of Mississippi Sound, U.S.A. *Nova Hedwigia*, 64: 241-249.

SUNDBACK, K. & GRANELI, W. 1988. Influence of microphytobentos on the nutrient flux between sediment and water: A laboratory study. *Mar. Ecol. Prog. Ser.*, 43: 63-69

TELL, G. 1972. Algas epífitas de las lagunas Chascomús, El Burro, Yalca y Vitel (Bs.As., Arg.). *Darwiniana*, 17: 552-567.

TELL, G. 1979. Chlorophyceae epífitas y epizoicas de agua dulce de Tierra del Fuego. *ECOSUR*, 6 (11): 1-23.

TOJA, J. 1980a. Limnología del embalse de La Minilla durante 1976. I. Ciclo del fitoplancton en relación con los factores del medio. *Oecologia aquatica*, 4: 71-88.

TOJA, J. 1980b. Limnología del embalse de La Minilla durante 1976. II. Distribución del zooplancton. *Oecologia aquatica*, 4 : 89-110.

TOJA, J. 1982. Control de la eutrofia en embalses por utilización selectiva de agua a distintas profundidades. *Revista Obras Públicas*, abril-mayo: 223-231.

TOJA, J. 1983. Zooplancton de los embalses de Aracena y La Minilla durante 1977. In: *Actas I Congreso Español de Limnología* (Prat, N., ed.): 105-114.

TOJA, J. 1984. *Limnología de los embalses de abastecimiento de agua a Sevilla*. CEDEX. MOPU. Madrid.

TOJA, J. 1990. Longitudinal differentiation of physico-chemical parameters and phytoplankton in La Minilla reservoir (S.W. Spain). *Arch. Hydrobiol. Beih. Ergebn. Limnol.* , 33:733-747

TOJA, J., GONZÁLEZ-RULL, J. & RAMOS, D. 1983. Phytoplankton succession in Aracena, La Minilla and El Gergal Reservoirs (Huelva-Sevilla, Spain). *Wat. Supply*, 1(1): 103-113.

UTERMÖHL, H. 1931. Neue Wege in der quantitativen Erfassung des Planktons. *Verh.Int. Verein. Limnol*, 5: 567-596.

VARELA, M. 1981. El problema de la determinación de clorofila a en el fitomicrobentos: discusión sobre la metodología. *Oecologia aquatica*, 5: 7-19.

WETZEL, R.G. 1960. A comparative study of the primary productivity of higher aquatic plants, periphyton and phytoplankton in a large, shallow lake. *Int. Rev. Ges. Hydrobiol.*, 49(1): 1-61.

WETZEL, R.G. 1963. Primary productivity of periphyton. *Nature*, 197: 4871.

WETZEL, R.G. 1965. Techniques and problems of primary productivity measurements in higher aquatic plants and periphyton. *Mem. Ist. Ital Idrobiol.*, 18 Suppl. : 249-271.

WETZEL, R.G. & Westlake, D.F. 1969. Periphyton. In: *A manual of methods for measuring primary production in aquatic environments* (Vollenweider, R.A., ed.): 33-40 IBP. Handbook, 12. Blackwell. Oxford.

Homage to Ramon Margalef; or, Why there is such pleasure in studying nature (J. D. Ros & N. Prat, eds.). 1991. *Oecologia aquatica*, 10: 77-94

THE RELATIONSHIP BETWEEN FUNCTIONAL ANATOMY OF LAKES AND PRIMARY PRODUCTION

JORDI CATALAN

Departament d'Ecologia. Facultat de Biologia. Universitat de Barcelona. Av. Diagonal, 645. 08028 Barcelona. Spain

Received: January 1990

SUMMARY

The morphology of a lake and exchanges of energy across its boundaries together constitute its functional anatomy. The amount and variability of primary production are related to this functional anatomy. This relationship involves preferential scales, intervals that can be characterized by non-linear functions, time delays and changes of phase. The larger the volume of reference and the sharper the boundaries, the more significant the link between functional anatomy and primary production. Dynamic physical and biological models must include the development of functions and estimation of parameters in certain characteristic intervals but must also forecast where (or when) qualitative or abrupt changes could appear. For research on intralake variability of primary production special attention should be paid to the relative duration of successive stable and unstable phases. In interlake studies, lakes having functional anatomy with features which determine sporadic ice-formation, meromixis, seasonal thermocline, etc. would show less predictable variability in primary production than lakes with a regular annual pattern. In extreme cases, low frequency fluctuations can lead to completely new biological systems.

KEY WORDS: Lake morphology, primary production, phytoplankton, hydrodynamics

INTRODUCTION

Pelagic ecosystems escape direct human perception. This has stimulated the development of conceptual aspects in the ecology of these aquatic systems to the detriment of descriptive details, which are more generously provided by benthic and terrestrial ecosystems. Human size is much smaller than lakes and oceans, and several orders of magnitude bigger than most organisms inhabiting them. As a consequence, our view of these ecosystems has been and is still highly conditioned by sampling methods, knowledge about them being based on imaginative re-creation from data given by the artificial senses of devices and analytical procedures. It is needless to go back to Ehrenberg's time; when POMEROY (1974), for instance, described the complicated marine microbial food web, the classical view of the ocean as an ecosystem started to change radically (GOLDMAN, 1988).

The main characteristic of pelagic ecosystem dynamics is the vertical segregation of the production elements. Particles sink and matter accumulates in the deepest part of the water body. Although organic matter is mineralized in the sediment or in the deepest part of the water column, nutrients remain distant from the light and from the primary producers. If the input of nutrients is not continuous in the photic zone, production is limited. Those nutrients in dark layers must be carried to the photic zone to be reincorporated with living biomass, and this transport requires energy. This mechanical energy (convective mixing, advective movement, etc.) is as important for ecosystem dynamics as energy incorporated by photosynthesis which flows internally through the food web.

Starting from this basis, the qualitative relationship between external energy and ecosystem primary production has been summarized as follows: "External energy stimulates primary production by increasing covariance between resources (light and nutrients) and primary producers (phytoplankton)" (MARGALEF, 1980). Nonetheless, it is very difficult to find a simple and general parametric equation of this relationship. MARGALEF (1978, 1980, 1989) proposed that primary production could be related to a certain power (less than unity, *ca.* 0.6) of the total available external energy. However, MARGALEF (1978) himself pointed out some limitations to the general application of such an expression.

On the one hand, the relationship has been found suitable in the ocean on annual scales (MARGALEF, 1978), but in lakes the volume of reference is much smaller, and so the role of boundaries becomes more significant or at least more apparent. Energy exchanges and morphology interact to constitute a functional anatomy of lakes. On the other hand, while the range of external energy is broad and changes according to the spatial and temporal scales, phytoplankton physiological and lifetime scales are short and stay fixed when the frame of reference is enlarged or decreased. As a consequence, the relationship between external energy and primary production may change its nature on different spatial and temporal scales.

The purpose of this paper is to present a view of lake ecosystems where functional anatomy is considered to be the main factor determining the non-anthropogenic intralake and interlake variability of primary production. The relationship involves preferential scales, time delays and non-linear relations with characteristic phase changes.

The lifetime of phytoplankton organisms is the appropriate reference for the time scales of the biological processes related to primary production of lakes (HARRIS, 1980). Physiological adaptation is the only possible answer to variable conditions in scales of less than hours, while community changes need days. Equivalence between spatial and temporal scales are determined by diffusion coefficients in the water mass. Diffusivity differs by two or three orders of magnitude between the horizontal plane and vertical dimension in lakes. With this set of scales in mind, let us consider the relationship between production and functional anatomy starting from the level of the microenvironment of a phytoplankton organism and moving on to the level of the whole lake.

THE MICROENVIRONMENT OF A PHYTOPLANKTON ORGANISM

It is very difficult for humans to imagine the microenvironment of phytoplankton and, probably as a consequence, these organisms have been considered trapped by the forces of water motion like inert bodies in a soup. But slowly a more structured and organism-dependent image of the phytoplankton world is emerging.

If we consider a time scale of seconds, we may assume that the light field during this period is nearly constant for algae and interaction between mixing and resources should involve nutrient uptake alone. The rate of nutrient uptake depends essentially on two kinds of processes: 1) those related to the uptake kinetics of each cell, which we will not consider here; and 2) those related with transport from the fluid to the cell wall. The transport depends on the nutrient gradient between the cell surface and the fluid, which generates a diffusion process.

In a turbulent environment, changes in fluid velocity occur above the Kolmogorov length (FRISCH & ORSZAY, 1990):

$$l = (\nu_k^3 / \varepsilon)^{1/4}$$

where ν_k is the kinematic viscosity (c. 10^{-6} m^2 s^{-1} at 20° C), and ε is the dissipation rate or the energy dissipated by unit mass (10^{-10}-10^{-5} J kg^{-1} s^{-1} = m^2 s^{-3}). Nowadays, the dissipation rate can be estimated directly from microestructure probes data. Its value, even during storms, rarely approaches 10^{-4} m^2 s^{-3} (DILLON & CALDWELL, 1980; IMBERGER, 1985). Thus, values of the Kolmogorov length are generally between 1 cm and 1 mm. Only in hot waters (40° C) with very strong mixing (10^{-4} m^2 s^{-3}) does one find figures of 200-500 µm.

Most phytoplankton organisms are much smaller than 1 mm, although some colonies are occasionally approximate to 1 mm (REYNOLDS *et al.*, 1987). Thus, the flow around the phytoplankton cells could be laminar, and changes in the field of the fluid velocities due to turbulent mixing may not directly affect diffusion around the cells. Only those organisms which move

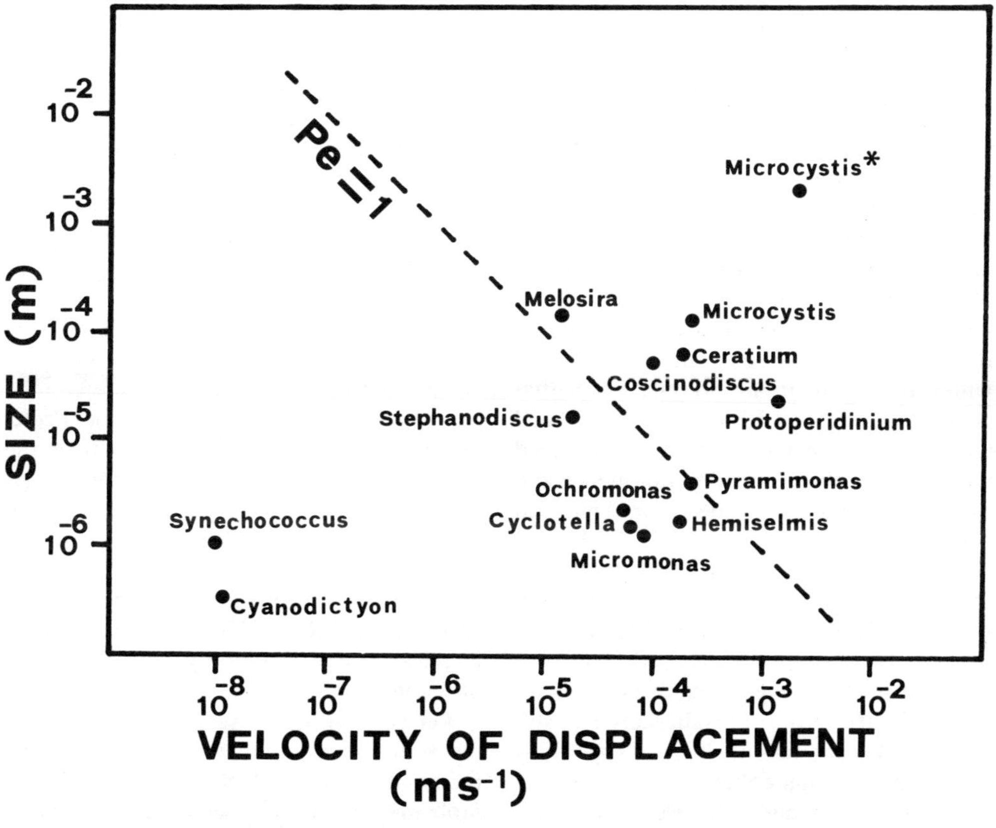

FIGURE 1. Size vs. maximum velocity of movement by sinking, floating, or swimming of some selected phytoplankton marine and freshwater species. Peclet number (Pe) = 1 divides phytoplankton into organisms depending on molecular diffusion and organisms whose movement enhances nutrient transport to the cell surface. *Synechococcus* sp., *Cyanodictyon* sp., *Stephanodiscus astraea*, *Ochromonas minima*, *Cyclotella nana*, *Micromonas pusilla*, *Hemiselmis simplex*, *Pyramimonas disomata*, *Melosira italica*, *Microcystis* sp. (*, from tropical lakes), *Ceratium tripos*, *Coscinodiscus wailesti*, *Protoperidinium quinquecorne*.

faster (large dinoflagellates, rapidly-sinking diatoms or colonies with significant excess of buoyancy) could jump from one eddy to another and, in this way, enlarge their potential dispersion or their capacity to explore space.

However, although there might be no direct effect, turbulence determines inertial delays of the organisms with respect to the fluid, implying an advective movement around the cell. Nowadays, it is not clear whether this process may be more relevant for nutrient uptake in some phytoplankton species than other ways of relative motion with respect to the fluid. The Peclet number (Pe) relates advection to molecular diffusion:

$$Pe = v\ d / D$$

where v is the organism velocity, d its size, and D the molecular diffusion (10^{-9} m^2 s^{-1}). When Pe >> 1, the movement increases nutrient transport to the cell surface although there is no turbulent boundary around it (Reynolds number v d / v_k < 1). Some large flagellates and colonies of cyanophytes and diatoms reach Peclet numbers higher than 1 by swimming, quick floating or sinking movements (Fig. 1). Unfortunately, there are no data for inertial delays. MUNK & RILEY (1952) calculated the ratio between mean slippage associated with velocity changes due to turbulence and Stokes' sinking velocities for spherical cells, and they gave values of 10^{-3}-10^{-2} in normal turbulence fields and 0.3 for strong mixing. Thus, from a theoretical viewpoint, inertial delays do not seem particularly important in nutrient uptake for spherical organisms. Nonetheless, many algal forms depart significantly from sphericity and still little is known about how morphology affects their inertial delay.

From size and motion velocities phytoplankton organisms can be divided into two main sets (Fig. 1): a) those trapped by viscosity depending on molecular diffusivity, including most of the freshwater species; and b) those which can enhance nutrient flow to the cell by their self-generated motion and are big enough to reach Pe >> 1, which comprise some species relevant in the productivity of lakes (*Microcystis*, *Ceratium*, *Melosira*). After CSANADY (1986) and SPIGEL & IMBERGER (1987), the nutrient flux (NF) per unit volume can be roughly characterized as follows for spherical cells:

for Pe < 1,

$$NF\ (mol\ m^{-3}\ s^{-1}) = 12\ \Delta N\ D\ d^{-2};$$

for Pe > 1,

$$NF = K\ \Delta N\ D^{2/3}\ v^{1/3}\ d^{-5/3},$$

ΔN being the nutrient difference (mol m^{-3}) between the cell surface and the environment, and K an adimensional constant whose value is about 6 for Pe > 1000, differing in less than one order of magnitude for smaller Peclet numbers. In any case, nutrient flux per unit volume depends mainly on size, decreasing with it. Therefore, when dealing with spherical cells, one expects to find smaller cells in oligotrophic waters. Organisms living at Peclet numbers < 1 have no preferential flux zones around the body. For Peclet numbers > 1, the higher the number the more localized the flux zones in some parts of the organisms. This phenomenon might bring about the development of a specialized zone for uptake or for heterotrophic nutrition (GAINES & ELBRÄCHTER, 1987), which may increase cell polarity or colony differentiation.

On the millimetre scale, although much research is required to gain a realistic image of the microplankton world, the most probable conclusion seems to be that a turbulent environment has little relevance in nutrient uptake, which depends more on the individual features of the organisms. The link between functional anatomy and primary production must be sought at a higher scale.

Recently, MITCHELL *et al.* (1990) report that millimetre-sized clusters of plankton can be produced by the shear of a laminar flow across the surface of a microplankter. Additionally, above critical concentrations of 1,000 to 10,000 cells ml^{-1} these clusters can became temporarily stable, sinking as streamers of plankton-laden water with radii of one or more millimetres. The streamers can increase migratory velocities by one order of magnitude. In turbulent waters, motion by vertical eddy diffusion is greater than the swimming and sinking velocity of most phytoplankton. Nonetheless, phototaxis, negative geotaxis or gyrotaxis (on the basis of which a mechanism generating spatial heterogeneity was proposed by MITCHELL *et al.*, 1990) can introduce biases in the general pattern generated by eddy diffusivity. The bias might be decisive in critical circumstances, like, for example, in phytoplankton being trapped or not on the thermocline, or as a way to accelerate phytoplankton circulation in the mixed layer which may be advantageous (MARGALEF *et al.*, 1979). But these are

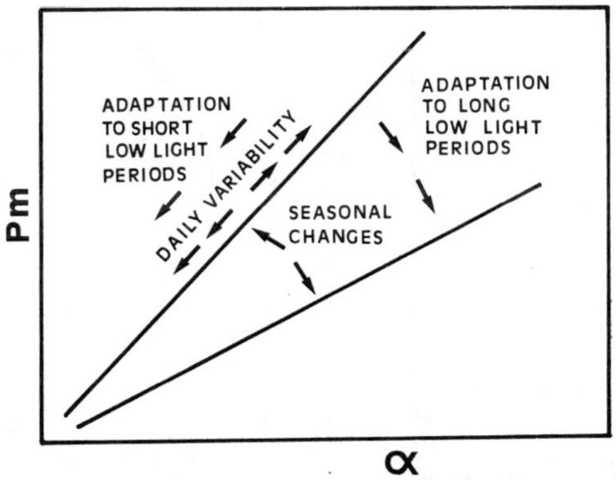

FIGURE 2. Variability in the relationship between the maximum production at light saturation (Pm) and the slope of the photosynthesis-irradiance curve in the light limited zone (α) associated with changes in light of different time scale. The light intensity that saturates the system (I_k = Pm / α) changes at larger time scales than Pm and α.

events which became relevant on time scales higher than those considered hitherto.

DIURNAL MIXED LAYER

At a scale of meters, minutes and hours, turbulent and advective processes in the water body can move phytoplankton populations up and down in a resource gradient. In a mixing layer, nutrient distribution will tend to be at least as homogenous as the distribution of organisms and therefore algae move up and down in an environment where light is the most variable resource. In this case, the link between external energy and primary production is related to the cost of adaptation of the phytoplankton to the light field.

Phytoplankton photoadaptation involves several processes which require different periods (HARRIS, 1980). Change in the fluorescence yield is the quickest response to changes in light intensity, taking minutes and serving as an emergency exit. On an hourly scale, photoadaptation results in changes in photosynthetic behaviour, which is commonly characterized by photosynthesis *vs.* irradiance curves (P-I). These curves have been characterized by several expressions (i.e. STEELE, 1962; HERRON & MAUZERAL, 1972; JASSBY & PLATT, 1976; PLATT *et al.*, 1980). Usual parameters are α, which is the slope of the curve in the light-limited zone; Pm, the maximum production at light saturation; and I_k, light intensity at Pm, which is equal to the quotient of Pm and α. Although α and Pm are apparently independent, it has been proved that α and Pm change proportionally to keep I_k constant during short periods of changes in irradiance and during diurnal light variations (HARDING *et. al.*, 1987). I_k changes seasonally or during prolonged low-light periods (Fig. 2). At the biochemical level, it represents a change in the size of the photosynthetic units: if I_k decreases, light-harvesting pigments increase more than reaction centres. Therefore, changes in pigment composition, or in pigments with respect to other cell components (carbon, carbohydrates, etc.), also take some days. Eventually, if physiological or structural cell changes cannot deal with the variation in resources, growth is modified.

The ratio between adaptation time and mixing time determines which type of adaptation is possible. If mixing time is higher than adaptation time, the organism must adapt to the mean value of the light field; if not, phytoplankton can adapt to each new light intensity.

The transition from autumn overturn to mid-winter in a high mountain lake, when the lake is covered by ice and snow, well illustrates the relation between mixing time and photosynthetical adaptation in several regimes of mixing and light levels. Figure 3 shows changes in photosynthetic parameters in Lake Redó (Pyrenees) during the winter (CAMARERO *et al.*, in prep.). During overturn the phytoplankton is distributed more or less homogeneously along the vertical axis (Fig. 3a). Nonetheless, there is a clear differentiation between Pm near the surface and Pm below a depth of 15 m. However I_k, which requires a longer adaptation time, is quite similar for phytoplankton populations at any depth, and its value corresponds to mid-depth light intensity (25 m). This means that during autumn overturn in Lake Redó, the mixing time of phytoplankton is shorter than the time of I_k adaptation and

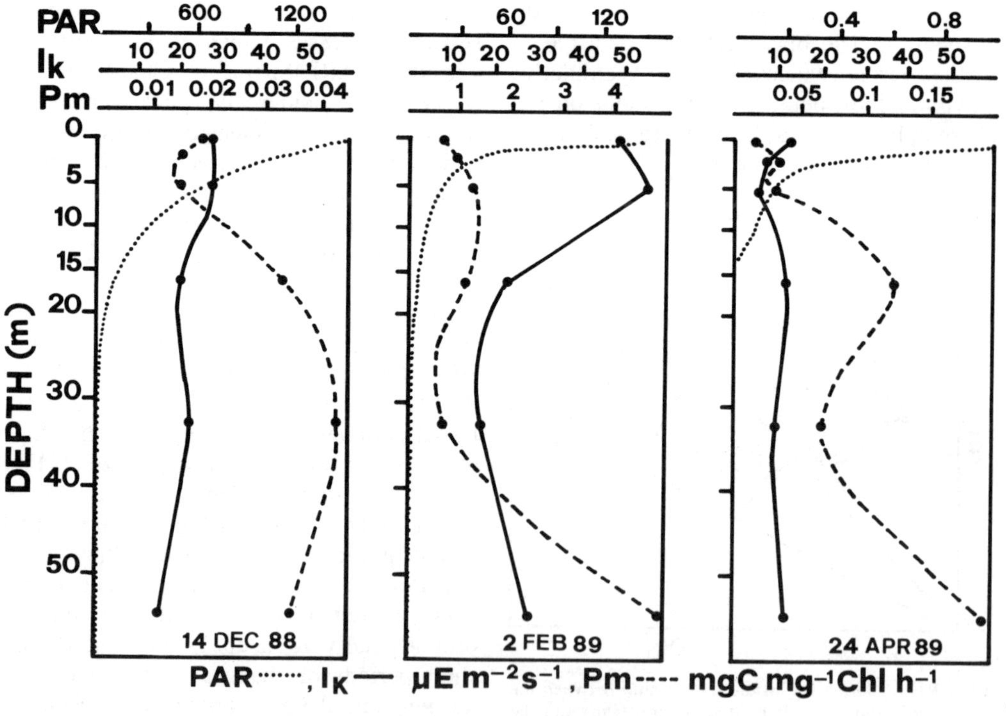

FIGURE 3. Photoadaptation in Lake Redó (Pyrenees) during 1988-89 winter. During the ice-free period (A) vertical mixing did not allow I_k differentiation and phytoplankton was adapted to mean depth irradiance. During the ice-covered period with little snow (B) I_k variability followed light values, time of mixing was larger than time of adaptation. When more snow was accumulated in the cover (C), phytoplankton was not able to adapt its I_k to the very low irradiance.

longer than the time of Pm adaptation; α changes parallel to Pm. During overturn, the mixing time of heat has been estimated to be less than an hour (CATALAN, 1988). For phytoplankton, it is likely to be longer, but less than a day. When ice forms on the lake surface, turbulent mixing reduces markedly and, as a consequence, Ik differentiation is possible (Fig. 3 b). Then snow accumulates and the light penetration goes down to very low levels (Fig. 3 c). Adaptation reaches its limit and I_k becomes homogeneous and very low throughout the water column again. I_k does not reach the values of light in the water column and so it seems impossible for these algae to be able to adapt to such low light levels.

Any characterization of phytoplankton photoresponse to a variable light field must involve a historical perspective. On a scale of days, the effect upon primary production of a sudden change of light (because of snow accumulation, or water rise or deepening, for instance) would differ depending on the phase of growth of the algae population. Not only is the mean light field important, but the variance of the light environment is also significant during the growth phase. Photosynthetic measurements showed that algae grown under steady illumination had higher photosynthetic efficiency (α) and capacity (Pm) when incubated under steady or little-fluctuating light (0.1 Hz). However, algae grown under highly-fluctuating light had higher photosynthetic efficiency and capacity than those grown under steady illumination, when incubated under 1.0 and 10.0 Hz light (QUÉGINER & LEGENDRE, 1986).

Although, as pointed out above, the relation between mixing time and adaptation time determines whether phytoplankton adapt to the mean light environment or follow light variation, it is not clear how photoadaptation relates to system net primary production. We may hold that an adaptation process would be energetically cheaper the longer it takes to happen, thus explaining the substitution of one photoadaptation mechanism for another. However, maximization of system primary production is not necessarily the main aim of phytoplankton cellular regulation.

Seasonal primary production dynamics may be understood as an accumulation of short-time events. At the scale of months, a statistical link between primary production and external energy may easily arise from the high weight that relatively few productive events, occurring when mixing breaks some nutricline, have in the statistical analysis. The coefficient of the power relations between primary production and external energy (MARGALEF, 1978) could be an indicator of the relative abundance of productive and unproductive mixing events. However, the relationship can appear confusing when we compare the spectra of mixing events from different lakes and the variability of primary production at each intensity.

There are different sources of mixing energy: wind action, inflows and outflows, heat exchange, etc. Each of them has particular characteristics: 1) wind has an immediate and local effect, followed by inertial arrangements when it ceases which may vary greatly from lake to lake, depending on morphology. 2) Inflow and outflow generate advection, which determines a clear directional spatial differentiation. Inflow differs from other external sources of energy in carrying nutrients and organic matter from outside the system. Inflow magnitude, and nutrient and matter concentration depend on the watershed features and thus the functional anatomy of lakes extends to the catchment area. 3) Finally, input of energy in the form of heat produces the mixing action with some delay - that is, when the heat is lost by the system sometime later. Its role has to be considered on a higher time scale. Hence, convection has a strong seasonal influence in most lakes. As an immediate effect, positive heat flux partially compensates for wind action.

Inputs of energy generally occur near the lake surface, so understanding the behaviour of the mixed layer near the surface is a first step in the matching of phytoplankton adaptation to physical processes on short-time scales. The daily behaviour of the mixed layer, integrated by rhythmic events and probabilistic fluctuations, determines the episodes of primary production, the integration of which will cause the seasonal and annual fluctuations.

The behaviour of the diurnal mixed layer can be typified by employing a series of dimensionless numbers and characteristic lengths (IMBERGER, 1985). For instance, a useful ratio showing functional anatomy at this scale is the Wedderburn number (PATTERSON *et al.*, 1984), whose magnitude determines the deepening regime of the mixed layer (SPIGEL & IMBERGER, 1980) (Fig. 4).

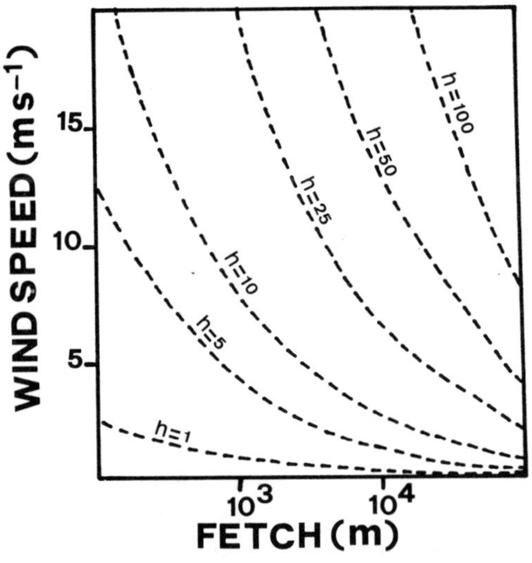

FIGURE 4. Given a particular density differential ($\Delta\sigma$) in a two layer water body, the efficiency of a wind action in altering the stratification depends on the geometry of the upper layer and time of response on the geometry of the basin (H,L). In the figure, wind speed (v), fetch (L), and thermocline depth (h) are plotted for Wedderburn number = 1, which means severe mixing, for a $\Delta\sigma$ = 0.08119 (that is, a 1° C temperature differential in the range of about 10° C). $W = (g\ \Delta\sigma\ h\ /\ (\sigma_a\ C_d\ v^2))\ /\ (h/L)$, where g is the acceleration due to gravity, σ_a the air density, and C_d a drag coefficient. Time of response is given by $t = (2\ L\ /\ (g\ \Delta\sigma/\sigma_0\ (H-h)/H)^{1/2})\ /\ 4)$.

Based on a two-layer rectangular cavity with densities of σ_0 and $\sigma_0 + \Delta\sigma$, the Wedderburn number represents the ratio between surface wind stress and the pressure gradient resulting from the slope of the interface (the maximum baroclinic pressure force). It is a Richardson number with an aspect ratio. W = 1 corresponds to the interface surfacing at the upwind end, where vertical mixing is enhanced, and the upwelled water is redistributed horizontally by advection. Rapid deepening accompanies the large interface displacements, and complete mixing may occur during a single wind episode. W > 10 indicates that mixing effects are much more relevant in the vertical dimension - there is no share production at the interface, and wind action and convection determine all mixing processes at the interface.

Another interesting property of the mixed layer is the ratio between wind-induced mixing velocity and penetrative convection velocity. This relationship has been traditionally expressed as the Monin-Obukov length (IMBERGER, 1985). For a net surface heat input, it represents the depth scale over which wind stirring approximately balances the rate of gain of potential energy from the surface heating. For a heat loss, it represents the depth of the mixed layer influenced by wind stirring.

The next step is to characterize how phytoplankton biology matches those patterns of mixing described by dimensionless numbers and characteristic length: how and when entrainment of nutrients or particles to the diurnal mixed layer occurs; what are the delays between the physical and biological processes; etc.

An early approach was the relation between primary production and the quotient between photic depth and mixing depth (SVERDRUP, 1953). Obviously, if phytoplankton remains outside the photic zone part of the time during mixing, primary production will decrease; although this is not always the case if the mixing depth represents the frontal zone where water richer in nutrients is entrained from a non-mixed layer. Both photic zone and mixed layer change during the day and from day to day, and they are not independent. The mixed layer depends on light absorption in the water column, and light absorption on phytoplankton growth, which in turn depends on the ratio between photic depth and mixing depth. The buoyancy characteristics of phytoplankton species involved and the pattern of daily variability of the mixing layer may determine competitive advantages for some species (SPIGEL & IMBERGER, 1987).

On the other hand, it is not easy to define the depth of the photic zone in relation to the light field. The depth of the photic zone has often been underestimated. One cannot talk of compensation depth in general, but for each cell. Some algae can grow at very low light levels. It is not rare for oligotrophic lakes to present production maxima at depths at which light is between 1 and 0.1% of the subsurface irradiance. Moreover, experiments carried out at very low light levels suggest that the energy required to sustain a cell may vary with growth rate (GIBSON, 1987). Thus, compensation depth requires a dynamic definition. In general, the models available require a shift to more dynamic approaches where both mixing processes and phytoplankton adaptability are compared at the appropriate scales. Since compensation depth is not general, but depends on the species and even on the status of the individuals, this is a mechanism for succession. At the base of the photic zone, the spatial segregation of species may increase and different species could develop successively.

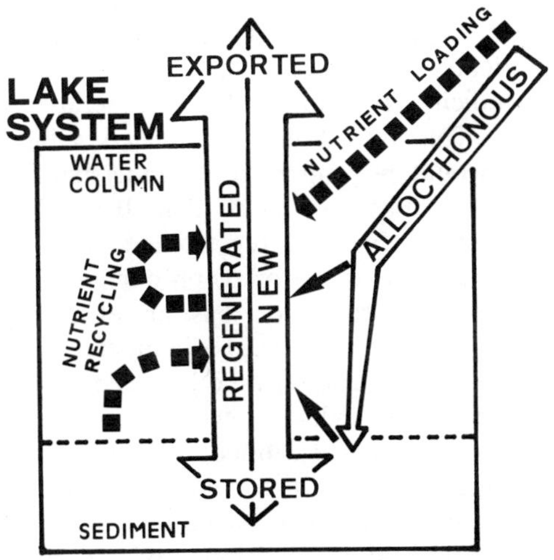

FIGURE 5. Lake production can be partitioned according to the inlake or outlake origin of nutrients (regenerated and new production, respectively), and according to its destination (exported or stored). For a given nutrient loading the relative importance of each fraction would depend on lake functional anatomy. Allocthonous (imported) production contributes to new production but with delay respect to its entrance to the system.

A compensation depth can be considered for the whole ecosystem. SMETACEK & PASSOW (1990) have pointed out the significance of grazing and heterotrophs in modifying the phototroph compensation depth in the oceans. Nonetheless, in many lakes, sediments are as relevant as open water biomass in determining the lake respiration centre of gravity (CATALAN, in press).

THE LAKE AS A SYSTEM

The primary production of lakes can be divided conceptually by considering the origin of the supporting nutrients (Fig. 5). Two components can be distinguished: new and regenerated production. The conceptual idea of DUGDALE & GOERING (1967) for the photic zone of the oceans can be stated for a lake as follows (AXLER *et al.*, 1982): new production is based on external loading of nutrients; and regenerated production is based on in-lake recycled nutrients. Sediments are considered part of the whole lake ecosystem. Notice that if this definition of new and regenerated production for the lakes were applied to the ocean considered as a "sole big lake", most of the new production measured today would have to be understood as regenerated. The partition between new and regenerated is always relative and depends on the definition of the "box" of reference.

Nutrients for new production can enter the system in a mineralized form or as part of imported organic matter. The latter portion determines the heterotrophic level of the system, and introduces a time delay in the use of elements brought in by the inflow. Delay depends both on the specific properties of the system and on the nature of the organic matter.

In a homogenous territory, nutrient loading is related to watershed size (SCHINDLER, 1971), and to catchment physiography. In a given watershed, external energy, in the form of precipitation and water flow, probably has less influence on average nutrient loading than on its seasonal variability.

On the other hand, matter produced in the system, after cycling in the water column several times - or with no cycling - is eventually stored in the sediments or exported through the outflow or through organism-mediated transport (for example, Chironomidae emergence). Work done by external energy is useful, as regenerated primary production, when it results in the transport of nutrients from a zone where stored production is mineralized to the photic layer. The functional anatomy of lakes can be viewed as an apparatus capturing energy, just as we conceive of the photosynthetic apparatus or of the plant morphology capturing photons. As in the photosynthetic process, the efficiency of the conversion of the energy reaching the system into production depends on a first component of physical nature (functional anatomy) and a second component of biochemical nature (metabolic biological processes which generate gradients and adaptative physiological processes). Functional anatomy determines the potential to produce, but biochemistry regulates the efficiency in the use of this potential. For instance, mixing events of the same magnitude which occur too often will have different results in incorporating nutrients to the productive zone if the periodicity is shorter than mineralizing time. This is a little investigated subject that would be important in understanding the variability of lakes that base their production in the regenerated portion. Recovery of the stored production into regenerated production has an efficiency that depends on the biogeochemistry in the sediments but also on time between storing and mixing episodes.

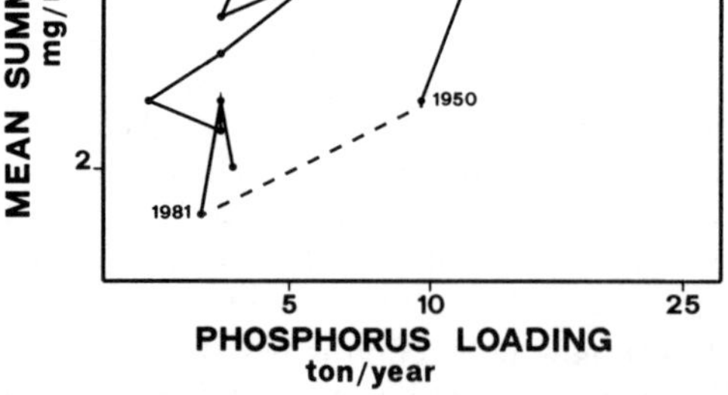

FIGURE 6. Trajectory followed by Lake Washington during its eutrophication and regeneration phases in the space defined by nutrient loading (intensive magnitude) and mean chlorophyll in surface samples for July and August (extensive magnitude). Area defined by the trajectory might depend on the functional anatomy of the lake (Data after EDMONDSON, 1985).

For theoretical limnology and lake management it is important to understand how the functional anatomy of lakes influences the division of primary production into the several types mentioned above. For instance, eutrophied lakes do not follow the same trajectory during recovery as during eutrophication, in terms of pairs of nutrient loading and phytoplankton biomass (Fig. 6). The shape and size of the area between the two trajectories depend on the functional anatomy of the particular lake. In theory, if more energy is available for mixing, production is kept at a higher level (ALONSO, 1989a,b) and the system loses and stores less material, hence the trajectories back and forth would be more similar.

Morphological characteristics of lakes and their biological production have long been related. THIENEMANN (1928) already pointed out that oligotrophy is associated with lake depth. FEE (1979) found that the variability of primary production in some lakes of the Experimental Lake Area (Canada) strongly correlated with the ratio between the area of epilimnetic sediments (Ae) and the volume of the epilimnion (Ve), but had little relation with the nutrient loading, except when the lakes had been artificially eutrophied. The ratio Ae/Ve is equivalent to the ratio between the probability of retention of a particle in the epilimnion and the mean depth of that epilimnion. In other words, the nearer a compartment where nutrient recycling occurs to the photic zone the higher the production. In general, the external loading to the lake is scaled to catchment size, the nutrients received are diluted according to the lake volume (SCHINDLER, 1971) and, finally, increased depth represents a higher dilution of nutrients recycled in the deepest part of the lake (MARGALEF, 1983).

This view of the relationship between production and lake morphology does not give sufficient emphasis to the role of mechanical energy as a necessary link between them. The relationship between morphology and production is usually established in rough terms on an annual time scale; the significance of external energy could be easily understood on a seasonal or even on a daily level.

The relationship between primary production and morphology and physical regimes of lakes is not monotone. A series of "switchers", "multipliers" and "accumulators" have their particular configuration in each lake. This determines the characteristics of the link between external energy driving the system and the dynamics of lake primary production on seasonal and annual scales. In the relation between temperature and water properties, two particular points represent qualitative changes in the relationship between the functional anatomy of lakes and primary production. These are the maximum density point (*ca.* 4° C in pure water) and the freezing point. When these temperatures are reached in a lake, there is a qualitative change in the effect of different external energy sources.

Radiation induces convective mixing until the water reaches 4° C, then the water increases in buoyancy and partially withstands wind action. Heat introduced in the lake represents a potential energy which will do its work later. Hence, going above 4° C will bring about a time delay in the relationship between external energy and primary production, but it also implies an accumulation of energy which will act in a shorter time. In large bodies like the ocean, this occurs spatially: energy accumulates in large areas and is given back to the atmosphere as sensible heat plus latent heat locally (upwelling zones). When the lake cools below 4° C, convective mixing by heat loss stops and radiation again has a direct effect as a mixing agent. Therefore, in a typology of lake variability the relative length of the periods below and above 4° C would be important.

Formation of surface ice has several implications for lake production and the external energy relationship. The best known are its insulating and light-reducing effects, especially if snow accumulates on the top (CATALAN, 1989). But there is another, less obvious effect of the formation of ice. Cover formation opens the possibility of integrating precipitation events for a long period and concentrate their effect on lake production when the cover thaws. Lakes receiving the same winter precipitation in their watershed might show different fluctuations in spring production depending on whether they freeze or not. An interesting example has recently been given by GOLDMAN *et al.* (1989), comparing primary production variability in Lake Tahoe and Castle Lake. The extreme production fluctuations in Castle Lake correlate with El Niño events but not the extreme fluctuations of Lake Tahoe, although they are both at a similar latitude. It seems that Castle Lake reflects El Niño-induced variability in precipitation because it freezes and, therefore, integrates in its cover the snowfalls throughout the winter, whereas Lake Tahoe because of its size does not freeze and the effect of the precipitation increase or decrease is diluted during the year.

In Castle Lake, the higher the snow accumulation during winter the lower the spring production. The exact mechanism involved in this relationship is unclear. In Lake Redó in the Pyrenees the same seems to occur, although data are much more scarce. In this lake, production depends on nutrient inputs from sediments situated below 25 m and particularly 40 m deep. The thicker the snowpack the slower the thawing of the lake cover and the shorter the period of spring isothermy. As a consequence, convective mixing action upon fine sediments is also shorter during spring, and nutrient input lower. A similar mechanism may occur in Castle Lake, because persistence of the cover during thaw implies a more stratified flow of melting water, and, thus, delays action upon deep fine sediment. This phenomenon introduces variability in the primary production series of these lakes because their total production depends more on the regenerated portion than on the new portion. Probably, the larger the nutrient loading, implying a relative increase in new production, the lower the relative effect of lake cover in annual production fluctuations, unless other mechanisms appear.

Solar heating introduces buoyancy stabilization of the water column. The kinetic energy introduced by wind action is usually insufficient to overcome the potential energy inherent in the stratification in some periods of the year. Episodic events and accumulation of results lead to the division of the summer water column into compartments. Particularly interesting is the relationship between the trophic state and this compartmentalization. Oligotrophic lakes present deep chlorophyll maxima which are usually below the thermocline, in the upper hypolimnion, where light is between 1 and 0.1 % of surface and where pseudo-eddy diffusivities incorporating the effects of all the mechanisms which result in vertical transport are higher than in the thermocline itself (QUAY *et al.*, 1980; CATALAN, 1988). Turbulence is active in the surface mixed layer, hypolimnion is largely quiescent with internal waves. Seiching and shear in the upper hypolimnion favour intrusions from the boundaries into the interior. These intrusions represent a nutrient transport which feeds the phytoplankton development far from the surface. Furthermore, waves and tilt of the thermocline can temporally increase light availability to the phytoplankton situated near the compensation depth. In these fertilization mechanisms of deep chlorophyll development, special attention should be paid to the relationship between the depth of the thermocline and the depth of sub-basins or low slope boundaries. Fine materials can accumulate here as sources of nutrients and of reductor power which may be horizontally transported.

A shift to new system characteristics during seasonal dynamics or a eutrophication process occurs when the deep chlorophyll maximum crosses the thermocline and the lake splits into a productive epilimnion and a respiratory hypolimnion.

The dynamics of lacustrine pelagic systems, probably due to the influence of open water oceanography, are usually treated with little attention to the boundaries: infinity is supposed to be quite near. However, boundaries are fundamental in lakes, because mixing upon them occurs at scales much closer to phytoplankton scales than in the ocean. The main nutrient deposits in lakes are sediments. Recycling of nutrients takes place in fine sediments (< 6 μm particles), where there is both more organic matter and higher water content. Therefore, factors determining in-lake sediment distribution are important elements of the functional anatomy of the lake. The main factors are slope, inflow and wind action.

Slope may cause sedimentary material to move. HÅKANSON (1977) established that fine sediments do not accumulate on slopes > 14%, and that below 4% slope has no effect. Therefore, the steeper the slopes and the deeper the lake, the higher the seasonal variability of regenerated production, because pumping of nutrients from the sediment is concentrated in shorter periods of the year or only occurs from year to year (meromixis).

River influence generally decreases logarithmically from the mouth, although it is difficult to establish where river action ceases to control bottom dynamics and wind action begins to dominate (HÅKANSON, 1982). Within areas dominated by wind action, three main zones can be distinguished (HÅKANSON, 1977): accumulation areas, where fine materials are continuously being deposited; transportation areas, where periods of accumulation are interrupted by shorter periods of transportation; and, finally, areas of erosion where there is no deposition of fine materials. These areas are highly determined by lake morphology. Only in shallow lakes can wind-induced waves transfer energy directly to the lake bed. HILTON (1985) proposes that wind energy causes redistribution of the semifluid material (gyttja, for instance) down to a depth of at least one wavelength of the wind-driven waves (SMITH, 1975). Then a boundary can be calculated delimiting lakes with quite a different productive response to wind.

Another mechanism causing sediment redistribution and implying a net transport towards the deepest parts of the lake is intermittent complete mixing (ICM; HILTON, 1985). During overturn, a fairly constant concentration of particles occurs in the water column. When quiescent conditions follow, the mass of particulate material that settles at any point in the lake is directly proportional to the depth of the overlying water. The turbulent levels reached in the water column determine the relevance of intermittent complete mixing. In large deep lakes, it is unlikely that an efficient mix of resuspended particles occurs and, therefore, the significance of the ICM is greater in medium size lakes (< 6 km^2). It is interesting to note the dual role of external energy in ICM lakes. On the one hand, it contributes to the concentration of materials, especially organic matter, in the deepest part of the lake, thus increasing the distance between the gravity centre of primary production and the gravity centre of respiration. On the other hand, it stimulates regenerated production by accelerating diffusion of nutrients. This dual role may contribute to the splitting of the annual cycle into periods of different contribution of new and regenerated production to global production; and, especially in oligotrophic lakes, it contributes to increasing seasonal variability in primary production and dependence from one productive period to the next.

Lake compartments and water density singularities determine the existence of interfaces (spatial and temporal) between phases of appreciably different turbulent kinetic energy. These sharp transitions from high to low levels of turbulence have been called "ergoclines" by LEGENDRE & DEMERS (1985). Processes near or around ergoclines are usually determinant of aquatic ecosystem dynamics. Functional anatomy can also be seen as a set of ergoclines.

In spatial ergoclines, both phases coexist and interact through the interface in the course of time. Examples are summer stratification, thermal bars, marine fronts, etc. In temporal ergoclines, the two phases do not coexist and the interface is ephemeral in relation to the duration of the phases, for instance, formation of the thermocline, lake freezing, etc. The whole ecosystem crosses the interface at the same time and therefore transport or exploitation processes from one phase to the other have an abstract sense.

It is important to consider the interface position relative to the light-gravity axis. Obviously, asymmetry between the two phases is higher when the interface is orthogonal to the light-gravity axis, which is the case of most spatial lacustrine ergoclines. However, in large lakes frontal systems parallel to the light-gravity axis may develop, and thermal bars may occur near the shore line even in small lakes (IVEY, 1983).

The interface between a mixed phase and a stable phase can represent merely a boundary between two states, or it may be an active front where turbulence erodes a nutricline whose origin is in the previous compartmentalization of lake biological activity. This latter case corresponds to the classical deepening of the thermocline or the resuspension of sediments, for instance, and results in a higher production of the mixed phase because of the exclusion of one of the reactants (light) from the stable phase.

When the interface does not represent an obvious rupture of a nutricline, biomass maxima usually occur on the stabilized side of the interface. This is intuitively understood when the ergocline is the boundary between water and a solid (rock, sediment, etc.). When the ergocline is

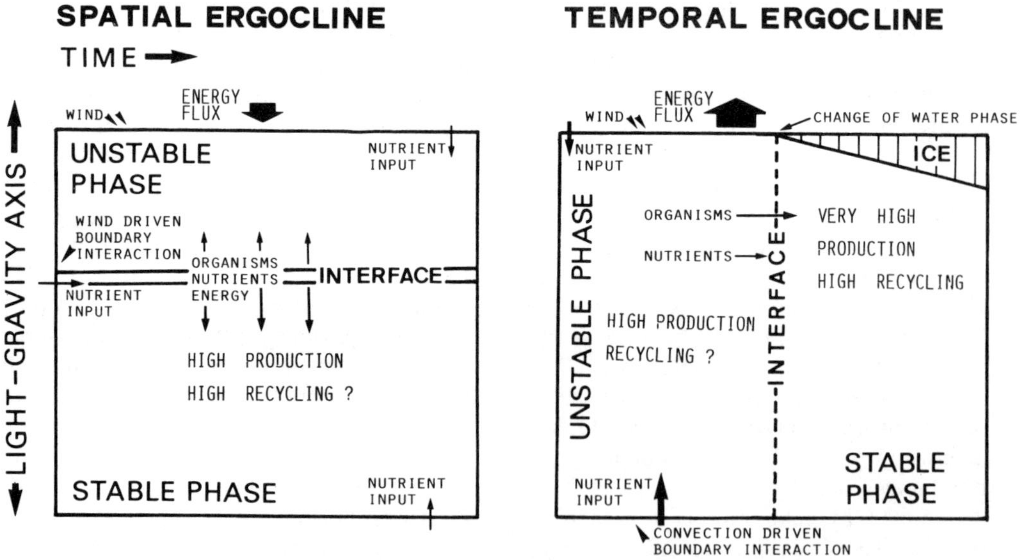

FIGURE 7. Many oligotrophic high-mountain lakes present two chlorophyll maxima associated with two different ergoclines: a summer deep chlorophyll maximum (SDCM) usually develops in the stable phase of the vertically stratified lake just below the thermocline; and a chlorophyll maximum under ice (CMUI) often develops when lake freezes. The temporal ergocline is thought to be an archetypal example (CATALAN & CAMARERO, 1991) for the case that maximum of phytoplankton biomass on the stable phase of the ergocline does not depend on an accumulation process but depends rather on an increase of productivity induced by the sudden change of phase of the system, from a turbulent to a stable environment, without any apparent parallel increase of nutrient input in the system, but with a more efficient and quick recycling.

between two water phases, biomass maximum on the stable phase may occur due to accumulation from production in the unstable phase or to the occurrence of a production maximum on the stabilized side itself. Unless the latter option is associated with a nutrient gradient, it is difficult to understand. Nonetheless, some oligotrophic lakes provide examples where higher production seems to occur in the stable phase. CATALAN & CAMARERO (1991) suggest that this is the case in the summer deep chlorophyll maximum and in the chlorophyll maximum under ice after freezing which some high-mountain lakes present (Fig. 7). Factors suggested as contributing to the higher production in the stable phase are: 1) more efficient phytoplankton growth in a dim and stable light regime in very low nutrient environments; and 2) higher *in situ* recycling in the low turbulence phase. This higher efficiency in recycling in the less turbulent environment could be an adaptative strategy developed among plankton to conserve essential elements in the system (SMETACEK, 1985; SMETACEK & POLLEHNE, 1986). On a microscale, greater recycling means more interaction between autotrophs and microheterotrophs, but it is difficult to imagine any relation between microorganism interactions and the microstructure of the environment as stated at the beginning of this paper. The higher production in the stable phase is relatively ephemeral because there is hardly any input of nutrients. Additionally, it may be insignificant in quantitative terms with respect to the ergoclines where mixing acts upon a nutrient gradient.

The succession of mixing and stratification periods in lakes has been recognized as one of the most useful criteria for lake typology (HUTCHINSON, 1957). Temporal transitions from stable to unstable phases represent changes in the lake compartments and the breaking of biologically induced gradients by external energy. Transitions range from annual to semidiurnal and sometimes shorter time scales. For instance, the classical seasonal changes of the structure of the water column of the dimictic lakes of the temperate zone (freezing, isothermy, stratification, isothermy) occur daily in tropical high mountain lakes. At first sight it seems that, the more frequent the succession of stable and unstable phases, the higher the variance of production and also the higher the mean annual production. However, frequency in phase succession probably has a range that better matches biological processes (LEGENDRE *et al.*, 1986). Higher frequencies would not allow enough time for the development of significant chemical and biological gradients, while shorter frequencies would favour the definitive exclusion of some elements from the system. In a dimictic lake, for instance, two productive periods (spring and autumn) and two mineralizing periods (summer and winter stratification) can usually be recognized. It will be interesting to see the results of a study of the relation between the relative duration of the stable and unstable phases and the magnitude of the productive periods from a long-term series of data.

CONCLUSION

Work done by external energy is more efficient when both anisotropy and sharpness of the boundaries of the volume of reference increase, and also when the time scale includes more variability in energy exchange across the air-water interface. Therefore, the relevance of external energy in primary production should be investigated at the level of the whole lake and on time scales much greater than the microorganisms' life span.

Understanding the link between functional anatomy and primary production in detail demands a parallel development of dynamic physical and biological models, which will probably take some time. Nonetheless, at annual scales and for inter-lake comparison, it is noteworthy that many characteristics of functional anatomy are related to lake size and profile. How primary production variability changes with size for a constant latitude, or how it changes with latitude for a constant size may be relatively easily investigated, both empirically, in district lakes such as in areas of Canada, and theoretically, by physical models including simple biological processes. In my view, in inter-system studies, special attention should be paid to localizing qualitative functional changes such as the characteristics which determine the ice formation, thermocline persistence, the appearance of Coriolis effect, and formation of frontal structures in the mixed layer (IMBERGER & PARKER, 1985), for instance. For intra-lake variability research, the relative duration of successive stable and unstable phases could be significant. The relationships between different components of the functional anatomy and primary production of lakes are non-linear. Therefore, the relations should be monotone in some intervals but points where the relationship changes qualitatively or abruptly should also appear. The recognition of these situations might be more significant than characterization of monotone intervals. Lakes with functional anatomy with characteristics near any of these singular points might present more unpredictable variability in primary production, and in extreme cases, low frequency fluctuations can conduct them to completely new biological systems (SCHINDLER & COMITA, 1972).

ACKNOWLEDGMENTS

Margalef has the quality to fascinate the audience in his lectures, because of his interdisciplinary views, encyclopedic knowledge and intuition, but also because of the enthusiasm and even tenderness with which he talks about organisms. I wonder how much of my addiction to plankton research and ecology responds to these irrational stimuli.

Several friends kindly read an early draft of this paper: Quico Sabater, Quique Ballesteros, Ramon Margalef, Josep Anton Morguí and Miquel Alonso; their suggestions were always useful.

REFERENCES

ALONSO, M. 1989a. Temperature increase effects on zooplankton structure in a cooling reservoir. *Arch. Hydrobiol. Beih. Ergebn. Limnol.*, 33: 503-511.

ALONSO, M. 1989b. Els efectes de les centrals nuclears sobre les aigües continentals. *Revista de Catalunya*, 26: 47-59.

AXLER, R. P., GERSBERG, R. M. & GOLDMAN, C. R. 1982. Inorganic nitrogen assimilation in a subalpine lake. *Limnol. Oceanogr.*, 27: 53-65.

CAMARERO, L., LATASA, M., JAUME, D. & CATALAN, J. in prep. Changes in photoadaptation during the formation of the ice and snow cover in a deep mountain lake.

CATALAN, J. 1988. Physical properties of the environment relevant to the pelagic ecosystem of a deep high-mountain lake (Estany Redó, Central Pyrenees). *Oecol. aquat.*, 9: 89-123.

CATALAN, J. 1989. The winter cover of a high-mountain Mediterranean lake (Estany Redó, Pyrenees). *Water Resour. Res.*, 25: 519-527.

CATALAN, J. in press. Evolution of dissolved and particulate matter during the ice-covered period in a deep high-mountain lake. *Can. J. Fish. Aquat. Sci.*

CATALAN, J. & CAMARERO, L. 1991. Ergoclines and biological processes in high-mountain lakes: similarities between the summer stratification and the ice-forming period in Lake Redó (Pyrenees). *Verh. Internat. Verein. Limnol.*, 24: 1011-1015

CSANADY, G. T. 1986. Mass transfer to and from small particles in the sea. *Limnol. Oceanogr.*, 31: 237-248.

DILLON, T. M. & CALDWELL, D. R. 1980. The Batchellor spectrum and dissipation in the upper ocean. *J. Geophys. Res.*, 85: 1910-1916.

DUGDALE, R. C. & GOERING, J. J. 1967. Uptake of new and regenerated forms of nitrogen in primary productivity. *Limnol. Oceanogr.*, 12: 196-206.

EDMONDSON, W. T. 1985. *Recovery of Lake Washington from eutrophication. Proceedings of the International Congress on Lakes pollution and recovery.* Rome 15th-18th April. European water pollution control association.

FEE, E. J. 1979. A relation between lake morphometry

and primary productivity and its use in interpreting whole lake eutrophication experiments. *Limnol. Oceanogr.*, 24: 401-406.

FRISCH, U. & ORSZAY, S. A. 1990. Turbulence: challenges for theory and experiment. *Physics today,* 43: 24-33.

GAINES, G. & ELBRÄCHTER, M. 1987. Heterotrophic nutrition. In: *The Biology of Dinoflagellates.* (F.J.R. Taylor, ed.): 168-224. Blackwell. Oxford.

GIBSON, C. E. 1987. Adaptations in *Oscillatoria redekei* at very slow growth rates - changes in growth efficiency and phycobilin complement. *Br. Phycol. J.;* 22: 187-191.

GOLDMAN, J. C. 1988. Spatial and temporal discontinuities of biological processes in pelagic surface waters. In: *Toward a theory on biological-physical interactions in the world ocean.* (B.J. Rothschild, ed.): 273-296. Kluwer. Dordrecht

GOLDMAN, C. R., JASSBY, A. & POWELL, T. 1989. Interannual fluctuation in primary production: meteorological forcing at two subalpine lakes. *Limnol. Oceanogr.*, 34: 310-323.

HÅKANSON, L. 1977. The influences of wind, fetch, and water depth on the distribution of sediments in Lake Vänern, Sweden. *Can. J. Earth. Sci.,* 14: 397-412.

HÅKANSON, L. 1982. Bottom dynamics in lakes. *Hydrobiologia,* 91: 9-22.

HARDING, L. W. Jr., FISHER, T. T., Jr. & TYLER, M. A. 1987. Adaptive responses of photosynthesis in phytoplankton: specificity to time-scale of change in light. *Biological Oceanogr.*, 4: 403-437.

HARRIS, G. P. 1980. Temporal and spatial scales in phytoplankton ecology. Mechanisms, methods, models and management. *Can. J. Fish. Aquat. Sci.,* 37: 877-900.

HERRON, H. A. & MAUZERAL, D. 1972. Development of photosynthesis in greening mutant of *Chlorella* and an analysis of the light saturation curve. *Plant Physiol.*, 50: 141-148.

HILTON, J. 1985. A conceptual framework for predicting the occurrence of sediment focusing and sediment redistribution in small lakes. *Limnol. Oceanogr.*, 30: 1131-1143.

HUTCHINSON, G. E. 1957. 1957. *A treatise on limnology.* Wiley. New York.

IMBERGER, J. 1985. The diurnal mixed layer. *Limnol. Oceanogr.*, 30: 737-770.

IMBERGER, J. & PARKER, G. 1985. Mixed layer dynamics in a lake exposed to a spatially variable wind field. *Limnol. Oceanogr.*, 30: 473-488.

IVEY, G.N. 1983. Experiments on convection near the temperature of maximum density and their application. In: *Proceedings of the 8th Australasian Fluid Mechanics Conference.* 2A4-2A6. University of Newcastle.

JASSBY, A. D. & PLATT, T. 1976. Mathematical formulation of the relationship between photosynthesis and light for phytoplankton. *Limnol. Oceanogr.*, 21: 540-547.

LEGENDRE, L. & DEMERS, S. 1985. Auxiliary energy, ergoclines and aquatic biological production. *Naturaliste can. (Rev. ecol. Syst.),* 112: 5-14.

LEGENDRE, L., DEMERS, S. & LEFAIVRE, D. 1986. Biological production at marine ergoclines. In: *Marine interfaces ecohydrodynamics.*: (J.C.J. Nihoul, ed.): 1-29. Elsevier. Amsterdam.

MARGALEF, R. 1978. Life-forms of phytoplankton as survival alternatives in an unstable environment. *Oceanol. Acta*, 1: 493-509.

MARGALEF, R. 1980. *La Biosfera. entre la termodinámica y el juego.* Omega. Barcelona.

MARGALEF, R. 1983. *Limnología.* Omega. Barcelona.

MARGALEF, R. 1989. Le plancton: survivre et s'organiser dans un fluide mobile. In: *Océanologie actualité et prospective.*: (M. Denis, ed.): 170-183. Centre d'Océanologie de Marseille. Marseille.

MARGALEF, R. ESTRADA, M. & BLASCO, D. 1979. Functional morphology of organisms involved in red tides, as adapted to decaying turbulence. In: *Toxic dinoflagellates blooms.* (D.L. Taylor & H.H. Seliger, eds.): 89-94. Elsevier. North Holland. New York.

MITCHELL, J. G., OKUBO, A., FUHRMAN, J. A. 1990. Gyrotaxis as a new mechanism for generating spatial heterogeneity and migration in microplankton. *Limnol. Oceanogr.*, 35: 123-129.

MUNK, W. H. & RILEY, G. A. 1952. Absorption of nutrients by aquatic plants. *J. mar. Res.* 11: 215-240.

PATTERSON, J.C., HAMBLIN, P.F. & IMBERGER, J. 1984. Classification and dynamic simulation of the vertical density structure of lakes. *Limnol. Oceanogr.*, 29: 845-861.

PLATT, T., GALLEGOS, C. L. & HARRISON, W. G. 1980. Photoinhibition of photosynthesis in natural assemblages of marine phytoplankton. *J. mar. Res.*, 38: 687-708.

POMEROY, L. R. 1974. The ocean's food web, a changing paradigm. *BioScience,* 24: 499-504.

QUAY, P. D., BROECKER, M. S., HESSLEIN, R. H. & SCHINDLER, D. W. 1980. Vertical diffusion rates determined by tritium tracer experiments in the thermocline and hypolimnion of two lakes. *Limnol. Oceanogr.*, 25: 201-218.

QUÉGUINER, B. & LEGENDRE, L. 1986. Phytoplankton photosynthetic adaptation to high frequency light fluctuations simulating those induced by sea surface waves. *Mar. Biol.*, 90: 483-491.

REYNOLDS, C. S. 1984. *The ecology of freshwater phytoplankton.* Cambridge University Press. Cambridge.

REYNOLDS, C. S., OLIVER, R. L. & WALSBY, A. E. 1987. Cyanobacterial dominance: the role of buoyancy regulation in dynamic lake environments. *N. Z. J. Mar. Freshwat. Res.*, 21: 379-390.

SCHINDLER, D. W. 1971. A hypothesis to explain differences and similarities among lakes in the Experimental Lake Area, Northwestern Ontario. *J. Fish. Res. Bd. Canada*, 28: 295-301.

SCHINDLER, D. W. & COMITA, G. W. 1972. The dependence of primary production upon physical and chemical factors in a small senescing lake, including the effects of complete winter oxygen depletion. *Arch. Hydrobiol.*, 69: 413-451.

SMETACEK, V. 1985. Role of sinking in diatom life-history cycles: ecological, evolutionary and geological significance. *Mar. Biol.*, 84: 239-251.

SMETACEK, V. & PASSOW, U. 1990. Spring bloom initiation and Sverdrup's critical-depth model. *Limnol. Oceanogr.*, 35: 228-234.

SMETACEK, V. & POLLEHNE, F. 1986. Nutrient cycling in pelagic systems: a reappraisal of the conceptual framework. *Ophelia*, 26: 401-428.

SPIGEL, R. H. & IMBERGER, J. 1980. The classification of mixed-layer dynamics in lakes of small to medium size. *J. Physic. Oceanogr.*, 10: 1104-1120.

SPIGEL, R. H. & IMBERGER, J. 1987. Mixing processes relevant to phytoplankton dynamics in lakes. *N. Z. J. Mar. Freshwat. Res.*, 21: 361-367.

SMITH, I. R. 1975. *Turbulence in lakes and rivers.* Freshwater Biological Association. Ambleside.

STEELE, J. H. 1962. Environmental control of photosynthesis in the sea. *Limnol. Oceanogr.*, 7: 137-150.

SVERDRUP, H. U. 1953. On conditions for the vernal blooming of phytoplankton. *J. Cons. Cons. Int. Explor. Mer.*, 18: 287-295.

THIENEMANN, A. 1928. Der Sauerstoff im eutrophen und oligotrophen See. *Die Binnengewaser*, 4. Schweizerbart. Stuttgart.

Homage to Ramon Margalef; or, Why there is such pleasure in studying nature (J. D. Ros & N. Prat, eds.). 1991. *Oecologia aquatica*, 10: 95-111

FACTORS CONTROLLING PHYTOPLANKTON COMMUNITY STRUCTURE IN AN ALKALINE VERSUS A SOFTWATER LAKE

DOLORS PLANAS

Département de Sciences Biologiques. GEOTOP. Université du Québec à Montréal. C.P. 8888, Succursale A. Montréal. Québec. H3C 3P8. Canada.

Received: February 1990.

SUMMARY

The objective of this paper is to compare nutrient resources that might control phytoplankton species composition and succession in two Spanish lakes located in very different geological areas but at the same latitude. The lakes, Banyoles and Sanabria, are respectively located on Eocene gypsum carbonate rocks and in a Pre-Cambrian region dominated by granitic rock. Their species compositions are quite different: the first lake is dominated by diatoms, the second by chroococcal cyanobacteria.

Our hypothesis is that the limiting nutrient in soft-water Lake Sanabria will be nitrogen, and in the alkaline Lake Banyoles, phosphorus (because of its strong precipitation with calcium or coprecipitation with $CaCO_3$).

In order to test this hypothesis we used principal component analyses with physico-chemical and biological variables on the one hand and species composition on the other. The results show nitrogen to be the more important nutrient, controlling species succession in both lakes. This element is relatively low in both ecosystems during the summer.

KEY WORDS: Phytoplankton, Lake Sanabria, Lake Banyoles, nutrient limitation.

INTRODUCTION

The factors responsible for the seasonal variation of phytoplankton as a whole and of its particular component species may be grouped roughly as: 1) partly independent physical factors - that is, temperature, illumination, and turbulence; 2) interdependent chemical factors - nutrients, accessory organic compounds (vitamins) and antibiotics; and 3) biological factors - predation, competition and parasitism (HUTCHINSON, 1967).

The chemical characteristics of waters - soft-water versus alkaline - could be important in characterizing regional differences in the specific composition of algal communities, as was demonstrated by SPARLING & NALEWAJKO (1970) in natural lakes or by MARGALEF *et al.* (1976) in the phytoplankton characterization of Spanish reservoirs. Species affinity to water type can be a direct effect, such as for example the need for Ca by some haptophyceae with coccolith deposits, or an indirect effect related to the pH-carbon dioxide-bicarbonate system (REYNOLDS, 1984) or to the control of macronutrients as is the case of calcium and phosphorus (MARGALEF, 1983; AVNIMELECH, 1984).

In relation to water chemistry and phytoplankton association, most research carried out during the first part of this century was based on the importance of the monovalent to divalent cation ratio. For example, communities dominated by diatoms were associated with a lower ratio, and desmid and chrysophyceae communities with a higher ratio (PEARSALL, 1922). Although cation concentrations in the water could be important to the regulation of the osmotic pressure in algal cells (MARGALEF, 1983) it is more probable that the difference in species composition in communities living in more or less mineralized waters is related more to nutrient availability than to the abundance of a specific cation or anion. This fact is clearly illustrated by African tropical lakes which show a wide range of chemical characteristics and resource limitations which vary accordingly (KALFF, 1983; KILHAM *et al.*, 1986). If the physical conditions of the lakes are similar, the response of the phytoplankton depends on changes in the N:P supply ratio (SUTTLE & HARRISON, 1988) which will govern the species succession.

The objective of our study is to compare the nutrient resources that might control phytoplankton species composition in two Spanish lakes located in very different geological areas but at the same latitude, ensuring that physical factors controlled by annual irradiation are very similar. The lakes are Banyoles and Sanabria.

In order to arrive at this objective we used principal component analyses with physico-chemical and biological variables on the one hand and species composition on the other. The data for Lake Banyoles were taken from the 1970-1971 cycle (PLANAS, 1973) and for Lake Sanabria from the 1973-1974 cycle studied by Planas in collaboration with the Comisaría de Aguas del Duero (Planas, unpublished data).

Our hypothesis is that the limiting nutrient in soft-water Lake Sanabria will be nitrogen whereas in the alkaline Lake Banyoles, phosphorus, because of its strong precipitation with calcium or coprecipitation with $CaCO_3$, will be the controlling factor. Nitrogen has been found to be the limiting nutrient in some low conductivity temperate lakes where phosphorus concentration is very low (GOLDMAN, 1981).

CHARACTERISTICS OF THE LAKES

Lake Sanabria and Lake Banyoles are the largest natural water bodies in Spain. Both are pre-alpine lakes and are located geographically at 42°1' N - 3°1' W and 42°7' N - 2°45' E, respectively. They are situated in very different geological regions and have very different origins. Lake Sanabria is located in a Pre-Cambrian region dominated by granitic rock. It occupies the valley excavated by two convergent glaciers closed by the Riss moraines (MARGALEF, 1955). Lake Banyoles is located in Eocene gypsum overlying carbonate rocks of the same geological period. It has a karstic origin and is the remainder of a much larger Tertiary lake (SOLÉ-SABARÍS, 1958).

Water inflow derives mainly from the surface (the Tera river) in Lake Sanabria and from the bottom in Lake Banyoles (PLANAS, 1973; in press).

The first lake is bigger: its surface area is three times, its volume six times and its mean depth twice that of Lake Banyoles. The renewal time of the water in the former is eight times faster that in Lake Banyoles due to its geographical location: Lake Sanabria is located in the wettest part of Spain with a yearly mean precipitation of 1446 mm, whereas Lake Banyoles receives only 837 mm of rain per year.

The main anion is sulphate in Banyoles and bicarbonate in Sanabria. The dominant cations are respectively calcium and magnesium.

According to their respective nutrient

concentrations, Lake Banyoles is considered to be oligo-mesotrophic (PLANAS, 1973) and Lake Sanabria oligotrophic. The major nutrient differences are in nitrate and silica concentrations. Nitrate concentrations are four times greater in Lake Banyoles than in Lake Sanabria. Silica, which is very abundant and shows strong seasonal variations (0.25 μM to 200 μM) in the epilimnion of the alkaline lake, has a concentration typical of a pre-Cambrian lake (40 to 74 μM) in the soft-water Lake Sanabria. Phosphorus concentration in Lake Banyoles is twice that in Lake Sanabria.

MATERIALS AND METHODS

Data used for the two principal component analyses (PCA) were integrated from samples taken at different depths of the euphotic zone of each lake. The thickness of this zone varied from 2.6-10 m in Lake Banyoles (LB) and from 8-16 m in Lake Sanabria (LS). The samples came from the deepest station of the euphotic zone of each lake: 40 m in Banyoles and 50 m in Sanabria.

Lake Sanabria sampling began at the end of July 1973 and finished in mid-October 1974. In Lake Banyoles sampling took place from the end of October 1969 to January 1971. Symbol identification for the various dates is given in Table I.

For the chemical and biological methods used in Lake Banyoles see PLANAS (1973); the same methods were used for Lake Sanabria. The first lake was sampled fortnightly, the second every three weeks during stratification and monthly during the winter period.

The variables used for the physico-chemical-biological principal component analysis (PC-PCA) were: the nutrients, phosphorus (PO_4) measured as total soluble reactive phosphorus and nitrogen (NO_3 = nitrates + nitrites); water transparency (SD); phytoplankton biomass, as calculated from chlorophyll-a concentrations (CHL); and cell numbers (CN). In Lake Banyoles, silica (Si) was also included as a nutrient, because the dominant species are diatoms, and silica, which presents a marked cycle, could act as a limiting factor. Since Si was analysed only from June onwards, the PCA for this lake comprises only the sampling data collected during the stratification period (June to October 1970). In Lake Sanabria we did not include this variable since, as mentioned above, it behaves as a rather conservative element. Another difference in the variables used in the PC-PCA of each lake is the primary production (PP). We included it in Lake Banyoles' PC-PCA in which case we had a complete cycle, but not in Lake Sanabria', where it was only measured 9 times out of 22.

For the species principal component analysis (S-PCA) we retained the species

TABLE I. Lakes Banyoles and Sanbria sample number identification. * Mixing period.

BANYOLES		SANABRIA	
NUMBER	DATE	NUMBER	DATE
1	14.02.70	1	27.07.73
2	28.03.70	2	10.08.73
3	14.03.70	3	11.09.73
4	28.03.70	4	27.09.73
5	11.04.70	5	19.10.73
6	25.04.70	6	31.10.73*
7	09.05.70	7	16.11.73*
8	25.05.70	8	10.12.73*
9	09.06.70	9	20.12.73*
10	20.06.70	10	07.02.74*
11	06.07.70	11	27.03.74*
12	20.07.70	12	25.04.74*
13	03.08.70	13	16.05.74*
14	18.08.70	14	30.05.74*
15	31.08.70	14b	16.06.74*
16	14.09.70	15	25.06.74*
17	28.09.70	16	11.07.74*
18	10.10.70	17	26.07.74
19	24.10.70	18	09.08.74
20	07.11.70	19	21.08.74
21	21.11.70	20	11.09.74
22	05.12.70	21	26.09.74
23	19.12.70	22	16.10.74
24	02.01.71		
25	16.01.71		
26	30.01.71		

which represented more than 10% of the total for at least one sampling, except if the species was observed only once, in which case it was excluded.

In discussing the distribution of the objects in the reduced space of the PCA, we considered only the axes with an eigenvalue near or above 1. We considered that descriptors with a component greater than or equal to 0.60 defined the axes.

For each PCA we determined object (in this case sampling date) clustering by applying a hierarchic classification (LEGENDRE & LEGENDRE, 1979).

RESULTS

The first three axes of the PC-PCA in Lake Banyoles (LB) contribute 82.63% of the variance, with Axis I explaining 37.45%. The loading variable which defines the positive part of the Axis is Si (0.70) and those for the negative part are CN (-0.73) and SD (-0.72). Axis II explains 26.34% of the variance. The descriptors that characterize this axis are NO_3 (-0.84) and PP (-0.83).

These two axes clearly separate the stratification period (Cluster 1) from the mixing period (Clusters 2 and 3). The separation of the clusters is highly significant (Fig. 1a). In general the stratification period is characterized by low nitrogen and silica concentrations, low primary production and relatively high transparency (PLANAS, 1990) and phytoplankton cell density (Fig. 2).

Axis III contributes 18.83% of the variance. The most important loading variables on this axis are nutrients. This axis separates a few dates of the stratification period from those of the mixing period (Fig. 1b).

The objects in the principal component analysis of Lake Banyoles using species as descriptors (S-PCA) are spatially distributed on five axes, which explain 71.72% of the variance. Axis I explains 24.78% of the variance. It is defined by the following species: *Rhodomonas minuta* (-0.91), *R. minuta nannoplanctonica*

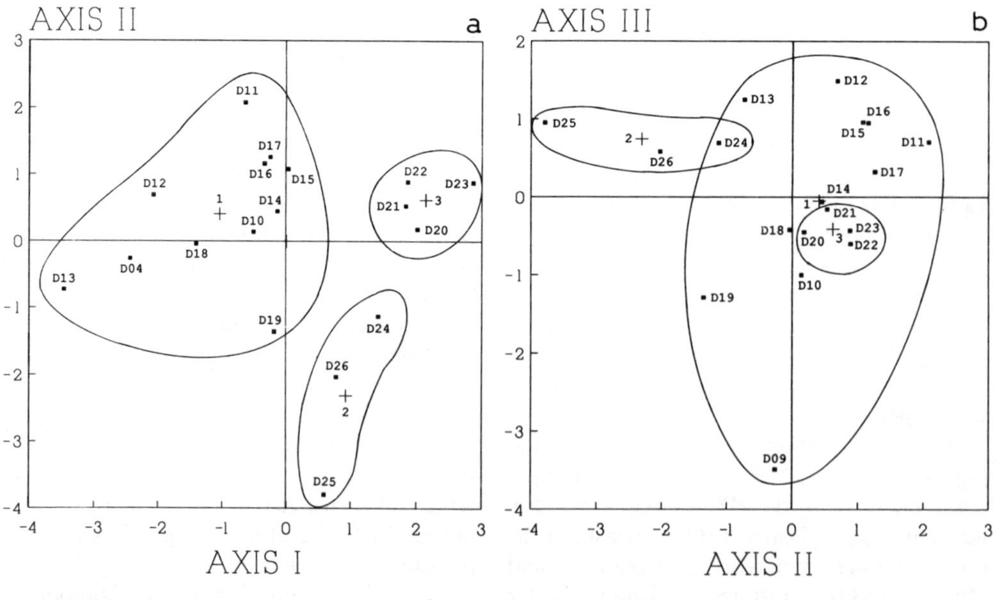

FIGURE 1. Lake Banyoles sample distribution using the physico-chemical and biological variables. (a) Axes I and II. (b) Axes II and III. For identification of object numbers, see Table I.

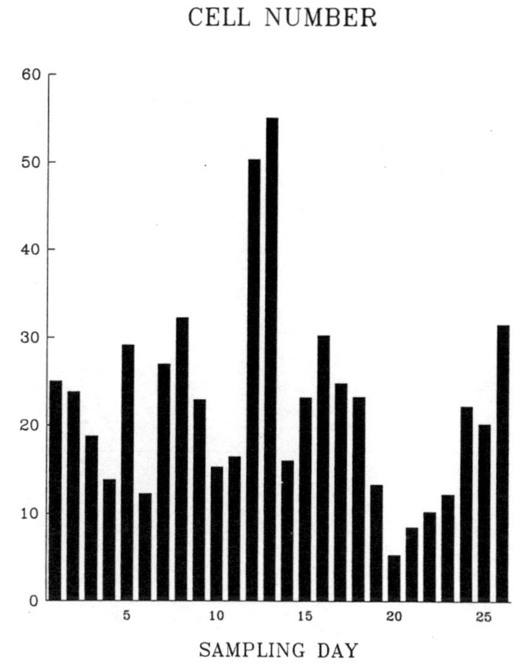

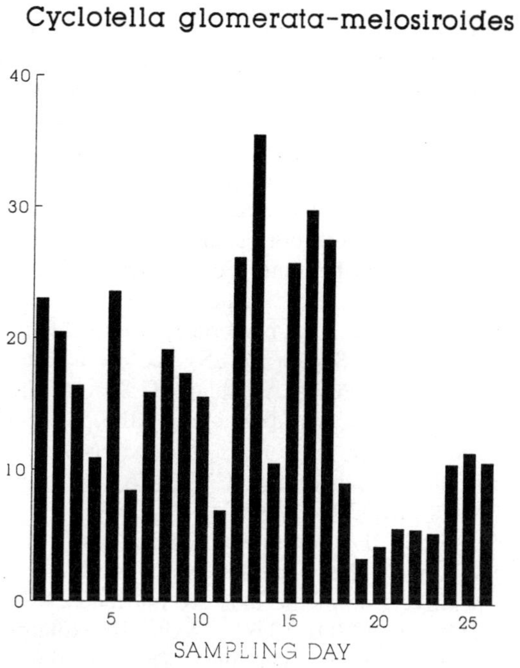

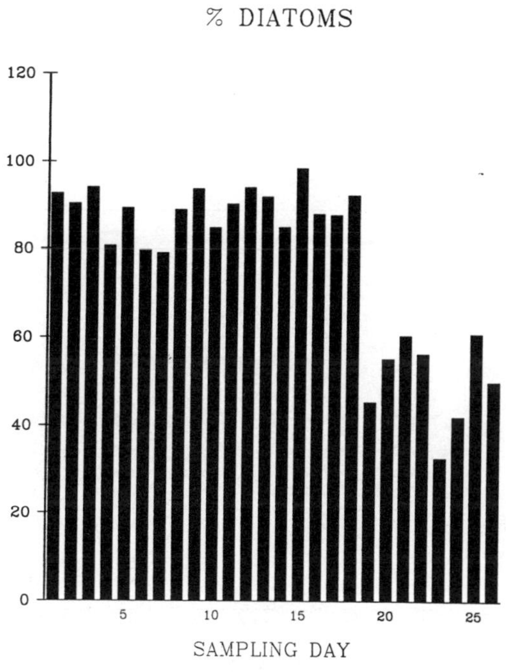

FIGURE 2. Temporal evolution of Lake Banyoles variables: Cell number (cell m^{-2} x 10^9), *Cyclotella glomerata-melosiroides* (cells m^{-2} x 10^9), and percentage of diatoms.

(-0.81), *Monoraphidium contortum* (-0.84) and *Dictyosphaerium pulchellum* (-0.83), all loading the negative part of the axis. Axis II contributes 17.16% of the variance; the species defining this axis are *Chrysoccocus rufescens* (-0.78) and *Chlamydomonas chlamydella* (-0.71) (Fig. 3a).

Axis III explains 12.70% of the variance, with the bacteria *Sphaerotilus* sp. (0.83) and *Cyclotella comta* (0.73) as the most important contributors, loading the positive part of the axis (Fig. 3b).

Axis I clearly separates the stratification and the mixing period into two clusters. The separation is highly significant. The overturn date is isolated from both groups with $p \leq 0.01$ (Fig. 3a). The mixing period is mainly characterized by the abundance of *Rhodomonas* sps. (Fig. 4) and to a lesser degree by *Dictyosphaerium pulchellum* and *Monoraphidium contortum* (Fig. 4).

Axis II separates the sampling period around overturn into two groups; the sampling dates before overturn, and those after it. This period is characterized by a relatively greater abundance of *Chrysoccocus rufescens* (Fig. 5a). The greater abundance of *Chlamydomonas chlamydella* (Fig. 5b) in the overturn period isolates this date from the other clusters.

Axis III separates the sampling dates where *Cyclotella comta* was more abundant (Fig. 5c). These correspond to the samples taken at the beginning of the stratification period. At that time there was a proliferation of the bacteria *Sphaerotilus* sp.

In Lake Sanabria (LS) the first three axes of the PC-PCA explain 81.86% of the variance. The variables loading the positive and negative part of the first axis (40.35% of the variance) are CN (0.76) and SD (-0.75) respectively.

Axis II contributes 25.16% of the variance. The variables which define the positive side of the axis are nutrients: NO_3 (0.72) and PO_4 (0.62). Axis III explains 16.35% of the variance. The descriptor highly associated with the negative part of this axis is CHL (-0.71).

The first axis in the PC-PCA of this lake (LS) mainly separates 1974 stratification dates from the mixing period. Axis II discriminates objects corresponding to the 1973 late summer period (Fig. 6a) when PO_4 was high (Fig. 7a) from winter samples with high NO_3 concentrations (Fig. 7b). Axis III discriminates objects with high CHL concentrations (Fig. 7d).

The spatial distribution of the objects in the three axes could be clustered into four groups. Statistically, Groups 1 and 2 are very highly significantly different; 3 and 4, highly significantly. The first cluster includes samples with high PO_4 and CHL. The sampling dates clustered in 2 are characterized by high transparency (Fig. 7c) or high NO_3. In Cluster 3 the sampling dates are characterized by high CN (Fig. 9a) and in Cluster 4 by elevated PO_4 concentration. If we eliminate Cluster 4, its objects are absorbed into Cluster 3, in the positive part of the axis (Fig. 6a). The difference between the three remaining groups thus becomes very highly significant.

In the principal component analysis of LS, using phytoplankton species composition to separate objects, the first 5 axes explain 75% of the variance. Axis I contributes 21% of the variance. The species loading the negative and positive

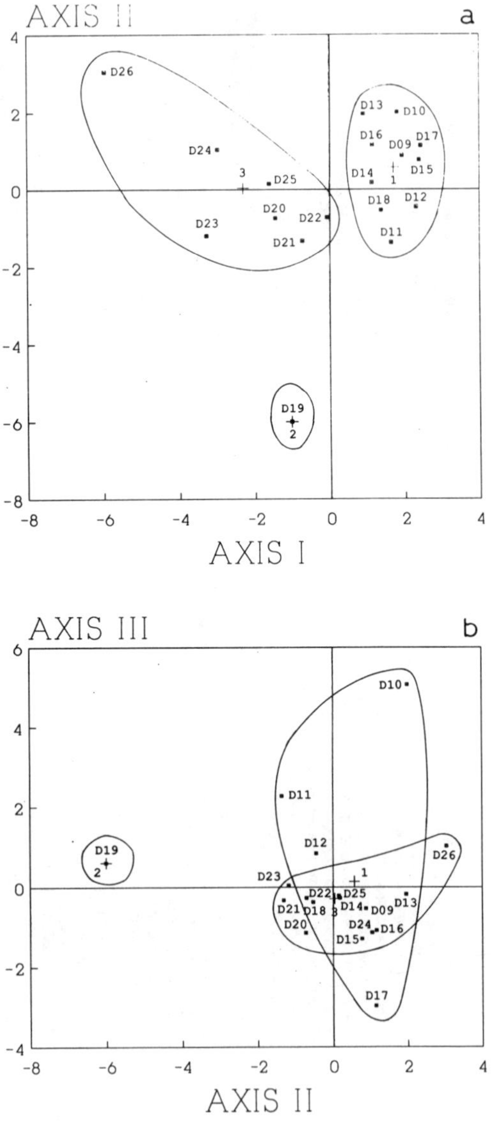

FIGURE 3. Lake Banyoles sample distribution using species composition as descriptors. (a) Axes I and II, (b) Axes II and III. For identification of object numbers, see Table I.

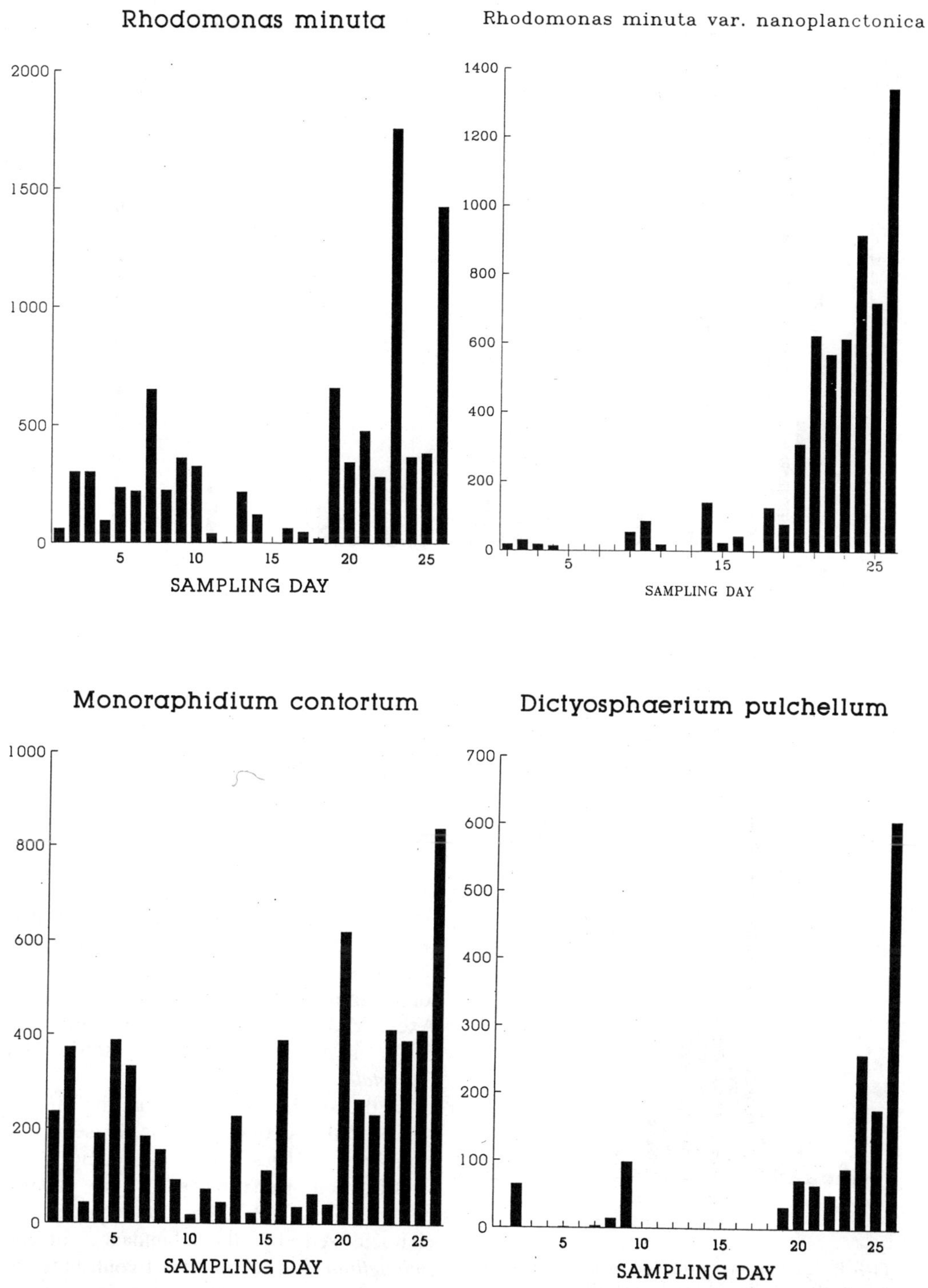

FIGURE 4. Temporal evolution of Lake Banyoles species loading the first axis (cells m^{-2} x 10^6): *Rhodomonas minuta*, *R. minuta nannoplanctonica*, *Monoraphidium contortum* and *Dictyosphaerium pulchellum*.

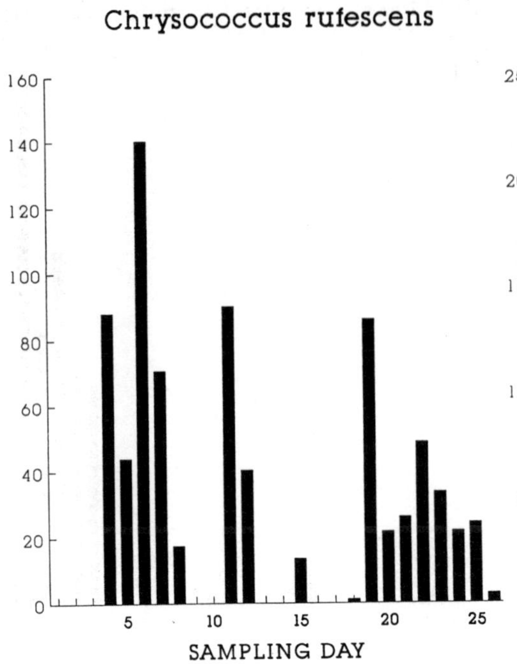

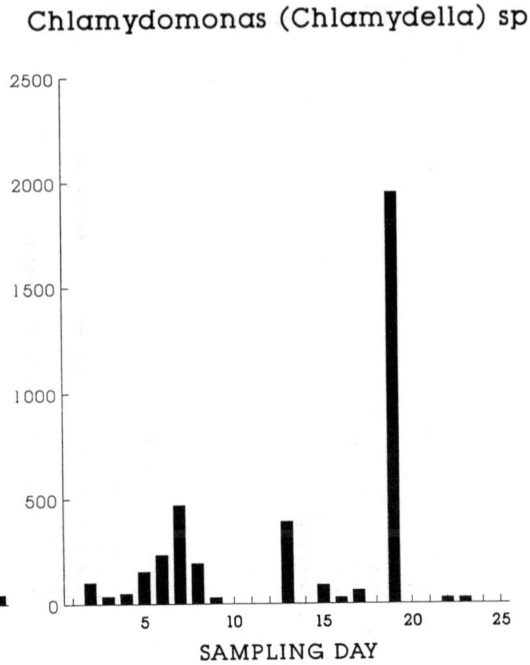

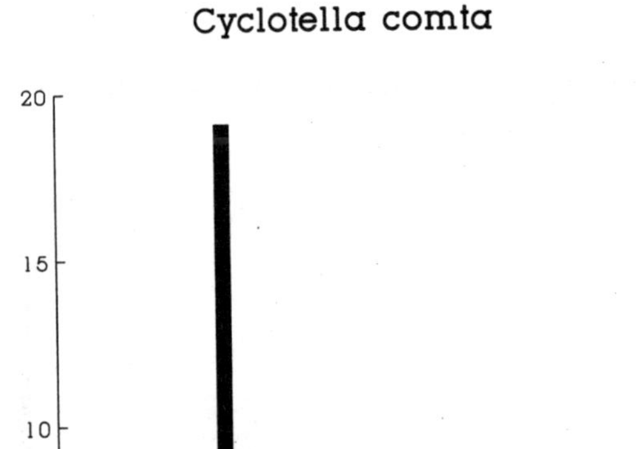

FIGURE 5. Temporal evolution of the species (cells m^{-2} x 10^6) loading Axes II and III: *Chrysoccocus rufescens*, *Chlamydomonas chlamydella*, and *Cyclotella comta*.

part of this axis are the chloroccocal green algae *Crucigenia tetrapedia* (-0.67) and *Monoraphidium contortum* (0.66), respectively.

Axis II explains 18.09% of the variance. The main species contributing to this axis are *D. pulchellum* (0.65) and *Chromulina rossanofii* (0.58). *C. pseudonebulosa* (-0.66) and *Merismopedia marssonii* (-0.66) define Axis III and explain 15% of the variance (Fig. 11a, 9c) respectively. The two remaining axes contribute only 10% of the total variance. The species contributing to Axes IV and V are respectively (Fig. 9b, 11b): *Aphanothece clathrata* (0.61) and *Cyclotella* sp. (0.81).

In the spatial distribution of the objects in the S-PCA of LS, Axis II separates sampling dates corresponding to stratification from those of the mixing period (Fig. 8). The first period is characterized by the abundance of *D. pulchellum* (Fig. 10c). Axis I contributes to separating the summer dates where *C. tetrapedia* presents its maximum density

(Fig. 10a), from samples where for the same period in the former year *Cyclotella* sp. was very abundant (Fig. 11b).

The spatial distribution of samples in Axes II and III (Fig. 8b) is determined by the absence of *C. pseudonebulosa* (Fig. 11a). Its cell density was greatest in the summer of 1974, especially from August to September.

DISCUSSION

The phytoplankton species richness in Lake Sanabria was higher than in Lake Banyoles: 232 and 77 species respectively. The first lake could be characterized as a myxophycean plankton lake, the second as a diatom plankton lake (HUTCHINSON, 1967). Cyanobacteria of the chroococcal order dominated Lake Sanabria throughout the year (Planas, unpublished data); the more abundant species were *Aphanothece clathrata* and *Merismopedia marssonii*. In Lake Banyoles the diatoms *Cyclotella glomerata* and *C. melosiroides* were dominant during the whole study period (PLANAS, 1973; Fig. 2b).

The dominance of diatoms in the silica-rich, alkaline Lake Banyoles (Fig. 2c) was to be expected, although diatoms could have been limited by other nutrients or by light (KILHAM *et al.*, 1986). HUTCHINSON (1967) characterizes oligotrophic diatom plankton as being present in nutrient-poor lakes with neutral or slightly alkaline water.

The dominance of cyanobacteria in Lake Sanabria is more striking. Some species of these algae are normally associated with waters low in nitrogen, as is the case of the nitrogen fixing species which dominate in eutrophic ecosystems where phosphorus concentrations are high (SCHINDLER, 1978). They are also very abundant at high phosphorus and nitrogen concentrations (VOLLENWEIDER, 1968). However, these trophic characteristics do not correspond to Lake Sanabria, which, as we mentioned before, might be classified as oligotrophic. The two chlorococcal species that dominate this lake are very abundant in the low conductivity waters of the Canadian Shield (Planas, personal observation), especially after the spring outburst when nitrogen and phosphorus are not detectable in the water. These Canadian shield lakes are naturally quite rich in dissolved organic matter, as is

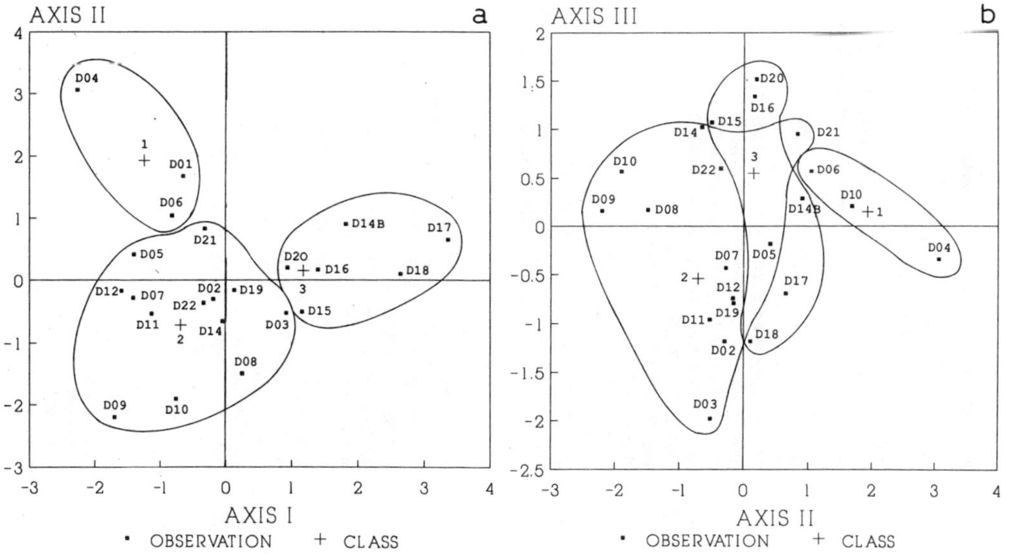

FIGURE 6. Lake Sanabria sample distribution using the physico-chemical and biological variables. (a) Axes I and II. (b) axes II and III. Points, observations; crosses, classes. For identification of object numbers, see Table I.

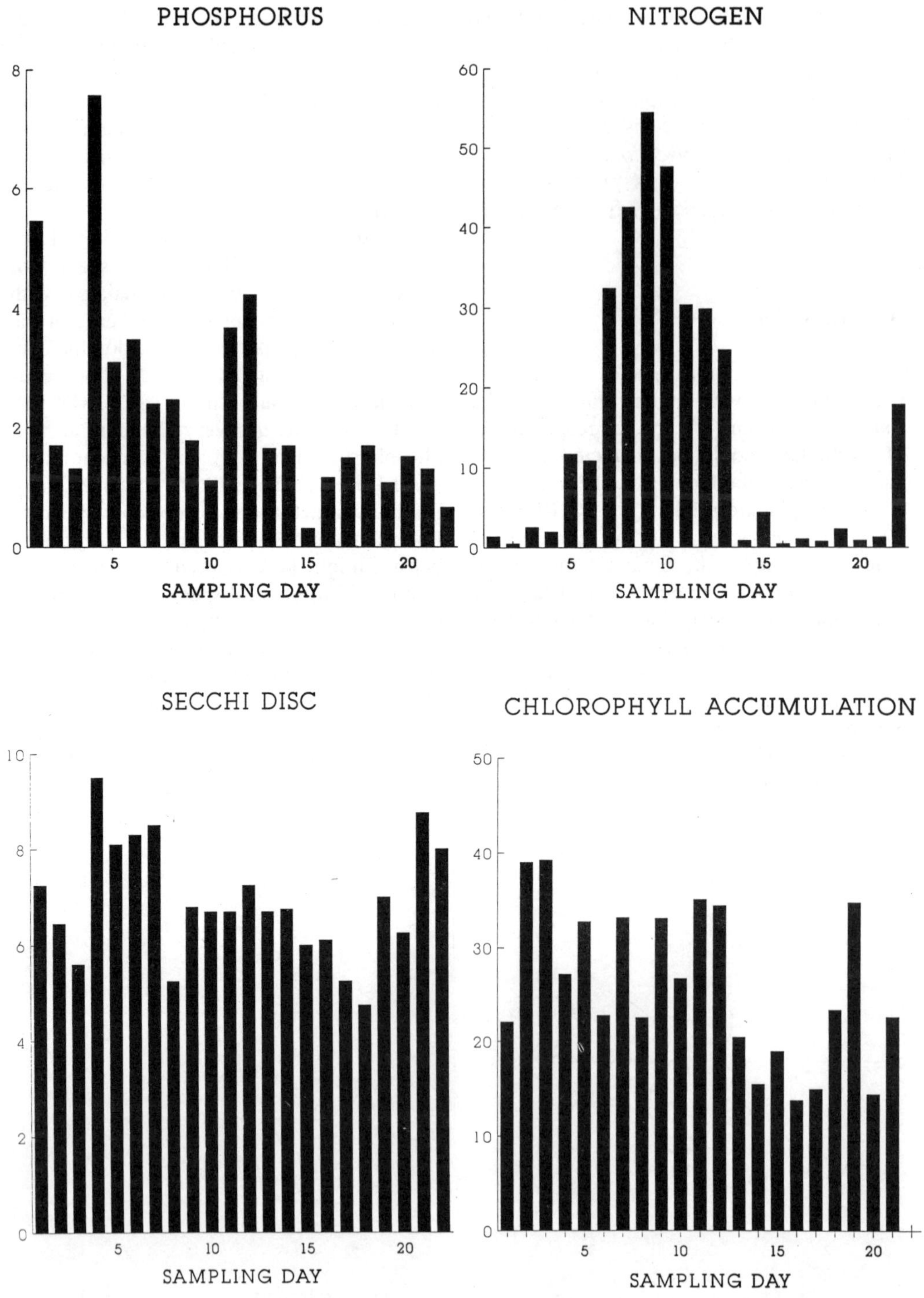

FIGURE 7. Temporal evolution of Lake Sanabria variables: phosphate (mM m^{-2}), nitrates + nitrites (mM m^{-2}), Secchi disc depth (m) and chlorophyll-a (mg m^{-2}).

the case of Lake Sanabria (Planas, unpublished data). It has been mentioned in the literature, on several occasions and for different parts of the world, that planktonic blue-greens are favoured by dissolved organic matter (FOGG *et al.*, 1973). This, however, does not imply that such compounds are needed by cyanobacteria, as the same authors show that no freshwater cyanobacteria species require exogenous organic growth factors.

For both lakes and particularly for Lake Banyoles, we cannot speak of true succession, at least not in reference to the dominant species, since they were present throughout the study period. Phytoplankton succession is a general phenomenon observed not only in temperate lakes (HUTCHINSON, 1967; REYNOLDS, 1984) but also in tropical and subtropical lakes, if one can rely on the little information that exists (GANF & VINER, 1973; MELACK, 1979; LEWIS, 1978 a, b; HECKY & KLING, 1981).

The two dominant cyanobacteria species of LS constitute more than 50% of the total cell counts during lake stratification and early mixing. *M. marssonii* was more abundant at the beginning of stratification than *A. clathrata* but the latter species attained its maximum in October 1974. HUTCHINSON (1967) mentions that *Aphanothece* sp. develops in productive lakes of temperate regions during the warmest month, usually when the combined nitrogen in the water is low and PO_4 indetectable. This does not correspond to the conditions found in LS during the summer, where the species was most abundant when PO_4^{3-} was either detectable or at its maximum. *Merismopedia* are generally considered to be oligotrophic indicators (DROUET & DAILY, 1956).

In LB the dominant species (Fig. 2b) constitute 90% of the population from early spring to overturn. Thereafter they decrease to 20-50% of the total cell number (Fig. 2b; PLANAS, 1973) but continue to be dominant. According to Grim's study of Lake Constance (GRIM, 1939) the same diatom species were dominant throughout the year. The author mentions that the only non-diatom species with a significant contribution was the cryptomonad *R. lacustris*, which was abundant towards the end of the winter when the temperature of

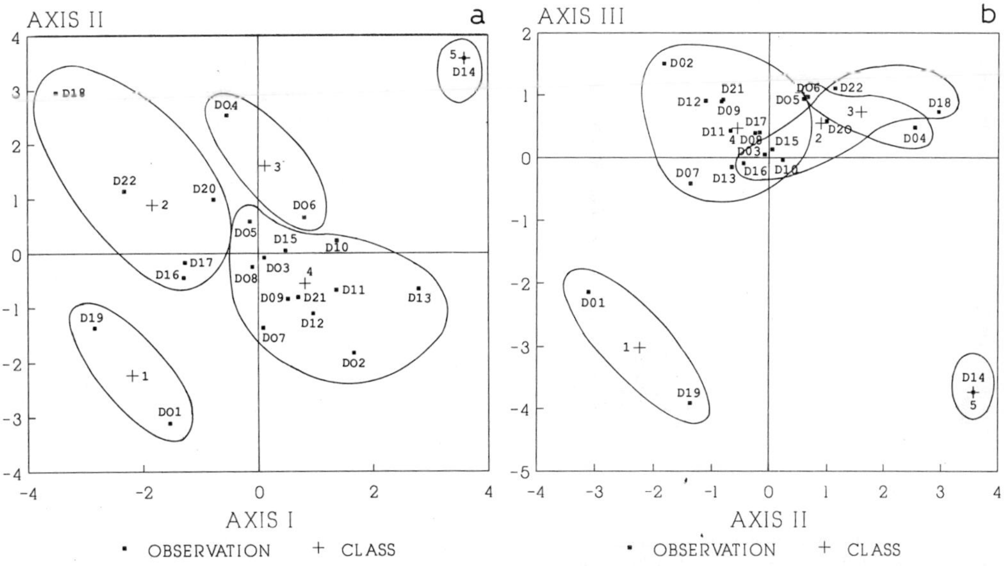

FIGURE 8. Lake Sanabria sample distribution using species composition as descriptors. (a) Axes I and II, (b) Axes II and III. Points, observations; crosses, classes. For identification of object numbers, see Table I.

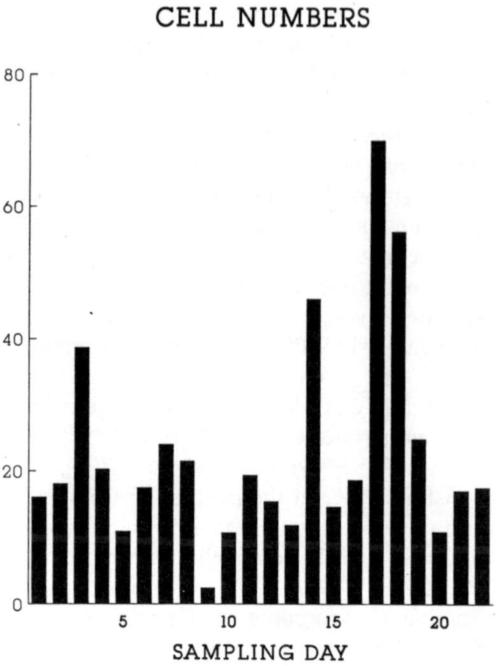

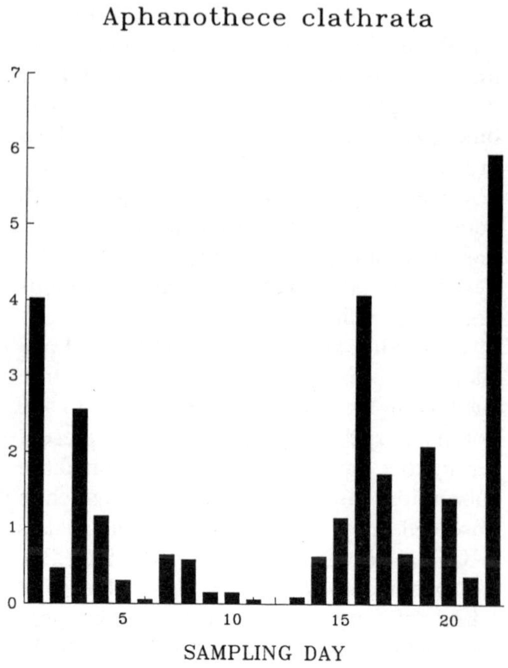

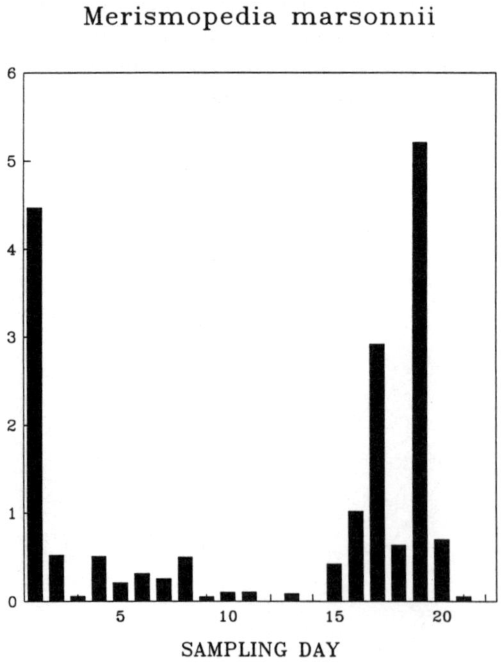

FIGURE 9. Temporal evolution of Lake Sanabria phytoplankton density (cell m^{-2} x 10^9) and dominant species: *A. chlathrata* (colonies m^{-2} x 10^9), and *Merismopedia marssonii* (colonies m^{-2} x 10^9).

the water was between 5-13 °C. As we will discuss later, Cryptophyceae also became abundant in LB during the winter.

The fact that the dominant species are present throughout the year explains why they do not contribute to the S-PCA axis formation and that when they do, as in the case of Lake Sanabria, they explain only a small percentage of the total variance.

In both lakes we observed a succession of the subdominant species. These species, which had a very marked annual cycle, contributed to the axis formation, together with sporadic species that became very abundant for short periods.

For both lakes, the PC-PCA shows that transparency and cell density are the main variables loading the first axis. In LB the mean transparency is lower than that in LS, 2.8m and 6.9m respectively. Its contribution in positioning the objects of the first lake is low. SD is more important in Lake Sanabria where it distributed the sampling dates around overturn. Overturn corresponds to the period of maximum

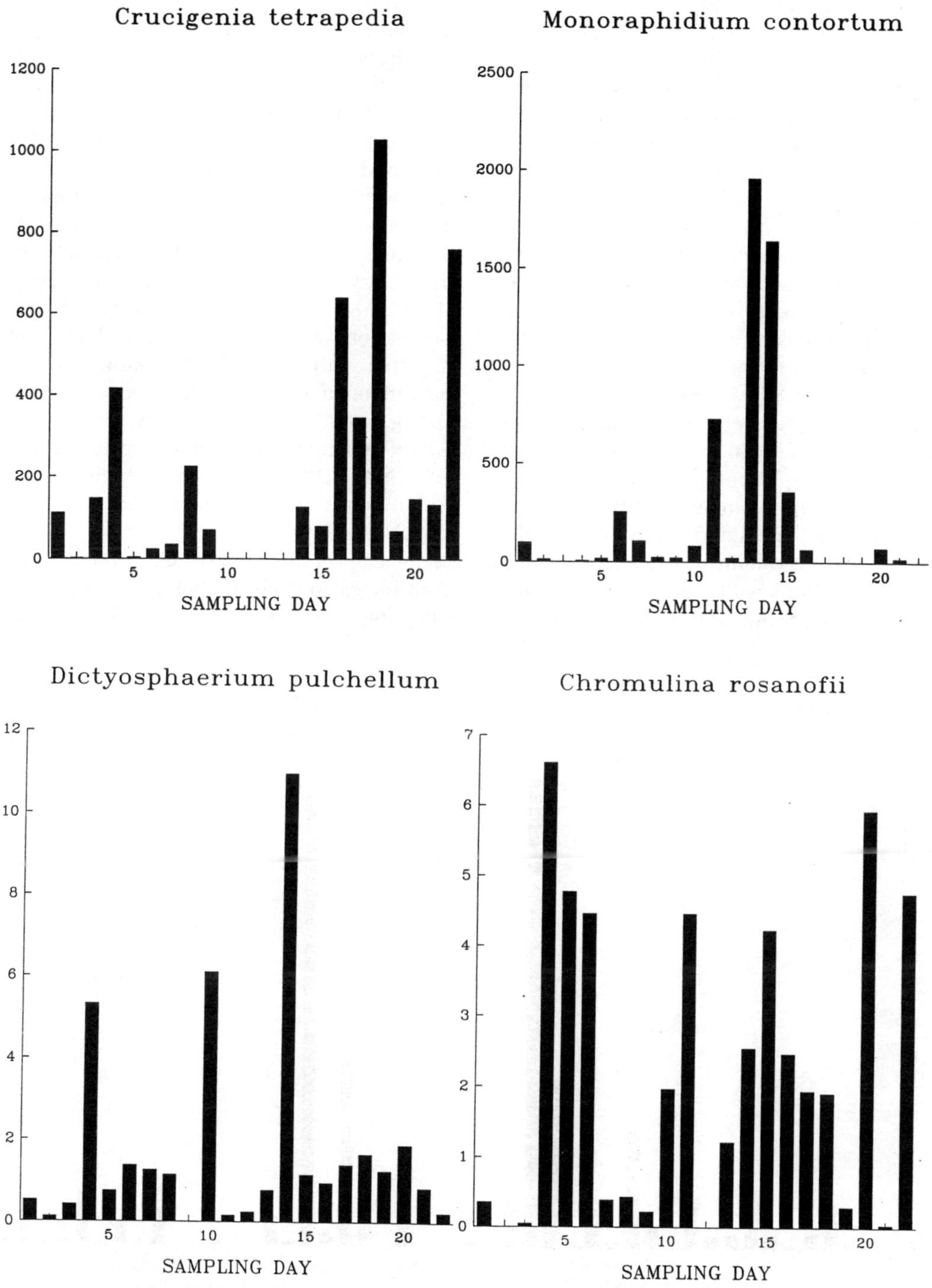

FIGURE 10. Temporal distribution of Lake Sanabria species loading the first two axes: *Crucigenia tetrapedia* (cells $m^{-2} \times 10^6$), *Monoraphidium contortum* (cells $m^{-2} \times 10^6$), *Dictyosphaerium pulchellum* (cells $m^{-2} \times 10^9$), and *Chromulina rossanofii* (cells $m^{-2} \times 10^9$).

transparency in this lake.

As stated above, cell numbers also contribute to the formation of Axis I. It is interesting to note that the two variables which contribute to Axis I have different signs in each lake. In LB both variables loaded the same sign of the axis, whereas in Sanabria SD loaded the negative part of the axis and CN the positive. The factors which contribute to light transmission in each lake could explain this difference. During our study cycle of LB, suspended matter increased after overturn due to a high underwater inflow during the autumn rains (PLANAS, 1973). The transparency then remained very low until the next summer stratification. In Lake Sanabria transparency seems related more to phytoplankton abundance, although not to biomass as measured by CHL. In both lakes, the maximum in CN occurred in the summer. The smaller size of the dominant species in LS probably induced more light scattering than the diatoms which dominated LB.

Nitrogen contributes to Axis II formation in both lakes. This nutrient was more abundant in winter than in summer, both in LB and LS. In the latter lake, NO_3 increased after overturn and remained relatively high until spring. In LB, although a slight increase occurred during the fall overturn, NO_3 became really abundant only in January when the lake mixed completely (PLANAS, 1973, 1990). In LB, primary production as well as NO_3 load the negative part of Axis II. Although the maximum of PP in this lake occurs in the spring, another maximum is present in winter concurrent with the nitrogen increase (PLANAS, 1990). Both variables thus seem to influence the positioning of the same sampling stations.

In LS, besides NO_3, phosphorus also contributes to Axis II, but in the opposite direction. It increases slightly during

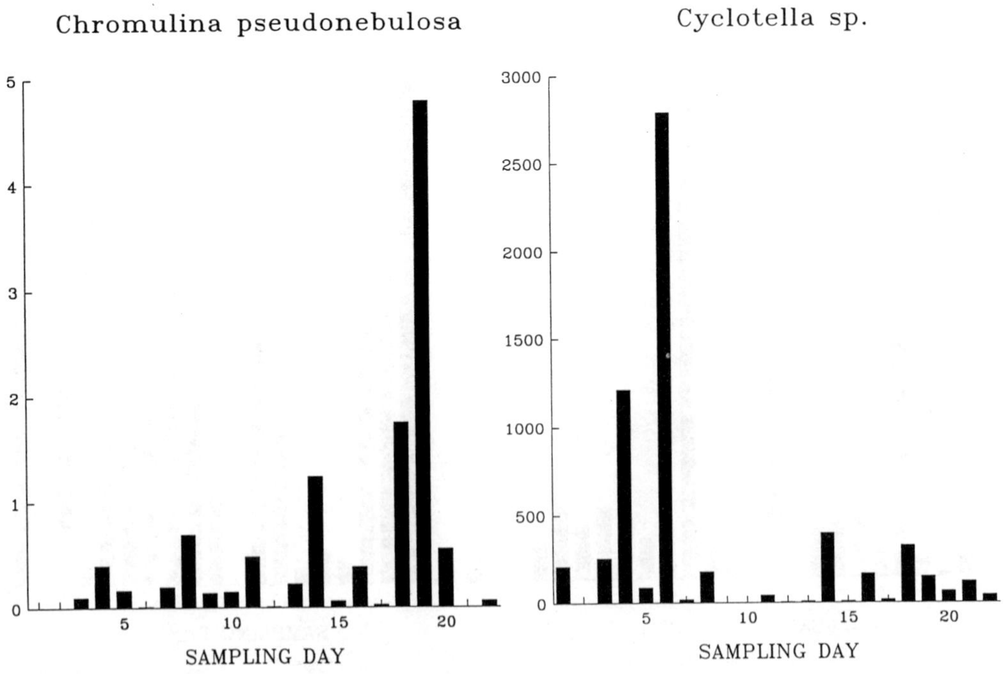

FIGURE. 11. Temporal distribution of Lake Sanabria species loading Axis III: *Chromulina pseudonebulosa* (cells $m^{-2} \times 10^9$), and Axis V: *Cyclotella* sp. (cells $m^{-2} \times 10^6$).

overturn but attains its maximum towards the end of the summer of 1973, which does not correspond at all to the nitrogen pattern.

In Lake Sanabria, Chlorophytes, which as a taxon contributed the greatest percentage of identified species (Planas, unpublished data), were the most important in defining Axes I and II. *Crucigenia tetrapedia* is a summer species more abundant in 1974 than in 1973. It explains the formation of Cluster III. The sampling data of this period were characterized by very low NO_3 and moderate PO_4 concentrations; the biomass as measured by chlorophyll-a was high.

Dictyosphaerium pulchellum was most abundant towards the end of the mixing period as well as at the end of the stratification. *Monoraphidium contortum* was abundant during the whole winter. These species position a few objects characterized by either higher PO_4 or NO_3 concentrations.

The Chrysophyceae and the Cyanobacteria contributed to the formation of Axis III in the S-PCA of SL. *Chromulina pseudonebulosa* had a pulse in the second part of the summer of 1974 when PO_4 and NO_3 concentrations were very low and biomass (measured as chlorophyll-a) was relatively high. *M. marssonii* became abundant at the same period. Its abundance during a period when both nutrients were low confirm its capacity for growth in poor waters (DROUET & DAILY, 1956).

Aphanothece clathrata defines the spring data with relatively low biomass, when NO_3 is becoming scarce and phosphorus less abundant. This species has been found in small productive lakes (REYNOLDS, 1984) but as mentioned before its presence is associated with low nutrient concentrations. This fact, as well as its abundance in ultra-oligotrophic lakes of the Canadian Shield, suggest that this species can develop in waters very low in phosphorus and nitrogen, independently of water temperature, contrary to the statement made by HUTCHINSON (1967).

In Lake Banyoles the subdominant species that characterize the mixing period are species very abundant in winter such as *Rhodomonas minuta*, *R. minuta nannoplanctonica* and the two green species also abundant in LS during this period, *D. pulchellum* and *M. contortium*. The two green species characterize samples clustered in Group III. They correspond to environmental conditions where Si and NO_3 are abundant, and PO_4 relatively low. Maximum density for the Cryptophyceae occurs when NO_3 concentration is at its peak and PO_4 is still low. As mentioned for Lake Constance, the *Rhodomonas* sps. are usually very abundant in winter, in oligotrophic lakes of temperate regions. We do not think that temperature is the variable that controls its growth but that nitrogen could be a more important factor. In the acidifying lakes of North America (YAN, 1978), where 1/3 of the acidity of rain is due to NO_x (LIKENS *et al.*, 1979), the Cryptophyceae generally increase during the summer. This would seem to confirm the importance of nitrogen in the proliferation of this taxon.

The sampling dates characterized by *Chrysococcus rufescens* and *Chlamydomonas chlamydella* correspond to a period of low productivity in LB. This low productivity was associated with low nitrogen concentrations (Planas, in press). The first species was more abundant at the beginning of stratification and the second at overturn.

The species that contributed to Axis III appeared sporadically at the beginning of stratification. *Sphaerotilus* sp. had a very short pulse; *Cyclotella comta* persisted longer, until the middle of stratification. These species characterize the sampling dates corresponding to a period of high transparency. They also become abundant in Lake Constance towards the end of the summer (GRIM, 1939). In this lake their outburst correspond to a decrease in the dominance of *C. melosiroides* and *C.*

glomerata, but this is not the case in LB where the maximum of *C. comta* occurred when the dominant species were very abundant, just prior to their peak.

In summary, the nutrient which seems to control the succession in species composition in the two lakes studied is nitrogen. This element is relatively low in both ecosystems during the summer and becomes more abundant at overturn. Phosphorus (measured as total dissolved reactive phosphorus) seems to have relatively less importance in the species composition evolution. In both lakes the concentration of this nutrient seems related more to recycling in the water column or its inflow by surface runoff, than to fluxes from the sediments or deep hypolimnetic waters. This would explain the relatively constant phosphorus concentrations, independent of the lakes' temperature regime.

Nitrogen, on the contrary, has a very marked cycle and its abundance in the euphotic zone seems more closely associated with water turnover. It is striking that in both lakes its low concentrations during the summer do not favor the nitrogen fixing cyanobacteria. It is probable that the affinity of these cyanobacteria to phosphorus is very low and consequently the concentration present in both lakes did not satisfy their needs. The abundance of phosphorus during the mixing period might possibly explain the appearance or increase of the same green algae species (*M. contortum*, *D. pulchellum*) in both lakes during the winter.

The chemical differences in the two lakes considered, an alkaline and a moderately acidic lake, seem to determine the type of phytoplankton at the dominant species level. In the alkaline lake, the availability of silica and possibly other minor ions determine its diatom dominance throughout the year.

The dominance of Cyanobacteria in Lake Sanabria is more difficult to interpret. In this lake, silica and phosphorus concentrations are lower than in Lake Banyoles. Silica, due to its relatively constant concentration, does not seem to be a determining factor in phytoplankton succession, but the low concentration of phosphorus could be. In Lake Sanabria phosphorus concentration was half that measured in Lake Banyoles. In studies carried out by SUTTLE & HARRISON (1988) on the natural phytoplankton populations of two ultraoligotrophic lakes of the Rocky Mountains with nutrient characteristics close to those of Lake Sanabria, a chroococcal cyanobacteria (*Synechococcus*) showed the highest affinity to phosphorus; this species could develop at very low phosphorus concentrations contrary to the diatoms, which are better competitors at low nitrogen levels.

These findings could explain the dominance of the blue-greens and the scarcity of diatoms in Lake Sanabria as well as the abundance of diatoms in Lake Banyoles but, as was mentioned before, the water alkalinity and the high abundance of silica could also be determining factors.

ACKNOWLEDGEMENTS

Many people participated in the sampling of the lakes discussed in this publication, and I would like to express my sincere gratitude to them all. I want to underline the contribution of Manuel Jiménez Espuelas, director of the Comisaría de Aguas del Duero, to the Sanabria study. His laboratory did practically all the sampling and most of the analyses. The ICONA delegation of Zamora at that time under the direction of Alfredo Prieto, assisted us in various stages of the sampling.

I also wish to thank the late Juan Jesús Torán who allowed us to include part of the Lake Sanabria sampling in the reservoir programme study, carried out between 1973 and 1975. Many thanks to C. de Montigny,

S. Paquet and L. Zhang for their help with the data analyses, and to P. Wickham for improvements to the English.

REFERENCES

AVNIMELECH; Y. 1984. Phosphorus and calcium carbonate solubilities in Lake Kineret. *Limnol. Oceanogr.*, 28: 640-645.

DROUET, F. & DAILY, W.A. 1956. Revision of the coccoid Myxophyceae. *Butler Univ. Bot. Stud.*, 12: 1-218.

FOGG, G.E. FAY, W.D.P.P. & WALSBY, A.E. 1973. *The blue green algae*. Academic Press. London.

GANF, G.G. & VINER, A.B. 1973. Ecological stability in a shallow equatorial lake (Lake George, Uganda). *Proc. R. Soc. Lond.*, Ser. B, 184: 321-346.

GOLDMAN, C.R. 1981. Lake Tahoe: two decades of change. *Verh. Internat. Verein. Limnol.*, 21: 45-70.

GOLDMAN, C.R. & AMSTRONG, R. 1969. Primary productivity studies in Lake Tahoe, California. *Vere. Internat. Verein. Limnol.*, 17: 49-71.

GRIM, J. 1939. Beobachtungen am Phytoplankton des Bodensees (Obersee) Sowie deren rechnerische Auswertung. *Int. Rev. ges. Hydrobiol. Hydrogr.*, 39: 193-315.

HECKY, R.E. & KLING, H.J. 1981. The phytoplankton and protozooplankton of the euphotic zone of Lake Tanganika. Species composition, biomass, chlorophyll content and spatio-temporal distribution. *Limnol. Oceanogr.*, 26: 548-564.

HUTCHINSON, E. 1967. *A treatise on Limnology*. V(II). *Introduction to lake biology and the limnoplankton.* John Wiley. New York.

KALFF, J. 1983. Phosphorus limitation in some tropical African lakes. *Hydrobiologia,* 100: 101-112.

KILHAM, P., KILHAM, S.S. & HECKY, R.E. 1986. Hypothesized resource relationships among African planktonic diatoms. *Limnol. Oceanogr.*, 31: 1169-1181.

LEGENDRE, L. & LEGENDRE, P. 1979. *Écologie numerique. 2. La structure des données écologiques.* Masson & Les Presses de l'Université du Québec. Paris.

LEWIS, W.M. 1978a. Analysis of succession in a tropical phytoplankton community and a new measure of succession rate. *Amer. Natur.*, 112: 401-414.

LEWIS, W.M. 1978b. A compositional, geographical analysis of the phytoplankton in a tropical lake; Lake Lanao, Philippines. *J. Ecol.*, 66: 849-880.

LIKENS, G.E., WRIGHT, R.F., GALLOWAY, J.N. & BUTLER, T.J. 1979. The burning of fossil fuels is increasing the acidity of precipitation over wide areas. *Scientific American*, 241(4); 43- 51.

MARGALEF, R. 1955. Comunidades dulceacuícolas del NW de España. *Pub. Inst. Biol. Apl.*, 21: 5-31.

MARGALEF, R. 1983. *Limnología.* Omega. Barcelona.

MARGALEF, R. PLANAS, D., ARMENGOL, J., VIDAL, A. PRAT, N. GUISET, A., TOJA; J. & ESTRADA, M. 1977. *Limnología de los embalses españoles.* Dirección General de Obras Públicas. Madrid.

MELACK, J.M., KILHAM, P. & FISHER, T.R. 1982. Responses of phytoplankton to experimental fertilization with ammonium and phosphate in an African soda lake. *Oecologia,* 52: 321-326.

PEARSALL, W.H. 1922. A suggestion as to factors influencing the distribution of free-floating vegetation. *J. Ecol.*, 9: 241-253.

PLANAS, M.D. 1973. Composición, ciclo y productividad del fitoplancton del lago de Banyoles. *Oecol. aquat.*, 1: 3-106.

PLANAS, D. 1990. Factores de control de la distribución espacial y temporal de la producción primaria del fitoplancton del lago de Banyoles. *Scientia gerundensis*., 16(2):193-204.

REYNOLDS, C.S. 1984. *The ecology of freshwater phytoplankton.* Cambridge University Press. Cambridge.

RHEE, G-YULL. 1978. Effects of N:P atomic ratios and nitrate limitation on algal growth, cell composition, and nitrate uptake. *Limnol. Oceanogr.*, 23: 478-486.

SCHINDLER, D.W. 1978. Factors regulating phytoplankton production and standing crop in the worlds' freshwaters. *Limnol. Oceanogr.*, 23: 478-486.

SPARLING, J.H. & NALEWAJKO, C. 1970. Chemical composition and phytoplankton of lakes of southern Ontario. *J. Fish. Res. Bd. Can.*, 27: 1405-1428.

SOLÉ SABARÍS, L. 1958. Les aigues subterrànies. In: *Geografia de Catalunya* (L. Solé Sabarís, ed.), 1: 215-219. Aedos. Barcelona.

SUTTLE, C.A. & HARRISON, P.J. 1988. Ammonium and phosphate uptake rates, N:P supply ratios, and evidence for N and P limitation in some oligotrophic lakes. *Limnol. Oceanogr.*, 33: 186-202.

VOLLENWEIDER, R. 1968. Scientific fundamentals of the eutrophication of lakes and flowing waters with particular reference to nitrogen and phosphorus as factors in eutrophication. *OCDE, Directorate for Scientific Affairs*. Paris.

YAN, N.D. & STOKES, P. 1978. Phytoplankton of an acidic lake, and its response to experimental alteration of pH. *Environ. Conserv.*, 5: 93-100.

Homage to Ramon Margalef; or, Why there is such pleasure in studying nature (J. D. Ros & N. Prat, eds.). 1991. *Oecologia aquatica*, 10: 113-125

SEDIMENTARY PHOSPHORUS DYNAMICS IN EPICONTINENTAL SYSTEMS: A LITERATURE REVIEW

PILAR LÓPEZ LASERAS

Departament d'Ecologia. Facultat de Biologia. Universitat de Barcelona. Av. Diagonal, 645. 08028 Barcelona. Spain

Received: June 1990

SUMMARY

Sedimentary phosphorus dynamics being a very complex process involving a large number of different mechanisms, literature data present a wide range of variations both from the quantitative and from the qualitative point of view. Three main aspects are considered in this paper: the range of sedimentary phosphorus observed in lakes worldwide and relationships with other lake characteristics such as trophic status, depth, sedimentary iron and organic matter; phosphorus release rates and factors controlling them, and forms in which phosphorus is bound in sediments and their ecological significance. Because of the complexity of the mechanisms involved, no general conclusions valid for any kind of lakes may be drawn on the basis of the information now available.

KEY WORDS: Sedimentay phosphorus, lake trophy, release rates, fractionation, NAI-P, AI-P, organic-P.

INTRODUCTION

There is now little doubt about the major role played by phosphorus in the eutrophication process of natural waters. From the classical papers of EINSELE (1936, 1937, 1938) and MORTIMER (1941, 1942), many workers have demonstrated the importance of the sediment as source or sink for water phosphorus. A great volume of investigation has focussed on the quality and quantity of sediment phosphorus and on the process affecting its exchange between bottom and water. A good example of this interest is the high proportion of studies involving aspects related to phosphorus presented at several International Congresses on Interactions between Sediment and Water (GOLTERMAN, 1977a; SLY, 1982). In spite of this, however, the variety and complexity of factors influencing sedimentary phosphorus behavior has not yet led to a global interpretation of its dynamics.

Many different approaches have been used in the study of phosphorus in sediments. Among the major ones, three may be considered:

A) Studies that mainly focus on vertical or horizontal variations of phosphorus concentrations in sediments of one or more lakes. Other lake or sediment characteristics are also studied simultaneously in order to establish relationships between different lake descriptors.

B) Studies on phosphorus release rates and the factors controlling them. This approach has received major attention in recent years because release from sediments seems to be one of the major processes

TABLE I. Total phosphorus concentrations in the sediment and trophic status of worldwide lakes. o: oligotrophic. om: oligomesotrophic; m: mesotrophic; me: meso-eutrophic; e: eutrophic; h: hypereutrophic. vv: vertical variation; hv: horizontal variation; lv: interlakes variation; tv: temporal variation. Values in mg P kg^{-1} dw. *, Poland, Italy, Finland, Canada, USA, Japan. **, Poland, Italy, Japan. ***, Poland, Italy, Finland, USA, Malawi, India, Japan.

Lake	Trophic status	Study	TP	Reference
Char, Arctic	o	hv	570- 620	26
E.L.A. (16), Canada	o	lv	1300-3300	11
Ennerdale, England	o	vv	600-2300	25
Weber, U.S.A.	o	vv	2710-3950	8
Mowich, U.S.A.	o	vv	750-2170	21
Florida (8), U.S.A.	o	lv	100-1100	13
Worldwide (9) *	o	lv	910-4060	16
Maggiore, Italy	om	-	1470	16
Wisconsin (2), U.S.A.	om	vv,lv	894-9940	8
Poland (2)	m	lv	1100-1400	38
Wuras Dam, South Africa	m	hv	5000-11000	19
Lucerne, Switzerland	m	vv	400-1700	5
Winnipeg, U.S.A.	m	vv,hv	380-740	1
Worldwide (3) **	m	lv	1010-1600	16
Mulwala reservoir, Australia	me	hv	350-1188	17
Little John, U.S.A.	me	vv	3880-12100	8
Australia (3)	e	lv,hv	378-1760	33
Austria (2)	e	lv	258-705	27
Wabamun, Canada	e	hv	1700-2300	30
Denmark (6)	e	lv	800-8000	3
Pluβsee, Germany	e	tv	1460-2260	15
Kinnereth, Israel	e	vp	1000-4500	32
Lugano, Italy	e	hv	1000-4500	29
Nakanoumi, Japan	e	vv	500-1200	39
Poland (3)	e	lv	1500-2700	38
Coastal lagoons, Spain	e	lv	300-1000	23
Spanish reservoirs (2)	e	lv	350-1190	22
Lilla Ullevifjarden, Sweden	e	hv	1000-2000	31
Rotsee, Switzerland	e	hv	1000-1200	5
Long Lake reservoir, U.S.A.	e	vv	1000-4700	35
Mendota, U.S.A.	e	vv	850-1630	7
Monona, U.S.A.	e	vv	950-2000	9
Potomac Estuary, U.S.A.	e	vv,hv	310-1860	12
Worldwide (10) ***	e	lv	900-3050	16
Esrom, Denmark	h	hv	2440-6380	18
Sobygaard, Denmark	h	vv	2000-12000	34
Poland (3)	h	lv	2200-3700	38
La Massona, Spain	h	vv	496-1022	24

involved in eutrophication, and may thus be taken into account in protection and restoration programs.

C) Studies on chemical fractions in which phosphorus is bound in sediments. The contradictory results observed, when comparisons between sediment or lake characteristics are made, and when factors controlling release rates are evaluated, have led to the conclusion that the chemical speciation of phosphorus is almost certainly among the more important factors regulating phosphorus dynamics.

These three approaches will be considered in the next sections.

TOTAL PHOSPHORUS IN SEDIMENT RELATED TO OTHER LAKE AND SEDIMENT CHARACTERISTICS

The concentrations of total phosphorus observed in the sediments of lakes around the world are highly variable, as can be observed in Table I, where values reported by different authors have been summarised. The range of variation has similar magnitude when concentrations are compared at various levels of a vertical profile, at various sites of one lake or at different lakes (see Table I for references). There is little information about temporal changes of sedimentary phosphorus, but the results observed at Lake Plußsee (GOLACHOWSKA, 1979) and in Spanish reservoirs (MORGUÍ & LÓPEZ, unpublished data) also show wide variation.

Numerous attempts have been made to relate the variation of sedimentary phosphorus to other lake or sediment characteristics, but until present no general relationships valid for any lake have been found.

Because of the relation between water phosphorus and eutrophication, a similar relation between sedimentary phosphorus and trophic status of lake could be suspected. Although some authors have reported higher concentrations of total phosphorus in eutrophic sediments than in oligotrophic sediments (FLANNERY *et al.*, 1982; WISNIEWSKY & PLANTER, 1985), a general relationship between these two variables cannot be established when lakes from different regions are considered. For example, GOLACHOWSKA (1984) found similar values of total phosphorus in the sediments of oligotrophic Lake Char (an Arctic lake) and in eutrophic Lake Chilwa (a tropical African lake). From the literature data compiled in Table I, no relation can be deduced.

In a few cases, sedimentary phosphorus has been related to water depth. Results are not conclusive, since both significant (RYDING & BORG, 1976) and non-significant (KAMP-NIELSEN & HARGRAVE, 1978; KEULDER, 1982) correlations have been found. Data from the literature has been plotted in figure 1, in order to evaluate this relationship in a wide variety of lakes. As can be observed, sedimentary phosphorus is not dependent on water depth, when different lakes are

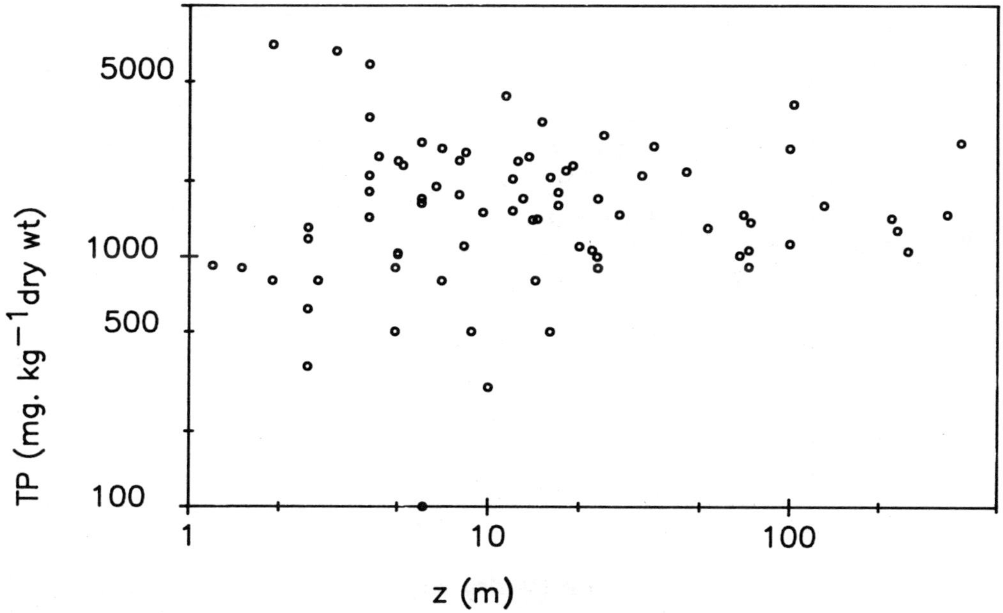

FIGURE 1. Relationship between total phosphorus concentration in sediment and water depth (z). (Data from BORTLESON & LEE, 1974, 1975; BRUNSKILL *et al.*, 1971; FLANNERY *et al.*, 1982; GOLACHOWSKA, 1984; KEMP *et al.*, 1978; KLAPWIJK *et al.*, 1982; SLATER & BOAG, 1978).

taken into account.

The concentrations of iron in sediment is among the variables most frequently studied in relation to sedimentary phosphorus, because of the chemical association between these two compounds revealed by EINSELE (1936, 1937, 1938) and MORTIMER (1941, 1942). Water phosphorus can precipitate associated to ferric hydroxides and then it can be released into solution under favorable conditions. Nevertheless, the results observed are very contradictory. High correlation between phosphorus and iron has been noticed for vertical variation in eutrophic lakes (BORTLESON & LEE, 1972, 1975; LÓPEZ, 1986b), for horizontal variations in oligo- and mesotrophic lakes (ALLAN & BRUNSKILL, 1977; KEMP *et al.*, 1978) and for interlake variations (BRUNSKILL *et al.*, 1971; SHUKLA *et al.*, 1971). In contrast, no significant correlation has been reported by other authors in similar conditions: for vertical profiles of both oligo- and eutrophic lakes (MACKERETH, 1966; BORTLESSON & LEE, 1974; THOMAS & SOLTERO; 1977; CROSS & RIGLER; 1983), horizontal variations (PREMAZZI & RAVERA, 1977) or interlake variations (FLANNERY *et al.*, 1982; LÓPEZ, 1986a). One cause of these contradictory results could derive from the fact that usually only small groups in lakes with similar characteristics are considered. OSTROFSKY (1987) studies this relationship in 66 very different lakes of North America, reporting a significant correlation by using a logarithmic transformation of phosphorus values ($r = 0.58$, $p < 0.01$). Literature data for 78 lakes (Ostrofsky values are not included) have been plotted in figure 2, where shallow (< 10 m) and deep (> 10 m) lakes are clearly distinguishable. A weak but significant correlation has been found for shallow lakes ($r = 0.36$, $p < 0.05$), whereas for deeper lakes correlation becomes non-significant ($r = 0.21$). This is in agreement with the observation made by SLY (1977) that in shallow sediments the

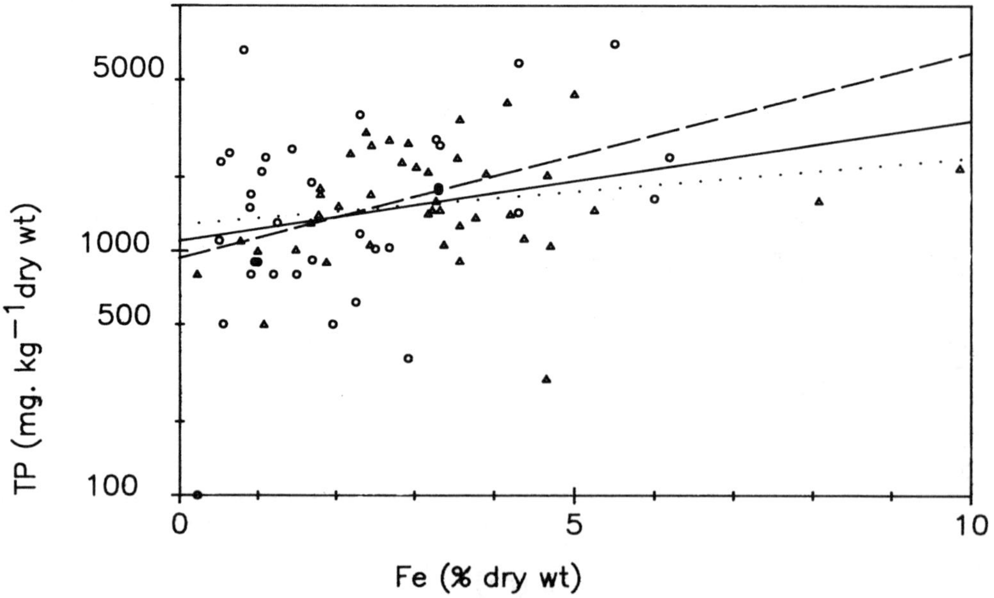

FIGURE 2. Relationship between total phosphorus and total iron concentrations in sediments. Data from the same sources as in figure 1. Circles and dashed line correspond to lakes with depth lower than 10 m; triangles and dotted line, to deeper lakes.

quotient Fe/P remains constant, but in deeper sediments phosphorus should appear closely related to organic carbon. However, it must be noted that correlation between sedimentary phosphorus and organic matter has been found both in shallow and deep lakes (BRUNSKILL *et al.*, 1971; HART *et al.*, 1976; PREMAZZI & RAVERA, 1977; LARSON *et al.*, 1976; REYNOLDSON & HAMILTON, 1982; FLANNERY *et al.*, 1982), non-significant values being reported less frequently (KAMP-NIELSEN & HARGRAVE, 1978; KEMP *et al.*, 1978).

Calcium is another element that can play an important role in phosphorus dynamics. Precipitation of phosphorus adsorbed on calcium carbonate surfaces has been demonstrated (KUO & LOTSE, 1974) and GOLTERMAN (1982a) has also suggested a direct precipitation as apatite. However, the relationship between calcium and sedimentary phosphorus has not been studied in terms of total concentrations but in terms of phosphorus species and will be discussed later.

PHOSPHORUS RELEASE

Phosphorus release (often called internal phosphorus loading) is defined by the net flux of phosphorus from sediment to lake waters, which occurs when solubilization and upwards transport exceed phosphorus sedimentation. Since the phosphorus concentration in sediment is much higher than in lake water, phosphorus release can enhance the eutrophication of lakes. So, internal phosphorus loading is an important factor that must be taken into account in the prediction models used to provide the basis of lake restoration programs.

Two strategies have been used to estimate phosphorus release. One is the experimental approach where fluxes are studied in the field or in the laboratory. With the other strategy, phosphorus release is determined by mass balance calculations in whole lakes.

The magnitude of phosphorus release has been related to lake trophic status. NURNBERG *et al.* (1986) have established

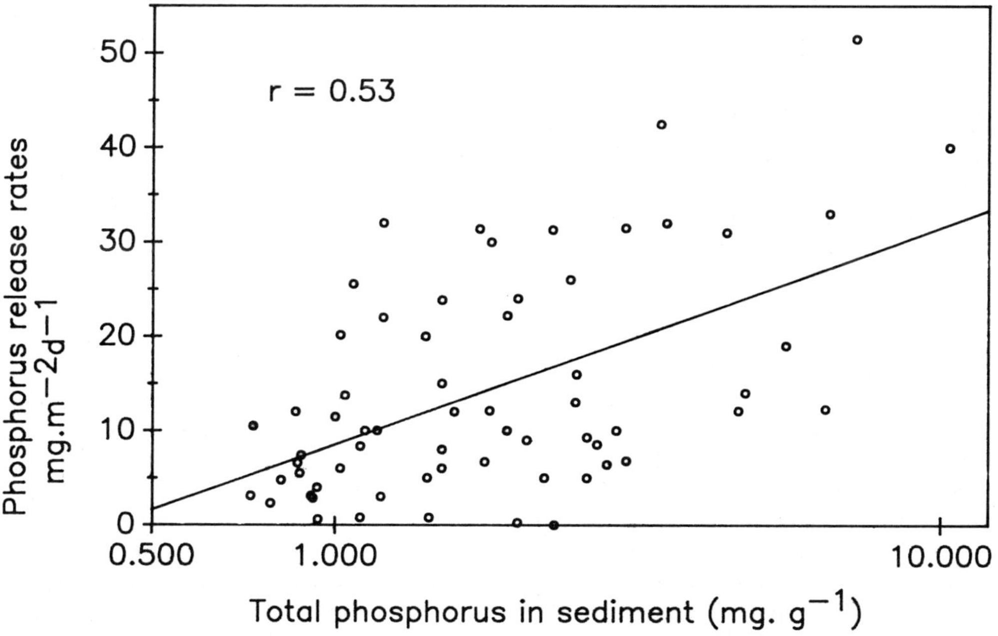

FIGURE 3. Relationship between total phosphorus concentration in sediment and phosphorus release rates. Data from NURNBERG (1988).

a dependence of phosphorus anoxic release rates on water phosphorus concentrations in 56 lakes from the literature (r = 0.50). Oligrotrophic lakes show anoxic release rates below 1 mg m^{-2} d^{-1}, whereas in eutrophic lakes values up to 50 mg m^{-2} d^{-1} can be observed (NURNBERG, 1988).

Several authors have also looked for a correlation between total phosphorus concentration in sediment and phosphorus release rates. BOSTROM (1984) did not found such a correlation, but NURNBERG (1988) reported a regression line PR= -5.59 + 0.0079 TP (r^2 = 0.53, n=14) for several North American lakes. In figure 3, literature data in lakes worldwide compiled by NURNBERG (1988) have been plotted in a semilogarithmic graph. A weak correlation can be observed using a logarithmic transformation of TP values, but the range of variation is very wide and the relation observed cannot be used for prediction purposes.

In the last decade, several articles have reviewed the phosphorus release concept (BOSTROM *et al.*, 1982; RYDING, 1985; NURNBERG, 1988; JANSSON *et al.*, 1988). Different mechanisms have been suggested as main factors influencing the rates of phosphorus release: redox conditions, pH, water temperature, mineralization, nitrate concentration, bioturbation, effects of macrophytes and periphytic biolayer, diffusion and wind turbulence and different sediment characteristics.

CHEMICAL FACTORS AFFECTING PHOSPHORUS RELEASE

The best documented mechanism of phosphorus release involves the chemical reduction of Fe(III) to Fe(II). The works of EINSELE (1936, 1937, 1938) and MORTIMER (1941, 1942) pioneered this interpretation. Since iron reduction occurs under anaerobic conditions, it was suggested that phosphorus release would be only significant in lakes with stagnant anaerobic bottom waters. On the basis of this, aerobic sediments had been regarded as exclusively phosphorus sinks (HAKANSON & JANSSON, 1983). In laboratory and/or field experiments, many authors have observed high rates of phosphorus release under anaerobic conditions, supporting this classical approach (for example, LI *et al.*, 1972; SERRUYA *et al.*, 1974; VINER, 1975; RILEY & PREPAS, 1984; NURNBERG, 1985).

However, the role of iron controlling the overall phosphorus dynamics in lakes was demonstrated to be much less general than previously thought (GOLTERMAN, 1977a; LEE *et al.*, 1977).

On the other hand, phosphorus release under aerobic conditions has been reported for both eutrophic and oligotrophic waters (STEVENS & GIBSON, 1977; LEE *et al.*, 1977; RYDING & FORSBERG, 1977; TWINCH & PETERS, 1984), whereas some lakes with hypolimnetic anoxia for much of the year showed very small internal phosphorus loading (LEVINE, 1986; SCHINDLER *et al.*, 1987).

Water pH can also affect phosphorus release rates. Several laboratory studies reveal that a drastic increase of phosphorus flux from sediments takes place when pH increases (ANDERSEN, 1975; RIPPEY, 1977; LIJKLEMA, 1977), this dependence being in part a function of the initial P:Fe ratio in the sediment (LIJKLEMA, 1977). So, it was believed that pH changes were the major factor controlling phosphorus release in shallow productive lakes (BOSTROM *et al.*, 1982).

Some field studies reporting closely related variations of pH and phosphorus (ANDERSEN, 1974; BOSTROM *et al.*, 1982) were interpreted as the result of a causal relationship between the two variables. However, high phosphorus levels may be a sign of increased primary production, which in turn increases pH. In fact, when other conditions are taken into account it is not clear that pH is the key factor to explain phosphorus release in these systems, and probably the role of the pH as regulator of phosphorus release under natural conditions has been overemphasized (RYDING, 1985).

The other major chemical factor influencing phosphorus release rates is the concentration of nitrate, for which two counteracting effects have been described. On the one hand, nitrate can decrease the oxygen consumption and, thus, prevent phosphorus release from anoxic sediments by maintaining a high redox potential at the sediment surface (RIPL, 1976; ANDERSEN, 1982). On the other hand, addition of nitrate to oxygen depleted sediments leads to a stimulation of the mineralization of organic matter during the denitrifying process (ANDERSEN, 1977), and thus organic-bound phosphorus may return to water (AHL, 1979; SORENSEN, 1982). TIREN & PETTERSSON (1985) concluded from different experiments that whether or not high nitrate loading will cause an increase of phosphorus release depends on several factors such as continuity of the nitrate supply, nitrate uptake rate and sediment properties. Moreover, the results observed by RYDING (1985), who found the relationship between nitrate and phosphorus release most significant over long intervals of time (3 or 6 years), indicate the need for long term studies in order to provide a reliable basis to understand nitrate effects.

BIOLOGICAL FACTORS AFFECTING PHOSPHORUS RELEASE

Activity of organisms is the other major controlling factor in phosphorus release. The widely observed relationship between water temperature and internal phosphorus loading (KELDERMAN, 1980; HOLDREN & ARMSTRONG, 1980; KELDERMAN & VAN DE REPPE, 1982; RILEY & PREPAS, 1984; BOERS, 1986) suggests that microbial activity can enhance phosphorus release rates. Such activity may result in a direct mobilization of phosphorus through degradation (i.e. mineralization) of organic matter, or it may lead to a lowering of Eh values due to O_2 and nitrate consumption. Results of some studies in which bacterial activity has been suppressed (by adding antibiotics or by γ-irradiation) in shallow eutrophic lakes are consistent with the theory that bacterial activity plays the key role in phosphorus release in aerobic sediments (VINER, 1975; BATES & NEAFUS, 1980; KAMP-NIELSEN, 1974; FILLOS, 1977; SINKA & CAPPENBERG, 1988). Moreover, it has been suggested that part of the anaerobically released phosphorus has its origin in bacterial cells (FLEISCHER, 1983).

Other organisms apart from bacteria can also affect the phosphorus exchange between sediments and water. The periphytic biolayer seems to reduce phosphorus release because of the biological uptake of phosphorus from sediments, although laboratory results may not be quantitatively transferred to natural conditions (HANSSON, 1989). Macrophytes also influence phosphorus release, since phosphorus is taken up by roots and transferred to plant tissues from which it can be released again. BARKO & SMART (1980) concluded that this process may be quite significant for phosphorus recycling.

The role of bioturbation (defined as the stirring of sediments by the activity of benthic burrowing organisms) on phosphorus release is not clear. DAVIES *et al.* (1975) reported a reduction of phosphorus concentrations in the uppermost interstitial water as a consequence of higher redox potential, caused by tubificid worms. On the other hand, GRANELI (1979) and WISNIEVSKY *et al.* (1985) showed an increased transport of phosphorus from sediments caused by benthic macro-invertebrates. In fact, bioturbation is a very complex process that probably depends on, among other factors, the chemical and physical characteristics of sediment in which it takes place (PETR, 1977).

TABLE II. Phosphorus species from the sediments of worldwide lakes. Species are described in text. Values in brackets correspond to percentages of total phosphorus. Data are in mg $P.kg^{-1}$ dw. # NAI-P fraction obtained by sum of NaOH and BD extracts. * NAI-P values obtained directly by NaOH extraction.

LAKE	ADS-P	NAI-P	AI-P	TP	References
Prairie Lakes (8) Canada	n.a.	120-1196 (22-62)#	205-307 (15-46)	542-1992	2
Balaton, Hungary	64 (10.5)	39 (6.4)*	284 (46.5)	610	36
Loosdrecht, Netherlands	30-200 (3- 6)	120-1550 (20-43)*	90-370 (6-19)	600-3620	6
Netherlands (4)	72-432 (9-39)	216-2232 (27-62)*	72-540 (8-39)	800-3600	20
La Minilla reservoir, Spain	n.a.	62-789 n.a.#	39-152 n.a.	n.a.	14
Sau reservoir, Spain	n.a.	310-787 (46-61)#	251-439 (19-59)	703-1388	4
Sweden (5)	6-42 (1-2)	44-375 (5-30)*	241-452 (19-37)	920-1670	36
Sweden (9)	6-250 (1-9)	44-4010 (5-68)*	240-710 (12-37)	917-5860	10
Eire, USA	n.a.	38-2329 (5-86)#	144-859 (13-86)	188-2863	37
USA (66)	0-872 (0-11)	140-5879 (9-68)*	43-737 (3-38)	1329-9212	28

PHOSPHORUS SPECIATION

Another approach to the study of sedimentary phosphorus dynamics is to analyze its fractional distribution, determining which fractions are potentially mobile within the natural range of variation for the main environmental factors.

Phosphorus is bound in the sediments to various chemical species which have different properties and behavior. The concentration of these forms depends on the whole characteristics of the lake. Moreover, since fractions differ in their behavior, the global dynamics of sedimentary phosphorus is greatly influenced by the fractional composition. For these reasons, quantitative and qualitative studies on phosphorus speciation should provide us with better knowledge and comprehension of the role of sedimentary phosphorus in water quality.

Different methods have been proposed for the analysis of phosphorus fractions. GOLTERMAN (1977b, 1982b, 1984) suggested the use of NTA extractions in order to evaluate the fraction available for algal growth. However further studies provided contradictory results for relationship between NTA-extractable and bioavailable phosphorus (KLAPWIJK *et al.*, 1982; GUNATILAVA *et al.*, 1988).

Other fractionation methods derive from the analysis of phosphorus forms in soils (CHANG & JACKSON, 1957) and go through sequential extractions with different solvents. Several procedures and solvents have been proposed, many of them by modification of those elaborated by WILLIAMS *et al.* (1971a, 1971b, 1976). Detailed discussion of the methods currently used are given by VAN ECK (1982) and PSENNER *et al.* (1988).

At present, the definition of the fractions is purely operational, four main forms being the most usually determined (PSENNER *et al.*, 1988):

A) Loosely-bound or adsorbed phosphorus, that is usually extracted with NH_4Cl (HIELTJES & LIJKLEMA, 1980) or distilled water (PSENNER, 1984). It gives an estimate of the immediately available phosphorus.

B) Non-apatite inorganic phosphorus (NAI-P), extracted with NaOH (HIELTJES & LIJKLEMA, 1980) or by successive addition of NaOH and buffered dithionite (BD) (PSENNER, 1984; PSENNER *et al.*, 1988). It corresponds to phosphorus associated with metal oxides and hydroxides. BD extracts reductant soluble phosphorus mainly from hydroxide surfaces (PSENNER & PUCSKO, 1988), although it has been suggested that in calcareous sediments BD could also extract some calcium-bound phosphorus (GABELLONE & GUISANDE, 1989). NaOH-P represents phosphorus which is exchangeable against OH^- and phosphorus compounds soluble in bases.

C) Inorganic apatite phosphorus (IA-P),

extracted with HCl. It represents phosphorus associated with calcium as carbonates or apatite (or similar compounds). BONZONGO *et al.* (1989) suggest that this fraction is usually overestimated by bacterial lysis.

D) Residual phosphorus. Obtained by the difference between total sedimentary phosphorus and fractions A,B and C. It includes organic and refractory phosphorus compounds.

Data from the literature about chemical fractionation of phosphorus in different lakes has been summarised in Table II. Some relations between fractions and other variables will be discussed next.

LOOSELY BOUND PHOSPHORUS

Adsorbed phosphorus is usually the smallest fraction, although in some cases it can rise to 39% of total phosphorus (Table II). It seems to be the most easily mobilized fraction. SONDEGAARD (1989) reports a reduction of 50-80% of this form in release experiments under aerobic conditions. However, decrease is restricted to the upper 2-4 cm of sediments, which agrees with the decrease observed in release rates with time (SINKE & CAPPENBERG, 1988) because diffusion from deeper sediments may be limited by distance. Because of its low percentage, adsorbed phosphorus has only small effects on global release rates. BOSTROM (1984) has observed that lakes in which loosely bound phosphorus is the main fraction mobilized have low release rates, which are not affected by redox or pH changes.

NON-APATITE INORGANIC PHOSPHORUS (NAI-P)

NAI-P is among the most variable fractions, wether the total concentration or percentage of total phosphorus is considered (Table II). It has usually been related to iron concentration (WILLIAMS *et al.*, 1976; OSTROFSKY, 1987) and it is associated with fine grained sediment (WILLIAMS *et al.*, 1976; GABELLONE & GUISANDE, 1989).

NAI-P seems to be the most important fraction for release and uptake mechanisms (PSENNER, 1984). Actually, phosphorus release has been related to NaOH-P in many studies (HOSOMI *et al.*, 1981; BOSTROM, 1984; TIREN & PETTERSSON, 1985; SONDEGAARD, 1989; OSTROFSKY *et al.*, 1989). The dithionite fraction shows a strong correlation with release rates in anoxic conditions (NURNBERG, 1988). Moreover, NAI-P can be related to water input of reactive soluble phosphorus (LOGAN, 1982; ARMENGOL *et al.*, 1986).

NAI-P has been assumed as the fraction that may be used for algal growth. Good agreement between NaOH-P and bioavailable phosphorus has been found by several authors (WILLIAMS *et al.*, 1980; YOUNG & DE PINTO, 1982), whereas in other cases results are not conclusive (KLAPWIJK *et al.*, 1982) or show no correlation (BOERS *et al.*, 1984).

APATITE PHOSPHORUS (AI-P)

Calcium-bound phosphorus has traditionally been considered as allogeneous or of detrital origin (WILLIAMS *et al.*, 1976; JONES & BOWSER, 1978) and it can be clearly related to water inflow and erosion rates (ARMENGOL & VIDAL, 1988). The narrow range of variation (especially when total concentrations rather than percentages of total phosphorus are considered) observed in Table II is in agreement with this. Because apatite does not dissolve at environmental conditions, AI-P is not related to release rates (HOSOMI *et al.*, 1981; BOSTROM, 1984) and it has been considered as non-available phosphorus (WILLIAMS *et al.*, 1976). However, it has been demonstrated that both algae and bacteria can grow with normally occurring apatite as the only phosphorus source (SMITH *et al.*, 1978). On the other hand,

GOLTERMAN (1982a) shows that precipitation of apatite is possible in hard waters and suggests that calcium-phosphorus rather than iron-phosphorus is the controlling exchange factor in hard waters. Unfortunately, there is very little information about calcareous sediments and no general conclusions may be drawn about the role of this fraction.

ORGANIC AND REFRACTORY PHOSPHORUS

Organic phosphorus may also be a significant fraction of sedimentary phosphorus, and is usually related to the organic carbon content (WILLIAMS *et al.*, 1976; ARMENGOL *et al.*, 1984; OSTROFSKY, 1987). Its role in phosphorus release is not clearly understood: several studies demonstrate a decrease of this fraction during release experiments (BOSTROM, 1984; PETTERSSON, 1984; TIREN & PETTERSSON, 1985; SONDEGAARD, 1989). The organic phosphorus mobilized may be partially transferred to bacterial biomass and then released to water or transferred to NAI-P (PETTERSSON, 1984) or it may be directly transferred to the HCl-P fraction (TIREN & PETTERSSON, 1985). On the other hand other authors (KAMP-NIELSEN, 1974; OSTROFSKY *et al.*, 1989) have not found any relation between the organic fraction and phosphorus release.

As a whole, sedimentary phosphorus dynamics is a very complicated process involving a large number of mechanisms, so partial descriptions or models which are valid for one or more lakes may not be fully applicable to other lakes in which one or several factors differ. General models in which all possible mechanisms are included are extremely difficult, if not impossible, to construct with data now available.

REFERENCES

AHL, T. 1979. Natural and human effects of trophic evolution. *Arch. Hydrobiol. Beih.*, 43: 259-279.

1. ALLAN, R.J. & BRUNSKILL, G.J. 1977. Relative atomic variation (RAV) of elements in lake sediments: Lake Winnipeg and other Canadian Lakes. In: *Interactions between sediments and freshwater* (H.L. Golterman, ed.): 108-120. Junk. The Hague.

2. ALLAN, R. J., WILLIAMS, J. D. H., JOSHI, S. R. & WARWICK, W. F. 1980. Historical changes and relationship to internal loading of sediment phosphorus forms in hypereutrophic prairie lakes. *J. Environ. Qual.*, 9: 199-206

3. ANDERSEN, J.M. 1974. Nitrogen and phosphorus budgets and the role of sediments in six shallow Danish lakes. *Arch. Hydrobiol.*, 74: 528-550.

ANDERSEN, J.M. 1975. Influence of pH on release of phosphorus from lake sediments. *Arch. Hydrobiol.*, 76: 17-32.

ANDERSEN, J.M. 1977. Importance of the denitrification process for the rate of degradation of organic matter in lake sediments. In: *Interactions between sediments and fresh water* (H.L. Golterman, ed.): 357-362. Junk. The Hague.

ANDERSEN, J.M. 1982. Effect of nitrate concentration in lake water on phosphate release from the sediment. *Water Res.*, 16: 1119-1126.

4. ARMENGOL, J., CRESPO, M. & MORGUÍ, J.A. 1984. Phosphorus compounds in the sediment of the Sau reservoir (Barcelona, NE Spain) throughout its twenty-year existence. *Verh. Internat. Verein. Limnol.*, 22: 1536-1540.

ARMENGOL, J., CRESPO, M., MORGUÍ, J.A. & VIDAL, A. 1986. Phosphorus budgets and forms of phosphorus in the Sau reservoir sediment: an interpretation of the limnological record. *Hydrobiologia*, 143: 331-336.

ARMENGOL, J. & VIDAL, A. 1988. The use of different phosphorus fractions for the estimation of the trophic evolution of the Sau reservoir. *Arch. Hydrobiol. Beih. Ergebn. Limnol.*, 30: 61-70.

BARKO, J.W. & SMART, M. 1980. Mobilization of sediment phosphorus by submersed fresh water macrophytes. *Freshw. Biol.*, 10: 229-238.

BATES, M.H. & NEAFUS, J.E. 1980. Phosphorus release from sediments from Lake Carl Blackwell, Oklahoma. *Water Res,* 14: 1477-1481.

5. BLOESCH, J. 1977. Sedimentation rates and sediment cores in two Swiss lakes of different trophic state. In: *Interactions between sediments and fresh water* (H.L. Golterman, ed.): 65-71. Junk. The Hague.

BOERS, P.C.M. 1986. Studying the phosphorus release from the Loodrescht lakes sediments using a continuous flow system. *Hydrobiol. Bull.,* 20: 17-29.

6. BOERS, P.C.M., BONGERS, J.W.T., WISSELO, A.G. & CAPPENBERG, T. E. 1984. Loodrescht lakes restoration project: sediment phosphorus distribution and release from the sediments. *Verh.*

Internat. Verein Limnol., 22: 842-847.

BONZONGO, J.C., BERTRU, G. & MARTIN, G. 1989. Les méthodes de speciation du phosphore dans les sediments: Critiques et propositions pour l'évaluation des fractions minérales et organiques. *Arch. Hydrobiol.*, 116: 61-69.

7. BORTLESON, G.C. & LEE, G.F. 1972. Recent sedimentary history of Lake Mendota. *Environmental*, 6: 799-808.

8. BORTLESON, G.C. & LEE, G.F. 1974. Phosphorus, iron and manganese distribution in sediment cores of six Wisconsin Lakes. *Limnol. Oceanog.*, 19: 794-801.

9. BORTLESON, G.C. & LEE, G.F. 1975. Recent sedimentary history of Lake Monona, Wisconsin. *Water Air Soil Pollut.*, 4: 89-98.

10. BOSTROM, B. 1984. Potential mobility of phosphorus in different types of lake sediments. *Int. Revue ges. Hydrobiol.*, 69: 457-474.

BOSTROM, B., JANSSON, M. & FORSBERG, C. 1982. Phosphorus release from lake sediments. *Arch. Hydrobiol. Beih. Ergebn. Limnol.*, 18: 5-59.

11. BRUNSKILL, G.J., POVOLEDO, D., GRAHAM, B.W. & STAINTON, M.P. 1971. Chemistry of surface sediments of sixteen lakes in E.L.A., Northwestern Ontario. *J. Fish. Res. Bd. Can.*, 28: 277-294.

12. CALLENDER, E. 1982. Benthic phosphorus regeneration in the Potomac River estuary. *Hydrobiologia*, 92: 431-446.

CHANG, S.C. & JACKSON, M.L. 1957. Fractionation of soil phosphorus. *Soil Sci.*, 84: 133-144.

CLEUGH, T.R. & HAUSER, B.W. 1971. Results of the initial survey of the Experimental Lakes Area, Northwestern Ontario. *J. Fish. Res. Bd. Can.*, 28: 129-137.

CROSS, P.M. & RIGLER, F.H. 1983. Phosphorus and iron retention in sediments measured by mass budget calculations and directly. *Can. J. Fish. Aquat. Sci.*, 40: 1589-1597.

DAVIES, R.B., THRURLOW, D.L. & BREWSTER, F.E. 1975. Effects of burrowing tubificid worms on the exchange of phosphorus between lake sediment and overlying water. *Verh. Internat. Verein. Limnol.*, 19: 382-394.

ECK, G.T.M. VAN. 1982. Forms of phosphorus in particulate matter from the Holland Diep/Haringuliet, The Netherlands. *Hydrobiologia*, 92: 665-681.

EINSELE, W. 1937. Physikalisch-chemische Betrachtung einiger Probleme des limnischen Mangan und Eisenkreislaufs. *Verh. Internat. Verein Limnol.*, 5: 69-84.

EINSELE, W. 1936. Über die Beziehungen des Eisenkreislaufs zum Phosphatkreislauf im eutrophen See. *Arch. Hydrobiol.*, 29: 664-686.

EINSELE, W. 1938. Über chemische und kolloidchemische Vorgange in Eisen-Phosphat-systemen unter limnochemischen und limnogeologischen Gesichtspunkten. *Arch. Hydrobiol.*, 33: 361-387.

FILLOS, J. 1977. Effects of sediments on the quality of the overlying water. In: *Interactions between sediments and fresh water* (H.L. Golterman, ed.): 266-271. Junk. The Hague.

13. FLANNERY, M.S., SNODGRASS, R.D. & WHITMORE, T.J. 1982. Deepwater sediments and trophic conditions in Florida lakes. *Hydrobiologia*, 92: 597-602.

FLEISCHER, S. 1983. Microbial phosphorus release during enhanced glycolysis. *Naturwiss.*, 70: 415.

14. GABELLONE, N.A. & GUISANDE, C. 1989. Relationship between texture and fractions of inorganic phosphorus in the surface sediments of a reservoir. *Aquat. Sci.*, 51: 306-316.

15. GOLACHOWSKA, J. 1979. Phosphorus forms and their seasonal changes in water and sediments of lake Plusssee. *Arch. Hydrobiol.*, 86: 217-241.

16. GOLACHOWSKA, J. 1984. Phosphorus in the bottom sediments of some lakes of the world. *Pol. Arch. Hydrobiol.*, 31: 175-205.

GOLTERMAN, H.L. 1977a. *Interactions between sediments and fresh water*. Junk. The Hague.

GOLTERMAN, H.L. 1977b. Sediment as a source of phosphate for algal growth. In: *Interactions between sediments and fresh water* (H.L. Golterman, ed.): 286-293. Junk. The Hague.

GOLTERMAN, H.L. 1982a. Loading concentration models for phosphate in shallow lakes. *Hydrobiologia*, 91: 169-174.

GOLTERMAN, H.L. 1982b. Differential extraction of sediment phosphate with NTA solutions. *Hydrobiologia*, 92: 638-687.

GOLTERMAN, H.L. 1984. Sediments, modifying and equilibrating factors in the chemistry of freshwater. *Verh. Internat. Verein Limnol.*, 22: 23-59.

GRANELI, W. 1979. The influence of *Chironomus plumosus* larvae on the oxygen uptake of sediment. *Arch. Hydrobiol.*, 87: 385-403.

GUNATILAVA, A., HERODEK, S., ISTANOVICS, V. & DOBOLYI, E. 1988. Biological availability of sediment phosphorus. *Arch. Hydrobiol. Beih. Ergebn. Limnol.*, 30: 93-98.

HAKANSON, L. & JANSSON, M. 1983. *Principles of lake sedimentology*. Springer. Berlin.

HANSSON, L.A. 1989. The influence of a periphytic biolayer on phosphorus exchange between sustrate and water. *Arch. Hydrobiol.*, 115: 21-26.

17. HART, B.T., MC, GREGOR, R.J. & PERRIMAN, W.S. 1976. Nutrient status of the sediments in lake Mulwala. I: Total phosphorus. *Aust. J. Mar. Freshwater Res.*, 27: 129-135.

HIELTJES, A.H. & LIJKLEMA, L. 1980. Fractionation of inorganic phosphates in calcareous sediments. *J. Environ. Qual.*, 9: 405-407.

HOLDREN, G.C. & ARMSTRONG, D.E. 1980. Factors affecting phosphorus release from intact lake sediment cores. *Environ. sci. Technol.*, 14: 79-86.

HOSOMI, M., OKADA, M. & SUDO, R. 1981. Release of phosphorus from sediments. *Verh. Internat. Verein Limnol.*, 21: 628-633.

JANSSON, M., ENELL, M., FLEISCHER, S., GACHTER, R., KELDERMAN, P., LOFGREN, S., NURNBERG, G., PROVINI, A. & SINKE, A. 1988. Phosphorus release from lake sediments. *Arch. Hydrobiol. Beih. Ergebn. Limnol.*, 30: 90-93.

JONES, B.F. & BOWSER, C.J. 1978. The mineralogy and related chemistry of lake sediments. In: *Lakes: chemistry, geology and physics.* (A. Lerman, ed.): 179-235. Springer. New York.

KAMP-NIELSEN, L. 1974. Mud-water exchange of

phosphate and other ions in undisturbed sediment cores and factors affecting the exchange rates. *Arch. Hydrobiol.*, 73: 218-237.

18. KAMP-NIELSEN, L. & HARGRAVE, B.T. 1978. Influence of bathymetry on sediment focussing in Lake Esrom. *Verh. Internat. Verein. Limnol.*, 20: 714-719.

KELDERMAN, P. 1980. Phosphate budget and sediment exchange in Lake Grevelingen (SW Netherlands). *Neth. J. Sea Res.*, 14: 229-236.

KELDERMAN, P. & VAN DE REPE, A.M. 1982. Temperature dependence of sediment-water exchange in Lake Grevelingen, SW Netherlands. *Hydrobiologia*, 92: 489-490.

KEMP, A.L.W., WILLIAMS, J.D.H., THOMAS, R.L. & GREGORY, M.L. 1978. Impact of man's activities on the chemical composition of the sediments of Lakes Superior and Huron. *Water Air Soil Pollut.*, 10: 381-402.

19. KEULDER, P.C. 1982. Particle size distribution and chemical parameters of the sediments of a shallow turbid impoundment. *Hydrobiologia*, 91: 341-353.

20. KLAPWIJK, S.P., KROON, M.W. & MEJLER, M.L. 1982. Available phosphorus in lake sediments in the Netherlands. *Hydrobiologia*, 92: 491-500.

KUO, S. & LOTSE, E.G. 1974. Kinetics of phosphate adsorption and desorption by lake sediments. *Soil Sci. Soc. Amer. Proc.*, 38: 50-54.

21. LARSON, G.L., ABERNATHY, A.R. & ALLISON, D.E. 1976. Chemical features of the recent sediments of a high-mountain lake. *Arch. Hydrobiol.*, 78: 456-467.

LEE, G.F., SONZOGNI, W.C. & SPEAR R.D. 1977. Significance of oxic vs anoxic conditions for lake Mendota sediment phosphorus release. In: *Interactions between sediments and fresh water* (H.L. Golterman, ed.): 294-306. Junk. The Hague.

LEVINE, S.N., STAINTON, M.P. & SCHINDLER, D.W. 1986. A radiotracer study of phosphorus cycling in lake 227, Northwestern Ontario. *Can. J. Fish. Aquat. Sci.*, 43: 366-378.

LI, W.C., ARMSTRONG, D.E., WILLIAMS, J.D.H., HARRIS, R.F. & SYERS, J.K. 1972. Rate and extent of inorganic phosphate exchange in lake sediments. *Soil Sci. Soc. Amer. Proc.*, 36: 279-285.

LIJKLEMA, L. 1977. The role of iron in the exchange of phosphate between water and sediments. In: *Interactions between sediments and fresh water* (H.L. Golterman, ed.): 313-317. Junk. The Hague.

LOGAN, T.J. 1982. Mechanisms for release of sediment-bound phosphate to water and the effects of agricultural land management on fluvial transport of particulate and dissolved phosphate. *Hydrobiologia*, 92: 519-530.

22. LÓPEZ, P. 1979. *Estudio de los parámetros químicos en el agua y el sedimento de dos embalses de la cuenca del Guadiana: Cíjara y García de Sola*. Master Thesis. University of Barcelona.

23. LÓPEZ, P. 1986a. Composición del sedimento en sistemas acuáticos del litoral mediterráneo español. *Limnetica*, 2: 11-18.

24. LÓPEZ, P. 1986b. Calcio, fósforo y metales del sedimento de la laguna de La Massona (Girona). *Butll. Inst. Cat. Hist. Nat.*, 53: 45-50.

25. MACKERETH, F.J.H. 1966. Some chemical observations on post-glacial lake sediments. *Phil. Trans. R. Soc.* (B), 250: 165-213.

26. MARCH, L. DE. 1978. Permanent sedimentation of nitrogen, phosphorus and organic carbon in a high arctic lake. *J. Fish. Res. Bd. Can.*, 35: 1089-1094.

MORTIMER, C.H. 1941. The exchange of dissolved substances between mud and water in lakes I. J. *Ecol.*, 29: 280-329.

MORTIMER, C.H. 1942. The exchange of dissolved substances between mud and water in lakes II. *J. Ecol.*, 30: 147-201.

27. NEWRKLA, P. & GUNATILAVA, A. 1982. Benthic community metabolism of three Austrian pre-alpine lakes of different trophic conditions and its oxygen dependency. *Hydrobiologia*, 92: 531-536.

NURNBERG, G.K. 1985. Availability of upwelling phosphorus from iron-rich anoxic hypolimnia. *Arch. Hydrobiol.*, 104: 459-476.

NURNBERG, G.K. 1988. Prediction of phosphorus release rates from total and reductant-soluble phosphorus in anoxic lake sediments. *Can. J. Fish. Aquat. Sci.*, 45: 453-462.

NURNBERG, G.K., SHAW, M., DILLON, P.J. & MCQUEEN, D.J. 1986. Internal phosphorus load in an oligotrophic Precambrian Shield Lake with an anoxic hypolimnion. *Can. J. Fish. Aquat. Sci.*, 43: 574-580.

28. OSTROFSKY, M.L. 1987. Phosphorus species in the surficial sediments of lakes of Eastern North America. *Can. J. Fish. Aquat. Sci.*, 44: 960-966.

OSTROFSKY, M.L., OSBORNE, D.A. & ZEBULSKE, T.J. 1989. Relationship between anaerobic sediment phosphorus release rates and sedimentary phosphorus species. *Can. J. Fish. Aquat. Sci.*, 46: 416-419.

PETR, T. 1977. Bioturbation and exchange of chemicals in the mud-water interface. In: *Interactions between sediments and fresh water* (H.L. Golterman, ed.): 216-226. Junk. The Hague.

PETTERSSON, K. 1984. Transfers between forms of sedimentary phosphorus induced by nitrate treatment. *Verh. Internat. Verein. Limnol.*, 22: 233-238.

29. PREMAZZI, G. & RAVERA, O. 1977. Chemical characteristics of Lake Lugano sediments. In: *Interactions between sediments and fresh water* (H. L. Golterman, ed.): 121-124. Junk. The Hague.

PSENNER, R. 1984. Phosphorus release patterns from sediments of a meromictic mesotrophic lake (Piburger See, Austria). *Verh. Internat. Verein. Limnol.*, 22: 219-228.

PSENNER, R., BOSTROM, B., DINKA, M., PETTERSSON, K., PUCSKO, R. & SAGER, M. 1988. Fractionation of phosphorus in suspended matter and sediment. *Arch. Hydrobiol. Beih. Ergebn. Limnol.*, 30: 98-112.

PSENNER, R. & PUCSKO, R. 1988. Phosphorus fractionation: Advantages and limits of the method for the study of sediment P origins and interactions. *Arch. Hydrobiol. Beih. Ergebn. Limnol.*, 30: 43-59.

30. REYNOLDSON, T.B. & HAMILTON, H.R. 1982. Spatial heterogeneity in whole lake sediments -

towards a loading estimate. *Hydrobiologia*, 91: 235-240.

RILEY, E.T. & PREPAS, E. 1984. Role of internal phosphorus loading in two shallow productive lakes in Alberta, Canada. *Can. J. Fish. Aquat. Sci.*, 41: 845-855.

RIPL, W. 1976. Biochemical oxidation of polluted lake sediments. A new restoration method. *Ambio*, 5: 132-135.

RIPPEY, B. 1977. The behaviour of phosphorus and silicon in undistrubed cores of Lough Neagh sediments. In: *Interactions between sediments and fresh water* (H. L. Golterman, ed.): 348-353. Junk. The Hague.

RYDING, S.O. 1985. Chemical and microbiological processes as regulators of the exchange of substances between sediments and water in shallow eutrophic lakes. *Int. Revue ges. Hydrobiol.,* 70: 657-702.

31. RYDING, S.O. & BORG, H. 1976. Chemical studies of superficial sediments in lake Lilla Ullevifjarden. Int. Revue ges. *Hydrobiol.*, 61: 581-593.

RYDING, S.O. & FORSBERG, C. 1977. Sediments as a nutrient source in shallow polluted lakes. In: *Interactions between sediments and fresh water* (H.L. Golterman, ed.): 227-234. Junk. The Hague.

SCHINDLER, D.W., HESSLEIN, R.H. & TURNER, M.A. 1987. Exchange of nutrients between sediments and water after 15 years of experimental eutrophication. *Can. J. Fish. Aquat. Sci.*, 44: 26-33.

32. SERRUYA, C. 1971. Lake Kinneret; the nutrient chemistry of the sediments. *Limnol. Oceanog.*, 16: 510-521

SERRUYA, C., EDELSTEIN, M., POLLINGHER, U. & SERRUYA, S. 1974. Lake Kinneret sediments: Nutrient composition of pore water and mud water exchanges. *Limnol. Oceanog.*, 19: 489-508.

SHUKLA, S.S., SYERS, J.K., WILLIAMS, J.D.H., ARMSTRONG, D.E. & HARRIS, R.F. 1971. Sorption of inorganic phosphate by lake sediments. *Soil Sci. Soc. Amer. Proc.*, 35: 244-249.

SINKE, A.J.C. & CAPPENBERG, T.E. 1988. Influence of bacterial processes on the phosphorus release from sediments in the eutrophic Loodsdrecht lakes, The Netherlands. *Arch. Hydrobiol. Beih. Ergebn. Limnol.*, 30: 5-13.

33. SLATER, S.E.J. & BOAG, A.J. 1978. The phosphorus status of the sediments of three eutrophic lakes in Victoria. *Aust. J. Mar. Freshwater Res.*, 29: 263-274.

SLY, P.G. 1977. Sedimentary environments in the Great Lakes. In: *Interactions between sediments and fresh water* (H.L. Golterman, ed.): 76-82. Junk. The Hague.

SLY, P.G. 1982. *Sediment/freshwater interaction.* Junk. The Hague.

SMITH, E.A., MAYFIELD, C.J. & WONG, P.T. 1978. Naturally occurring apatite as a source of ortophosphate for growth of bacteria and algae. *Microbiol. Ecol.*, 4: 105-117.

34. SONDERGAARD, M. 1989. Phosphorus release from a hypereutrophic lake sediment: Experiments with intact sediments cores in a continuous flow system. *Arch. Hydrobiol.*, 116: 45-59.

SORENSEN, J. 1982. Reduction of ferric iron in anaerobic marine sediment and interaction with reduction of nitrate and sulfate. *Appl. Envir. Microbiol.*, 43: 319-324.

STEVENS, R.J. & GIBSON, C.E. 1977. Sediment release of phosphorus in Lough Neagh, Northern Ireland. In: *Interactions between sediments anad fresh water* (H.L. Golterman, ed.): 343-347. Junk. The Hague.

35. THOMAS, S.R. & SOLTERO, R.A. 1977. Recent sedimentary history of a eutrophic reservoir: Long Lake, Washington. *J. Fish. Res. Bd. Can.*, 34: 669-676.

36. TIREN, T. & PETTERSSON, K. 1985. The influence of nitrate on the phosphorus flux to and from oxygen depleted lake sediments. *Hydrobiologia,* 120: 207-223.

TWINCH, A.J. & PETERS, R.H. 1984. Phosphate exchange between littoral sediments and overlying water in an oligotrophic North-temperate lake. *Can. J. Fish. Aquat. Sci.*, 41: 1609-1617.

VINER, A.B. 1975. The sediments of Lake George (Uganda) II: Release of ammonia and phosphate from an undisturbed mud surface. *Arch. Hydrobiol.*, 76: 368-378.

37. WILLIAMS, J.D.H., JAQUET, J.M. & THOMAS, R.L. 1976. Forms of phosphorus in the surficial sediments of Lake Eire. *J. Fish. Res. Bd. Can.*, 33: 413-429.

WILLIAMS, J.D.H., SHEAR, H. & THOMAS, R.L. 1980. Availability to *Scenedesmus quadricauda* of different forms of phosphorus in sedimentary materials from the Great Lakes. *Limnol. Oceanog.*, 25: 1-11.

WILLIAMS, J.D.H., SYERS, J.K., ARMSTRONG, D.E. & HARRIS, R.F. 1971a. Characterization of inorganic phosphate in non-calcareous lake sediments. *Soil Sci. Soc. Amer. Proc.*, 35: 556-561.

WILLIAMS, J.D.H., SYERS, J.K., HARRIS, R.F. & ARMSTRONG, D.E. 1971b. Fractionation of inorganic phosphate in calcareous lake sediments. *Soil. Sci. Soc. Amer. Proc.*, 35: 250-255.

38. WISNIEWSKI, R.J. & PLANTER, M. 1985. Exchange of phosphorus across sediment-water interface (with special attention to the influence of biotic factors) in several lakes of different trophic status. *Verh. Internat. Verein Limnol.*, 22: 3345-3349.

39. YOSHIDA, T. 1980. Mathematical model of phosphorus release from lake sediment. *Verh. Internat. Verein. Limnol.*, 21: 21-25.

YOUNG, T.C. & DEPINTO, J.V. 1982. Algal availability of particulate phosphorus from diffuse and point sources in the lower Great Lakes basin. *Hydrobiologia,* 91: 111-119.

(J. D. Ros & N. Prat, eds.). 1991. *Oecologia aquatica*, 10: 127-135

FLUCTUATIONS, THE KEY ASPECT FOR THE ECOLOGICAL INTERPRETATION OF SALINE LAKE ECOSYSTEMS*

FRANCISCO A. COMÍN[1], RAMON JULIÀ[2] & PILAR COMÍN[1]

[1] Departament d'Ecologia. Facultat de Biologia. Universitat de Barcelona. Av. Diagonal, 645. 08028 Barcelona. Spain.
[2] Institut Jaume Almera (CSIC). Martí i Franquès, s/n. 08028 Barcelona. Spain.

Received: February 1991

SUMMARY

In Lake Gallocanta (Aragón, NE Spain) changes of the biological communities occur at different scales of time in accordance with physical and chemical characteristics of the water. Different stages of the lake can be defined by the values of abiotic and biotic variables. However, processes occurring at any time influence the characteristics of later stages, even after dry periods. Fluctuations of physical, geochemical and biological characteristics in Lake Gallocanta make clear this idea. Definition of the intensity and frequency of the fluctuations and their relationships with climatological and geomorphological characteristics of the watershed are the key aspects in understanding the ecological functioning of saline wetlands.

KEY WORDS: Saline lake ecosystems, fluctuations.

INTRODUCTION

Interpretation of the ecological characteristics of aquatic ecosystems requires definition of time and space scales because the significance of the data depends on the scale of observation (TURNER *et al.*, 1989). This is particularly clear for ecosystems, such as saline wetlands, where salinity changes are related to water volume in a very wide range. The structure and functioning of the biological community changes in accordance with the intensity and frequency of the external disturbances (HARRIS, 1980; MARGALEF, 1986). The most conspicuous case is represented by ephemeral saline lakes.

The community fluctuations are strongly related to physical and chemical changes of the water, and these to climatological fluctuations (COMÍN *et al.*, 1983). The succession resumes after every desiccated event. However, fluctuations of different lengths of time can take place between dry episodes, and during longer periods.

This paper presents an example of such a type of ecosystems: Lake Gallocanta, where some chemical and biological characteristics have been studied over a period of years, although not continuously (COMÍN *et al.*, 1990). The major point

* Research supported by Diputación General de Aragón, Servicio de Conservación del Medio Natural.

here is to make it clear that observation of components and processes at different scales is necessary for any interpretation or applied interest of this type of ecosystems. Or, as previously mentioned by MARGALEF (1987), these ecosystems are better characterized by the type (intensity and frequency) of the fluctuations than by the values observed at any precise time and place.

GEOGRAPHICAL SETTING

Lake Gallocanta (40° 50' N and 2° 11' W) is the hydrological terminus of a 520 km^2 watershed in the Iberian Range (western limit of the Zaragoza and Teruel provinces, Aragón region). The geomorphological origin of the lake is a tectonic depression. Accumulation of eroded salts from the watershed and concentration by evaporation in this endorheic area is the origin of the present saline character of Lake Gallocanta.

The interesting point here is that Lake Gallocanta and its watershed are located at the boundary between semiarid and semihumid weather (COMÍN & ALONSO, 1988). The average annual rainfall varies widely from north to south in the Iberian Peninsula. This, with evaporation, determines wide spatial differences in the net annual rainfall-evaporation (r-e) budget. Because of the strong negative net annual r-e, every saline lake in Spain but one is ephemeral. Lake Chiprana is the only permanent saline lake in Spain. It is located very close to many other ephemeral lakes, under the same climatic conditions. The difference lies on the fact that Lake Chiprana is at relatively low altitude (150 m a.s.l.) and very close to the phreatic water table compared to other lakes.

Between the ephemeral Spanish saline lakes, Gallocanta is noteworthy because its water level shows very distinct seasonal and longer period fluctuation (Fig. 1), while the other saline lakes in Spain dry every year in response to very strong negative rainfall- evaporation budgets (COMÍN & ALONSO, 1988).

The peculiar location of Lake Gallocanta explains the water level fluctuations. Interannual differences of the rainfall cause mid-term trends in the water level fluctuations superimposed on the seasonal fluctuations. Fluctuations at any organizational level of the ecosystem are supposed to follow those of water level in saline lakes because salinity changes are wide enough to affect physical, chemical and biological characteristics. This is the uniqueness of Lake Gallocanta. Very few other examples like this are available. In saline lakes which dry every year and in permanent saline lakes the biological components do not show such clear

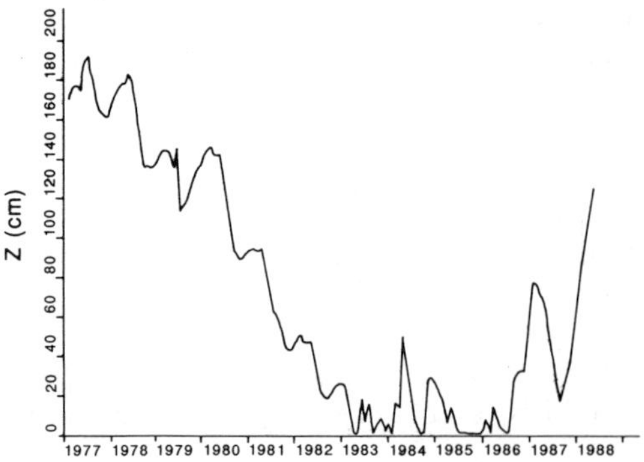

FIGURE 1. Recorded changes of the water level in Lake Gallocanta

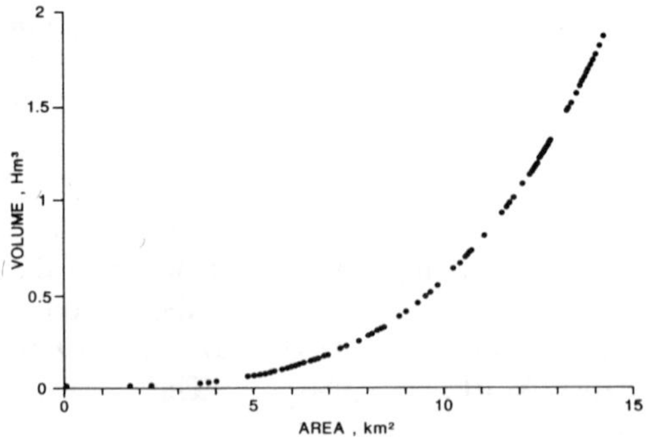

FIGURE 2. Relationship between area covered by the water and water volume in Lake Gallocanta.

fluctuations in water level or in related characteristics (physical, chemical and biological).

The average annual rainfall in Gallocanta is 488 mm. This is the average of annual rainfall measured every day at five meteorological stations located in the watershed. No significant differences were observed between the five stations. A rough calculation of the net annual evaporation from the lake, taking into account changes in water volume during months with no rainfall at all, gives a figure of 750 mm. The volumes of different vertical sections of the lake were calculated as the product of the water height by the simultaneous change of area. The areas were estimated from aerial photographs at precise dates in 1977 and 1983, when lake water was at different levels, and these were recorded. The relationship between water volume and area covered by the water is quite linear for a part of the graph (Fig. 2). However, water level and rainfall are far from a direct relationship.

The water volume in the lake is the result of the balance inputs-outputs. Rainfall on the lake, run-off and groundwater constitute the inputs. Outputs consist, basically, of evaporation from the lake. Most of the Gallocanta watershed is occupied by cereal fields. Considering a long enough period of time, inputs equal outputs. In this steady-state condition, $V_i = V_0$, V_i and V_0 being the volumes of water entering and going out of the lake. Following BOWLER (1981), in an endorheic area this expression can be written as:

$$A_l P + (A_c - A_l) P_f = A_l E \qquad (1)$$

where A_l and A_c are, respectively, the areas (m^2) of the lake and the watershed. P is the rainfall ($l\ m^{-2}$), E the evaporation ($l\ m^{-2}$) and P_f is the water volume entering to the lake per square meter of watershed.

Reordering the elements in (1),

$$A_c/A_l = [(E-P)/P_f] + 1 \qquad (2)$$

This equation includes morphometric parameters in the first part and climatic parameters in the second part. In this equation $P_f = P\ c$, where c is the equivalent to a run-off coefficient or a coefficient to transform from rainfall in the watershed to water entering the lake.

Equation (2) is useful to define the conditions for the hydrological balance of a terminal lake in a closed basin. In the case of Gallocanta,

$$520/20 = [(720-500)/\ 500\ c] + 1$$

From this, a value of 0.019 is obtained. This is the value of the run-off coefficient which equals the second part of the equation, named the climatic function, to the first part, the morphometric function.

Then, under the present conditions, the water level of the lake will remain in steady state when the value for both functions is 26.

This model can be applied only to closed basins. In the case of Gallocanta, some corrections should be introduced if direct measurement of the hydrological balancc gives support to the idea that groundwater relation exists with other close aquifers (CRUZ *et al.*, 1979).

Another important criticism of the model proposed refers to the run-off coefficient, which is estimated from the other parameters in the equation and not measured directly. The value of the coefficient can change at least one order of magnitude depending on the rainfall intensity and the land state, among other factors (BOWLER, 1981). So, it can change during the period of time of a fluctuation in the lake water level. However, for the period of time between identical water levels, an average could be used if direct estimates are not available.

The combination of the climatic and the morphometric parameters define the state of any saline lake (Fig. 3). For those lakes where great changes of water volume occur, an oscillating area, rather than a point in that space, could be defined.

In the case of Lake Gallocanta, the value of A_c/A_l was 26 in 1981, when the water level was relatively high. Depending on the parameters in the climatic function, the lake will change from an ephemeral condition to either of the other two possibilities in figure 3. For example, a rainfall decrease to 250 mm will transform the lake into an evaporitic saltern, that is, dry or with huge amounts of salt precipitates. This change would be forced if outgoings from the watershed (e.g., use of groundwater for irrigation, extraction from connected watersheds) increase. Just a 6.6% increase of the evaporation, combined with a low rainfall, gives an $Fc = [(800- 250)/(250 \times 0.02)] + 1 = 111$, characteristics with which the lake would change to the evaporitic stage (Fig. 3).

THE RESPONSE TIME OF THE LAKE TO CLIMATIC CHANGES

The water volume in the lake does not change immediately after changes in the parameter which constitute inputs and outputs from the watershed. So, changes

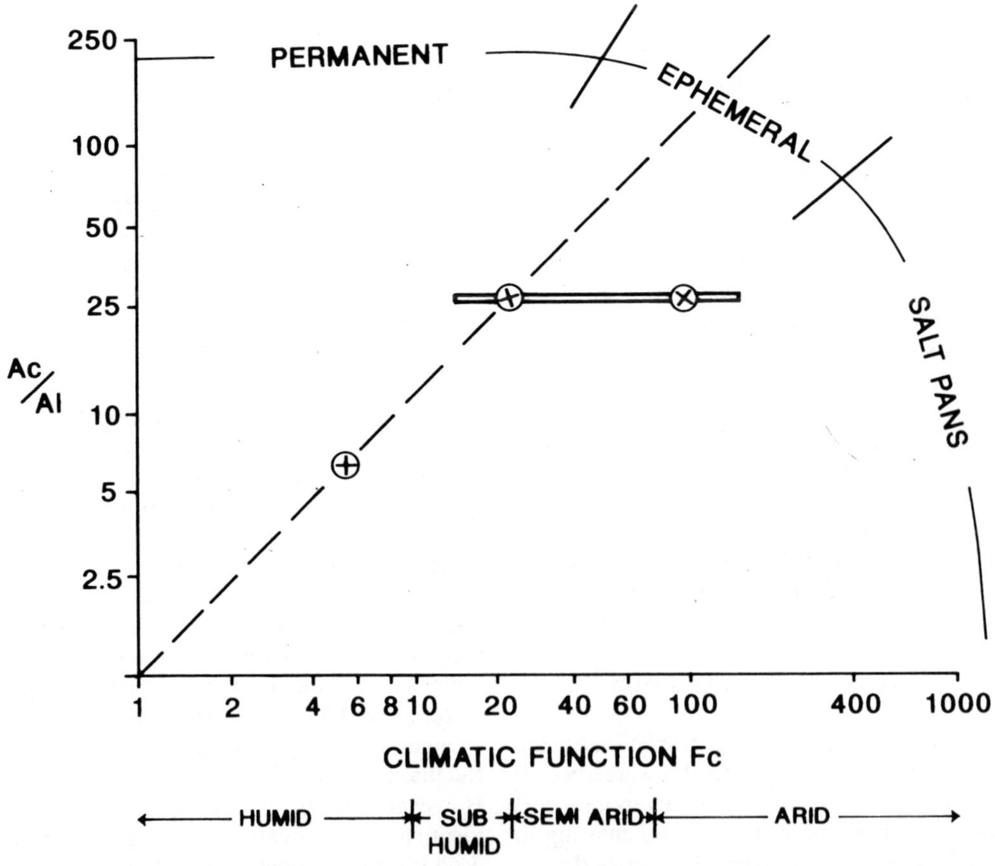

FIGURE 3. Change in Lake Gallocanta as a consequence of increased water extraction and decreased rainfall in the space defined by the relationship between the morphometric function (A_c/Al) and the climatic function (Fc).

TABLE I. Values of the response time (years) of different saline lakes (Data from LANGBEIN, 1961, except for those of Gallocanta).

Lake	k, year^{-1}
Pyramid	65
Redberry	50
Dead Sea	35
Basin	25
Quill	20
Great Salt Lake	9
Eyre	1.5
Elton	1
Gallocanta	2.4-0.3

from year to year in the water inputs and outputs are not paralleled by changes of the same volume in the lake water. In some terminal lakes, water level hardly changes in spite of great year to year changes of rainfall. In other lakes, great changes of water level are observed from one year to the next, although rainfall hardly changes. The point is that the geomorphological structure of the basin deviates the relationship between the input-output balance of the watershed and lake water volume from linearity.

The response time of the lake water volume as affected by changes in the water inputs to the watershed can be calculated following LANGBEIN (1961). A change of the lake water volume (V), will occur if there is a difference between inputs (I) and outputs (O):

$$\Delta V = I - O \quad (3).$$

In a terminal lake the water outflow

$$Q = EA_l = (V + b) / k,$$

where E is the net evaporation, A_l the lake area, V the lake water volume, and b and k are constants for every lake. Then, in equation (3)

$$\Delta V = I - (V/k) - (b/k).$$

If the time intervals are short enough,

$$V_t - V_{t-1} = I_t - (V_t/k) - (b/k) \quad (4)$$

where t and t-1 refer here to consecutive years. After n years,

$$V_n = I_n \ K/(1+k) + I_{(n-1)} \ k/(1+k) + I_{(n-2)} \ k/(1+k) + ...$$

This equation indicates that the volume of water in the lake is a weighted average of the inflows during the preceding years. The effect of the inflows of a determinate year on lake volume decreases geometrically with time.

The values of k, the solutions of equation (4), are the response times of the lake water volume to the inflows of the previous year. For Gallocanta, these response times for consecutive years during the period 1976-87 fluctuate between 0.3 and 2.4 years. That is, the lake water volume takes between 3.6 and 28.8 months to change in response to the changes in the water volume inflowing to the lake from one year to the next which occurred in Gallocanta during that decade.

Table I lists the response times, k, of several saline lakes. In Gallocanta, the lowest values of k correspond to great changes in the water level between consecutive years (1980-1981, 1981-1982, 1986-1987, 1987-1988). The highest values of the response time correspond to the lowest differences in water level between two consecutive years (1977-78, 1979-80).

GEOCHEMICAL RELATED BIOLOGICAL FLUCTUATIONS

The model of salinity versus water volume changes proposed by LANGBEIN (1961) is common for every ephemeral saline lake (Fig. 4). The time necessary for a lake to show the complete fluctuation will depend on the particular climatic circumstances. For Gallocanta in recent times, at least twelve years have been necessary lately (COMÍN *et al.*, 1990). The lake has also dried up on at least three other occasions during the present century (GUIRAL, 1981). HERNÁNDEZ & ARANEGUI (1926) referred to another dry period at the end of nineteenth century during which cartloads of salt were collected.

The physical, chemical and biological characteristics of the lake change as the water level fluctuates within the area defined in figure 4. Several phases can be distinguished bearing in mind geochemical and biological features observed in Lake Gallocanta from 1980 to 1987. During 1980-81, at a relatively high water level, a seasonal fluctuation was observed, e.g.: small changes in the amounts of different dissolved ions and biological populations occurred (COMÍN *et al.*, 1983). From mid 1981 to 1983 a dominant water level decrease and salt concentration increase took place. The saturation points were reached for most of the ion couples and precipitation of salts proceeded intensively (Fig. 5). The lake dried up after a period of brines. During the years 1983-86 the lake remained dry or with very low water levels.

A refilling period was initiated in 1986. The water level increased and, since 1988,

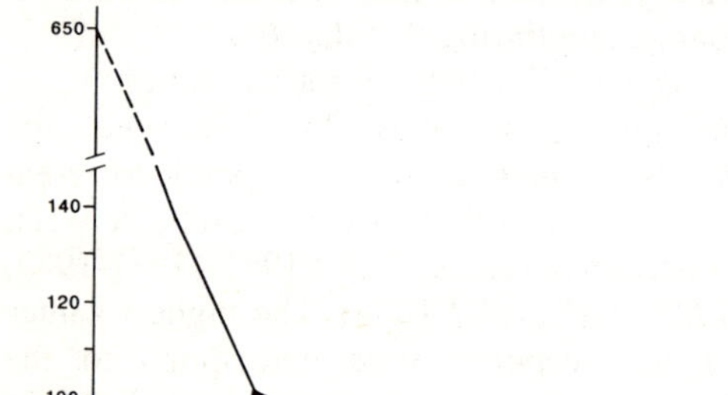

FIGURE 4. Salinity versus water volume in Lake Gallocanta, 1986-87. The dashed lines complete an hypothetical emptying-drying- refilling fluctuation.

TABLE II. Amounts of salts lost from the water of Lake Gallocanta during the dry period 1983-86.

Ion	tm
Cl^-	66,435
SO_4^{2-}	35,965
$CO_3^{2-} + HCO_3^-$	1,941
Na^+	20,811
Mg^{2+}	9,134
Ca^{2+}	2,192
K^+	539

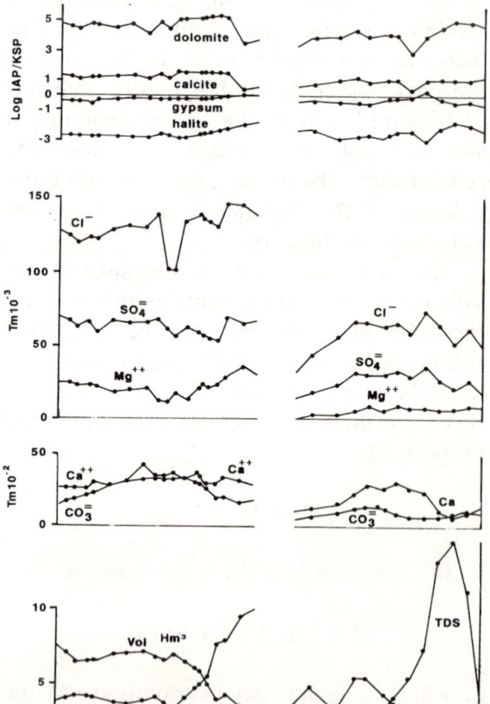

FIGURE 5. Changes in total dissolved solids (g l^{-1}), water volume (hm^3), absolute amounts of the most abundant anions and cations (tm) and the saturation coefficient (lg AP/kt) of the most abundant salts in the water of Lake Gallocanta during two annual cycles (1980-81 and 1986-87). A dry period occurred in between (1983-86).

TABLE III. Outstanding processes and components of the biological facies which occur in Lake Gallocanta at different phases.

	DESICCATION	BRINE	DRY	FLOODING	REFILLING	HIGH LEVEL
TDS (g l^{-1})	60-200	200-600		35-50	40-20	15
WATER LEVEL (cm)	50-20	20-0	0	0-75	75-100	150
PRECIPITATING SALTS	Calcite Dolomite Gypsum Halite Bischofite					
BIOGEOCHEMICAL PROCESSES	Denitrification Fermentation	Sulfate reduction Methane formation		Nitrification	Aerobic respiration Sulfide oxidation Mn, Fe oxidation	
PHYSICAL PROCESSES	Washing margins		Wind deflation			
PHYTOPLANKTON	*Dunaliella salina* *Nannochloris* *Spirulina* Floating mats		Cysts	*Lobocystis dichotoma* *Oocystis solitaria* *Tetraselmis* *Gymnodinium*		*Rhodomonas* *Monoraphidium*
ZOOPLANKTON	*Arctodiaptomus* *Brachionus* *Fabrea*		Resting eggs	*Daphnia mediterranea* *Moina mongolica* *Branchinella*		*D. dolicocephala*
MACROPHYTES	Benthic microbial mats (decomposing)		Resting seeds (drying)	*Ruppia drepanensis* *Lamprothamnion papulosum*	*Cladophora*	*Chara galioides*
OTHER FACIES	Insects colonizing margins Anoxic sediments		Terrestrial plants	Sediment disturbance		Flooded margins Highly dense bird populations

it has remained fluctuating around 100 cm. Huge amounts of salts were lost from the lake during the dry and low level periods (Table II) as can be seen by comparing the total concentration of dissolved solids in the same water volume at different times at the drying phase in 1981 and the refilling phase in 1987. Therefore the amount of salts during the latter are much lower than during the former. In addition to this, the molar ratios of the major ions changed, and ions were lost from the lake waters selectively by different processes.

Precipitation of salts is very important. The sediments of the lake contain different mineral components (COMÍN *et al.*, 1990), mainly carbonates (aragonite, high and low magnesium, calcite, dolomite) and sulphates (gypsum). In order to determine the relationship between the ionic composition of the water and the saturation point, after which precipitation takes place, the IAP/KSP ratio (the product of the free activities of the ions divided by the constant of the solubility product at 25 °C) for each mineral form was calculated using a computer program (KHARAKA *et al.*, 1988).

In Lake Gallocanta the pattern of change of ionic combinations over the saturation point during 1980-1981 is similar to that occurring in hypersaline marine environments while water evaporates. Carbonates precipitated first (10 times oversaturation for calcite and aragonite, and 100,000 times oversaturation for dolomite was observed during most of the 1980-81 period). The ratio for aragonite is very similar to calcite and is not represented in figure 5. Gypsum and halite increased the ratio IAP/KSP by mid-1981, paralleling the water volume decrease and TDS increase. Gypsum reached the equilibrium point at the end of 1981 while halite precipitated later on (COMÍN *et al.*, 1990).

The high dolomite oversaturation is related to the high magnesium concentration (200-800 meq l^{-1}). It is probably related to gypsum precipitation, which favors dolomite precipitation because this mineral salt needs a high nuclear energy and is controlled by kinetics inhibition mechanisms (BERNER, 1971). Dolomite, calcite and gypsum hardly redissolve during the refilling period. Hence, the mineral composition of the water is extremely influenced by water level fluctuations.

Eolic deflation contributes directly to the export of salts from the lake. Strong winds affect Gallocanta. Evaporation is increased by the wind. The water mass is displaced in the wind direction, sometimes increasing the water level in the leeward shore by 20 cm, which means flooding of a large land area where water is adsorbed onto clays.

Biogeochemical processes occurring in the sediments of wetlands include aerobic and anaerobic processes (i.e.: denitrification, fermentation, sulphate reduction, methanogenesis) which affect the carbonate and sulphate concentrations in the water (Table III). For example, a sulphate reduction rate of 9.5 mmol SO_4^{2-} m^{-2} d^{-1} was estimated in a coastal marine sediment (JØRGENSEN, 1977). If this, quite acceptable activity took place in Lake Gallocanta, 20 tm SO_4^{2-} per day could be reduced in the sediments during the low water level periods.

The specific composition of the biological communities also changed during the same period. Populations characteristic of the different phases can be proposed (Table III). However, it is not possible to define rigid limits between phases. Moreover, the main point we want to make here is the idea of continuity in saline lakes; over a period of time very different geological and biological facies can appear at successive times of a fluctuation.

The great spatial heterogeneity described before (DE DECKKER, 1988) is also a substantial characteristic of saline lakes. For example, the branchiopod *Branchinella spinosa* is present in the waters of Gallocanta after a dry period and under a combination of salinity, temperature and probably other factors. During the other phases, this population is also present in the lake sediment, like many other animals and plants, in the form of resistance organs (eggs, ephippia, cysts, seeds). Those species which are not able to persist under unfavorable conditions migrate (e.g.: birds) or die. Migration or exportation of organic matter, also of salts, has a buffer effect on the ecology of saline lakes. Accumulation of salts and dead organic matter has a stress effect.

MARGALEF (1987) proposed a mathematical model which considers the biological production depending on the change of both the inputs of external energy and the covariance of the production factors. He derived the original expression

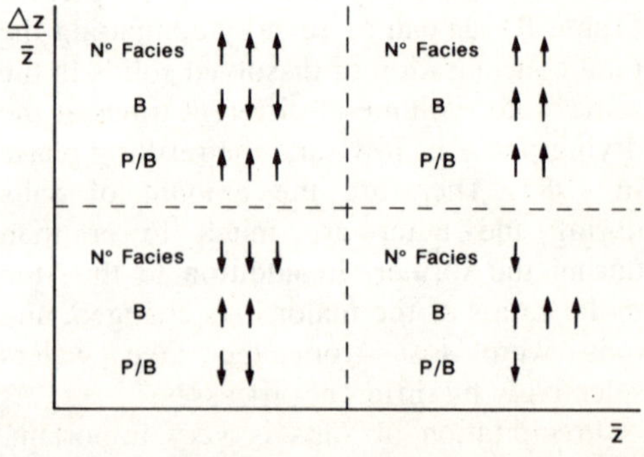

FIGURE 6. Relative intensities of synthetic characteristics in saline wetlands at different stages defined by water level and changes of the water level (Δ Z/z).

dB/dt = A C (where B is biomass, t is time, A is external energy and C is covariance of production factors) into

$$d^2B / dt^2 = C (dA/dt) + A (dC/dt)$$

and named the three terms, respectively, deceleration, energy degradation and ecological segregation in reference to general trends occurring during the ecological succession.

The application of models of this type to saline lakes is not possible yet because of the lack of data. However, endorheic wetlands areas like Gallocanta, where the water level fluctuates at different time scales, seem promising for providing sufficient reliable data to test some of the theoretical models proposed for fluctuating ecosystems.

A simplified graphic representation can relate the changes in the relationship between salinity and water volume to the community changes through time and space. In the case of most saline wetlands the change in time of external energy inputs can be represented by water level fluctuations (changes through time, vertical axis in figure 6). The ecological segregation, mostly horizontal heterogeneity in wetlands, can be represented by the water level (changes through space, horizontal axis in figure 6). Validation of this type of relationship requires data at different time scales (LIKENS, 1987), which will provide tools for a better knowledge of the way aquatic ecosystems work (MARGALEF, 1986).

REFERENCES

BERNER, R.A. 1971. *Principles of chemical sedimentology*. Mc-Graw Hill. London

BOWLER, J.M. 1981. Australian salt lakes, a paleohydrologic approach. *Hydrobiologia,* 82: 431-444.

COMÍN, F.A. & ALONSO, M. 1988. Spanish salt lakes, their chemistry and biota. *Hydrobiologia,* 158: 237-245.

COMÍN, F.A., ALONSO, M., LÓPEZ, P. & COMELLES, M. 1983. Limnology of Gallocanta Lake, Aragón, Northeastern Spain. *Hydrobiologia,* 197: 51-66

COMÍN, F.A., JULIÀ, R., COMÍN, M.P. & PLANA, F. 1990. Hydrogeochemistry of Lake Gallocanta (Aragón, NE Spain). *Hydrobiologia,* 197: 51-66.

CRUZ, J., DOBLAS, J.G. & GARRIDO, M. 1979. Hidrogeología de la laguna de Gallocanta. *Actas 2º Symp. Nac. Hidrogeol. Secc. 1ª Hidrogeol.*, 4: 431-446.

DE DECKKER, P. 1988. Biological and sedimentary facies of Australian salt lakes. *Palaeogeography, Palaeoclimatology, Palaeoecology,* 62: 237-270.

GUIRAL, J. 1981. Introducción al conocimiento del clima de la cuenca de la laguna de Gallocanta. In: *Estudio de la biocenosis de la laguna de Gallocanta y su cuenca*: 119-138. (Proyex, S.A., ed.). CEOTMA. MOPU. Madrid.

HARRIS, G.P.. 1980. Temporal and spacial scales in phytoplankton ecology: mechanisms, methods, models and management. *Can. J. Fish. Aquat. Research,* 37: 877-900.

HERNÁNDEZ, E. & ARANEGUI, P. 1926, La Laguna de Gallocanta y la geología de sus alrededores. *Bol. R. Soc. Esp. Hist. Nat.*, 26: 419-429.

JØRGENSEN, B.B. 1977. The sulfur cycle of a coastal marine sediment. *Limnol. Oceanogr.*, 22: 814-832.

KHARAKA, Y.K., GUNTER, W.D., AGGARWAL, P.K., PERKINS, E.H. & DEBRAAR, J.D. 1988. *Solmineq 88, a computer program code for geochemical modelling of water-rock interactions.* U.S. Geol. Survey. Menlo Park. Cal.

LANGBEIN, W.B. 1961. *Salinity and hydrology of closed lakes*. U.S.G.S. Prof. Paper 412. U.S. Gort. print Office. Washington.

LIKENS, G.E. (ed.) 1987. *Long-term studies in ecology*. Springer. New York.

MARGALEF, R. 1986. Reset successions and suspected chaos in models of marine populations. *Proc. Int. Symp. Long Term Changes Mar. Fis. Pop.*: 321-343. Vigo.

MARGALEF, R. 1987. Teoría y modelado de los ecosistemas fluctuantes. In: *Bases científicas para la protección de los humedales en España*: 31-41. Real Academia Ciencias Exactas, Físicas y Naturales. Madrid.

TURNER, M.G., O'NEILL, R.W., GARDNER, R.H. & MILNE, B.T. 1989. Effects of changing spatial scale on the analysis of landscape pattern. *Landscape ecology*, 3: 153-162

Homage to Ramon Margalef; or, Why there is such pleasure in studying nature 137
(J. D. Ros & N. Prat, eds.). 1991. *Oecologia aquatica*, 10: 137-144

EXTERNAL ENERGY AND PLANKTON: NEW INSIGHTS ON THE ROLE OF SMALL-SCALE TURBULENCE ON ZOOPLANKTON ECOLOGY

MIQUEL ALCARAZ & ENRIC SAIZ

Institut de Ciències del Mar (CSIC), Pg. Nacional, s/n. 08039 Barcelona. Spain.

Received: May 1990.

SUMMARY

Some of the already known effects of small scale turbulence on zooplankton populations maintained in laboratory microcosms (higher development rates, changes in the demographic composition of populations, higher ammonia concentration) seemed to be a consequence of increased energy expenditure driven by behavioural changes. Higher frequency of escape reactions and fast swimming in copepods could be the main reasons for it, although direct measurements of metabolic rates under turbulence were not available. Here we discuss the effects of small scale turbulence on two parameters directly related to metabolic activity: heart-beat rate on pond cladocerans, and NH_4-N and PO_4-P excretion rates on marine copepods.

KEY WORDS: Turbulence, zooplankton, metabolism, heart-beat rate, excretion rate.

INTRODUCTION

Mechanical (exosomatic) energy has a significant influence on plankton distribution (MACKAS *et al.*, 1984; TETT & EDWARDS, 1984) and on the structural and functional properties of pelagic ecosystems. The external supply of turbulent energy not only seems to modulate primary production, acting as auxiliary energy (MARGALEF, 1974, 1978, 1982; LEGENDRE, 1981) but also can contribute to the selection of phytoplankton life-forms (MARGALEF, 1978).

In general, these effects are a consequence of the combination of turbulence plus advection, and cannot easily be distinguished from those caused by co-variant factors such as light or nutrients (MARRASÉ, 1986; ESTRADA *et al.*, 1987 a). In the case of small scale turbulence, which can be considered to have a null transporting effect, its study in natural systems is hindered by, among other things, the lack of replicability, the absence of well-defined boundaries and the impossibility of isolating factors (ESTRADA *et al.*, 1987 a, b; ALCARAZ *et al.*, 1988, 1989).

These problems can be partially solved through laboratory experiments in which planktonic populations are exposed to the desired conditions in enclosures (HARTE *et al.*, 1980; PILSON & NIXON, 1980; PÉREZ *et al.*, 1977; MARRASÉ, 1986; ESTRADA *et al.*, 1987 a, b; ALCARAZ *et al.*, 1988). Experiments involving time scales from several days to months allow us to study the effects of different factors (turbulence, light, nutrient availability, presence of predators, etc.) on state variables (biomass, population structure, etc.), while their effect on rate processes (physiology), which respond faster, can be deduced either from their cumulative effects on state variables, or by direct measurement of instantaneous rates.

Most information on the influence of turbulence on planktonic populations refers to phytoplankton. Data on zooplankton are scarce and less conclusive, probably because of the dependence of herbivorous zooplankton on phytoplankton (itself affected by turbulence; LEGENDRE & DEMERS, 1984; KIORBØE *et al.*, 1990).

Microcosm experiments suggest that the consequences of small scale turbulence on state variables of zooplanktonic populations (the higher concentration of ammonia in microcosms in which zoo- and phytoplankton populations were exposed to turbulence, or the increased development rates of copepod populations, ALCARAZ *et al.*, 1988) are a consequence of an acceleration of their metabolic activity, similar to that induced by increasing temperatures. However, although circumstantial evidence such as increased encounter rates between predators and prey (ROTHSCHILD & OSBORN, 1988; EVANS, 1989) or changes in swimming behaviour and in the frequence of "fast swimming" and escape reactions of copepods exposed to turbulence (COSTELLO *et al.*, 1990) suggests enhanced energy consumption (MARRASÉ *et al.*, 1990), it is as yet not conclusive.

In this paper we evaluate the effect of small scale turbulence on the heart-beat rate of pond cladocerans (females of *Daphnia pulex*, which have differentiated hearts) and on excretion rates of calanoid marine copepods (*Acartia clausi* and *A. margalefi*). Heart-beat rates are considered to be sensitive descriptors of metabolic activity in cladocerans (MAC ARTHUR & BAILLIE, 1929; INGLE *et al.*, 1937), marine copepods (PAVLOVA & MINKINA, 1983) and decapod crustaceans (SPAARGAREN, 1977; SPAARGAREN & ACHITUV, 1977). Heart-beat rates are also appropriate descriptors of metabolic rates because they show a rapid response to changes in temperature and other environmental factors affecting metabolic activity (MAC ARTHUR & BAILLIE, 1929). Because the excretion rates require long incubations (from 12 to 24 h), they reflect the integrated effect of turbulence rather than instantaneous responses. The changes due to turbulence on these indicators of metabolic activity and their relation with the mentioned effects on state variables of zooplankton have been discussed.

MATERIAL AND METHODS

HEART-BEAT RATES

The heart-beat rates of *D. pulex* exposed to turbulent or calm conditions were quantified by direct observation through an horizontally mounted dissecting microscope (Fig. 1). Individuals (obtained from a long-established culture) were kept in focus tethered to a thin hair (ALCARAZ *et al.*, 1981; PAFFENHÖFER *et al.*, 1982; MARRASÉ *et al.*, 1990; COSTELLO *et*

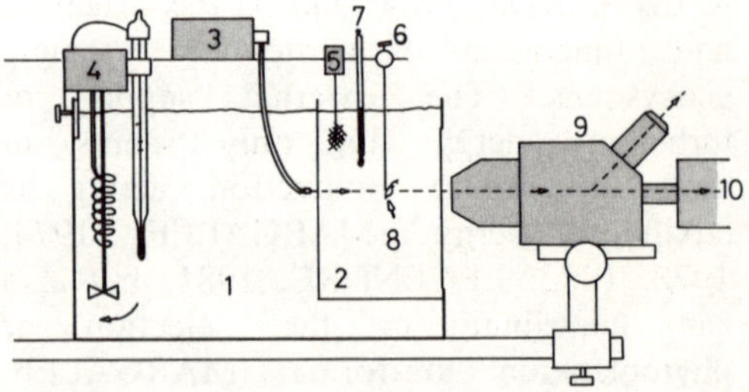

FIGURE 1. Diagrammatic description of the observation chamber for heart rate estimations. 1: Thermostatic bath. 2: Observation aquarium. 3: Light source (light fibre). 4: Heater-cooler with thermostat and homogenizer. 5: Turbulence generator. 6: Manipulator for positioning the organisms. 7: Thermometer. 8: Tethered organism. 9: Stereo microscope. 10: Video camera.

al., 1990). The observation chamber (Fig. 1) consisted of a small aquarium (10x5x5 cm) maintained in a thermostatic bath at the selected temperature (8 to 18 ± 0.1 °C). In order to avoid any disturbance by light intensity, measurements were made under dim red light.

Turbulence was generated by the vibration of a 1.5 x 1.5 cm netlon grid (0.5 cm mesh-size) attached to a speed-regulated vibration motor, similar to that described by MARRASÉ *et al.* (1990) and COSTELLO *et al.* (1990). We performed 10 to 20 replicated measurements with a stop watch of the time spent for series of 50 or 100 heart beats at each temperature and turbulence condition. The time lag between successive measurements was 1 min, and when changing from turbulence to calm conditions the organisms were allowed to acclimatize for 15 min. Despite the high rate of heart-beat in *Daphnia* (INGLE *et al.*, 1937) the counting error was negligible, and differences between independent counts by different observers were not significant.

EXCRETION RATES

Experimental organisms were pelagic copepods of the genus *Acartia* (*A. clausi* and *A. margalefi*), captured by short (5 min) horizontal net tows made in the harbours of Barcelona and Masnou, on the Catalan coast (Western Mediterranean). They were transported to the laboratory in plastic carboys filled with *in situ* water. Adult females were sorted and placed in 10 l carboys filled with 60 µm-filtered *in situ* water and allowed to acclimatize in a temperature-controlled chamber under dim light (30 - 50 µE m^{-2} sec^{-1}) and 12 h light: 12 h dark photoperiod. A phytoplankton suspension (*Isochrysis*, *Thalassiosira*) was added periodically to ensure abundant food supply. Excretion rates were measured by the differences in concentrations of ammonia and soluble reactive phosphorus in control and experimental flasks after 24 h incubation in darkness. The flasks consisted of 25 or 50 ml screw-cap plastic vials 3/4 filled with air-saturated filtered sea water (GF/F glass fibre filters). The concentration of organisms in the experimental flasks was about 1 ind ml^{-1}. Turbulence was created by placing the flasks in a shaker oscillating at 100 strokes min^{-1}, with an amplitude of 2.5 cm. Non-turbulent flasks were kept undisturbed. Temperature was the same for the two experimental conditions (17 ± 1.0 °C). Initial and final ammonia concentrations were analysed according to KOROLEFF (1970), and phosphorus determinations followed the procedure of GRASSHOFF (1976). The results were expressed as µg-at NH_4-N or µg-at PO_4-P ind^{-1} h^{-1}. Statistical tests were made following SOKAL & ROHLF (1969).

RESULTS

Heart-beat rates corresponding to a single female of *Daphnia pulex* (1.8 mm length without terminal spine) as a function of environmental temperature for agitated

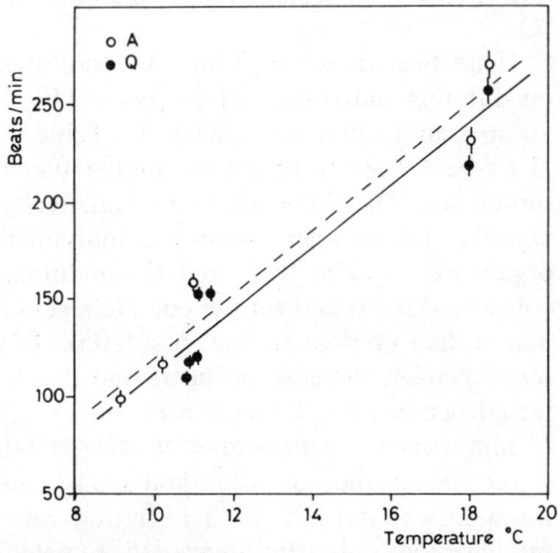

FIGURE 2. Heart rates of a *Daphnia pulex* female (1.8 mm length without spine) in A (turbulent, open dots and dashed line) and Q (non-turbulent, full dots and continuous line) conditions as function of temperature.

TABLE I. Heart rates (beats min^{-1}) corresponding to different specimens of *D. pulex* under Q (non-turbulent) and A (turbulent) conditions. Mean, standard deviation, number of measurements and percent increase of A versus Q conditions.

t °C increase	Q			A			%
	mean	s.d	n	mean	s.d.	n	
9.7	136.9	4.1	10	154.7	6.3	10	13.0
10	98.1	2.4	10	109.3	3.0	20	11.5
15.7	86.5	0.7	10	96.3	2.1	10	11.9
16.0	241.5	24.6	10	283.5	23.8	10	17.3
19.0	143.3	1.3	20	161.9	2.0	20	12.9
*11.1	153.6	1.5	10	158.3	2.1	10	3.0
*18.0	218.3	3.0	10	230.0	6.5	10	5.5
*18.5	257.5	5.0	10	269.7	6.0	10	4.6

* measurements made on the same individual.

(A) and calm (Q) conditions are represented in figure 2. The estimated value of Q_{10} according to the Belerhadek equation for pooled (Q and A) data is $Q_{10} = 2.34$. The regression equations relating temperature and heart rate for turbulent (A) and calm (Q) conditions (Fig. 2) are respectively

$$H = 16.13\ T - 41.51,\ r^2 = 0.948\ \ (A) \quad \text{and}$$

$$H = 15.89\ T - 46.64,\ r^2 = 0.904\ \ (Q),$$

where H represents heart-beat rate in beats min^{-1}, and T environmental temperature in °C.

Heart-beat rates for Q and A conditions on different individuals of *D. pulex* (10 to 20 measurements) are shown in Table I. Heart-beat rates were always higher under turbulence. The differences are statistically significant both when comparing individual organisms exposed to A and Q conditions (t-test, $p < 0.05$) and for the complete set of paired data (Wilcoxon test, $p < 0.05$). The mean percent increase on heart rate due to turbulence is 9.9 ± 3.4 %, n = 8).

Simultaneous estimations of NH_4-N (as µg-at NH_4-N $ind^{-1}\ h^{-1} 10^{-6}$) and PO_4-P (as µg-at PO_4-P $ind^{-1}\ h^{-1} 10^{-6}$) excretion rates on copepods (*Acartia margalefi* females) for turbulent and calm conditions (Fig. 3) were significantly correlated (r=0.70, n=14). The corresponding regression equation was

$$N = 21.1\ P + 3.46$$

where N represents µg-at NH_4-N $ind^{-1}\ h^{-1} 10^{-6}$, and P = µg-at PO_4-P $ind^{-1}\ h^{-1} 10^{-6}$.

The average increase in ammonia excretion due to turbulence was 32 %, while phosphorus excretion increased 65 % (Table II). N:P excretion ratios were higher under calm conditions, and in both conditions were well above the theoretical ones (REDFIELD *et al.*, 1963).

The ammonia excretion rates for *A. clausi* are indicated in Table III. In this species, the increase in the excretion rates

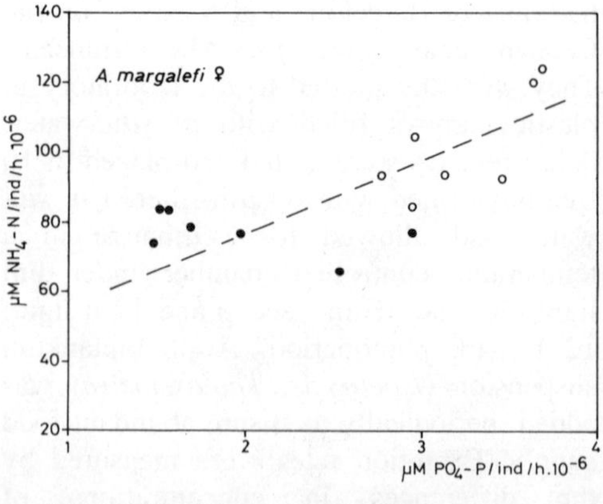

FIGURE 3. Simultaneous measurements of phosphorus (abcissae) and ammonia (ordinate) excretion rates under A (turbulent, open dots) and Q (non-turbulent, full dots) conditions for *Acartia margalefi*.

TABLE II. Mean excretion rates corresponding to turbulent (A) and non-turbulent (Q) conditions for adult females of *A. margalefi*. N = µg-at. NH_4-N ind^{-1} $h^{-1} 10^{-6}$; P = µg-at. PO_4-P ind^{-1} $h^{-1} 10^{-6}$. N:P in atoms. The differences between Q and A are statistically significant (t-test, $p < 0.05$).

	Q			A		
	N	P	N:P	N	P	N:P
Mean	76.9	1.7	42.2	101.7	3.22	31.5
s.d	5.7	0.5	11.3	13.8	0.3	2.8
n	7	7	7	7	7	7
A increase (%)				32	65	

TABLE III. Ammonia excretion rates for *A. clausi* under non-turbulent (Q) and turbulent (A) conditions. N = µg-at. NH_4-N ind^{-1} h^{-1} 10^{-6}

	Q	A
Mean	142.0	180.0
s.d.	20.0	14.9
n	5	5
A increase (%)		26

due to turbulence is 26 %, slightly lower than for *A. margalefi*.

DISCUSSION

Changes in state variables of copepod populations maintained under turbulence in laboratory microcosms include enhanced development rates, changes in the population dynamics and higher ammonia concentration (ALCARAZ *et al*., 1988, 1989). These effects are attributable to the increased metabolic activity and energy expenditure resulting from the higher frequency of escape reactions or "fast swimming", which require up to 400 times more energy consumption than normal swimming (STRICKLER, 1975).

Recent estimates of time budget of swimming and feeding behaviour of copepods support this assumption, for there is a significant increase in the frequency of fast-swimming under turbulence (COSTELLO *et al.*, 1990), and estimations obtained from these data about the relative energy gains and losses of copepods under calm and turbulent conditions on high and low food concentrations (MARRASÉ *et al.*, 1990) indicate energy advantages under turbulent and low food concentration conditions. These results, however, correspond to the interaction of zooplankton with food particles, and it is difficult to differentiate the effects of turbulence alone on metabolism (due to the higher frequency of escape reactions) from those derived from changes in feeding behaviour (probably triggered by the higher encounter rates between copepods and their food). Thus, the two parameters chosen to estimate the effects of turbulence on metabolism (heart rate and ammonia and phosphorus excretion rates) have been measured here avoiding the interference of food particles.

Despite the scarcity of data, it is evident that heart-beat rates are highly temperature dependent. The value of Q_{10} for *D. pulex* is similar to that observed by MAC ARTHUR & BAILLIE (1929) for *D. magna* for a similar temperature range. Regarding the differences between A and Q conditions for the temperatures tested (Fig. 2), the enhancement of heart rates due to turbulence seems to be constant and independent of temperature. The percent increase observed when exposing the organisms to turbulence is therefore equivalent to a rise of 1.0 °C in the environmental temperature. In cladocerans, environmental temperature rises lead to accelerated development rates and reduced life-span, for longevity and heart rates do vary inversely (MAC ARTHUR & BAILLIE, 1929; SHAW & BERCAW, 1962). This is in agreement with the enhanced development rates of copepods in laboratory microcosms under turbulence (ALCARAZ *et al*., 1988), although no direct data on copepod heart-beat rate or longevity are available.

Excretion rates for both species of *Acartia* indicate turbulence-enhanced ammonia excretion rates. This coincides with observations on other calanoid species from natural systems (mixed and frontal zones in comparison to stratified ones; HARRIS & MALEJ, 1986), and could explain the differences in ammonia concentration between turbulence and non-turbulence observed in laboratory microcosms (ALCARAZ *et al.*, 1988). The response to turbulence appears more important in *A. margalefi* than in *A. clausi* (32 % against 26 % increase), and probably are the consequence of differences in the scale of turbulence perceived due to the different size of both species. Specific excretion rates for *A. clausi* are similar to those reported by MAYZAUD (1973, 1976) and NIVAL (1974) for the same species and similar temperature conditions.

Phosphorus and ammonia excretion rates were well correlated, although phosphorus appeared to be more affected by turbulence than is ammonia. Despite their higher relative increase (about twice that of ammonia), absolute values of phosphorus excretion rates in A and Q conditions were low in comparison with the N:P excretion ratios observed by ALCARAZ (1988) in mixed zooplankton.

The results indicate that turbulence accelerates metabolism. The organismal response must be very rapid, for heart rate responds to changes from calm to turbulence conditions and vice-versa in 15 min at most, the acclimatation time between changes on turbulence conditions. Similar rapid responses to successive switching from turbulence to calm conditions have been observed in the frequency of fast-swimming events, a high-energy consuming behaviour, in *Centropages* (COSTELLO *et al.*, 1990).

Whatever the behavioural or physiological mechanism affected by turbulence and responsible for the metabolic enhancement, the perceiving structures could be e-type receptors (BARRIENTOS, 1980; PAFFENHÖFER & STEARNS, 1988; GILL & CRISP, 1985), which correspond to the scale range at which random turbulent velocity fluctuations can have hydrodynamical importance (LAZIER & MANN, 1989). Apart from differences in the scale of turbulence generated by the different stirring methods (oscillation of grids, shakers, vibration), the relative scarcity of this type of receptors in *Daphnia* could be the reason for the relative lower metabolic enhancement in cladocerans as compared to copepods.

Our results point towards an accelerating effect of turbulence on metabolism comparable to that of increased temperatures. Effective measurements of turbulence and more data regarding its behavioural and physiological effects are needed in order to understand better the global importance of the physical-biological interactions in aquatic ecosystems.

ACKNOWLEDGEMENTS

The ideas and comments of Prof. R. Margalef have been a constant source of inspiration and encouragement. This study is dedicated to him. The authors are indebted to M. Estrada and C. Duarte for their critical reading of the manuscript, and to J. Armengol, who provided the *Daphnia* specimens. This work was supported by grants PR84-0061 and MAR-88-0252 of the CICYT and the Institut de Ciències del Mar, CSIC.

REFERENCES

ALCARAZ, M. 1988. Summer zooplankton metabolism and its relation to primary production in the Western Mediterranean. In: *Océanographie pélagique méditerranéenne*. (H.J. Minas & P. NivaL, eds.) *Oceanol. Acta*, N. sp. 9 : 185-191.

ALCARAZ, M., PAFFENHÖFER, G.A., & STRICKLER, J.R. 1980. Catching the algae: a first account of visual observations of filter-feeding calanoids. In: *The evolution and ecology of zooplankton communities*. (W.C. Kerfoot, ed.):

241-248. Univ. Press of New England.

ALCARAZ, M., SÁIZ, E., MARRASÉ, C. & VAQUÉ, D. 1988. Effects of turbulence on the development of zooplankton biomass and copepod populations in marine microcosms. *Mar. Ecol. Prog. Ser.,* 49: 117-125.

ALCARAZ, M., ESTRADA, M., & MARRASÉ, C. 1989. Interaction between turbulence and zooplankton in laboratory microcosms. In: *Proceedings 21st. E.M.B.S.* (R.Z. Klewoski, E. Styczynska-Jurewicz & L. Falkowski, eds.): 191-204. Ossolineum. Wroclaw.

BARRIENTOS, Y. 1980. *Ultrastructure of sensory units of the first antennae of copepods.* M. Sc. thesis. Univ. of Ottawa.

COSTELLO, J.H., STRICKLER, J.R., MARRASÉ, C., TRAGER, G., ZELLER, R. & FREISE, A.J. 1990. Grazing in a turbulent environment: behavioral response of a calanoid copepod, *Centropages hamatus. Proc. Natl. Acad. Sci. USA.* 87: 1648-1652.

ESTRADA, M., ALCARAZ, M. & MARRASÉ, C. 1987 a. Effects of turbulence on the composition of phytoplankton assemblages in marine microcosms. *Mar. Ecol. Prog. Ser.,* 38: 267-281.

ESTRADA, M., ALCARAZ, M. & MARRASÉ, C. 1987 b. Effect of reversed light gradients on the phytoplankton composition in marine microcosms. *Inv Pesq.,* 51: 443-458.

EVANS, G.T. 1989. The encounter speed of moving predator and prey. *J. Plankton Res.,* 11: 415-417.

GILL, C.W. & CRISP, D.J. 1985. Sensitivity of intact and antennule amputated copepods to water disturbance. *Mar. Ecol. Prog. Ser.,* 21: 221-227.

GRASSHOFF, K., EHRHARDT, M. & KREMLING, K. (eds.). 1983. *Methods of seawater analysis.* Verlag Chemie. Weinheim.

HARRIS, R.P., MALEJ, A. 1986. Diel pattern of ammonium excretion and grazing rhythms in *Calanus helgolandicus* in surface stratified waters. *Mar. Ecol. Prog. Ser.,* 31: 75-85.

HARTE, J., LEVY, D., REES, J. & SAEGEBATH, E. 1980. Making microcosms an effective assessment tool. In: *Microcosms in ecological research.* (J.P. Giesy, ed.): 105-137. Technical Information Center, US Dept. of Energy, Springfield, Virginia

INGLE, L. WOOD, T.R. & BANTA, A.M. 1937. A study of longevity, growth, reproduction and heart rate in *Daphnia longispina* as influenced by limitations in quantity of food. *J. Exp. Zool.,* 76: 325-352.

KIØRBOE, T., KASS, H., KRUSE, B., MOHLENBERG, F., TISELIUS, P. & AERTEBJERG, G. 1990. The structure of the pelagic food web in relation to water column structure in the Skagerrak. *Mar. Ecol. Prog. Ser.,* 59: 19-32.

KOROLEFF, F. 1970. Direct determination of ammonia in natural waters as indophenol blue. *Inter. Counc. Explor. Sea, CM papers and Reprints, C9: Information on techniques and methods for sea water analysis,* Interlab. Rep. 3: 19-22.

LAZIER, J.R.N. & MANN, K.H. 1989. Turbulence and the diffusive layers around small organisms. *Deep Sea Res.,* 36: 1721-1733.

LEGENDRE, L. 1981. Hydrodynamic control of marine phytoplankton production: the paradox of stability. In: *Ecohydrodynamics.* (J.C.J. Nihoul, ed.): 19-207. Elsevier. London

LEGENDRE, L. & DEMERS, S. 1984. Towards dynamic biological oceanography and limnology. *Can J. Fish. Aquat. Sci.,* 41: 2-19.

MAC ARTHUR, J.W. & BAILLIE, W.H.T. 1929. Metabolic activity and duration of life. *J. Exp. Zool.,* 53: 243-268.

MACKAS, D.L., DENMAN, K.L., & ABBOT, M.R. 1985. Plankton patchiness: biology in the physical vernacular. *Bull. mar. Sci.,* 37: 652-674.

MARGALEF, R. 1974. *Ecología.* Omega. Barcelona.

MARGALEF, R. 1978. Life-forms of phytoplankton as survival alternatives in an unstable environment. *Oceanol. Acta,* 1: 493-509.

MARGALEF, R., 1982. Instabilities in ecology. In: *Stability of thermodynamic systems.* (J. Casas-Vázquez, J. & G. Lebon, eds.): 295-306. Springer. New York

MARRASÉ, C., 1986. *Experimentos multifactoriales con plancton marino en microcosmos.* Ph. D. Thesis, Univ. of Barcelona.

MARRASÉ, C., COSTELLO, J.H., GRANATA, T. & STRICKLER, J.R. 1990. Grazing in a turbulent environment: Energy dissipation, encounter rates, and efficacy of feeding currents in *Centropages hamatus. Proc. Natl. Acad. Sci USA,* 87: 1653-1657.

MAYZAUD, P. 1973. Respiration and nitrogen excretion of zooplankton II. Studies on the metabolic characteristics of starved animals. *Mar. Biol.,* 21: 19-28.

MAYZAUD, P. 1976. Respiration and nitrogen excretion of zooplankton IV. The influence of starvation on the metabolism and the biochemical composition of some species. *Mar. Biol.,* 37: 47-58.

NIVAL, P., MALARA, G., CHARRA, R., PALAZZOLI, I. & NIVAL, S. 1974. Étude de la respiration et de l'excrétion de quelques copépodes planctoniques (Crustacea) dans la zone de remontée d'eau profonde des côtes marocaines. *J. exp. mar. Biol. Ecol.,* 15: 231-260.

PAFFENHÖFER, G.A., & STEARNS, D.E. 1988. Why is *Acartia tonsa* (Copepoda: Calanoida) restricted to nearshore environments? *Mar. Ecol.,* 42: 33-38.

PAFFENHÖFER, G.A., STRICKLER, J.R. & ALCARAZ, M. 1982. Suspension feeding by herbivorous calanoid copepods: A cinematographic study. *Mar. Biol.,* 67: 193-199.

PAVLOVA, E.V. & MINKINA, N.I. 1983. Evaluation of basal energy metabolism in marine copepods. *Doklady Biol. Sci.,* 265: 406-408 (in russian).

PÉREZ, K.T., MORRISON, G.M., LACKIE, N.F., OVIATT, C.A., NIXON, S.W., BUCKLEY, B.A. & HELTSHE, J.F. 1977. The importance of physical and biotic scaling to the experimental simulation of a coastal marine ecosystem. *Helgolander Wiss. Meeresunters.,* 30: 14-162.

PILSON, M.E.Q. & NIXON, S.W. 1980. *Marine microcosms in ecological research.* Technical Information Center, US Dept. of Energy, Springfield, Virginia:724-741.

REDFIELD, A.C., KETCHUM, B.H. & RICHARDS, F.A. 1963. The influence of organisms on the composition of seawater. In: *The Sea,* 2. (M.N. Hill, ed.): 26-77. Wiley. New York.

ROTHSCHILD, B.J. & OSBORN, T.R. 1988. Small-scale turbulence and plankton contact rates. *J. Plankton Res.*, 10: 465-474.

SHAW, R.F. & BERCAW, B.L. 1962. Temperature and lifespan in poikilotherm animals. *Nature,* Lond., 196: 454-457.

SOKAL, R.R. & ROHLF, F.J. 1969. *Biometry.* Freeman. San Francisco.

SPAARGAREN, D.H. 1977. On the metabolic adaptation of *Carcinus maenas* to reduced oxygen tensions in the environment. *Nether. J. Sea Res.*, 11: 325-333.

SPAARGAREN, D.H. & ACHITUV, J. 1977. On the heart rate response to rapid temperature changes in various marine and brackish water crustaceans. *Nether. J. Sea Res.*, 11: 107-117.

STRICKLER, J.R. 1975. Swimming of planktonic *Cyclops* species (Copepoda, Crustacea): pattern, movements and their control. In: *Swimming and flying in nature,* 2. (T.Y.T. Wu, C.J. Brokaw & C. Brennen, eds.): 599-613. Plenum. New York.

TETT, P. & EDWARDS, A. 1984. Mixing and plankton: an interdisciplinary theme in oceanography. *Oceanogr. Mar. Biol. Ann. Rev.,* 22: 99-123.

Homage to Ramon Margalef; or, Why there is such pleasure in studying nature (J. D. Ros & N. Prat, eds.). 1991. *Oecologia aquatica*, 10: 145-152

PRODUCTION-CONSUMPTION BUDGET IN AN ESTUARINE BAY: HOW ANOXIA IS PREVENTED IN A FORCED SYSTEM

JORDI CAMP[1], JAVIER ROMERO[2], MARTA PÉREZ[2], MONTSE VIDAL[1], MAXIMINO DELGADO[1] & AMPARO MARTÍNEZ[1]

[1] Institut de Ciències del Mar (CSIC). Pg. Nacional s/n. 08039 Barcelona. Spain.
[2] Departament d'Ecologia. Facultat de Biologia. Universitat de Barcelona. Av. Diagonal, 645. 08028 Barcelona. Spain.

Received: December 1990.

SUMMARY

Althought the generation of anoxic situations is a well known fact in estuarine bays, in the southern embayment of the Ebro delta (NW Mediterranean), the Alfacs Bay, oxygen depletion is a rare event. In this paper we examine the processes preventing anoxia in the system through a consumption/production budget. From calculations based upon annual mean values, we conclude that the bay is slightly heterotrophic, that is, that total consumption exceeds autochtonous production by only 10%.

But heterogeneity in both space and time results in a high variance around the mean used in our budget. We have examined the main components of this variance (stratification of the water column, horizontal heterogeneity, phytoplankton production in pulses, etc.) and some of them have been incorporated in a simple box-flux model. From this model, we conclude that only the deep water layer is strongly heterotrophic and that its sources of oxygen are vertical mixing and renovation (50%), diapycnal eddy diffusion (25%) and local production (25%).

The generation of anoxia must be driven by the co-occurrence of events associated with oxygen depletion and/or limitation of oxygen transport (i.e., calm weather period after a phytoplankton pulse; lack of estuarine circulation during a long calm weather period, etc.), what we call the covariance between processes. Such covariance is in general low, and in consequence anoxia must be considered a low probability event.

Comparing Alfacs Bay with other estuaries in the framework of the proposed conceptual model, three features emerge that may be relevant in order to prevent anoxia: low organic loading (produced plus imported); active estuarine circulation forced by high wind frequency; degradation of the organic matter mainly by benthic consumers.

KEY WORDS: Estuaries, eutrophication, anoxia, oxygen budget.

INTRODUCTION

The estuaries of developed countries have suffered dramatic changes in the last few decades. While they preserve the characteristics of transit systems between a continent that contributes and an ocean that dilutes, the processes have accelerated. Human action increases energy and matter flows, upsets the natural equilibrium and leads to situations that may compromise the existence of the estuarine ecosystem in its current form.

Each estuary has its own features that determine its functioning and define the extent of the possible fluctuations in the system's variables of state; beyond certain limits of these variables, important qualitative changes take place. Standing out amongst these are the processes of eutrophication, implying local or overall oxygen depletion, with consequences both for the general functioning of the system (loss of biological diversity, fall in production, etc.), and for activities of economic interest (fishing, aquaculture).

The current study is the first attempt at a functional synthesis, undertaken after more than eight years of research in the Alfacs Bay (Ebro delta, Tarragona, N.E., Spain), and is based on the results of diverse studies either published or in preparation, and on data from doctoral theses either already completed or in the process of elaboration (DELGADO, 1986; PÉREZ, 1989; MARTÍNEZ, 1990; VIDAL, 1991; CAMP, 1991).

This database represents precise analytical and quantitative knowledge of the most relevant processes in the ecological dynamics of the bay (primary production, nutrient cycling, etc.) and, in spite of the gaps for a complete and comprehensive general model, it permits a functional approach to the ecosystem of the bay.

The current study proposes a simple model that describes the overall ecological dynamics of the bay in terms of a budget between autotrophic and heterotrophic processes, using dissolved oxygen concentration as the main system descriptor. The model explains the series of field values obtained from repeated measurements of this variable, gives an indication of the processes that maintain the oxygen concentrations relatively high and allows some speculative discussion about the possible limits of these mechanisms and some peculiarities of interest in the bay under study.

THE STUDY AREA

The zone is a semi-confined mass of sea water, with a surface area of 49 km^2 and an average depth of 3.13 m ; the mouth of the bay is 2 km width (Fig. 1). The edges are surrounded by a shelf or platform (18 km^2), which falls in a gentle slope from 0 to 1.5 m, with an average depth of 0.64 m and which connects via a more pronounced slope to the muddy central basin, with an average/maximum depth of 4.16/6.5 m and an area of 31 km^2. The northern part of the shelf occupies 6 km^2, has a silty sediment and is almost totally covered (90 %) by dense beds of macrophytes, mainly *Cymodocea nodosa*; sediments of the eastern and southern sides of the shelf, which together extend over 12 km^2, are sandy, with a vegetation cover of 10 % (PÉREZ & CAMP, 1986).

The bay receives freshwater inputs from the northern edge of about $275x10^6$ m^3 per year between April and October, and much less during the rest of the year, a seasonality caused by the needs of rice cultivation. This water brings considerable quantities of inorganic nutrients and organic matter into the bay (CAMP & DELGADO, 1987; PRAT *et al.*, 1988; MUÑOZ, 1989).

The hydrographic structure of the bay habitually presents a salinity dominated stratification, with a superficial layer (0 to 2-3 m depth) of low salinity (30-35 ‰) and a more or less pronounced outward movement, and a deep, salty (36-38 ‰) layer with an inward movement. Total

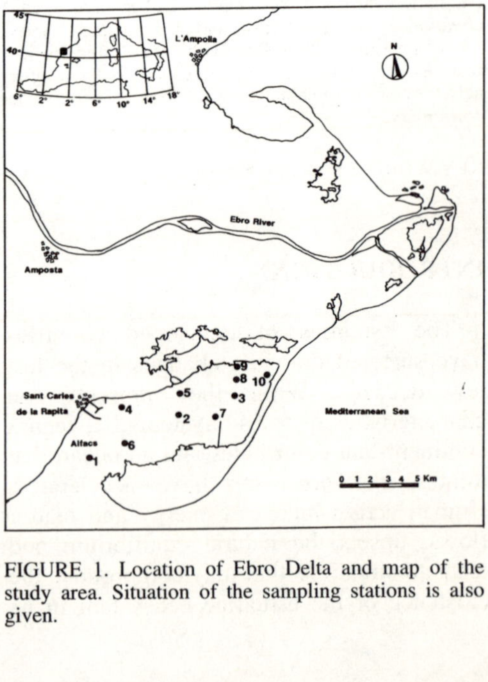

FIGURE 1. Location of Ebro Delta and map of the study area. Situation of the sampling stations is also given.

TABLE I. Some morphological and hydrological features of the Alfacs Bay relevant to the present work.

SURFACES (km^2)	
Total	49
Platforms	18
Northern	6
Southern and Eastern	12
Central basin	31
VOLUMES OF WATER (m^3)	
Total of the bay	153×10^6
Annual freshwater inputs	275×10^6
DEPTHS (m)	
Average of the bay	3.13
Maximum of the bay	6.50
Average of the platforms	0.64
Average of the central basin	4.17
HYDROLOGICAL FEATURES	
Water temperature (annual mean)	18.2 °C
Water temperature (April-November)	22.0 °C
Salinity 0-2 m (annual mean)	35 $g.l^{-1}$
Salinity 2-4 m (annual mean)	37 $g.l^{-1}$

mixing of both layers is rare and occurs in the case of very strong north-western winds, which blow with an average frequency of 12 days and a duration of 2-3 days (FONT, 1986). Re-stratification occurs within 48 hours after the wind dropping. This implies an export of the mixed water, which is substituted by sea water from outside the bay. In the absence of strong wind there is still an exchange between the bay and the open sea, either following a sequence of dilution with a periodicity of ca. 10 days (CAMP & DELGADO, 1987) or through a typical estuarine circulation with a volume exchange of $5\text{-}15 \times 10^6$ m^3 day^{-1}. Some of the general features of the bay are summarized in Table I.

FIELD DATA FOR THE MODEL

Primary production in the bay (Table II) is mainly due to phytoplankton, the activity of which does not show apparent seasonality. This production is higher in the upper, diluted nutrient-rich and well illuminated layer (DELGADO, 1986). Microphytobenthos production is more intense in the shallow water of the shelf and does not show any marked seasonality

TABLE II. Organic carbon inputs to the Alfacs Bay. The data are expressed on a surface basis (in mg C m^{-2} day^{-1}, averaged throughout the year) and as total inputs to the bay (in Tm C y^{-1}). From DELGADO (1986); PÉREZ (1989), MARTÍNEZ (1990) and unpublished data.

	mg C m^{-2} day^{-1}	Tm C y^{-1}
PHYTOPLANKTON		
Upper layer (0-2 m)	273	4874
Bottom layer (2-4 m)	64	725
MICROPHYTOBENTHOS		
Shallow waters (platforms)	96	631
Deep waters (central basin)	48	543
MACROPHYTES		
Seagrasses	1080	1643
Macroalgae	610	493
ALLOCHTHONOUS INPUTS		
Discharge channels (only POC)	62	1100

(DELGADO, 1989). Among macrophytes, the phanerogams show production maxima between March and October (PÉREZ, 1989), while macroalgae growth peaks between January and June (MARTÍNEZ, 1990). Other significant organic carbon inputs come through the water from irrigation channels (Table II).

An important part of the total input (produced plus imported) is consumed in the sediment, as reflected in its oxygen uptake rates (VIDAL *et al.*, 1989; see Table III).

The data for dissolved oxygen concentration come from diverse sources. One data set was obtained from chemical analysis (Winkler) and the other from polarographic sensors (Orbisphere models 2608-2609, Syland Simplair and WTW OXI-196).

In consequence, comparisons in absolute terms are difficult; but the several Winkler/sensor intercalibration experiments performed showed that although differences in absolute values are evident, the concentration gradients (e.g. following a vertical profile) are faithfully reproduced independently of the method used or the calibration of the sensor. This justifies our preference for emphasizing differences rather than absolute values, especially when data from different authors are presented

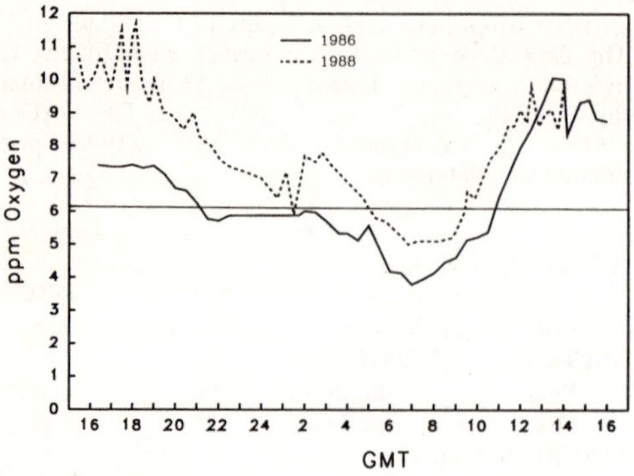

FIGURE 2. Daily evolution of oxygen concentration in the water over the shallow, northern platform, corresponding to two different years. Horizontal line represents an approximate value for oxygen saturation.

together.

Data on dissolved oxygen concentration can be grouped in the three following categories:

a) two daily cycles on the shallow shelves with continuous measurements by sensor (Station 10 in figure 1).

b) vertical profiles distributed throughout the bay and measured in three consecutive years by different authors (Fig. 2; Table IV). The frequency distribution of the surface/bottom differences (that is, the difference in oxygen concentration between 0 and 4 metres) appears in figure 3.

c) a smaller set of chemically determined values, which serve as a reference for items (a) and (b) (see Table IV).

There is a lack of systematic and reliable data for consumption in the water column (BOD). The only available information indicates that the BOD_5 values are in general lower than 2 ppm (C. Lleti, pers. comm. and J. Camp, unpublished data).

TABLE III. Oxygen consumption in the sediment, in g m^{-2} day^{-1}. From VIDAL *et al.* (1990) and VIDAL (1991).

	Minimum	Maximum	Average
Northern platform	1.76	2.30	1.80
Southern platform	0.23	1.02	0.61
Central basin	1.23	1.84	1.54
Average for the bay			1.53

Likewise, the data from deep water consumption in short incubation experiments (3 hours) never indicates significant oxygen uptake (VIDAL, 1991), which, given the precision of the method, results in a maximum consumption of less than 1 ppm day^{-1}.

THE MEAN: AN ANNUAL OVERALL BUDGET

The first approach to this model is in terms of a production/consumption balance on an annual basis.

The primary production in the bay can be estimated at 8,900 Tm of C, while the external contribution from channels is 1,100 Tm of C in particulate form (Tables I and II). To these must be added the dissolved organic carbon (DOC), the magnitude of which, estimated according to sporadic analysis (MUÑOZ, 1990), oscillates between 1,500 and 3,500 Tm. Therefore, the total input (produced plus imported) of organic carbon oscillates between 11,500 and 13,500 Tm C y^{-1}.

The oxygen consumed in the sediment is 27,364 Tm O_2 y^{-1}. This can represent, in terms of organic matter consumed, between 7,885 and 10,287 Tm C y^{-1}, depending on the oxidation pathways involved, aerobic (106 C:138 O_2) or anaerobic (basically

TABLE IV. Data of oxygen concentration (ppm) at 0 and 4 m depth, averaged throughout complete studies for different stations, numbered as in figure 1. For Stations 1, 2 and 3, data from an annual cycle (12 months) by DELGADO (1986) obtained through Winkler titration. For Stations 9, 8 and 3, data from an annual cycle (16 months) by CAMP (1991), obtained by polarographic sensor. For Stations 4, 5, 6 and 7, data from an annual cycle (12 months) obtained by a polarographic sensor (unpublished data from Departament d'Agricultura, Ramaderia i Pesca). Vertical gradient (difference) is also given.

Station	1	2	3	9	8	3	4	5	6	7
0m	8.12	8.20	8.24	9.64	7.80	7.83	6.2	6.4	7.3	6.9
4m	7.88	7.52	7.27	-	6.60	6.82	5.5	5.4	6.7	6.3
Difference	0.24	0.69	0.97	-	1.20	1.01	0.7	1.0	0.6	0.6

sulphate reduction, in which case the molar ratio is close to 1 C:1 O_2).

These data show that the primary production of the bay and the carbon consumption in the sediment are of the same order of magnitude and give a positive balance between inputs (+) and outputs (-) ranging from 1,000 to 5,500 Tm C y^{-1}, the indetermination mostly being due to the scarcity of values of DOC. The excess of organic carbon may be exported towards the exterior of the bay, accumulated in the sediment or consumed in the water column. Even in the case of a total consumption of the maximum value of 5,500 Tm C in the water column, the average BOD_5 would not be over 1.5 ppm, which is consistent with the existing data.

The exportation depends on the exchange rate with the open sea. Calculations from successive balances of salt in distinct periods of time (CAMP & DELGADO, 1987) give an average exchange rate of 60 x 10^6 m^3 every ten days. Assuming that the concentration of total organic carbon (dissolved plus particulated, TOC), is close to 3.6 mg l^{-1}, and that the coastal sea water has a TOC concentration of the order of 1.5 mg l^{-1}, a net annual export around 4,000 Tm C can be estimated. The maximum surplus carbon

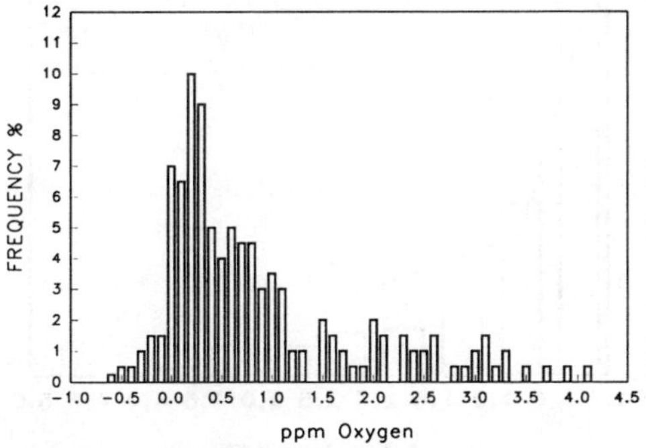

FIGURE 3. Frequency distribution of the oxygen concentration depth gradient observed in data from several years. The gradient is assimilated to the difference between the oxygen concentration at 0.5 m and the one at 4 m. Total number of paired observations, n= 188.

loading to the bay is therefore of 1,600 Tm, which can either be consumed in the water column (yielding an average BOD_5 of 0.5 ppm) or increase the carbon pool in the sediment. In an ideal situation in which daily production and consumption values would be equal to the respective average values throughout the year, net loss of oxygen to the atmosphere would not be expected, and therefore the creation of oxygen deficits would only be related to the balance between input by channels and export to the open sea. As shown, this balance yields a maximum surplus of 500 Tm C inside the bay, the consumption of which would give daily oxygen deficits of the order of 0.05 ppm, easily balanced by influx from the atmosphere. In conclusion, and regarding the mean, the situation reveals very slight heterotrophy, if any.

THE VARIANCE: SPACE-RELATED COMPONENTS

Obviously, the balance presented gives a very global picture but with a scarce predictive value, as the processes involved show strong heterogeneity in both space and time. To incorporate both horizontal and vertical spatial heterogeneity into our model, the budget will be recalculated dividing the bay into subcompartments:

(1) shallow northern platform, with a high macrophytic cover;

(2) shallow southern platform, with a low macrophytic cover;

(3) central basin:

(3.1) upper layer, mostly productive due to phytoplankton activity;

(3.2) lower layer, mostly degradative due to sediment activity;

The two layers are separated by a pycnocline, mostly a halocline, situated at 2 m depth.

The water over the platforms is supposed to be in conditions close to the annual average (water temperature 20 ^oC, salinity 35 ‰, photoperiod: 12 hours, average

TABLE V. Daily oxygen budget in the shallow platforms of the bay. Unless otherwise stated, data are in mg O_2 m^{-2}

	Northern platform		Southern platform	
	Day	Night	Day	Night
Macrophyte production	3340	0	370	0
Microphytobenthos prod.	362	0	362	0
Phytoplankton production	340	0	340	0
Macrophyte respiration	-250	-250	-28	-28
Sediment uptake	-880	-880	-305	-305
Daily change in [O2], ppm	+4.55	-1.77	+1.15	-0.52
Daily budget, ppm	+2.78		+0.63	

TABLE VI. Daily oxygen budget in the central basin. Oxygen consumption in the water column has not been considered. Data, unless otherwise stated, are in mg O_2 m^{-2}

	Day	Night
Upper layer (0-2 m)		
Phytoplankton production	1050	0
Daily change (ppm)	+0.5	0
Deep layer (2-4 m)		
Phytoplankton production	249	0
Microphytobenthos production	181	0
Sediment uptake	-770	-770
Daily change (ppm)	-0.17	-0.38
Net daily change	-0.55	

values of phytoplankton and benthic activities). Production data are taken from Table II; respiration of macrophytes is 15% of its production (PÉREZ, 1989). Consumption in the water column is not considered, neither is exchange with the atmosphere nor exchange with the central water mass. The resulting budget is shown in Table V. Regarding the central basin, the surface (0-2 m) water mass is assumed to be in conditions similar to those of the water over the platforms. The deep (2-4 m) layer has a salinity of 37 ‰. Consumption/ production values are the yearly averages (see Tables II and V). The budgets of oxygen per m^2 of basin surface are shown in Table VI.

The shelf budget is clearly positive, and the excess of oxygen is likely to be exported to the atmosphere. The iteration of the basin balance presented in Table VI will lead to the exhaustion of the oxygen in the deep zone in 12-13 days if the consumption in the sediment is independent of its concentration in the water column, or slightly more if we consider that the consumption is depressed below 3 ppm (VIDAL, 1991). Since generalized anoxic situations have never been detected (Fig. 3), an alternative oxygen supply to the deep

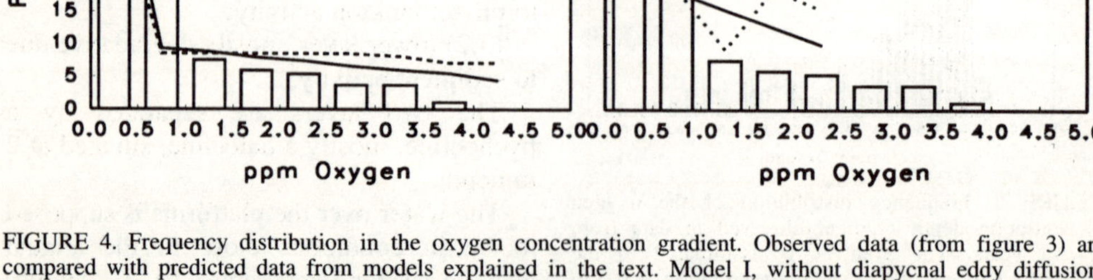

FIGURE 4. Frequency distribution in the oxygen concentration gradient. Observed data (from figure 3) are compared with predicted data from models explained in the text. Model I, without diapycnal eddy diffusion; Model II, including diapycnal eddy diffusion. A certain variance (high, dashed line; low, continuous line) has been assumed for the strong winds frequency.

zone, mainly through hydrodynamic processes, must exist. Two possibilities can be considered: i) water column mixing and/or exchange with the open sea, and ii) eddy diffusion through the pycnocline.

The bay/open sea exchange is controlled by two processes: dilution by fresh water (period around 10 days) and forced vertical mixing due to strong winds (period 10-12 days, duration 2 days). Although the first preserves stratification, the layer of deep (2-4 m) water is substituted by oxygen-saturated surface seawater. The second process implies that the water column loses its stratification, giving vertical profiles of oxygen concentration close to saturation values.

A simple model based on consumption data of Table VI was prepared to predict the oxygen concentration gradient frequency under these two hypotheses:

i) dilution and vertical mixing operate with exactly the same frequency throughtout the year. This means that a sequence of 4 days of vertical homogeneity (2 days of disturbance plus 2 days of re-stratification) followed by a fall of 0.6 ppm day^{-1} during 8 more days would be obtained.

ii) there is a certain variance in the time lag between two consecutive vertical mixing episodes, which means that the dilution cycle can also act.

The results are shown in figure 4a; these predictions clearly do not agree with the observed values, which suggest transport throught the pycnocline.

The magnitude of this process can be summarized by the coefficient of diapycnal eddy diffusion, k, but its estimation raises serious difficulties. k is related to the vertical density gradient and to the energy dissipation rate. LEDWELL & WATSON (1988) discuss an allometric relationship between N (the Brunt-Väisälä frequency) and k, whose parameters bear an error of at least one order of magnitude. From our field values of N, we obtain values of k ranging from 0.1 to 0.01 $cm^2\ s^{-1}$.

A different approach may be followed based on the salt balance of the bay. The average salinity of the deep layer is 0.8 ‰ less than the water from the open sea and 2‰ less than the surface water (average of 26 profiles in a stratified situation). This implies an exchange of the deep layer with the surface layer of 0.5 $m^3\ m^{-2}$ over an average distance of 2 meters every 8-12 days, which can be translated into a diffusion coefficient k of 0.011-0.015 $cm^2\ s^{-1}$. A realistic figure for k must be situated between 10^{-1} and 10^{-2} $cm^2\ s^{-1}$, coinciding with the very general magnitude proposed by several authors of 0.2 $cm^2\ s^{-1}$. Such values are obviously higher than the coefficients of passive diffusion, near 10^{-5} $cm^2\ s^{-1}$.

Oxygen concentration gradient frequencies were recalculated including the diapycnic eddy diffusion. As shown in figure 4b, the expected values become much closer to the field measurements.

Consequently, a tentative summary of the sources of oxygen to fit the deep layer demand can be summarized as in figure 5. As shown there, the deep layer is strongly

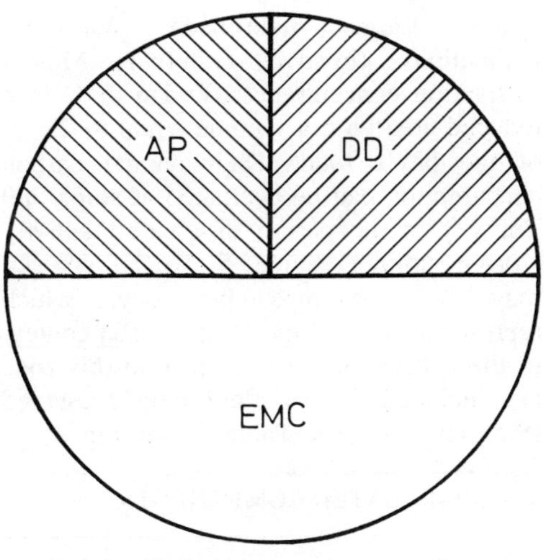

FIGURE 5. Relative contribution of the different processes oxygenating the deep water of the bay. AP, autochtonous production; DD, dyapicnal diffusion; EMC, exchange/mixing cycle.

heterotrophic (75 %) and the main oxygen supply pathway is supported by the mixing/exchange cycle.

Regarding the shelves, the day is not a good time scale for testing the model, as there are no net increases in oxygen concentration at the end of a 24 hour cycle (Fig. 2), due to the loss of oxygen to the atmosphere, which can be estimated at 1,800 mg O_2 m^{-2} per day for the northern shelf and 403 mg O_2 m^{-2} day^{-1} for the southern. The model should be tested on an hourly basis, but in fact, as we will discuss later, the oxygen dynamics of the shelves is scarcely relevant in the overall oxygen dynamics of the bay, due to the close interaction with the atmosphere. Moreover, some of these aspects, mostly concerning seagrass beds, have already been discussed in another study (PÉREZ & ROMERO, in press).

OTHER COMPONENTS OF THE VARIANCE

The relative agreement between the theoretical values obtained under the assumptions described and those observed in the field seems to indicate that the system works, at least in general trends, in accordance with our hypothesis. Even so, since other components of the variance have been neglected, certain doubts may persist on the validity of the model. Some of these possibilities are examined below.

a) SPACE-RELATED COMPONENTS

a1) Spatial heterogeneity in oxygen turbulent advection due to the hydrodynamic and morphological characteristics of the bay. In general, vertical mixing due to NW winds is more intense in the eastern part, but, conversely, estuarine circulation brings more oxygen to the western part of the bay.

a2) Important horizontal differences in planktonic production. Such differences indeed exist (DELGADO, 1986), but this fact is not reflected in the oxygen consumption of the sediment of the central basin (VIDAL, 1991). There is probably a certain homogeneisation of the phytoplankton biomass before its sedimentation.

a3) Spatial segregation of the production and the degradation of macrophytes. Zones of detritus accumulation have been not detected while in the seagrass beds stocks of necromass are built up to 50 % of the foliar biomass in September (M.A. Mateo, pers. comm.). Conversely, dead macroalgae accumulate on the shallow shelves, in which anoxia has been detected in a fine layer (a few cm) close to the sediment; the rest of the water column is oxygenated from the atmosphere. Likewise, these accumulations of dead thalli have also been found at the base of the slope of the northern shelf, although in relatively small areas.

a4) Localised allochthonous contributions. The sedimentation, precipitation and flocculation of materials take place in the northern shelf, which is reflected in the greater oxygen consumption of the sediment there.

a5) Large man-made consumer populations (bivalve cultures). Mussel cultivation concentrates 100 Tm of C in a reduced area with a volume equal to 2.2 % of the bay's total. Their annual carbon consumption can be evaluated at some 700 Tm y^{-1}. In the period of greatest activity (spring-summer) some 5 Tm of oxygen may be consumed per day, which represents 17 % of the total oxygen content of the cultivation area. Data from this zone are included in the field data presented (Stations 4 and 5 in Table IV and Fig. 1).

b) TIME-RELATED COMPONENTS

b1) Time-dependence of allochthonous organic matter inputs. The irrigation channels are open from April to October, inducing higher consumption rates in the water column and, finally, in the sediment.

b2) Pulses of phytoplankton production. Plankton production is likely to be coupled with the passing of meteorological fronts. Strong winds induce both water mixing and sediment resuspension and presumably a massive release of phosphorus (VIDAL, 1991), a nutrient which appears to have a limiting action. The expected result is a phytoplankton growth pulse a few days after the event, ending due to the exhaustion of the limiting nutrient or due to dilution/exportation forced by the following mixing and exchange event. The pulse results in a more active oxygen depletion in the deep zone due both to an enhanced particulated matter sedimentation and to a minimization of autochthonous oxygen production caused by the shadowing of the phytoplankton of the upper layer.

b3) Variability of the exchange frequency with the open sea, especially during the period in which the dilution cycle is not operating due to the closure of the irrigation channels (from October to April).

b4) Variability in the diapycnic diffusion rate, which depends on a wide number of factors ranging from the most local (wind conditions, intensity of stratification) to the most general (internal waves, etc.).

GENESIS OF EXTREME SITUATIONS: SOME REFLECTIONS ON COVARIANCE

The appearance of situations of anoxia will be linked to the degree of coupling amongst the variance components, that is the coincidence in space and/or time of processes enhancing oxygen consumption above the mean. Some of the possible coincidences are briefly discussed below.

In the case of the shelves, in early autumn oxygen budgets can be strongly negative due to: (i) high water temperatures; (ii) scarce macrophyte production; (iii) large amounts of dead leaves (30 g C m^{-2}) decomposing at a rate of 0.1 g C day^{-1} (PÉREZ, 1989); (iv) lack of phytoplankton production; (v) massive contributions from the channels leading to BOD_5 up to 5 ppm. Under these assumptions, daily consumption on the shelves could reach an extreme value of 4.3 ppm. Nevertheless, oxygen concentration does not reach low values since diffusion from the atmosphere, a process estimated at 3.5×10^5 mg O_2 m^{-2} day^{-1} atm^{-1} (HUTCHINSON, 1957) can largely balance the oxygen demand. Moreover, oxygen transport from the atmosphere is accelerated by the high frequency of soft winds (breezes) in the zone. For the same reason, oxygen depletion is not likely to occur at the superficial layer of the central basin, even in extreme conditions (without primary production and with a high BOD_5).

In the deep water layer the situation is far more complex. According to the proposed model, and using the minimum diapycnal diffusion coefficient, it would take 7 windless days of stratification to reach values less than 3 ppm, and 11-12 days to fall to 1 ppm, a value around which the parameter would stabilise. Using a maximum coefficient, 25-30 days of calm would be necessary to reach 3 ppm. Thus, and bearing in mind the "stand-by" times duration of the disturbance + restratification events, the 14-day cycle can lower the oxygen concentration to 50 % of saturation value, while the 10-day cycle can produce values of 60-70%. It seems likely, considering the superposition of both, that there is only a low probability of values of oxygen concentration below 60 % of saturation. Moreover, when the frequency of mixing events is minimal due to calm weather (summer), the channel flows are at their maxima, and the dilution cycle and/or estuarine circulation accelerates; conversely, when the channel flows are minimum (winter) the frequency of strong winds is high. The maximum probability of a lasting stability in the water column would correspond to winter anticyclonic situations

TABLE VII. Variance-covariance matrix between processes affecting oxygen depletion in the deep zone (see explanation in text). Positive covariances are represented by + symbols, negative covariances by - symbols. The number of symbols represents the magnitude of the variance (or covariance).
1) Sediment uptake
2) Organic allochthonous inputs
3) *In situ* production of phytoplankton (pulses)
4) Low frequency of deep water renewal through dilution cycle and/or estuarine circulation
5) Low frequency of vertical mixing due to the absence of strong winds
6) Low rate of dyapycnical diffusion
7) Absence of primary production in the deep zone

	1	2	3	4	5	6	7
1	+	+	+	-	.	.	.
2		++	+	---	.	+	.
3			+++	-	---	.	+
4				++	.	---	.
5					+	.	-
6						?	.
7							+

(typical in January-February). But this situation rarely lasts more than 15 days, and, in addition, the low water temperatures (8-10 ^{o}C) in this period would minimize the oxygen consumption rate in the water column.

Consumption in the sediment was not seen to be particulary variable (VIDAL, 1991; see Table III); therefore, departures from the mean in the consumption rates can only be expected in the water column. For a consumption rate of 1 ppm day^{-1} for 5 days an easily degradable organic carbon content of 2 mg l^{-1}, obviously corresponding to much higher TOC contents, would be required. Bearing in mind that the maximum POC values measured in the water of the bay were never higher than 1.2 mg l^{-1}, the above mentioned consumption is unlikely to be reached, except in clearly-defined situations. One of these could be due to a massive resuspension of sediment by a heavy storm, but this would represent a short-lived phenomenon, and besides would bring agitation and oxygenation. Another would correspond to a phytopalnkton bloom, but given that these blooms take place 5-7 days after the passing of the front and implying an oxygen increase in both surface and deep layers, before the enhancement of the consumption rates, the high oxygen demand induced by these blooms will only be relevant in a small part of the mixing/dilution cycles.

All these speculations can be formalised a little more, if necessary, by accepting that the degree of coupling or uncoupling between two processes, on which the oxygen exhaustion depends, can be expressed in terms of covariance. The functioning of the system can be summarised using a variance/covariance matrix, like the one tentatively proposed in Table VII. Since data are not available for a detailed quantification, this matrix represents a purely conceptual and qualitative construction. However, we can accept that if there is a set of processes linked by high and positive values of mutual covariances and independent of the rest, anoxia events are likely to occur, since deviation from the mean will be maximum. Conversely, if such subsets do not exist, the values will be kept closer to the average situations, through compensation among diverging processes.

Accepting the values proposed for the matrix in Table VII, the prediction for Alfacs Bay appears to be that anoxia is an unlikely event, given that the highest covariances are negative. The maximum probability comes from an intense phytoplankton pulse followed by a long calm period and minimal freshwater inputs, but this combination appears unlikely and, in any case, very localised in time.

An analogous matrix can be constructed paying attention to the space variance/covariance of the processes involved. The most positive covariances would appear between the allochthonous carbon inputs, the chlorophyll values (estimation of phytoplankton activity) and the stock of consumer organisms (mussel beds). These three variables are at their maxima in the northern zone of the bay. Since the shallow shelves are oxygenated from the atmosphere, the anoxia events are most likely to occur at the base of the slope and adjacent deep zones where materials from the shelves (including macroalgal and other organic detritus) can accumulate and where the mussel beds are installed. This is the place where anoxia situations can be generated at particular times of the year, which again takes us back to the time problem and leads us to consider the possibility of a three-dimensional variance/covariance matrix which integrates spatial and temporal elements.

CONCLUSION

The preceding discussion has shown how anoxia is prevented in the system under study. From a more general point of view and apart from emphasizing the close coupling between physical and biological processes, a central question arises: why are situations of anoxia more frequent in other estuaries? An answer based on three properties of the Alfacs Bay can be tentatively proposed; these are:

a) a relatively low rate of external input, from 40 to 70 g C m^{-2} y^{-1}, and a similarly low primary production (180 g C m^{-2} y^{-1});

b) a high wind frequency along with a surface to volume ratio high enough to transform efficiently the wind energy into estuarine circulation;

c) a pathway of carbon degradation and oxygen consumption mainly in the sediment, i.e., dominance of benthos over plankton in the catabolic processes.

These singularities are clearly reflected in the data compiled by NIXON (1981), summarized in figure 6. Considerations a) and b) refer above all to the relatively reduced size of the channel through which the energy of the ecosystem flows, which gives an average situation in which anoxia can be prevented by physical oxygenation processes. c) refers to properties of the subsystem controlling the heterotrophic

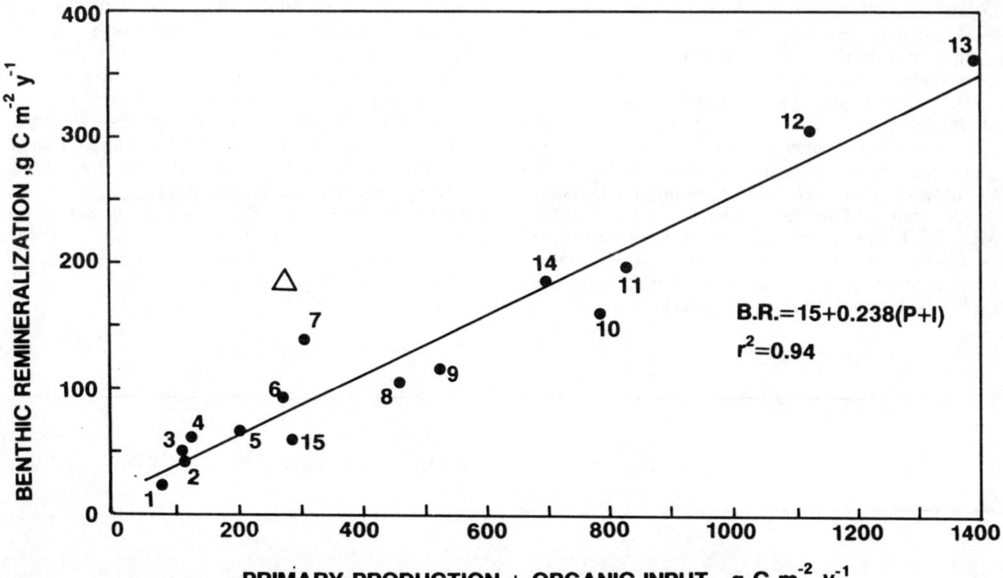

FIGURE 6. Benthic remineralization plotted against total organic carbon input for a series of different estuaries (from 1 to 15; data compiled by NIXON, 1981). Alfacs Bay is also represented (Delta).

aspects. Benthos can be considered as a subsystem of greater inertia and memory, which is able to make the oxygen consumption independent of environmental fluctuations, at least to a certain extent. This in turn produces a lowering in both the variance and the covariance, allowing the system to persist at values not excessively far from the average.

ACKNOWLEDGEMENTS

This work has benefited from CICYT grant MAR89-0556 and CE89-0017, and financial support from DARP-Generalitat de Catalunya (Conveni CSIC-IEC).

REFERENCES

CAMP, J. 1991. *Aproximación a la dinámica ecológica de una bahía estuárica mediterránea.* PhD Thesis. University of Barcelona. In prep.

CAMP, J. & DELGADO, M. 1987. Hidrografía de las bahías del Delta del Ebro. *Inv. Pesq.*, 51(3): 351-369.

DELGADO, M. 1986. *Ecología de las diatomeas costeras. Relación plancton-bentos.* PhD Thesis. University of Barcelona.

DELGADO, M. 1989. Abundance and distribution of microphytobenthos in the bays of Ebro Delta (Spain). *Estuar. Coast. Shelf Sci.*, 29: 183-194.

FONT, J. 1986. *La circulació general a la Mar Catalana.* PhD Thesis. University of Barcelona.

HUTCHINSON, G.E., 1957. *A treatise on Limnology.* Wiley. New York.

LEDWELL, J.R. & WATSON, A.J. 1988. The use of deliberately injected tracers for the study of diapycnal mixing in the ocean. In: *Small scale turbulence and mixing in the ocean* (J.J. Nihoul & B.M. Jamart, eds.): 11-21. Elsevier. Amsterdam.

MARTÍNEZ, M.A. 1990. *Estudio ecológico de algas efemerofíceas: papel en los flujos de materia y energía en un sistema estuárico (bahía de los Alfaques, Delta del Ebro, España).* Ph D Thesis. University of Barcelona.

MUÑOZ, I. 1990. *Limnologia de la part baixa del riu Ebre i els canals de reg: els factor fisico-químics, el fitoplàncton i els macroinvertebrats bentònics.* Ph. D. Thesis. University of Barcelona.

NIXON, S.W. 1981. Remineralization and nutrient cycling in coastal marine ecosystems. In: *Estuaries and Nutrients* (B.J. Neilson & E.L. Cronin, eds.):111-138. Hermana Press. Clifton. New Jersey.

PÉREZ M. 1989. *Fanerógamas marinas en sistemas estuáricos: producción, factores limitantes y algunos aspectos del ciclo de nutrientes.* Ph D Thesis. University of Barcelona.

PÉREZ, M. & CAMP, J. 1986. Distribución espacial y biomasa de las fanerógamas marinas de las bahías del Delta del Ebro. *Inv. Pesq.*, 50 (4): 519-530.

PÉREZ, M. & ROMERO, J. in press. Photosynthetic response to light and temperature of seagrass *Cymodocea nodosa* and its use in a preliminary carbon budget. *Aquatic Botany.*

PRAT, N., MUÑOZ, I., CAMP, J., COMÍN, F.A., LUCENA, J.R., ROMERO, J. & VIDAL, M., 1988. Seasonal changes in particulate organic carbon and nitrogen in the river and drainage channels of the Ebro Delta (NE Spain). *Verh. Internat. Verein. Limnol.*, 23:1344-1349.

VIDAL, M., ROMERO, J. & CAMP, J. 1989. Sediment-water nutrient fluxes: Preliminary results of in situ measurements in Alfaques Bay, Ebro River Delta. *Scientia Marina*, 53(2-3): 505-511.

VIDAL, M. 1991. *Estudios dobre la dinámica de nutrientes en sistemas estuáricos: flujos sedimento-agua de nutrientes y oxígeno.* Ph. D. Thesis. University of Barcelona.

Homage to Ramon Margalef; or, Why there is such pleasure in studying nature 157
(J. D. Ros & N. Prat, eds.). 1991. *Oecologia aquatica,* 10: 157-185.

PHYTOPLANKTON ASSEMBLAGES ACROSS A NW MEDITERRANEAN FRONT: CHANGES FROM WINTER MIXING TO SPRING STRATIFICATION

MARTA ESTRADA

Institut de Ciències del Mar (CSIC). Pg. Nacional, s/n. 08039 Barcelona. Spain.

Received: September 1990

SUMMARY

A series of stations located along the transect between Barcelona and the Balearic Islands were sampled to study the composition of the phytoplankton community, during June 1984, March 1985, June 1985 and June 1986. In March, the concentration of phytoplankton cells presented a maximum in the upper part of the water column. The Catalan front, near the continental slope of the Catalan coast side, separated a diatom-dominated assemblage in the coastal zone from a flagellate-dominated assemblage offshore. In the spring cruises, the most characteristic feature of the phytoplankton distribution was the presence of a deep chlorophyll maximum (DCM) which corresponded to increased phytoplankton cell numbers. The influence of the front on the phytoplankton distribution was well marked in early June (1984), but in the cruises carried out later in the season, warm waters extended over the frontal zone. The distribution of the phytoplankton assemblages showed some regularities: during the spring cruises, diatom assemblages with *Chaetoceros* spp., *Nitzschia* spp. and *Thalassiosira* spp. formed patches in the DCM layer, while a flagellate and dinoflagellate assemblage occupied the upper part of the water column. In May 1984 and May 1986 a third assemblage with *Coscinodiscus radiatus*, *Dictyocha fibula* and *Oxytoxum margalefi* formed a background population in the DCM level. Thus, in spite of high environmental variability, persistent groupings of species could be recognized.

KEY WORDS: Phytoplankton, hydrographic front, NW Mediterranean

INTRODUCTION

Two major hydrographic features of the Catalano-balearic Sea are the Catalan and the Balearic frontal systems (ESTRADA & MARGALEF, 1988; ESTRADA & SALAT, 1989; LA VIOLETTE *et al.,* 1990, TINTORÉ *et al.,* 1990). The Catalan front follows the continental slope on the peninsular side, and its main marker is a salinity gradient between low salinity waters on the shelf and higher salinities offshore. The Balearic front, on the Balearic Islands side, appears to be connected to the front off the northern coast of Corsica; it is associated with temperature and salinity gradients, but during the summer months its surface signature does not correspond to the deeper structures, due to the extension, over the upper layers of the water column, of relatively warm and low salinity waters of Atlantic origin.

During the stratification period, which lasts approximately from April to November (MARGALEF & BALLESTER, 1967), the phytoplankton distribution typically presents a deep chlorophyll maximum (DCM) and the central zone between the Catalan and Balearic fronts shows a persistent doming of the isopycnals which is often associated with increased phytoplankton biomass in the DCM (ESTRADA, 1985a). This doming has been considered as a divergence (OVCHINNIKOV, 1966; FONT, 1986), although recent studies have challenged this interpretation. The central zone appears to have lower stability than the marginal sides and may be strongly affected by instabilities from both frontal regions (WANG *et al.*, 1990).

The distribution of oceanographic parameters across the main fronts of the Catalano-balearic Sea has been studied in a series of oceanographic cruises covering a transect from the Catalan coast to the channel between Majorca and Minorca. Information on nanoplankton composition and primary production during the first two cruises, carried out from 12 to 28 July 1982 and from 30 June to 17 July 1983, can be found in ESTRADA (1985a, b). The microplankton distribution during these cruises and the following one, carried out from 17 to 30 May 1984, was studied by MARGALEF (1985) and MARGALEF & ESTRADA (1987). These authors showed that the phytoplankton distribution during summer was strongly conditioned by the vertical stratification patterns of the water column. High phytoplankton and chlorophyll concentrations were found near the coast and in a deep chlorophyll maximum which extended across the frontal zones, at depths of 60-80 m in summer and 40-70 m in May. The present study is based on data from May 1984 and from the three following cruises, carried out in late winter and spring 1985 and in spring 1986 (Table I), and attempts to draw a generalized picture of the main changes in the distribution of the nanoplankton communities during the transition from the mixing to the stratification period. Details concerning winter 1985 and spring 1984 cruises have been given by ESTRADA & MARGALEF (1988). The basic data for all the cruises can be found in PEPS (1986) and MASÓ & PEPS (1988).

MATERIAL AND METHODS

The distribution of the stations referred to in this paper is given in figure 1. Other details concerning the cruises are given in Table I. In general, the transects were completed in less than three days, except for the PEP 86 cruise and for the second transect of the FRONTS 3-85 cruise. In these cases, the stations were mainly devoted to biological experimentation work and each one was occupied for 24 hours; therefore, the resulting distributions should be interpreted with caution. Previous surveys had shown that the small-scale details of the phytoplankton biomass distribution presented great heterogeneity (ESTRADA, 1985a). However, the emphasis of the present work is on global patterns, and these tended to be persistent within the duration of each cruise.

Water samples were obtained by means of a Rosette sampler attached to a CTD system, during the first part of the FRONTS 85 cruise, and using 5-litre Niskin oceanographic bottles in the remaining cruises. Water was taken from nine depths at 10-20 m intervals between surface and 100 m depth in FRONTS 85 and at 10 m intervals between the surface and 110 m depth in the other cruises. Additional samples were taken at greater intervals down to 200 m (phytoplankton analyses) or 400 m (physico-chemical determinations). Information on the physico-chemical methodology is given in PEPS (1986) and MASÓ & PEPS (1988). Chlorophyll *a* concentration was determined fluoro- metrically (YENTSCH & MENZEL, 1963).

Subsamples of 120 ml of water for phytoplankton counts were fixed with concentrated Lugol's solution without acetic acid (MARGALEF, 1972) and stored in hermetically capped Pyrex bottles. 100 ml of water were sedimented in a composite chamber and examined using the inverted microscope technique. One transect of the chamber (approximately 1 ml of sample) was observed at 500 X to enumerate the smaller and more frequent organisms. An additional transect was examined at 125 X to count cells of intermediate size and the entire chamber bottom was scanned at 125 X to count the larger forms. Many organisms could not be classified at the level of genus or species and had to be lumped in categories such as "flagellates" (which also included small forms in which flagella could not be seen) and "unidentified dinoflagellates" (mainly

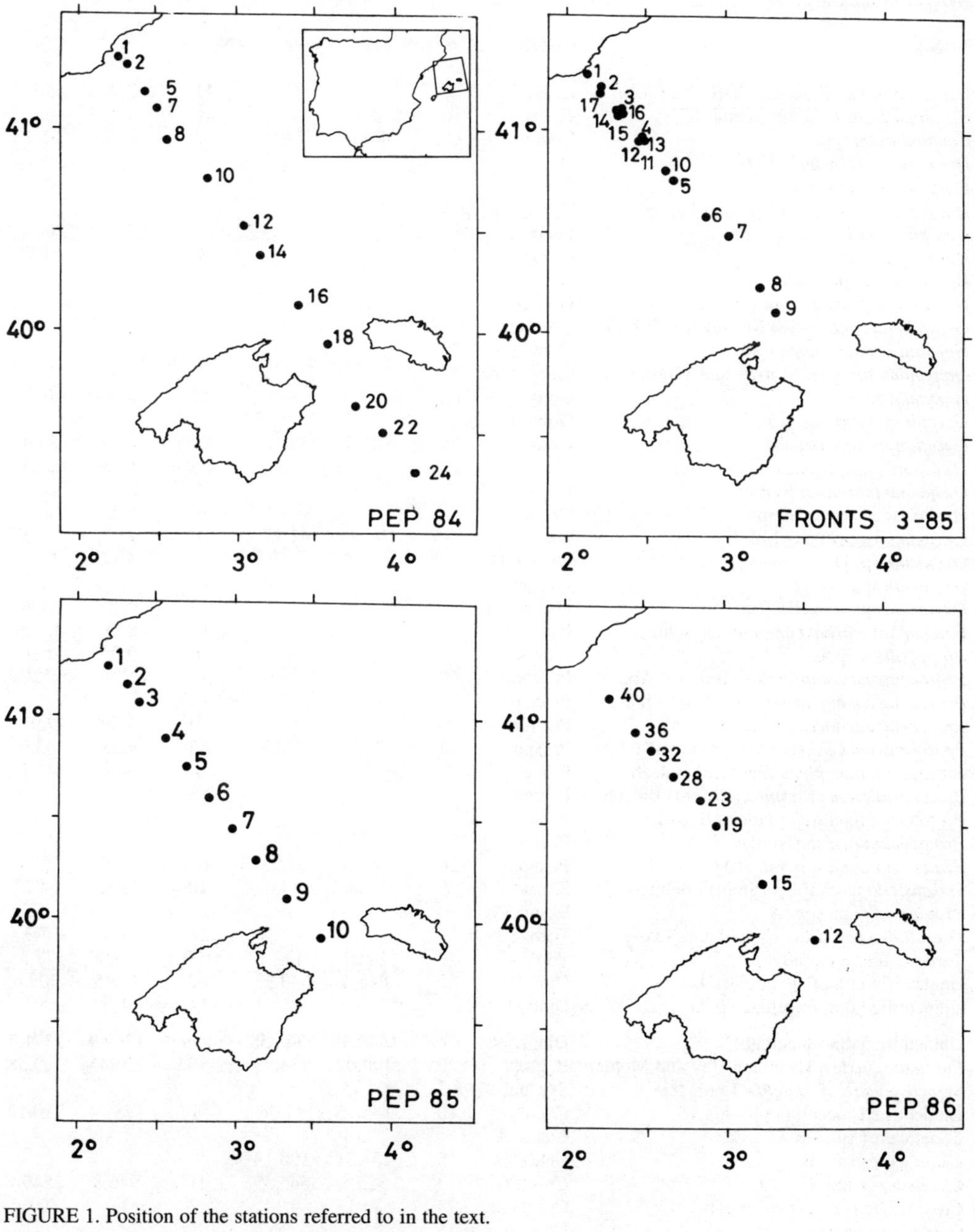

FIGURE 1. Position of the stations referred to in the text.

TABLE I. Sampling period and statistical data of the taxa used in the multivariate analyses corresponding to each cruise: m, mean number in cells ml^{-1}; s, standard deviation, computed for the number of occurrences (N. obs.). Code gives the key for the names used in figure 12.

CRUISE		FRONTS 3-85			PEP 84		
Transect (first-last station)	1-9		10-17			1-24	
Sampling period	15-16/3/85		24-26/3/85			16-19/5/84	
Number of samples			131			200	
Number of taxa selected			34			49	
TAXA	CODE	N. obs.	m	s	N. obs.	m	s
Ceratium furca (Ehrenb.) Clap. and Lacm.	C. fur.				11	2.9	5.8
Ceratium fusus (Ehrenb.) Dujardin	C. fus.				8	1.2	0.4
Cochlodinium spp.	Cochlo.	11	4.68	7.81	62	13.7	36.1
Dinophysis cf. *ovum* Schütt	Din. ov.						
Dinophysis acuta Ehrenb.	Din. acu.						
Dinophysis rotundata Clap. and Lachm.	Din. rot.						
Dinophysis (*Phalachroma*) spp.	Dinop.1				7	1.3	0.7
Dinophysis spp.	Dinop. 2				8	1.1	0.3
Dissodium (*Diplopsalis*) spp.	Dis.						
Goniodoma polyedricum (Pouchet) Jörg.	Gon. pol.						
Goniodoma sphaericum Murray and Whitting	Gon. sph.						
Gonyaulax polygramma Stein	Gony. pol.						
Gonyaulax turbynei Murray and Whitting	Gony. tur.						
Gonyaulax spp.	Gony.				22	1.4	0.7
Gyrodinium spp. (large)	Gony. l						
Gyrodinium spp. (small)	Gony. s	18	18.45	35.42	85	16.0	25.4
Mesoporos perforatus Gran (Lillick)	Mes. perf	19	18.29	25.98	23	10.6	23.9
Oxytoxum margalefi Rampi	Ox. mar.				7	20.9	40.4
Oxytoxum scolopax Stein	Ox. sco.				9	3.5	6.4
Oxytoxum variabile Schiller	Ox. var.	7	101.94	41.27	10	97.7	64.3
Oxytoxum sp. D	Ox. sp. D				9	3.44	6.22
Oxytoxum spp. (large)	Ox. spp. l						
Oxytoxum spp. (small)	Ox. spp. s				13	50.88	53.96
Palaeophalachroma unicinctum Schiller ?	Pal. uni.				11	8.78	13.29
Pronoctiluca sp.	Pronoc.				10	9.05	9.86
Prorocentrum compressum (Bailey) Abe	Pr. comp.				27	3.0	5.2
Prorocentrum maximum (Gourret) Schiller	Pr. max.						
Prorocentrum micans Ehrenb.	Pr. mic.				7	3.89	7.07
Prorocentrum (*Exuviaella*) spp. (small)	Pr. spp.	29	10.91	22.53	63	46.8	63.6
Protoperidinium bipes (Paulsen) Balech	P. bip.				8	34.8	47.7
Protoperidinium crassipes (Kofoid) Balech	P. cras.						
Protoperidinium steini (Jörg.) Balech	P. ste.						
Protoperidinium spp. (large)	P. spp. l						
Protoperidinium spp. (small)	P. spp. s	26	14.75	31.28	61	12.9	28.4
Scrippsiella trochoidea (Stein) Loeblich	S. troc.	7	11.04	14.6	10	18.82	34.27
Protoperidinium spp. A	P. spp. A						
Torodinium robustum Kofoid and Swezy	Toro. rob.				25	4.8	7.3
Torodinium sp. (small)	Toro. s	47	24.97	48.84	100	45.8	77.5
Unidentified dinoflagellate sp. B	Dino B	68	846.4	643.23	62	413.5	309.6
Unidentified dinoflagellate sp. D	Dino D				14	4.2	5.5
Unidentified dinoflagellates	Dinos.	130	1640.99	1348.58	195	1309.9	1371.7
Emiliania huxleyi Lohmann Hay and Mohler	E. hux.	22	196.02	134.83	13	223.45	309.58
Syracosphaera cf. *pulchra* Lohmann	Syr. pul.						
Unidentified coccolithophorids	Cocol.	10	155.75	71.06	8	175.24	210.12
Chaetoceros sp. A	Chaet. A						
Chaetoceros spp. B	Chaet. B	15	101.31	100.14			
Chaetoceros spp. C	Chaet. C	13	355.5	545.59	19	946.8	2545.1
Coscinodiscus radiatus Ehrenb.	Cosc. rad.	23	2.01	4.14	13	1.3	0.6
Coscinodiscus spp.	Cosc. spp	15	2.75	5			

TABLE I (Cont.)

Diploneis sp.	Diplo.	7	6.81	9.19			
Ditylum brightwelli Grunow	Dit. bri.	7	1.29	0.7			
Hemidiscus cuneiformis Wallich	Hem. cun.	12	1.5	0.65	11	1.3	0.5
Leptocylindrus danicus Cleve	Lep. dan.				7	25.6	51.0
Nitzschia (*Pseudonitzschia*) spp. (small)	Nitz. s						
Nitzschia (*Pseudonitzschia*) spp. (large)	Nitz. spp	14	39.73	66.21	14	87.2	178.3
Pleurosigma spp.	Pleuro.	31	6.04	20.15	7	6.77	9.13
Rhizosolenia alata Cleve	Rh. al.				7	13.1	18.6
Rhizosolenia imbricata Brightwell	Rh. imb.	11	8.16	12.24			
Rhizosolenia spp.	Rh. spp.				15	3.8	7.0
Schröderella delicatula (Per.) Pavillard	Sch. del.				8	10.35	13.65
Thalassionema nitzschioides Grunow	Thaln. nit.						
Thalassiosira spp. (small)	Thal. s	53	347	517.8	17	111.99	90.08
Thalassiosira spp. (medium)	Thal. m	36	33.72	87.59	21	21.7	39.8
Thalassiosira spp. (large)	Thal. l						
Thalassiothrix frauenfeldii (Grunow) Cleve and Möller	Thalx. fr.						
Unidentified centric diatoms	Diats.						
Pennate diatoms (small)	Pen. s	38	98.71	122.51	9	76.86	75.14
Pennate diatoms (large)	Pen. l	17	10.56	15.44			
Unidentified pennate diatom sp. A	Pen. A	42	14.7	36.03	52	24.75	65.39
Dictyocha fibula Ehrenb.	Dic. fib.	97	55.65	81.72	68	39.4	50.1
Dictyocha speculum	Dic. spec.						
Flagellates	Flag.	131	6450.4	15981.02	193	1975.9	3096.1
Calycomonas spp.	Calyc.	25	127.21	37.74	44	196.8	132.8
Phaeocystis sp.	Phaeo.	9	288.94	490.21			
Cryptomonads	Cryptos.	58	595.91	1060.62	42	483.1	1103.3
Leucocryptos?	Leuco.	36	170.29	113.36	77	401.4	445.9
Solenicola setigera	Sol. set.				9	2127.4	4780.6
Mesodinium sp.	Mesod.				12	7.8	12.7
Ciliates	Cili.	101	130.29	135.87			

CRUISE	PEP 85	PEP 86
Transect (first-last station)	1-10	12-40
Sampling period	1-2/6/85	17-24/6/86
Number of samples	129	107
Number of taxa selected	47	50

TAXA	N. Obs.	m	s	N. obs.	m	s
Ceratium furca (Ehrenb.) Clap. and Lacm.	11	1.36	2	9	2.78	3.08
Ceratium fusus (Ehrenb.) Dujardin	12	2	1.6			
Cochlodinium spp.	54	25.7	55.29	73	21.96	57.57
Dinophysis cf. *ovum* Schütt	8	3.75	6.53			
Dinophysis acuta Ehrenb.				10	1.5	0.92
Dinophysis rotundata Clap. and Lachm.	12	5.12	7.38	32	1.5	0.83
Dinophysis (*Phalachroma*) spp.				12	2.8	5.68
Dinophysis spp.				16	1.5	1.06
Dissodium (*Diplopsalis*) spp.				8	1.75	1.98
Goniodoma polyedricum (Pouchet) Jörg.				8	1.75	1.98
Goniodoma sphaericum Murray & Whitting				22	1.23	0.6
Gonyaulax polygramma Stein				8	1.25	0.43
Gonyaulax turbynei Murray & Whitting				9	12.73	26
Gonyaulax spp.				8	18.01	37.58
Gyrodinium spp. (large)				19	1.37	0.67
Gyrodinium spp. (small)	54	21.82	46.79	59	8.58	21.69
Mesoporos perforatus Gran (Lillick)	18	10.59	11.82	44	13.55	37.74
Oxytoxum margalefi Rampi	8	25.74	37.12	27	33.93	55.42
Oxytoxum scolopax Stein				10	3.05	6.15
Oxytoxum variabile Schiller	10	118.78	45.61	8	130.71	91.36

TABLE I (Cont.)

Oxytoxum sp. D						
Oxytoxum spp. (large)	9	16.01	35.87			
Oxytoxum spp. (small)	17	57	69.82			
Palaeophalachroma unicinctum Schiller ?						
Pronoctiluca sp.						
Prorocentrum compressum (Bailey) Abe				46	2.7	2.29
Prorocentrum maximum (Gourret) Schiller				9	26.86	72.42
Prorocentrum micans Ehrenb.						
Prorocentrum (*Exuviaella*) spp. (small)	63	117.99	367.04	53	9.35	68.19
Protoperidinium bipes (Paulsen) Balech	8	94.2	74.55			
Protoperidinium crassipes (Kofoid) Balech				12	1	0
Protoperidinium steini (Jörg.) Balech				9	1.56	0.5
Protoperidinium spp. (large)	8	1.25	0.43	11	1.18	0.57
Protoperidinium spp. (small)	59	61.72	90.48	64	5.93	15.66
Scrippsiella trochoidea (Stein) Loeblich	8	81.25	74.14			
Protoperidinium spp. A	13	165.41	180.83			
Torodinium robustum Kofoid and Swezy	26	5	7.13	51	12.03	61.8
Torodinium sp. (small)	35	37.61	62.28	37	18.7	30.9
Unidentified dinoflagellate sp. B	30	251.96	225.38	62	325.75	229.9
Unidentified dinoflagellate sp. D	10	7.94	8.75	21	5.27	6.52
Unidentified dinoflagellate sp. F	18	21.85	35.15	31	14.91	9.56
Unidentified dinoflagellates	126	2862.57	2442.46	107	2513.13	1981.89
Emiliania huxleyi Lohmann Hay and Mohler						
Syracosphaera cf. *pulchra* Lohmann	8	79.59	78.35			
Unidentified coccolithophorids						
Chaetoceros sp. A	14	8008.9	9094.06			
Chaetoceros spp. B	7	106.37	111.77	10	744.07	1323.27
Chaetoceros spp. C	17	995.76	1611.15	8	1168.61	1546.39
Coscinodiscus radiatus Ehrenb.	9	1	0	29	2.17	1.42
Coscinodiscus spp.						
Diploneis sp.						
Ditylum brightwelli Grunow						
Hemidiscus cuneiformis Wallich				39	4.35	5.87
Leptocylindrus danicus Cleve				12	866.2	2338.27
Nitzschia (*Pseudonitzschia*) spp. (small)	15	103.46	157.96			
Nitzschia (*Pseudonitzschia*) spp. (large)	10	82.22	70.84	15	477.87	841.35
Pleurosigma spp.						
Rhizosolenia alata Cleve	8	16.2	38.71	28	49.27	92.79
Rhizosolenia imbricata Brightwell						
Rhizosolenia spp.	15	78.09	89.07	18	66.74	170.85
Schröderella delicatula (Per.) Pavillard						
Thalassionema nitzschioides Grunow				11	8.7	13.16
Thalassiosira spp. (small)	13	454.72	1071.33			
Thalassiosira spp. (medium)	8	50.18	115.04			
Thalassiosira spp. (large)	10	3.42	5.96			
Thalassiothrix frauenfeldii (Grunow) Cleve and Möller	8	7.43	13.29			
Unidentified centric diatoms	12	133.4	95.39			
Pennate diatoms (small)	51	82.1	99.32			
Pennate diatoms (large)	11	5.53	11.85	33	16.84	21.3
Unidentified pennate diatom sp. A						
Dictyocha fibula Ehrenb.	30	73.6	79.75	39	123.61	171.15
Dictyocha speculum Ehrenb.				10	51.71	53.99
Flagellates	127	4811.9	7329.26	107	5742.16	5787.35
Calycomonas spp.	26	728.72	1154.74	10	128.13	36.45
Phaeocystis sp.				16	761.98	821.9
Cryptomonads	22	230.64	134.34	8	145.89	51.58
Leucocryptos?	33	373.07	311.6	42	203.26	121.48
Solenicola setigera	19	1256.17	1306.27	31	1516.35	1650.65
Mesodinium sp.	24	8.91	11.66	15	33.03	65.31
Ciliates	88	179.36	194.77			

small naked gymnodinioids). In general, these two groups were numerically the most abundant. It should be noted, however, that many small cells deteriorate in fixed samples, so that the "flagellate" concentration is likely to be underestimated. Many of the problems associated with the inverted microscope technique and with phytoplankton counts in general have been discussed in several papers (MARGALEF, 1972; 1978; TRAVERS, 1972; VENRICK, 1978). An important consideration in this context is that each technique of phytoplankton enumeration is only able to give a more or less distorted appreciation of a segment of the phytoplankton community. For example, one of the limitations of the inverted microscope technique is its unsuitability for the examination of the organisms in the picoplankton size group. An additional problem, especially in oligotrophic waters such as those studied here, is that the concentrations of most of the relatively large and readily identifiable forms are low and subject to high counting errors in 100 ml samples. This introduces much noise when dealing with the distribution of individual species. However, even considering these limitations, the data should be useful for comparison purposes, provided that the same examination criteria have been used for all the samples.

The characteristics of the data sets used in this paper are summarized in Table I. Mean chlorophyll a and phytoplankton concentration values for the upper 100 m (FRONTS 3-85) or 110 m (in the other cruises) of the water column were calculated according to the expression:

$$\overline{X} = \sum_{i=1}^{k} (D_{i+1}-D_i) * (C_{i+1}+C_i)/2/D_k$$

where Ci is the chlorophyll *a* or phytoplankton concentration at depth D_i, is the number of sampling depths and D_k the deeper sampling depth.

Principal component analysis (PCA) was used to summarize the information contained in the phytoplankton lists. A description of the mathematical techniques involved can be found in COOLEY & LOHNES (1971) and LEGENDRE & LEGENDRE (1983). The taxa were selected for the analyses according to their frequency of appearance in the samples (some taxa were excluded because their identification was considered inconsistent). Separate analyses were performed on the correlation matrix among the logarithm-transformed abundance data of the selected species for each cruise. The computer programs were implemented from the IBM Scientific Subroutine Package (IBM, 1969).

RESULTS

The sections of temperature, salinity and chlorophyll along the studied transects are shown in figures 2-5. The distribution of σ_t (not represented) closely followed that of temperature. During FRONTS 3-85, the hydrographical conditions were typical of the mixing period, and temperature and salinity showed little vertical variation. The Catalan front was marked by sharp temperature, salinity and silicate (Fig. 6) gradients in the FRONTS 3-85 and PEP 84 cruises. In the other surveys, carried out later in the season, the upper part of the water column was covered by a layer of warm water; apart from their signature in the salinity distribution, the presence of the fronts and the doming of the isolines in the central zone were apparent only at depth. The distributions (not shown) of nitrate and phosphate (the concentrations of this nutrient were not available for PEP 86) were similar to those of silicate, except that nitrate and phosphate concentrations in the spring cruises were generally undetectable in the surface layers.

During FRONTS 3-85, the highest chlorophyll concentrations were found offshore of the front, in the upper 40-60 m of the water column (Fig. 2). By mid-May

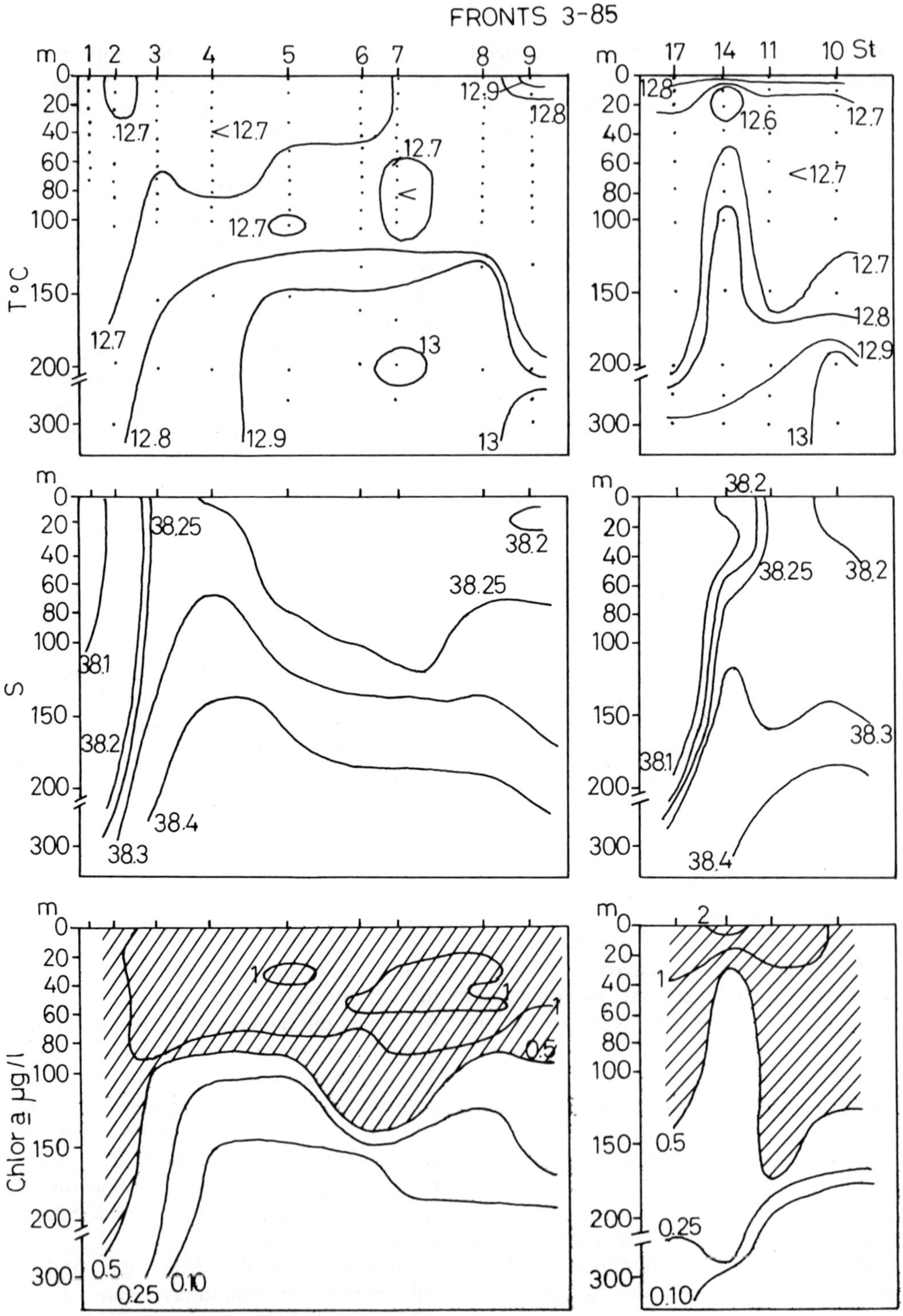

FIGURE 2. Distribution of temperature (T °C), salinity (S) and chlorophyll *a* (chlor. *a*, μg l^{-1}) along the two FRONTS 3-85 transects. The numbers on top on the graph indicate the stations.

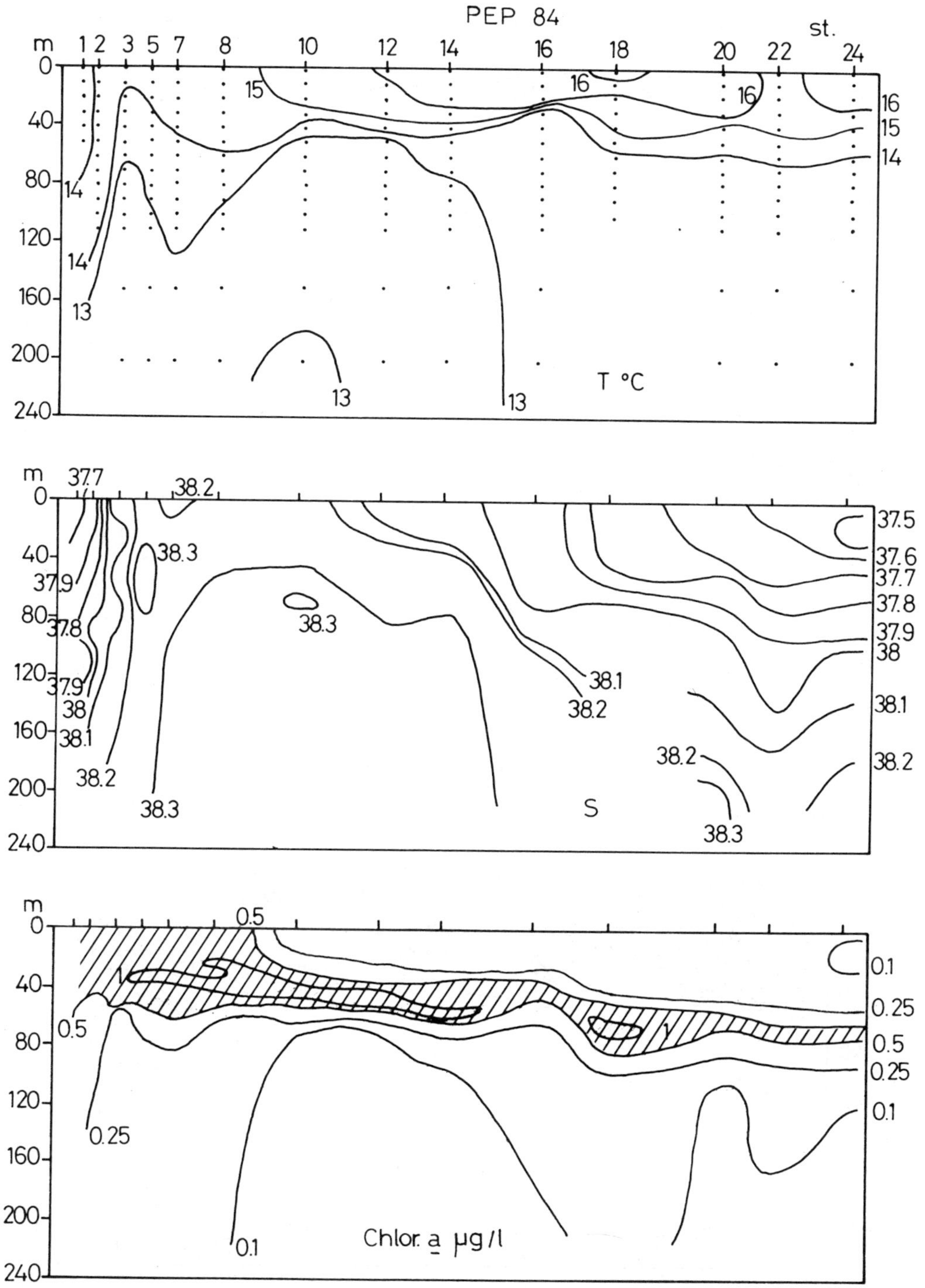

FIGURE 3. Distribution of temperature (T °C), salinity (S) and chlorophyll (chlor *a*, µg l^{-1}) along the PEP 84 transect.

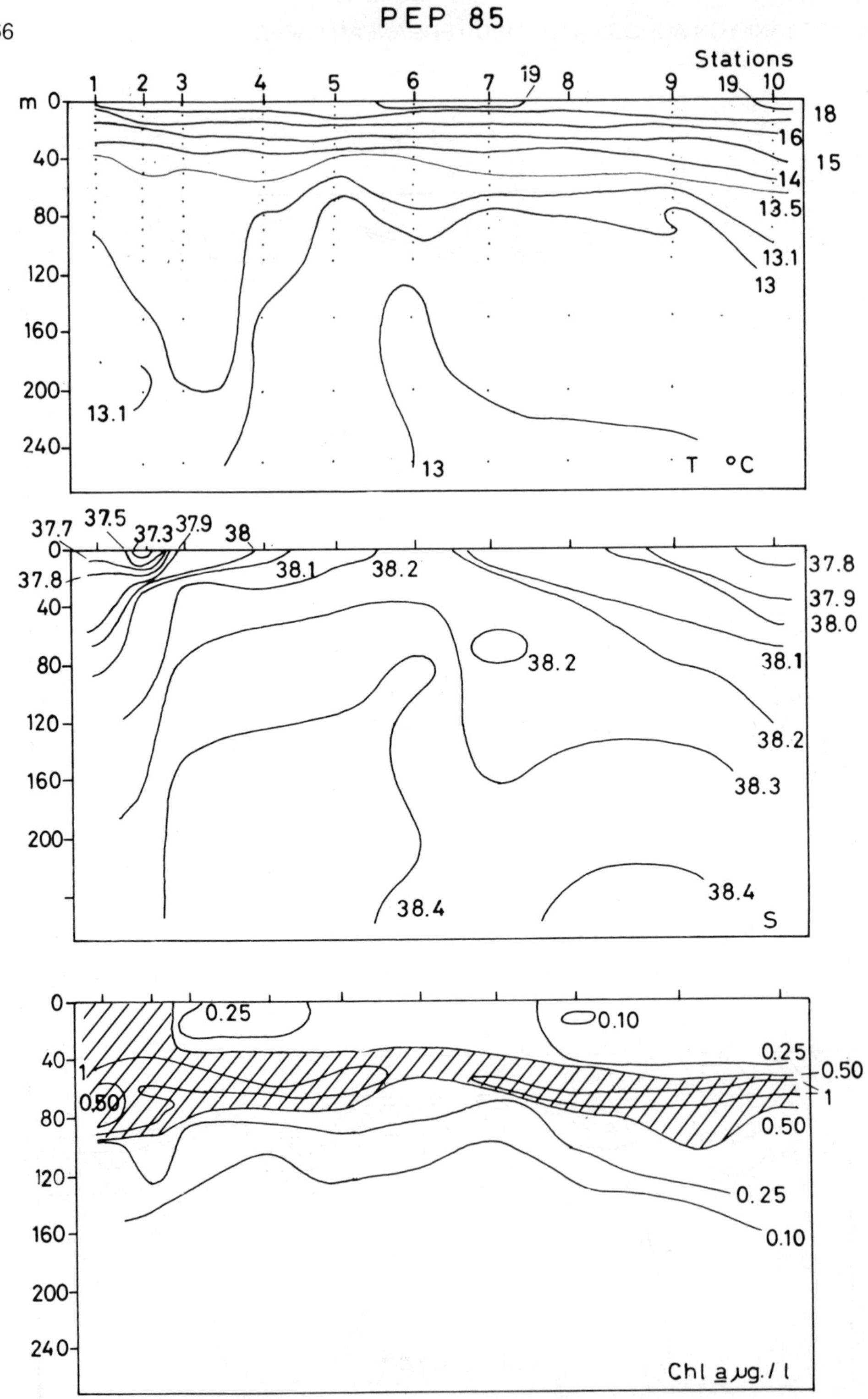

FIGURE 4. Distribution of temperature (T °C), salinity (S) and chlorophyll (chlor. *a*, µg l^{-1}), along the PEP 85 transect.

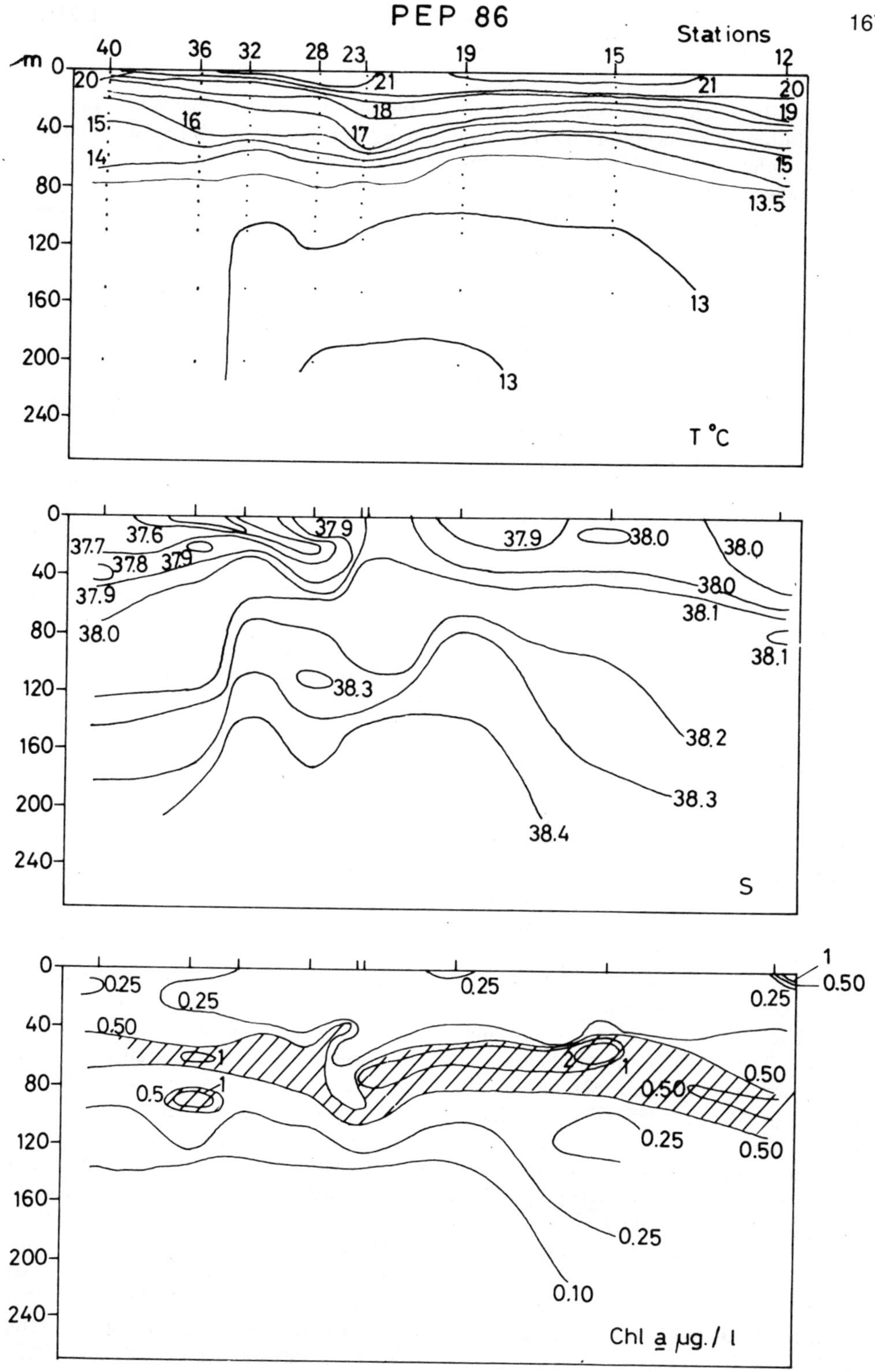

FIGURE 5. Distribution of temperature (T °C), salinity (S) and chlrophyll (chlor *a* µg l^{-1}), along the PEP 86 transect.

(PEP 84), the DCM was well established, with chlorophyll concentrations in some spots exceeding 2 mg m^{-3} (Fig. 3). In this and the other spring cruises (Figs. 4 and 5), the DCM was associated with nitrite and oxygen maxima (not shown) and presented similar global features, although there was a deepening of the chlorophyll maximum with the advance of the season. During PEP 84 and PEP 85, relatively high chlorophyll values (> 0.5 mg m^{-3}) reached the surface, inshore of the Catalan front. A similar situation, although with the front closer to the coast, was found in the first transect of PEP 86 (not shown; see MASÓ & PEPS, 1988); Station 40, in the transect presented here, was too far from the coast to show this feature. No comparable increases of surface chlorophyll concentration were found in the Balearic frontal zone, although high chlorophyll patches were often present at the DCM.

The total phytoplankton abundance showed maxima in the upper euphotic zone, offshore of the Catalan front, in FRONTS 3-85 (Fig. 7). During PEP 84, the highest phytoplankton densities were found offshore, at surface, and in the DCM, at depth (Fig. 8). In the other cruises (Figs. 9 and 10), the higher phytoplankton abundances were found nearshore and at the DCM level. The distributions of flagellates (not shown), which were numerically the dominant group, and of small dinoflagellates tended to show similar features. The distribution of diatoms was much more patchy than that of the other groups. In March 1985 (Fig. 7), diatoms formed the bulk of the phytoplankton community inshore of the Catalan front, while flagellates with a large proportion of haptophytes (including *Phaeocystis*) dominated offshore. In the other cruises later in the season, the diatoms were restricted to the DCM zone, where they formed localized high concentration patches.

The mean chlorophyll *a* concentration for the upper 100 or 110 m of the water column (Fig. 11) exceeded 1 µg l^{-1} for many stations of FRONTS 3-85. During the spring cruises, the mean chlorophyll *a* value was generally lower than 0.6 µg l^{-1}; the higher mean concentrations tended to be found in the frontal zones. The relationship between mean chlorophyll *a* and phytoplankton concentrations (Fig. 11) was very variable, as could be expected from the different size and physiological conditions of the organisms involved; this relation was highest during FRONTS 3-85 and tended to decrease from mid to late spring.

Some statistical data for the most frequent and/or abundant taxa in the four cruises are given in Table I. These lists should be compared with caution, since vaguely defined taxa may represent different species or forms from one cruise to another. As stated before, flagellates and small dinoflagellates were the numerically dominant forms in all cruises. The most common among the thecate dinoflagellate forms were species of *Protoperidinium, Prorocentrum, Oxytoxum* and *Ceratium.* Coccolithophorids were scarce in all the cruises; it appears that they may be important only at particular times of the year (ESTRADA, 1985a; ESTRADA & SALAT, 1989; and unpublished data). The silicoflagellate *Dictyocha fibula* was present in all cruises but, except in March 1985, was restricted to the lower layers of the euphotic zone, a typical feature of its distribution during the stratification period in the Mediterranean (NIVAL, 1965; TRAVERS & TRAVERS, 1968; ESTRADA, 1985a). *D. speculum* was only appreciably frequent in the PEP 86 cruise. The most abundant diatom genera in all cruises were *Chaetoceros, Nitzschia, Rhizosolenia* and *Thalassiosira. Ditylum brightwelli* was well represented in coastal waters during the FRONTS 3-85 cruise; it was common in the microplankton samples from the July cruises of 1982 and 1983, but had only been found rarely in earlier surveys of the zone (MARGALEF &

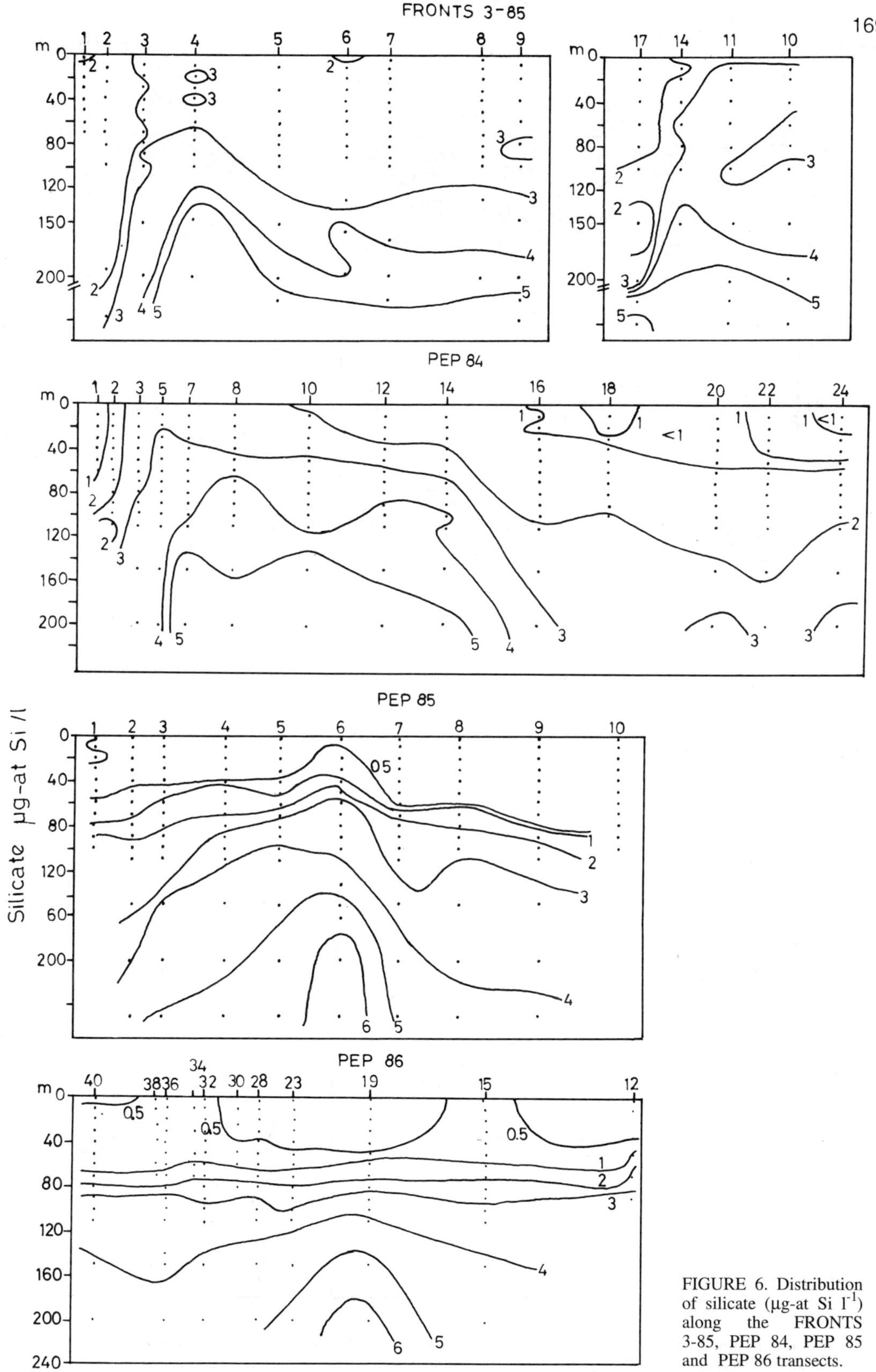

FIGURE 6. Distribution of silicate (µg-at Si l^{-1}) along the FRONTS 3-85, PEP 84, PEP 85 and PEP 86 transects.

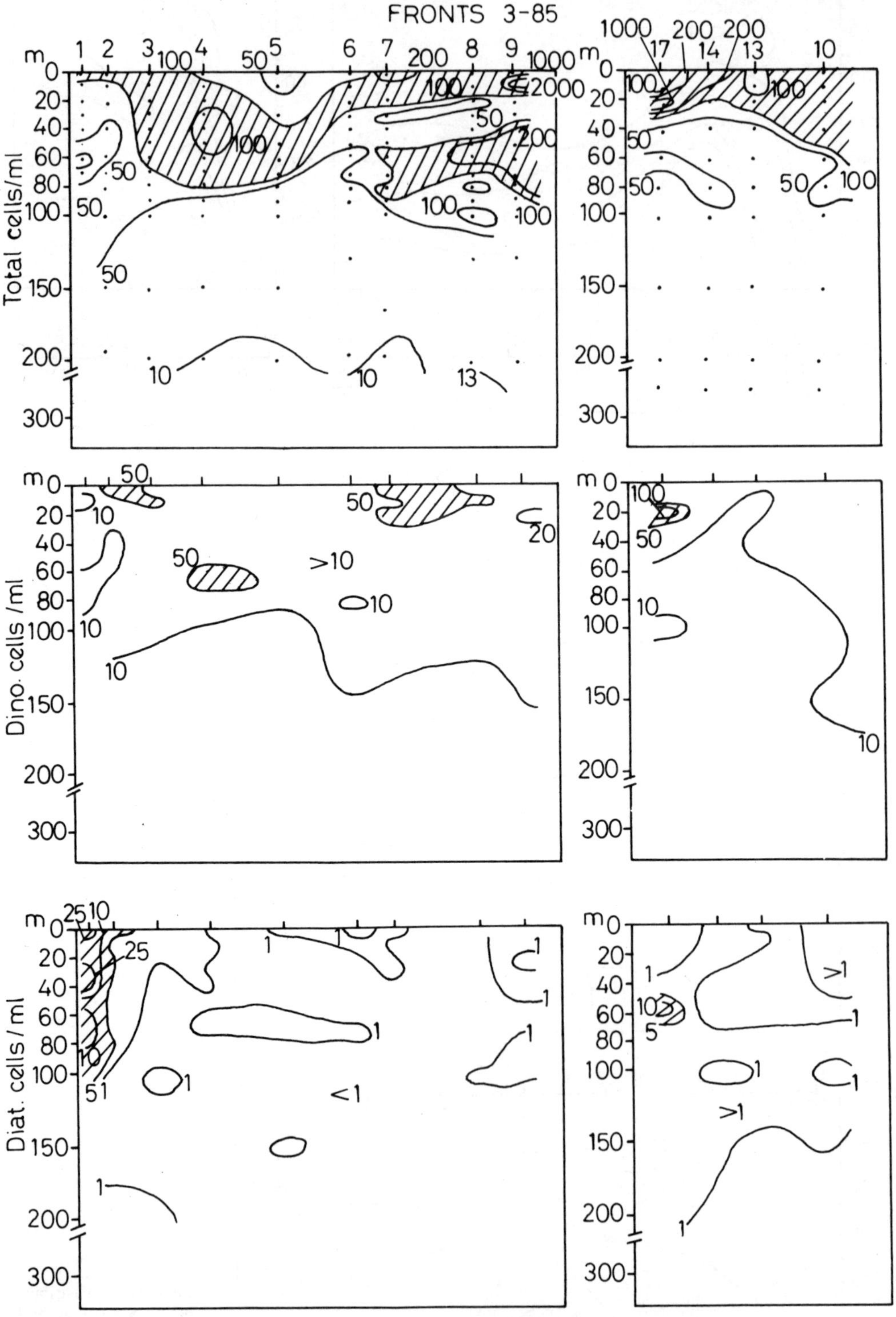

FIGURE 7. Distribution of the population density (cells ml^{-1}) of total phytoplankton, dinoflagellates and diatoms, along the FRONTS 3-85 transects.

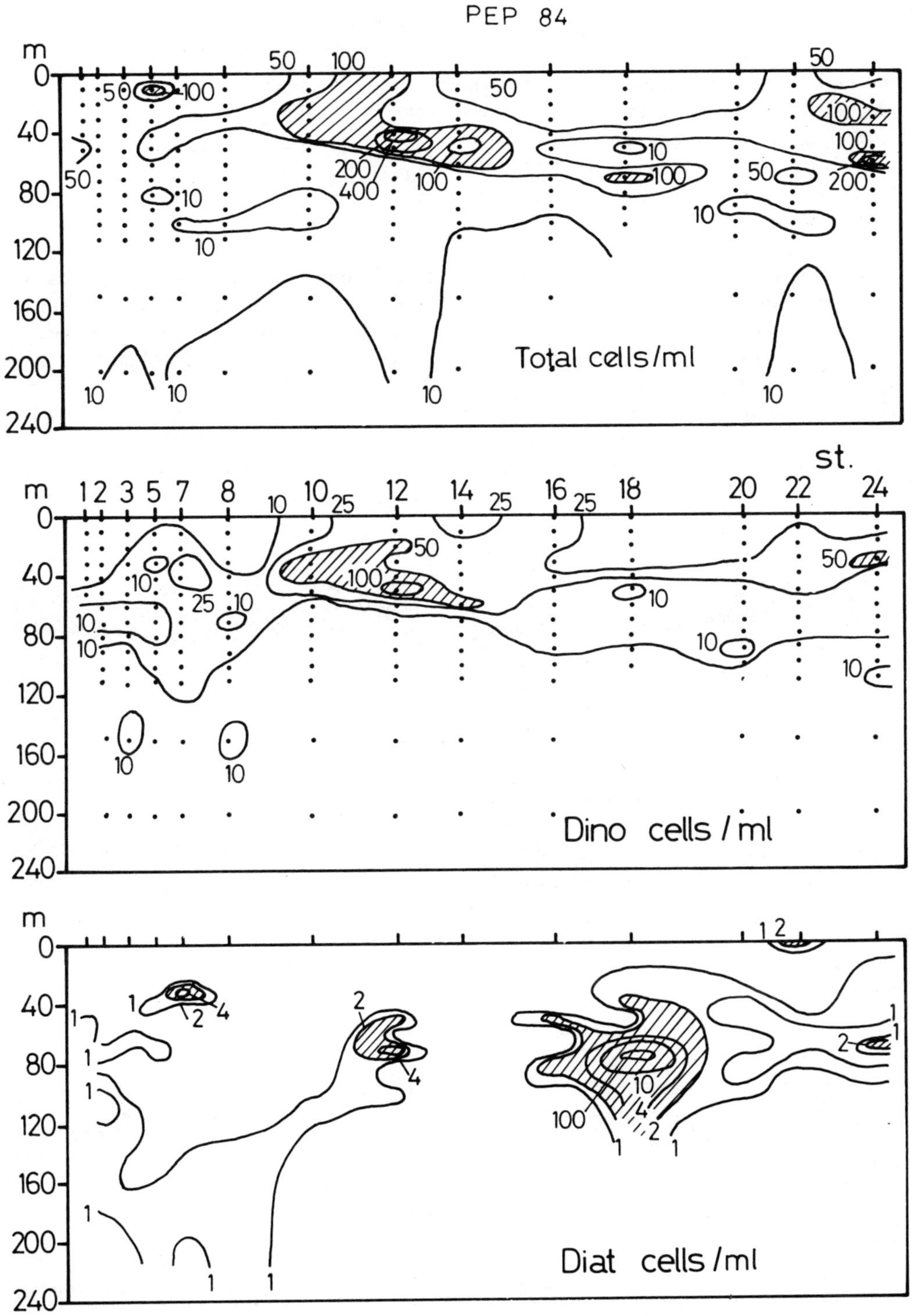

FIGURE 8. Distribution of the population density (cells ml^{-1}) of total phytoplankton, dinoflagellates and diatoms, along the PEP 84 transect.

ESTRADA, 1987). Weakly silicified *Chaetoceros* forms generally contributed the bulk of the high diatom concentration patches in the DCM. In the spring cruises, *Coscinodiscus radiatus* and *Hemidiscus cuneiformis* were typically distributed in the lower part of the euphotic zone; they were found at shallower depths in the winter cruise.

The variance explained by the first three principal components (PC) in all the analyses ranged from 29% to 33% (Table II). Examination of the ordination of the species according to their correlations (loadings) with the components (Table III and Fig. 12) showed several regularities. PC2 from FRONTS 85 and PC1 from the remaining cruises presented positive correlations with most of the species and could be considered as generalized "abundance-richness of species" components. Their distribution (Figs. 13-16) was comparable with that of total cell abundances and chlorophyll concentration; this reflects the fact that most taxa tend to coincide in occupying the richer areas and is a general finding in

TABLE II. Percentage of variance explained by the first three principal components (PC1, PC2, PC3) of the analyses.

cruise	PC1	PC2	PC3
PEP 84	11.1	7.9	5.3
FRONTS 3-85	10.7	9.2	6.6
PEP 85	11.6	6.8	5.4
PEP 86	11.3	9.3	6.9

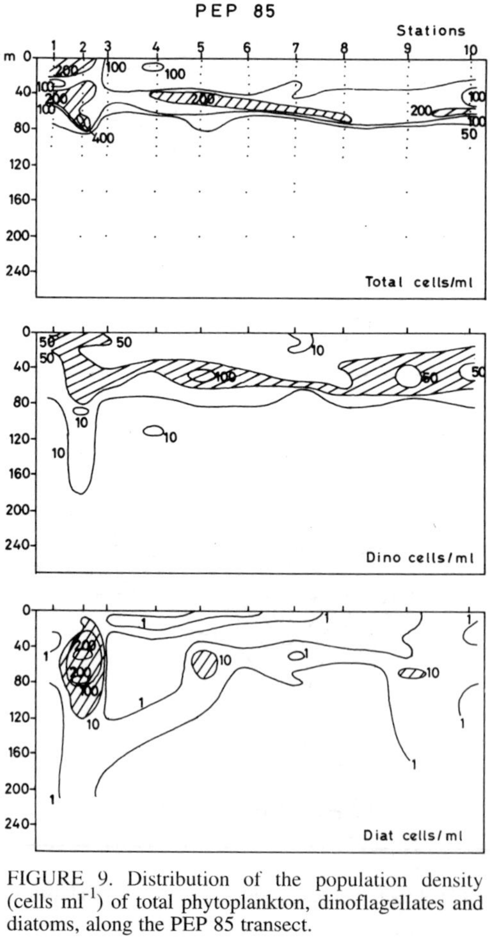

FIGURE 9. Distribution of the population density (cells ml^{-1}) of total phytoplankton, dinoflagellates and diatoms, along the PEP 85 transect.

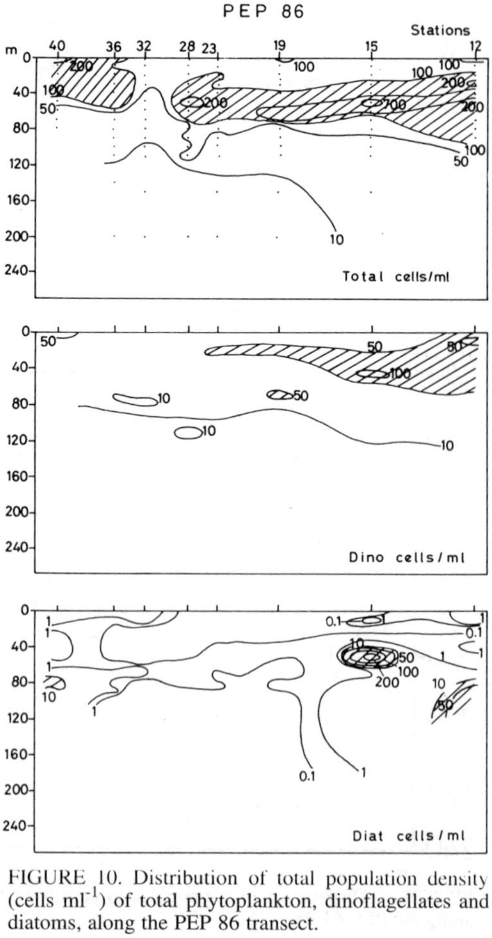

FIGURE 10. Distribution of total population density (cells ml^{-1}) of total phytoplankton, dinoflagellates and diatoms, along the PEP 86 transect.

multivariate analyses concerned with distributions of organisms when there are contrasts in fertility within the area studied.

A comparison of the loading ordinations of PC1 for FRONTS 85 and PC2 for the spring cruises (Fig. 12) showed that these components coincided in expressing a contrast between major taxonomic groups, with diatoms on one side and dinoflagellates and other groups on the other. An estimation of the similarities among these ordinations was carried out by calculating the Spearman rank correlation coefficients (Rs) among the positions of the n common taxa in each pair of ordinations. The results indicated significant positive correlations between the PC2 ordinations of the spring cruises (Rs =0.46 between PEP 84 and PEP 85 for n=32 common taxa; Rs =0.58 between PEP 85 and PEP 86 for n=30, and Rs =0.60 between PEP 84 and PEP 86 for n=33). The distributions of the components (Figs. 13-16) presented a clearcut boundary between the phytoplankton populations in the coastal and oceanic side of the Catalan front during the FRONTS 85 and PEP 84 cruises, and reflected the importance of diatoms in the DCM.

The distributions of PC3 in PEP 84 and PEP 86 (Figs. 14-16) consistently showed a marked gradient between the upper and the lower part of the euphotic zone. In both cruises, a group of taxa including *Oxitoxum margalefi*, *Coscinodiscus radiatus*, *Hemidiscus cuneiformis, Dictyocha fibula,* and an unidentified dinoflagellate (sp. D), which were typically distributed in the lower parts of the euphotic zone, characterized one of the extremes of the ordination list, while the highest correlations on the other side corresponded basically to dinoflagellates and flagellates, but also to some diatom taxa. The

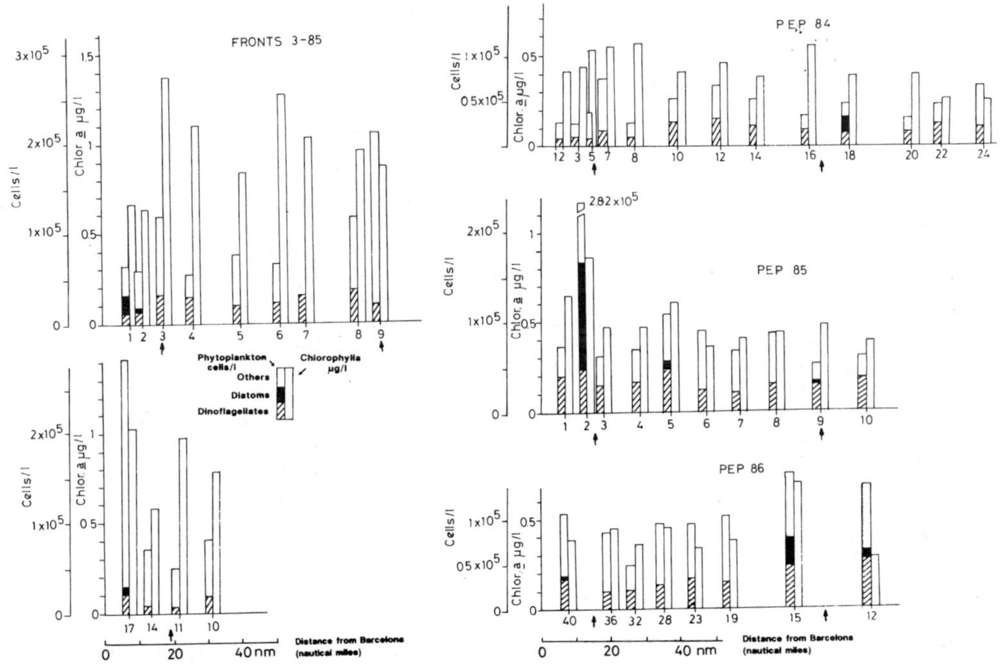

FIGURE 11. Mean values of chlorophyll *a* (chlor. *a* µg l^{-1}) and population densities (cells ml^{-1}) of total phytoplankton, dinoflagellates and diatoms for the upper 100 m (FRONTS 3-85) or 110 m (PEP 84, PEP 85 and PEP 86) of the water column.

TABLE III. Loadings of the selected taxa on the "biomass" component (first principal component, C1, of the analyses of PEP 84, PEP 85 and PEP 86 and second principal component, C2, of the analysis of FRONTS 3-85). Taxa ordained according to decreasing value of the loading

PEP 84

TAXA	Loading on C1
Unidentified dino. sp. B	0.73
Prorocentrum spp. (small)	0.62
Unidentif. pennate diat. sp. A	0.59
Dictyocha fibula	0.54
Nitzschia spp.	0.53
Gonyaulax spp.	0.50
Schröderella delicatula	0.49
Pleurosigma spp.	0.47
Chaetoceros spp. C	0.47
Solenicola setigera	0.47
Cochlodinium spp.	0.43
Unidentified dinoflagellates	0.42
Prorocentrum compressum	0.40
Gyrodinium spp.	0.40
Flagellates	0.38
Oxytoxum scolopax	0.38
Oxytoxum spp. (small)	0.38
Pennate diatoms (small)	0.36
Oxytoxum sp. D	0.35
Mesoporos perforatus	0.34
Leptocylindrus danicus	0.34
Protoperidinium spp. (small)	0.33
Dinophysis spp.	0.31
Unidentified dino. sp. F	0.28
Unidentified dino. sp. D	0.26
Palaeophalachroma unicinctum	0.26
Pronoctiluca sp.	0.25
Rhizosolenia alata	0.23
Torodinium robustum	0.22
Scrippsiella trochoidea	0.21
Oxytoxum variabile	0.17
Oxytoxum margalefi	0.16
Dinophysis (Phalachr.) spp.	0.16
Ceratium fusus	0.15
Calycomonas spp.	0.15
Thalassiosira spp. (small)	0.13
Ceratium furca	0.12
Protoperidinium bipes	0.07
Prorocentrum micans	0.05
Rhizosolenia spp.	-0.00
Unidentified coccolithophorids	-0.00
Mesodinium sp.	-0.03
Hemidiscus cuneiformis	-0.05
Emiliania huxleyi	-0.05
Coscinodiscus radiatus	-0.08
Thalassiosira spp. (medium)	-0.08
Leucocryptos?	-0.09
Torodinium spp.	-0.11

PEP 85

TAXA	Loading on C1
Flagellates	0.75
Unidentif. small dinof.	0.72
Calycomonas spp.	0.59
Gyrodinium spp.(small)	0.56
Prorocentrum spp. (small)	0.56
Protoperidinium spp. (small)	0.50
Leucocryptos?	0.46
Protoperidinium sp. A	0.46
Cochlodinium spp.	0.44
Torodinium robustum	0.44
Ceratium fusus	0.43
Cryptomonads	0.42
Unidentif. centric diatoms	0.40
Syracosphaera pulchra	0.40
Mesoporos perforatus	0.40
Ciliates	0.39
Chaetoceros spp. C	0.36
Unidentified dino sp. B	0.35
Solenicola setigera	0.32
Dinophysis ovum	0.29
Unidentified dino sp. F	0.28
Ceratium furca	0.26
Nitzschia spp. (small)	0.26
Thalassiosira spp. (small)	0.25
Scrippsiella trochoidea	0.24
Rhizosolenia spp.	0.24
Torodinium spp (small)	0.23
Chaetoceros sp. A	0.21
Rhizosolenia alata	0.21
Unidentified dino. sp. D	0.20
Protoperidinium bipes	0.19
Pennate diatoms (small)	0.18
Oxytoxum spp. (small)	0.18
Oxytoxum variabile	0.17
Thalassiosira spp. (medium)	0.16
Dinophysis rotundata	0.16
Nitzschia spp. (large)	0.15
Dictyocha fibula	0.14
Pennate diatoms (large)	0.13
Chaetoceros spp. B	0.13
Oxytoxum margalefi	0.12
Oxytoxum spp. (large)	0.04
Thalassiothrix frauenfeldii	0.02
Mesodinium sp.	0.01
Thalassiosira spp. (large)	-0.01
Protoperidinium spp. (large)	-0.16
Coscinodiscus radiatus	-0.24

Spearman rank correlation coefficient among the ordinations of the 33 common taxa (Rs =0.43) was significant at the 5% level. The PC3 of the FRONTS 85 analysis was more difficult to interpret, but it also appeared to express a contrast in vertical distributions and its positive side was related to the presence of several taxa with

TABLE III (Cont.)

PEP 86

TAXA	Loading on C1
Unidentified dinoflagellates	0.74
Flagellates	0.72
Rhizosolenia alata	0.59
Chaetoceros spp. B	0.57
Chaetoceros spp. C	0.56
Pennate diatoms (large)	0.53
Leptocylindrus danicus	0.49
Solenicola setigera	0.48
Prorocentrum spp. (small)	0.46
Unidentified dino. sp. B	0.45
Rhizosolenia spp.	0.45
Phaeocystis sp.	0.42
Leucocryptos	0.41
Unidentified dino. sp. D	0.41
Goniodoma sphaericum	0.40
Gonyaulax turbynei	0.39
Dinophysis acuta	0.35
Mesodinium sp.	0.35
Protoperidinium spp. (small)	0.35
Gyrodinium spp. (small)	0.30
Goniodoma polyedricum	0.28
Dissodium (*Diplopsalis*) spp.	0.26
Dinophysis rotundata	0.25
Prorocentrum compressum	0.25
Mesoporos perforatus	0.25
Gyrodinium spp. (large)	0.25
Protoperidinium crassipes	0.22
Torodinium robustum	0.22
Cochlodinium spp.	0.20
Protoperidinium spp. (large)	0.19
Ceratium fusus	0.14
Oxytoxum scolopax	0.14
Prorocentrum maximum	0.13
Oxytoxum variabile	0.12
Cryptomonads	0.12
Dictyocha speculum	0.12
Hemidiscus cuneiformis	0.12
Thalassionema nitzschioides	0.12
Oxytoxum margalefi	0.1
Coscinodiscus radiatus	0.09
Dinophysis spp.	0.09
Gonyaulax spp.	0.06
Dinophysis (*Phalachr.*) spp.	0.05
Unidentified dino. sp. F	0.05
Torodinium spp. (small)	0.04
Gonyaulax polygramma	0.02
Calycomonas spp.	0.02
Nitzschia spp.	0.02
Protoperidinium steini	0.00
Dictyocha fibula	-0.02

FRONTS 3-85

TAXA	Loading on C2
Flagellates	0.54
Rhizosolenia imbricata	0.53
Thalassiosira spp. (small)	0.47
Chaetoceros spp. B	0.47
Ditylum brightwelli	0.46
Chaetoceros spp. C	0.45
Nitzschia spp.	0.44
Unidentified dinoflagellates	0.43
Scrippsiella trochoidea	0.41
Pennate diatoms (small)	0.37
Pennate diatoms (large)	0.35
Dictyocha fibula	0.33
Mesoporos perforatus	0.32
Thalassiosira spp. (medium)	0.31
Unidentif. pennate diat. sp. A	0.31
Cryptomonads	0.30
Prorocentrum spp. (small)	0.28
Coscinodiscus radiatus	0.25
Unidentified dino. sp. B	0.22
Ciliates	0.21
Pleurosigma spp.	0.19
Cochlodinium spp.	0.11
Protoperidium spp. (small)	0.08
Torodinium sp. (small)	0.03
Oxytoxum variabile	0.01
Unidentified coccolithophorids	0
Coscinodiscus spp.	-0.01
Phaeocystis sp.	-0.03
Hemidiscus cuneiformis	-0.04
Leucocryptos?	-0.05
Calycomonas spp.	-0.10
Emiliania huxleyi	-0.18
Diploneis sp.	-0.20
Gyrodinium spp. (small)	-0.25

benthic affinities (large pennate diatoms, *Coscinodiscus, Pleurosigma* spp.). These taxa with relatively deep distributions tended to present the lowest correlations with the "abundance-richness of taxa" components, a finding which presumably reflects the segregation of their distribution from the high biomass zones. In contrast, the distribution of the PC3 of PEP-85 appeared mainly to express a difference between the communities neighbouring the Balearic zone and those of the rest of the transect. The taxa showing the highest correlation with this component were several small thecate dinoflagellates.

For simplicity, in the following comments, PC2 of FRONTS 85 and PC1 of the remaining cruises will be referred to as the "biomass component"; PC1 of FRONTS 85 and PC2 of the other cruises will be referred to as the "diatom component" and PC3 of PEP 84 and PEP 86 as the "vertical gradient component".

Considering the loadings of the taxa on the first three components, three main assemblages can be loosely distinguished.

A) A group occupying the upper euphotic zone during stratified conditions, comprising basically dinoflagellates and flagellates presenting negative correlations with the "diatom component", positive correlations with the "vertical gradient component" and more or less positive correlations with the "biomass component". Typical of this assemblage are taxa such as *Torodinium robustum*, *Ceratium furca*, *Ceratium fusus*, *Prorocentrum* spp. (small) and the flagellates.

B) A group of taxa with negative loadings on the "biomass" and the "vertical gradient" components, occupying the lower parts of the euphotic zone, within the

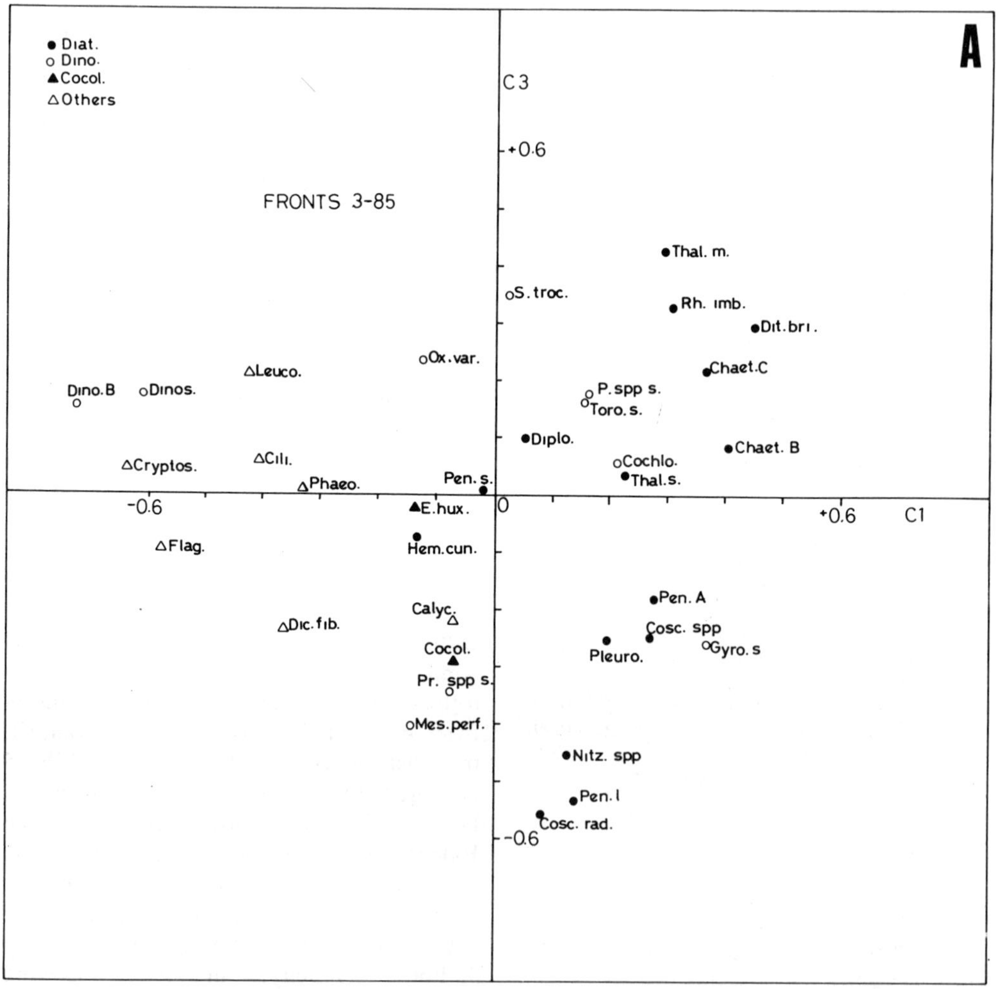

FIGURE 12. Position of the extremes of the taxa vectors in the space determined by the "diatom" component (first principal component, C1, of FRONTS 3-85 and second principal component, C2, of PEP 84, PEP 85 and PEP 86) and the third principal component, C3 (which represents the "vertical gradient" component for PEP 84 and PEP 86). The scale of the third principal component of PEP 85 has been reversed to facilitate comparison with the other graphs. The key for the codes is given in Table I.

DCM, but not being important contributors to the high chlorophyll patches in it. Characteristic of this group are *Oxytoxum margalefi*, *Coscinodiscus radiatus*, *Dictyocha fibula*, *Hemidiscus cuneiformis* and the unidentified dinoflagellate sp. D.

C) Practically all the diatoms (the exceptions being mainly *C. radiatus* and *H. cuneiformis*), characterized by positive loadings on the "diatom component" and high or medium loadings on the "biomass component", were distributed as an assemblage in each cruise, in the coastal zone inside the Catalan front in winter, and in the DCM in spring. However, given the differences in the diatom species present in the different cruises (especially when comparing winter and spring surveys) and the fact that many of the taxa identified to the genus level could include different species in each cruise, it is not possible to ascertain precisely the similarities between the diatom assemblages present in each cruise.

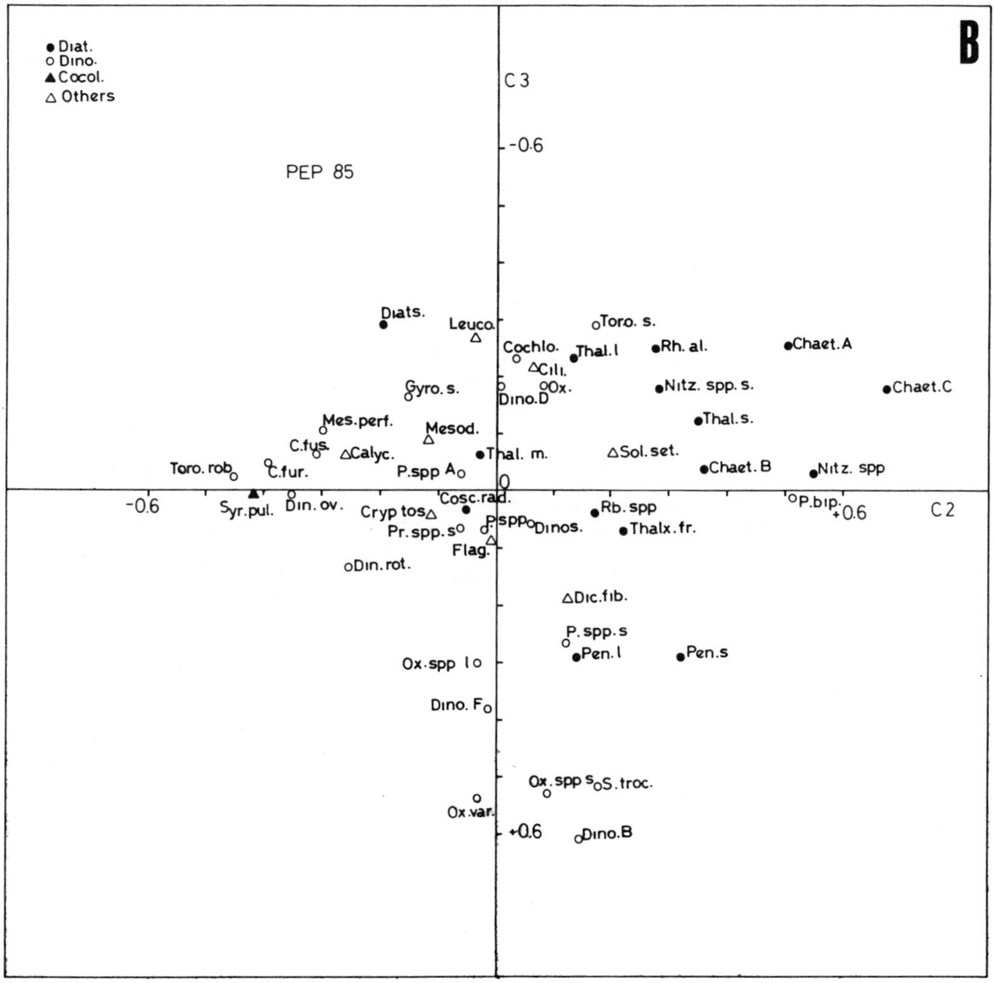

FIGURE 12. (Cont.).

DISCUSSION

Comparison of the phytoplankton and chlorophyll distributions showed overall similarities, although there were some exceptions, for example in the upper 30 m of the water column in the stations of the Catalan coast side of the PEP 84 cruise, where relatively high chlorophyll *a* concentrations corresponded to relatively low phytoplankton densities. Apart from the possibility of certain artifacts, such as differential deterioration of cells in the fixed samples, this could be due, for example, to a lower proportion of heterotrophic forms, larger average cell volume or increased chlorophyll content per cell (which could be a result of the higher nutrient concentrations) in the coastal zone.

During March 1985, although diatoms were more abundant in the coastal zone inshore of the Catalan front, chlorophyll concentrations were higher offshore, coinciding with high flagellate concentrations. This finding stresses the importance of the contribution of frontal instabilities to the fertility of the central zone of the Catalan Sea. Both the coastal

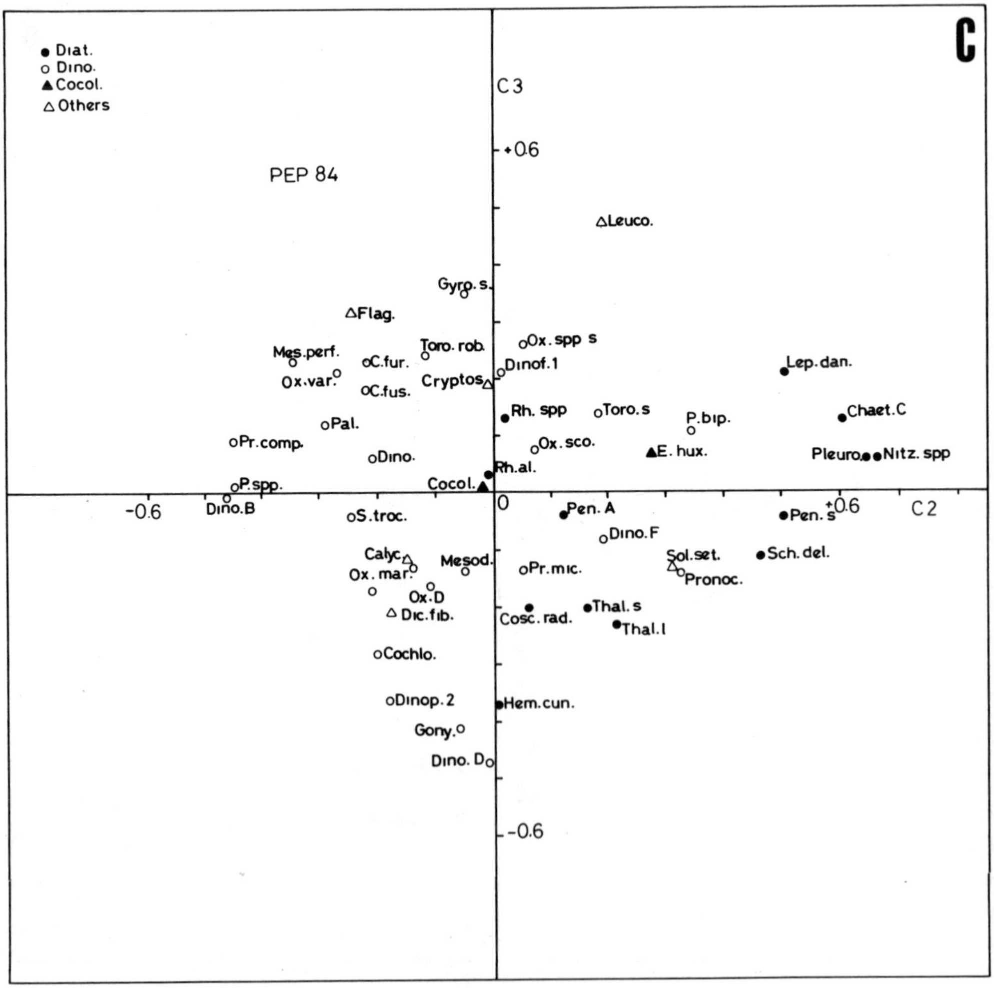

FIGURE 12. (Cont.).

and offshore assemblages included taxa cited by MARGALEF (1969) as typical of the winter-spring bloom in coastal waters of the Catalano-Balearic sea. MARGALEF & BALLESTER (1967) and MARGALEF & CASTELLVÍ (1967) had pointed out the variable contribution of diatoms and flagellates (including a large number of organisms now included within the haptophytes) to the spring bloom; however, as earlier data were restricted to littoral stations (depths < 1000 m), it is difficult to make comparisons with the inshore and offshore assemblages found in the present study.

The global features of the phytoplankton distribution during the cruises studied showed a major change from winter to spring, in correspondence with the hydrographical transition from a mixing to a stratified situation. All the spring cruises showed the presence of a well established DCM. In general, this DCM was associated with a peak in cell numbers and there was good correspondence between high chlorophyll spots at the DCM and patches of high diatom abundance. However, as shown by other studies (ESTRADA, 1985a; VANDEVELDE *et al.*, 1987; GOULD, 1988), comparison of chlorophyll and cell

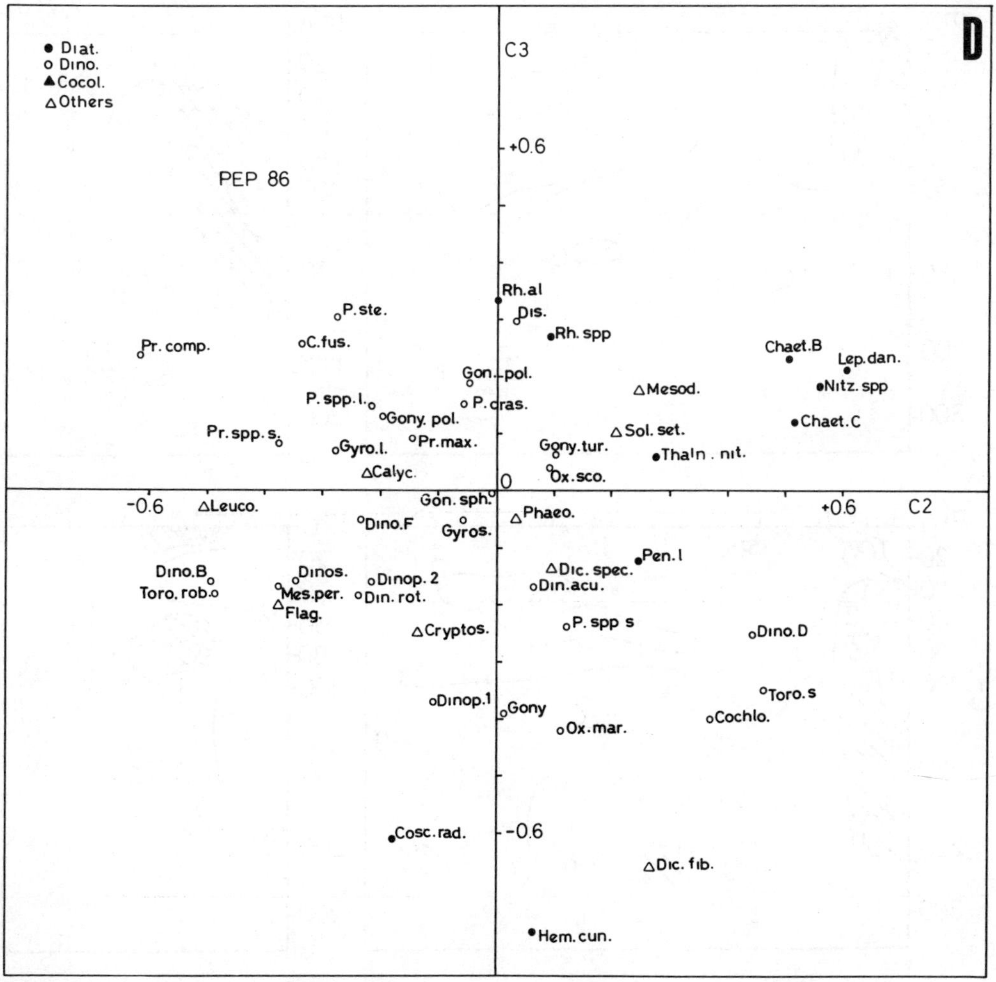

FIGURE 12. (Cont.).

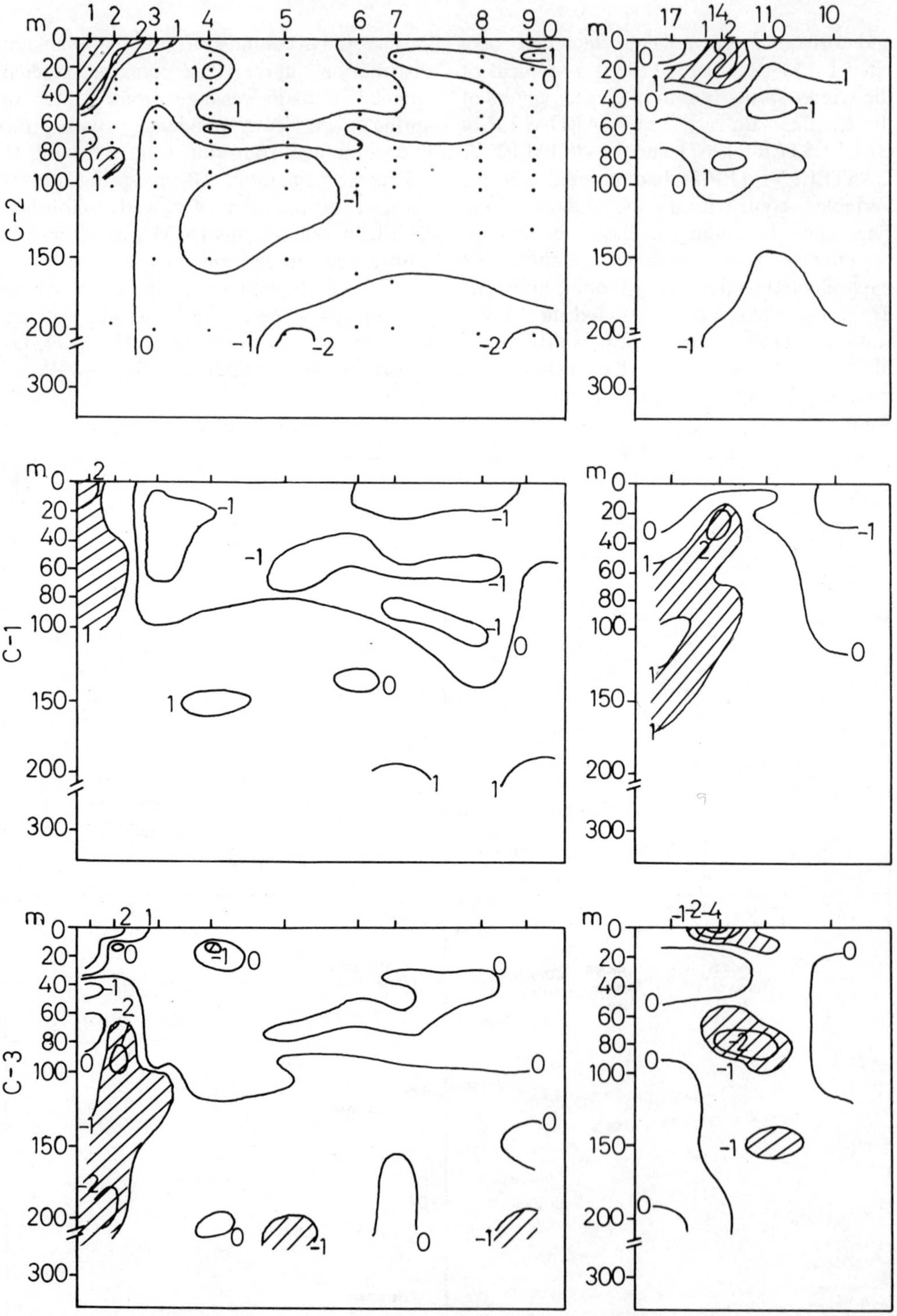

FIGURE 13. Distribution of the first three principal components (C2, C1 and C3) along the FRONTS 3-85 transects. C2 represents the "biomass" component for this cruise and is shown in the upper graph.

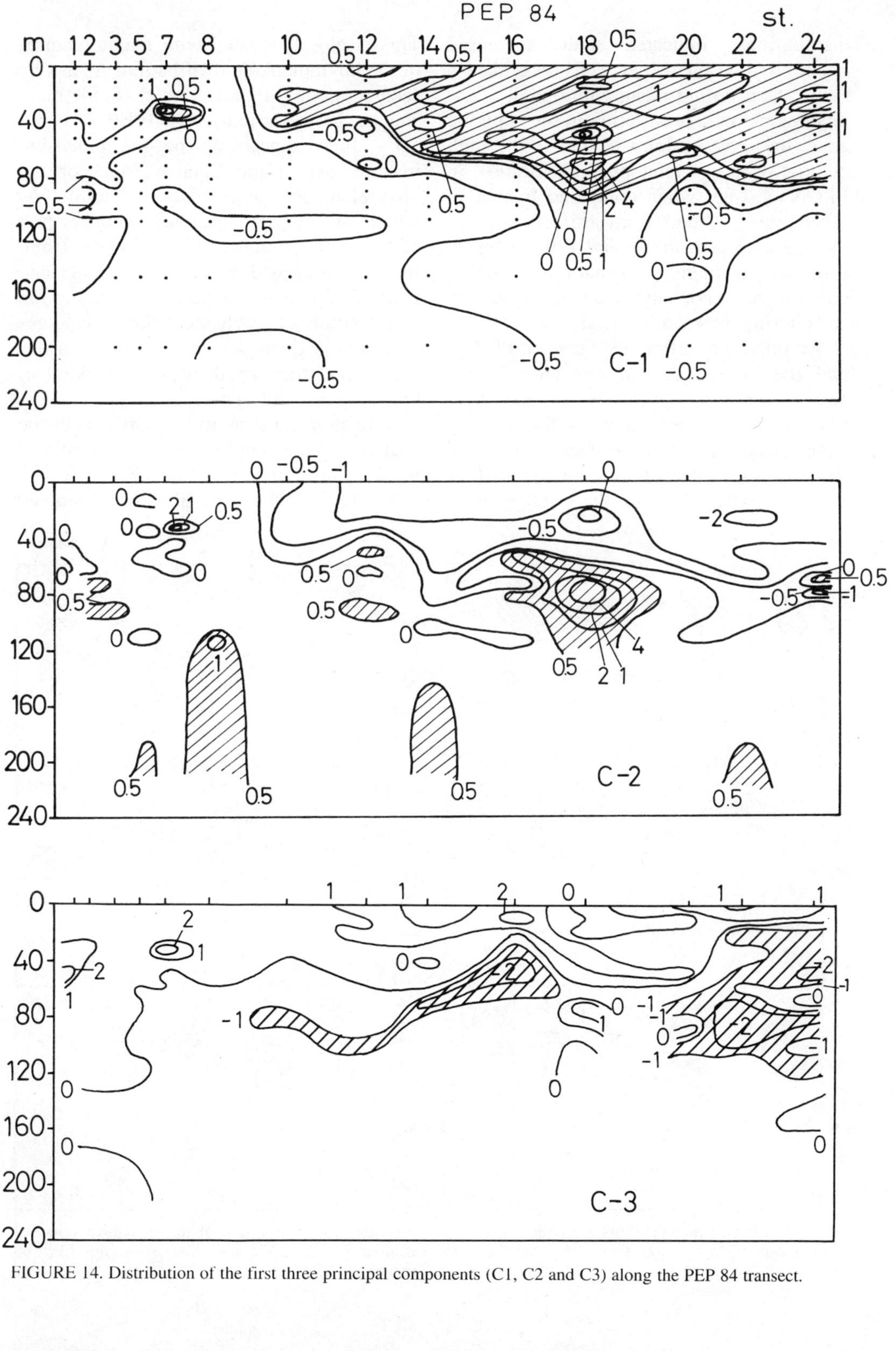

FIGURE 14. Distribution of the first three principal components (C1, C2 and C3) along the PEP 84 transect.

distributions indicated that increased chlorophyll content per cell was also a factor. Concerning mean chlorophyll a and phytoplankton concentration in the euphotic zone, the most striking feature is the high chlorophyll to phytoplankton ratio in FRONTS 3-85, when compared to that of other cruises. Apart from artifacts such as those mentioned above and differences in cell size, a likely explanation is higher chlorophyll *a* content of actively growing cells during the bloom period.

An important difference between PEP 84 and the cruises carried out later in the season was the presence, in the later cruises, of a surface layer of low density water which covered the coastal zone and the disappearance of the strong surface signatures characterizing the Catalan front. This is closely related with major features of the phytoplankton distribution. As can be seen comparing figures 15 and 16, the main change between mid May (PEP 84) and early June appears to be the extension inshore, across the Catalan front, of the phytoplankton assemblage (Group A) which, during May, occupied the upper part of the euphotic zone, offshore of the front. In the two early June cruises, PEP 85 and PEP 86, this assemblage extends along the entire transect. Whether these changes represent expansion of offshore surface waters and their resident phytoplankton or the response of phytoplankton to local stratification remains to be clarified. In the first case, the phenomenon could be related to the spring invasion of surface waters from the east which had already been noted

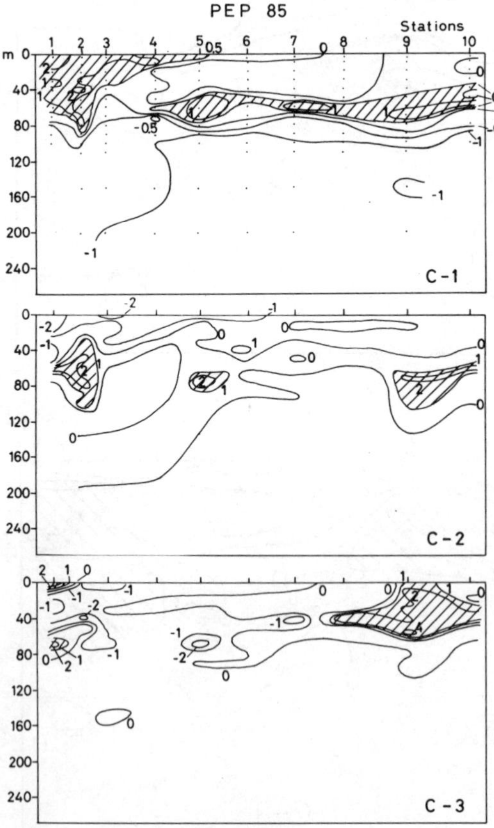

FIGURE 15. Distribution of the first three principal components (C1, C2 and C3) along the PEP 85 transect.

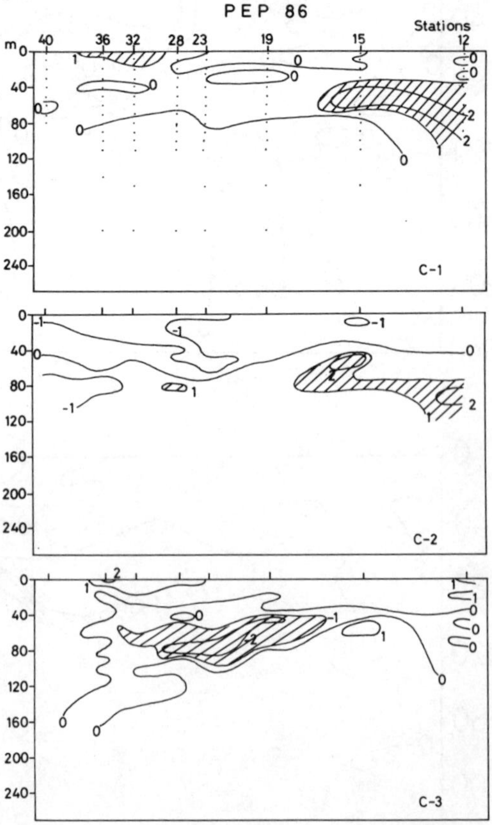

FIGURE 16. Distribution of the first three principal components (C1, C2 and C3) along the PEP 86 transect.

by MARGALEF (1963) and whose consequences were reflected in local fishing traditions.

With respect to the deepening of the front, the diatom assemblages (C), which in winter occupied the upper part of the water column in the nearshore waters, become restricted to patches within the DCM, in the lower part of the euphotic zone. At the same time, another assemblage (Group B) tends to constitute a background population within the DCM.

The distribution of the assemblages can be related to different trophic conditions within the water column. The high phytoplankton abundances in March 1985 were associated with the supply of nutrients to the upper layers caused by winter mixing. This situation led to diatom dominance inshore and flagellate dominance offshore. These differences may have been related to hydrological differences or to a different timing in the initiation of the bloom inshore and offshore; the lack of an adequate temporal sequence of sampling prevents clarification. During a recent cruise in February 1990, in the same area (Estrada, unpublished data), it was noted that diatom blooms appear offshore in response to hydrographic perturbations, but they may be relatively short-lived.

During the stratification period, the community of the upper euphotic layer, where nutrients are generally undetectable, is probably associated with a predominance of regenerated production. Irradiance does not constitute a limitation in this zone, and assimilation numbers may be relatively high, although biomasses remain low, perhaps due to a close equilibrium between production and predation. It is important to note in this context that many flagellates and dinoflagellates typical of this zone are likely to be heterotrophic.

The presence of localized high population densities of diatoms in the DCM suggests that they are the result of *in situ* growth and low diffusion losses, as had been noted in previous studies (ESTRADA, 1985a). Diatoms are typically associated with situations of high "new" production (MARGALEF, 1978; CUSHING, 1989) and the association of the Mediterranean DCM with the nutricline (ESTRADA, 1985a, b) indicates the importance of the supply of "new" nutrients. However, irradiance levels at the DCM (1-3% of surface irradiance during the spring cruises studied here) are generally too low to allow for high productivity indices (ESTRADA, 1985b). Increased production at the DCM could occur in pulses, when hydrographic perturbations expose nutrient rich water bodies to higher irradiances in shallower levels. The presence of the fronts and the lower stability of the central zone are likely to be very important for the origin of these perturbations. This would explain why the high biomass patches of the DCM tend to appear in the vicinity of the frontal zones.

In contrast to the distribution of most diatoms, the taxa of Group B are widely distributed within the DCM, but generally in low concentration. They could belong to a "shade flora" adapted to low irradiance conditions. However, these taxa were present in upper water levels during the winter cruise; as noted by SOURNIA (1982), it should be taken into account that the occurrence of algae at depth can be related to factors such as nutrient availability and water circulation in addition to light conditions. It is possible that the Group B taxa are part of a successional sequence following blooms occuring at depths above or within the DCM.

Comparison of these results with those from the previous summer cruises (PEP 82 and PEP 83) shows similar patterns, but also some differences (ESTRADA, 1985a; ESTRADA & SALAT, 1989). Both an upper euphotic zone assemblage (A) and a diatom assemblage were also present during the summer cruises, but the contribution of Group B was less clear. In July 1983, a tongue with high chlorophyll concentration (up to 4 mg m^{-3} of chlorophyll), probably due to local enrichment by coastal runoff, occupied the waters of the coastal zone. The corresponding phytoplankton community presented an important contribution of large dinoflagellates (such as *Ceratium furca*, *C. fusus* and *Protoperidinium brochi*) typical of stratified situations with adequate nutrient supply (ESTRADA & SALAT, 1989). A similar assemblage appeared to be present throughout the PEP 82 section, although the transect did not cross any high chlorophyll tongues near the coast (ESTRADA, 1985a). As shown by these studies and those of MARGALEF (1985) and MARGALEF & ESTRADA (1987) concerning the microplankton distribution during the PEP 82, PEP 83 and PEP 84 cruises, an assemblage of large thecate dinoflagellates is typical of the summer conditions in the top 10-20 m of the water column. This suggests that a segregation of phytoplankton also occurs within the upper part of the euphotic zone. The analysis by MARGALEF & ESTRADA (1987) also indicated a contrast between the upper layers and those of the lower and more productive levels and noted the presence of a "shade flora", although in this case it was characterized by large taxa such as *Ceratium platycorne*, *Amphisolenia globifera* and *Planktoniella sol*, which were not adequately sampled with the techniques used here.

Although fragmentary, the available data on the Northwestern Mediterranean ecosystem illustrate how fronts and gradients provide opportunities for enhanced biological activity and play a role in maintaining ecological boundaries. Superimposed on the general effects of the seasonal changes, the interplay of physical and chemical factors produces a mosaic of environmental situations which allows the coexistence of phytoplankton assemblages in different successional phases. In this context, it is interesting to note that repeatable patterns in the composition of phytoplankton assemblages can be found in different years, in spite of high environmental variability. However, more information is needed to complete a basic picture of the seasonal changes and to evaluate regularities and patterns of fluctuation at different temporal and spatial scales.

ACKNOWLEDGEMENTS

This work was supported by the Consejo Superior de Investigaciones Científicas, by grants no. 2311308 and PQ84-0067 of the Comisión Asesora de Investigación Científica y Técnica and by grant no. CCA8411054 of the Comité Conjunto Hispano-Norteamericano para la Cooperación Científica y Tecnológica. I am very grateful to all the participants in the cruises who cooperated in the collection of the samples. Mss. A. Cruz and M Vélez and Mr. J. Biosca provided technical assistance.

REFERENCES

COOLEY, W. W. & LOHNES, P. R. 1971. *Multivariate Data Analysis*. Wiley. New York.

CUSHING, D. H. 1989. A difference in structure between ecosystems in strongly stratified waters and in those that are only weakly stratified. *J. Plankton Res.*, 11: 1-13.

ESTRADA, M. 1985a. Deep phytoplankton and chlorophyll maxima in the Western Mediterranean. In: *Mediterranean Marine Ecosystems* (Moraitou-Apostolopoulou, M. & Kiortsis, V., eds.): 247-277. Plenum Press. New York.

ESTRADA, M. 1985b. Primary production at the deep chlorophyll maximum in the Western Mediterranean. In: *Proceedings of the 19th European Marine Biology Symposium* (Gibbs, P.E., ed.): 109-121. Cambridge University Press. Cambridge.

ESTRADA, M. & MARGALEF, R. 1988. Supply of nutrients to the Mediterranean photic zone along a persistent front. *Oceanol. Acta.* spec. iss. 9: 133-142.

ESTRADA, M. & SALAT, J. 1989. Phytoplankton

assemblages of deep and surface water layers in a Mediterranean frontal zone. *Scientia Marina,* 53(2-3): 203-214.

FONT, J. 1986. *La circulació general a la Mar Catalana.* Ph. D. Thesis. University of Barcelona.

GOULD, R. W. 1988. Net phytoplankton in a Gulf Stream warm-core ring: species composition, relative abundance, and the chlorophyll maximum layer. *Deep-Sea Res.*, 35:1595-1614.

IBM. 1969. *1130 Scientific Subroutine Package (1130-CM-02X).* Programmer's Manual. IBM Corporation.

LA VIOLETTE, P. E., TINTORÉ, J. & FONT, J. 1990. The surface circulation of the Balearic Sea. *J. Geophys. Res.*, 95(C2): 1559-1568.

LEGENDRE, L. & LEGENDRE, P. 1983. *Numerical Ecology.* Elsevier. Amsterdam.

MARGALEF, R. 1963. El ecosistema pelágico de un área costera del Mediterráneo Occidental. *Mem. R. Acad. Ciencias Artes Barcelona*, 35(1): 1-46.

MARGALEF, R. 1969. Composición específica del fitoplancton de la costa catalano-levantina (Mediterráneo occidental) en 1962-1967. *Inv. Pesq.*, 33: 345-380.

MARGALEF, R. 1972. Fitoplancton marino de la región de afloramiento del NW de Africa. *Res. Exp. Cient. B/O Cornide,* 2: 65-94.

MARGALEF, R. 1978. Life-forms of phytoplankton as survival alternatives in an unstable environment. *Oceanol. Acta,* 1: 493-509.

MARGALEF, R. 1985. Environmental control of the mesoscale distribution of primary producers and its bearing to primary production in the Western Mediterranean. In: *Mediterranean Marine Ecosystems* (Moraitou-Apostolopoulou, M. & Kiortsis, V., eds.): 213-229. Plenum Press. New York.

MARGALEF, R. & BALLESTER, A. 1967. Fitoplancton y producción primaria de la costa catalana de junio de 1965 a junio de 1966. *Inv. Pesq.*, 31: 165-182.

MARGALEF, R. & CASTELLVÍ, J. 1967. Fitoplancton y producción primaria de la costa catalana, de julio de 1966 a julio de 1967. *Inv. Pesq.*, 31: 491-502.

MARGALEF, R. & ESTRADA, M. 1987. Synoptic distribution of summer microplankton (Algae and Protozoa) across the principal front in the Western Mediterranean. *Inv. Pesq.,* 51: 121-140.

MASÓ, M. & PEPS. 1988. Datos oceanográficos básicos de las campañas FRONTS-3-85, FRONTS-6-85, PEP 86, FRONTS-11-86 y PEP 87 en el mar Catalán. *Datos informativos*, 24: 1-83.

NIVAL, P. 1965. Sur le cycle de *Dictyocha fibula* Ehrenberg dans les eaux de surface de Villefranche-sur-Mer. *Cah. Biologie Mar.*, 6: 67-82.

OVCHINNIKOV, I. M. 1966. Circulation in the surface and intermediate layers in the Mediterranean. *Okeanologiya*, 6: 48-5.

PEPS. 1986. Datos oceanográficos básicos de las campañas PEP-82, PEP-83 y PEP-84 en el mar Catalán. *Datos informativos*, 19: 1-100.

SOURNIA, L. 1982. Is there a shade flora in the marine plankton? *J. Plankton Res.*, 4: 391-399.

TINTORÉ, J., WANG, D. P. & LA VIOLETTE, P. E. 1990. Eddies and termohaline intrusions of the shelf/slope front off the Northeast Spanish coast. *J. Geophys. Res*, 95(C2): 1627- 1633.

TRAVERS, M. 1972. Le microplancton du golfe de Marseille. Méthodes d'étude microscopique des organismes. *Téthys,* 4: 535-558.

TRAVERS, A. & TRAVERS, M. 1968. Les Silico flagellés du Golfe de Marseille. *Mar. Biol.*, 1: 285-288.

VANDEVELDE, T., LEGENDRE, L., THERRIAULT, C., DEMERS, S & BAH, A. 1987. Subsurface chlorophyll maximum and hydrodynamics of the water column. *J. Mar. res.*, 45: 377-396.

VENRICK, E. L. 1978. The implications of subsampling. In: *Phytoplankton Manual* (Sournia, A., ed.): 75-87. UNESCO. Paris.

WANG, D. P., VIEIRA, M., SALAT, J. & TINTORÉ, J. 1988. A shelf/slope filament off the Northeast Spanish coast. *J. Mar. Res*,. 46: 321-332.

YENTSCH, C. S. & MENZEL, D. W. 1963. A method for the determination of phytoplankton chlorophyll and phaeophytin by fluorescence. *Deep-Sea Res.*, 10: 221- 231.

NERITIC ZOOPLANKTON SUCCESSION STUDIED WITH STATISTICAL ENTROPY MEASURES: THE ROLE OF SPACE AND THE HIERARCHICAL MATCHING OF THE ENVIRONMENT

JORDI FLOS[1] & AINA CARBONELL[2]

[1] Departament d'Ecologia. Facultat de Biologia. Universitat de Barcelona. Av. Diagonal, 645. 08028 Barcelona. Spain
[2] Instituto Español de Oceanografía. Centro Costero. Ap. Correos, 130. La Coruña. Spain.

Received: July 1990

SUMMARY

Succession in neritic zooplankton is studied, and the role of space and hierarchic adaptation to the environment assessed. Four stations located at the vertices of a square with sides of 1.5 miles, off Tarragona (Western Mediterranean), at depths between 20 and 50 m, were sampled from October 1984 to October 1985. Hydrographical data were obtained and zooplankton was caught by oblique hauls; copepoda were classified to the species level and the individuals were counted, as well as those belonging to other groups.

Two disturbances (perturbations) were identified in March and June, which gave rise to various successional events. The process has been studied by computing entropies, taking into account the space (conditional probabilities),which allowed us to calculate the contribution of space to total entropy, and the mutual information between space and populations. These calculations were performed separately for the copepoda populations and for the rest of zooplanktonic groups.

Evidence is given that during a successional episode initiated by a perturbation, relative mutual information between the space and the populations rises abruptly at the beginning and decreases slowly afterwards, until the next disturbance, while the maximum total entropy is obtained approximately two months after the beginning of the process. There is a time lag between the maximum of relative mutual information calculated for copepoda and that calculated for the groups of zooplankton. Taken together, this suggests that during succession, plankton structures itself hierarchically, it being possible to identify two steps in each episode: the first one is directed by external energy, the second is directed from the system itself. During each successional episode, the specific composition of the populations of copepoda changes continuously, but at a decreasing rate, while higher order taxonomic groups remain fixed in the first step but change in the second.

KEY WORDS: Zooplankton, succession, entropy, Western Mediterranean.

INTRODUCTION

MARGALEF (1958) described the general trends of marine phytoplankton succession and unified some concepts by relating them to terrestrial succession (MARGALEF, 1963). He also proposed the selection of phytoplankton life-forms during succession in an unstable environment (MARGALEF, 1978) and recently (MARGALEF, 1985), he put more emphasis on the spectra of disturbances that generate pulses of successional events, which might generate chaos in any model of marine populations (MARGALEF, 1986).

According to MARGALEF (1985, 1986), energy in ecosystems is continuously being converted into organization and information. External energy enters the system in an irregular way, at more or less disjointed centers and with different intensities. Each input of energy or disturbance starts an ecological succession and stops or resets the one that was in action. This resetting from outside remains unpredictable from inside the system and imposes small or large retrogressions in the degree of organization. As energy from an external input decays, spatial organization of the system develops, and although the system is governed and punctuated by inputs of external energy, it is expected that the process of ecological succession at any given place may follow predictable regularities.

Among the characteristics which change during succession (MARGALEF, 1968, 1974; ODUM, 1971), most attention has been given to the variation in species composition. It is also the only effect that can be studied without hypothesizing about the holistic behavior of the system; and so it is the minimum definition of ecological succession.

MARGALEF (1957) introduced the use of statistical entropy in ecology as a tool for measuring species richness or diversity, assuming that the set of species is a communication channel (SHANNON & WEAVER, 1949), and that its capacity in bits per individual may be calculated on the basis of species frequencies in the samples. In this way, the information per individual gives an idea of the width of the biological communication channel (MARGALEF, 1961), which depends on the system of codification used by the biologist (most of the time, species are used and individuals are considered to be interchangeable within the same species). The Shannon diversity measure (mostly in a spectral formulation) has been used since then in many ecological problems to study the spatial complexity of a system or to compare maturity stages within the successional process (MARGALEF, 1961, 1962, 1969a, 1969b, 1974, 1975; PIELOU, 1969, 1975). In the last decade the theory of communication and the concept of statistical entropy have been widely used in the study of systems complexity and self-organization, as well as in marine ecology (ATLAN, 1985; MARGALEF, 1985; CONRAD, 1983, 1985; ULANOWICZ, 1986). However, this does not prevent Margalef writing that "In my experience there is a widespread and noticeable dislike for the concepts of diversity and succession, expressed by ecologists who are otherwise fond of statistical approaches and claim to be the genuine reductionists" (MARGALEF, 1989).

In the present paper we analyse the temporal seasonal evolution of statistical entropy of zooplankton and the contribution of space in total entropy, from the winter-spring phytoplankton bloom to the end of the summer, in a neritic Mediterranean area, on the NE coast of Spain. Two disturbances or inputs of energy were identified that were followed by a process of succession. During succession, energy decays and is transmitted from the lower levels (primary producers), through the herbivores, to the higher levels in the zooplankton. This flux of energy helps the organization of the biological system made up of organisms of different types and species. From the point of view of quantity, energy input is followed by an increase of biomass. First of all, there is an increase in chlorophyll, which may be viewed as the actual disturbance for herbivores. Animal biomass may then be considered as the input for higher order trophic levels. The decay of energy gives way to an increase of species diversity and spatial structuring if the physical environment is stable enough. In consequence, there is an adherence between environment and plankton that is transmitted hierarchically (MARGALEF,

1961; LEVASSEUR *et al.*, 1984). In the present study we have not attempted to measure anything like a statistical entropy of environment (ATLAN, 1985), which could have helped a close analysis of the hierarchical transmission of spatial structures from the environment to the biological sphere. However, we have compared the evolution of the statistical entropy calculated in the group of copepoda (classified into species) and the whole zooplankton (classified into higher order taxonomic groups), looking especially at the contribution of space. Mutual information between space and taxonomic composition of organisms is displayed and interpreted.

MATERIAL AND METHODS

THE DATA

Hydrographic and zooplankton data used here were obtained in a series of cruises (PENTA) carried out by the Instituto Español de Oceanografía. Twelve stations covering an area of some 30 km^2 on the Catalan continental shelf (Western Mediterranean, NE coast of Spain), were sampled from October 1984 to October 1985. Four of the stations (at depths between 20 and 50 m) were sampled eleven times, while the others were sampled every two or three months. The frequency of sampling was higher from March to the end of the summer, and lower in winter (the longest gap was between December and March). In this study we present the analysis of the data obtained in these four stations, which were situated in the vertices of a square with a side of 1.5 miles (Fig. 1).

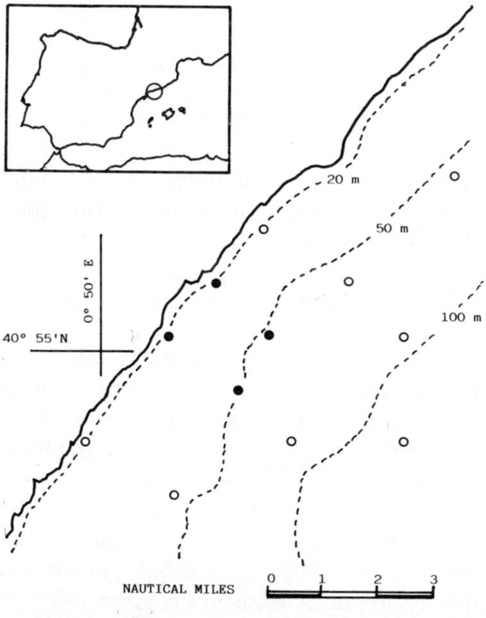

FIGURE 1. Geographical situation on the NE Spanish coast of the four stations analysed in the present paper (black dots), in the frame of the whole sampled grid (open circles).

Zooplankton was caught by oblique hauls, using nets of 250 µm mesh. Copepods were identified at the level of species, when possible, and counted (87 species or items). Other major groups were classified at higher taxonomic levels (18 groups), and counted (Tables I and II).

The small scale (1-10 km) structure of phytoplankton in the neritic waters of the NE Spanish coast has been studied by MARGALEF (1969a) and ESTRADA (1979), but there are no comparable studies of zooplankton in the same area. The most complete study of the neritic zooplankton in the area is the one by VIVES (1966). A good summary of the annual zooplankton cycle in the neritic waters of the NE coast of Spain can be found in ESTRADA *et al.* (1985).

Previous studies concerning zooplankton in the NW Mediterranean that implied a temporal high frequency sampling did not take space into account (e.g. IBANEZ & DALLOT, 1969) and a recent work that implied high frequency sampling in space could not resolve intermediate scales between daily and seasonal variability (BOUCHER *et al.*, 1987). In fact, plankton dynamics is now being more often studied in relation to hydrodynamics (LEGENDRE & DEMERS, 1984; ROTHSCHILD, 1988;

TABLE I. Mean number of individuals per cubic meter in the four stations studied of those species of copepoda that represented more than 1 % of the copepoda population, at some time during the sampling period. Those species indicated by sp. were not adult individuals. The number of indetermined copepodites are also shown.

Copepoda species	1984 Oct	Dec	1985 Mar	Apr	May	May	Jun	Jul	Aug	Sep	Oct
Calanus helgolandicus	0	1	0	11	15	41	48	13	0	0	0
Calanus sp.	11	3	2	3	12	0	0	0	1	0	1
Mecynocera clausi	0	0	0	0	0	10	0	0	0	0	0
Paracalanus parvus	96	12	26	68	42	125	439	209	12	21	10
Clausocalanus arcuicornis	76	11	0	0	4	7	10	10	2	1	3
Clausocalanus furcatus	3	2	0	0	1	0	5	0	7	1	0
Clausocalanus pergens	0	0	0	1	79	8	65	4	45	57	94
Clausocalanus sp.	9	1	271	2	10	9	24	9	25	7	7
Ctenocalanus vanus	1	14	170	3	36	2	9	2	0	0	
Diaixis hibernica	1	3	0	0	2	0	0	0	0	0	0
Temora stylifera	185	72	0	0	0	0	1	25	63	263	209
Pleuromamma sp.	0	10	3	0	0	0	0	0	0	0	0
Centropages typicus	6	0	24	4	25	72	197	15	0	0	1
Centropages violaceus	6	0	0	0	0	0	0	0	0	1	1
Centropages kroyeri	0	0	0	0	0	0	0	22	6	29	0
Centropages sp.	6	0	0	0	0	0	0	0	0	0	0
Isias clavipes	9	1	0	0	0	0	0	0	0	5	2
Acartia clausi	12	28	32	23	13	33	10	100	5	21	12
Acartia latisetosa	0	0	0	0	0	0	0	3	4	1	2
Oithona plumifera	60	45	8	4	7	1	2	3	9	36	96
Oithona helgolandica	3	1	11	24	29	108	58	48	0	1	0
Oithona nana	1	0	0	0	0	0	14	244	1	0	
Oithona sp.	17	3	2	0	0	0	0	0	0	0	0
Oncaea sp.	17	13	2	0	0	0	0	0	0	0	0
Corycaeus flaccus	0	3	0	0	0	0	0	0	0	0	1
Corycaeus latus	3	6	0	0	0	0	0	1	14	12	10
Corycaeus brehmi	0	2	5	1	4	0	0	1	2	3	1
Corycaeus furcifer	0	3	0	0	0	0	0	0	0	0	0
Corycaella carinata	12	2	0	0	0	0	0	0	0	0	0
copepodites indet.	16	7	15	6	9	9	14	10	4	8	5

LEGENDRE & LE FÈVRE, 1989), but space-time scale problems are difficult to solve in practice.

In the sampling region, currents along the coast at 20 to 50 m depth are of the order of 10 cm s^{-1}, changing their direction irregularly. The net flow ranges from 5 cm s^{-1} on the surface to 1 cm s^{-1} at 50 m depth (FONT, 1983). Sampling on the vertices of a square with a side of 1.5 miles ensures that one is not sampling the same body of water in the interval of a few hours, while gathering the spatial variability linked to internal and inertial waves and horizontal turbulent patches (e.g. DICKEY, 1988). On the other hand, the temporal sampling intervals (less than one month on average between March and October 1985) seem appropriate for examining the seasonal successional evolution of zooplankton (e.g. STEELE, 1988).

MEASUREMENT OF ENTROPIES

The basic expression for the measurement of entropies in each sampling or cruise considers the species (or zooplanktonic groups) and the geographical positions as the two sources of total statistical entropy, so that total and marginal frequencies are used in the calculations. The relationship among the different entropy measures is given by:

$$HT = HP + HS/P = HS + HP/S,$$

TABLE II. Mean number of individuals per cubic meter in the four stations studied of the more frequent zooplanktonic groups.

Zooplankton group	1984 Oct	Dec	1985 Mar	Apr	May	May	Jun	Jul	Aug	Sep	Oct
Medusae	3	5	23	56	12	9	1	2	2	1	1
Siphonophora	26	56	23	785	75	136	153	187	13	37	27
Copepoda	552	254	582	154	299	428	904	735	211	472	474
Cladocera	424	12	4	3689	231	1646	231	4167	216	155	38
Ostracoda	1	11	0	2	1	1	1	1	0	0	0
Chaetognatha	32	52	0	3	5	3	5	61	8	30	29
Pteropoda	5	3	0	11	21	19	1	0	1	0	13
Appendicularia	71	98	44	387	90	71	145	76	24	99	65
Salpae	1	6	0	0	0	0	0	0	0	0	1
Doliolidae	4	17	0	0	0	0	9	58	35	24	9
larv. Echinodermata	2	8	7	10	2	0	0	1	0	1	1
larv. Crustacea	7	13	14	37	62	43	30	31	10	11	4

where HT is the total entropy, HP (P stands for population) is that of species (or groups) joining the four samples of the same cruise, HS/P is the mean entropy of stations (S stands for station or space) weighted by the populations (species or groups), HS is the entropy of stations computed with the frequencies obtained with the total number of individuals per station, and HP/S is the mean population entropy weighted by the stations.

It is important to note, and to bear in mind in all that follows, that the two sets of entropies (one calculated from the specific composition of copepods and the other from the groups of zooplankton) are not completely comparable. In the first case, individuals of the same species can be considered interchangeable, but in the second, individuals of the same taxonomic group are not interchangeable. So, interpretation must be very cautious and must take into account the principle that has been used in the division of the set into separate items (MARGALEF, 1989). Diversity is a function of the organization of the ecosystem and is expressed and should be recognized at all levels of observation. In order to estimate the distribution of overall diversities, MARGALEF (1969b) recommended the use of a taxonomic group that covers a broad ecological spectrum, and not to use groups that are ecologically more specialized. In the marine environment, copepods are present throughout the year, and cover a wide spectrum of trophic positions. The measures of diversity made on the set of zooplanktonic groups has a different meaning as we shall stress later in the text.

If instead of considering the global system containing individuals and space as components we consider populations and space as two interacting systems, one of the SHANNON & WEAVER (1949) formulae allows us to compute the transmitted information between S and P

$$T(S,P) = HP - HP/S = HS - HS/P$$

where HP and HS are the information content of the output in each of the two interacting systems, and HP/S and HS/P are the corresponding ambiguity-functions, using the terminology of the theory of communication in a channel with noise.

We can compute redundancy or relative mutual information between space and population:

$$R = (Hmax - HT)/Hmax,$$

where Hmax = HP + HS.

RESULTS AND DISCUSSION

DISTURBANCES

In order to show the general evolution of the system and the moments when the main inputs of external energy took place, a synthesis of the hydrographic data (temperature, salinity, nutrients and chlorophyll) obtained during the sampled period is presented in figures 2-4. Two perturbations can be recognized in March and June. The first is typical at the end of winter, when the environment shifts from a high nutrient and well mixed situation to a more stable one, with growing incoming irradiance. In fact, the March sampling shows high and uniform salinity in the water column, but nutrients, although they are relatively high, have already dropped to one third of those found in December, while chlorophyll is more than 2 mg m^{-3}. So, the physical perturbation occurred before the March sampling, and phytoplankton had already reacted in March (response time of phytoplankton is of the order of 10 days or less). It is the phytoplankton bloom that must be considered to be the actual input of energy from the point of view of zooplankton.

The second disturbance occurred between the last sampling in May and that in June, and was due to an input of nutrients through the surface. It must be remembered that the stations are close to the coast (between depths of 20 and 50 m) and that they may be subjected to continental inputs as well as to perturbations coming from the shelf or the shelf break zones. In June, high levels of nutrients were detected at several stations. Some of them presented nutrient ratios characteristic of deep water (well compensated, from below the thermocline) while at least in one station, relatively high proportions of organic nitrogen and ammonia clearly showed a sewage outfall influence. Phytoplankton response to this perturbation is detected in the sampling of July, so that, from the point of view of zooplankton, the disturbance shows two aspects: a) an input of waters from different

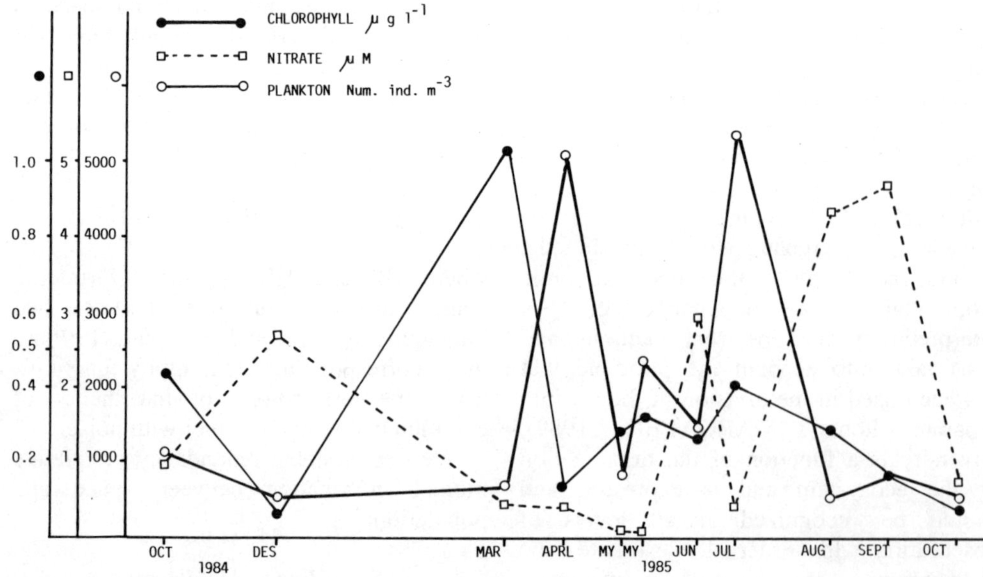

FIGURE 2. Mean values for the upper 20 m and the four stations studied of chlorophyll (µg l^{-1}) and nitrate (µM). Zooplankton is the mean value integrating from bottom to surface (20 or 50 m, depending on the station). Time of sampling is indicated on the horizontal axis.

sources in June with a general increase of nutrients, with an almost perfect vertical thermal homogeneity from the surface to at least 20 m depth; followed by b) an increase of exploitable phytoplanktonic biomass in July.

A maximum of nutrients was also observed in August, but it was followed neither by an increase of phytoplankton, nor by an increase of zooplanktonic biomass. So, this rise in nutrients cannot be considered a disturbance from the point of view of the plankton.

A first and simple approximation to the response of zooplankton to environmental disturbances is the distribution of the total number of individuals over time (Fig. 2). Copepods are always present during the annual cycle and their population most often accounts for 40 to 80 per cent of the total number of zooplanktonic organisms (Table II). However, they never exhibit numbers as high as those presented by cladocerans (maxima in April and July, when copepoda represent only 3 and 14 per cent of total organisms respectively). The higher numbers for total zooplankton are usually due to dense populations of cladocerans. In the first peak of zooplankton (Fig. 2), the density of cladocerans is very different from one station to another, while in the others they are more uniformly distributed in space. Other taxonomic groups contribute significantly to the relative maxima at the beginning of May. At the end of May the zooplankton is rather diverse, with cladocera, copepoda, siphonophora, appendicularia, medusae, pteropoda and chaetognatha larvae as main groups. In June, copepoda increase and cladocera decrease. In July, cladocera increase again, doliolids become important and chaetognatha increase. The disturbance of June was not uniform in space and neither was the response of plankton.

ZOOPLANKTON SUCCESSION

In what follows, it is important not to forget that entropy may be understood as uncertainty, so that a higher entropy means a more homogeneous set of frequencies. When entropy refers to populations (calculated on the frequencies of the different species or groups) high entropies

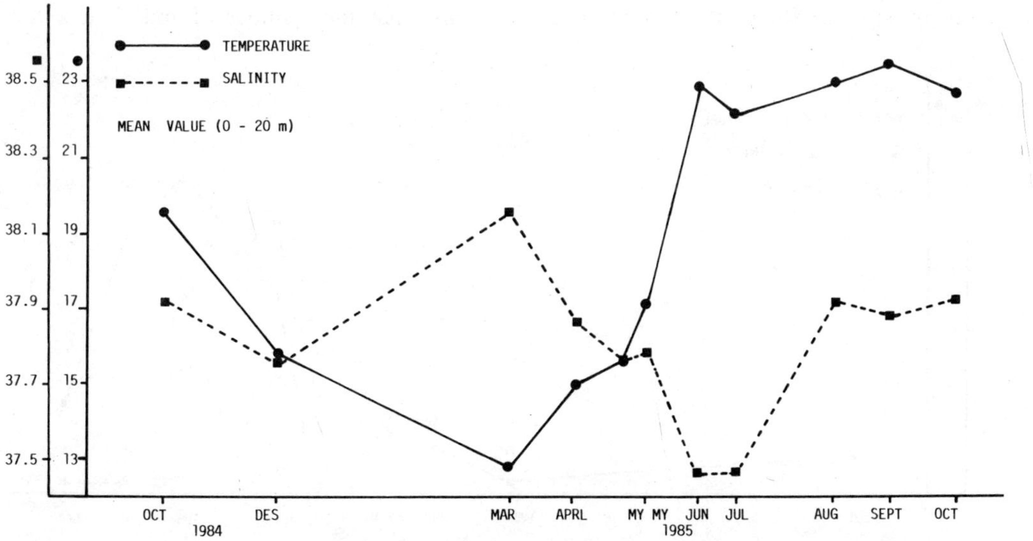

FIGURE 3. Temporal evolution of the mean values of temperature (°C) and salinity for the upper 20 m and the four stations studied.

mean uniformity in the frequencies of species (high diversity) and low entropies mean that some species (or groups) are clearly more abundant than the rest. When the entropy refers to space (calculated on the stations, using the relative abundance of individuals in each station as frequency), high entropies mean uniformity in the total number of individuals per station, and low entropies mean that one or two of the stations present a much higher number of individuals than the others (there is heterogeneity among stations in terms of the total amount of individuals or biomass).

Absolute entropy values obtained for the populations of copepods are usually higher than those obtained for the zooplanktonic groups, because there are more species of copepoda than groups of zooplankton considered. On the other hand, there are always only four stations, so that the entropy of space is of the same order irrespective of the way in which the frequencies are calculated. In consequence, the relative contribution of space in the total entropy is higher when the groups of zooplankton are considered.

In figure 5 the evolution of total entropy is presented over a time span. The general trend is similar for both copepods and groups of zooplankton, with some differences that will be discussed later. The first input of energy in March is reflected by the lowest total entropy. At that time, almost eighty percent of zooplanktonic individuals are copepods, but the total number of individuals is already rather low (less than 1,000 individuals per cubic meter, Table II). In April, cladocera increase in numbers and dominate the zooplankton, while copepoda decrease. Entropy of zooplanktonic groups remains low, but copepoda entropy increases a little. The highest entropy is reached in May, for both copepoda and groups of zooplankton. The second input of energy is also followed by an increase of cladocera (July) which is reflected by a low total entropy calculated for zooplankton groups, while the total maximum entropy is reached for both sets in August.

Maximum total entropy occurs in May and August, approximately two months after the inputs of energy. However, looking at the contribution of space in total entropy (Fig. 6) we realize that the absolute contribution when calculated for copepods rises from March to June, decreases in July and rises again during the summer. This means that the number of individuals for

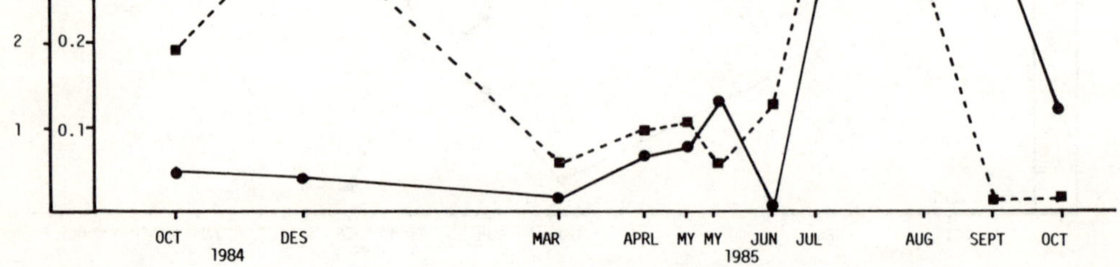

FIGURE 4. Temporal evolution of the mean value of the vertical standard deviation (in the upper 20 m) of temperature and salinity in the four stations studied.

the different species becomes relatively more uniform in space during succession. The entropy of stations calculated on the zooplankton groups has its minimum in April (when there is the maximum of cladocera) and rises in May, maintaining thereafter a similar absolute contribution until the end of the sampling period in October.

In terms of relative contribution, that is, the fraction of total entropy (HT) accounted for by HS/P, we realize that it is maximum, both for copepoda and zooplanktonic groups, in March, at the beginning of the successional process. The minimum is found at the time the total entropy is maximum (May and August), and rises at the end of the successional sequence. On the one hand, this U-shaped pattern follows approximately the inverse of the total entropy, which is logical because total entropy enters the calculations in the denominator (HS/P / HT). However, the mutual information between space and species (or zooplanktonic groups) gives us a clue to the interpretation (Fig. 7). Mutual information (or redundancy in relative terms), is minimum at the time of the first input of energy for both copepods and zooplanktonic groups. For copepods, it rises to a maximum in the following sampling, to decrease steadily afterwards until the next input. For zooplanktonic groups, the rise is somewhat delayed, as is the second minimum after June.

In March, 6 species out of 22 account for more than 90% of the total number of copepoda, the same in every station. It seems that the input of energy represents a spatial equalization of biomass (high relative contribution of space) and a simplification of the composition of communities (low diversity). In April, the number of species of copepoda present in the samples decreases to 17 and only the biomass of a few species increases markedly in some spots, while the most abundant species present in March disappears from the scenario (because of decrease of relative contribution of space, increase of redundancy, diversity does not increase significantly). Once the first bloom

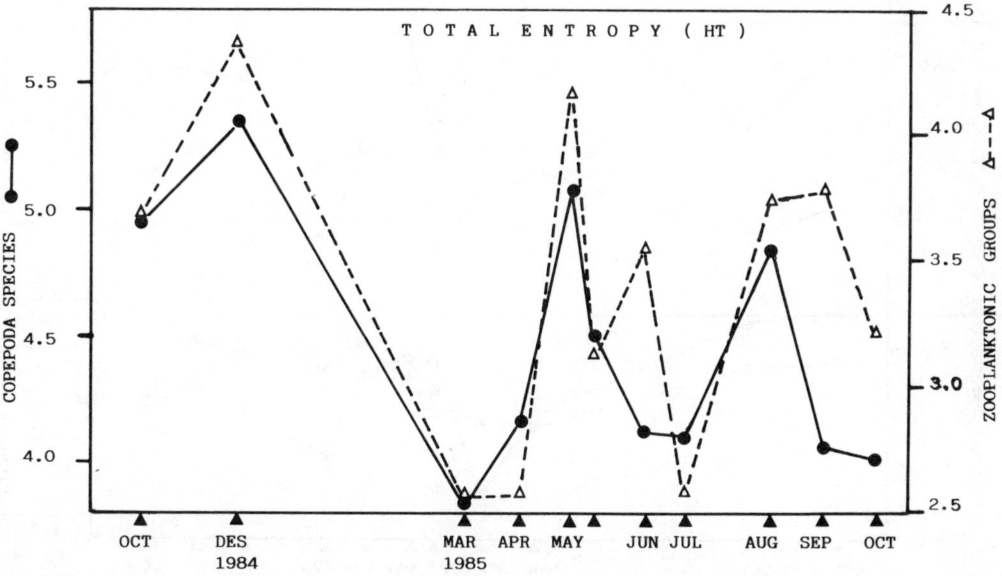

FIGURE 5. Temporal evolution of total entropy (HT) calculated for the population of copepoda and the groups of zooplankton.

of algal biomass has been incorporated in the trophic web, communities begin to organise, and the flux of energy per unit of standing biomass decreases considerably. The fact that in May chlorophyll is a little higher than in April, indicates that a flux of energy is maintained, helping the diversification of herbivores, but the presence of animals and presumably an increase of detritus and micro-heterotrophic organisms gives a more diverse environment from a trophic point of view (with a maximum of diversity for copepods and zooplanktonic groups in May and August). However, this moment of maximum specific diversity corresponds to a decreasing trend for redundancy in copepods and maximum redundancy in zooplankton groups. This moment corresponds to the minimum of relative contribution of space to total entropy and means that a maximum of specific diversification is reached, but without uniformization in space. From this point, entropy is transferred to space, or in other words, spatial complexity reduces its grain, which is compatible with a reduced flux of energy and an increase of environmental physical stability. The decreasing trend in redundancy means a decrease in the degree of communication between space and species from the point of view of the biologist who is taking the samples. In fact, populations become more independent of space, and have the capacity to respond without great changes to the small disturbances occurring in an increasingly stable environment. The decrease in diversity with increasing stability in later successional stages was already pointed out by MARGALEF (1961).

In March and June we find the maximum concentrations of copepoda, and in April and July those of cladocera. The first response of zooplankton to each input of auxiliary energy ends with maximum

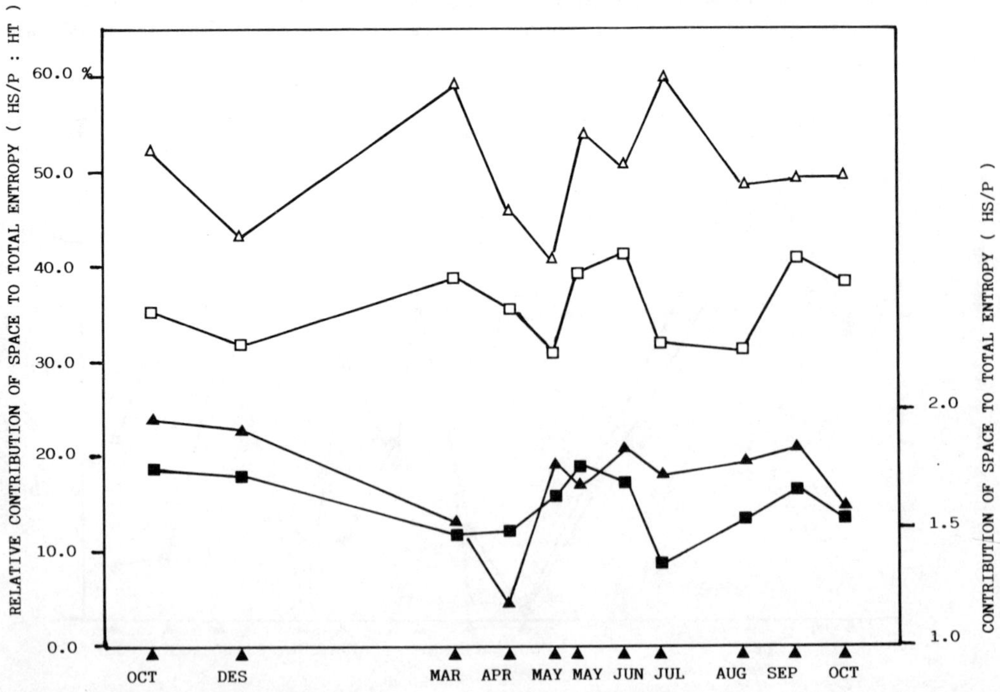

FIGURE 6. Absolute (right) and relative (left) contribution of space to total entropy for the population of copepoda (squares) and the groups of zooplankton (triangles).

diversity (in May and August). Changes of diversity are the outcome of the interference between mechanisms that regulate the numbers of species and those that are operative in relation to the total biomass or abundance. Diversity decreases when it is easier to add a species than to add an individual (MARGALEF, 1989). In other words, diversity drops when scarcity begins to play a role, and species are compelled to rework the biological matter that has entered the system. Then species become more connected. There are no means for populations to grow almost free from competition, and while energy is transferred to higher trophic levels, connectivity among species increases, bringing a concomitant decrease in diversity (MARGALEF & GUTIÉRREZ, 1983).

Comparing the total entropy measurements of the zooplankton groups and of copepods it may be seen that entropy of zooplankton groups remains relatively high while that of copepoda decreases (May-June, August-September). On the other hand, at least after the first input of energy, total entropy for zooplanktonic groups increases more slowly (or later) than the entropy for copepods. These facts, together with the delay presented by the maximum redundancy of zooplanktonic groups with respect to that of copepoda, indicates that succession proceeds in a hierarchical way (ALLEN & STARR, 1982; O'NEIL *et al.*, 1986).

Finally, we computed the quotient between the diversity of species (or zooplanktonic groups) calculated from the sample obtained joining the eight samples from two consecutive cruises (the same four stations), and the mean of the diversities calculated for each sample. This quotient reflects the relative change of species (or zooplanktonic groups) with time. A quotient greater than one indicates change of species or groups, and of less

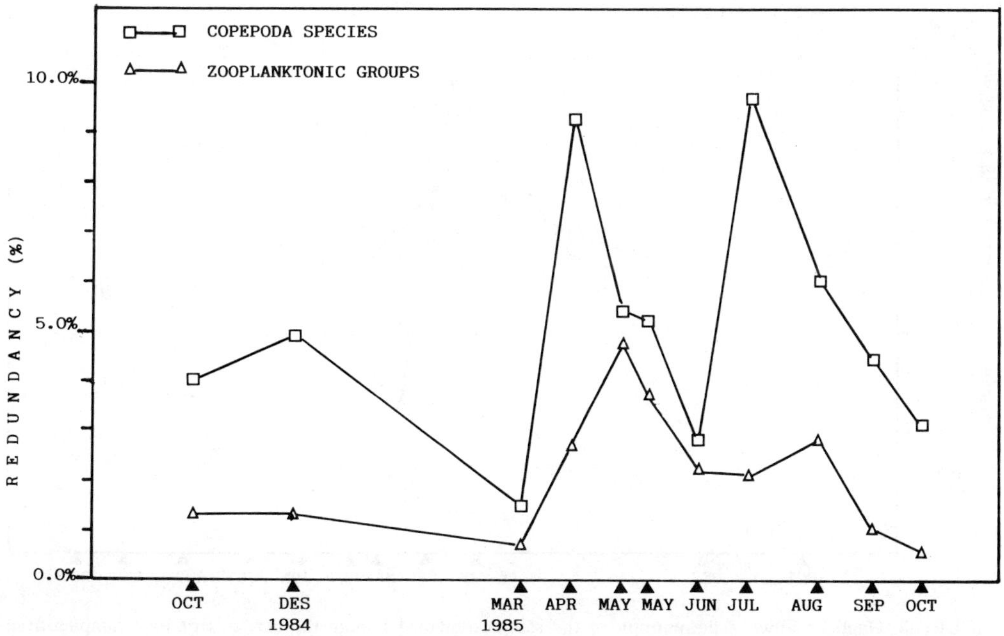

FIGURE 7. Relative mutual information (redundancy) between space and populations calculated for copepoda and for groups of zooplankton.

than one indicates a reinforcement of the present species or groups (Fig. 8). It can be observed that the greater change takes place between March and April, after the first input of energy, both for groups of zooplankton and for copepoda. Throughout the succession process, the specific composition of copepoda is changing, but with decreasing intensity, while zooplanktonic groups are reinforced between April and May.

At the end of the first successional sequence, zooplanktonic groups change again, but less than in March, and the rate of specific change of copepoda increases after the June input. From June to the end of summer, copepoda follow the same trend observed in spring, and zooplanktonic groups also show the reinforcement in mid-succession.

The unity of each successional event, which is initiated by a pulse of auxiliary energy, progresses in an increasingly stable environment and ends in another disturbance, is recognized by the trends of two parameters calculated from the populations of copepoda: redundancy (Fig. 7) and the rate of change of the specific composition (Fig. 8). However, measurements made on the groups of zooplankton allow us to recognize two parts in the successional event.

The first culminates with the maximum of copepod diversity (May and August), coinciding with the minimum contribution of space to total entropy, and reinforcement of zooplanktonic groups. The second part sees the diversity of copepoda decrease while zooplanktonic groups change and their diversity increases (June and September). Maxima of redundancy reflect the maximum of communication or matching between environment and populations. The initial increase of redundancy measured in the population of copepods (April and July) indicates this

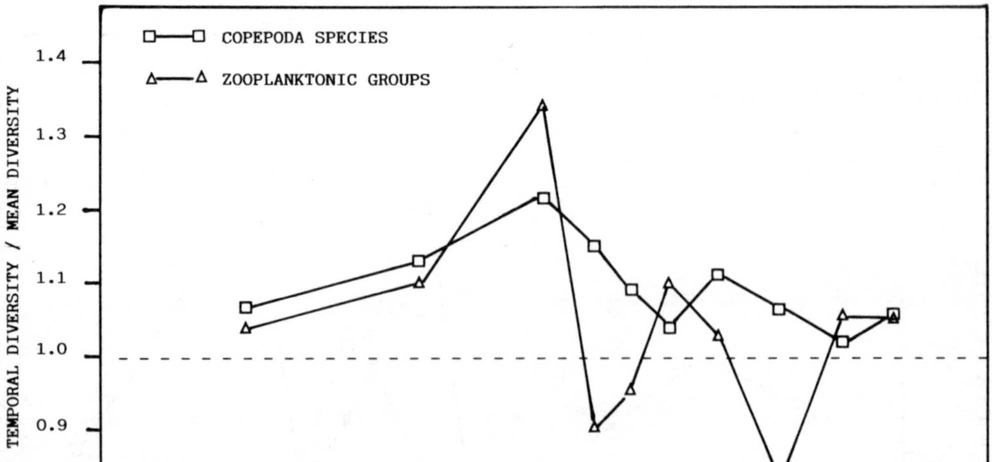

FIGURE 8. Quotient between the entropy of the sample obtained joining the two sets of four samples from successive cruises and the mean of the entropies of populations calculated from each one of the eight samples. The quotient was computed for copepoda and zooplanktonic groups. Values greater than one mean change and less than one mean reinforcement of previous biological composition. See text.

matching between chlorophyll and herbivores (represented in this case by copepoda), while the maximum of redundancy measured in the groups of zooplankton, indicates the matching of the first established populations that incorporated the input of energy, and the secondary populations that follow and feed on the organic matter accumulated in the environment.

The reinforcement of zooplanktonic groups, that is, their persistence over time, indicates their effective adaptation to the environment and accompanies their specific diversification (as is the case of copepoda). Once the initial fuel is consumed, a microbial food loop (POMEROY, 1974) develops and secondary phytoplankton communities emerge, which represent an environmental change to be coped with by zooplankton. The maximum redundancy of zooplanktonic groups indicates a match between what must then be considered the environment and zooplankton. Later, the rate of change of zooplanktonic groups increases. If the first steps in succession are governed by the degree of matching between temporal characteristics of physical phenomena and those of biotic responses (LEGENDRE & LE FÈVRE, 1989), in later stages, with increasing environmental physical stability, it is the biological composition and its chemical and physical implications that drive the system. Time lags appear between the first driving physical event and the resulting changes in pelagic food chains, but also between the first changes in pelagic communities and the following biological responses. We see here all the elements for a hierarchical organization of zooplankton during succession (O'NEIL *et al.*, 1986).

These results are compatible with the increase of the mean size of organisms during succession (RODRÍGUEZ *et al.*, 1987) as most bigger animals belong to some of the non-copepod groups. A shift in the mean size of organisms to larger forms must give a decrease of species diversity measured within a taxonomic group (IGLESIAS, 1988), but diversity of groups can be maintained or even increased.

The trends observed give support to the idea that the communities selected or established in the first stages of succession, until diversity is maximum, become organized thereafter, increasing the number of effective connections in the food web, which is only possible if species diversity decreases. On the other hand, small disturbances occurring during the later stages of succession are assimilated by the populations. Their specific composition changes continuously though with decreasing rates, so that there is no equilibrium and we cannot speak of a persistent and steady state composition of species, although we may consider that this is the actual form of increasing stability in planktonic ecosystems. Manna models (WANGERSKY & WANGERSKY, 1980, 1983) predict that in stable conditions the total diversity can be constant but the species composition changes constantly.

We hold that the increase of spatial uniformity (mean entropy of space weighted by species) reflects the miniaturization of structure in very stable environments. Since our samples are integrated vertically through oblique hauls, we are dealing with horizontal spatial structuring. Patterns at scales of less than 1000 m, which are more closely linked to vertical structures, fall outside our observation window.

CONCLUSIONS

Zooplanktonic succession begins and is reset by inputs of energy. The two successional episodes observed in the present study are slightly different: the first disturbance represents a very large environmental change (vertical stabilization, increasing irradiance and temperature), so that many of the present species, adapted to a low flux of energy in an unstable and cold medium, cannot persist. In terms of primary and secondary succession, this first input triggers what we could call a seasonal primary successional process in the neritic ecosystem. When the second input takes place, the environment is rich in biological types of zooplankton organisms, already adapted to high temperatures and vertical stability, and the microbial food loop is well developed. Response times are shorter and chlorophyll does not accumulate in the environment. During the summer inputs are not important enough to override the system that has been established in a physically stable environment, and the system adapts to disturbances without drastic changes of the biological composition of zooplankton. It must be said that the weather in October 1985 was much more a continuation of summer than the beginning of fall, so that we could not detect any interruption of the second successional process before our last sampling.

We conclude that decreasing mutual information reflects the internalization of environmental changes or disturbances. The shift from a physically driven system to a biologically governed one, although continuous, is marked by the maximum redundancy between space and zooplanktonic groups, and is reflected by a secondary increase in total entropy of zooplanktonic groups. During succession, there is no stabilization of the taxonomic composition, but a decreasing rate of its change, as well as miniaturization of spatial structuration. It is possible to look at these results as a case of a shift from a Lotka-Volterra to a manna type of community growth as conditions go from excess food evenly distributed to scarce food patchily distributed.

During succession, specific diversity measured on traditional samples (vertical or oblique hauls) reaches its maximum halfway through the process, while there is still a high enough externally driven flux of energy to allow a reduced degree of inter and intra-specific interactions. The accumulation of matter in life-forms reduces the energy flux per unit of biomass and brings scarcity into play. The reworking of matter forces the increase of connectivity among species and the decrease of specific diversity. One of the effects of this would be a hierarchical shift of control to those species in plankton with a broader activity range (bigger, with a higher motility). These species would help the spatial organization of plankton and lengthen the trophic or behavioral paths connecting species. This would represent a decrease of connectivity as some potential relations could be cut, so that a hierarchy or "chain of command" structure (MARGALEF, 1989) could develop.

Further research is needed to elucidate whether there is actually a shift in the scale of spatial structuring in the system (miniaturization) that might help to relieve effective connections among species in a higher order spatial scale, which could preserve a relatively high overall specific diversity; this diversity would otherwise be much lower.

ACKNOWLEDGEMENTS

This study was possible thanks to a contract between the IEO, INTECSA and FECSA. Thanks are also due to all the people participating in the PENTA cruises. The comments of Peter Wangersky are gratefully appreciated.

REFERENCES

ALLEN, T.F.H. & STARR, T.B. 1982. *Hierarchy: perspectives for ecological complexity*. University of Chicago Press. Chicago.

ATLAN, H. 1985. Information Theory and Self-Organization in Ecosystems. *Can. Bull. Fish. Aquat. Sci.*, 213: 187-199.

BOUCHER, J., IBÁÑEZ, F. & PRIEUR, L. 1987. Daily and seasonal variations in the spatial distribution of zooplankton populations in relation to the physical structure in the Ligurian Sea Front. *J. Mar. Res.*, 45: 133-173.

CONRAD, M. 1983. *Adaptability: The significance of variability from molecule to ecosystem.* Plenum Press. New York.

CONRAD, M. 1985. The statistical basis of ecological potentiality. *Can. Bull. Fish. Aquat. Sci.*, 213:179-186.

ESTRADA, M. 1979. Observaciones sobre la heterogeneidad del fitoplancton en una zona costera del Mar Catalán. *Inv. Pesq.*, 43: 637-666.

ESTRADA, M., VIVES, F. & ALCARAZ, M., 1985. Life and productivity of the open sea. In: *The Western Mediterranean* (R. Margalef, ed.): 149-197. Pergamon Press. Oxford.

FONT, J. 1983. Corrientes permanentes en el borde de la plataforma continental frente al delta del Ebro. *Estudio oceanográfico de la plataforma continental. Seminario Científico.* Cádiz 15-18 marzo, 1983:230-248.

IBÁÑEZ, F. & DALLOT, S. 1969. Étude du cycle annuel des chaetognates planctoniques de la rade de Villefranche par la méthode d'analyse des composantes principales. *Marine Biology*, 3:11-17.

IGLESIAS, R. 1988. *Diversidad taxonómica y ataxonómica en poblaciones de insectos: un ejemplo del ecosistema restinga.* Ph. D. Thesis. University of Barcelona.

LEGENDRE, L. & DEMERS, S. 1984. Towards dynamic biological oceanography and limnology. *Can. J. Fish. Aquat. Sci.*, 41:2-19.

LEGENDRE, L. & LE FÈVRE, J. 1989. Hydrodynamical singularities as controls of recycled versus exported production in oceans. In: *Productivity of the ocean: present and past* (W.H. Berger, W.S. Smetacek & G. Wefer, eds.): 49-64. J. Wiley. New York.

LEVASSEUR, M., THERRIAULT, J.C. & LEGENDRE, L. 1984. Hierarchical control of phytoplankton succession by physical factors. *Mar. Ecol. Prog. Ser.*, 19:211-222.

MARGALEF, R. 1957. La teoría de la información en ecología. *Mem. real. Acad. Ciencias Artes Barcelona*, 32:373-499.

MARGALEF, R. 1958. Temporal succession and spatial heterogeneity in natural phytoplankton. In: *Perspectives in Marine Biology* (A.A. Buzzati-Traverso, ed.): 323-349. University of California Press. Berkeley.

MARGALEF, R. 1961. Communication of structure in planktonic populations. *Limnol. Oceanogr.*, 6:124-128.

MARGALEF, R. 1962 Succession in marine populations. *Adv. Frontiers Plant Sci.*, 2:137-188.

MARGALEF, R. 1963. On certain unifying principles in ecology. *Am. Nat.*, 97:357-374.

MARGALEF, R. 1968. *Perpectives in ecological theory.* University of Chicago Press. Chicago.

MARGALEF, R. 1969 a. A small scale distribution of phytoplankton in the Western Mediterranean at the end of July. *Publ. Staz. Zool. Napoli*, 37 suppl.:40-61.

MARGALEF, R. 1969 b. Diversity and stability: a practical proposal and a model of interdependence. In: *Diversity and stability in ecological systems,* Brookhaven Symposia in Biology, 22:25-37.

MARGALEF, R. 1974. *Ecología.* Omega. Barcelona.

MARGALEF, R. 1975. Diversity, stability and maturity in natural ecosystems. In: *Unifying concepts in ecology* (W.H. Van Dobben & R.H. Lowe-McConnell, eds.): 151-160. Junk. The Hague.

MARGALEF. R. 1978. Life-forms of phytoplankton as survival alternatives in an unstable environment. *Oceanol. Acta*, 1: 493-509.

MARGALEF, R. 1985. From hydrodynamic processes to structure (information) and from information to process. *Can. Bull. Fish. Aquat. Sci.*, 213: 200-220.

MARGALEF, R. 1986. Reset successions and suspected chaos in models of marine populations. *Int. Symp. Long Term Changes Mar. Fish Pop.*, Vigo: 321-343.

MARGALEF, R. 1989. On diversity and connectivity as historical expressions of ecosystems. *Coenoses*, 4: 121-126.

MARGALEF, R. & GUTIÉRREZ, E. 1983. How to introduce connectance in the frame of an expression for diversity. *Am. Nat.*, 121: 601-607.

ODUM, E.P. 1971. *Fundamentals of ecology.* Saunders. Philadelphia.

O'NEIL, R.V., DEANGELIS, D.L., WAIDE, J.B. & ALLEN, T.F.H. 1986. *A hierarchical concept of ecosystems.* Princeton University Press. Princeton.

PIELOU, E.C. 1969. *An introduction to mathematical ecology.* Wiley. New York.

PIELOU, E.C. 1975. *Ecological diversity.* Wiley. London.

POMEROY, L.R. 1974. The ocean's food web, a changing paradigm. *Bioscience*, 24: 499-504.

RODRÍGUEZ, J., JIMÉNEZ, F., BAUTISTA, B. & RODRÍGUEZ, V. 1987. Planktonic biomass spectra dynamics during a winter production pulse in the Mediterranean coastal waters. *J. Plankton Res.*, 9: 1183-1194.

ROTHSCHILD, B.J.(Ed.) 1988. *Toward a theory on biological-physical interactions in the world ocean.* Kluwer-Academic Press. Dordrecht.

SHANNON, C.E. & WEAVER, W. 1949. *The mathematical theory of communication.* University of Illinois Press. Urbana.

STEELE, J.H. 1988. Scale selection for biodynamic theories. In: *Toward a theory on biological-physical interactions in the world ocean* (B.J. Rothschild, ed.): 513-526. Kluwer-Academic Press. Dordrecht.

ULANOWICZ, R.E. 1986. *Growth and development: ecosystems phenomenology.* Springer. New York.

VIVES, F. 1966. Zooplancton nerítico de las aguas de Castellón (Mediterráneo Occidental). *Inv. Pes.*, 30:49-166.

WANGERSKY, P.J. & WANGERSKY, C.P. 1980. The mana effect: A model of phytoplankton patchiness in a regenerative system. *Int. Revue ges. Hydrobiol.*, 65: 681-690.

WANGERSKY P.J. & WANGERSKY, C.P. 1983. The manna effect: Paradox of the plankton. *Int. Revue ges. Hydrobiol.*, 68: 327-338.

(J. D. Ros & N. Prat, eds.). 1991. *Oecologia aquatica*, 10: 203-213

PROCESSES OCCURRING DURING FISH LARVAE STAGES AS DETERMINANTS OF YEAR-CLASS STRENGTH IN FISH POPULATIONS IN THE NORTH-WESTERN MEDITERRANEAN

ANA SABATÉS & JOSEP-MARIA GILI

Institut de Ciències del Mar (CSIC). Passeig Nacional s/n. 08039 Barcelona. Spain.

Received: April 1990.

SUMMARY

The spatial and temporal mesoscale distribution pattern of fish larvae on the Catalan coast is closely linked both to mesoscale hydrodynamic mechanisms in the area, and to the distribution patterns of the remainder of zooplankton, including their predators and potential prey. Spatial coincidence with their most likely prey and the non-coincidence in time with their predators or potential competitors for the same prey are considered to be the mechanisms that most contribute to the survival of fish larvae in the area studied. The set of processes described which occur during the larval fish stages on the Catalan coast is one of the factors that may explain the fact that the number of adults captured has remained constant over recent years, despite the fact that the fishing effort has suffered few variations, that fishing power has increased (a greater number of ships with more efficient techniques), and that fishing techniques have diversified (there is now a tendency to exploit a greater number of species).

KEY WORDS: Fish larvae, spatio-temporal distribution, potential prey and predators, North-Western Mediterranean.

INTRODUCTION

Over the last few decades, fish populations along the Catalan coast have been intensely exploited by the local fishing fleets. However, recent studies in which stocks are evaluated on the basis of hauls and of size distributions have led to the hypothesis that despite the high intensity of fishing activities, stocks have reached a state of equilibrium (MARTÍN, 1989). The representativity of those studies, taking into account the fishing statistics for a given area in order to evaluate the evolution of the actual fish populations, is a factor that has been discussed at length (CADDY & GARCÍA, 1983; GARCÍA, 1984). The two main aspects argued are that the studies concentrate on those species of greatest commercial interest, and that the catches should be considered as representative only during the period of the fish life cycle, when the individuals can be caught using fishing techniques. Nevertheless, other information on fish populations, different or complementary to that offered by fisheries, is scarce or inaccurate. In the Mediterranean especially, the fishery is multispecific: a considerable number of species are exploited and fishing techniques have become diversified. This has meant that information on the evaluation of stocks is very incomplete, which is an obstacle to the attainment of a global vision, a vision which today is available only for those species of greater commercial interest.

The fishing techniques used off the Catalan coast are very diversified, ranging from traditional fishing gear to others of greater impact on fish populations such as trawling and purse seining. This has led to a situation in which the different types of fishing techniques tend to select different species and different fractions of the same population (LLEONART, 1990). Despite these limitations to evaluate adult populations, the evolution of fisheries on the Catalan coast gives us a fairly accurate idea of how the populations of the commonest fish species in the area have evolved. If the intensity of fishing activities has varied little (especially trawling, although there has been an increase in traditional methods), the level of hauls has remained constant (even among the fish first-year class; MARTÍN, 1989) and, furthermore, immature specimens represent a major proportion of the total catch. There must, then, be mechanisms or phenomena within the life cycle of the fish, and specifically in the area studied, which keep the stocks constant. Many pelagic and some demersal species have a high fecundity, which means that the number of eggs and yolk-sack larvae in the sea will initially be very high; however, if there are no mechanisms at work to maximize the survival of the larvae until they reach the juvenile stage, only with great difficulty will these stocks of adults be maintained at a constant level.

Recently, a great number of authors seem to coincide in the description of the most general mechanisms that ensure that the larvae of a given species survive in sufficient numbers to guarantee the recruitment and development of the fish first-year class. These mechanisms can be summarized into two general theories which, despite an initial discrepancy as regards the importance that different authors attach to them, must be considered together, though at different scales of space and time.

In the first place, the need to have sufficient food available is essential for the development of any planktonic organism, such as fish larvae. These, like all plankton, are distributed in a wide three-dimensional space where they develop more or less together in groups or patches. The time and place of encounter between larvae and their prey is particularly crucial during the critical period (HJORT, 1926) when larvae come to depend on external resources once they have absorbed the yolk. When this moment is reached, the availability of sufficient and adequate prey is a limiting factor in larval survival, which led CUSHING (1975) to postulate his theory of match-mismatch. More recently, and gathering the opinions of other authors, BAILEY & HOUDE (1989) have placed predation on larvae by other planktonic and nektonic organisms at the same level of importance as the availability of food when it comes to guaranteeing success in the recruitment of a given species.

In the second place, the close relationship between hydrodynamic mechanisms and the dispersion-concentration of zooplankton organisms, led ILES & SINCLAIR (1982) to postulate their theory of hydrographic retention areas. These authors link the survival of larvae to the possibility of developing, in geographically limited areas, particular hydrodynamic features that ensure the integrity of larval populations. Subsequently, and with the idea of widening the scope of this theory, SINCLAIR (1988) renamed it the member-vagrant theory, in which it is considered to be more advantageous for larvae to be grouped together in a more or less delimitated area than to be wandering in the open sea as isolated individuals.

The aim of the present study is to describe the spatial and temporal distribution patterns of fish larvae on the Catalan coast in relation to the mesoscale hydrodynamic mechanisms that originate them, as well as to their potential predators and prey. The hypothesis we put forward here is that the set of mesoscale processes occurring during planktonic stages of fishes in the area contributes towards an increase in larval survival, and may be one of the factors which have greatest influence on the fact that the adult stocks of many species remain constant, at least in their first year class.

THE SPATIAL AND TEMPORAL MESOSCALE DISTRIBUTION PATTERN OF FISH LARVAE ON THE CATALAN COAST IN RELATION TO THE HYDROGRAPHICAL PATTERN

The mesoscale spatial distribution pattern of fish larvae on the Catalan coast follows two general tendencies. Within the overall pattern, however, the species present small variations related to the spatial distribution of the adults and their spawning areas and periods. Thus, for example, the fact that the structure of the Catalan continental shelf is very heterogeneous, and therefore the distribution of the types of bottoms as well (GILI *et al.*, 1987a), gives rise to a greater number of species where the shelf is wider than where it is narrower.

The first tendency observed is a greater abundance of larvae on the edge of the continental shelf, the result of spatial coincidence between larvae of mesopelagic and shelf species (Fig. 1; SABATÉS, 1990a). This concentration has been observed to be linked to the presence of a hydrographic shelf-slope front on the edge of the shelf, associated to the so-called Catalan current (FONT, 1987). Though the situation of this front is variable, it is always present and remains between the 200 and 1000 m isobaths (WANG *et al.*, 1988; FONT *et al.*, 1988). If this front is assumed to be a continuation of that of the Ligurian Sea with a similar hydrographic structure, then, according to BOUCHER *et al.* (1987), it presents a continuous series of convergence/divergence cells.

SABATÉS (1990a) and SABATÉS & MASÓ (1990) argue that this front could act as a barrier to the dispersal of the coastal and shelf species offshore and, at the same time, would favour the transport of oceanic and mesopelagic fish larvae over the shelf. The instability of the shelf-slope front gives rise to the penetration of slope waters onto the shelf, especially in the area to the north of the Ebro Delta (FONT *et al.*, in press) and of the Blanes canyon (MASÓ, 1989), due largely to the effect on the circulation of the structure of the continental shelf. With this, high concentrations of mesopelagic fish larvae, associated with waters related to the front, can be observed near the coast in certain zones and at certain periods of the year.

The second general tendency in the spatial distribution of larvae is related to the year-round thermal regime on the Catalan coast. The temperature of the water rises from north to south along the coast. This thermal gradient between the northern and southern parts of the area is especially marked in spring and autumn. This gives rise to the fact that in some species - e.g. *Cepola macrophthalma* (L.) or anchovy, *Engraulis encrasicolus* (L.) - the spawning begins in the southern zone, when temperature increases, and progressively extends northward (SABATÉS, 1988; PALOMERA & SABATÉS, 1990). Thus, in these species, the spawning is more protracted in the southern than in the northern zone. In contrast, in the species which reproduce at the beginning of autumn, in colder waters, the spawning begins in the northern zone and later spreads southwards. These thermal variations are less accentuated in areas further away from the coast than in nearer areas, where the greatest fluctuations in salinity and temperature have been observed (MASÓ & DUARTE, 1989).

From the seasonal point of view, it was observed that the concentration of larvae increases in mid-spring, reaching maximum densities in June and July (Fig. 2), coinciding with the spawning period of most of the species, anchovy being the dominant species (SABATÉS, 1988; PALOMERA, 1989). This seasonal phenomenon is observable in the whole of the Western Mediterranean (ABOUSSOUAN, 1964; MARINARO,

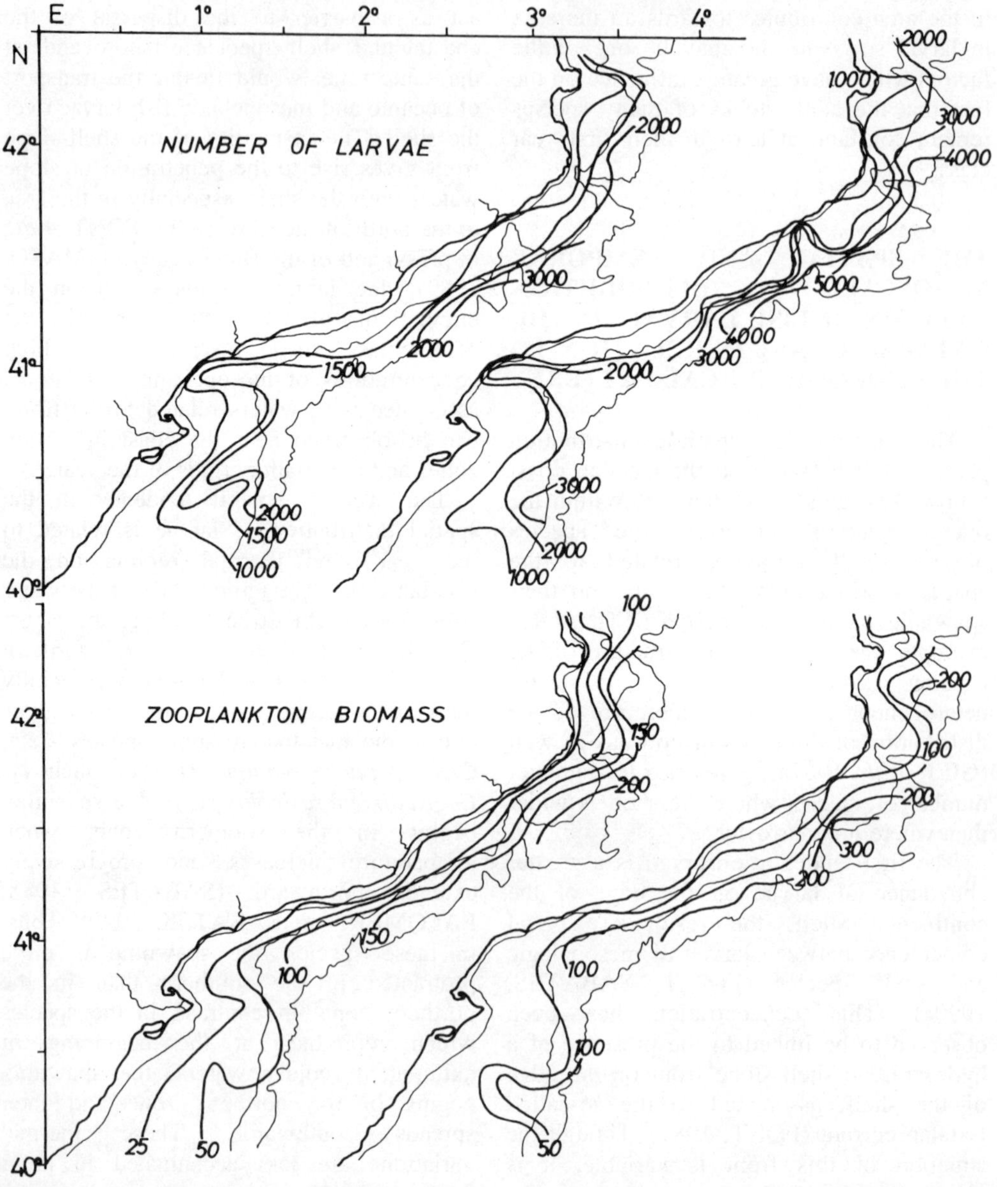

FIGURE 1. Spatial distribution of average number of larvae x 10 m^{-2} (left) and standard deviation (right), and zooplankton biomass (= displacement volume, ml 100 m^{-3}) along the Catalan coast. Data from SABATÉS (1990a) and SABATÉS *et al.* (1989).

1971), although the spawning lasts from one or few months in some species to several months in others. In the Western Mediterranean, as opposed to other geographical areas such as the North and South Atlantic, some mesopelagic fish species show a marked seasonality in their spawning (SABATÉS & OLIVAR, 1989) and, in many cases, this coincides with the aforementioned period of time.

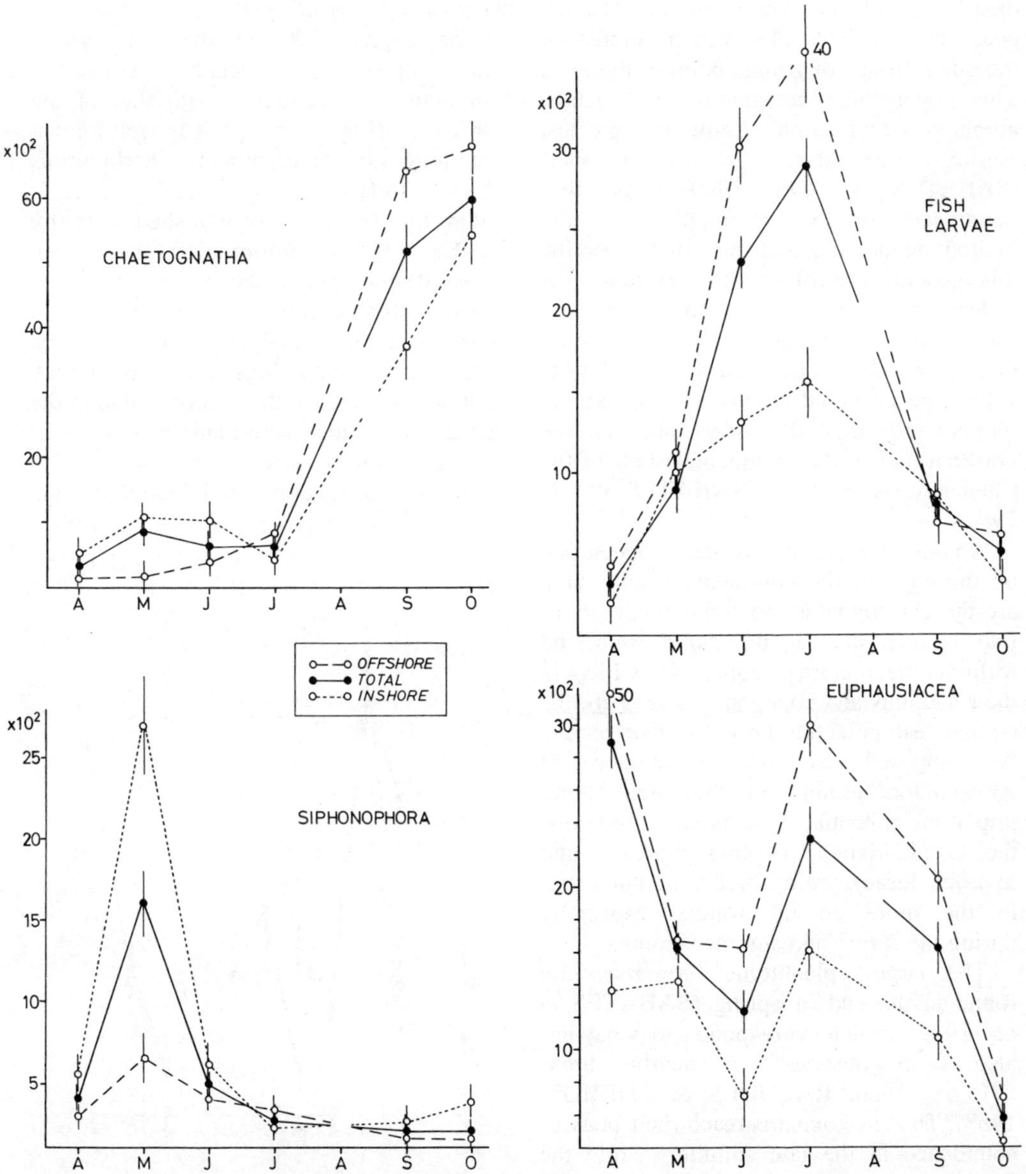

FIGURE 2. Seasonal distribution of fish larvae concentration and other zooplankton groups (especially main potential predators) along the Catalan coast. Vertical bars show standard error. Data from SABATÉS (1988).

THE SPATIAL AND TEMPORAL MESOSCALE DISTRIBUTION PATTERN OF POTENTIAL PREY AND PREDATORS OF FISH LARVAE ON THE CATALAN COAST

The general tendency observed in the distribution of fish larvae on the Catalan coast has also been observed in almost all the other groups of zooplankton in the area. Thus copepods, chaetognaths, euphausiids, amphipods and salpids show the greatest densities near the edge of the shelf (SABATÉS *et al.*, 1989). As was mentioned for the ichthyoplankton, the hydrodynamic mechanisms that describe this general pattern are the existence of waters associated with the persistent shelf-slope front and the north-south thermal gradient in the area. Overall, higher values are observed for the zooplanktonic biomass towards the edge and in the northern half of the continental shelf of the Catalan coast (Fig. 1; SABATÉS *et al.*, 1989).

Among the organisms that concentrate on the edge of the continental shelf, there are the chaetognaths, potential predators of fish larvae, and, at the same time and forming the majority, copepods, which in their nauplius and copepodite stages are the commonest potential prey for fish larvae. As opposed to the remainder of zooplankton groups in the area, other important potential fish larvae predators, the cnidarians (siphonophores and hydromedusae), are at their most numerous in the more coastal waters, especially during the spring and summer months.

The largest planktonic biomasses are found at the end of spring (SABATÉS *et al.*, 1989), which corresponds to what has also been observed in nearby areas (VIVES, 1966; RAZOULS & THIRIOT, 1968). The chaetognaths reach their peak of abundance in the late summer, while the cnidarians (and specifically the siphonophores) reach their peak in spring (Fig. 2). In the area of the Catalan coast a second peak of abundance was observed for cnidarians (siphonophores and medusae), although of lesser intensity, at the height of summer (RIERA *et al.*, 1986; GILI *et al.*, 1987b); a clear tendency was also observed to find the commonest species near the coast during this second abundance peak in the area (GILI *et al.*, 1987c).

The euphausiids are the only group whose peak of density coincides, temporally and spatially, with that of the fish larvae (Fig.2), though it is well known that they are predominantly herbivorous (RAYMONT, 1983).

On the basis of unpublished data by VIVES (1985), information has been obtained concerning the variation in the concentration of copepodites in the Catalan coast, over an annual cycle (Fig. 3). The densities of copepodites observed, though high, are lower than the real ones due to the fact that the hauls were made using nets of 200 μm mesh size, not adequate in retaining copepodites and nauplius, the

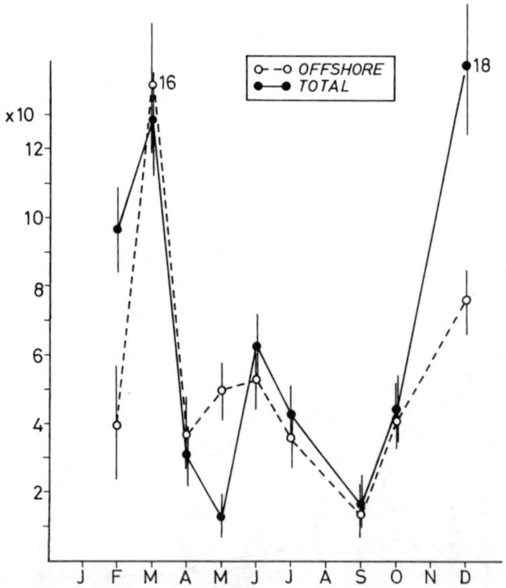

FIGURE 3. Temporal variation of copepodite concentration on the Catalan coast from unpublished data in VIVES (1985). Vertical bars show standard error.

common prey of fish larvae. Two peaks of maximum density of copepodites are observed at the beginning of spring and of autumn, besides a third peak of lower density at the end of the spring and beginning of the summer. As regards the data mentioned in VIVES (1985), if we consider only the stations furthest from the coast, the same pattern of seasonal variation may be observed and the concentrations are very similar to those of the coastal zone (Fig. 3).

THE MAINTENANCE OF FISH FIRST YEAR CLASS THANKS TO THE PROCESSES WHICH ENHANCE THE LARVAL SURVIVAL: A HYPOTHETICAL APPROACH

The mesoscale distribution patterns hitherto described for both fish larvae and zooplankton have a clear persistence. At the same time, the heterogeneity that the aforementioned hydrodynamic mechanisms are able to generate for the zooplanktonic populations can be understood only on this scale (MACKAS *et al.*, 1985; HAURY & PIEPER, 1988), as consequence of a coupling between both biological and hydrodynamic processes in coastal areas (DENMAN & POWELL, 1984). Unusual hydrodynamic events that might take place on a smaller scale may affect the general patterns observed in the larval populations. In this respect SABATÉS (1990b) has observed that the effect of accidental phenomena, such as the unusually high outflow of continental waters, might affect the survival of larvae in a particular area, but after a time the mesoscale distribution pattern and concentration of individuals becomes quite similar to the initial pattern.

In the more coastal zones of the area studied, the level of environmental heterogeneity is greater than towards the edge of the shelf. Together with this hydrographic heterogeneity the phytoplanktonic biomass is much greater nearer the coast (MASÓ & DUARTE, 1989) and may generate a greater concentration of zooplankton (LEGENDRE & DEMERS, 1984). Nevertheless, it seems that such fluctuating environmental conditions do not favour the mesoscale persistence of zooplankton populations. As SABATÉS *et al.* (1989) reported for the mesozooplankton, the less fluctuating hydrographical conditions at the end of the shelf allow the development of denser and more permanent populations, thanks to their conservative-type strategy.

The first process that must occur to enhance fish larvae survival is the synchronization between the period of larval development and the maximum density of its potential resources (HJORT, 1914; CUSHING, 1972). In the area studied, if we consider the density of copepodites, which probably coincides with a high number of nauplius, when the fish larvae concentrations are highest, we can deduce that the resources are sufficient for the estimated larval population. Densities of copepodites such as those observed, especially in the stations situated on the edge of the continental shelf (Fig. 3), are more than sufficient if we consider the rates of capture and consumption of prey observed by different authors both in the laboratory (e.g. LASKER *et al.*, 1970; HOUDE, 1975; FRANK & LEGGETT, 1986) and *in situ* (e.g. LASKER & ZWEIFEL, 1978; HUNTER, 1981; JENKINS, 1987). Furthermore, one of the commonest prey observed in some species of the area studied, together with the nauplius of copepods, are cladocerans (Fortuño, personal communication), organisms which are extraordinarily common in the area at the end of the spring and beginning of the summer in surface waters (VIVES, 1966).

Spatial coincidence of fish larvae and their potential prey, especially copepods, increases the probability of survival of the former. Frontal areas like those observed on the Catalan coast have been shown by several authors to be very productive (e.g. KIØRBOE & JOHANSEN, 1986), and also support important concentrations of fish eggs and larvae (e.g. NAKATA, 1989), thanks to the presence of high numbers of a potential food supply (KIØRBOE *et al.*, 1988; IWATSUKI *et al.*, 1989). Furthermore, on the Catalan coast the populations of fish larvae associated with waters of the density front are highly multispecific (SABATÉS, 1988). This could lead to a problem of denso-dependence between the different species when it comes to sharing the available resources. Furthermore, recent studies (e.g. KANE, 1984; FORTIER & HARRIS, 1989) have brought to light the fact that there are quite specific prey selection mechanisms between larvae of different species within the same population, which favours their coexistence. These trophic selection mechanisms involve both a selection of prey belonging to different species (e.g. JENKINS, 1987) and prey of different sizes of the same species (e.g. HEATH *et al.*, 1989).

While LEGGET (1986) points out that food supply is as important as predation in the survival of the larvae, BAILEY & HOUDE (1989) consider that the action of the predators is one of the most important mechanisms in the control of the recruitment of fish. Although a wide range of predators are known, some of these, such as chaetognaths and cnidarians, can at certain times produce a high death toll (e.g. BREWER *et al.*, 1984; PURCELL, 1985). Recently, some authors have considered that the relationship between fish larvae and their potential predators is established on the level of competition for the same prey rather than as direct predation upon the fish larvae (FORTIER & HARRIS, 1989; BAILEY & HOUDE, 1989). One of the mechanisms observed as being more effective in avoiding these predators or competitors is to minimize the temporal overlap of their maximum densities (e.g. BOLLENS *et al.*, 1989). A similar situation has been observed on the Catalan coast (Fig. 2). Both cnidarians (especially siphonophores) and chaetognaths show their blooms just before and after those of the larvae. Furthermore, the high concentrations of both groups of predators coincide with the blooms of copepodites in the area (Fig. 3). In this way, although the concentration of prey is lower during the period of maximum concentration of fish larvae, these can thus avoid their predators and the competition with them for the same prey. In any case, in multispecific populations in which there is always a dominant species (such as the anchovy on the Catalan coast), predation rates would be reduced since the dominant species saturate the predator field (FRANK & LEGGETT, 1983). According to these authors, this is a common situation in temperate and boreal regions. In the Catalan coast, PALOMERA & LLEONART (1989) observed that the mortality rate of anchovy larvae was highest during the period of maximal abundance of the species, although it coincided with important concentrations of their potential food.

FRANK & LEGGETT (1985), on observing the inverse correlations between fish larvae and their predators described by other authors, affirmed that they were due more to a spatial distribution of both collectives of organisms in different water masses than to direct action on the part of predators, which would drastically reduce the population of fish larvae. A similar situation may be observed on the Catalan coast, where the cnidarians, as potential predators or competitors, are distributed in the more coastal waters, even in those periods when the densities of larvae are high (GILI *et al.*, 1988).

In conclusion, observations made on the mesoscale spatial and temporal distribution pattern of fish larvae on the Catalan coast in terms of population, leads us to postulate a spatial and temporal synchronization between the populations of larvae and those of the remainder of zooplankton.

Meanwhile, the distribution of the whole range of zooplanktonic organisms depends upon the mesoscale hydrodynamic mechanisms in the area. Both phenomena together make possible a level of survival sufficient to determine the year-class strength of many species in the area, probably giving rise in turn to a level of recruitment of many species that would corroborate our initial hypothesis. The processes that occur during the larval stages of fishes on the Catalan coast may be one of the causes of the constancy of fishing stocks in the area.

REFERENCES

ABOUSSOUAN, A. 1964. Contribution à l'étude des oeufs et larves pélagiques des poissons téléostéens dans le golfe de Marseille. *Recl. Trav. Stn. mar. Endoume*, 32 (48): 87-171.

BAILEY, K.M. & HOUDE, E.D. 1989. Predation on eggs and larvae of marine fishes and the recruitment problem. *Adv. Mar. Biol.*, 25: 1-83.

BOLLENS, S.M., FROST, B.W. & SCHWANINGER, H.R. 1988. Seasonal abundance patterns in the ichthyoplankton and zooplankton of a temperate fjord, with a discussion of selective forces molding the timing of spawning in marine fishes. *ICES, 1988 Early Life History Symposium*, 23: 10.

BOUCHER, J., IBANEZ, F. & PRIEUR, L. 1987. Daily and seasonal variations in the spatial distribution of zooplankton populations in relation to the physical structure in the Ligurian Sea Front. *J. mar. Res.*, 45: 133-173.

BREWER, G.D., KLEPPEL, G.S. & DEMPSEY, M. 1984. Apparent predation on ichthyoplankton by zooplankton and fishes in nearshore waters of southern California. *Mar. Biol.*, 80: 17-28.

CADDY, J.F. & GARCÍA, S. 1983. Production modelling without long data series. *FAO Fish. Rep.*, 278 (Suppl.): 309-313.

CUSHING, D.H. 1972. The production cycle and numbers of marine fisheries. *Symp. Zool. Soc. London*, 29: 213-232.

CUSHING, D.H. 1975. *Marine ecology and fisheries*. Cambridge University Press. London.

DENMAN, K.L. & POWELL, T.M. 1984. Effects of physical processes on planktonic ecosystems in the coastal ocean. *Oceanogr. mar. Biol.*, 22: 125-168.

FONT, J. 1987. The path of the Levantine intermediate water to the Alborán Sea. *Deep-Sea Res.*, 10: 1745-1755.

FONT, J., SALAT, J. & TINTORÉ, J. 1988. Permanent features in the circulation of the Catalan Sea. In: *Oceanologica Acta*, Special issue, *Océanographie pelagique méditerranéenne*, (H.J. Minas & P. Nival, eds.), 9: 51-57.

FONT, J., SALAT, J. & JULIÀ, A. in press. Marine circulation along the Ebro continental margin. *Mar. Geol.*

FORTIER, L. & HARRIS, R.P. 1989. Optimal foraging and density-dependent competition in marine fish larvae. *Mar. Ecol. Progr. Ser.*, 51: 19-33.

FRANK, K.T. & LEGGETT, W.C. 1983. Multispecies larval fish associations: accident or adaptation? *Can. J. Fish. Aquat. Sci.*, 40: 754-762.

FRANK, K.T. & LEGGETT, W.C. 1985. Reciprocal oscillations in densities of larval fish and potential predators: a reflection of present or past predation? *Can. J. Fish. Aquat. Sci.*, 42: 1841-1849.

FRANK, K.T. & LEGGETT, W.C. 1986. Effect of prey abundance and size on the growth and survival of larval fish: an experimental study employing large volume enclosures. *Mar. Ecol. Progr. Ser.*, 34: 11-22.

GARCÍA, S. 1984. Un exemple d'utilisation des modèles de production composites en Meditérranée espagnole. *FAO Fish. Rep.*, 305: 97-105.

GILI, J.M., ROS, J.D. & PAGÈS, F. 1987a. Types of bottoms and benthic cnidaria from the trawling grounds (littoral and bathyal) off Catalonia (NE Spain). *Vie Milieu*, 37: 85-98.

GILI, J.M., PAGÈS, F. & VIVES, F. 1987b. Distribution and ecology of a population of planktonic cnidarians in the western Mediterranean. In: *Modern trends in the systematics, ecology, and evolution of hydroids and hydromedusas* (J. Bouillon, F. Boero, F. Cicogna & P.F.S. Cornelius, eds.): 157-170. Oxford University Press. Oxford.

GILI, J.M., PAGÈS, F. & RIERA, T. 1987c. Distribución de las especies más frecuentes de sifonóforos calicóforos en la zona norte del Mediterráneo occidental. *Inv. Pesq.*, 51: 323-338.

GILI, J.M., PAGÈS, F., SABATÉS, A. & ROS, J.D. 1988. Small-scale distribution of a cnidarian population in the Western Mediterranean. *J. Plankton Res.*, 10: 385-401.

HAURY, L.R. & PIEPER, R.E. 1988. Zooplankton: scales of biological and physical events. In: *Marine organisms as indicators* (D.F. Soule & G. S. Kleppel, eds.): 35-72. Springer. Berlin.

HEATH, M., LEAVER, M., MATTHEWS, A. & NICOLL, N. 1989. Dispersion and feeding of larval Herring (*Clupea harengus* L.) in the Moray Firth during September 1985. *Estuar. Coast. Shelf Sci.*, 28: 549-566.

HJORT, J. 1914. Fluctuations in the great fisheries of Northern Europe. *Rapp. P.-v. Réun. Cons. Int. Explor. Mer.*, 20: 1-228.

HJORT, J. 1926. Fluctuations of the year-classes of important food fishes. *J. Cons. Int. Explor. Mer.*, 1: 1- 38.

HOUDE, E.C. 1975. Effects of stocking density and food density on survival, growth and yield of laboratory-reared larvae of sea bream *Archosargus rhomboidalis* (L.) (Sparidae). *J. Fish. Biol.*, 7:

115-127.

HUNTER, J.R. 1981. Feeding ecology and predation of marine larvae. In: *Marine larvae. Morphology, ecology and relation to fisheries*. (R. Lasker, ed.): 33-77. University of Washington Press. Seattle.

ILES, T.D. & SINCLAIR, M. 1982. Atlantic herring: stock discreteness and abundance. *Science*, 215: 627-633.

IWATSUKI, Y., NAKATA, H. & HIRANO, R. 1989. The thermohaline front in relation to fish larvae. *Rapp. P.-v. Réun. Cons. int. Explor. Mer.*, 191: 119-126.

JENKINS, G.P. 1987. Comparative diets, prey selection, and predatory impact of co-occurring larvae of two flounder species. *J. Exp. Mar. Biol. Ecol.*, 110: 147-170.

KANE, J. 1984. The feeding habits of co-occurring cod and haddock larvae from Georges Bank. *Mar. Ecol. Progr. Ser.*, 16: 9-20.

KIØRBOE, T. & JOHANSEN, K. 1986. Studies of larval herring (*Clupea harengus* L.) patch in the Buchan area. IV. Zooplankton distribution and productivity in relation to hydrographic features. *Dana*, 6: 37-51.

KIØRBOE, T., MUNK, P., RICHARDSON, K., CHRISTENSEN, V. & PAULSEN, H. 1988. Plankton dynamics and larval herring growth, drift and survival in a frontal area. *Mar. Ecol. Progr. Ser.*, 44: 205-219.

LASKER, R., FEDER, H.M., THEILACKER, G.H. & MAY, R.C. 1970. Feeding, growth, and survival of *Engraulis mordax* larvae reared in the laboratory. *Mar. Biol.*, 5: 345-353.

LASKER, R. & ZWEIFEL, J.R. 1978. Growth and survival of first-feeding anchovy larvae (*Engraulis mordax*) in patches containing different proportions of large and small prey.In: *Spatial pattern in plankton communities*. (J.H. Steele, ed.): 329-354. Plenum. New York.

LEGENDRE, L. & DEMERS, S. 1984. Towards dynamic biological oceanography and limnology. *Can. J. Fish. Aquat. Sci.*, 41: 2-19.

LEGGETT, W.C. 1986. The dependence of fish larval survival on food and predator densities. In: *The Role of Freshwater Outflows in Coastal Marine Ecosystems*. (S. Skreslet, ed.): 117-137. Springer. Berlin.

LLEONART, J. 1990. *La pesquería de Cataluña: descripción global y planteamiento de las bases para su seguimiento*. Final Report of C.E.E. Program. General Direction XIV- B-1. Brussels (unpublished manuscript).

MACKAS, D.L., DENMAN, K.L. & ABBOT, M.R. 1985. Plankton patchiness: biology in the physical vernacular. *Bull. mar. Sci.*, 37: 652-674.

MARINARO, J.Y. 1971. Contribution à l'étude des oeufs et larves pélagiques de poissons méditérranéens. V Oeufs pélagiques de la baie d'Alger. *Pelagos*, 3 (1): 1-115.

MARTÍN, P. 1989. *Dinámica de la pesquería de arrastre en Cataluña*. Ph. D. Thesis. University of Barcelona.

MASÓ, M. 1989. *Variabilidad espacio-temporal de las características oceanográficas de la zona costera y su relación con el sistema planctónico*. Ph. D. Thesis. University of Barcelona.

MASÓ, M. & DUARTE, C. 1989. The spatial and temporal structure of hydrographic and phytoplankton biomass heterogeneity along the Catalan coast (NW Mediterranean). *J. mar. Res.*, 47: 813-827.

NAKATA, H. 1989. Transport and distribution of fish eggs and larvae in the vicinity of coastal fronts. *Rapp. P.-v. Réun. Cons. int. Explor. Mer.*, 191: 153-159.

PALOMERA, I. 1989. *Primeras fases del desarrollo de la anchoa (Engraulis encrasicolus) en la costa catalana*. Ph.D. Thesis. University of Barcelona.

PALOMERA, I. & LLEONART, J. 1989. Field mortality estimates of anchovy larvae, *Engraulis encrasicolus*, in the western Mediterranean. *J. Fish. Biol.*, 35 (Supplement A): 133-138.

PALOMERA, I. & SABATÉS, A. 1990. Co-occurrence of *Engraulis encrasicolus* and *Sardinella aurita* eggs and larvae in the Northwestern Mediterranean. *Scient. Mar.*, 54 (1): 61-67.

PURCELL, J.E. 1985. Predation on fish eggs and larvae by pelagic cnidarians and ctenophores. *Bull. mar. Sci.*, 37: 739-755.

RAYMONT, J.E.G. 1983. *Plankton and productivity in the oceans. 2. Zooplankton*. Pergamon. Oxford.

RAZOULS, S. & THIRIOT, A. 1968. Le macrozooplancton de la région de Banyuls-sur-Mer (Golfe du Lion). *Vie Milieu*, 19: 133-184.

RIERA, T., GILI, J.M. & PAGÈS, F. 1986. Estudio cuantitativo y estacional de dos poblaciones de cnidarios planctónicos frente a las costas de Barcelona (Mediterráneo occidental): ciclos entre 1966-67 y 1982-83. *Misc. Zool.*, 10:23-32.

SABATÉS, A. 1988. *Sistemática y distribución espacio-temporal del ictioplancton en la costa Catalana*. Ph.D. Thesis. University of Barcelona.

SABATÉS, A. 1990a. Distribution pattern of larval fish populations in the Northwestern Mediterranean. *Mar. Ecol. Prog. Ser.*, 59: 75-82.

SABATÉS, A. 1990b. Changes in the heterogeneity of mesoscale distribution patterns of larval fish associated with a shallow coastal haline front. *Estuar. Coast. Shelf Sci.*, 30: 131-140.

SABATÉS, A., GILI, J.M. & PAGÈS, F. 1989. Relationship between zooplankton distribution, geographic characteristicsand hydrographic patterns off the Catalan coast (Western Mediterranean). *Mar. Biol.*, 103: 153-159.

SABATÉS, A. & MASÓ, M. 1990. Effect of a shelf-slope front on the spatial distribution of mesopelagic fish larvae in the Western Mediterranean. *Deep-Sea Res.*, 37: 1085-1098

SABATÉS, A. & OLIVAR, M.P. 1989. Comparative spawning strategies of mesopelagic fishes in two marine systems with different productivity. *Rapp. P.-v. Réun. Cons. int. Explor. Mer.*, 191: 27-33.

SINCLAIR, M. 1988. *Marine populations: an essay on population regulation and speciation in the oceans*. Sea Grant/University of Washington, Seattle (unpublished manuscript).

VIVES, F. 1966. Zooplancton nerítico de las aguas de Castellón (Mediterráneao occidental). *Inv. Pesq.*, 30: 49-166.

VIVES, F. 1985. *El ecosistema pelágico y bentónico de la costa NE española (Cataluña) en relación*

con la hidrografía local. Final Report C.A.I.C.Y.T. Madrid (unpublished manuscript).
WANG, D. P., VIEIRA, M., SALAT, J., TINTORÉ, J. & LA VIOLETTE,P.E. 1988. A shelf-slope front filament off Northeast Spanish coast. *J. mar. Res.*, 46: 321-332.

(J. D. Ros & N. Prat, eds.). 1991. *Oecologia aquatica*, 10: 215-222.

BACTERIA AS TRANSFORMERS OF THE ARENA OF PLAY OF OTHER ORGANISMS

FERRAN VALLESPINÓS & SÍLVIA MALLO

Institut de Ciències del Mar (CSIC). Passeig Nacional, s/n. 08039 Barcelona. Spain.

Received: May 1990

SUMMARY

The following parameters were determined from a series of 70 samples taken from different estuary systems in the Iberian peninsula: oxygen uptake of the superficial sediment, number of viable heterotrophic bacteria, Eh, concentration of organic matter and percentage of fine-grained materials. The existence of positive and significant correlations between organic matter content, aerobic respiration and bacterial biomass is shown. Between 35% and 45% of the accumulated matter is mineralized annually by aerobic metabolism, and bacteria are the principal agents of this transformation. The greater organic load entails, however, conditions which are increasingly negative in redox potential, due to the demand of oxidants and the accumulation of reduced chemical species, which means that the circulation of the system's matter and energy through the bacterial compartment is gradually more important, with the installation of conditions that prevent the continuity of the eukaryote metabolism (with the exception of certain yeasts).

KEY WORDS: Bacteria, estuaries, respiration

INTRODUCTION

Coastal zones display a relatively higher concentration of organic matter (resulting from *in situ* primary production plus continental contributions), which increases the possibilities of heterotrophic metabolism. The functioning of estuary systems is clearly influenced by the metabolism of the bacterial populations which affects the cycles of matter and fluxes of energy. So there is an association between bacterial and particulate matter, which is seen most clearly in the processes of mineralization and the production of nutrients.

The most quantitatively significant part of these processes takes place at a sedimentary level, and the importance of the joining of the benthic and pelagic communities in coastal waters is recognised (NIXON, 1981). The consumption of organic matter is therefore associated with an important flow of nutrient towards the water column, in such a way that the rates of respiration in the sediment are indicators of the trophic levels of the system (EDMONDSON, 1966). At the same time, the activity of bacteria in the sediment is the principal factor responsible for the establishment of reducing conditions that call for the use, in the respiratory chains, of end acceptors of electrons different from molecular oxygen (nitrates, sulphates and carbon dioxide), in a sequence which depends on its relative energy (WATSON *et al.*, 1985), and which is itself accompanied by a diminishing of the Eh of the system, and consequently of the entire chemistry of both the sediments and the water column.

The participation of the various bacterial metabolisms (both aerobic and anaerobic) in the global balances of C and N features both a seasonal and spatial variability (MALLO & *et al.*, in press) associated with the availability of organic material, the thermic conditions and the value of the redox potential. In any case, the metabolism of the bacterial populations gradually establishes certain environmental conditions which make impossible the activity of the majority of the decomposing eukaryotic organisms (due to its chemical inadequacy), with the exception of the yeasts, whose importance is practically unknown. The final result is that virtually the entire flux of energy and matter ends up circulating through the bacterial compartment of the system, in a relation directly proportional to the existing quantity of organic material.

This study attempts to discuss some aspects of this model of the progressive "dominance" of the medium by the bacteria through the analysis of the information obtained in a series of coastal media of distinctive characteristics, in relation both to the quantity and quality of the inputs received and to the type of dynamics to which they are subjected.

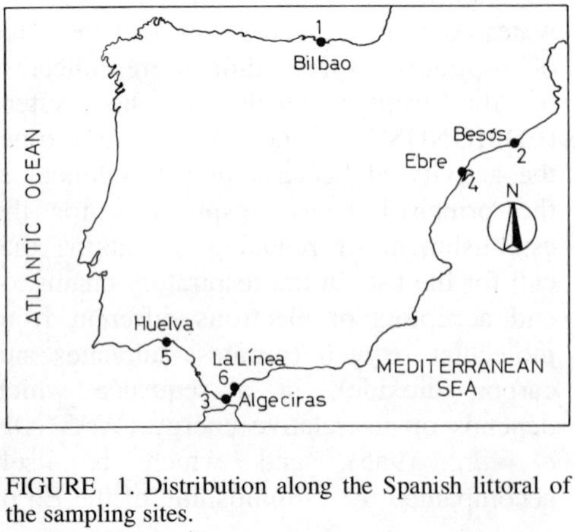

FIGURE 1. Distribution along the Spanish littoral of the sampling sites.

MATERIAL AND METHODS

Seventy samples were taken from six sites distributed throughout the Iberian peninsula (Fig. 1), with environmental characteristics embracing a wide variety of conditions (in relation to organic matter content, percentage of fine-grained materials and redox potential). In every case the stations are coastal zones affected by continental contributions. The samples were taken in summer or pre-summer conditions. In the Besòs river and the Ebro river delta, a seasonal study was carried out, and winter samples were also available.

The samples were obtained by hand (on occasion with the help of scuba divers) using polystyrene cylinders 4.5 cm in diameter. In the Besòs area a box-corer was used. In each case the temperature was determined *in situ* with a portable probe (ORION 917001), along with the redox potential, using a platinum electrode (ORION RESEARCH s500 C-ORP) connected to a portable millivoltmeter (ORION 231).

The sediment was transferred to the laboratory under refrigeration, with its structure maintained unaltered. The analysis was always begun within four hours of the sample being taken. A dilution bank was prepared from a suspension of sediment (approximately 10% w:v) in sterile sea water obtained from the same station, for the counting of the total number of viable heterotrophic bacteria (aerobic and facultative), on Marine agar plates (DIFCO 0791-01). The samples were incubated at 30 °C for a maximum period of a week.

The levels of aerobic respiration were determined in 25 ml vials (with rubber septum and hermetically sealed) which contained suspensions of sediment and seawater. The potential oxygen uptake was measured by gas chromatography in relation to the changes in the composition of the gases in the head space of the vials, which were also incubated at 30 °C for 12 hours. In the conditions of the analysis, the

TABLE I. Principal characteristics of the sampled zones. A variable number of samples was obtained from each zone (70 in total), with which 16 groups of stations of equivalent significance were defined.

LOCATION	ORG. INPUT ($Tm\ C\ d^{-1}$)	STATION GROUP	DEPTH (m)	WATER . TEMP (°C)	ORGANIC MATTER (%)	FINE-GRAINED MATERIALS (%)	Eh (mV)
RÍA DE BILBAO	20	1.1	6-10	20.0	12.1-18.4	30.0-80.0	-150/-420
BESÒS DELTA	120	2.1	11-52	13.6-21.1	3.1-7.7	25.4-84.4	-180/-401
		2.2	52	14.0	4.7	77.4	-220
		2.3	62-63	13.0-14.0	1.8-2.5	24.5-79.5	-158/-218
		2.4	11-52	13.0-13.2	3.6-5.3	25.4-84.4	-126/-387
		2.5	52	13.1	3.6	77.4	161
		2.6	62-63	13.5-13.9	1.6-2.5	24.5-79.5	-210/-304
ALGECIRAS BAY	-	3.1	15	16.0	0.7-2.9	8.0-84.1	-
EBRO DELTA	30	4.1	1	3.5	0.4	3.0	179
		4.2	1	3.0	5.6	33.0	-106
		4.3	1-2	26.3-29.3	0.5-1.2	0.01-4.0	16/-280
		4.4	1-4	23.0-29.1	1.9-7.8	24.5-97.9	-73/-347
		4.5	4-6	24.0-29.0	2.4-6.2	22.5-72.5	-128/-290
RÍA DE HUELVA	5	5.1	10	15.2	0.3-0.8	3.1-3.9	180/210
		5.2	10	15.1	4.3-4.7	0.1-1.7	53/80
LA LÍNEA	-	6.1	10	15.0	0.3-1.4	0.4-5.7	-

greatest part of this uptake (always above 85%) corresponds to bacterial metabolism, as the results obtained with the addition of cyclohexymide (inhibitor of eukaryotic respiration) show.

A detailed decription of this technique and those used to determine the other accompanying parameters (granulometry, organic matter, etc.) can be found in MALLO (1989).

Information about all the aforementioned parameters is available for the majority of samples, although in some cases there are no data for counts of heterotrophic bacteria or of redox potential.

RESULTS

Taking the sediment studied as a whole, the concentration of organic matter oscillates between 0.25 and 18.4%; the proportion of fine-grained materials, between 0.10 and 97.9%, and in general there is a relation between the two parameters. The number of viable heterotrophic bacteria varies between 2 x 10^5 and 4 x 10^7 cell g^{-1}. The most negative redox potential value determined was -420 mV. This information is summarised in Table I, in which a series of groups is established for each zone that include stations of equivalent characteristics. The grouping was obtained in each case by the application of a multivariate analysis (principal components) to the group of chemical and biological parameters considered (up to 15), which include those used in this study (MALLO, 1989; VALLESPINÓS *et al.*, 1989).

Although the samples include a variety of types of coastal marine systems, a strong lineal and positive relation (r= 0.94) is found between the amount of organic matter accumulated in the surface sediment (and which includes both the matter produced *in situ* and the imported matter) and the mineralised organic matter, measured by the oxygen uptake according to the technique described (Fig. 2).

In those cases in which the quantity of organic matter apported to the system is known, it has been related to the influenced area and divided by an average value of oxygen uptake. The results are shown in Table II, and it can be observed that between a third and a half of the contributions are mineralized by aerobic means by the bacterial populations of the superficial sediment. Moreover, the results seem to point to an inverse relation between this proportion and the total inputs, as will be discussed below.

In relating the concentration of organic matter of the sediment to the bacterial populations (Fig. 3) it is observed that an increase in the total number of heterotrophic bacteria (aerobic and facultative) is produced as a result of the enrichment in the organic matter content (r= 0.62). The greater availability of carbon and nitrogen therefore enhances the presence of richer bacterial flora which participates in the degradation of the organic matter accumulated in the superficial sediment. This implies the consumption of an equivalent quantity of oxydants (oxygen, free or incorporated in mineral molecules) and the corresponding accumulation of reduced chemical species. These processes taken as a whole necessarily involve a diminishing of the redox potential of the sediments, a diminishing that is the result of the combination of bacterial metabolisms (flux of electrons generated by the organotrophic processes).

TABLE II. Percentage of benthic remineralization in relation to the total of organic matter deposited in each system.

LOCATION	ORGANIC INPUT ($g\ C\ m^{-2}\ y^{-1}$)	BENTHIC REMINERAL. ($g\ C\ m^{-2}\ y^{-1}$)	%
RÍA DE BILBAO	3,600	1,280	35
BESOS DELTA	1,700	580	34
EBRO DELTA	250	100	40
RÍA DE HUELVA	110	50	45

The relation between the consumption of organic matter and the redox potential of the sediments summarizes a set of complex thermodynamic balances, but it is evident that the highest rates of mineralization of the excess of accumulated organic matter is translated into a diminishing of the Eh value, as shown in figure 4, with a value of r= 0.49.

The quotient between the concentration of organic matter and the uptake of oxygen (OM/OU) measures, to an extent, the turnover of material incorporated in the superficial sediment. If this value is referred to the bacterial biomass (Fig. 5), a significant (r= 0.73) and positive lineal relation is found, which implies that, contrary to expectations, the sediments with a population more abundant in heterotrophic bacteria (aerobic and

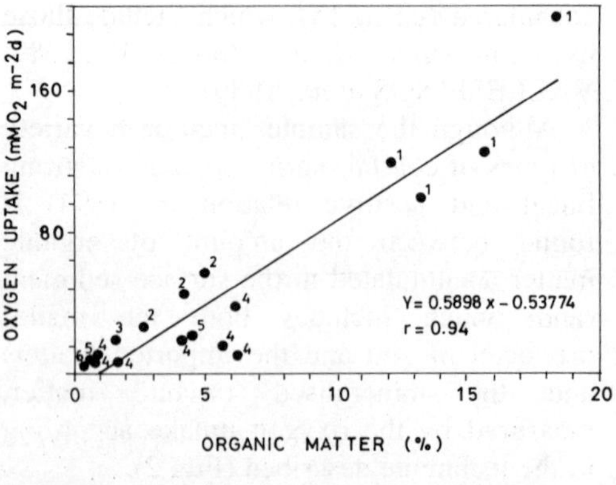

FIGURE 2. Relation between the rate of mineralization (measured as oxygen uptake) and the content of organic matter in the sediment. It corresponds to average values for distinct groups of stations (1: Ría de Bilbao; 2: Besòs Delta; 3: Algeciras Bay; 4: Ebro Delta; 5: Ría de Huelva; 6: La Línea).

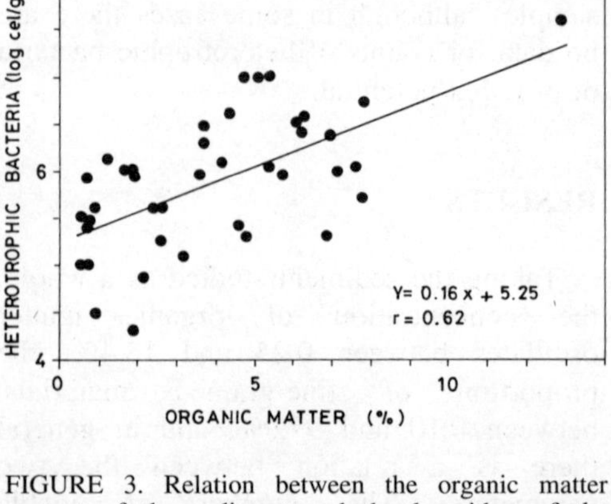

FIGURE 3. Relation between the organic matter content of the sediment and the logarithm of the number of viable heterotrophic bacteria (aerobic and facultative).

facultative) remineralize the organic matter in a relatively slower fashion.

DISCUSSION

In the conditions of the analysis, the uptake of oxygen is a measure integrated in all the oxidative processes that take place in the sediment. The comparison between the data presented in this study (which range from 0.16 to 200 mM m^{-2} d^{-1}) and those published by other authors who have studied the communities of the sediments of shallow water is difficult, given the wide disparity both between methods of measurement and presentation of results. However, all the uptakes found are similar to those described for other estuarine environments (NOWICKI & NIXON, 1985), even in the case of the highest values of oxygen demand, which correspond to the Ría de Bilbao (BOYNTON *et al.*, 1980; HOPKINSON & WETZEL, 1982; VAN ES, 1982).

Where it was possible to follow the temporal variation (Ebro Delta), it was seen that a high variability exists and that the oxygen is consumed more slowly in winter (9.5 mM m^{-2} d^{-1}) than in summer (36.8 mM m^{-2} d^{-1}). This seasonal difference in oxygen flow is similar to that described in other coastal sediments of equivalent characteristics (BALZER, 1984; HALL *et al.*, 1989), and corresponds to a Q_{10} of 1.9. The proximity of this value to that assigned to marine bacteria (JOHNSON, 1936) in the interval of sediment temperatures discovered during the entire cycle (8.2-29.8°C) and the results of the experiments in which cyclohexymide was added, suggest that the greater part of this demand is exercised by the populations of heterotrophic bacteria involved in the degradation of the organic matter, which use the oxygen as a final acceptor of electrons in their respiratory chains. In fact, a relation is found in the body of available data (Fig. 3) between the organic content of the sediment and an index of bacterial biomass (in this study the count of viable heterotrophic bacteria on plates).

In comparing the results obtained in this study from distinct types of sediment, with a variable quantity of detritic matter (HARGRAVE, 1972) and which also feature an unknown refractory fraction to bacterial attack (RILEY & CHESTER, 1971), everything seems to indicate that the determinations of oxygen uptake rates are useful as a measure of total biological activity (on the whole bacterial) in the sediments (PAMATMAT, 1971). This relation is illustrated in figure 1. On the other hand, in comparing the benthic metabolism of oxygen uptake, transformed in terms of carbon (for which the quotient

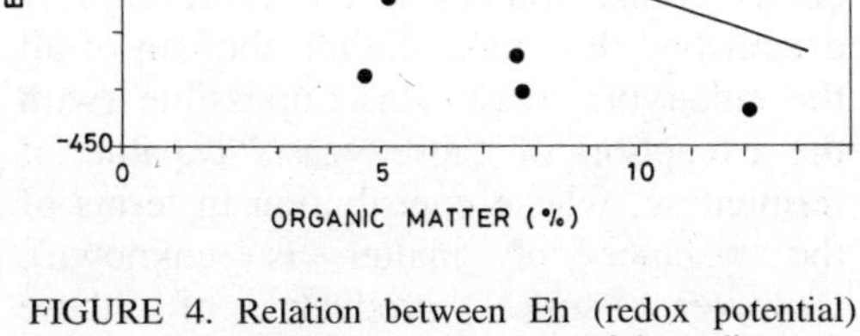

FIGURE 4. Relation between Eh (redox potential) value and the organic matter content of the sediment.

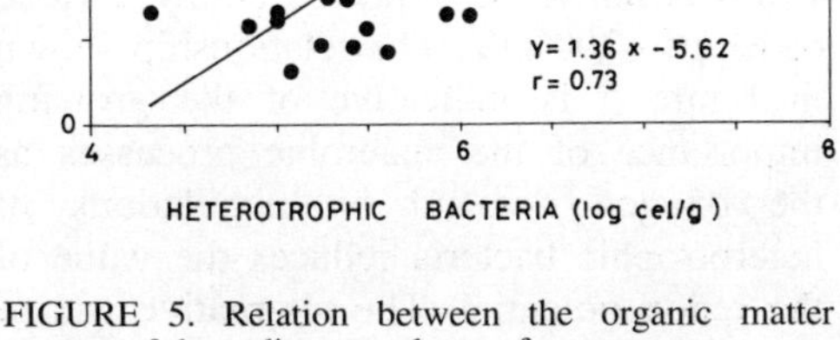

FIGURE 5. Relation between the organic matter content of the sediment and rate of turnover.

RQ = 1 has been used; NIXON 1980) with the primary production total (P, measured or estimated) in each one of the stations considered, it is evident that in estuarine systems an external contribution of organic matter (I) is required to allow the adjustment of budgets (in terms of carbon and nitrogen).

In the sampled stations as a whole, the situation is usually I >> P. In spite of the organic matter input having in some cases a "natural" origin (drainage from cultivated fields in the Ebro Delta), and in others stemming from processes of pollution by waste waters (Ría de Bilbao or Besòs Delta), the final result is not very different, in that the bacterial activity in the sediment is always carried out in a fairly equivalent manner, so that by aerobic means 35-45% of the organic matter accumulated in the sediment is remineralized annually (Table II). Finally, although the data currently available does not allow an absolute statistical validity, the relative weight of the aerobic processes in the overall bacterial activity appears to be less in sediments with higher organic matter content (and with a lower redox potential).

This is also supported by the fact that the lower the turnover of matter accumulated in the sediment (estimated using the relation MO/OU), the higher its organic content. It must be borne in mind that these sediments correspond to the most polluted zones (Ría de Bilbao and Besòs Delta); the presence of high concentrations of micropollutants (principally heavy metals; Borràs, pers. comm.) partially inhibits the bacterial activity. Furthermore, they are sediments with the most negative redox potential (see Table I), and the use of molecular oxygen in the respiratory chains (WATSON *et al.*, 1985) is not to be expected below a value of Eh of +200 mV. The relationship shown in figure 5 is indicative of the growing importance of the anaerobic processes as the oxygen demand by populations of heterotrophic bacteria reduces the value of the redox potential. The alternative use of other electron acceptors also affects the balance of the consumption of organic carbon accumulated in the sediment (SØRENSEN *et al.*, 1979), for as the nitrates and sulphates oxidize 1.25 and 2 moles of organic carbon respectively, the oxygen acts in an equimolecular relation.

The degradation of organic matter is accompanied by a consumption of oxidants and an accumulation of reduced chemical species, the effect being a diminishing of the redox potential. The estuarine systems in which I >> P are especially suitable for studying the dynamics of this process, in terms of both time and space, amongst other reasons because they usually represent an unbalanced medium, in which the greater availability of material liable to be oxidized by the bacterial populations exercises a demand for oxidants that rapidly reduces the value of the redox potential, as shown in figure 4. Once the availability of oxygen is exhausted, the decomposition of organic matter continues through the use of NO^{-}_{3}, Fe^{+3}, SO^{2-}_{4}, CO_2 and anaerobic fermentations. The process occurs precisely in this sequence due to respiration involving reactions of phosphorylation tied to transport of electrons, and thus with oxide reduction reactions that depend on the value of the Eh of the medium.

In agreement with the data presented, which refers to a wide variety of estuaries, the bacterial populations of the sediment work as active transformers of the quality of the medium, with a high oxygen demand for the metabolisation of the available organic matter, which causes its reduction and rapid exhaustion. The intensity of this process is directly related to the quantity of accumulated organic matter, in such a way that in those systems with greater external contributions conditions are more rapidly established that make the metabolism of all the eukaryotic organisms impossible (with the exception of those yeasts capable of fermenting, whose contribution in terms of the balance of matter is unknown). However, some products of these alternative respiratory mechanisms (e.g. H_2S) are toxic for the majority of organisms.

It must thus be concluded that the overall tendency of these processes is a gradual control of the medium by the bacterial populations of the sediment, in the sense that the percentage of matter and energy that circulates through them is higher as more strictly reducing conditions are established. Moreover, these conditions are themselves the result of the activity of the same bacterial populations in the initial phases of aerobic oxidation of the organic matter. The final result is that the bacterial compartment of the sediment is only the agent in the mineralization and production of nutrients, which in turn affects the conditions of the water column. The comparison between samples taken from systems with different characteristics shows that the availability of organic matter constitutes the principal factor in explaining the temporal and spatial variability of the activity of the bacterial populations.

ACKNOWLEDGEMENTS

We gratefully thank the people who gave their help in the collection of samples, and P. Sacristán for help in the composition of the figures.

The greater part of the given data were obtained during the development of the following programmes: "Integrated study of the Ebro Delta. Determining factors of biological production" (financed by CAYCIT) and "Integrated project for the study of the effect of submarine mud deposits of the River Besòs area on the coastal zone off Barcelona" (financed by MOPU-Corporació Metropolitana).

REFERENCES

BALZER, W. 1984. Organic matter degradation and biogenic element cycling in a nearshore sediment (Kiel Bight). *Limnol. Oceanography*, 29: 1231-1246.

BOYNTON, W.R., KEMP, W.M. & OSBORNE, C. 1980. Nutrient fluxes across the sediment-water interface in the turbid zone of a coastal plain estuary. In: *Estuarine perspectives* (V. Kennedy, ed.): 93-110. Academic Press. New York.

EDMONSON, W.T. 1966. Changes in oxygen deficit of Lake Washington. *Ver. intern. Verein. theoret. angew. Limnologie*, 16: 153-158.

HALL, P.O.J., ANDERSON, L.G., VAN DER LOEFF, M.M.R., SUNDBY, B. & WESTERLUND, S.F.G. 1989. Oxygen uptake kinetics in the benthic boundary layer. *Limnol. Oceanography*, 34: 734-746.

HARDGRAVE, B.T. 1972. Oxidation-reduction potentials, oxygen concentration and oxygen uptake of profundal sediments in a eutrophic lake. *Oikos*, 23: 163-177.

HOPKINSON, C.S. & WETZEL, R.L. 1982. *In situ* measurements of nutrient and oxygen fluxes in a coastal marine benthic community. *Marine Ecology Progr. Ser.*, 10: 29-35.

JOHNSON, F.H. 1936. The oxygen uptake of marine bacteria. *J. Bacteriology*, 31: 547-556.

JONES, J.G. 1982. Activities of aerobic and anaerobic bacteria in lake sediments and their effect on the water column. In: *Sediment Microbiology* (D.B. Nedwell & C.M. Brown, eds.): 107-145. Academic Press. London.

MALLO, S. 1989. *Estudio de la variación espaciotemporal de la desnitrificación bacteriana en sedimentos marinos litorales y su contribución al balnace de nitrógeno en los sistemas*. Ph. D. Thesis. University of Barcelona.

MALLO, S., VALLESPINÓS, F., FERRER, S. & VAQUÉ, D. In press. Microbial activities (Ebro Delta, Spain) influenced by organic matter influx. *Estuarine Coastal Shelf. Sci.*

NIXON, S.W. 1980. Between coastal marshes and coastal waters -a review of twenty years of speculation and research on the role of salt marshes in estuarine productivity and water chemistry. In: *Estuarine and wetlands processes* (P. Hamilton & K. Macdonald, eds.): 437-525. Plenum Press. New York.

NIXON, S.W. 1981. Remineralization and nutrient cycling in coastal marine ecosystems. In: *Estuaries and Nutrients* (B.J. Nielson & L.E. Cronin, eds.): 111-138. Humana Press. New Jersey.

NOWICKI, B.L. & NIXON, S.W. 1985. Benthic community metabolism in a coastal lagoon ecosystem. *Marine Ecology Prog. Ser.*, 22: 21-30.

PAMATMAT, M.M. 1971. Oxygen consumption by the seabed. IV Shipboard and laboratory experiments. *Limnol. Oceanography*, 16: 536-550.

RILEY, J.P. & CHESTER, R. 1971. *Introduction to marine chemistry*. Academic Press. New York.

SØRENSEN, J., SØRENSEN, B.B. & REVSBECH, N.P. 1979. A comparision of oxygen, nitrate and sulphate respiration in coastal marine sediments. *Microbial Ecology*, 5: 105-115.

VALLESPINÓS, F. AMENGUAL, P. BORRÀS, G., FERRER, S., MALLO, S. & VAQUÉ, D. 1989. *Proyecto integrado para el estudio del efecto del depósito submarino de lodos del rio Besòs sobre la zona costera de Barcelona*. Technical Report (MOPU-CMB).

VAN ES, F.B. 1982. Community metabolisms of intertidal flats in the Ems-Dollard Estuary. *Marine Biology*, 66: 95-108.

WATSON, P.G., FRICKERS, P.E. & GOODCHILD, C.M. 1985. Spatial and seasonal variations in the chemistry of sediment interstitial waters in the Tamar estuary. *Estuarine Coastal Shelf Sci.*, 21: 105-119.

(J. D. Ros & N. Prat, eds.). 1991. *Oecologia aquatica*, 10: 223-242

STRUCTURE AND DYNAMICS OF NORTH-WESTERN MEDITERRANEAN PHYTOBENTHIC COMMUNITIES: A CONCEPTUAL MODEL

ENRIC BALLESTEROS

Centre d'Estudis Avançats (CSIC). Camí de Santa Bàrbara, s/n. 17300 Blanes. Spain.

Received: January 1990

SUMMARY

This paper describes the structural (species richness, species diversity, pattern-diversity, biomass, coverage, specific distribution, homogeneity) and functional (production, P/B ratio) changes observed during the annual cycles of eight phytobenthic Mediterranean communities by using PCA analysis. The first axis obtained discriminated between variables related to pattern-diversity and those related to production and coverage (production- diversification or P-D axis). The second axis was mainly related to biomass and homogeneity (biomass-heterogeneity or b-h axis). Shallow-water communities were positioned in the quadrant of the plane associated with high production while deep-water communities were located in the quadrant characterized by high diversification values. The transit of communities through this plane during their annual cycles suggests their idealization as geometric figures revolving clockwise along the plane defined by the P-D and the b-h axes. These changes allow the distinction of two phases and two stages in the annual cycle of the communities, according to the sequence: diversified community (high heterogeneity)→ production phase (high production)→ developed community (high biomass)→ diversification phase (high pattern-diversity). Peak production occurs in spring for the mediolittoral communities and is delayed towards summer and autumn at increasing depths. This time lag affects the whole annual cycle described above and appears attributable to the adaptation of the different species and communities to the seasonal cycle of the main factors limiting their growth (BALLESTEROS, 1989b).

Although production and diversification are inversely related, P/B ratio is not negatively correlated with diversity because of the different mechanisms generating diversity in natural communities: disturbance and evolution. Short-time diversity increases in periods of intense physical disturbance (diversification phases) due to lack of dominance within the community and the growth of opportunistic species. Thus, short-term diversity and productivity are not inversely correlated. Long-term diversity increases along evolutionary time under environmental stability, explaining the greater diversity of deep-water communities compared to shallow ones, where an inverse correlation between productivity and diversity is observed.

An interannual model of the structural and functional performances of phytobenthic Mediterranean communities can be depicted over the plane defined by the P-D axis and the b-h axis by introducing time as a third axis. In general, the dynamics of persistent communities can be depicted as helicoidal trajectories whose turns represent annual cycles. This representation allows the adequate definition of concepts such as persistence, resistance and resilience of communities, and creates an adequate framework for describing successional processes and the effects of all kind of disturbances. The hypothesized existence of multiple stable points in natural communities closely agrees with the model if different degrees of non-punctual, biotic, disturbances are considered. In our opinion this model could also be extended to other temperate seas.

KEY WORDS: Phytobenthos, structure, dynamics, ecological modelling.

STRUCTURE AND DYNAMICS OF BENTHIC MARINE COMMUNITIES: A REVIEW

Some of the concepts central to modern ecological theory (e.g., stability and predictability, succession and pattern formation, predation, competition and disturbance in community organization and dynamics, etc.) have been developed largely from the study of benthic ecosystems. This suggests that the structure and dynamics of benthic marine communities must be well understood. We do however, contend, together with SCHIEL & FOSTER (1986) that this is not the case, and base our assertion on the acute imbalance of our present knowledge of benthic ecosystems, which derives predominantly from the study of a few systems (e.g., kelp beds and fouling communities), and focuses mainly on the consequences of environmental forcing.

Most studies of benthic communities focus on intertidal systems (e.g DAYTON, 1971, 1975; PAINE, 1974; LUBCHENKO & MENGE, 1978; SOUSA, 1979; NIELL, 1979; PAINE & LEVIN, 1981). Consequently, our knowledge of benthic ecosystems is biased towards the peculiarities of these systems. Moreover, whenever subtidal communities are studied, these are often kelp beds (DAYTON *et al.*, 1984; JOHNSON & MANN, 1988) or fouling communities (e.g. SUTHERLAND, 1974; OSMAN, 1977, 1978). Kelp beds are among the most productive ecosystems (MANN, 1972; JACKSON, 1977) and, thereby, support an enhanced energy flow and can undergo radical changes because of their low diversity and the high connectivity between the main components of these systems. On the other hand, fouling communities are essentially unstable and immature, so they are not representative of the more permanent communities that dominate the marine benthos. Thus, there is no guarantee that the results obtained in these systems will be transferable to other benthic systems, given the great differences that exist among benthic communities both worldwide (ZABALA & BALLESTEROS, 1989) and in particular localities. In fact, the comparison of Atlantic and Pacific kelp beds has demonstrated significant differences in the dynamics of community organization and in the relative importance of the environmental factors that impinge upon them (DAYTON *et al.*, 1984; JOHNSON & MANN, 1988).

The majority of studies in benthic ecology have focused on the factors that control community organization (e.g., physical disturbances, competition, predation) and not on the organization (or structure) itself. As a result, the structural evolution of the communities has been inferred from experimentally induced changes or from changes caused by extreme natural disturbances. The different stages these perturbed communities leave behind have been used to describe the natural pattern of undisturbed communities and has led to the formulation of the multiple stable point hypothesis as the model which best describes the structure and function of highly dynamic systems such as fouling communities (SUTHERLAND & KARLSSON, 1977), communities growing over sandy bottoms (GRAY, 1977) and kelp beds (JOHNSON & MANN, 1988). As a result of the historical development of benthic ecology, most benthic ecosystems are depicted as intrinsically unstable, contrasting with the interannual constancy described for the epibiota of many areas (HISCOCK & MITCHELL, 1980). Moreover, the degree of stability or instability of benthic communities depends on the temporal and spatial scales at which they are examined, the relevant scale of observation depending on the research goals.

There is a paucity of studies of the dynamics of relatively undisturbed benthic communities. Despite extensive use of the term normal-structured communities to refer to investigated communities, the reality is often that the structural and dynamic changes that the benthic communities of temperate seas experience during the year are generally unknown. Seasonal variations in the Mediterranean (FELDMANN, 1937; BALLESTEROS, 1984, 1987, 1988a, 1988b; VERLAQUE, 1987), as well as in other temperate (NEUSHUL, 1967; NIELL, 1975; HRUBY, 1975; LITTLER, 1980; FERNÁNDEZ *et al.*, 1983; CHOCK & MATHIESON, 1983; MATHIESON & PENNIMAN, 1986; MURRAY & HORN, 1989) and tropical seas (LAWSON, 1957; JOHN *et al.*, 1977; PRICE, 1989) often involve significant changes in community structure. Seasonal changes also involve changes in community structure and organization (HRUBY, 1975; SEAPY & LITTLER, 1982; FERNÁNDEZ *et al.*, 1983; BALLESTEROS, 1984, 1987, 1988a, 1988b), and in functional processes (MANN, 1973; GAGNE *et al.*, 1982; WHEELER & DRUEHL, 1986; BALLESTEROS,1989b; TUSSENBROEK, 1989; KAIN, 1989). Some of this research focuses on specific attributes of the community (e.g., biomass, diversity, production, P/B ratios) and, while a model has been proposed to explain the seasonal changes in intertidal systems from temperate seas (NIELL, 1975, 1977), an accurate description of the annual cycles for the subtidal vegetation and for a large number of phytobenthic communitites is as yet unavailable.

Our goal here is to develop a conceptual model of the dynamics of North-western Mediterranean phytobenthic communities. To achieve this goal we shall use data derived from the study of different communities from the phytal system (mediolittoral, infralittoral and circalittoral levels). The dynamics of these communities have been described within the framework of seasonal changes in structure and production. In addition to this seasonal model, we propose a model to account for the long-term dynamics of benthic communities. We believe that the model derived for Mediterranean communities can easily be extended to phytobenthic communities, both subtidal and intertidal, in other temperate seas.

SEASONALITY OF ENVIRONMENTAL FACTORS IN THE NORTH-WESTERN MEDITERRANEAN

Seasonal changes in the structure and function of phytobenthic communities derive from the seasonal nature of the dominant environmental factors. The seasonal variability of environmental factors in the NW Mediterranean Sea is, in general, similar to that observed in most other temperate seas. Potential differences, however, include lower variability, greater underwater irradiance and temperature, and lower nutrient concentrations than those usually observed in oceanic waters elsewhere in the temperate zone (Fig. 1). In addition, seasonal environmental variability differs across a trophic gradient, from the communities in the relatively eutrophic Gulf of Lions to those in oligotrophic insular areas (ZABALA & BALLESTEROS, 1989), although the extreme oligotrophy is found around the Balearic Islands which are traditionally considered as Central Mediterranean. In this paper, because our study is largely devoted to the phytobenthic communities of the North-western Mediterranean (mainly the Gulf of Lions and the Catalan coast), we focus on the annual cycle of environmental factors from this area (see Table I).

PHYTOBENTHIC COMMUNITIES IN THE NORTH-WESTERN MEDITERRANEAN

The seasonal variability of environmental factors in the Mediterranean

TABLE I. Mean seasonal values of some physical and chemical parameters that affect phytobenthic communities in coastal waters from a north-western Mediterranean locality (Tossa de Mar, NE Spain, year 1983). The extinction coefficient (K_{480}) corresponds to a light with a wavelength of 480 nm. Nitrate and phosphate concentrations are expressed in micromols l^{-1} and temperature is expressed in degrees centigrade. Data from BALLESTEROS (1984)

Season	K_{480}	NO_4^{3-}	PO_4^{3-}	T	Physical disturbances
Winter	0.120	0.5-2.0	< 0.08	12-14	Violent storms (N winds) Becalmed waters (January)
Spring	0.114	< 1.0	< 0.06	15-20	High hydrodynamism (April) Low sea level (May, June)
Summer	0.090	< 0.1	< 0.03	21-25	High irradiance
Autumn	0.140	< 1.0	< 0.08	15-21	Violent storms (E winds) High hydrodynamism Heavy rains

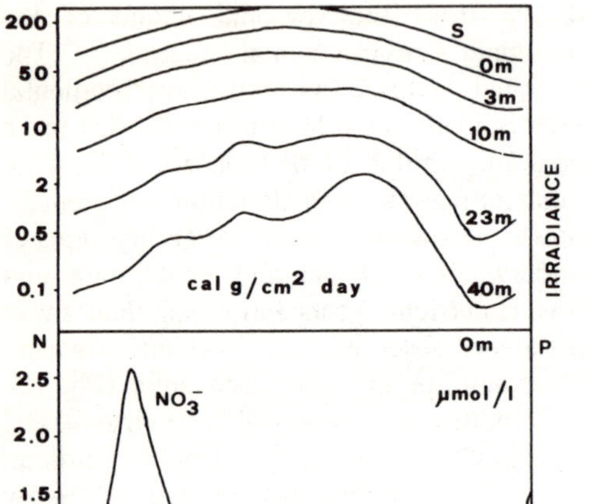

FIGURE 1. Annual cycle of irradiance, nutrient concentration and water temperature from a coastal locality of the North-western Mediterranean (Tossa de Mar, NE Spain). Data from BALLESTEROS (1984).

is similar to that in other temperate seas, but average values tend to be closer to those in tropical and subtropical seas. Kelps and corals have failed to master these phytal systems (ZABALA & BALLESTEROS, 1989), which are dominated by small, erect frondose algae in shallow waters (photophilic communities) and calcareous algae in localities with high densities of sea urchins (AUGIER & BOUDOURESQUE, 1974; VERLAQUE, 1987) and deep waters (sciaphilic communities, coralligenous; cf. ROS *et al.*, 1985). Because of the intensive studies of these communities, using phytosociological and other descriptive methods (FELDMANN, 1937; PÉRÈS & PICARD, 1964; BOUDOURESQUE, 1970, 1985; GIACCONE & BRUNI, 1972; COPPEJANS, 1977; BALLESTEROS, 1984; VERLAQUE, 1987; BOISSET, 1987), current knowledge of their floristic composition is, perhaps, greater than that for any other subtidal communities. One of the salient features of Mediterranean phytobenthos is the tendency towards miniaturization (COPPEJANS, 1980; BOUDOURESQUE *et al.* 1980; BALLESTEROS, 1984) that allows the study of the communities using very small sample sizes. The high level of knowledge and the extreme miniaturization of phytobenthic communities renders the Mediterranean a convenient location in which to examine the seasonal variations of phytobenthic communities.

Zonation patterns, which are well-documented for the Mediterranean, are apparent despite the lack of significant tides (BALLESTEROS & ROMERO, 1988). FELDMANN (1937) established the zonation sequence for rocky bottoms in the Alberes Coast, which was found to adequately describe zonation patterns elsewhere in the Gulf of Lions and the Catalan Coast (ROS *et al.*, 1985). In moderately exposed areas of Tossa de Mar (NE Spain), the gradation of communities from the upper mediolittoral to the lower circalittoral (2 to 60 meters) comprises at least ten well-differentiated seaweed communities (Fig. 2). Their species composition, structure, dynamics, and the forcing environmental factors have been extensively studied by BALLESTEROS (1984, 1988a, 1988b, 1989a, 1990a, 1990b, 1991, in press a, b).

A METHODOLOGICAL APPROACH TO THE SYNTHETIC DESCRIPTION OF NATURAL COMMUNITIES

Natural communities are generally described through their species composition, implemented with indicative values of their abundance. These values are often derived by using previously standardized scales (BRAUN BLANQUET, 1979) or by quantitative descriptors of abundance, such as the number of individuals, biomass or coverage (MARGALEF, 1974). Biomass (generally expressed as g dw m^{-2}) and coverage (generally expressed as the percentage of surface covered) measurements have been widely used to characterize phytobenthic communities (BOUDOURESQUE, 1971; NIELL, 1974; ROMERO, 1984) because they are excellent indicators of abundance, each emphasizing different properties of benthic organisms (BALLESTEROS, 1986).

In addition, benthic communities are often described by providing indication of their degree of structure, which is scale-dependent. This characteristic of benthic communities can be described through three main spectral parameters, namely species richness, diversity and homogeneity. Because of their spectral nature, the choice of the area in which to measure these parameters has important implications. Species/area relationships are exponential (GLEASON, 1925; HAWKINS & HARTNOLL, 1980), whereas diversity/area and homogeneity/area curves are asymptotic (NIELL, 1974; NÉDELEC, 1979; BALLESTEROS, 1986). Thus, it is often assumed that the value of each one of these parameters characterizing a particular community is the value achieved when its variation for successive increments in area is minimal. In practice, this minimal variation is assessed using subjective criteria, often consisting in the choice of the

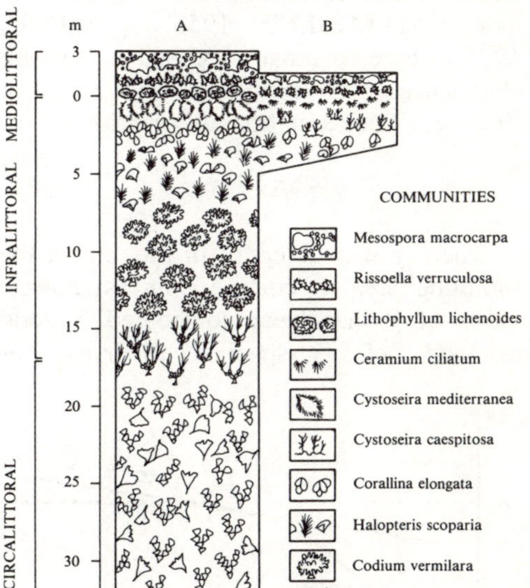

FIGURE 2. Diagrammatic representation of a typical transect in a moderately exposed place (A) and a sheltered place (B) from the Costa Brava (North-western Mediterranean)

abscissa where the curve achieves a particular slope as the relevant area. By using the same criteria to study different communities, the values obtained should be comparable. Obviously, the construction of the parameter/area relationship requires the sampling of much greater areas than the relevant area (GOUNOT, 1969). The area relevant to study the communities in Table II was estimated by sampling from areas nested within larger sampling areas (BALLESTEROS, 1986) and fitting species/area curves to a semilogarithmic function,

$$y = a \ln x + b \qquad (1)$$

where **y** was the number of species and **x** the sampling area in cm^2. Species richness (**R**) was estimated as the number of species corresponding to the Calleja point 5×10^{-2} (BALLESTEROS, 1986). Diversity was calculated with the Shannon index (MARGALEF, 1958; PIELOU, 1975), where coverage was used to measure abundance. Diversity/area curves were fitted to a Michaelis-Menten function,

$$y = Ax / B+x \qquad (2)$$

where **y** was diversity in bits and **x** the sampling area in cm^2. **A**, an asymptotic value of **y**, has been considered a good estimate of species diversity or

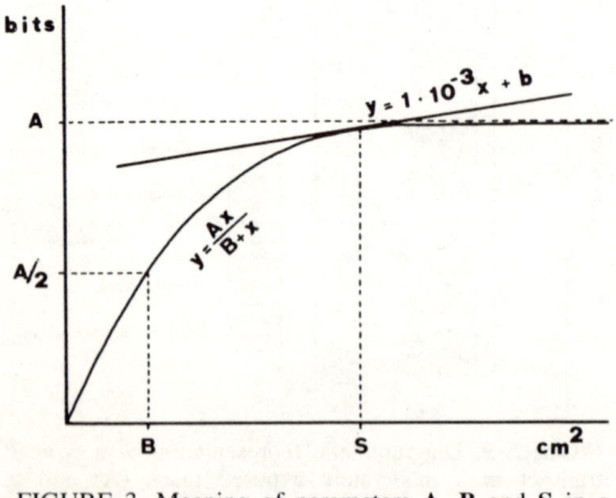

FIGURE 3. Meaning of parameters **A**, **B** and **S** in a diversity/area curve fitted to a Michaelis-Menten function.

alpha-diversity in the sense of PIELOU (1977) (Fig. 3). The value of the Kulczynski index (BOUDOURESQUE, 1971; KRONBERG, 1987) obtained at half of the largest area sampled, where it always reached the asymptotic value, was used to estimate homogeneity.

Moreover, the shape of these curves reflects important structural features of the communities because it is related to their stability or persistence (MARGALEF, 1969). In fact, the same diversity indexes can be obtained with very different shapes of the diversity/area curve. The extreme shapes of the curves, flat or rectangular spectra and step or diagonal spectra (MARGALEF, 1980, 1985), are characteristic of communities with low and high pattern-diversities, respectively. We have attempted to estimate pattern-diversity (also known as spatial-diversity; PIELOU, 1966) by using two different parameters: (1) the parameter **B**, derived from equation 2, which indicates the size of the area at which the diversity index equals half the species diversity of the entire community (Fig. 3), and presents minimum and maximum values for communities with rectangular and diagonal spectra, respectively; and (2) the parameter **S**, defined as the area corresponding to the Calleja point 1×10^{-3} in the diversity/area curve fitted to equation 2 (Fig. 3). This parameter corresponds to the area at which diversity is practically stabilized (structural minimal area).

In addition, the shape of the species/area curve can be described by the parameter **k**, referred to as specific distribution (BALLESTEROS, 1986), which describe the change in the slope of the species/area curve at different areas. It is calculated as

$$k = e^{-b/a} \qquad (3)$$

where **a** and **b** are derived from the semilogarithmic equation 1; it also corresponds to the value of **k** in the potential equation

$$x = k\,e^{ay} \qquad (4)$$

derived from equation 1. Large **k** values correspond to communities where large sampling areas are necessary to obtain a representative number of the species in the community (BALLESTEROS, 1986).

From a functional perspective, a community can be described from its primary production, understood as the synthesis of organic matter by the plants (BALLESTEROS, 1989b). Thus, production is measured in units of biomass (g dw) produced per unit of time (days) and surface (m^2). Another suitable measure is the ratio of primary production to total biomass (i.e., P/B ratio). The P/B ratio, or productivity, has often been related to the energy flow inside an ecosystem and, as such, is negatively correlated with diversity and with other measures of ecosystem organization (MARGALEF, 1968, 1969; NIELL, 1981; NIELL & VARELA, 1984). Production and P/B ratios are used here to describe the dynamics of phytobenthic communities.

SEASONAL CHANGES IN PARADIGMATIC COMMUNITIES: PRODUCTION AND DIVERSIFICATION PHASES

Although some authors have described seasonal variations in the species composition and the biomass of phytobenthic communities (e.g. MANN, 1972; JOSSELYN & WEST, 1985; MCQUAID, 1985; MURRAY & HORN, 1989), the description of significant changes in structural parameters such as the species/area curve (HRUBY, 1975) or diversity (NIELL, 1975; SEAPY & LITTLER, 1982; FERNÁNDEZ *et al.*, 1983) is rare, and the existence of persistent annual rhythms remains uncertain. However, seasonal phenomena are likely to influence dynamic processes from year to year (PARKE, 1947; MANN, 1972).

We can depart from the description of the changes in some structural (species richness, species diversity, pattern diversity, coverage) and dynamic parameters (production, P/B ratio) in three thoroughly studied phytobenthic communities: the *Rissoella verruculosa* community (mediolittoral zone), the *Cystoseira mediterranea* community (infralittoral zone) and the *Halimeda tuna* community (circalittoral zone) (Fig. 4). Because a detailed description of the observed changes has been provided elsewhere (BALLESTEROS, 1984; 1988b), we focus here on the salient features and relationships within (A) and between (B) communities.

(A) WITHIN COMMUNITIES

1. Seasonal changes in community structure and function are evident in all three communities. Similar to the findings of NIELL (1977) for intertidal communities in the temperate Atlantic, the annual change can be separated into a diversification phase, characterized by a low production and P/B ratio, and a production phase, characterized by high production and P/B ratio. Community structure varies with these dynamic changes. Two opposite stages can be recognized as sequences in a continuum: a developed stage and a diversified stage. The developed stage is characterized by communities with high coverage (and biomass) and, often, low pattern-diversity. Conversely, the diversified stage is characterized by low coverage (and biomass) and high pattern-diversity. Species diversity and species richness do not show any distinct trends in the different stages.

2. An inverse relationship between the descriptors of structure (species diversity, pattern diversity, species richness), and coverage and production is apparent for the *Rissoella* and the *Cystoseira* communities, but not for the *Halimeda*, community. In

addition, the postulated negative correlation between P/B ratio and diversity (MARGALEF, 1969, 1980) is not apparent here.

(B) BETWEEN COMMUNITIES

1. The seasonality of the changes in the different parameters decrease with depth,

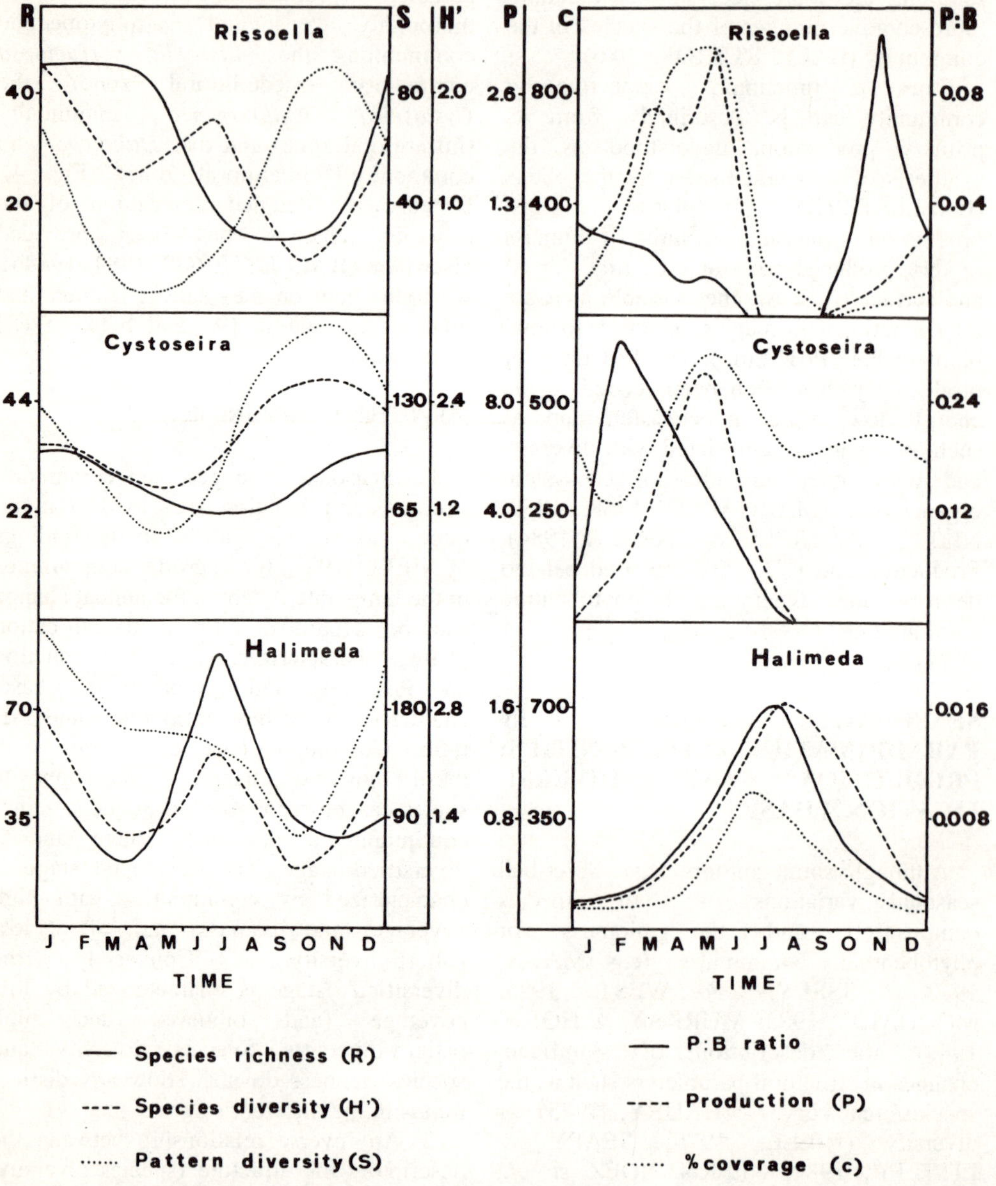

FIGURE 4. Annual variation in species richness, species diversity (in bits per individual), pattern diversity (in cm^2), P/B ratio (in days^{-1}), production (in g C m^{-2} day^{-1}) and % coverage in three selected communities from different bathymetric levels in the North-western Mediterranean: *Rissoella verruculosa*, *Cystoseira mediterranea* and *Halimeda tuna*. Data from BALLESTEROS (1984).

suggesting a lower turnover in deep water communities, a feature already suggested by some authors (ROS *et al.*, 1985; BALLESTEROS, 1989c).

2. There is a temporal lag in the time of peak P/B ratios from shallow communities, which show peak P/B ratios in autumn, and deep communities, that present peak P/B ratios in summer. This lag is similar to that in peak production, which occurs in spring in shallow communities and in summer for deep communities. Moreover, there is also a lag in the structural parameters; peak pattern-diversity occurs in October in shallow communities (*Rissoella* community), but it is displaced to January in deep communities (*Halimeda* community).

3. Productivity and structure appear inversely correlated.

AN ALTERNATIVE APPROACH: MULTIVARIATE ANALYSIS

To summarize relationships between the different parameters, and to describe patterns in the structural and functional affinities shown by the different communities, we applied PCA analysis to the data (LEGENDRE & LEGENDRE, 1979). Each community was described, for each season, using ten variables: species richness (**R**), specific distribution (**k**), biomass (**b**), coverage (**c**), homogeneity (**h**), species diversity (**H'**), pattern-diversity (**B** and **S**), production (**P**) and productivity (**P/B**). Unfortunately, missing values of production and P/B ratios for some communities prevented the analysis for all communities shown in figure 2. However, the analysis did include eight communities representative of all the different levels of the phytal system in the North-western Mediterranean (Fig. 2).

The correlation matrix obtained (Table II) reveals a cluster of positive correlated parameters (**H'**, **k**, **B** and **S**), all of which are related to pattern-diversity and attaining

TABLE II. Significative correlations between the different parameters used to characterize the structure and dynamics of eight phytobenthic communities from the North-western Mediterranean (Pearson coefficient). Parameters are: species richness (**R**), species diversity (**H'**), pattern diversity (**B** and **S**), specific distribution (**k**), homogeneity (**h**), production (**P**), % coverage (**c**), P/B ratio and biomass (**b**).

-, + p < 0.05
-- -, ++ p <0.01
- - -, +++ p <0.001

	R									
H'		H'								
B		+	B							
S		+++	+++	S						
k			++	++	k					
h		- -		- -		h				
P		-	-	- -			P			
c				-			+	c		
P/B							+		P/B	
b		-							-	b

peak values during the diversification phase. These parameters, and particularly pattern-diversity, are negatively correlated with the parameters showing peak values in the production phase (homogeneity, production and coverage). Biomass is poorly related to all of these variables, probably because of the influence of calcareous algae (BALLESTEROS, 1984). P/B ratios are, as expected, positively correlated to production and negatively correlated to biomass.

The rooted and ordered weight factor matrix is shown in Table III, and the projection of the original variables over the plane defined by the first and the second factorial axes is shown in figure 5. The first factor explains 37.9% of the variance and clearly discriminates the parameters related to pattern-diversity (**B**, **S** and **k**) from production and coverage. The second axis explains 17.3% of the variance and is mainly related to biomass. Species diversity and homogeneity are represented by opposite vectors with high weight factors over the first and the second axis, respectively. The third axis accounts for only 13.7% of the variance and is related to species richness. Because the first two axes encompass more than 50% of the variance, we believe that they suffice to describe adequately the spatial and seasonal variability of Mediterranean phytobenthic

TABLE III. Rooted and ordered weight factors matrix.

Pattern diversity-B	0.913	-0.054	0.119
Pattern diversity-S	0.885	-0.283	0.203
Species diversity	0.665	-0.523	0.371
Specific distribution	0.551	0.007	-0.050
Production	-0.682	-0.040	0.033
Coverage	-0.604	-0.098	0.625
Biomass	0.092	0.785	-0.158
Homogeneity	-0.421	0.716	-0.030
Species richness	0.075	-0.187	0.745
P/B ratio	-0.375	-0.559	-0.601

communities. Because of their relationship to different variables, we shall refer hereafter to the axes as production-diversification axis (**P-D** axis) and biomass-heterogeneity axis (**b-h** axis), respectively.

The distribution of the communities studied and their seasonal variations in the plane defined by **P-D** and **b-h** axes is shown in figure 6. The following aspects may be noted:

(1) Communities are ordered along the first axis in relation to their diversification. Temporal variation in their structure appears to be similar for all communities. Ordering along the second axis reflects a biomass increment from low to high values.

(2) Annual cycles of communities are reflected in rotational clockwise paths along the plane. The position of the communities at the different seasons reflects their structural and functional characteristics. The extremes of each axis can be considered as representative of the two phases and the two stages described before. Further, irregularities in the annual cycle are well reflected in the analysis. For instance, the *Ceramium ciliatum* community, which lacks annual periodicity, has an inverse rotation sense, but this may be peculiar to the year studied. Similarly, temporal discontinuities in structure and dynamics are well reflected in the analysis.

(3) Seasonal variation in the structure and dynamics of each community is proportional to the area comprised within the polygon that describes the seasonal path of the community. Thus, seasonal variations are greatest in *Cystoseira* species communities and are lowest in the *Lithophyllum tortuosum* community.

(4) Periods of peak production - which coincide with a minimum diversification - are attained in spring for the mediolittoral communities, and are progressively displaced to summer and autumn for increasing depths.

Thus, the patterns and trends suggested by the detailed analysis of three selected communities are confirmed when studying a greater number of communities. Thus, we think that these structural and functional patterns can be generalized to other Mediterranean phytobenthic communities.

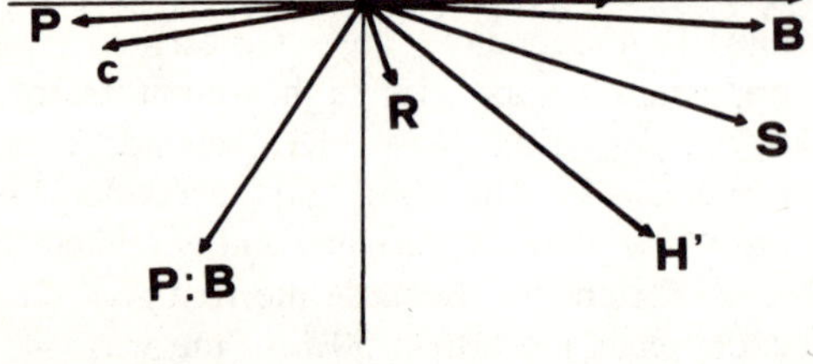

FIGURE 5. Projection of the original variables over the plane defined by the first and the second factorial axes. Symbols as in Table II.

A MODEL OF STRUCTURAL AND FUNCTIONAL DYNAMICS IN PHYTOBENTHIC COMMUNITIES FROM THE MEDITERRANEAN SEA

The model we propose to represent the seasonal variations in the Mediterranean phytobenthic communities is probably applicable to other temperate seas with a distinct growing season. The model considers that the majority of communities belonging to advanced successional stages have closed annual cycles that allow for the distinction between a production phase, when the dominant species has an intense growth, and a diversification phase, when a net biomass loss of the dominant species is detected. Both phases end in two stages: the developed community and the diversified community. The developed community has high biomass and coverage values, a high vertical structuration and a

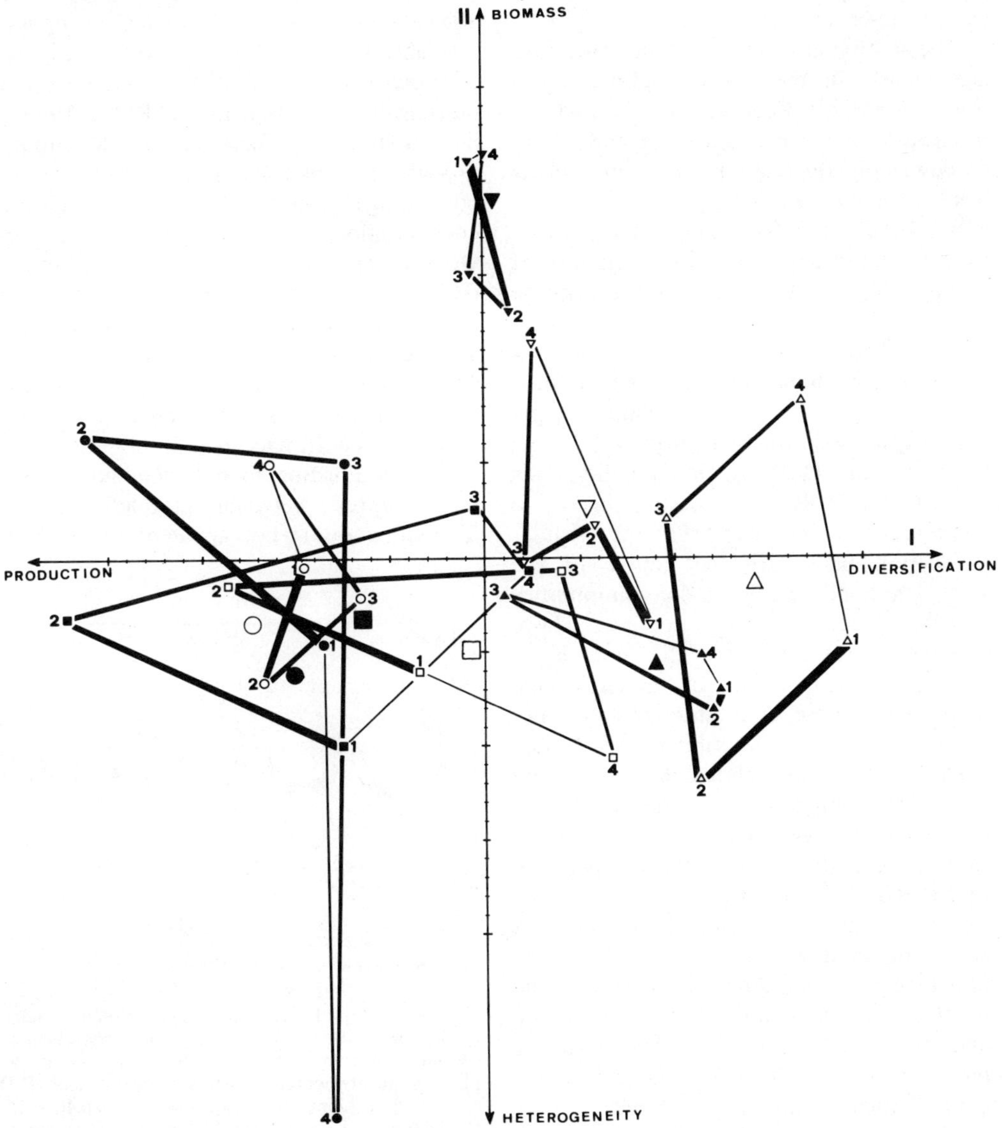

FIGURE 6. Communities representation over the plane defined by the first and the second factorial axes. Each community is represented by its position at each season (1-winter, 2-spring, 3-summer, 4-autumn) and by its average value. Communities are symbolized as follows: *Rissoella verruculosa* (full circles), *Lithophyllum lichenoides* (full reverse triangles), *Ceramium ciliatum* (open circles), *Cystoseira mediterranea* (full squares), *Cystoseira caespitosa* (open squares), *Codium vermilara* (full triangles), *Cystoseira zosteroides* (open triangles), *Halimeda tuna* (open reverse triangles).

high homogeneity, which causes a low diversification or pattern-diversity. The diversified community, in turn, has low values of biomass and coverage, low vertical structuration and significant horizontal heterogeneity, caused by the highly patchy distribution of encrusting and turf algae.

These structural and dynamic variations are evident in the plane defined by a production-diversification axis and a biomass-heterogeneity axis. The possibility of describing fluctuations in such a plane has recently been pointed out by MARGALEF (1989). The annual cycles may be represented by closed figures of diverse shapes which turn in a clockwise direction. The phases and stages described are arranged at the extremes of the two axes in communities with well defined annual cycles. Opportunistic algal communities also fluctuate along this plane, but their paths lack any directionality. This pattern can also be extrapolated to communities elsewhere showing similar production phases in the course of the year. The annual cycle of the communities may be idealized as circumferences whose diameter increases with the degree of seasonality (Fig. 7). In general, the diameter of these circumferences or other geometric figures describing the annual cycle of the communities decreases with increasing complexity. Therefore, it may be considered a good estimate of the energy flow through the community, as suggested by MARGALEF (1989).

The ordering of different communities according to different zonation levels can easily be discerned along the **P-D** axis and the **b-h** axis in relation to their structural and functional properties. The ordering basically follows the **P-D** axis and all the communities are distributed along this axis. Shallow-water communities, dominated by frondose algae, that have little structure and high P/B ratios, are distributed in the portion of the plane defined by high production. On the opposite side of the axis, deep-water communities, characterized by their high diversification, predominate (Fig. 7).

The delay in peak production towards late summer at increasing depths is a general feature of phytobenthic Mediterranean systems (BALLESTEROS, 1989b) because of the typical temporal availability of light and nutrients. Shallow-water communities are often nutrient-limited (BALLESTEROS, 1989b) and, as such, have their period of maximum growth (production phase) in spring, coinciding with higher nitrate concentrations in seawater (Fig. 1). Deep-water communities are light-limited (BALLESTEROS, 1989b), and their annual cycle is accommodated to exploit the high summer irradiances in order to grow (Fig. 1). However, extrapolation of this depth-dependent delay to communities elsewhere still needs confirmation.

It is interesting to note that an inverse relationship between productivity and diversification is not apparent (Table II,

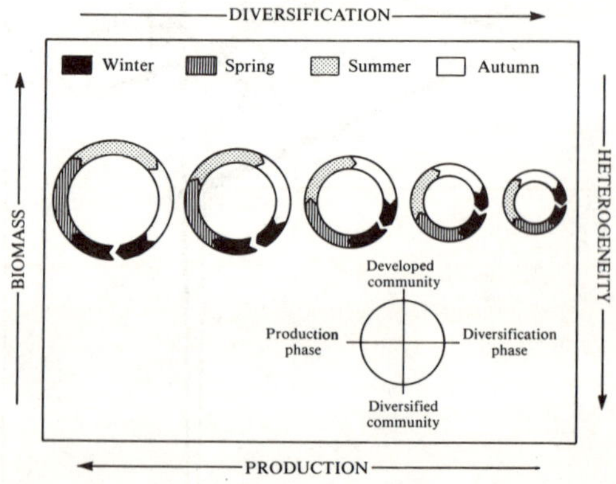

FIGURE 7. Hypothetical representation of the annual cycle of different Mediterranean phytobenthic communities from different bathymetric levels in the plane defined by a production-diversification axis (**P-D** axis) and a biomass-heterogeneity axis (**b-h** axis). Production and diversification phases correspond to the extremes of the **P-D** axis and developed community and diversified community stages are situated at the extremes of the **b-h** axis. A time lag in the different phases and stages from spring to summer and from shallow to deep water communities has been represented

Fig. 5). Nevertheless, in our opinion, the term diversification differs when applied to the annual cycle of a certain community or when referred to different communities. Within communities, the concept of diversification is not related to the classical meaning of structure, but to patchiness. High diversities only occur in diversification phases and, therefore, high pattern-diversities result from the spatial heterogeneity in the distribution of species within the community. In a developed community the diversity is usually minimal because of dominance of the main species in the community. On the other hand, the meaning of diversification among-communities corresponds to structure and organization, and is closer to the traditional interpretation (MARGALEF, 1974).

These observations allow us to reconsider the concepts of short-term (non-equilibrium) diversity and long-term (evolutionary) diversity suggested by SLOBODKIN & SANDERS (1969) and formulated by GRASSLE & SANDERS (1973). Short-term diversity reflects local effects and acts on a short time scale. Its variability is induced by small disturbances such as those involved in the annual cycle of temperate seas. The main increase in diversity within a community during an annual cycle results from a sudden increase in diversity at the end of the diversification phase, caused both by the lack of dominance inside the community and by the appearance of some new species. The new species can grow in this stage because of the reduced spatial competition, as predicted by the niche pre-emption model (WHITTAKER, 1975). This increase in diversity is often associated with periods of relatively intense physical disturbance caused by autumnal storms, corresponding to a maximum energy flow through the community. The transition to the production phase and to the developed community involves increased competition and subsequent competitive exclusion of species, yielding - in most Mediterranean communities - communities dominated by a single species, with the associated reduction in diversity (but not necessarily in structure). Clearly, short-term diversity and productivity need not be inversely correlated, as observed by NIELL (1977) and FERNÁNDEZ *et al.* (1983). Instead, disturbances may be considered as mechanisms leading to increased species richness and diversity (DAYTON & HESSLER, 1972; ABELE, 1976; OSMAN, 1977; CONNELL, 1978). On the other hand, long-term diversity refers to the development of diversity through evolution, and it appears to increase with environmental stability. Long-term diversity agrees with the stability-time hypothesis of SANDERS (1968, 1969). Deep communities are more diverse than shallow communities (Fig. 6), so there is an inverse correlation between productivity and diversity on a between-communities scale.

Seasonal changes generate moderate environmental instabilities or disturbances which allow short-term diversity to operate. At another level, communities and their annual cycles are subject to long-term diversity changes, which result from an evolutionary tendency towards greater species richness and community organization in stable environments (Fig. 8). Indeed, mature communities (coralligenous) are diverse, and immature ones (mediolittoral zone) have low diversity (Fig. 7). Physical and biological disturbances - absorbed or not absorbed by the community - increase the P/B ratio (NIELL, 1981). This effect is apparent in the annual cycle of real communities (Fig. 6). The consequences of this process in our model would be the displacement of the communities - in the case of total destruction of the community - towards areas of the plane defined by low diversification and biomass values. Conversely, succession should transfer the communities to the portion of the plane corresponding to high values of diversification and biomass (Fig. 8).

Up to this point, we have referred community dynamics to the annual cycle, without considering interannual dynamics. Interannual dynamics can be incorporated into the model by considering time as a third axis (Fig. 9). In this implemented model, annual cycles are represented by a 360-degree turn around an ascending helicoidal trajectory that reflects the development of structural and functional community characteristics (Fig. 9). This ascending helix should have some irregularities which represent specific responses to disturbances or to deviations from the average annual behaviour of the environmental factors. Each turn of the helix is different from the others because of the unrepeatability of each and every one of the processes which determine the situation of each community at each moment.

Although the term stability is vague in ecology, it has been defined through three complementary properties (DAYTON *et al.*, 1984): persistence, resistance and resilience. Persistence is here associated to the height of the helix. Even though the community has some capacity to buffer disturbances, which are reflected in modifications of the helix, this capacity is

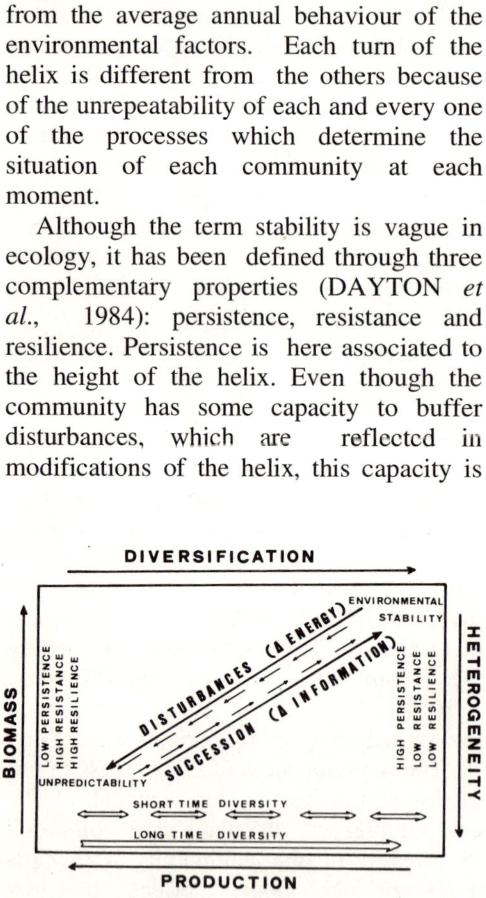

FIGURE 8. Tendencies of change caused by disturbances and succession in the plane defined by the **P-D** axis and the **b-h** axis. The role of short-term diversity and long-time diversity and the main characteristics of community stability at the two extremes of the **P-D** axis are also represented.

limited, and it is a function of the community resistance (amplitude stability in the sense of ORIANS, 1975). So, resistance can be figured as a cylinder, (which may be of irregular shape) including the helix (Fig. 9). If disturbance is sufficiently strong, the helix can break and go beyond the limits of the cylinder (Fig. 9). This breakage represents the disappearance of the community and it is always linked to a change in the species assemblage. The new trajectory moves towards lower values of biomass and diversification (Figs. 8 and 9). If disturbance is punctual, the new assemblage may evolve following a successional sequence, perhaps restoring the original community. The time necessary to complete this sequence should be inversely related to its resilience (Fig. 9). The process of succession would be ruled by different phases (HUVÉ, 1974), depending on the intensity of the disturbance. Each phase has its own cycle which may have an annual or lower periodicity, and it moves along the plane defined by the **P-D** and the **b-h** axes. FERNÁNDEZ *et al.* (1981) distinguish different phases, each one with two stages named development moments and transient moments. The energy flow through the community is minimized and the community structure is improved during the development moments, while the transient moments correspond to surges of energy flow (high P/B ratios) through the community. Although this pattern has been described to explain succession in rocky intertidal systems from the temperate Atlantic, it may adequately describe the successional sequence of sublittoral communities and can be easily introduced into our model. Development moments have a very similar performance to the production phase, and transient moments have similar properties to the diversification phase. Succession phases reproduce in shorter periods of time and, perhaps more importantly, the annual dynamics of normally-constituted phytobenthic communities.

If the disturbance acts for a long period of time (e.g. an increase of herbivory, some human modifications), the community cannot return to its original state and, thus, it would be stationed at another position on the plane, with its own annual cycle and its own stability properties. This scenario supports the theory that predicts the existence of multiple stable points in natural communities (SUTHERLAND, 1974). At each depth different communities (or stable points) coexist and this determines the patchy aspect that characterizes the rocky bottoms in temperate seas. The coexistence of different multiple stable points at the same depth and in places subject to similar mean inputs of exosomatic energy, can only be explained if different degrees of biotic or abiotic disturbance are considered. The most usual and well studied situation derives from an excessive grazing pressure by sea-urchins, a situation which has been described in oceanic communities (MANN, 1977) as well as in Mediterranean ones (VERLAQUE, 1987).

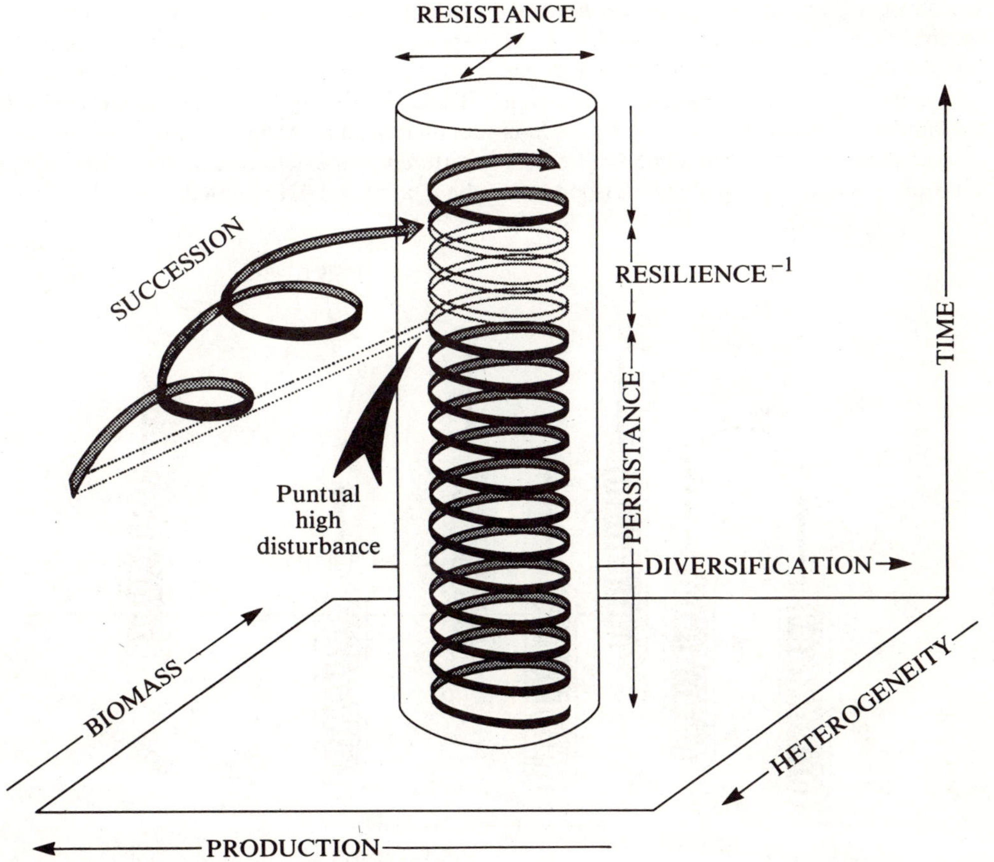

FIGURE 9. Three-dimensional representation of the community dynamics over a long period of time. Each turn of the helix corresponds to an annual cycle. Persistence, resistance and resilience of the community are easily defined in the figure. High disturbances that cannot be absorbed by the community, and the subsequent succession pattern that the new assemblage undergoes after the disturbance, are also schematically represented

In general, the dynamics of the persistent communities from temperate seas can be depicted as helixes whose turns represent annual cycles along the plane defined by a **P-D** axis and a **b-h** axis (Fig. 10). The transfer from one community to another is usually rare (i.e., the probability of such a change occurring is very low) because the community structure and dynamics correspond largely to mean values of the physical environment. As an example, a *Cystoseira mediterranea* community will never turn into an *Halimeda tuna* community. However, the existence of some different communities at, for example, the same depth, may be related to catastrophic events or to disturbances attributed to biotic factors. Under these conditions, community interconversion has a certain, time-dependent, non-zero probability of occurring. A good example is provided by the different communities (or multiple stable points) described in Mediterranean photophilic algae communities according to predation intensity by sea-urchins (VERLAQUE, 1987).

When the model is extrapolated to evolutionary time scales, it is obvious that communities would not persist as rigid cylinders, but would oscillate. The extent of oscillation may be a function of the change in qualitative and quantitative species composition, or in species characteristics. These oscillations may also be interpreted as changes in the interconversion probabilities between different stable points. However, natural selection operates over the individuals belonging to some species, and evolutionary changes in species modify, in turn, the structural and functional performances of communities. Thus, it is difficult to introduce the role of evolution into the model proposed. From a different perspective, two main forces can be assumed to modulate long-term

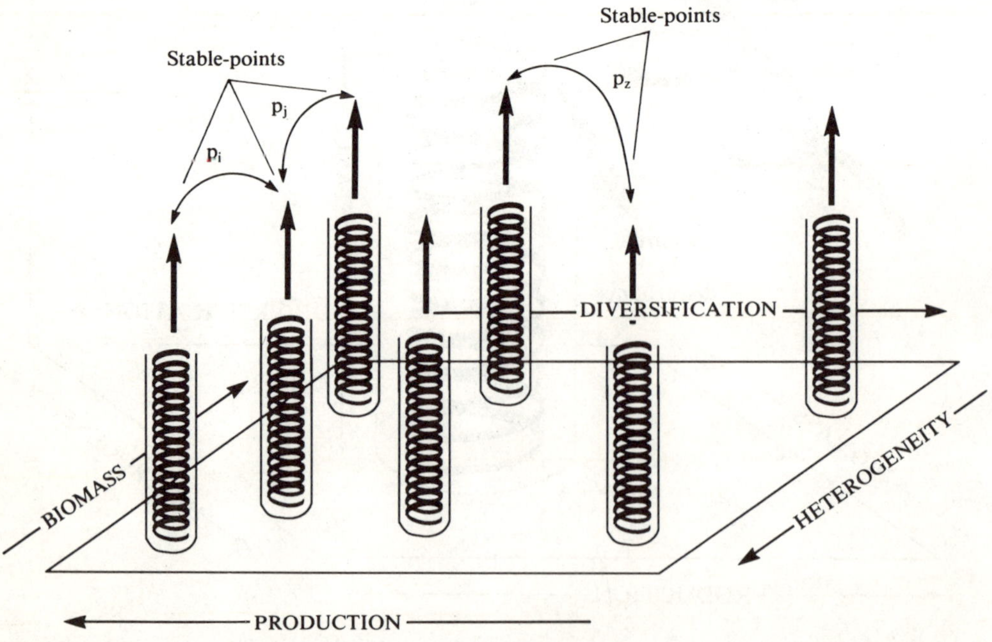

FIGURE 10. Dynamics of some different phytobenthic communities in the space defined by a production-diversification axis, a biomass-heterogeneity axis, and time. Conversion from one community to another is only possible if they are successionally related or they belong to a bathymetric level with the same energy inputs. Interconversion probabilities may change along evolutionary time.

ecosystem dynamics: environmental stability and disturbances (CONNELL, 1978; WALBRAN *et al.*, 1989). Under stable conditions, communities would tend to increase diversity, accumulate information and use the energy to create persistent structures, minimizing the diameter of the helixes and, sometimes, their resistance. Systems would become more mature, in the classic sense of MARGALEF (1963, 1975). Disturbances would act in the opposite sense. The result is the aspect of the submerged world, scattered into different communities, each one adapted to particular environmental conditions and determined by a series of unavoidable historical factors.

ACKNOWLEDGEMENTS

Firstly, R. Margalef and J. Romero suggested applying a PCA analysis to data obtained from different phytobenthic communities in the course of my Ph. D. thesis. Help in the numerical treatment was provided by J. Romero and O. Delgado.

M. Zabala and A. García were present in a stage in the island of Cabrera, when the conceptual model was established; creative discussions with them served to elaborate the model. J. Romero, C.M. Duarte, J. Camp, J. Flos, M. Zabala, M. J. Uriz and other members of the Centre d'Estudis Avançats, the Department of Ecology of the University of Barcelona and the Institut de Ciències del Mar, helpfully commented on the model and gave me some ideas to be exploited in the future.

C.M. Duarte kindly revised the original text and thoroughly corrected and improved the English version. J. Corbera and my father carefully drew the most difficult figures. The author is grateful to all for their assistance.

REFERENCES

ABELE, L.G. 1976. Comparative species richness in fluctuating and constant environments: coral-associated decapod crustaceans. *Science*, 192: 461-463.

AUGIER, H. & BOUDOURESQUE, C.F. 1974. Dix ans de recherches dans la zone marine du parc national de Port-Cross (France). Deuxième partie. *Ann. Soc. Sci. nat. Archeol. Toulon. Var.*, 26: 119-150.

BALLESTEROS, E. 1984. *Els vegetals i la zonació litoral: espècies, comunitats i factors que influeixen en la seva distribució.* Ph. D. Thesis. University of Barcelona.

BALLESTEROS, E. 1986. Métodos de análisis estructural en comunidades naturales, en particular del fitobentos. *Oecol. Aquat.*, 8: 117- 131.

BALLESTEROS, E. 1987. Estructura i dinàmica del poblament algal de les fulles de *Posidonia oceanica* (L.) Delile als herbeis de Tossa de Mar (Girona). *Butll. Inst. Cat. Hist. Nat.*, 54 (Sec. Bot., 6): 13-30.

BALLESTEROS, E. 1988a. Composición y estructura de la comunidad infralitoral de *Corallina elongata* Ellis & Solander de la Costa Brava (Mediterráneo Occidental). *Inv. Pesq.*, 52(1): 135-151.

BALLESTEROS, E. 1988b. Estructura y dinámica de la comunidad de *Cystoseira mediterranea* Sauvageau en el Mediterráneo Noroccidental. *Inv. Pesq.*, 52(3): 313-334.

BALLESTEROS, E. 1989a. Estructura y dinámica de la comunidad infralitoral de *Codium vermilara* (Olivi) Delle Chiaje de la Costa Brava (Mediterráneo Occidental). *Anales de Biología*, 19 (B. Ambiental, 5): 73-90.

BALLESTEROS, E. 1989b. Production of seaweeds in North-western Mediterranean marine communities: its relation with environmental factors. *Topics in Marine Biology*. (*Sci. Mar.*, 53, 2-3: 357-364).

BALLESTEROS, E. 1989c. El bentos coster: estructura i procés. *Revista de Catalunya*, 30: 48-55.

BALLESTEROS, E. 1990 a. Structure and dynamics of the community of *Cystoseira caespitosa* Sauvageau (Fucales, Phaeophyceae) in the North-western Mediterranean. *Sci. Mar.*, 54(2): 155-168.

BALLESTEROS, E. 1990 b. Structure and dynamics of the community of *Cystoseira zosteroides* (Turner) C. Agardh (Fucales, Phaeophyceae) in the North-western Mediterranean. *Sci. Mar.*, 54(3): 217-229.

BALLESTEROS, E. 1991. Structure and dynamics of the community of *Rissoella verruculosa* (Bertolini) J. Agardh (Gigartinales, Rhodophyceae) in the North-western Mediterranean. *Sci. Mar.*, 55(2): 439-451.

BALLESTEROS, E. in press a. Growth, production and dynamics of a deep-water population of *Halimeda tuna* (Chlorophyceae, Caulerpales) in the North-western Mediterranean. *Bot. Mar.*

BALLESTEROS, E. in press b. Structure of a deep-water population of *Halimeda tuna* (Chlorophyceae, Caulerpales) in the North-western Mediterranean. *Collect. Bot.*

BALLESTEROS, E. & ROMERO, J. 1988. Zonation patterns in tideless environments (Northwestern Mediterranean): looking for discontinuities in species distributions. *Inv. Pesq.*, 52(4): 595- 616.

BOISSET, F. 1987. *Estudio del fitobentos esciáfilo infralitoral de sustratos duros en el litoral valenciano (España). Flora y vegetación. Tesis Doctoral. Universidad de Valencia.*

BOUDOURESQUE, C.F. 1970. *Recherches de bionomie analytique, structurale et expérimentale sur les peuplements benthiques superficiels en Méditerranée Occidentale (fraction algale).* Thèse. Université d'Aix-Marseille II.

BOUDOURESQUE, C.F. 1971. Méthodes d'étude qualitative et quantitative du benthos (en particulier du phytobenthos). *Téthys*, 3(1): 79- 104.

BOUDOURESQUE, C.F. 1985. Groupes écologiques d'algues marines et phytocoenoses benthiques en Méditerranée Occidentale: une revue. *Giorn. Bot. ital.*, 118 (suppl.): 7-42.

BOUDOURESQUE, C.F., MARCOT-COQUEUGNIOT, J., NÉDELEC, H., PERRET-BOUDOURESQUE, M. & BELSHER, T. 1980. Le phytobenthos d'un port méditerranéen, Port-Vendres. *Compt. rend. 105 cong. Nat. Soc. Sav. Caen Sci.*, 3: 261-273.

BRAUN-BLANQUET, J. 1979. *Fitosociología.* Blume. Madrid.

CHOCK, J.S. & MATHIESON, A.C. 1983. Variations of New England estuarine seaweed biomass. *Bot. Mar.*, 26: 87-97.

CONNELL, J.H. 1978. Diversity in tropical rain forests and coral reefs. *Science*, 199: 1033-1034.

COPPEJANS, E. 1977. *Bijdrage tot de studie van der wierpopulaties (Chlorophyceae, Phaeophyceae, Rhodophyceae) van het fotofiel infralittoraal in het noordwestejelik mediterranean bekken.* Doktoraat Thesis. Rijksuniversiteit Gent.

COPPEJANS, E. 1980. Phytosociological studies on mediterranean algal vegetation: rocky surfaces of the photophilic infralittoral zone. In: *The shore environment. Ecosystems.* (J.H. PRICE,D.E.G. IRVINE & W.F. FARNHAM, eds.): 371-393. Academic Press. London.

DAYTON, P.K. 1971. Competition, disturbance and community organization: the provision and subsequent utilization of space in a rocky intertidal community. *Ecol. Monogr.*, 41: 351- 389.

DAYTON, P.K. 1975. Experimental evaluation of ecological dominance in a rocky intertidal algal community. *Ecol. Monogr.*, 45: 137-159.

DAYTON, P.K. & HESSLER, R.R. 1972. The role of disturbance in the maintenance of deep-sea diversity. *Deep Sea. Res.*, 19: 199-208.

DAYTON, P.K., CURRIE, V., GERRODETTE, T., KELLER, B.D., ROSENTHAL, R. & TRESCA, D.V. 1984. Patch dynamics and stability of some California kelp communities. *Ecol. Monogr.*, 54(3): 253- 289.

FELDMANN, J. 1937. Recherches sur la végétation marine de la Méditerranée: la côte des Albères. *Rev. Algol.*, 10: 1- 339.

FERNÁNDEZ, M.C., NIELL, F.X. & FUENTES, J.M. 1981. Remarks on succession in rocky intertidal systems. *Proceed. Xth Int. Seaweed Symp.*: 283-288.Walter de Gruyter. Berlin.

FERNÁNDEZ, M.C., NIELL, F.X. & ANADÓN, R. 1983. Comparación de dos comunidades de horizontes intermareales con abundancia de *Bifurcaria bifurcata* Ros. en las costas N y NO de España. *Inv. Pesq.*, 47(3): 435-455.

GAGNÉ, J.A., MANN, K.H. & CHAPMAN, A.R.O. 1982. Seasonal patterns of growth and storage in *Laminaria longicruris* in relation to differing patterns of availability of nitrogen in the water. *Mar. Biol.*, 69: 91-101.

GIACCONE, G. & BRUNI, A. 1972. Le cistoseire e la vegetazione sommersa del Mediterraneo. *Att. Inst. Venetto Sci. Lett. Art.*, 81: 59-103.

GLEASON, H.A. 1925. Species and area. *Ecology*, 6: 66-74.

GOUNOT, M. 1969. *Méthodes d'étude quantitative de la végétation.* Masson. Paris.

GRASSLE, J.F. & SANDERS, H.L. 1973. Life histories and the role of disturbance. *Deep Sea Res.*, 20: 643-659.

GRAY, J.S. 1977. The stability of benthic ecosystems. *Helgol. wiss. Meeres.*, 30: 427-444.

HAWKINS, S.J. & HARTNOLL, R.G. 1980. A study of the small-scale relationship between species number and area on a rocky shore. *Estuar. Coast. Mar. Sci.*, 10: 201-214.

HISCOCK, K. & MITCHELL, R. 1980. The description and classification of sublittoral epibenthic ecosystems. In: *The shore environment. Ecosystems.* (J.H. PRICE, D.E.G. IRVINE & W.F. FARNHAM, eds.): 323-370. Academic Press. London.

HRUBY, T. 1975. Seasonal changes in two algal populations from the coastal waters of Washington state. *J. Ecol.*, 63: 881-889.

HUVÉ, P. 1974. Considérations sur la genèse de quelques peuplements algaux benthiques de Méditerranée Occidentale: les étapes, stades, phases. *Bull. Soc. Phycol. Fr.*, 19: 101- 120.

JACKSON, G.A. 1977. Nutrients and production of giant kelp *Macrocystis pyrifera* of southern California. *Limnol. Oceanogr.*, 22: 979-995.

JOHN, D.M., LIEBERMAN, D. & LIEBERMAN, M. 1977. A quantitative study of the structure and dynamics of benthic subtidal algal vegetation in Ghana (Tropical West Africa). *J. Ecol.*, 65: 497-521.

JOHNSON, G.R. & MANN, K.H. 1988. Diversity, patterns of adaptation, and stability of Nova Scotian kelp beds. *Ecol. Monogr.*, 58(2): 129-154.

JOSSELYN, M.N. & WEST, J.A. 1985. The distribution and temporal dynamics of the estuarine macroalgal community of San Francisco Bay. *Microbiol.*, 129: 139-152.

KAIN, J. 1989. The seasons in the subtidal. *Br. Phycol. J.*, 24: 203-215.

KRONBERG, I. 1987. Accuracy of species and abundance minimal areas determined by similarity area curves. *Mar. Biol.*, 96: 555- 561.

LAWSON, G.W. 1957. Seasonal variation of intertidal zonation on the coast of Ghana in relation to tidal factors. *J. Ecol.*, 45: 831- 860.

LEGENDRE, L. & LEGENDRE, P. 1979. *Écologie numérique*. Masson. Paris.

LITTLER, M.M. 1980. Southern California rocky intertidal ecosystems: methods, community structure and variability. In: *The shore environment. Ecosystems*. (J.H. PRICE, D.E.G. IRVING & W.F. FARNHAM eds.): 565-608. Academic Press. London.

LUBCHENCO, J. & MENGE, B.A. 1978. Community development and persistence in a low rocky intertidal zone. *Ecol. Monogr.*, 59: 67- 94.

MANN, K.H. 1972. Ecological energetics of the seaweed zone in a marine bay on the Atlantic coast of Canada. II. Productivity of the seaweeds. *Mar. Biol.*, 14: 199-209.

MANN, K.H. 1973. Seaweeds: their productivity and strategy for growth. *Science*, 182: 975-981.

MANN, K.H. 1977. Destruction of kelp beds by sea-urchins: a cyclical phenomenon or irreversible degradation? *Helgol. wiss. Meeres.*, 30: 455-467.

MARGALEF, R. 1958. Information theory in ecology. *Gen. Syst.*, 3: 36-71.

MARGALEF, R. 1963. On certain unifying principles in ecology. *Am. Nat.*, 97: 357-374.

MARGALEF, R. 1968. *Perspectives in ecological theory*. Chicago University Press. Chicago.

MARGALEF, R. 1969. Diversity and stability: a practical proposal and a model of interdependence. *Brookh. Symp. Biol.*, 22: 25-37.

MARGALEF, R. 1974. *Ecología*. Omega. Barcelona.

MARGALEF, R. 1975. Diversity, stability and maturity in natural ecosystems. In: *Unifying concepts in ecology*. (W.H. VAN DOBBEN & R.H. LOWE-MC CONNELL, eds.): 139-150. Junk. The Hague.

MARGALEF, R. 1980. *La Biosfera, entre la termodinámica y el juego*. Omega. Barcelona.

MARGALEF, R. 1985. From hydrodynamic processes to structure (information) and from information to process. *Can. Bull. Fish. Aquat. Sci.*, 213: 200-220.

MARGALEF, R. 1989. Reflexiones sobre la diversidad y significado de su expresión cuantitativa. *Simposio sobre la Diversidad Biológica*. Madrid.

MATHIESON, A.C. & PENNIMAN, C.A. 1986. Species composition and seasonality of New England seaweeds along an open coastal-estuarine gradient. *Bot. Mar.*, 29: 161-176.

MC QUAID, C.D. 1985. Seasonal variation in biomass and zonation of nine intertidal algae in relation to changes in radiation, sea temperature and tidal regime. *Bot. Mar.*, 28: 539-544.

MURRAY, S.N. & HORN, M.H. 1989. Seasonal dynamics of macrophyte populations from an Eastern North Pacific rocky intertidal habitat. *Bot. Mar.*, 32: 457-473.

NÉDELEC, H. 1979. *Étude structural et problèmes d'échantillonnage dans une phytocoenose portuaire*. Mém. Univ. P. et M. Curie. Marseille-Luminy.

NEUSHUL, M. 1967. Studies of subtidal vegetation in Western Washington. *Ecology*, 48: 83-94.

NIELL, F.X. 1974. Les applications de l'index de Shannon à l'étude de la végétation intertidale. *Soc. Phycol. Fr. Bull.*, 19: 238- 254.

NIELL, F.X. 1975. *Estudios sobre la estructura, dinámica y producción del fitobentos intermareal (facies rocosa) de la ría de Vigo*. Tesis Doctoral. Universidad de Barcelona.

NIELL, F.X. 1977. Rocky intertidal benthic systems in temperate seas: a synthesis of their functional performances. *Helgol. wiss. Meeres.*, 30: 315-333.

NIELL, F.X. 1979. Structure and succession in rocky algal communities of a temperate intertidal system. *J. Exp. Mar. Biol. Ecol.*, 36(2): 185-200.

NIELL, F.X. 1981. Estructuras disipativas en la organización de los sistemas bentónicos. *Oecol. Aquat.*, 5: 239- 245.

NIELL, F.X. & VARELA, M. 1984. Initial colonization stages on rocky coastal substrates. *Mar. Ecol.*, 5(1): 45-56.

ORIANS, G.H. 1975. Diversity, stability and maturity in natural ecosystems. In: *Unifying concepts in ecology*. (W.H. VAN DOBBEN & R.H. LOWE-MC CONNELL, eds.): 139-150. Junk. The Hague.

OSMAN, R.W. 1977. The establishment and development of a marine epifaunal community. *Ecol. Monogr.*, 47: 5763.

OSMAN, R.W. 1978. The influence of seasonality and stability on the species equilibrium. *Ecology*, 59(2): 383-399.

PAINE, R.T. 1974. Intertidal community structure. Experimental studies on the relationship between a dominant competitor and its principal predator. *Oecologia*, 15: 93-120.

PAINE, R.T & LEVIN, S.A. 1981. Intertidal landscapes: disturbance and the dynamics of pattern. *Ecol. Monogr.*, 51: 145-178.

PARKE, M. 1947. Studies on British Laminariaceae. I. Growth in *Laminaria saccharina* (L.) Lamour. *J. Mar. Biol. Ass. U.K.*, 27: 651-709.

PÉRÈS, J.M. & PICARD, J. 1964. Nouveau manuel de bionomie benthique de la mer Méditerranée. *Rec. Trav. St. Mar. Endoume*, 31(47): 1-137.

PIELOU, E.C. 1966. Species diversity and pattern diversity in the study of ecological succession. *Theor. Biol.*, 10: 370-383.

PIELOU, E.C. 1975. *Ecological diversity*. John Wiley. New York.

PIELOU, E.C. 1977. *Mathematical ecology*. John Wiley. New York.

PRICE, I.R. 1989. Seaweed phenology in a tropical Australian locality (Townswille, North Queensland). *Bot. Mar.*, 32: 399-406.

ROMERO, J. 1984. Relaciones entre unidades de volumen y unidades de biomasa en distintas especies de algas bentónicas. Aplicaciones a evaluaciones de biomasa del fitobentos. *Oecol. Aquat.*, 7: 37-42.

ROS, J., ROMERO, J., BALLESTEROS, E. & GILI, J.M. 1985. Diving in blue water: the benthos. In: *Western Mediterranean* (Margalef, R., ed.): 233-295. Pergamon. Oxford.

SANDERS, H.L. 1968. Marine benthic diversity: a comparative study. *Am. Nat.*, 102: 243-282.

SANDERS, H.L. 1969. Benthic marine diversity and the stability-time hypothesis. *Brookh. Symp. Biol.*, 22: 71-80.

SCHIEL, D.S. & FOSTER, M.S. 1986. The structure of subtidal algal stands in temperate waters. *Oceanogr. Mar. Biol. Ann. Rev.*, 24: 265-307.

SEAPY, R.R. & LITTLER, M.M. 1982. Population and species diversity fluctuations in a rocky intertidal community relative to severe aerial exposure and sediment burial. *Mar. Biol.*, 71: 87-96.

SLOBODKIN, L.B. & SANDERS, H.L. 1969. On the contribution of environmental predictability to species diversity.*Brookh. Symp. Biol.*, 22: 82-92.

SOUSA, W.P. 1979. Experimental investigations of disturbance and ecological succession in a rocky intertidal algal community. *Ecol. Monogr.*, 49(3): 227-254.

SUTHERLAND, J.P. 1974. Multiple stable points in natural communities. *Am. Nat.*, 108: 859-873.

SUTHERLAND, J.P. & KARLSSON, R.H. 1977. Development and stability of the fouling community at Beaufort, North Carolina. *Ecol. Monogr.*, 47: 425-446.

TUSSENBROEK, B.I. 1989. Seasonal growth and competition of fronds of *Macrocystis pyrifera* in the Falkland islands. *Mar. Biol.*, 100: 419-430.

VERLAQUE, M. 1987. *Contribution à l'étude du phytobenthos d'un ecosystème photophile thermophile marin en Méditerranée Occidentale.* Thése. Université d'Aix-Marseille.

WALBRAN, P.D., HENDERSON, R.A., TIMOTHY JULL, A.J. & JOHN HEAD, M. 1989. Evidence from sediments of long-term *Acanthaster planci* predation on corals of the Great Barrier Reef. *Science*, 245: 847- 850.

WHEELER, W.N. & DRUEHL, L.D. 1986. Seasonal growth and productivity of *Macrocystis integrifolia* in British Columbia, Canada. *Mar. Biol.*, 90: 181-186.

WHITTAKER, R.M. 1975. *Communities and ecosystems.* Mc Millan. New York.

ZABALA, M. & BALLESTEROS, E. 1989. Surface-dependent strategies and energy flux in benthic marine communities, or why corals do not exist in the Mediterranean. *Sci. Mar.*, 53(1): 3- 17.

Homage to Ramon Margalef; or, Why there is such pleasure in studying nature 243
(J. D. Ros & N. Prat, eds.). 1991. *Oecologia aquatica*, 10: 243-254

STRUCTURE OF CNIDARIAN POPULATIONS IN MEDITERRANEAN SUBLITTORAL BENTHIC COMMUNITIES AS A RESULT OF ADAPTATION TO DIFFERENT ENVIRONMENTAL CONDITIONS

JOSEP-MARIA GILI[1] & ENRIC BALLESTEROS[2]

[1] Institut de Ciències del Mar (CSIC). Passeig Nacional, s/n. 08039 Barcelona. Spain
[2] Centre d'Estudis Avançats de Blanes (CSIC). Camí de Santa Bàrbara, s/n. 17300 Blanes. Spain

Received: March 1990

SUMMARY

The presence and abundance of Cnidarians, extremely common in Mediterranean sublittoral benthic communities, exerts a profound effect on the entire structure of such communities, although the roles of Anthozoans and Hydrozoans differ in accordance with their different biological characteristics. Structural differences in benthic cnidarian populations were observed in three sublittoral communities located within 100 m of each other yet differing considerably in light intensity and water movement. Species-area and Shannon's diversity-area curves were computed for each community from reticulate samples constituted by 18 subsamples of 289 cm^2. Species-area curves were fitted to semilogarithmic functions, while diversity-area curves were fitted to Michaelis-Menten functions by the method of least squares. Species richness, species distribution, alpha-diversity, and pattern diversity were estimated from the fitted curves. Patch size for Anthozoans was smaller than that for Hydrozoans in all three communities. Diversity was greatest in the communities with the lowest and with the highest light intensity and water flow levels, but species richness was highest in the intermediate community. The most structured community (with the most heterogeneous species distribution pattern) had the largest organisms with the longest life spans. It is suggested that the difference of water flow conditions in these three communities leads to the development of different spatial patterns of individuals and colonies. The size, shape and small-scale distribution of these organisms are probably correlated to their ability of enhancing prey capture under different water flow conditions.

KEY WORDS: Cnidarians, benthos, population structure, diversity-area curves, Western Mediterranean.

INTRODUCTION

Cnidarians are extremely abundant in nearly all Mediterranean marine benthic communities. However, the highest number of species is found on rocky sublittoral bottoms, where cnidarians make up a substantial proportion of the biomass of such communities (TRUE, 1970; GILI & ROS, 1985a). Cnidarians constitute one of the dominant group of benthic organisms in sciaphilic communities where algal growth is limited by low light irradiance. In such places, different communities develop in relation to light intensity. Strong changes in irradiance can occur in short spaces where holes and caves open on rocky walls. Light attenuation from outside to inside the caves involves a gradient in species composition along an horizontal axis, a gradient which is enhanced by the decrease in water hydrodynamism along the same axis

(RIEDL, 1966). This gradual change in species composition has led Mediterranean benthic biologists to define three different communities (PÉRÈS & PICARD, 1964; ROS *et al.*, 1985; GILI & ROS, 1985b) named (from outside to inside) coralligenous community, semidark cave community and dark cave community.

According to previous observations made in sublittoral communities (RIEDL, 1966; GILI *et al.*, 1986; ZABALA *et al.*, 1989), the water flow is intense and bidirectional in the coralligenous, moderate and unidirectional at the entrance of the caves and reduced even though constant inside the cave. This decrease in water flow intensity involves a parallel zooplankton depletion (RIEDL, 1966; M. Zabala, personal communication).

Passive organisms, such as hydroids and anthozoans, are particularly dependent on adequate degrees of water movement (RIEDL, 1971). Many benthic cnidarians are passive suspension feeders and depend upon water current to carry suspended food across their food-capturing surfaces. The ability and distribution of benthic cnidarians is enhanced to maximize the volume of food-laden water they encounter per unit time (WAINWRIGHT & KOEHL, 1976).

The shape, size and spatial distribution of benthic cnidarians affects the water flow around them. Individuals or colonies can survive in rapid or moderate flow if they are small, aggregate or if they adjust their body shape and size to surrounding objects or organisms, (KOEHL, 1984). These organisms, whose structures affect the water flow, operate like obstacles that decrease fluid velocity and increase the turbulence in their surface layer (VÖGEL, 1981; LaBARBERA, 1984) and, therefore, enhance the possibilities of prey retention and catch. Under fast flow conditions, feeding ability could be enhanced by the presence of neighbours with high body size (SEBENS, 1984). However, under moderate flow velocities, populations of small colonies or less aggregated individuals alter the flow over the substratum, turbulence and pattern of particle transport (ECKMAN, 1983; OKAMURA, 1984). Our hypothesis is that the different hydrodinamical regimes in these three benthic communities, partly because of resulting differences in food supply, do condition the spatial distribution of the species, and therefore the population structure and patch size.

The structure of a community is determined by the spatial distribution of the species within it. Studies designed to find a representative sampling area, based on species-area curves, diversity-area curves, and spatial variations in biomass, provide a considerable amount of information on community structure for calculating patch size. Thus, structure and minimal area are closely related and, even though minimal sampling area may vary according to the criteria applied by different researchers (e.g., BOUDOURESQUE & BELSHER, 1979; HAWKINS & HARTNOLL, 1980; BALLESTEROS, 1986), a series of numerical parameters can be obtained for comparing the structural complexity of communities. The interest in obtaining such parameters is obvious since topics such as patch-size can be quantified and different communities can easily be compared (BALLESTEROS, 1986).

The object of the present study is to examine differences in the structure of three benthic cnidarian populations occupying a short but sharp light-intensity and water-flow gradient. The consequence of such environmental gradients would be that colonies and individuals would have a different spatial pattern in each community which lead to different patch-sizes. Cnidarians make up a taxocoenosis that can be quite representative of the community as a whole (WEINBERG, 1978; GILI *et al.*, 1989a) and for this reason they may be a good indicator of the structural pattern within each benthic community.

MATERIALS AND METHODS

Samples were collected off the northeastern coast of the Medes Islands (Costa Brava, Northwestern Mediterranean, UTM 31TEG15), where transition between the coralligenous community and the dark cave community occurs in a reduced space (Fig. 1).

Three samples were obtained by denudation (BOUDOURESQUE, 1971), scraping a surface consisting of 3 x 6 squares measuring 17 x 17 cm each. So, each sample consisted of 18 squares of 289 cm^2, making a total surface area of 5,202 cm^2 (Fig. 1). The squares (subsamples) were separated in the laboratory, and the hydrozoans and anthozoans were quantified in terms of:

- total biomass, measured as dry weight

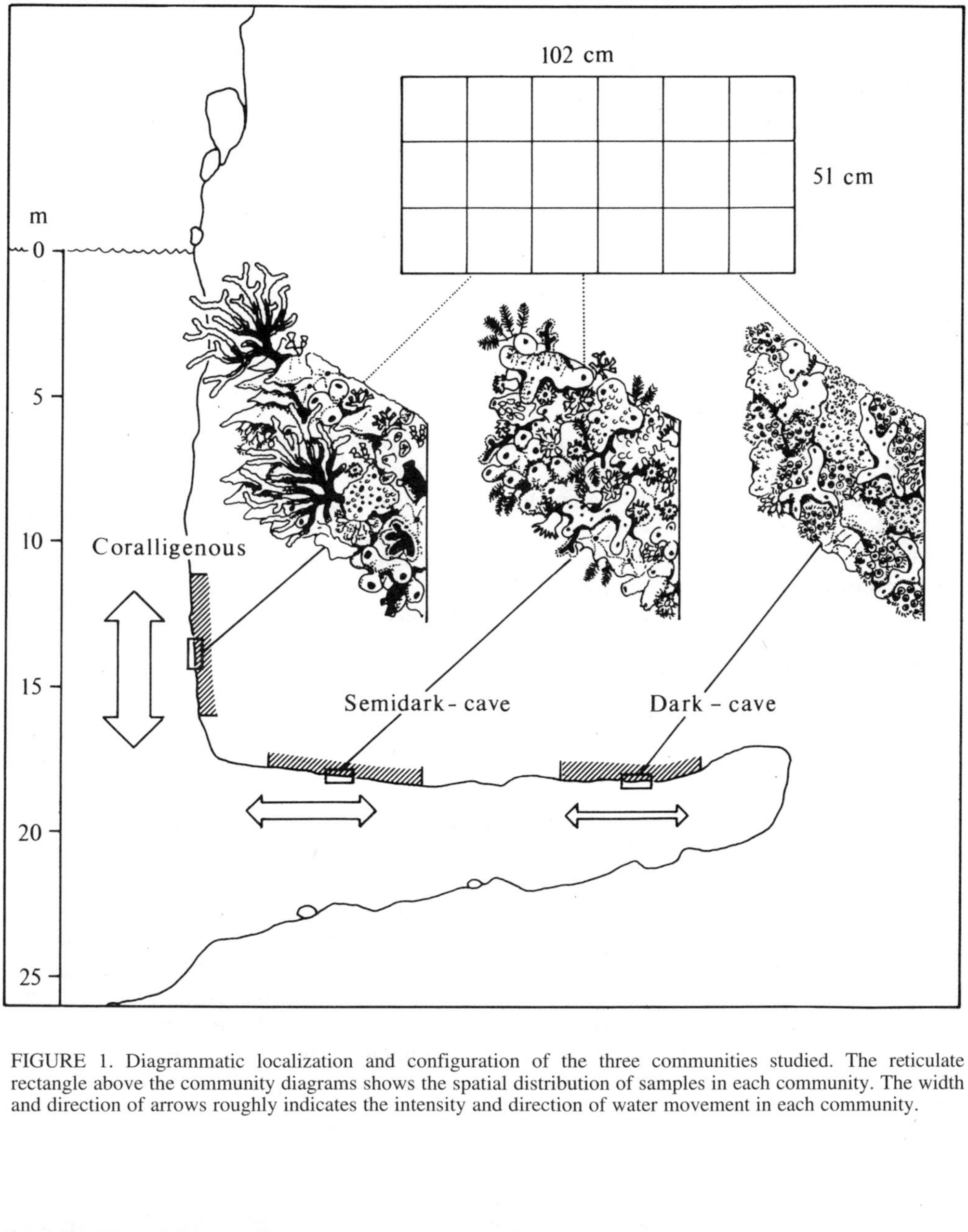

FIGURE 1. Diagrammatic localization and configuration of the three communities studied. The reticulate rectangle above the community diagrams shows the spatial distribution of samples in each community. The width and direction of arrows roughly indicates the intensity and direction of water movement in each community.

in grams, after drying each species at 110 °C for 24 h;

- organic matter, i.e., ash-free dry weight; ashes were prepared by burning samples for 6 h at 500 °C, in order to minimize the importance of accreted skeletons.

Two data matrices were thereby obtained for the samples, in each of which x_{ij} was the biomass (or organic matter) for species i in subsample j. Species-area and diversity-area curves were derived from these data matrices for total cnidarians, anthozoans, and hydrozoans. The species-area curves were fitted to a semilogarithmic function of the form

$$y = a \ln x + b,$$

where **y** was the number of species and **x** was the surface area in cm^2. The diversity-area curves were fitted to a Michaelis-Menten function:

$$y = Ax / (B + x)$$

were **y** was the diversity in bits and **x** the surface area in cm^2.

The following structural parameters (BALLESTEROS, 1986) were then calculated:

1) Number of species per sample.

2) Species distribution (**k**): the value of **k** in the fitted species-area curve; the semilogarithmic equation thus obtained can be expressed as a power equation:

$$x = k.e^{\lambda y}$$

The parameter **k** is a good index for defining the shape of the curve, indicating changes in slope for different patch sizes. The higher the value of **k**, the larger the surface area required to obtain a representative number of species for the community.

3) The parameter **a**, related to species richness, equal to the slope of the line fitting the species-area curve on the semilogarithmic coordinates.

4) The Molinier 20/5 point (NÉDÉLEC, 1979), corresponding to the point on the species-area curve at which a 20 % increase in surface area yields a 5 % increase in the number of species; this point can be regarded as the minimum species patch size value (BALLESTEROS, 1986).

5) Species diversity (**A**), the asymptotic value of diversity when the diversity-area curve is fitted to a Michaelis-Menten function.

6) Pattern diversity (**S**), the value of surface area corresponding to the Calleja $1 \cdot 10^{-3}$ point on the diversity-area curve when fitted to a Michaelis-Menten function; this parameter also furnishes the area at which diversity practically stabilizes (minimal structural area; NIELL, 1974).

All these parameters were used as indicators in evaluating the structure of the benthic cnidarian populations found.

RESULTS

A total of 39 cnidarian species were recorded in the three communities (Table I). *Paramuricea clavata* was the dominant species in the coralligenous community, although other species of solitary madreporarians, zoantharians, and alcyonarians formed small but dense colonies that contributed a substantial portion to the total biomass of the community. Unlike the other communities, some species of hydrozoans were present, forming dense colonies that made up a significant share of the total biomass. The semidark-cave community presented the largest number of species. Small colonies of *Hoplangia durotrix* predominated in this community, although other solitary madreporarians also had high biomass values. An important number of species shared in the hydrozoan biomass, and colonies were not large. In the dark-cave community four madreporarian species account for a great part of the total

TABLE I. Biomass (n) in grams (dry weight) per m^2, and frequency (f) in 18 samples for all species in the three (coralligenous, semidark-cave, and dark-cave) communities considered.

Species	Community					
	Coralligenous		Semi-dark cave		Dark-cave	
	n	f	n	f	n	f
HYDROZOA						
Bougainvillia ramosa	0.0124	1	0.9838	5	-	-
Perigonymus repens	0.0986	1	-	-	-	-
Clava multicornis	0.0880	1	-	-	-	-
Eudendrium motzkossowskae	0.0098	1	0.4912	4	-	-
Eudendrium capillare	0.1406	2	0.4402	3	-	-
Halecium tenellum	3.5226	5	0.0408	1	-	-
Halecium labrosum	1.3380	4	0.8912	5	-	-
Clytia haemisphaerica	0.8790	4	0.4246	10	0.0064	1
Obelia dichotoma	1.3814	8	0.3316	12	-	-
Hebella scandens	0.0078	1	0.0394	5	-	-
Sertularella gaudichaudi	16.7862	4	0.1028	1	-	-
Antennella secundaria	0.1624	2	-	-	0.0846	1
Aglaophenia octodonta	0.0158	1	-	-	-	-
Coryne pusilla	-	-	0.0060	1	-	-
Eudendrium racemosum	-	-	2.3782	11	1.1588	2
Halecium mediterraneum	-	-	0.0100	1	-	-
Phialella quadrata	-	-	0.0676	2	-	-
Laomedea pelagica	-	-	0.0368	2	-	-
Synthecium evansi	-	-	0.2806	3	-	-
Plumularia setacea	-	-	0.0400	2	-	-
Nemertesia antennina	-	-	0.0582	4	0.0080	1
Aglaophenia pluma	-	-	0.0002	1	0.0040	1
Eudendrium rameum	-	-	-	-	1.7194	3
Laomedea angulata	-	-	-	-	0.0032	1
SCYPHOZOA						
Stephanoscyphus eumedusoides	-	-	-	-	0.0580	6
ANTHOZOA						
Alcyonium acaule	8.6820	5	-	-	-	-
Parerythropodium coralloides	8.1902	12	0.1192	3	0.0098	1
Paramuricea clavata	266.9146	9	-	-	-	-
Epizoanthus arenaceus	0.1934	1	-	-	-	-
Parazoanthus axinellae	14.8068	8	-	-	-	-
Hoplangia durotrix	1.3972	1	162.5834	13	183.7603	17
Caryophyllia smithi	11.4878	9	6.0980	5	1.9984	4
Caryophyllia inornata	14.8224	6	41.9614	8	279.0548	7
Leptopsammia pruvoti	83.3340	12	33.9306	7	327.0448	17
Corallium rubrum	-	-	1.8490	1	58.5484	6
Cornularia cornucopiae	-	-	0.0354	3	-	-
Rolandia rosea	-	-	11.0832	13	0.7326	9
Corynactis viridis	-	-	6.0132	12	0.1672	4
Polycyathus muellerae	-	-	-	-	289.4380	3

biomass. Two of these species formed dense colonies, moderate (*H. durotrix*) or large (*P. muellerae*) in size. Unlike the other two communities, the solitary madreporarians formed facies, while just two species accounted for nearly the total hydrozoan biomass (Table I).

Figure 2 sets out in more general and summarized form the percentage of species in each of the three communities according to three categories, namely, average life span, biomass (dry weight), and polyp size (taken as mouth size). Each of these three categories has been divided into five subdivisions. In the coralligenous community the highest percentages corresponded to the lower and upper subdivisions in all three categories; in the

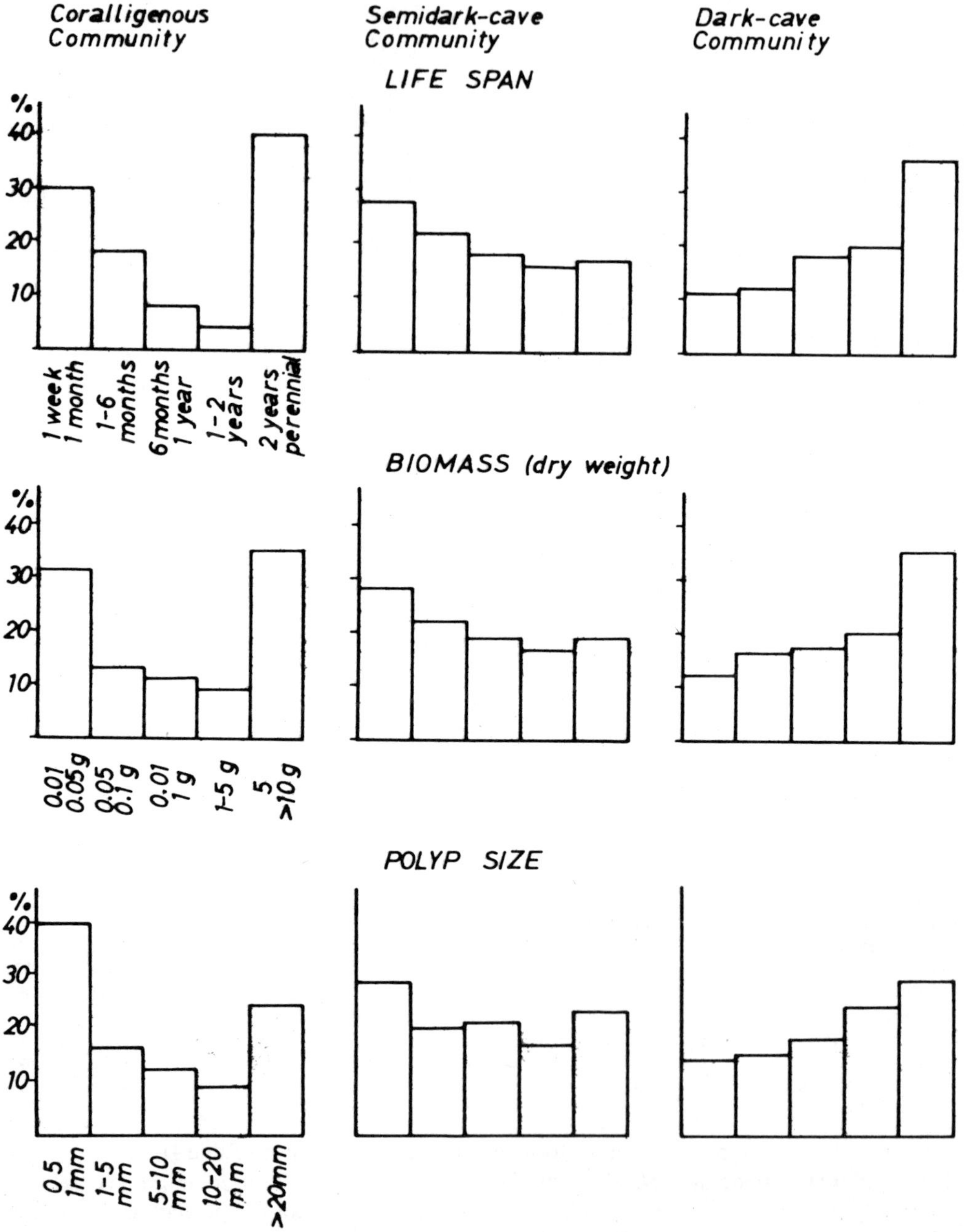

FIGURE 2. Percentage number of species in each of the three (coralligenous, semidark-cave, dark-cave) communities studied in terms of life-span, biomass (dry weight), and polyp size (based on mouth size). Each category has been divided into five subdivisions.

TABLE II. Parameters calculated from the species-area curves (the number of species, N, refers to the total number collected in each community; the Molinier 20/5 point is a measure of patch size in cm^2).

	k	a	r	N	$M_{20/5}$
TOTAL CNIDARIANS					
Coralligenous	129.9	6.245	0.999	23	4976
Semidark-cave	100.4	6.793	0.999	27	3846
Dark-cave	113.9	4.577	0.999	17	4636
ANTHOZOANS					
Coralligenous	38.4	1.857	0.993	9	1472
Semidark-cave	32.4	1.812	0.983	9	1240
Dark-cave	42.8	2.180	0.993	10	1640
HYDROZOANS					
Coralligenous	176.0	4.075	0.988	14	6742
Semidark-cave	150.9	4.980	0.995	18	5779
Dark-cave	275.7	2.398	0.985	7	10559

TABLE III. Parameters calculated from the diversity-area curves (S is a measure of patch size in cm^2). -, not significant data.

	A	S	r
TOTAL CNIDARIANS			
Coralligenous			
dry weight	2.08	508	0.997
ash-free dry weight	1.49	320	0.963
Semidark-cave			
dry weight	2.02	317	0.787
ash-free dry weight	3.05	575	0.997
Dark-cave			
dry weight	2.25	565	0.959
ash-free dry weight	2.47	617	0.971
ANTHOZOANS			
Coralligenous			
dry weight	1.77	441	0.999
ash-free dry weight	1.10	228	0.911
Semidark-cave			
dry weight	1.89	424	0.936
ash-free dry weight	2.49	565	0.997
Dark-cave			
dry weight	2.24	559	0.961
ash-free dry weight	2.28	571	0.969
HYDROZOANS			
Coralligenous			
dry weight	2.34	555	0.974
ash-free dry weight	2.46	585	0.977
Semidark-cave			
dry weight	3.26	790	0.998
ash-free dry weight	3.08	742	0.998
Dark-cave			
dry weight	-	-	-
ash-free dry weight	-	-	-

semidark-cave community the percentage of species in each subdivision within each category was approximately equal; and in the dark-cave community the percentage of species became progressively larger in the upper part of the range for each category.

The fits of the species-area curves presented higher values of **k**, associated with species distribution over surface area, in the coralligenous and dark-cave communities (Table II). On the other hand, the highest value of **a**, associated with species richness, was attained in the semidark-cave community. The trends for these two parameters were very similar for the anthozoans in the first two communities, whereas there were substantial differences for the hydrozoans: the value of **a** was highest in the semidark-cave community, while the dark-cave community yielded the highest value of **k** (Table II).

The fits of the diversity-area curves gave similar values for species diversity in all three communities. However, there were appreciable differences between the value of **A** for biomass as dry weight and for biomass as ash-free dry weight, particularly in the semidark cave community. The value of **A** for anthozoans followed the same trend as for the total cnidarians, whereas for hydrozoans the highest value was for the semidark-cave community. The low biomass of hydrozoans in the dark-cave community gave rise to non-significant fits of the diversity-area curves to the Michaelis-Menten function for this community. For biomass including ashes, the pattern-diversity (**S**) value for total cnidarians was highest for the dark-cave community and lowest for the semidark-cave community. In contrast, for ash-free biomass, the value of **S** increased from the coralligenous community to the dark-cave community (Table III). The values of **S** for anthozoans alone followed a similar trend to those for total cnidarians, but for the hydrozoans the value was lower in the coralligenous community both for ash-free biomass and biomass including ashes.

The behaviour of the two types of curve considered (see the shapes of the curves in figure 3) suggests that the values of **k** and **S** furnish very similar indications of patch size in the three communities. In any case, patch size for total cnidarians was slightly higher qualitatively in the coralligenous community and slightly higher quantitatively in the dark-cave community. Patch size was smallest for the semidark-cave community. For anthozoans patch size was highest in the dark-cave community, lowest in the semidark-cave community. In contrast, for hydrozoans patch size was qualitatively highest in the dark-cave and coralligenous communities but quantitatively highest in the semidark-cave community. For ash-free biomass, however, patch size was smallest in the coralligenous community for both anthozoans and total cnidarians.

DISCUSSION

Patch size for total cnidarians based on the species-area curves was somewhat larger in the coralligenous community than in the cave communities because of the distribution of the dominant species (*P. clavata*), in the former community, whose larger and long lived colonies affect the settlement and development of the other species, which grew scattered in the basal stratum. On the other hand, the lower level of spatial competition in the cave communities probably enables certain madreporarians to form large colonies by budding as well as dense aggregations of small colonies or individuals through larval recruitment, resulting in a smaller

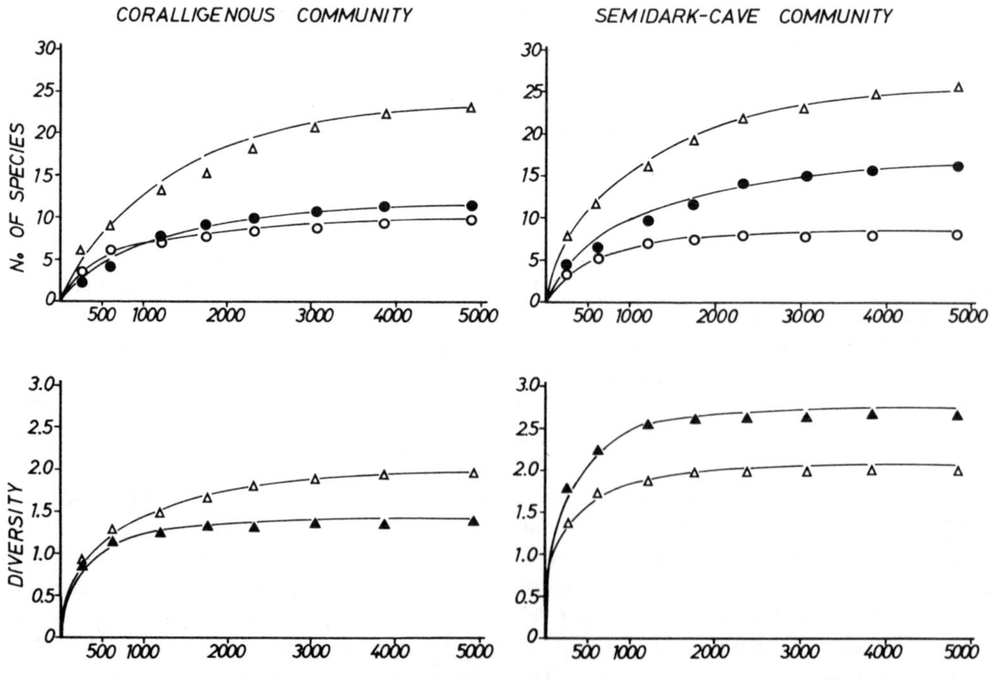

FIGURE 3. Species-area and diversity-area curves for the three (coralligenous, semidark-cave, and dark-cave) communities for each of the three groups (total cnidarians, hydrozoans, and anthozoans) considered, with

patch-size.

From a quantitative point of view, the larger patch size for cnidarians in the dark-cave and coralligenous communities, compared with that in the semidark-cave community, is caused by the presence of larger species, which attain high biomass levels, in the first two. These species left little room for other species, which were therefore confined to smaller patches. As a consequence, in the coralligenous community a high percentage of larger species with high biomasses coexisted with many other much smaller species (Fig. 3). This was still more evident in the dark-cave community, in which patch size was largest thanks to the patchy distribution of madreporarian colonies.

The lowest patch size occurred in the semidark-cave community, reflecting the occupation of the available space by smaller species. Individuals or colonies of these, in turn, occupied smaller areas, thereby allowing a larger number of species to coexist. Both solitary anthozoan species and small colonies of anthozoans and hydrozoans contributed to this. Unlike the hydrozoans in the other two communities, the hydrozoans in the semidark-cave community enjoyed a greater number of potential sites to colonize among the anthozoan individuals and colonies.

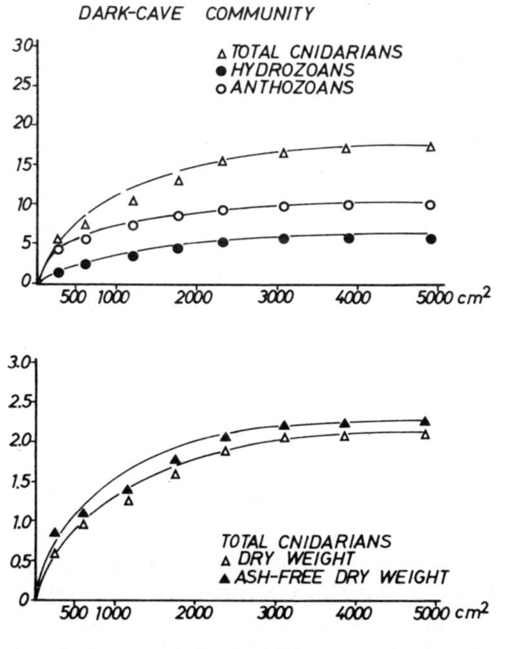

quantitative curves for both biomass measures (dry weight and ash-free dry weight) used.

Quantitatively, the difference in patch size as measured in terms of total dry weight and ash-free dry weight is indicative of the importance of the skeletal carbonates of certain anthozoans in determining the structural patterns in the communities. This was quite clear in the case of the coralligenous community, in which the difference between *P. clavata* and all the other species increased, and, as a result, patch size and diversity decreased, when the skeletal weight contribution was disregarded. In contrast, in the semidark-cave community the converse held true, while in the dark-cave community there was hardly any change. Due to the predominance of madreporarians in these last two communities, which, compared to gorgonians have high organic matter contents despite their calcium skeletons (GILI & ROS, 1985b), differences between species biomass measured as dry weight and as ash-free dry weight were minimal in the dark-cave community and decreased (i.e., diversity increased) in the semidark-cave community. Dominance of a community by a single species, termed a facies, also exerts an important effect on patch size (ROS *et al.*, 1985). This was true in the case of *P. clavata* in the coralligenous community and, in part, of *P. muellerae* in the dark-cave community. Such dominance caused total cnidarian biomass to be higher in these two communities (GILI & ROS, 1985b). In contrast, in a study of similar communities off Marseilles by TRUE (1970), the highest biomass was recorded for a semidark-cave community, occupied by a facies of *Corallium rubrum*; for this reason, patch size was quite similar to that in a coralligenous community, also occupied by a facies of *P. clavata*.

Benthic cnidarians are suspension feeders and as such are dependent upon the external availability of food transported by the water flow. In the context of a flow gradient like that existing in the three communities studied, it would seem reasonable to expect potential food sources to be most abundant in the coralligenous community and least abundant in the dark-cave community. Greater food availability and greater colonizing ability would be associated with a larger number of species and, therefore, with a larger patch-size (WRIGHT, 1983; KEOUGH, 1984). However, the community situated in between the two extremes had the smallest patch size and the highest number of species. The largest patch size found in the coralligenous community is consistent with the observations by WRIGHT (1983) and would be caused by the large area covered by *P. clavata*, which has a colonial habit well adapted to capturing prey in high water flows (LEVERSEE, 1976), and a certain gregarious nature that apparently facilitates the capture of prey in certain anthozoan species (McFADDEN, 1986).

Sessile cnidarians found in slowly-moving water, in the dark-cave community, tend to have morphological and physiological adaptations which compensate for the lack of external transport (WAINWRIGHT & KOEHL, 1976), among them: tolerance of low oxygen levels; feeding on motile prey in lower concentrations; aperiodical appearance of potential prey. At the same time, scleractinians in dark-caves may have a lower energy demand than other benthic cnidarians in temperate waters (SEBENS, 1987). In the less accessible parts of caves the amount of water movement is not adequate to cover the demands for the development of dense populations like those found in the entrance of the cave or in the coralligenous community (RIEDL, 1971). The minimum range of water movement allows the zooplankton and oxygen concentrations near the organisms to fall and, thus, become intolerable to most sessile cnidarians, like hydroids or some octocorals, with high food requirements. In the coralligenous community again, under conditions of high water movement, the zooplankton availability increases and allows the development of a diversified population of benthic cnidarians; large colonies (gorgonians) close to others of moderate (e.g. alcyonarians) or small size (e.g. hydroids). However, the lower availability of food in the dark-cave community would act, probably, to reduce competition so that the species were able to grow and occupy a large space (SEBENS, 1982).

The moderate water movement in semidark-cave community enables the establishment of an intermediate spatial distribution pattern as a consequence of the development of colonies or individuals of moderate and small size. Under these moderate flow conditions, feeding ability could be enhanced by the presence of neighbours, or by forming aggregations of small and clonal colonies (ECKMAN, 1983; McFADDEN, 1986).

In consequence, as shown in figure 2, the coralligenous and dark-cave communities contained higher percentages of longer-lived species that capture larger prey items than did the semidark-cave community. In this type of species, which have adopted more of a K strategy (*sensu* MacARTHUR & WILSON, 1967; see GILI & ROS, 1985a), the level of competition with other species similar in habits and biology is higher (CONNELL, 1975), which tends to favour coexistence by more dissimilar species, like the r-strategists or opportunistic ones (CONNELL & SLATYER, 1977; GILI & ROS, 1985a), which are smaller, shorter-lived, and more specialized in the capture of smaller prey. In the semidark-cave community the lower, yet nonetheless sufficient, level of food availability enabled colonies and individuals to develop but not to attain large sizes. This, in turn, enabled a larger number of species to coexist, but the patch-size would decrease because of the lack of compartmentation caused by spatial-structuring species such as *P. clavata* in the coralligenous community and the total quantitative dominance of *Hoplangia durotrix.* The qualitative dominance by hydroids and by a few octocorals in the semidark-cave community is indicative of the ecotone character of this community, dominated as it is by r-strategists.

The seasonality observed in the major part of the Mediterranean hydroids (e.g. BOERO & FRESI, 1986; LLOBET *et al.*, 1991) could change the spatial pattern observed in communities such as semidark-caves, where these organisms play an important structural role. But their great capacity for continuous growth and recruitment enhance their presence and similar distribution in this community (GILI *et al.*, 1989b).

In conclusion, in communities in which the growth of certain biomass-storing, structure-creating species is enhanced, there is a tendency towards the contagious distribution of individuals of a single species, resulting in larger patch-sizes. This occurs in a situation of large food availability as is the case in the coralligenous community with *P. clavata* or in highly food-limited environments such as in the dark-cave communities, where species with low food requirements can survive. Nevertheless, if there are few biomass-storers, or if there are none, specially adapted to a given environment, species richness could be increased by the appearance of opportunistic forms (e.g. hydroids) due to lower competitive levels. Also, patch-size decreases because of the lack of a rigid structural pattern and low spatial heterogeneity.

Spatial arrangements of particular populations can probably be correlated with the ways in which certain colonies and individuals modify water currents (e.g., HISCOCK, 1983) and affect the flow forces and feeding currents. This suggestion reveals the importance of size, shape and patch distribution of these animals to their ability to utilize moving water, catch suspension preys and successfully colonize sublittoral benthic communities.

ACKNOWLEDGEMENTS

We wish to thank Dr. J.D. Ros for his useful comments and criticism of the manuscript and Mr. R. Sacks for helping with the English version.

REFERENCES

BALLESTEROS, E. 1986. Métodos de análisis estructural en comunidades naturales, en particular del fitobentos. *Oecol. aquat.*, 8: 117-131.

BOERO, F. & FRESI, E. 1986. Zonation and evolution of a rocky bottom hydroid community. *P.S.Z.N. I: Mar. Biol.*, 7: 123-150.

BOUDOURESQUE, C.F. 1971. Méthodes d'étude qualitative et quantitative du benthos (en particulier du phytobenthos). *Téthys.* 3: 79-104.

BOUDOURESQUE, C.F. & BELSHER, T. 1979. Le peuplement algal du port de Port-Vendres: recherches sur l'aire minimale qualitative. *Cah. Biol. Mar.* 20: 259-269.

CONNELL, J.H. 1975. Some mechanisms producing structure in natural communities: a model and evidence from field experiments. In: *Ecology and Evolution of communities.* (M. Cody & J. Diamond, eds.): 460-490. Harvard University Press. Cambridge, Mass.

CONNELL, J.H. & SLATYER, R.O. 1977. Mechanisms of succession in natural communities and their role in community stability and organization. *Am. Nat.* 111: 1119-1144.

ECKMAN, J.E. 1983. Hydrodynamic processes affecting benthic recruitment. *Limnol. Oceanogr.*, 28: 241-257.

GILI, J.M., MURILLO J. & ROS, J.D. 1989a. The distribution pattern of benthic Cnidarians in the Western Mediterranean. *Scient. Mar.*, 53: 19-35.

GILI, J.M., RIERA, T. & ZABALA, M. 1986. Physical and biological gradients in a submarine cave on the Western Mediterranean coast (NE Spain). *Mar. Biol.*, 90: 291-297.

GILI, J.M. & ROS,J.D. 1985a. Estudio cuantitativo de

tres poblaciones infralitorales de cnidarios bentónicos. *Inv. Pesq.*, 49: 323-352.

GILI, J.M. & ROS, J.D. 1985b. Study and Cartography of the benthic communities of Medes Islands (NE Spain). *P.S.Z.N. I Mar. Ecol.*, 6: 219-238.

GILI, J.M., ROS, J.D. & ROMERO, J. 1889b. The structural role of Cnidaria in Mediterranean benthic communities: a simulation-like approach. In: *Proceedings of the 21st European Marine Biology Symposium (*Klekowski, R., Styczynska- Jurewicz, E. & Falkowski, L., eds.): 279-289. Ossolineum. Wroclaw.

HAWKINS, S.J. & HARTNOLL, R.G. 1980. A study of the small-scale relationship between species number and area on a rocky shore. *Estuar. Coast. Mar. Sci.*, 10: 201-214.

HISCOCK, K. 1983. Water movement. In: *Sublittoral ecology. The ecology of the shallow sublittoral benthos.* (R. Earll & D.G. Erwin, eds.): 58-96. Clarendon. Oxford.

KEOUGH, M.J. 1984. Effects of patch size on the abundance of sessile marine invertebrates. *Ecology*, 65: 423-437.

KOEHL, M.A.R. 1984. How do benthic organisms withstand moving water? *Amer. Zool.*, 24: 57-70.

LaBARBERA, M. 1984. Feeding currents and particle capture mechanisms in suspension feeding animals. *Am. Zool.*, 24: 71-84.

LEVERSEE, G.J. 1976. Flow and feeding in fan-shaped colonies of the gorgonian coral, *Leptogorgia. Biol. Bull.*, 151: 344-356.

LLOBET, I., GILI, J.M. & HUGHES, R.G. 1991. Horizontal, vertical and seasonal distribution of epiphytic hydrozoa on the alga *Halimeda tuna* in the North-western Mediterranean Sea. *Mar. Biol.*, 110: 151-159.

MacARTHUR, R.H. & WILSON, E.O. 1967. *The theory of island biogeography.* Princenton University Press. Princenton, N.J.

McFADDEN, C.S. 1986. Colony fission increases particle capture rates of a soft coral: advantages of being a small colony. *J. Exp. Mar. Biol. Ecol.*, 103: 1-20.

NÉDÉLEC, C. 1979. *Étude structurale et problèmes d'échantillonage dans une phytocoenose portuaire.* Mem. Univ. P. et M. Curie. Marseille-Luminy

NIELL, F.X. 1974. Les applications de l'index de Shannon à l'étude de la végétation intertidale. *Soc. Phycol. Fr. Bull.*, 19: 238-254.

OKAMURA, B. 1984. The effects of ambient flow velocity, colony size, and upstream colonies on the feeding success of Bryozoa. I. *Bugula stolonifera* Ryland, and arborescent species. *J. Exp. Mar. Biol. Ecol.*, 83: 179-193.

PÉRÈS, J.M. & PICARD, J. 1964. Nouveau manuel de bionomie benthique de la Mer Méditerranée. *Rec. Trav. St. mar. Endoume.*, 31: 5-137.

RIEDL, R. 1966. *Biologie der Meereshöhlen.* Paul Parey. Berlin.

RIEDL, R. 1971. Water movement. In: *Marine Ecology, 1, Environmental factors.* (O. Kinne, ed.): 1123-1156. Wiley. London.

ROS, J.D., ROMERO, J, BALLESTEROS, E. & GILI, J.M. 1985. Diving in blue water. The benthos. In: *Western Mediterranean* (R. Margalef, ed.): 233-295. Pergamon. London.

SEBENS, K.P. 1982. Competition for space: growth rate, reproductive output and escape in size. *Am. Nat.* 120: 189-197.

SEBENS, K.P. 1984. Water flow and coral colony size; Inter-habitat comparisons of the octocoral *Alcyonium siderium. Proc. Natl. Acad. Sci. USA*, 81: 5473-5477.

SEBENS, K.P. 1987. Coelenterata. In: *Animal Energetics*. 1. (T.J. Pandian & F.J. Vernberg, eds.): 55-120. Academic Press. San Diego.

TRUE, M.A. 1970. Étude quantitative de quatre peuplements sciaphiles sur substrat rocheux dans la région marseillaise. *Bull. Inst. Océanogr. Monaco.* 1410: 1-48.

VÖGEL, S. 1981. *Life in moving fluids.* Willard Grant. Boston.

WAINWRIGHT, S.A. & KOEHL, M.A.R. 1976. The nature of flow and the reaction of benthic Cnidaria to it. In: *Coelenterate Ecology and Behavior*. (G.O. Mackie, ed.): 5-21. Plenum. New York.

WEINBERG, S. 1978. The minimal area problem in invertebrate communities of Mediterranean rocky substrata. *Mar. Biol.*, 49: 3-40.

WRIGHT, D.H. 1983. Species-energy theory: an extension of species-area theory. *Oikos,.* 41: 496-506.

ZABALA, M., RIERA, T., GILI, J.M., BARANGÉ, M., LOBO, A. & PEÑUELAS, J. 1989. Water flow, trophic depletion, and benthic macrofauna impoverishment in a submarine cave from the Western Mediterranean. *P.S.Z.N. I: Mar. Ecol.,* 10: 271-287.

Homage to Ramon Margalef; or, Why there is such pleasure in studying nature
(J. D. Ros & N. Prat, eds.). 1991. *Oecologia aquatica*, 10: 255-270

PRODUCTION DYNAMICS IN A MACROPHYTE-DOMINATED ECOSYSTEM: THE MAR MENOR COASTAL LAGOON (SE SPAIN)

JORGE TERRADOS[1] & JOANDOMÈNEC ROS[2]

1 Departamento de Biología Animal y Ecología. Universidad de Murcia. Campus Universitario de Espinardo. 30100 Murcia. Spain.
2 Departament d'Ecologia. Facultat de Biologia. Universitat de Barcelona. Av. Diagonal, 645. 08028 Barcelona. Spain.

Received: August 1991

SUMMARY

The greater part of the bottom of the Mar Menor, a hypersaline coastal lagoon in the south-east of the Iberian Peninsula, is covered by meadows of *Caulerpa prolifera* and *Cymodocea nodosa*. Both species show marked seasonality in their biomass cycles, reaching their highest annual values in summer-autumn and their lowest in winter-spring. The annual production of the *C. prolifera* mud-based meadows is estimated at 630-707 g dw m^{-2} y^{-1}. Leaf production of the *C. nodosa* meadows varies between 162 and 427 g dw m^{-2} y^{-1}, while rhizome production varies between 45 and 78 g dw m^{-2} y^{-1}. *C. prolifera* growth seems to be limited by the availability of nitrogen in the water column, while *C. nodosa* growth is limited by the availability of nitrogen and phosphorus. Both species show adaptation in their photosynthetic characteristics (I_c, I_k, P_{max}) to seasonal variation in the water temperature of the lagoon. The values of these parameters for *C. prolifera* suggest that this species is adapted to life in habitats of less irradiance than *C. nodosa*. The distribution of *C. nodosa* meadows in the lagoon is controlled by the availability of light at the bottom. Overall primary production of the Mar Menor can be estimated at roughly 41270 Tm C y^{-1} (some 300 g C m^{-2} y^{-1}); 64% of the lagoon's primary production is produced by the macrophytobenthos, 11% by the microphytobenthos and 25% by the phytoplankton.

KEY WORDS: Macrophytobenthos, primary production, *Caulerpa prolifera*, *Cymodocea nodosa*, Mar Menor lagoon.

INTRODUCTION

The Mar Menor is a coastal lagoon situated in the south-east of the Iberian peninsula (Fig. 1). Its surface area is 135 km^2 and its maximum depth is not more than 7 m; average depth is roughly 4 m. The bottoms are mainly soft substrates, more muddy in the central basin of the lagoon and sandy in the shallow parts, mainly of La Manga (ROS, 1987; PÉREZ RUZAFA *et al.*, 1989). In areas near the bank with low hydrodynamism the substrate is also muddy. The dryness of the climate (subdesert Mediterranean dominium, CAPEL, 1981) and the absence of freshwater inputs mean that the Mar Menor, in terms of its water exchange regime with the Mediterranean Sea, functions as an antiestuarine lagoon (ROS *et al.*, 1985). Now water salinity varies between 42 ‰ and 47 ‰, compared with 36-38 ‰ in the Mediterranean. Water temperature varies between 12° C in winter and 30° C in summer; at the shallow banks more extreme temperatures can be reached

(ROS & MIRACLE, 1984; TERRADOS, 1986, 1991; PÉREZ RUZAFA, A., 1989). The waters of the lagoon are always well oxygenated (TERRADOS, 1986) and nutrient concentration is low (< 1 µM NO_3^-, < 0.03 µM PO_4^{3-}; TERRADOS, 1991).

Benthic vegetation in the lagoon is predominantly made up by macrophyte meadows. The two most important species, in terms of area occupied, in the functioning of the lagoon ecosystem are the chlorophycean *Caulerpa prolifera* (Forsskal) Lamouroux and the seagrass *Cymodocea nodosa* (Ucria) Ascherson (TERRADOS, 1991). At certain shallow points with negligible hydrodynamism small patches of *Ruppia cirrhosa* (Petagna) Grande appear. In Las Encañizadas and in Santiago de la Ribera *Chaetomorpha linum* (O. F. Müll.) Kütz. may attain considerable development. Populations of photophilic algae on rocks (inner islands, El Estacio) reach high levels of biomass and local production (TERRADOS, 1986), but their contribution is small in the lagoon overall.

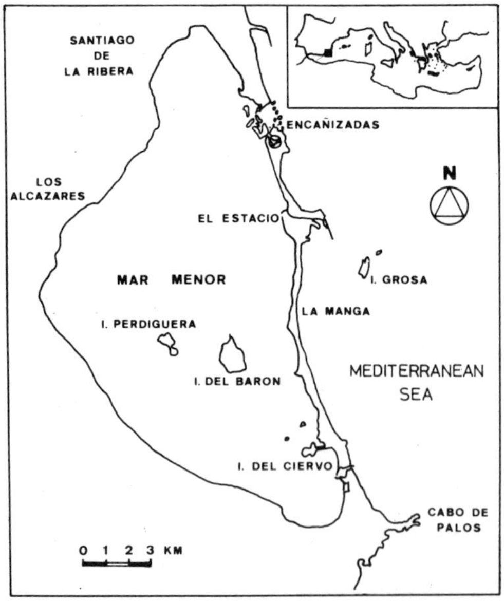

FIGURE 1. The Mar Menor lagoon and its location on the Spanish Mediterranean coast.

The great development of macrophytobenthos in the lagoon and the low phytoplanktonic density (ROS & MIRACLE, 1984) indicate that benthic primary production is more important than planktonic production in the lagoon economy (ROS, 1987). The nutrient input rate in the water mass and the extension of environments favourable to the development of submerged macrophytic vegetation (substrate stability, low hydrodynamism, sufficient light) are the main factors which determine how the primary production of the lagoon ecosystem is shared out between these two components. The present work summarizes the most important aspects of the production ecology of the two species dominant in the macrophytobenthos of the Mar Menor - *Caulerpa prolifera* and *Cymodocea nodosa*. These aspects are discussed in relation to the general functioning of the lagoon ecosystem.

BIOMASS AND PRIMARY PRODUCTION

CAULERPA PROLIFERA

The biomass of the mud-based *Caulerpa prolifera* meadow varies between 30-50 g dw m^{-2} and 150-280 g dw m^{-2} (BALLESTER, 1985; TERRADOS, 1986, 1991; PÉREZ RUZAFA, I.M., 1989). With small differences between areas, the meadow biomass reaches its lowest values in April-June and its highest in August-September, which are then held until the end of autumn (TERRADOS, 1986, 1991). On sand, *Caulerpa prolifera* meadows reach maximum annual biomass (97 g dw m^{-2}) at the end of spring; it gradually falls afterwards and is low (36-41 g dw m^{-2}) from January to April (TERRADOS, 1986).

The seasonal variation in the biomass of the *Caulerpa prolifera* meadow is also reflected in a variation in its structure

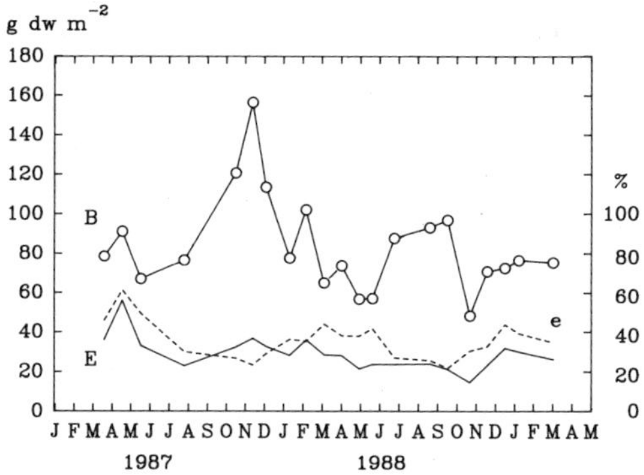

FIGURE 2. Annual variation of total biomass (B, solid line with open dots), stolon biomass (E, solid line) and the percentage of total biomass corresponding to stolons (e, dashed line; frond percentage: 100-e) in a *Caulerpa prolifera* meadow at the Mar Menor central basin (TERRADOS, 1991).

(TERRADOS, 1991). The development of the meadow is basically the consequence of the development of the fronds. The percentage of the total meadow biomass corresponding to the fronds (positive correlation, $r = 0.438$; $p < 0.05$, between the total meadow biomass and the percentage of the meadow corresponding to fronds; Fig. 2), the percentage of proliferations (negative correlation, $r = -0.848$; $p < 0.01$, between the percentages of total meadow biomass corresponding to stolons and proliferations; Figs. 2 and 3) and frond dimensions (Fig. 4) all increase. In this way the leaf area index (LAI) varies between 0.6 m^2 m^{-2} in April and 2.9 m^2 m^{-2} in October (TERRADOS, 1991).

From the results obtained in different laboratory experiments in which the influence of temperature on growth rates of *Caulerpa prolifera* fragments and the *in situ* annual variation in water temperature were studied (Fig. 5), TERRADOS (1991) constructed a growth model which permits estimation of the annual primary production. The annual primary production of mud-based *C. prolifera* meadows in the central part of the lagoon is estimated at 630-707 g dw m^{-2} y^{-1} (roughly 212-237 g C m^{-2} y^{-1}).

The lower limit of *C. prolifera*'s survival temperature varies from 10 ^{o}C in February to 20 ^{o}C in June; the higher limit remains constant at around 35 ^{o}C, while optimum growth temperature is 26-28 ^{o}C (TERRADOS, 1991). These results suggest that growth of *C. prolifera* is practically nil during the winter, and that from the middle of May until the middle of September the water temperature allows almost optimum growth of the alga.

Caulerpa prolifera also grows on sand,

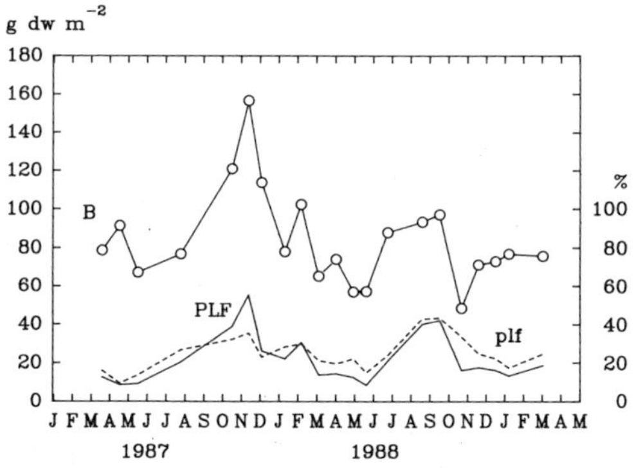

FIGURE 3. Annual variation of total biomass (B, solid line with open dots), proliferation frond biomass (PLF, solid line) and the percentage of total biomass corresponding to proliferations (fronds inserted on other fronds; plf, dashed line) in a *Caulerpa prolifera* meadow at the Mar Menor central basin (TERRADOS, 1991).

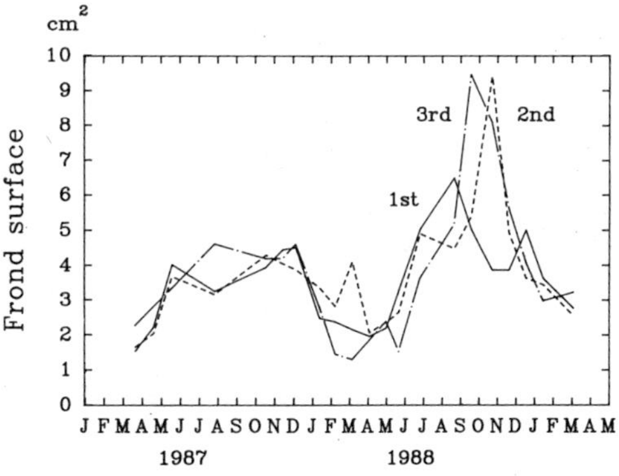

FIGURE 4. Annual variation of the surface of primary fronds (1st, solid line), secondary fronds (2nd, dashed line) and tertiary fronds (3rd, dotted-dashed line) in a *Caulerpa prolifera* meadow at the Mar Menor central basin (TERRADOS, 1991).

clay or rock, but the levels of biomass reached are much lower (TERRADOS, 1986), probably because of the lower availability of nutrients in these substrates (TERRADOS, 1991).

CYMODOCEA NODOSA

Cymodocea nodosa meadows present an unimodal annual cycle of leaf biomass (Fig. 6; TERRADOS, 1986; TERRADOS & ROS, in press a). Leaf growth is a continuous process (Fig. 7): the highest growth rates are registered from May to July, with the result that the maximum leaf biomass values (54 g dw m^{-2} in meadows on sand; 180 g dw m^{-2} in meadows on mud) are recorded in July and August. Leaf biomass and leaf growth rates are minimal in January and February. Maximum total biomass registered by different *C. nodosa* meadows in the lagoon varies between 86 and 169 g dw m^{-2} (BALLESTER, 1985). Shoot density in the meadow shows similar seasonal variations, with maximum values (1700-1900 shoots m^{-2}) in summer and minimum (1000 shoots $\cdot m^{-2}$) at the end of winter. Maximum annual shoot density of *C. nodosa* meadows in the Mar Menor varies between 312 and 2314 shoots m^{-2} (TERRADOS & ROS, in press a). The LAI varies between 0.14-0.20 m^2 m^{-2} in January-February and 1.2-1.8 m^2 m^{-2} in July-August; from May to September the photosynthetic surface is the same as or greater than the substrate surface.

Rhizome growth shows a seasonal variation rhythm different from that of the leaves; rhizomes grow from February to October, and the highest rates are found in May and June; growth ceases from October to January (TERRADOS & ROS, in press a).

There is a positive correlation ($p < 0.01$) between different parameters of the meadow and water temperature and incoming irradiance (Table I). "Extensive" variables, related to the quantity of photosynthetic biomass present (leaf biomass, LAI, shoot density) are linked more to water temperature. "Intensive" variables, instantaneous process rates (leaf growth, leaf P/B ratio, meadow production) are more linked to irradiance. This different degree of linking may be due to the different nature of processes involved. Intensive variables refer to short periods of

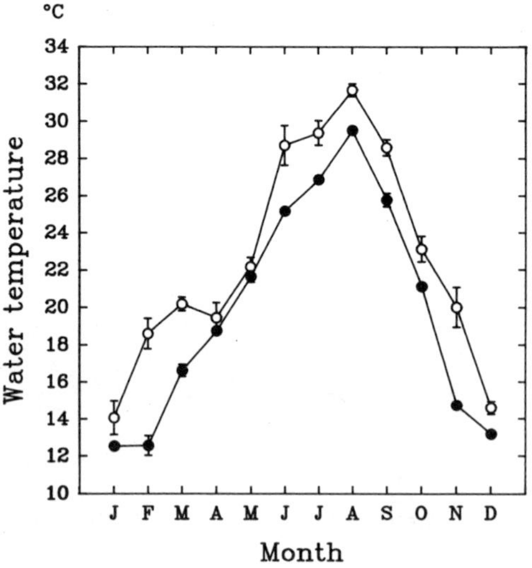

FIGURE 5. Monthly means of water temperature in the Mar Menor lagoon for the 1987/1990 period: open dots, shallow sites; full dots, central basin (TERRADOS, 1991). Vertical bars represent the standard error of mean.

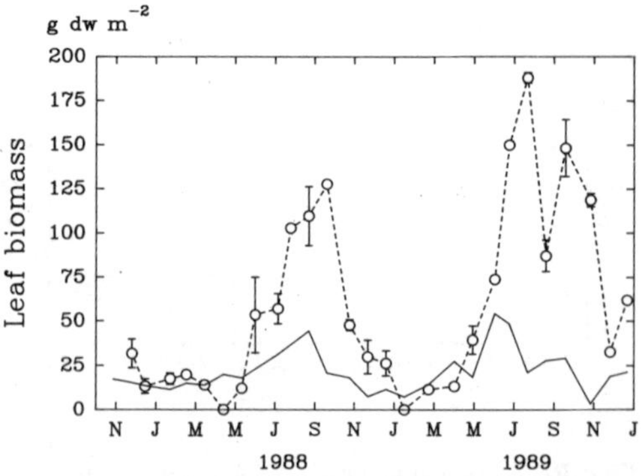

FIGURE 6. Leaf biomass annual variation of two *Cymodocea nodosa* meadows on sand (solid line) and on mud (dashed line; vertical bars, standard error of mean) in the Mar Menor (TERRADOS & ROS, in press a).

TABLE I. Values of linear correlation coefficient of Pearson (r) between some *Cymodocea nodosa* meadow variables and irradiance and water temperature (all significant at P< 0.01 level; TERRADOS, 1991).

Variable	Irrad.	Temp.
Shoot density (shoots m^{-2})	0.592	0.714
Leaf biomass (g dw m^{-2})	0.776	0.815
LAI (m^2 m^{-2})	0.805	0.863
Leaf growth (g dw $shoot^{-1}$ day^{-1})	0.866	0.698
Shoot P/B (day^{-1})	0.841	0.599
Meadow leaf production (g dw m^{-2} day^{-1})	0.849	0.741

time and depend heavily on the energy that maintains the process. Extensive variables are the organism's integrated response to the variability of the instantaneous rates, and here temperature takes on an important role, as it affects directly all the physiological processes of the organism.

Annual leaf production in meadows on sand is estimated at roughly 162 g dw m^{-2} y^{-1} (annual P/B: 2.46), and between 290 and 247 g dw m^{-2} y^{-1} in meadows on mud. A shoot produces 11-12 leaves in a year (TERRADOS & ROS, in press a). The annual rhizome production in meadows on sand is estimated at roughly 45-50 g dw m^{-2} y^{-1} (annual P/B: 0.29), which is equivalent to a growth in length of 39 m m^{-2} y^{-1}. In meadows on mud, rhizome production can be estimated at roughly

TABLE II. P-I relationship parameters of *Caulerpa prolifera* and *Cymodocea nodosa* throughout the year in the Mar Menor. Units: α, in mg O_2 g dw^{-1} h^{-1} μmol^{-1} m^{-2} s^{-1}; P_{max} and R, in mg O_2 g dw^{-1} h^{-1}; I_c and I_k, in µmol m^{-2} s^{-1} (TERRADOS & ROS, in press b, c).

	November	February	May	August
Caulerpa prolifera				
α	0.311	0.023	0.176	0.297
P_{max}	3.48	1.21	3.38	5.63
I_c	3	7	5	3
I_k	11	52	19	19
R	1.06	0.16	0.90	0.78
Cymodocea nodosa				
α	0.025	0.019	0.024	0.063
P_{max}	1.94	1.32	9.32	14.02
I_c	20	28	20	26
I_k	78	68	387	223
R	0.50	0.55	0.49	1.63

66-78 g dw m^{-2} y^{-1} (TERRADOS & ROS, in press a).

PHOTOSYNTHETIC FEATURES

CAULERPA PROLIFERA

Compensation irradiance (I_c) remains constant between 3 and 7 µmol m^{-2} s^{-1} while saturation irradiance (I_k) is higher in February (52 µmol m^{-2} s^{-1}) than in the rest

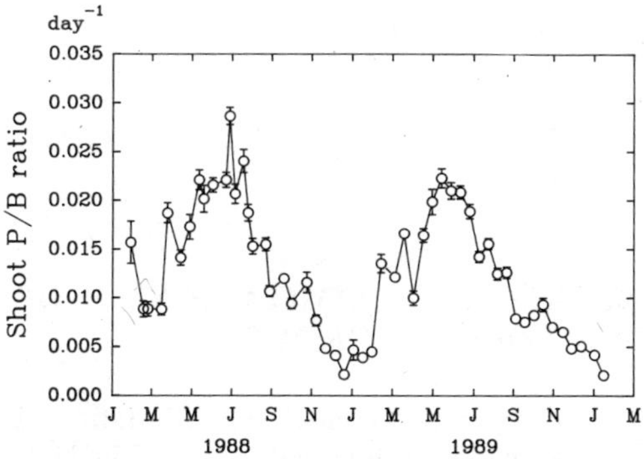

FIGURE 7. Shoot P/B ratio annual variation of a *Cymodocea nodosa* meadow on sand in the Mar Menor (vertical bars, standard error of mean) (TERRADOS & ROS, in press a).

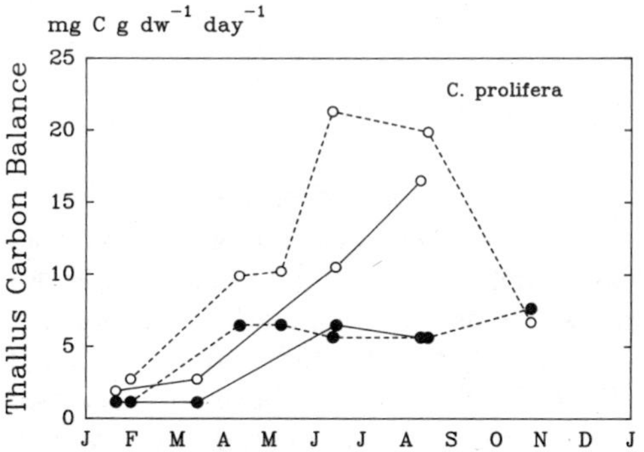

FIGURE 8. Daily carbon balances for *Caulerpa prolifera* at the central basin (solid lines) and shallow sites (dashed lines) in the Mar Menor lagoon. Net photosynthesis, open dots; dark respiration, full dots. (TERRADOS & ROS, in press b).

of the year (11-19 μmol m^{-2} s^{-1}; TERRADOS & ROS, in press b). The light-saturated net photosynthesis rate (P_{max}) goes up from February (1.21 mg O_2 g dw^{-1} h^{-1}) until August (5.63 mg O_2 g dw^{-1} h^{-1}), and goes down in November (3.48 mg O_2 g dw^{-1} h^{-1}); photosynthetic efficiency at non-saturating irradiances (α) is minimum in February and maximum in November (Table II).

P_{max} and dark respiration rate are linearly correlated with water temperature between 10 and 30°C and 10 to 35°C respectively (TERRADOS & ROS, in press b). The effect of temperature on R was smaller in May (Q_{10} = 1.39) and August (Q_{10} = 1.68) than in November (Q_{10} = 2.54) and February (Q_{10} = 2.03). In a similar way, the effect of temperature on P_{max} was higher in February (Q_{10} = 1.86) than in August (Q_{10} = 1.46).

The time during the day that the plant receives irradiances above I_k (H_{sat}, DENNISON & ALBERTE, 1982, 1985) varies between 7h 30' and 13h 15' in the central basin (depths of more than 4m) and between 9h 30' and 14h in the shallow areas by the bank (depths of less than 1 m; TERRADOS & ROS, in press b). Daily carbon balance (DENNISON, 1987) is always positive, and increases from January to July both in the central basin (1.9-16.5 mg O_2 g dw^{-1} day^{-1}) and in the shallow areas (2.7-21.3 mg O_2 g dw^{-1} day^{-1}; Fig. 8; TERRADOS & ROS, in press b).

CYMODOCEA NODOSA

I_c remains constant between 20 and 28 μmol m^{-2} s^{-1}, whereas I_k is greater in May (167 μmol m^{-2} s^{-1}) and August (223 μmol m^{-2} s^{-1}) than in November (78 μmol m^{-2} s^{-1}) and February (68 μmol m^{-2} s^{-1}; TERRADOS & ROS, in press c). P_{max} rises from February (1.32 mg O_2 dw^{-1} h^{-1}) to August (14.02 mg O_2 dw^{-1} h^{-1}) and falls in November (1.94 mg O_2 g dw^{-1} h^{-1}); α is minimum in February, maximum in August and intermediate in May and November

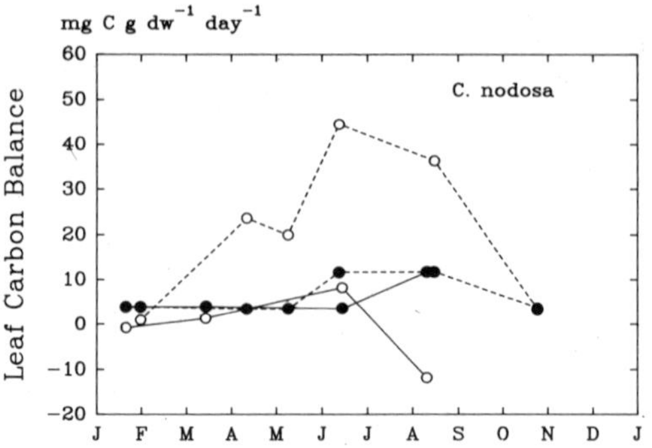

FIGURE 9. Daily leaf carbon balances for *Cymodocea nodosa* at the central basin (solid lines) and shallow sites (dashed lines) in the Mar Menor lagoon; net photosynthesis, open dots; respiration, full dots. (TERRADOS & ROS, in press c).

(Table II).

P_{max} and R are linearly correlated with water temperature between 10 and 30°C, and 10 and 35°C respectively (TERRADOS & ROS, in press c). The effect of temperature on R is smaller in August (Q_{10} = 2.07) than in November (Q_{10} = 2.21) and February (Q_{10} = 2.46). In contrast, the effect of temperature on P_{max} is higher in August (Q_{10} = 2.33) than in February (Q_{10} = 2.02). The P_{max}/R relationship shows a shifting of the maximum from 15°C in February to 30°C in August.

H_{sat} varies between 0h and 9h 30' in the central basin, and between 8h and 12h in shallow areas. The daily carbon balance in the leaves is positive throughout the year in the shallow areas, whereas it is near zero or negative in January and August in the central basin (Fig. 9; TERRADOS & ROS, in press c).

NUTRIENT LIMITATION OF MACROPHYTE GROWTH

The average annual C:N:P ratio of thallus in *Caulerpa prolifera* is 354:21:1. The carbon, nitrogen and phosphorus content of the fronds does not show defined seasonal variation over the year

TABLE III. Growth rates, (P/B)100 day^{-1} (mean, standard error of mean), of thallus fragments of *Caulerpa prolifera* (n=30) measured in aquaria with different nutrient addition treatments. NO_3^- was used as nitrogen source in Experiment 1; NH_4^+, in Experiment 2. PO_4^{3-} was used as phosphorus source in both experiments (TERRADOS, 1991).

	Experiment 1	Experiment 2
Control	3.39 (0.40)	5.24 (0.29)
N	6.27 (0.87)	6.51 (0.48)
N+P	6.23 (0.91)	7.26 (0.49)
P	2.42 (0.37)	4.61 (0.36)

(TERRADOS, 1991). Several fertilization experiments carried out in the laboratory (TERRADOS, 1991) indicate that the growth of *C. prolifera* is limited by the availability of nitrogen in the water column (Table III).

Nitrogen and phosphorus content of *Cymodocea nodosa* leaves is higher than that of the rhizomes and roots. The average annual C:N:P ratio of the leaves is 441:22:1, 672:17:1 in rhizomes, and 635:19:1 in roots (TERRADOS, 1991). Experimental additions of nitrogen and/or phosphorus to the sediment in meadows on sand indicate that the growth of *C. nodosa* is limited by the availability of nutrientes. Biomass and the growth rates of the shoots in the quadrats where phosphorus had been added were higher ($p < 0.05$) than in

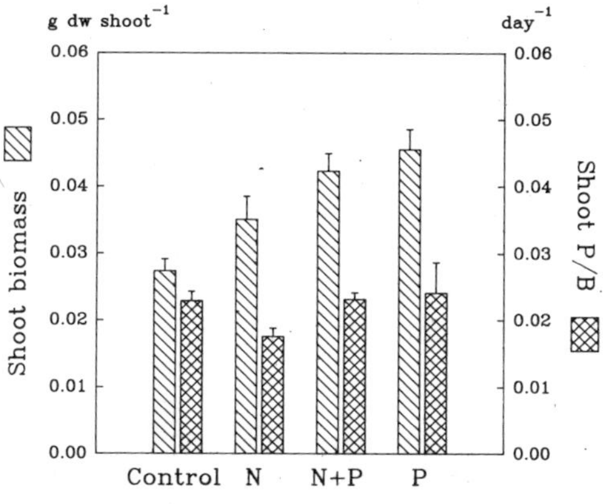

FIGURE 10. Biomass and P/B ratio of *Cymodocea nodosa* shoots after long-term fertilization of the sediment (sand). Vertical bars, standard error of mean. (TERRADOS & ROS, in press d).

TABLE IV. Effect of the long-term fertilization (monthly nutrient additions from April to July) of the sediment with nitrogen (NH_4^+) and/or phosphorus (PO_4^{3-}) on the shoot density of a *Cymodocea nodosa* meadow on sand. Mean (standard error of mean) of the number of shoots in 25 permanent quadrats (100 cm^2) before and after treatments (TERRADOS & ROS, in press d).

Treatment	Number of shoots per quadrat Initial	Final
Control	18.2 (0.9)	37.7 (1.4)
N	16.6 (0.9)	27.2 (1.2)
N+P	19.6 (0.8)	60.1 (2.7)
P	17.6 (0.7)	35.3 (1.2)

quadrats fertilized only with nitrogen and control (Fig. 10), whereas the P/B ratio did not change during the various treatments. The increase in shoot density in a meadow involves the growth of both leaves and rhizomes; shoot density only increased in the quadrat fertilized with both nitrogen and phosphorus (Table IV; TERRADOS & ROS, in press d).

TOTAL LAGOON PRIMARY PRODUCTION ESTIMATE

Following the bionomic mapping made by PÉREZ RUZAFA *et al.* (1989) and by studying aerial colour photographs taken on a flight in 1986, we have calculated that of the 135 km^2 of the lagoon's surface area approximately 85% of the bottom is muddy, 14% sandy and 1% rocky (TERRADOS, 1986, 1991). The shallow sandy bottom is covered by vegetation in approximately 18% of its area, making up 2.5% of the lagoon total. Similarly the area of the *Caulerpa prolifera* meadow on mud can be estimated at roughly 115 km^2, and the macrophyte meadows on sand at 3.5 km^2. The presence of communities of photophilic algae on rock is not important in the ecosystem as a whole.

With these data, supposing that half the meadows of macrophytes on sand are made up by *Caulerpa prolifera* and the other half by *Cymodocea nodosa* and with the

TABLE V. Biomass, primary production and surface covered by different plant communities in the Mar Menor lagoon (Data from BALLESTER, 1985; TERRADOS, 1986, 1991; PÉREZ RUZAFA,A., 1989 and Gilabert, pers. comm.). [1] isla del Ciervo, [2] El Estacio.,

	Biomass g dw m^{-2}	Annual production g dw m^{-2} y^{-1}	Annual production g C m^{-2} y^{-1}	Surface km^2	Total prod. Tm C	Production % of total
Caulerpa prolifera (mud)	31-157	630-707	211.7-237.5	115	24343-27318	63.18
Caulerpa prolifera (sand)	16-97	388-446	130.4-149.8	1.75	228-262	
Cymodocea nodosa, leaves	3-170	166-360	49.8-108.0	1.75	87-189	0.42
rhizomes	65-157	47-72	15.5-23.8	1.75	27-42	
Photophilic algae[1]	20-152	389	74	0.67	50	0.12
Photophilic algae[2]	0-956	1040	71	0.67	48	0.12
Microphytobenthos (mud)			36	115	4117	10.00
Microphytobenthos (sand)			23.5-44.6	20	470-892	1.62
Phytoplankton			75	135	10125	24.53
			TOTAL		41269.3 Tm C y^{-1}	
					≈ 306 g C m^{-2} y^{-1}	

estimates for annual production of the meadows of *C. prolifera* (TERRADOS, 1991), *C. nodosa* (TERRADOS & ROS, in press a), and the communities of photophilic algae, it is possible in turn to estimate the total primary production of macrophytobenthos in the lagoon (Table V).

The estimates for primary production of the phytoplankton and microphytobenthos are approximate and are based on carbon uptake rates measured at midday, using the ^{14}C method, during incubation periods of two hours, by Gilabert (pers. comm.) and PÉREZ RUZAFA, A. (1989), respectively. These authors took these measurements during the second half of 1987 and along all 1988 on a weekly basis (Gilabert, pers. comm.) or on a monthly basis (PÉREZ RUZAFA, A., 1989).

These estimates were calculated by multiplying the assimilation rates measured in one day by the number of hours of daylight on that particular day, and considering the daily assimilation as representative of the time period between sampling (one week for phytoplankton; one month for microphytobenthos). The annual estimate is the sum of the weekly or monthly estimates. As a result, and with all the limitations of the extrapolation method used, the production values obtained for these two compartments are considered maximum estimates.

Overall primary production of the Mar Menor lagoon can be estimated to be at least 41,269 Tm of carbon a year, which is equivalent to about 300 g C m^{-2} y^{-1}. Macrophytobenthos production is two to three times higher than phytoplankton production (Fig. 11). Microphytobenthos production makes up 11% of total lagoon

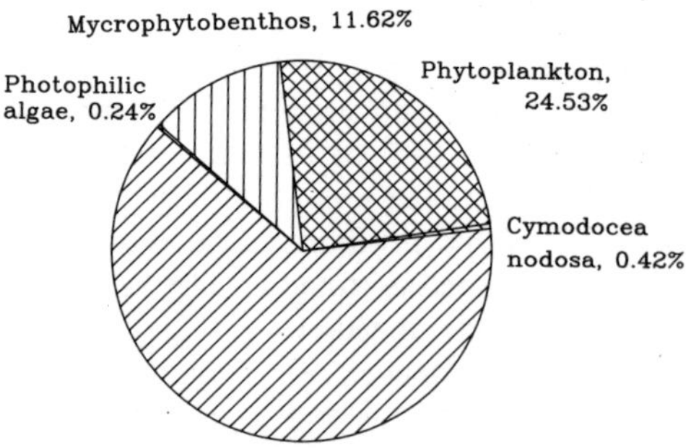

FIGURE 11. Percentage of annual primary production achieved by the different plant communities of the Mar Menor lagoon.

production, whereas macrophytobenthos production makes up slightly less than 64%.

ENVIRONMENTAL CONTROL OF MACROPHYTE DISTRIBUTION IN THE LAGOON BY LIGHT AVAILABILITY

Secchi disk depth values (D_{sd}) measured by Gilabert (pers. comm.) on a weekly basis at a point situated in the northern part of the central basin in 1988 can be used to calculate the light extinction coefficient (k) through the relation $k = 1.7 / D_{sd}$ (POOLE & ATKINS, 1929; GIESEN *et al.*, 1990).

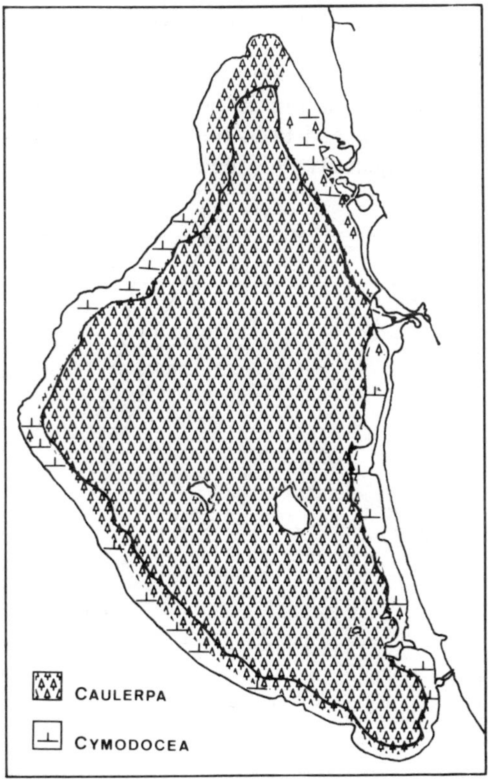

FIGURE 12. Distribution of *Caulerpa prolifera* and *Cymodocea nodosa* meadows in the Mar Menor lagoon, according to all information available. Compare with the map in PÉREZ RUZAFA *et al.* (1989); the 3.3 m isobath is shown (TERRADOS, 1991).

The average annual value obtained is k = 0.35, but during August and September k may reach 0.57. In these conditions the quantity of light reaching the bottom of the lagoon is small and in fact irradiances measured at the bottom of the Mar Menor at midday in summer (< 200 $\mu mol\ m^{-2}\ s^{-1}$) are lower than those measured during the rest of the year, even winter (TERRADOS, 1991).

DENNISON (1987) used H_{comp} values (the period of time during the day that the plant receives irradiances of more than I_c; DENNISON & ALBERTE, 1982, 1985) to predict the depth limit of distribution of *Zostera marina* in Woods Hole, and obtained the empirical relation between the light extinction coefficient and depth limit of distribution $Z_c = 1.62/k$, similar to that found by NIELSEN *et al.* (1989; in DUARTE, in prep.) in Danish estuaries ($Z_c = 1.53/k$). VICENTE & RIBERA (1982; in DUARTE, in prep.) obtained the relation $Z_c = 1.36/k$ for *Thalassia testudinum* on Puerto Rican coasts. Starting from a thorough bibliographical study DUARTE (in prep.), calculates an average relation for marine seagrasses of $Z_c = 1.86/k$, which is equivalent to the depth reached by 11% of the light incident on the surface.

If we apply this last relation to k values calculated for 1988 in the Mar Menor (Gilabert, pers. comm.) we find the annual variation of critical depth (Z_c) of seagrasses vertical distribution which, for the months of August and September, varies between 3.3 and 5.5 m.

Currently *Cymodocea nodosa* only forms dense meadows in the outer areas - the shallower areas - of the lagoon; at depths of more than 4 m only isolated shoots are observed, never a meadow (TERRADOS, 1991). Therefore it is necessary to modify the bionomic charting of the lagoon made by PÉREZ RUZAFA *et al.* (1989); we should extend the distribution of the monospecific meadow of *Caulerpa prolifera* to the area occupied by the *C. nodosa-C. prolifera* mixed meadow (Fig.

12). The distribution of the *C. prolifera* monospecific meadow in the lagoon coincides almost perfectly with the 3.3 m isobath and with the results of a dredging campaign carried out in April 1987 (PÉREZ-RUZAFA, unpublished data), in which the dominance of *C. prolifera* throughout the lagoon is shown.

Daily carbon balances calculated for *Caulerpa prolifera* and *Cymodocea nodosa* in different months of the year (Figs. 8 and 9) indicate that during certain periods of winter and summer not enough light reaches the bottom of the central basin to ensure a positive daily balance for *C. nodosa*. This is caused by the shorter day and the smaller angle of incidence of the sun's rays in winter, and mainly by the extinction of light in the water column in summer. *C. prolifera*, in contrast, maintains a positive carbon balance at the bottom of the central basin througout the year.

All these findings suggest that insufficient light reaches the bottom of the central basin of the Mar Menor in August and September for meadows of *Cymodocea nodosa* to develop. Light availability at depth appears therefore to be the main factor in determining the distribution of meadows of *C. nodosa* in the lagoon.

CYMODOCEA NODOSA REGRESSION

PÉREZ RUZAFA *et al.* (1987, 1989) and ROS (1987) indicate that after dredging and enlargement of the El Estacio channel at the beginning of the seventies, *Caulerpa prolifera* began to colonize the Mar Menor; as a result *Cymodocea nodosa* meadows were gradually replaced by *Caulerpa prolifera* meadows. If we check the biomass data for both species obtained by dredging in 1982/3 (BALLESTER, 1985) and 1987 (PÉREZ RUZAFA *et al.*, unpublished data) this replacement is not clearly seen - rather, we see the consistency over the past few years of a basically monospecific meadow of *C. prolifera*.

The only charted information available about macrophytobenthos prior to the enlargement of the El Estacio channel is the work carried out by LOZANO (1954; Fig. 13). The places where this author found "very abundant Zostera" (probably meadows of *Cymodocea nodosa*) coincide with the areas where today we find meadows of *C. nodosa* (shallow sandy bottoms), while the places where this author found "negligible Zostera" (in the central part of the northern basin) are also areas where today we find isolated shoots of *C. nodosa* (TERRADOS, 1991). Although we do not have information about all the areas of the lagoon, the comparison of these data with the situation today suggests that the distribution of *C. nodosa* in the lagoon has not changed significantly since 1953.

The colonization of the lagoon by *Caulerpa prolifera* is beyond all possible doubt; if it had been present before in its current quantity, this alga would have been

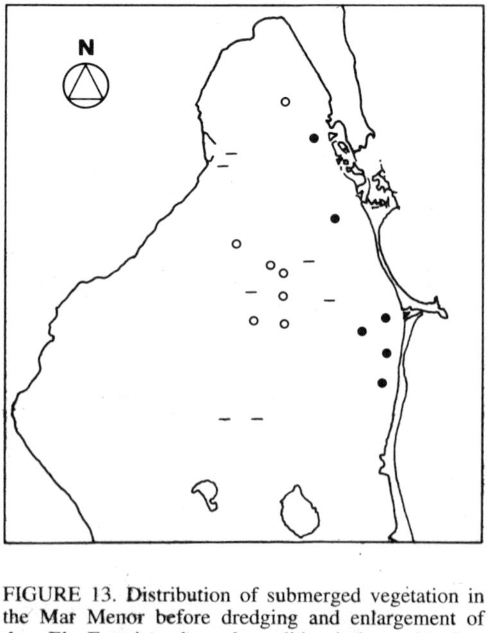

FIGURE 13. Distribution of submerged vegetation in the Mar Menor before dredging and enlargement of the El Estacio channel: solid circles, abundant "Zostera" (likely, *Cymodocea nodosa*); hollow circles, scarce "Zostera"; dashes, no vegetation (redrawn from LOZANO, 1954).

mentioned in the studies prior to the dredging of the El Estacio channel (LOZANO, 1954; SIMMONEAU, 1973). As we have said above, it is not altogether certain that the recent colonization by *C. prolifera* has meant that *C. nodosa* has been displaced.

Changes in salinity - a fall from 50-60‰ in the sixties to 45‰ today - and in water temperature - in the sense of damping off the extreme values - caused by the dredging and enlargement of the El Estacio channel, have not had a negative effect on the viability of the seagrass inside the lagoon - indeed, its development has been favoured. The increase in organic material in the sediment of the bottom areas colonized by *Caulerpa prolifera* and consequently the increase in the levels of the sediment's anoxia (PÉREZ RUZAFA *et al.*, 1987) is not in itself harmful to *C. nodosa* since seagrasses are capable of supplying enough oxygen to the rhizomes and roots through the lagunar air system to keep up their activity (PENHALE & WETZEL, 1983; DeLAUNE *et al.*, 1984; SMITH *et al.*, 1984; SORREL & DROMGOOLE, 1988; LARKUM *et al.*, 1990). Furthermore, the higher level of organic material in the sediment ensures greater availability of nutrients for the plant (TERRADOS, 1991).

There does not seem to be a direct interaction between *Caulerpa prolifera* and *Cymodocea nodosa*. The colonization of the lagoon by *C. prolifera* may simply have been the occupation of an empty niche in the ecosystem after the changes in temperature and salinity caused by the works in El Estacio, and not a process of competitive exclusion at the expense of *C. nodosa*.

Whereas the colonization of muddy substrates, with low hydrodynamism, by *Caulerpa prolifera* is a relatively fast process (ROS *et al.*, 1988), the consistency of some of the meadows of this alga on sand since 1985 (TERRADOS, 1991) leads us to suppose that the muddying and the formation of a meadow of high biomass in relatively shallow sandy substrates (less than 2 m deep) is a much slower process, indeed, a dense meadow on sandy sediment may not finally be created. We may put forward the hypothesis that the lower availability of nutrients in the sandy bottoms hampers the development of a dense meadow of *C. prolifera* (TERRADOS, 1991).

DISCUSSION

The maximum biomass attained by the meadows of *Caulerpa prolifera* in the Mar Menor is equivalent to the maximum leaf biomass attained by meadows of *Cymodocea nodosa* in shallow muddy areas; it is higher than the leaf biomass of the meadows of *C. nodosa* on sand, which are the most extended in the lagoon. Obviously, the total biomass attained by *C. nodosa*, including rhizomes and roots, is greater than that of *C. prolifera.*

The biomass cycle of both species show a certain time lag. *Caulerpa prolifera* reaches maximum biomass in summer and maintains it until November-December; *Cymodocea nodosa* also reaches maximum biomass in summer, but in autumn it already begins to fall. *C. nodosa* reaches minimum biomass in winter, while *C. prolifera* reaches its minimum biomass in spring.

The meadows of *Caulerpa prolifera* have an average annual LAI between 1.40 and 1.88 m^2 m^{-2}, higher than that of the meadows of *Cymodocea nodosa* on sand (0.71 m^2 m^{-2}). The presence of *C. prolifera* in the lagoon would seem to suppose an increase in the surface of the substrate available for the settlement of epiphytes. However, field observations carried out indicate that the presence of macroepiphytes on *C. prolifera* is negligible; the old leaves of *C. nodosa* usually have a fair number of epiphytes. This may be due to the higher rate of renovation of the biomass of *C. prolifera* compared with *C. nodosa*'s leaf biomass, or to the alga's use of defensive substances (PAUL & FENICAL, 1986).

In accordance with the estimates we have made, both annual primary production and the annual P/B ratio are higher in *Caulerpa prolifera* (211.7-237.5 g C m^{-2} y^{-1}; P/B = 4) than in *Cymodocea nodosa* (leaves: 49.8-108 g C m^{-2} y^{-1}; P/B = 2.5; rhizomes: 15.5-23.8 g C m^{-2} y^{-1}; P/B = 0.29). Light and temperature appear to be the main environmental factors which control meadow production of macrophytes, for the amounts of nutrient available.

N and P content in *Caulerpa prolifera* is higher than in *Cymodocea nodosa*. Both species have a similar N:P ratio. *C prolifera* growth seems to be limited by the availability of nitrogen, whereas *C. nodosa* growth seems to be limited by the availability of both nitrogen and phosphorus.

Caulerpa prolifera presents lower I_c and I_k figures than *Cymodocea nodosa.* α is higher in *C. prolifera* than in *C. nodosa*, whereas the P_{max} reached by the seagrass, especially in spring and summer, was higher than that of the chlorophycean. All this suggests that *C. prolifera* can live in conditions of less irradiance than *C. nodosa*, as shown by the daily carbon balances calculated for both species throughout the year. *C. nodosa* has a larger fraction of non-photosynthetic biomass than *C. prolifera*; therefore energy requirements for the maintenance of the plant are higher. Water temperature exerts a linear influence on the P_{max} and R rates in both species between 10 and 30 °C and 10 and 35 °C, respectively. Both species show adaptation in their photosynthetic characteristics to the seasonal variation of the water temperature.

The capacity for vegetative expansion and dispersion of *Caulerpa prolifera* is higher than that of *Cymodocea nodosa* (TERRADOS, 1991). After the dredging and enlargement of the El Estacio channel in 1972, *C. prolifera* carried out a spectacular colonization process in the Mar Menor, and today it can be found all over the lagoon. It is capable of implanting itself on rock (El Ciervo island and Los Alcázares), but where it reaches its greatest biomass is on mud and sand. Biological features such as almost continuous growth throughout the year or the capacity of generating vegetatively a new thallus from any fragment swept away by the water have enabled it to spread out rapidly throughout the lagoon.

The colonization of the Mar Menor by *Caulerpa prolifera* has brought about an increase in macrophytobenthic primary production and thus in the lagoon's primary production. On the one hand, the annual production of the meadow of *C. prolifera* on mud (212-235 g C m^{-2} y^{-1}) is higher than that of meadows of *Cymodocea nodosa* (65-130 g C m^{-2} y^{-1}); on the other hand, the presence of macrophytes in the lagoon is now much higher than before the opening of the El Estacio channel. The importance of the change in the Mar Menor caused by the enlargement of the El Estacio channel is obvious, since a species that colonized the lagoon after the project now accounts for 63% of the autotrophic incorporation of carbon into the ecosystem.

The colonization has, predictably, had other effects.

The establishment of a meadow of *Caulerpa prolifera* protects the sediment, increasing sedimentation rates and reducing sediment resuspension rates in situations of high hydrodynamism, especially in shallow bodies of water such as the Mar Menor; therefore water turbidity is probably now less. This process has been identified in other systems where there is significant macrophyte presence (SHORT & SHORT, 1984).

The decomposition rates of *Caulerpa prolifera* are higher than those of *Cymodocea nodosa*: experiments on decomposition of *C. nodosa* leaves carried out by PÉREZ (1989) show that after 140 days 40% of the initial dead material remains, while with *Caulerpa cupressoides*

(WILLIAMS, 1984a), which must have decomposition rates similar to those of *C. prolifera*, after seven days 50% of the initial dead material is lost. Although these differences are probably less with relation to nitrogen and phosphorus, the matter cycling in the macrophytobenthos is faster now that it is dominated by *C. prolifera* than when the dominant macrophyte was *C. nodosa*. However, the lower biomass turnover of macrophytes in comparison with phytoplankton causes the inflow of energy into the ecosystem to be slower now that the macrophytobenthos is dominant. For the lagunar ecosystem, the presence of macrophytes has meant the acquisition of compartments which have greater inertia and are stabilizing factors of both primary and secondary production (HARRISON & MANN, 1975; FENCHEL, 1977; KENWORTHY & THAYER, 1984).

As they are fixed to the substrate, aquatic macrophytes are capable of using the movement of the water to acquire dissolved inorganic nutrients more efficiently than phytoplankton (MANN, 1982). A moderate degree of hydrodynamism increases the volume of water used by the macrophytes and breaks the difussion gradients, facilitating the assimilation of nutrients (LOBBAN *et al.*, 1985). Furthermore, if they are on soft substrates, they can use the nutrients present in the sediment (McROY & McMILLAN, 1977; THURSBY & HARLIN, 1982; SHORT & McROY, 1984; BRIX & LYNGBY, 1985).

The relative importance of root and leaf assimilation depends on the level of availability of nutrients in the water or in the sediment (McROY & McMILLAN, 1977; THURSBY & HARLIN, 1982; BRIX & LYNGBY, 1985; BORUM *et al.*, 1989; HILLMAN *et al.*, 1990). It is generally accepted that the sediment is the main source of nutrients both for freshwater macrophytes (BARKO & SMART, 1980, 1981; DUARTE & KALFF, 1987; CHAMBERS *et al.*, 1988) and for marine phanerogames (SHORT, 1983 a, b, 1987; SHORT & McROY, 194; THURSBY & HARLIN, 1984; SHORT *et al.*, 1985; WILLIAMS, 1987, 1990; HILLMAN *et al.*, 1990).

The dominance of *Caulerpa prolifera* and *Cymodocea nodosa* in the lagoon determines, on the one hand, the consumption of the nutrients present in the sediment (which cannot be utilized directly by phytoplankton); and on the other hand, the competition with the phytoplankton in the acquisition of nutrients, as ROS & MIRACLE (1984) and ROS (1987) have suggested.

Although the concentrations of inorganic compounds of nitrogen and phosphorus dissolved in the water of the Mar Menor are very low, *Caulerpa prolifera* is probably capable of meeting its requirements of nitrogen and phosphorus from the nutrients available in the sediment. WILLIAMS (1984 b) showed that *Caulerpa cupressoides* was capable of assimilating NH_4^+ through the rhizoids and transporting it to the fronds. The values for the semisaturation constant for the absorption of NH_4^+ that this author obtained indicate that *C. cupressoides* is better adapted to use sediment than water as a nitrogen source. The experiments in sediment fertilization carried out in the Mar Menor (TERRADOS & ROS, in press d) and in other places (PÉREZ *et al.*, 1991) indicate that *Cymodocea nodosa* is capable of using nutrients present in the sediment.

As we have seen, nitrogen and phosphorus content in *Caulerpa prolifera* is greater than in *Cymodocea nodosa*. Because of this, the presence of the alga in the lagoon has probably increased the capacity of the macrophytobenthos to absorb the nutrients that enter the lagoon, in competition with the phytoplanktonic populations. This - together with the lower I_c and I_k of *C. prolifera*, the proposed reduction in water turbidity in the lagoon due to the greater presence of macrophytes and the channelling of the nutrient input towards the macrophytes - make us suppose that the colonization by the chlorophycean has probably increased the Mar Menor's resistance to eutrophication processes.

The participation of macrophytobenthos in the overall production of the Mar Menor is high compared with that in other places, such as the bays of the Ebro delta (PÉREZ, 1989). These bays constitute an estuarine environment where the inflow of freshwater and nutrients is important and where phytoplanktonic production predominates. In contrast, in the Mar Menor there is no marked inflow of freshwater, the interchange with the Mediterranean Sea is less, and although the amount of nutrient input from land sources has not been evaluated, it must be less than in the bays of the Ebro Delta. In these conditions of relative scarcity of nutrients in the water column, the light reaches the bottom in sufficient quantities to allow the development of submerged vegetation, capable of using the nutrients present in the sediment.

Summer seems to be a critical time of year for the functioning of the Mar Menor. It is then that development of the phytoplanktonic populations is at its highest (Gilabert, pers. comm.); therefore light extinction in the water column increases in such a way that at the bottom of the central basin there is insufficient light for *Cymodocea nodosa* to grow. The high water temperatures cause the metabolic cost of maintaining the vegetation to be higher. Furthermore, it has been suggested that the increase in the resident human population on the banks of the lagoon at this time of year, and the shortcomings of the sanitation network in the villages around the lagoon probably cause more nutrients to enter (TERRADOS, 1986), which may be one of the reasons for the proliferation of phytoplankton in summer.

This greater development of phytoplankton in summer causes a reduction in the quantity of light that reaches the bottom. If this reduction increased, the development of *Caulerpa prolifera* and other types of submerged vegetation would be hindered, thus triggering a process of eutrophication which would alter completely the functioning and environmental characteristics of the lagunar ecosystem. The relations between phytoplankton and macrophytes as far as the nutrient cycle and production-consumption of detritus are concerned require special attention in the future, because of their implications for the overall functioning of the Mar Menor lagoon ecosystem.

ACKNOWLEDGEMENTS

The authors are pleased to dedicate this paper to Professor Ramon Margalef, from whom they have tried to learn, among many other things, how to integrate apparently disperse data to obtain a coherent image of aquatic ecosystems. The work was made possible by a PFPI grant from the MEC to J. Terrados.

REFERENCES

BALLESTER R. 1985. Biomasa, estacionalidad y distribución de tres macrófitos: *Ruppia cirrhosa*, *Cymodocea nodosa* y *Caulerpa prolifera* en el Mar Menor (Murcia, SE de España). *Anales de Biología*, 4: 31-36.

BARKO, J.W. & SMART, R.M. 1980. Mobilization of sediment phosphorus by submersed freshwater macrophytes. *Freshwater Biology*, 10: 229-238.

BARKO, J.W. & SMART, R.M. 1981. Sediment-based nutrition of submersed macrophytes. *Aquat. Botany*, 10: 339-352.

BORUM, J., MURRAY, L. & KEMP, W.M. 1989. Aspects of nitrogen acquisition and conservation in eelgrass plants. *Aquat. Botany*, 35: 289-300.

BRIX, H. & LYNGBY, J.E. 1985. Uptake and translocation of phosphorus in eelgrass (*Zostera marina*). *Mar. Biol.*, 90: 111-116.

CAPEL MOLINA, J. 1981. *Los climas de España*. Oikos-Tau. Barcelona

CHAMBERS, P.A., PREPAS, E.E., BOTHWELL,

M.L. & HAMILTON, H.R. 1989. Roots versus shoots in nutrient uptake by aquatic macrophytes in flowing waters. *Can. J. Fish. Aquat. Sci.,* 46: 435-439.

DeLAUNE, R.D., SMITH, C.J. & TOLLEY, M.D. 1984. The effect of sediment redox potential on nitrogen uptake, anaerobic root respiration and growth of *Spartina alterniflora* Loisel. *Aquat. Botany,* 18: 223-230.

DENNISON, W.C. 1987. Effects of light on seagrass photosynthesis, growth and depth distribution. *Aquat. Botany,* 27: 15-26.

DENNISON, W.C. & ALBERTE, R.S. 1982. Photosynthetic responses of *Zostera marina* L. (eelgrass) to *in situ* manipulations of light intensity. *Oecologia,* 55: 137-144.

DENNISON, W.C. & ALBERTE, R.S. 1985. Role of daily light period in the depth distribution of *Zostera marina* (eelgrass). *Mar. Ecol. Prog. Ser.,* 25: 51-61.

DUARTE, C.M. & KALFF, J. 1988. Influence of lake morphometry on the response of submerged macrophytes to sediment fertilization. *Can. J. Fish. Aquat. Sci.,* 45: 216-221.

DUARTE, C.M. in prep. Seagrass depth limits.

FENCHEL, T. 1977. Aspects of the decomposition of seagrasses. In: *Seagrass ecosystems. A scientific perspective* (McRoy, C.P. & Helfferich, C., eds.): 123-145. Marcel Dekker. New York.

GIESEN, W.B.J.T., VAN KATWIJK, M.M. & DEN HARTOG, C. 1990. Eelgrass condition and turbidity in the Dutch Wadden Sea. *Aquat. Botany,* 37: 71-85.

HARLIN, M.M. & THORNE-MILLER, B. 1981. Nutrient enrichment of seagrass beds in a Rhode Island coastal lagoon. *Mar. Biol.,* 65: 221-229.

HARRISON, P.G. & MANN, K.H. 1975. Detritus formation from eelgrass (*Zostera marina* L.): The relative effects of fragmentation, leaching and decay. *Limnol. & Oceanogr.,* 5: 924-934.

HILLMAN, K., WALKER, D.I., LARKUM, A.W.D. & MCCOMB, A.J. 1990. Productivity and nutrient limitation. In: *Biology of Seagrasses* (Larkum, A.W.D., McComb, A.J. & Shepherd, S.A., eds.): 635-685. Elsevier. Amsterdam.

KENWORTHY, W.J. & THAYER, G.W. 1984. Production and decomposition of the roots and rhizomes of seagrasses, *Zostera marina* and *Thalassia testudinum,* in temperate and subtropical marine ecosystems. *Bull. Mar. Sci.,* 35: 364-379.

LARKUM, A.W.D., ROBERTS, G., KUO, J. & STROTHER,S., 1990. Gaseous movement in seagrasses. In: *Biology of Seagrasses* (Larkum, A.W.D., McComb, A.J. & Shepherd, S.A., eds.): 686-722. Elsevier. Amsterdam.

LOBBAN, C.S., HARRISON, P.J. & DUNCAN, M.J. 1985. *The physiological ecology of seaweeds.* Cambridge University Press. Cambridge.

LOZANO CABO, F. 1954. Una campaña de prospección pesquera en el Mar Menor (Murcia). *Bol. Inst. Esp. Oceanogr.,* 66: 1-34.

MANN, K.H. 1982. *Ecology of coastal waters. A systems approach.* Blackwell. Oxford.

McROY, C.P. & MCMILLAN, C. 1977. Production ecology and physiology of seagrasses. In: *Seagrass ecosystems: a scientific perspective* (McRoy, C.P. & Helfferich, C., eds.): 53-87. Marcel Dekker. New York.

PAUL, V.J. & FENICAL, W. 1986. Chemical defense in tropical green algae, order Caulerpales. *Mar. Ecol. Prog. Ser.,* 34: 157-169.

PENHALE, P.A. & WETZEL, R.G. 1983. Structural and functional adaptations of eelgrass (*Zostera marina* L.) to the anaerobic sediment environment. *Can. J. Bot.,* 61: 1421-1428.

PÉREZ RUZAFA, A. 1989. *Estudio ecológico y bionómico de los poblamientos bentónicos del Mar Menor (Murcia, SE de España).* Ph. D. Thesis. University of Murcia.

PÉREZ RUZAFA, A., MARCOS, C., PÉREZ RUZAFA, I.M. & ROS, J.D. 1987. Evolución de las características ambientales y de los poblamientos del Mar Menor (Murcia, SE de España). *Anales de Biología,* 12: 53-65.

PÉREZ RUZAFA, A., ROS, J.D., MARCOS, C., BALLESTER, R. & PÉREZ RUZAFA, I.M. 1989. Distribution and biomass of the macrophyte beds in a hypersaline coastal lagoon (the Mar Menor, SE Spain), and its recent evolution following major environmental changes. In: *International Workshop on Posidonia Beds.* (Boudouresque, C.F., Meinesz, A., Fresi, E. & Gravez, V., eds.), 2: 49-62. GIS Posidonie. Marseille.

PÉREZ RUZAFA, I.M. 1989. *Fitobentos de una laguna costera. El Mar Menor.* Ph. D. Thesis. University of Murcia.

PÉREZ, M. 1989. *Fanerógamas marinas en sistemas estuáricos: producción, factores limitantes y algunos aspectos del ciclo de nutrientes.* Ph. D. Thesis. University of Barcelona.

PÉREZ, M., ROMERO, J., DUARTE, C.M. & SAND-JENSEN, K. 1991. Phosphorus limitation of *Cymodocea nodosa* growth. *Mar. Biol.,* 109: 129-133.

POOLE, H.H. & ATKINS, W.R.G. 1929. Photo-electric measurements of submarine illumination throughout the year. *J. Mar. Biol. Assoc. U.K.,* 16: 297-324.

POWELL, G.V.N., KENWORTHY, W.J. & FOURQUREAN, J.W. 1989. Experimental evidence for nutrient limitation of seagrass growth in a tropical estuary with restricted circulation. *Bull. Mar. Sci.,* 44: 324-340.

ROS, J.D. 1987. El Mar Menor: fisiografía y biota acuática de una laguna hipersalina y su evolución reciente. In: *Bases científicas para la protección de los humedales en España*: 237-256. Real Academia de Ciencias Exactas, Fisicas y Naturales. Madrid.

ROS, J.D., PÉREZ RUZAFA, A. & MARCOS, C. 1985. *Posibilidad de aumento de la producción pesquera del Mar Menor.* Rapport for the Cámara de Comercio, Industria y Navegación de Murcia. Murcia.

ROS, J.D., GILABERT, J., GUILLÉN, J.E., MARCOS, C., MARÍN, A., MARTÍNEZ INGLES, A.M., PÉREZ RUZAFA, A. & TERRADOS, J. 1988. *Caracterización física y biológica de los fondos del Mar Menor con vistas a una posible utilización de sedimentos en la regeneración de sus playas, y seguimiento de los cambios que dicha utilización pueda producir en las comunidades bentónicas.* Unpublished final rapport.

Departamento de Biología Animal y Ecología. Universidad de Murcia. Murcia.

ROS, M. & MIRACLE, M.R. 1984. Variación estacional del fitoplancton del Mar Menor y su relación con la de un punto próximo en el Mediterráneo. *Limnetica*, 1: 32-42.

SHORT, F.T., 1983a. The seagrass *Zostera marina* L.: plant morphology and bed structure in relation to sediment ammonium in Izembek Lagoon, Alaska. *Aquat. Botany*, 16: 149-161.

SHORT, F.T. 1983b. The response of interstitial ammonium in eelgrass (*Zostera marina* L.) beds to environmental perturbations. *J. Exp. Mar. Biol. Ecol.*, 68: 195-208.

SHORT, F.T. 1987. Effects of sediment nutrients on seagrasses: literature review and mesocosm experiment. *Aquat. Botany*, 27: 41-57.

SHORT, F.T. & McROY, C.P. 1984. Nitrogen uptake by leaves and roots of the seagrass *Zostera marina* L. *Botanica Marina*, 27: 547-555.

SHORT, F.T. & SHORT, C.A. 1984. The Seagrass Filter: Purification of Estuarine and Coastal Waters. In: *The Estuary as a Filter* (Kennedy, V.S., ed.): 395-413. Academic Press. New York.

SHORT, F.T., DAVIS, M.W., GIBSON, R.A. & ZIMMERMANN, C.F. 1985. Evidence for phosphorus limitation in carbonate sediments of the seagrass *Syringodium filiforme. Est. Coast. Shelf Sci.*, 20: 419-430.

SIMONNEAU, J. 1973. *Mar Menor. Évolution sédimentologique et geochimique recente du remplissage.* Ph. D. Thesis. Paul Sabatier University. Toulouse.

SMITH, R.D., DENNISON, W.C. & ALBERTE, R.S. 1984. Role of seagrass photosynthesis in root aerobic processes. *Plant Physiol.*, 74: 1055-1058.

SORREL, B.K. & DROMGOOLE, F.I. 1988. Oxygen transport in the submerged freshwater macrophyte *Egeria densa* Planch. II. Role of lacunar gas pressures. *Aquat. Bot.,* 31: 93-106.

TERRADOS, J. 1986. *Pigmentos fotosintéticos y producción primaria de las comunidades bentónicas del Mar Menor, Murcia.* Master Thesis. University of Murcia.

TERRADOS, J. 1991. *Crecimiento y producción de las praderas de macrófitos del Mar Menor.* Ph. D. Thesis. University of Murcia.

TERRADOS, J. & ROS, J.D. in press a. Growth and primary production of *Cymodocea nodosa* (Ucria) Ascherson in a Mediterranean coastal lagoon: the Mar Menor (SE Spain). *Aquat. Botany.*

TERRADOS, J. & ROS, J.D. in press b. The influence of temperature on seasonal variation of *Caulerpa prolifera* (Forsskal) Lamouroux photosynthesis and respiration. *J. Exp. Mar. Biol. Ecol.*

TERRADOS, J. & ROS, J.D. in press c. Effects of temperature on seasonal variation of photosynthesis and respiration of *Cymodocea nodosa* (Ucria) Ascherson in a Mediterranean coastal lagoon: the Mar Menor (SE Spain). *Mar. Biol.*

TERRADOS, J. & ROS, J.D. in press d. Limitación por nutrientes del crecimiento de *Cymodocea nodosa* (Ucria) Ascherson en el Mar Menor, Murcia, SE España. *Trab. Inst. Esp. Oceanogr.*

THURSBY, G.B. & HARLIN, M.M. 1982. Leaf-Root interaction in the uptake of ammonia by *Zostera marina. Mar. Biol.,* 72: 109-112.

THURSBY, G.B. & HARLIN, M.M. 1984. Interaction of leaves and roots of *Ruppia maritima* in the uptake of phosphate, ammonia and nitrate. *Mar. Biol.,* 83: 61-67.

WILLIAMS, S.L. 1984a. Decomposition of the tropical macroalga *Caulerpa cupressoides* (West) C.Agardh: field and laboratory studies. *J. Exp. Mar. Biol. Ecol.,* 80: 109-124.

WILLIAMS, S.L. 1984b. Uptake of sediment ammonium and translocation in a marine green macroalga *Caulerpa cupressoides. Limnol. & Oceanogr.,* 29: 374-379.

WILLIAMS, S.L. 1987. Competition between the seagrasses *Thalassia testudinum* and *Syringodium filiforme* in a Caribbean lagoon. *Mar. Ecol. Prog. Ser.,* 35: 91-98.

WILLIAMS, S.L., 1990. Experimental studies of Caribbean seagrass bed development. *Ecol. Monogr.,* 60: 449-469.

ADAPTIVE ADVANTAGES OF THE "SYMBIOSIS" BETWEEN ALGAL CHLOROPLASTS AND SACOGLOSSAN MOLLUSCS

JOANDOMÈNEC ROS * & ARNALDO MARÍN **

* Departament d'Ecologia. Facultat de Biologia. Universitat de Barcelona. Av. Diagonal, 645. 08028 Barcelona. Spain
** Departamento de Biología Animal y Ecología. Facultad de Biología. Universidad de Murcia. Campus de Espinardo. 30100 Murcia. Spain

Received: September 1991.

SUMMARY

All available information is reviewed on the characteristics and the way the "symbiosis" between algal chromoplasts and sacoglossan molluscs works, using the existing bibliography and the authors' own studies. Emphasis is placed on the importance of the food subsidy received by the mollusc, which is one of the various benefits derived from the association, and on the possible benefit obtained from this consortium by the algal species involved and not by the retained chloroplasts, which are, after all, non-autonomic cell organelles. After making various theoretical considerations, and on the basis of the population dynamics of some pairs of alga-sacoglossan species, especially an *Acetabularia acetabulum* and *Elysia timida* population thoroughly studied by the authors, the following conclusions are drawn:

a) The establishment of a symbiotic relationship between sacoglossans and chromoplasts and the different levels of mutual adaptation and efficiency reached, are a direct function of the difficulty (seasonal or otherwise) experienced by the mollusc in obtaining food.

b) Depending on the moment of the biological cycle in which the animal receives it, the energy subsidy provided by the chloroplasts to the host mollusc acts either as a dietary complement, a dietary supplement or a partial substitute of the normally obtained algal food.

c) This subsidy is invested, successively and/or alternatively, in the mollusc's growth (Pg), in the shifting of its reproductive pattern from a lecithotrophic larval development to a direct one (Pr), or in increased exploratory and foraging activity (ΔR). All these entries represent a benefit for the mollusc, in allowing it to be more active, live longer, reproduce more efficiently and/or explore a greater area of the sea bed.

d) When the subsidy involves an energy complement for the sacoglossan species, it is plausible that the mollusc reduces its grazing rate on the alga species, in the same proportion at which it receives the subsidy. In this case the alga species also benefits, which would allow the relation between the alga and the sacoglossan to be classified as mutualistic. In all the other cases the benefit is exclusively for the mollusc.

KEY WORDS: Sacoglossa, chloroplasts, adaptation, symbiosis, *Elysia timida, Acetabularia acetabulum.*

"The close examination of any ecosystem reveals an infinity of comparable [trophic colateral] relations. The most obvious result of its evolution is that, in the end, a certain species forms a very strict relation with another defined species. The process of selection which has led to the establishment and maintenance of such relations has to be explained by some energy economy; in practice, however, the specificity of the relations between different organisms is more notable or more apparent than any general energy consideration, sometimes because it is unknown, and in most cases because once known, it is difficult to evaluate. In any case, the result is that traditional biology has always given more importance to the specificity of the relation than to the transfer or saving of energy that it implies". (MARGALEF, 1974: 543)

"Certain organisms can obtain "second hand" algae from their prey.... Stranger still is the survival, not of whole cells, but of Codium *and* Bryopsis *plastids in* Elysia, Tridachia *and* Placida, *and that these plastids continue to have a functional capacity...". (MARGALEF, 1974:538).*

INTRODUCTION

The Opisthobranchs are a group of marine gastropods which provide numerous examples of molluscs making use of morphological elements belonging to their prey species, such as plant and animal pigments (melanin, carotenoids, pterine and purine granules, etc.), sponge spicules, cnidarian nematocysts, etc. (EDMUNDS, 1966, 1974; THOMPSON, 1960, 1976; HAEFELFINGER, 1969; ROS, 1976, 1977; TODD, 1981; GREENWOOD & MARISCAL, 1984; CATTANEO, 1990; etc.)

The Sacoglossa (=Ascoglossa, Monostichoglossa) are an order of opisthobranchs of benthic habits. They are mainly herbivorous, and some species retain the chloroplasts of the algae that they consume (KAWAGUTI & YAMASU, 1965; TAYLOR, 1967, 1973; GREENE, 1970a; TRENCH, 1975; HINDE, 1983; MARÍN & ROS, 1988; MARÍN, 1991; etc.). The chloroplasts are not immediately digested, but remain active during weeks or even months in the cells of the digestive gland. The mollusc often houses so many chloroplasts that its colour perfectly matches that of the food alga (green, occasionally red), although sometimes these colours come from the ingesta and are seen by transparency, regardless of whether they correspond or not to functional chloroplasts. They may also be due to the animal's own pigments.

It has been known for over a century that some sacoglossans possess chlorophyllic pigments. DE NEGRI & DE NEGRI (1876) observed their presence in *Elysia viridis*, and BRANDT (1885) later described chloroplasts in the same species. Zooxanthellae were later observed in nudibranchs (NAVILLE, 1926, in *Aeolidiella alderi*), and YONGE & NICHOLAS (1940) described them in the sacoglossan *Tridachia crispata* (although they were really dealing with chloroplasts).

In the sixties, the electron microscope enabled easy identification of these organelles in *Plakobranchus ocellatus, Elysia atroviridis* and other sacoglossans (KAWAGUTI, 1941; KAWAGUTI & YAMASU, 1965; KAWAGUTI *et al.*, 1965; etc). Physiological and biochemical studies of various species (which included the measurement of oxygen produced and of the carbon fixed by photosynthesis) began shortly afterwards: TAYLOR (1967), GREENE (1970a, 1970b), TRENCH & SMITH (1970), GREENE & MUSCATINE (1972); etc.

These studies led to the first reviews and theoretical syntheses, which discussed the role of the chloroplast-sacoglossan relationship (TAYLOR, 1973, 1981; GREENE, 1974; TRENCH, 1975; TAYLOR 1973, 1981; HINDE, 1983; etc.). A relatively exhaustive list of the main recent works on the chloroplast-sacoglossan relationship can be found in the articles mentioned above and in WILLIAMS (1986), MARÍN (1988), MARÍN & ROS (1989, in press a, b). The synthesis by HINDE (1983) is particularly thorough.

The presence of plastids in their tissues has led these animals to be called cryptogamic molluscs and "leaves that crawl" (TRENCH, 1975). Many of the anatomical and behavioural characteristics of chloroplast-retaining sacoglossans cannot be explained if not as adaptations to this peculiar association: "Specialised tissues such as the large flat dorsal mantle of *Elysia [viridis]* which contains neat agricultural rows of chloroplasts, salvaged from *Codium*, are... a consequence of the importance of the symbiotic association" (DYER, 1989).

The functionality of the plant organelles enabled the term symbiosis to be used (as mentioned above) by the first researchers to look into this alga-animal relation (KAWAGUTI & YAMASU, 1965; GREENE, 1970a, 1970b; TRENCH, 1975), which has sometimes been qualified as chloroplast symbiosis (HINDE & SMITH,

1974; TRENCH, 1975). Such a symbiosis has always been interpreted as mutualistic (MARGALEF, 1974; AHMADJIAN & PARACER, 1986; DOUGLAS & SMITH, 1989), that is, with mutual benefit to the two species involved.

It has been effectively shown that the photosynthates produced by the chloroplasts are used, at least partly, by the molluscs (GREENE, 1970b; TRENCH *et al.*, 1970; GREENE & MUSCATINE, 1972; HINDE & SMITH, 1975; CLARK *et al.*, 1981; MARÍN & ROS, 1989; etc.). Individuals from different species of chloroplast-retaining sacoglossans live longer without eating than equivalent, non-retaining species, or grow faster if exposed to light than if they live in darkness; the animal consorts thus benefit from this unusual association.

The same cannot be said of the plant consorts, the chloroplasts. They are organelles separated from the algal cells and are, therefore, incapable of reproduction and of increasing their Darwinian fitness (which is one of the premises that has to be fulfilled in a mutualistic symbiosis). Then, strictly speaking, the relationship cannot be qualified as symbiotic (HINDE, 1983).

The same does not occur with the relationship established between cnidarian-eating nudibranchs and their symbiotic zooxanthellae, which are incorporated into the tissues of the predator (NAVILLE, 1926; KEMPF, 1984, 1991; MARÍN & ROS, 1991).

TRENCH (1980) proposed the term "foreign organelle retention" for the sacoglossan-chloroplast association (BLACKBOURN *et al.*, 1973), and CLARK & DeFREESE (1987) suggested "kleptoplasty", both terms in which the mutual benefit of the association is not implicit. "Chloroplast symbiosis" has also been used by some authors (HINDE & SMITH, 1974; CLARK & BUSACCA, 1978; CLARK *et al.*, 1981; WILLIAMS, 1986, etc.), while others use "symbiosis" (with quotation marks: HINDE & SMITH, 1974; MARÍN & ROS, 1989; etc.).

This paper, based on studies carried out by the authors on a score of sacoglossan species from the Spanish Mediterranean coast, contributes data and arguments which demonstrate that the symbiosis (the mutual association from which both partners benefit) is not between the mollusc and the plant organelle, but rather between the mollusc and the food alga.

At least in some situations, the benefit for the algal species lies in the reduction of its predation by the mollusc, in the same proportion in which the retained chloroplasts provide the mollusc with an energy subsidy based on photosynthates which are additional to the food consumed by the latter through the normal herbivory procedure. A very preliminary draft of the present arguments was advanced in MARÍN & ROS (1989).

SACOGLOSSANS AND THEIR ASSOCIATION WITH ALGAL CHLOROPLASTS

The Sacoglossans can be divided into three main morphological types (MORTON, 1979; THOMPSON, 1976; SCHMEKEL & PORTMANN, 1982, GASCOIGNE, 1985; see JENSEN, 1991). The most primitive are the testaceous, shelled forms of the suborders Juliacea, Volvatellacea and Oxynoacea, which apparently do not retain active chloroplasts.

There are two main types of more advanced morphological forms, naked and shell-less, of the suborders Elysiacea and Platyhedylacea, which can be slug-like, with parapodia and without dorsal appendages (families Elysiidae, Plakobranchidae, Boselliidae and Limapontiidae), or with cerata (dorsal papillae; families Polybranchiidae, Hermaeidae, Stiligeridae, Alderiidae, Platyhedylidae). Many of the species from these last two groups house functional

TABLE I. Systematic arrangement of the order Sacoglossa; the two suborders, the five superfamilies and the fourteen generally accepted families (GASCOIGNE, 1985; THOMPSON, 1976; MARCUS, 1982; JENSEN, 1983; LONG, 1983) are shown, along with the main world genera. Those genera are indicated in which chloroplast retention has been reported (full dots) and those in which retention does not exist (open dots, from the species studied). The species listed in standard type are those which, studied by different authors, have been shown to retain chloroplasts. The species listed in italics have been studied or are being studied by the authors (ROS & RODRÍGUEZ, 1985; MARÍN, 1988, 1991; MARÍN & ROS, 1988, 1989, in press a, b) and the preceding symbols indicate that they retain functional chloroplasts (•), that they do not retain them (°) or that it is not known whether they retain them or not (+; also see Table II). Note that many genera have not yet been prospected, and neither have the most primitive species (Juliacea, part of the Oxynoacea), or the most advanced ones, some of them egg-eaters (Alderiidae, Oleidae and part of the Stiligeridae).

SACOGLOSSA Bergh, 1876 (=ASCOGLOSSA v. Ihering, 1876; MONOSTICHOGLOSSA Pagenstecher, 1875)	
CONCHOIDEA Gascoigne, 1985	
Juliacea Boettger, 1963	
Juliidae E. A. Smith, 1885	
Julia Gould, 1862	
° Berthelinia Crosse, 1875	
Edenttellina Burn, 1960	
Tamanovalva Kawaguti & Baba, 1959	
Volvatellacea Marcus 1982	
Volvatellidae Pilsbry, 1895	
Volvatella Pease, 1860	
Cylindrobullidae Thiele, 1931	
Cylindrobulla P. Fisher, 1857	
Ascobulla Marcus, 1972	+ *A. fragilis*
Oxynoacea Boettger, 1963	
Oxynoidae Fischer, 1883	
Oxynoe Rafinesque, 1819	O. antillarum, *O. azuropunctata,* *O. olivacea,* O. panamensis
Lobigeridae Pruvot-Fol, 1954	
° Lobiger Krohn, 1847	° *L. serradifalci*
Lophopleurella Zilch, 1956	
Roburnella Marcus, 1982	
ACONCHOIDEA Gascoigne, 1985	
Elysiacea Bergh, 1872	
Elysiidae H. & A. Adams, 1854	
• Elysia Risso, 1818	E. atroviridis, E. australis, E. cauze, E. chlorotica, • *E. flava,* E. furvacauda • *E. gordanae,* E. hedgpethi, E. maoria, E. subornata, • *E. timida,* *E. translucens,* E. tuca, • *E. viridis*
Achillea Gascoigne, 1985	
Elysiobranchus Pruvot-Fol, 1930	
Elysiella Bergh, 1872	
Elysiopterus Marcus, 1957	
• Thuridilla Bergh, 1872	• *T. hopei*
• Tridachia Deshayes, 1857	T. crispata
• Tridachiella MacFarland, 1924	T. diomedea
Pattyclaya Marcus, 1982	
Plakobranchidae Rang, 1829	
• Plakobranchus Van Hasselt, 1824	P. ianthobapsus, P. ocellatus
Boselliidae, Marcus, 1982	
• Bosellia Trinchese, 1890-91	• *B. mimetica*
Polybranchiidae H. & A. Adams, 1854	
Caliphylla A. Costa, 1865(69)	
Mourgona Marcus, 1970	
Lobifera Pease, 1871	
Polybranchia Pease, 1860	
Cyerce Bergh, 1871	
Hermaeidae H. & A. Adams, 1854	
• Hermaea Lovén, 1844	• *H. bifida,* + *H. paucicirra*
• Costasiella Pruvot-Fol, 1951	C. lilianae
Aplysiopsis Deshayes, 1839-53	
Hermaeina Trinchese, 1874	
Stiligeridae Iredale & O'Donoghue, 1923	
Calliopaea D'Orbigny, 1837	° *C. bellula*
Stiliger Ehrenberg, 1831	
• Placida Trinchese, 1876	+ *P. cremoniana,* • *P. dendritica,* *P. viridis,* + *P. verticillata*
Ercolania Trinchese, 1872	+ *E. caerulea,* • *E. funerea*
° Olea Agersborg, 1923	
Limapontiidae Gray, 1842	
• Limapontia Johnston, 1836	° *L. capitata,* L. depressa
Alderiopsis Baba, 1968	
Alderiidae	
° Alderia Allman, 1846	
Alderella Odhner in Franc, 1968	
Platyhedylacea Rankin, 1979	
Platyhedylidae Salvini-Plawen, 1973	
Platyhedyle Salvini-Plawen, 1973	

chloroplasts from the food algae (Table I).

Almost all the Sacoglossan species are phytophagous suctorial feeders, that ingest suctorially the cytoplasm of caulerpalean algae and of other siphonalean chlorophyta and rhodophyta. For this reason the term chromoplast (coloured plastid) is preferred to chloroplast (green plastid), as some of the organelles retained are from red algae (rhodoplastids; TAYLOR, 1971; MARÍN & ROS, 1988). The term kleptoplast (stolen plastid) has also been used (WAUGH & CLARK, 1986; MARÍN & ROS, 1989).

Besides the morphology of the bucal apparatus, adapted for the perforation of the algal cells and the ulterior suction of their contents (THOMPSON, 1976; CLARK & BUSACCA, 1978; JENSEN, 1980, 1981, 1991; CLARK & DeFREESE, 1987; MARIN, 1988), these molluscs possess anatomical, physiological and behavioural adaptations designed for the efficient exploitation of the retained chloroplasts (Fig. 1). These adaptations range from the structure of the radula and the substitution mechanism of the only row of radular teeth, to the ramification of the digestive gland in the mantle (whose light-catching and gas- and nutrient-exchanging surface is increased by the presence of flaps or by cerata-like dorsal appendages; TAYLOR, 1981); and from the preference for certain intensities of light, to the synchronisation of the life cycle with that of the food alga (MARÍN & ROS, in press a), and to the animal behaviour, including syncopated ventilation movements of the mantle cavity that ensure the rapid renewal of the water

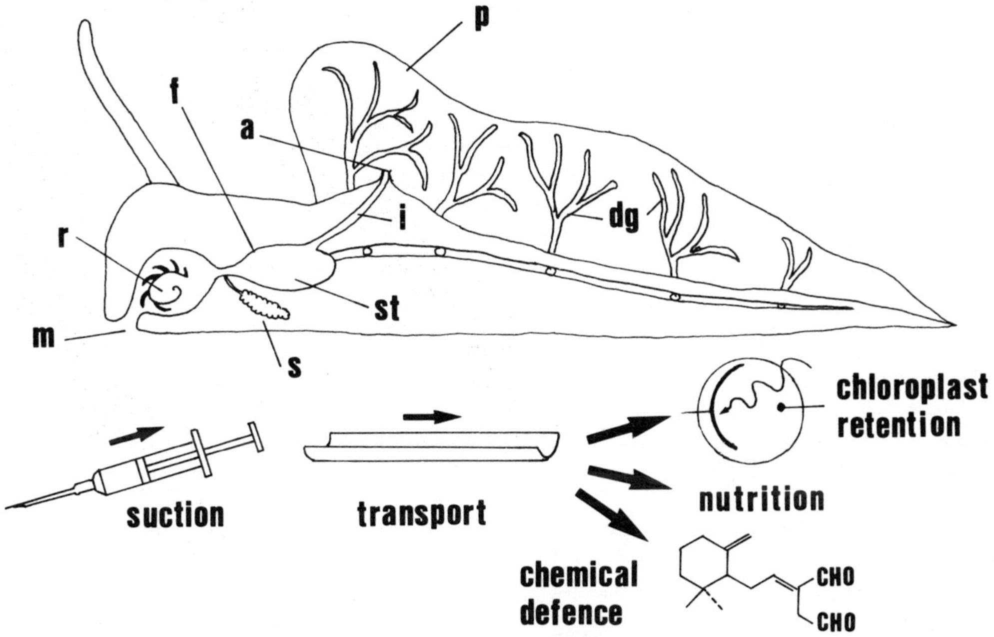

FIGURE I. Diagrammatic representation of the workings of a sacoglossan. Above, parts of the sacoglossan's digestive tube. Below, a functional and symbolic representation of the digestive tube. When it eats, the animal holds the food plant with its lips (m) and perforates the plant cells with the radula (r), while the pharynx (f) works as a suction pump for the stomach (st); also shown are the salival glands (s). The whole system works like a syringe (below). The plant cytoplasm ingested is distributed through the diverticles of the digestive gland (dg) (transport canal below). In the cells of the aforementioned digestive gland, the plant organelles and/or the molecules involved in the defence of the algae are retained intact, to be used later by the mollusc (screen to capture solar radiation and organic molecule below). The plant sap ingested is likewise used as food. The material which is not used is returned to the stomach, from where it is moved to the anus (a) through the intestine (i). This diagram shows a sacoglossan (an Elyisiidae, for example) with lateral parapodia (p), which in the Polybranchiidae and Stiligeridae are substituted for dorsal appendages (cerata).

TABLE II. Sacoglossan species from the south east of Spain studied by the authors, species of algae used as food and functional chloroplast retention in the molluscs. The symbols in the last column indicate that the existence of functional chloroplasts has been confirmed (+), that the mollusc does not retain them (-), or that it has not yet been possible to test for the presence or absence of such organelles (?). No attempt has been made to distinguish different degrees of kleptoplasty (in the sense of CLARK *et al.*, 1990). The references are identified in the bibliography; 0, this paper.

SACOGLOSSAN	ALGAL FOOD	CHLOROPLAST RETENTION
Ascobulla fragilis (Jeffreys, 1856)	Caulerpa prolifera (14, 15)	?
Oxynoe olivacea Rafinesque, 1819	Caulerpa prolifera (14, 15)	- (14, 15)
Lobiger serradifalci (Calcara, 1840)	Caulerpa prolifera (14, 15)	- (14, 15)
Elysia flava Verrill, 1901	Cladophora sp. ? (14, 15), Halimeda simulans (5)	+ (14, 15)
Elysia gordanae Thompson & Jaklin, 1988	Cladophora sp. (14, 15)	+ (14, 15)
Elysia timida (Risso, 1818)	Acetabularia acetabulum (14, 15, 21), A. calyculus (14, 15)	+ (14, 15, 20, 21)
Elysia translucens Pruvot-Fol, 1957	Udotea petiolata (14, 15)	+ (14, 15)
Elysia viridis (Montagu, 1804)	Codium fragile (12), C. bursa, C. tomentosum (2)	+ (12, 15, 26)
Thuridilla hopei (Vérany, 1853)	Cladophora vagabunda (14, 15), Pseudochlorodesmis furcellata, Derbesia tenuissima (0)	+ (14, 15)
Bosellia mimetica Trinchese, 1890	Halimeda tuna (14, 15, 19), H. simulans (5)	+ (14, 15)
Hermaea bifida (Montagu, 1815)	Griffithsia flosculosa (27), Lophosiphonia sp. (14, 15)	+ (14, 15, 27)
Hermaea paucicirra (Pruvot-Fol, 1954)	Codium fragile (22), C. tomentosum (2), Lophosiphonia sp. (14, 15)	+ (14, 15)
Placida cremoniana (Trinchese, 1892)	Unknown	?
Placida dendritica (Alder & Hancock, 1843)	Codium tomentosum, C. fragile (9), Bryopsis plumosa, C. vermilara (14, 15)	+ (10, 14, 15, 26)
Placida verticillata Ortea, 1982	Codium tomentosum, C. adhaerens (17), Bryopsis sp. (14, 15)	?
Placida viridis (Trinchese, 1873)	Bryopsis sp. (14, 15, 23)	+ (14, 15)
Calliopaea bellula D'Orbigny, 1873	Chaetomorpha linum (16)	- (14, 15)
Ercolania caerulea Trinchese, 1892	Valonia utricularis (28)	?
Ercolania funerea (A. Costa, 1867)	Chaetomorpha aerea, C. linum (14, 15), Cladophoropsis (5)	+ (14, 15)
Limapontia capitata (Müller, 1774)	Cladophora rupestris, Enteromorpha sp. (13), C. sericea (7); E. intestinalis, (8), C. arcta, C. vagabunda (14, 15)	- (9, 12, 14, 15)

that comes in contact with it (RAHAT & MONSELISE, 1979; CLARK *et al.*, 1981; ROS & RODRÍGUEZ, 1985; MARÍN & ROS, 1989),

The Sacoglossans are also characterised by a high trophic specialisation (Table II), which turns them into perhaps the most stenophagous of marine herbivores (CLARK & DeFREESE, 1987; MARÍN, 1988; MARÍN & ROS, 1988, JENSEN, 1991; MARÍN *et al.*, 1991), although the most highly evolved forms are egg-eaters. Their development is also characterised by a high proportion of lecithotrophic larval development and encapsulated metamorphosis (CLARK & GOETZFRIED, 1978; CLARK & JENSEN, 1981; ROS, 1981; MARÍN, 1988; MARÍN & ROS, in

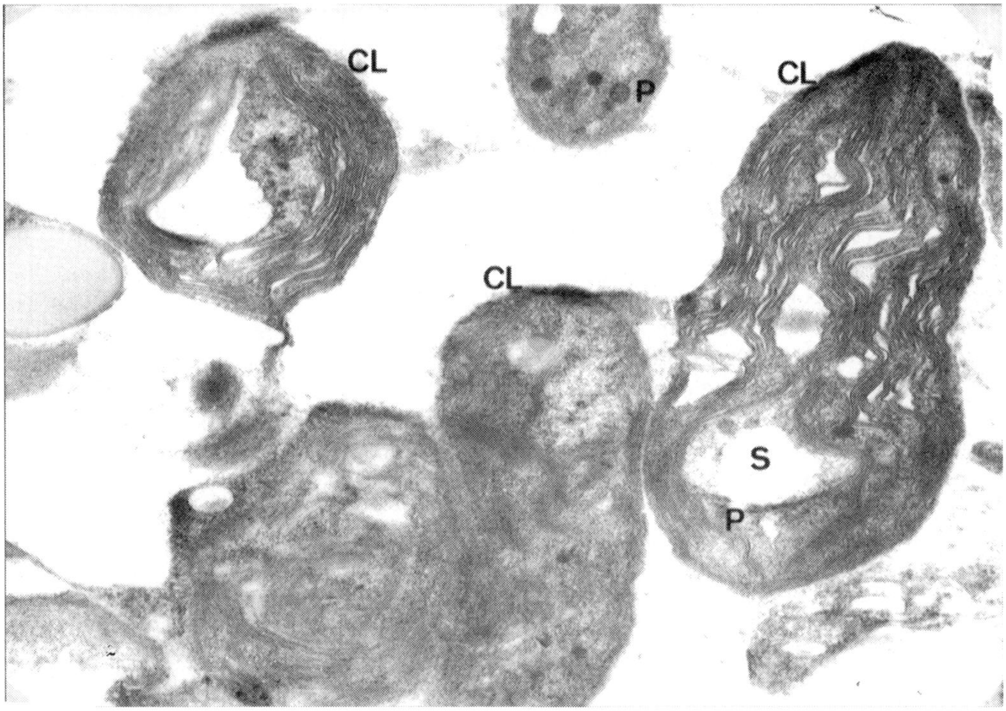

FIGURE 2. Chloroplasts of *Acetabularia acetabulum* (x6000). CL chloroplast; P, pyrenoid; S, starch.

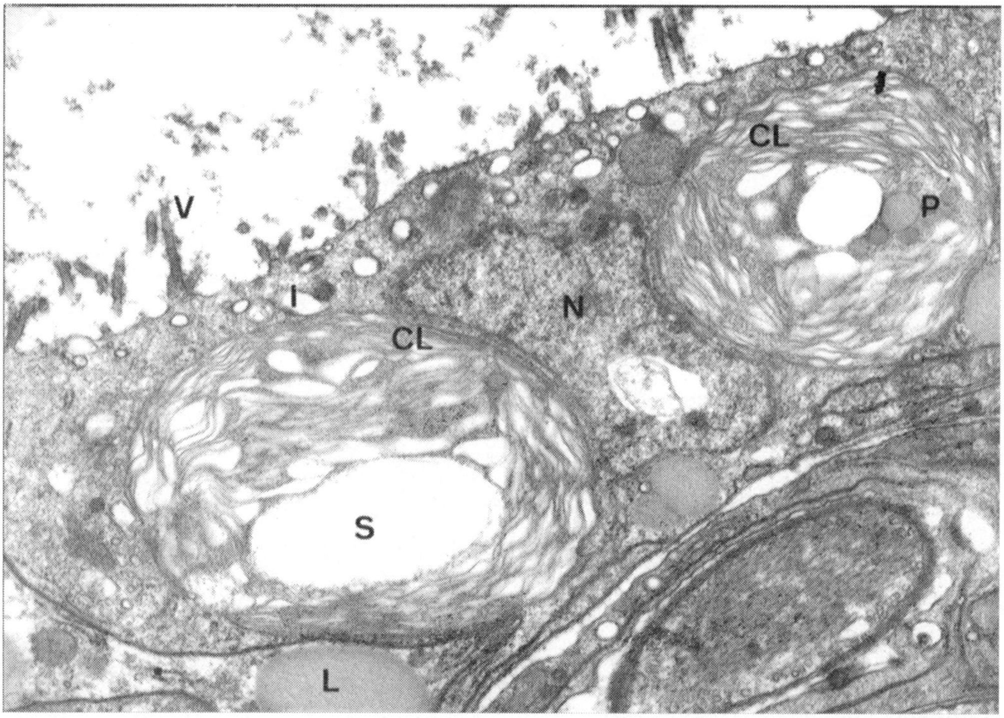

FIGURE 3. Digestive tube of *Elysia timida* showing a digestive cell with intact chloroplasts (x 6000). CL, chloroplast; I, vesicles along luminal border; L, lipid inclusions; N, nucleus; P, pyrenoids; V, microvilli.

press a).

In spite of appearing (like most nudibranch species) in relatively low densities in their natural habitats, as they exploit plant species which are chemically relatively well defended (PAUL & FENICAL, 1986), or calcified (CLARK & DeFREESE, 1987; MARÍN & ROS, in press a), the Sacoglossans perform a key role in the coastal ecosystems in which they live (CLARK & DeFREESE, 1987; MARÍN, 1988), as they make a fraction of the primary production of the marine and lagoon macrophytes (which is otherwise hard to obtain) available to the following trophic levels.

The digestive cells of the mollusc phagocyte the chromoplasts in the algal cytoplasm (Fig. 2; TRENCH, 1975; HINDE, 1973), which either remain free in the cytoplasm of the animal cells or are found inside a vacuole (MARÍN, 1988). The algal organelles and the cytoplasm which they carry stuck to them are surrounded by a host membrane (HAWES, 1979) in whose interior the chloroplasts remain undamaged (HAWES & COBB, 1980). The chromoplasts thus situated (Figs. 3 and 4) remain functional and continue to photosynthesise within the animal cell, in some species for at least as long as three months (HINDE & SMITH, 1974; MARÍN & ROS, 1989; CLARK *et al.*, 1990).

The duration of the photosynthetic capacity of the retained chromoplasts is directly correlated to the difficulty in obtaining food, or in other words, to the degree of calcification of the food alga (Table III; CLARK & DeFREESE, 1987; MARÍN & ROS, 1989). The photosynthetic capacity of the kleptoplasts (generally measured as amount of radioactive carbon fixed per unit of chlorophyll) is usually

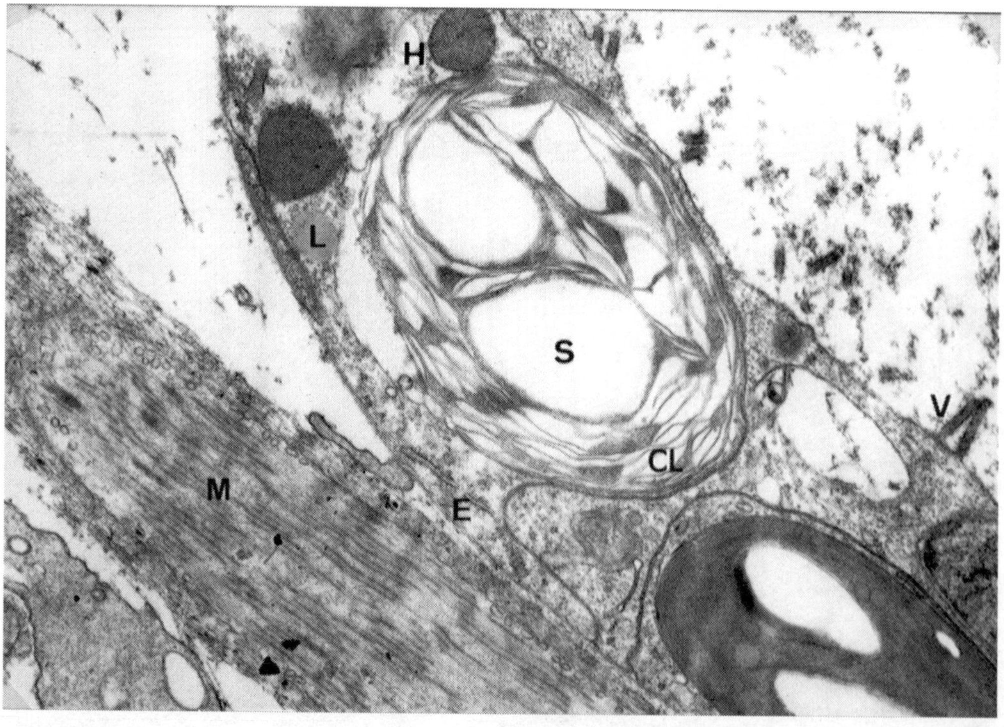

FIGURE 4. Digestive diverticula of *Elysia timida* showing single digestive cells containing intact chloroplasts (x 6000). CL, chloroplast; E, rough endoplasmic reticulum; H, mitochondria; M, smooth muscle fibre; S, starch; V, microvilli.

TABLE III. Ash content and organic matter content of the algal species eaten by the Iberian Sacoglossan species studied by the authors (ROS & RODRÍGUEZ, 1985; MARÍN, 1988; MARÍN & ROS, 1988, 1989). some figures have been estimated by the author of the reference (*) or by us (**). When there are two figures, they correspond to young and old plants, respectively. Note that, in general, the higher the ash content (or the lower the OM content) of the food alga, the longer are functional chromoplasts retained by the sacoglossan species. The references are identified in the bibliography. (Modified from MARÍN & ROS, 1989).

Algal species (eaten by sacoglossan species)	Ash, %	OM, %	Ref.
Caulerpa prolifera (Ascobulla fragilis, Oxyone olivacea, Lobiger serradifalci)	14.9	85.1	5
Griffithsia flosculosa (*Hermaea bifida*)	30*	70*	1
Chaetomorpha aerea (*Ercolania funerea*)	30**	70**	
Chaetomorpha linum (*Calliopaea bellula, E. funerea*)	30**	70**	
Cladophora vagabunda (*Limapontia capitata, Thuridilla hopei*)	31.7*	68.3*	1
Cladophora sericea (*L. capitata, Elysia flava*)	32*	68*	1
Bryopsis plumosa (*Placida dendritica, P. viridis*)	32.4*	67.6*	1
Pseudochlorodesmis furcellata (*Thuridilla hopei*)	32.4*	67.6*	1
Derbesia tenuissima (*T. hopei*)	32.4**	67.6**	1
Enteromorpha intestinalis (*Limapontia capitata*)	39*	61*	1
Cladophora arcta (L. capitata)	40*	60*	1
Cladophora rupestris (*L. capitata, Elysia gordanae*)	40*	60*	1
Codium fragile (Elysia viridis, Placida dendritica, Hermaea paucicirra)	42.0	58.0	18
Acetabularia acetabulum (Elysia timida)	19.2/ /73.1	26.9/ /80.8	4
Codium vermilara (E. viridis, H. paucicirra, P. dendritica, P. verticillata)	47.1/ /52.3	47.7/ /52.9	1
Udotea petiolata (Elysia translucens)	50.0	50.0	1
Codium bursa (E. viridis)	54.7	45.3	1
Codium adhaerens (Placida verticillata)	55*	45*	1
Valonia utricularis (Ercolania coerulea)	64.7	35.3	1
Halimeda tuna (Bosellia mimetica)	55.1/ /80.2	19.8/ /44.9	1
Halimeda simulans (B. mimetica, Elysia flava)	88.0	12.0	5

TABLE IV. Carbon fixation rates in several sacoglossan species, in relation to chlorophyll content and illumination. The references are identified in the bibliography (From MARÍN & ROS, 1989).

	Carbon fixation, $\mu gC\ mgChl^{-1}\ h^{-1}$	Illumination, $\mu E\ m^{-2}\ s^{-1}$	Ref.
A) Mediterranean species			
Elysia timida	144	200	15
Elysia viridis	95	162	11
Elysia translucens	68	100	15
Thuridilla hopei	40	100	15
Bosellia mimetica	45	100	15
Calliopaea bellula	0	-	14
Limapontia capitata	0	-	25
B) Non-Mediterranean species			
Costasiella lilianae	200-300	500	6
Elysia tuca	60	-	24

high (Table IV), sometimes as high as that of the food alga itself (BALLESTEROS, 1984; MARÍN, 1988; MARÍN & ROS, 1989; TERRADOS & ROS, in press).

The oxygen and the photosynthates produced by the chloroplasts (glucose and other carbohydrates, along with lipids and proteins) are released into the animal tissues (TRENCH *et al.*, 1973b; HINDE & SMITH, 1975; HINDE, 1978, 1983; ROS & RODRÍGUEZ, 1985; MARÍN & ROS, 1989). In this way, the mollusc can satisfy some of its metabolic requirements with this organic matter subsidy of autotrophic origin. (Some species have also been shown to absorb amino acids dissolved in water; DeFREESE & CLARK, 1991). In short, the mucus produced in abundance by these gastropods contains as much as 36% of the ^{14}C fixed by the chloroplasts (TRENCH *et al.*, 1970, 1972; HINDE, 1983; MARÍN & ROS, 1989), in the form of mucopolysaccharides or simple glucides such as galactose. The translocation of glucides, proteins and lipids from the chloroplast to the mollusc, and the fact that in some species up to a third of the carbon fixed photosynthetically is metabolised by the animal cells means that photosynthesis is an important source of organic nutrients for the sacoglossans with retained organelles.

The photosynthetic production of these molluscs is of the same order as that of the algae they eat (TRENCH *et al.*, 1973b; MARÍN, 1988; MARÍN & ROS, 1989). The sacoglossans, however, are not completely autotrophic, as in none of the cases studied is the algal herbivory completely substituted by the photosynthesis of the kleptoplasts; the degree of dependence varies according to the species. Through conventional feeding the molluscs also constantly manage to replace the chloroplasts that decay and become inactive. The chromoplasts cannot divide inside the mollusc because they are incapable of synthesising chlorophyll without the enzymes present in the algal cytoplasm. A sacoglossan is constantly renewing its supply of chromoplasts by feeding on new algae, but the feeding frequency seems to be above that which would be necessary if its only objective were the substitution of the non-functional chloroplasts. The sacoglossans, then, depend on photosynthesis for only a part of their metabolic requirements.

Other additional qualitative metabolic benefits have been suggested, apart from the simple quantitative contribution of nutrients. Thus, the molluscs could subsist on an unbalanced or incomplete diet (that of the algal sap or cytoplasm, which lacks the carbohydrates of the algal cell walls and thus constitutes a food input with a low

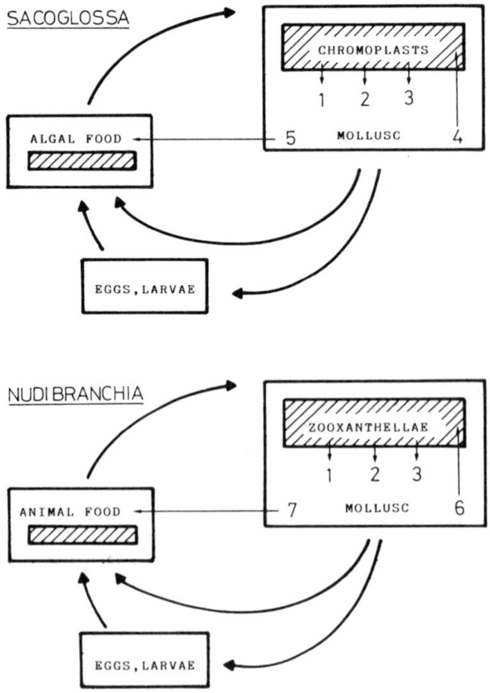

FIGURE 5. Diagrammatic representation of the benefits brought about by the association between algal chromoplasts and Sacoglossans (above) and between zooxanthellae and Nudibranchs (below). The organelles or the plant cells, depending on the case, are incorporated in the mollusc through the feeding process; the animal must eat continuously to ensure their replacement, although in the better established consortia the retention of plastids and/or cells may last weeks and even months. The eggs do not carry chromoplasts or zooxantellae and the larvae or the young acquire them when they first feed. The arrows going from the plant pseudosymbiont to the mollusc (above) or from the symbiont alga to the mollusc (below) indicate different types of benefits for the sacoglossan and the nudibranch, respectively: 1) Defensive (the animal is cryptic on its food species and on occasions the latter's defensive systems are used to the same ends by the mollusc); 2) Translocation of photosynthates (the passage of glucides, proteins and lipids has been demonstrated); 3) Other benefits, more speculative and not proven (which include the passage of oligoelements and vitamins, fixation of nitrogen by the algal consort and facilitation of the calcification processes in the mollusc). In the mollusc-algal consort direction there are no benefits (4) for the organelles, while the zooxanthellae, as well as obtaining a certain protection against external dangers, are dispersed (6), and with this can complete their biological cycle (see MARÍN & ROS, 1991). The benefit for the food species, on the other hand, is more apparent; the algae upon which the sacoglossans prey experience reduced grazing pressure in the same proportion in which its chromoplasts provide the mollusc with photosynthates (5). It has not been shown that the same (7) happens with the animal prey (cnidaria) which have symbiont zooxanthellae, on which dendronotacean and eolidacean feed and obtain, at the same time, secondary symbionts. From this diagram the conclusion can be drawn that only two of these four consortia are true mutualistic symbioses (alga-sacoglossan and zooxanthellae-nudibranch), the association between chromoplasts and sacoglossan and the one between cnidaria and nudibranch not fulfilling the requisite of mutual benefits.

C:N ratio), as the supplementary carbohydrates (in the form of polysaccharides necessary for the foot mucus and the jelly of the egg masses), would be contributed by the chloroplasts (HINDE, 1983). Epidermal uptake of dissolved organic matter has also been demonstrated (DeFREESE & CLARK, 1991), and together with nutritional transfer from kleptoplasts it may confer selective advantages to such specialized herbivores.

The possible passage of oligoelements and vitamins from plastids to host tissues (TAYLOR, 1973) and the fixation of nitrogen have also been mentioned (HINDE, 1983). Like other opisthobranchs, some species of sacoglossans are chemically defended, in many cases by products derived from substances produced by the food plant (GONOR, 1961; ROS, 1976; FAULKNER & GHISELIN, 1983; JENSEN, 1984; PAUL & FENICAL, 1986; CIMINO & SODANO, 1989; CIMINO *et al.*, 1990, 1991, etc.). It is possible - although this has never been proved - that the secondary plant metabolites, precursors of the defensive animal substances, are also produced by the symbiotic chromoplasts.

Finally, it has been suggested that, similarly to what happens in the madrepore-zooxanthella interaction, the sacoglossan-chromoplast symbiosis would facilitate the calcification of the gastropod. While the supposed benefits mentioned above are plausible, even though they have not yet been demonstrated, this last one appears to be highly improbable, as testaceous sacoglossans (the only ones for which calcification would have any significance) do not appear to retain chloroplasts (Table I), and those which retain them lack calcified structures, with the exception of the radular teeth.

DOUGLAS & SMITH (1989) indicate the difficulty of proving the supposed mutual benefits that a "classical" endosymbiotic relation (between two complete species) involves for the two consorts. Likewise, the conceptual difficulty of accepting a benefit for an organelle has been mentioned (HINDE, 1983), while numerous studies exist which demonstrate the benefit for the animal host in the chloroplast-mollusc pseudosymbiosis or parasymbiosis. This is illustrated in figure 5, which shows the three main types of benefits which, at least in some cases, the sacoglossan can obtain by retaining functional chloroplasts in its tissues. For comparison, the benefits of the zooxanthellae-nudibranch symbiosis are also shown.

Firstly, there is the benefit of camouflage: the sacoglossan is cryptic on its food alga (PORTMANN, 1958; RIEDL, 1967; ROS, 1976, 1977, 1978; HINDE, 1983). Secondly, it uses either completely or partially the photosynthates produced by the plastids, which thus represent a complement to its herbivorous diet (GREENE, 1970b; GREENE & MUSCATINE, 1972; TRENCH *et al.*, 1970, 1973a, 1973b; HINDE & SMITH, 1975; HINDE, 1978, 1983; TAYLOR, 1981; MARÍN & ROS, 1989; etc.). Thirdly, benefits have been drawn which, although plausible, have yet to be proved.

As regards the benefits that the algal consort may possibly obtain from the association, we shall first look at some theoretical aspects and then at a specific case which applies to the subject under discussion.

SIGNIFICANCE OF THE FOOD SUBSIDY FROM THE CHLOROPLASTS TO THE SACOGLOSSAN: FOUR POSSIBLE THEORETICAL SITUATIONS

The energy flow associated with the metabolism of a heterotrophic organism may be represented by means of a series of terms and symbols proposed by the IBP (PETRUSEWICZ & MACFADYEN, 1970; CRISP, 1984). We are interested in seeing how the production (the part of the assimilated food that is retained and transformed in growth and/or reproduction) differs in an organism receiving an energy subsidy, such as a sacoglossan that receives additional food inputs, as opposed to a typical, unsubsidised herbivore.

For a typical herbivore (a sacoglossan without kleptoplasts, for example), the following equations can be written (for an explanation of the symbols see Table V):

$$C = R + FU + P;$$

given that

$$P = Pg + Pr,$$

can be written:

$$C = R + FU + Pg + Pr,$$

which gives:

$$P = C - R - FU = Pg + Pr \qquad \text{(Eq. 1)}$$

For the case of the subsidised sacoglossan with functional kleptoplasts, the terms of the previous equations can be broken down as follows (see Table V):

$$C = Ca + Cc, \qquad \text{(Eq.2)}$$

$$R = Rs + Rc + \Delta Rs,$$

$$FU = FUs + FUc,$$

$$P = Pg + Pr + \Delta P.$$

Equation 1 then takes the following form:

$$P = Ca + Cc - Rs - Rc - \Delta Rs - FUs - FUc = \\ = Pg + Pr + \Delta P \qquad \text{(Eq. 3)}$$

Rc is a loss which affects the chloroplast's photosynthetic efficiency, but not the mollusc, so it can therefore be ignored; the same goes, as regards this demonstration, for FUc, although these exudates may become part of Cc (in the form of simple compounds or oligoelements; see Fig. 5). ΔRs, or the metabolic cost of maintaining the chloroplasts, may also be negligible (if, as may be assumed, the organelle-sacoglossan symbiosis is old, both consorts have co-evolved and have experienced mutual adaptations; HINDE, 1983). Furthermore,

TABLE V. Meaning of the terms of the energy flux equation used in the text. (Based on PETRUSEWICZ & MACFADYEN, 1970; see also CRISP, 1984).

C	Consumption	Total intake of food
Ca	Alga consumption	Algal food obtained by herbivory or DOM uptake
Cc	Chromoplast consumption	Nutrient input from retained chromoplasts to mollusc
FU	Rejecta	Egesta of faeces (F) plus excreta (U): urine and other secreted material, such as exudates, mucus, etc.
FUc	Chromoplast "rejecta"	Rejecta of the healthy chromoplasts retained in the mollusc
FUs	Sacoglossan rejecta	Rejecta of the mollusc, including healthy chromoplasts or (more frequently) decaying ones
P	Production	That part of the assimilated food that is retained and incorporated in the biomass of the organism (Pg) or invested in offspring (Pr)
Pg	Growth production	That part of the neat production the animal invests in increasing its own biomass (growth)
Pr	Reproduction production	That part of the neat production the animal invests in offspring
R	Respiration	That part of the assimilated food that is metabolically converted into heat and lost
Rc	Chromoplast respiration	Metabolic losses by the chromoplasts
Rs	Sacoglossan respiration	Metabolic losses by the mollusc
ΔP	Production increase	That part of the production brought about by the additional input of nutrients from the chromoplasts
ΔRs	Respiration increase	That part of the respiration of the mollusc devoted to the chromoplasts maintenance (that is, metabolic investment in "cryptogamic" behaviour by the mollusc; see text)

there is no way of estimating the importance of ΔRs, as those sacoglossan species which are comparable from a metabolic viewpoint are all either retainers of chromoplasts or non-retainers; besides, no species of sacoglossan has been reported which throughout its adult life alternates periods of chromoplast retention with other periods without retention (see below). For this demonstration, then, it can be considered that $\Delta Rs = 0$ or, alternatively, that $\Delta Rs \ll Rs$; in both cases it is negligible.

Equation 3 then becomes:

$$P = Ca + Cc - Rs - FUs = Pg + Pr + \Delta P \qquad \text{(Eq. 4)}$$

If we compare Eq. 4 with Eq. 1 it is easy to see that $Cc = \Delta P$, that is, the energy subsidy that the chloroplasts contribute is transformed by the mollusc into an increase in production, which may be directed towards Pg, Pr, both items, or partly to subsidise Rs (including the hypothetical or negligible ΔRs).

There exist, however, four theoretically possible situations that affect the balance of Equation 2.

i) Cc AS A COMPLEMENT

In the theoretical case proposed, effectively,

$$C = Ca + Cc,$$

which means that C, the mollusc's total consumption (be it estimated either at the individual or at the population level), is provided for by means of the joint input of Ca (algal food obtained by herbivory) and Cc (nutritious contribution or "subsidy" from the chromoplasts; Fig. 6.1). Cc is a *complement* of Ca, so Ca may be less than C (the total energy that the animal must consume for its maintenance). In fact, the greater Cc, the less Ca will be. The relative proportions between Ca and Cc may vary during the life of the same species and from one species to another.

The sacoglossan species that may be assumed to be most adapted to chloroplast retention (those which retain them the

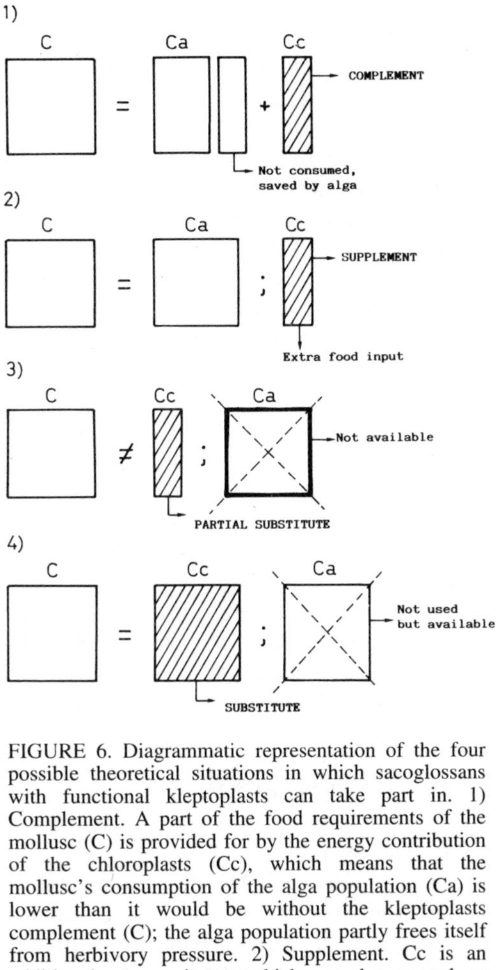

FIGURE 6. Diagrammatic representation of the four possible theoretical situations in which sacoglossans with functional kleptoplasts can take part in. 1) Complement. A part of the food requirements of the mollusc (C) is provided for by the energy contribution of the chloroplasts (Cc), which means that the mollusc's consumption of the alga population (Ca) is lower than it would be without the kleptoplasts complement (C); the alga population partly frees itself from herbivory pressure. 2) Supplement. Cc is an additional energy input, which can be wasted or invested in other metabolic needs; the basic metabolic requirements (C) are totally met by Ca. 3) Partial substitute. When Ca is not available (calcification, seasonal disappearance), the sacoglossan can thrive solely on Cc. As these energy inputs are less than C, the mollusc can only survive for a short time. At least in theory, it would be possible for Cc to provide for the greater part or all of the mollusc's energy requirements for a short period of several months (4. Total substitute), as happens in some cases of true symbiosis with zooxanthellae. No example of this kind has ever been reported for sacoglossans. See the explanation in the text.

longest, and those in which the plant organelles feature the highest production) will have a notable decrease in Ca with respect to C. This is the case of *Elysia viridis* (HINDE & SMITH, 1974), *E. timida* (MARÍN & ROS, 1989) and *Costasiella lilianae* (CLARK *et al.*, 1981), for example (Table IV).

In this case there is a clear benefit for the food alga, which is consumed in smaller quantities (since C > Ca, and a fraction of the potential tithe is saved by the alga). The mollusc, in turn, obtains part of its food without the corresponding foraging effort. It must be assumed that this energy saving is not compensated for by ΔR, the cost of keeping the chloroplasts active (convulsive movements, exposure of the mantle to the sun, migration in search of the light environment more suitable for the chloroplasts, etc.). This means that Rs - the respiration of a subsidised sacoglossan - is not much greater than R - the respiration of a non-subsidised sacoglossan; also applicable here are the considerations about ΔRs and Rs.

Given that the benefit is mutual when Cc is a complement to the diet, the mollusc-alga association in this case can clearly be qualified as a mutualistic symbiosis.

ii) Cc AS A SUPPLEMENT

It can also be supposed that in some situations in which the alga-mollusc association is more recent in evolutionary terms, or is for some reason less efficient (or is in the first evolutionary stages of the establishment of an association such as that described in i), it can be described thus:

$$C = Ca,$$

which means that Cc is a *supplement* to Ca, a subsidy that is obtained in addition to the energy intake necessary for the adequate functioning of the species (Fig. 6.2). This supplement may be "wasted" (in the same way that certain opportunistic, r-strategist species, both plant and animal, excrete an important part of the nutrients that they consume, or they ignore a considerable part of the food that they kill). Alternatively, it may be invested in P or R. In the first case, the energy supplement obtained by the mollusc may be the metabolic spur necessary in order for the sacoglossan species to evolve into a larger form (ΔP is invested in growth, Pg), with greater offspring production or with lecithotrophic or direct development (ΔP is invested in reproduction, Pr), or it may be invested in both entries. It has already been described how the sacoglossans are characterised by a high investment in reproductive output, often with intra- or extracapsular yolk (CLARK & GOETZFRIED, 1978; CLARK *et al.*, 1979; CLARK & JENSEN, 1981; ROS, 1981; MARÍN, 1988). In the second case, Cc may provide for an increase in subsistence metabolism (an increase directed, for example, toward increasing the area explored when food is hard to find or difficult to eat; see below).

When Cc is a supplement the benefit is exclusively for the mollusc, which can grow more, reproduce more or forage more; there is no reduction in the herbivory pressure on the algal species (in fact, in almost all the situations this pressure would increase; larger animals consume more, as do those which forage over a larger area, etc.). In this way the sacoglossan species can, for example, evolve from a strategy which is more r to one which is more K (ROS, 1981, 1982; see below).

iii) Cc AS A PARTIAL SUBSTITUTE

The third possibility is that

$$Ca = 0, \text{ and } C > Cc,$$

which means that the mollusc only has available to it the energy input from the retained chromoplasts, either because the food alga has disappeared (due to its vegetative season having come to an end) or because it is unavailable to the mollusc (due to calcification of the thallus, for example; Fig. 6.3). In both cases, Cc is always less than C, and acts as a *partial substitute* for Ca; the animal subsists for some time in unfavourable conditions thanks to a partial energy subsidy.

There also exists here a benefit for the mollusc. As shown by the multiple survival experiments in starved conditions (HINDE & SMITH, 1975; CLARK *et al.*, 1979; MARÍN & ROS, 1989, etc.), the sacoglossan can survive for an extra period of time (up to three months) with a reduced, but vital, substitute subsidy (Cc).

iv) Cc AS A TOTAL SUBSTITUTE

Finally, a situation would be theoretically possible in which

$$C = Cc, \text{ being } Ca \neq 0,$$

which means that the mollusc would depend entirely on the food intake from the kleptoplasts (which would then represent a *total substitute*), in spite of having food available (Fig. 6.4). This would be the final theoretical stage of the symbiosis, similar to the situation which appears in more advanced consortia (for example, in hermatypic corals and in other cnidaria, or in molluscs such as *Tridacna*, in which, however, the animal consort complements the energy contribution of the symbiotic algae by occasionally ingesting zooplankton or necton; YONGE, 1957, 1963; MARGALEF, 1974).

No species of sacoglossan has been reported with such an independence from the algal food, and given the short life of the isolated chloroplasts and the impossibility of their reproducing inside the mollusc, it may be that such a high level of dependence on the kleptoplasts would be wholly unviable. This could be noted, however, as a theoretical possibility which deserves to be explored in further studies, on particular because *Elysia timida* may be in the way to reaching this fully autonomous state (see below).

THE EXPLOITATION OF SYMBIONT CHLOROPLASTS, A QUESTION OF SPACE AND TIME

Except for certain marginal references, the previous section looked at an alga-sacoglossan relationship invariable in space and time. Although it is not infrequent in ecology to take a classical Newtonian approach to the study of populations, in which the systems of reference are felt to be inertial, the ecosystems are not fixed systems and they undergo changes in space and time (MARGALEF, 1974).

Likewise, the changes observed in the sacoglossan populations that retain chloroplasts, the characteristics of their life cycles, and the degree of chloroplast exploitation all depend on space and time. Examples are given below of how the spatio-temporal scale determines the degree of "symbiosis" of the mollusc with the algal organelles.

i) SPACE

Herbivores, and predators in general, move around to find their food in a space whose size varies according to the characteristics of prey and predator and the abundance of food. The optimum forage area for each predator is that in which the energy used in foraging is not more than the energy obtained from the capture of food. The exploited surface will depend on this equilibrium between energy benefit and cost.

Any factor affecting the success of foraging will have a direct effect on the foraging area; if the success rate falls, the predator will need to move further to make the same number of captures (or perhaps even further still, to make up for the additional energy used in exploring the additional surface area). The final result is, therefore, an increase in the exploration and forage area, which can be described as a bending of space (Fig. 7).

The example of *Elysia timida* and *Acetabularia acetabulum,* which will be explained in greater detail below, fits especially well with this situation. As the alga calcifies, the mollusc's forage area increases; at some point between May (in which the forage area is ten times more than normal) and June (in which this area is virtually infinite), the equilibrium between benefit and energy expenditure gets out of order.

The population of a non-subsidised mollusc would disappear at this moment; the subsidised sacoglossan still survives for a supplementary period.

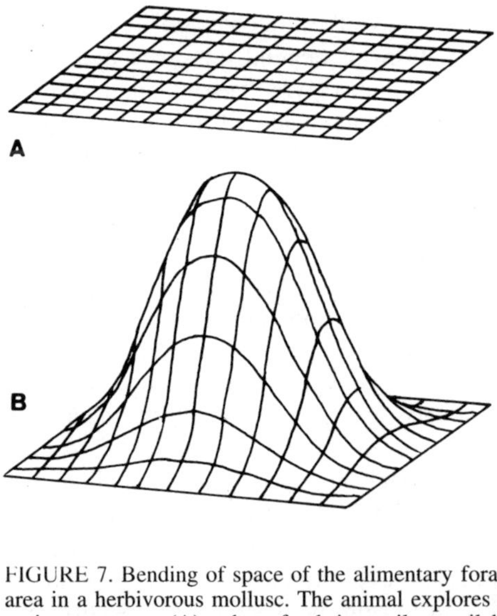

FIGURE 7. Bending of space of the alimentary forage area in a herbivorous mollusc. The animal explores an optimum space (A) when food is easily available and/or uniformly spread out. If, on the other hand, there is a factor which affects successful food capture (the food is not easily obtainable, for example), the effect is of an apparent deformation of the forage area, as the exploration surface necessary to obtain the same amount of food increases (B). If the food disappears entirely, then the exploration surface will become infinite (see Fig. 9D).

ii) TIME

The Mediterranean Sacoglossans feed on annual algae, or on algae which are distinctly seasonal in their production (BALLESTEROS, 1984, 1991; TERRADOS, 1986, 1991; TERRADOS & ROS, in press). The life cycle of these opisthobranchs is adapted to this seasonality of the food. The larval phases of the gastropods, according to the size and number of eggs and the importance of the yolk supply, adopt one of three feeding strategies. Some molluscs produce many small eggs that yield numerous larvae, which then feed for extended periods on the plankton (planktotrophy) and which are usually widely dispersed by ocean currents (THORSON, 1950; THOMPSON, 1976; ROS, 1981; TODD, 1981). Others produce fewer, larger eggs with more yolk - sometimes extracapsular - that develop as yolk-eating larvae, with limited dispersal abilities, either due to a short planktonic phase (lecithotrophy) or to a generally shorter benthic phase (direct development; THORSON, 1950; THOMPSON, 1976;

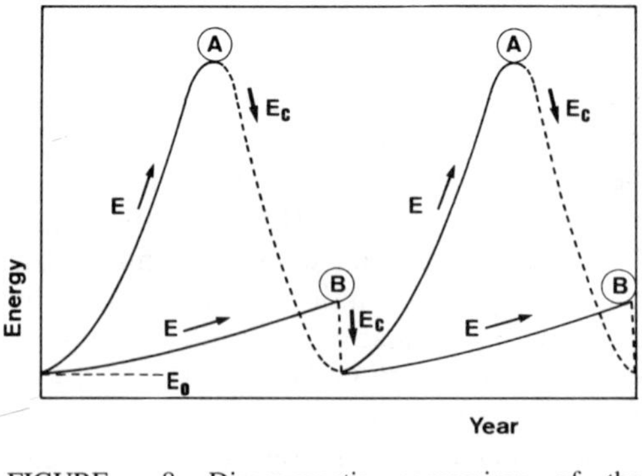

FIGURE 8. Diagrammatic comparison of the reproductive strategies in the sacoglossans, and implication of the use of chloroplast symbionts. A) Case of a typical herbivore, in which there is no energy subsidy from symbiosis (e.g *Oxynoe olivacea*). B) Situation in which the mollusc receives an energy subsidy from the algal symbiont (in this case, chloroplasts, e.g. *Elysia timida*). The continuous line represents the adult's benthic phase, and the broken line the embryonic and larval phases. See the explanation in the text.

ROS, 1981; TODD, 1981).

To sum up, we found some states of dispersion which feed on the means of dispersion (planktotrophic), while others depend on their reserves and do not feed on the means of dispersion (lecithotrophic and direct). The type of embryonary development shown by each sacoglossan species will be greatly influenced by the dynamics of the algal species used as food by the molluscs. The time that the larval phase has to remain in the plankton will depend on the spatial separation between the different food alga populations and on the period of time that these populations are functionally active.

We will now look at which theoretical situations are possible in terms of just how demanding the energy investment is which the mollusc dedicates to larval dispersal. Simplified, it can be expressed thus:

$$E = Eo + Ec,$$

in which E the is energy intake due to the food algae, Eo the threshold energy of each individual sacoglossan species and Ec the energy dedicated to larval dispersal. It is easy to see that the notation from the previous section could be used (E = C; Eo = R + FU + Pg + Pr [partially]; Ec = Pr [partially]), which we have not done so as not to complicate the reasoning that follows below.

It is known that the reproductive output depends directly on the food or energy input; for this case, there are two possible relationships between the energy dedicated by the mollusc to larval dispersal (Ec) and the total energy input (E).

iii) WHEN THERE IS PLENTY OF FOOD, AT LEAST SEASONALLY

When the food input is not limited, although it may have a distinct seasonal character, then E>> Eo, which means that investment in Ec may be important (Fig. 8). This situation corresponds to sacoglossan species that do not retain chloroplasts (like *Oxynoe olivacea*), or that retain them only for periods as short as a few hours, and so do not contribute to the energy economy of the mollusc (such as *Ercolania funerea*; MARÍN, 1988; MARÍN & ROS, 1988). *E. funerea* is a species distributed throughout the world, to be found in estuarine and coastal lagoon environments, where it feeds on the filamentous alga *Chaetomorpha linum*, whose high production is typical of the unstable environments in which it lives. The retention of *C. linum* chloroplasts by *E. funerea* never lasts longer than 24 hours (MARÍN & ROS, 1988). *Oxynoe olivacea*, which has a similarly extensive distribution, feeds on *Caulerpa prolifera* (JENSEN 1980; MARÍN & ROS, 1988), a perennial alga with a large biomass and production (TERRADOS & ROS, in press).

Both sacoglossan species have planktotrophic development, which at first seems to contradict the abundance of food. The environments in which *E. funerea* and *C. linum* live are very fluctuating and unstable, however, which results in a high larval mortality of the mollusc. *O. olivacea* feeds on the fronds of *C. prolifera*, which suffer a seasonal degradation cycle. The fronds develop in spring, then decay and fall in autumn. The mollusc's food disappears for 4-5 months; and this lack of food for such a long period is equivalent to a spatial separation of several kilometres. (If we assume that in one day a sacoglossan larva can move, say, 500m, the potential distance travelled in five months is some 75 km). When the alga used as food by a sacoglossan suffers bursts of high production and then suddenly vanishes, the herbivore anticipates the disappearance of its food species by producing a large quantity of planktonic larvae, capable of remaining in the plankton for prolonged periods. At the same time, the great abundance of the algal food, when present, makes the exploitation of symbiotic chloroplasts unnecessary.

iv) WHEN FOOD IS SCARCE OR DIFFICULT TO OBTAIN

When the food input is limited in adults (by calcification of the cell walls of the algae, for example), then E ≥ Eo, which means that the mollusc needs the active use of the organelles retained to obtain the energy which it must dedicate to Ec (Fig. 8). The type of larval development will depend in these cases on the seasonality of the food algae, which in turn will determine the greater or lesser diffusion of the larval phase. *Bosellia mimetica* feeds on *Halimeda tuna*, an alga which undergoes seasonal calcification (PORTMANN, 1958; MARÍN & ROS, 1988, 1989; BALLESTEROS, 1984). *B. mimetica* retains photosynthetically active chloroplasts and features planktotrophic development, allowing it to use food which is low in organic matter content, but of wide geographical dispersion and markedly seasonal.

An extreme case in this group can be illustrated by *Elysia timida*, which feeds on *Acetabularia acetabulum*, a seasonally calcified algae, low in organic matter content, but also low in geographical dispersion. *E. timida* alternates lecithotrophic development (when the food is scarce) with direct development (when food is abundant), in such a way that it substitutes a high dispersal phase for another of low dispersal, but capable of using abundant local or seasonal food (see below).

In *E. timida*, E has a value which is very close to that of Eo (Fig. 8). The "symbiosis" with chloroplasts allows energy input to be modulated and ensures the minimum number of offspring to colonise the alga's next vegetative cycle. Alternatively, exploitation of the retained chloroplasts prolongs the sacoglossans' benthic phase, which diminishes the uncertainty of the larvae's planktonic dispersion while also allowing greater energy investment in egg size, intra- or extra-capsular yolk and protection of the egg mass, which will ensure reproductive success.

A CASE STUDY: *ACETABULARIA ACETABULUM* AND *ELYSIA TIMIDA*

The theoretical situations explained above could have been successively described for a single species, during a prolonged process of alga-sacoglossan coevolution, up to the present, with different degrees of progressive "perfecting" of the consortium.

It is plausible that the typical herbivory relationship in a primitive sacoglossan could pass without too much difficulty (sturdiness of the plastids, transparency of the gastropod's body) through a first accidental phase of occasional use (as a supplement) of the products synthesised by the captive chloroplasts - it is reasonable to think that a fraction of them could continue photosynthesising during their stay in the digestive tract of the mollusc - and later towards the substitution of a part of the algal food intake for the complement represented by the photosynthates. This would have been encouraged by the progressive alimentary radiation of the group, from eaters of *Caulerpa* and other easily accessible algae, to exploiters of algal species which are more and more chemically defended or calcified, as mentioned above (CLARK & BUSACCA, 1978; CLARK & DeFREESE, 1987; MARÍN & ROS, 1989). It has already been mentioned that the final theoretical situation (in which the photosynthates of the chloroplasts would completely replace the algal food) has never been reported. On the other hand, this situation of total substitution is difficult to reach due to the limitations of the association itself, especially as it is impossible for the chloroplasts to self-reproduce.

What effectively happens is that distinct current sacoglossan species feature different levels of adaptation to the chloroplast symbiosis, and that this diversity seems linked to the taxonomic diversity of the group, as mentioned above (see for example CLARK *et al.*, 1990). Different sacoglossan species which feed on algal species of different energy content, variable perforation difficulty, diverse levels of calcification or defensive substances content, and which obtain photosynthates from them in quantities which are also variable, can coexist in the same environment (BURN, 1966; CLARK & BUSACCA, 1978; TAYLOR, 1981; HINDE, 1983; GASCOIGNE, 1985; MURILLO *et al.*, 1986; CLARK & DeFREESE, 1987; MARÍN & ROS, 1988, 1989; MARÍN, 1988, 1991; CLARK *et al.*, 1990; De FREESE & CLARK, 1991).

The detailed study of a sacoglossan species from the Iberian Mediterranean (MARÍN, 1988; MARÍN & ROS, in press a) has revealed that, throughout the year, the use that the mollusc makes of the algal food and the food input from its kleptoplasts passes through the three phases described here as supplement, complement and partial substitute. Besides, certain biological characteristics of the two species, sacoglossan and alga, can seemingly only be explained as an adaptation to a life in common and to the chloroplast symbiosis.

i) THE SETTING AND THE PLAYERS

The sacoglossan *Elysia timida* is an endemic Mediterranean mollusc that feeds on the green alga *Acetabularia acetabulum* (RAHAT, 1976; ROS & RODRÍGUEZ, 1985; MARÍN & ROS, 1988). The two species, the herbivorous mollusc and its algal food, were studied in their natural environment, which are the shallow, well-lit bottoms of the Mediterranean coast in south-east Spain (Mazarrón Bay), throughout an annual cycle (MARÍN & ROS, in press a). The object of this study was to investigate the main population parameters in natural conditions of both the mollusc and algal species: individual growth, population density and biomass seasonal pattern in *E. timida*, and phenology and calcification in *A. acetabulum*. Also studied were the herbivory dynamics, that is, the effect of the feeding by *Elysia* on the *Acetabularia* population, along with factors affecting photosynthesis in both algae and molluscs and the annual photosynthetic cycle. Furthermore, the development pattern of *E. timida* as a function of food abundance was established (MARÍN & ROS, in press b). Below is a summary of the results found (MARÍN & ROS, in press a).

The dasycladal *Acetabularia acetabulum* is a single-celled green alga widely used in morphogenetic studies (HÄMMERLING, 1930, 1963; BRACHET & BONOTTO, 1966). In Mazarrón Bay, the life cycle of the alga begins in October, when the young single-celled plant has two growing apices, one giving rise to the rhizoidal system that attaches the plant to the substrate, while the other growing apex becomes the erect thallus. From January, the cell walls of the thalli undergo progressive calcification; by April the base of the thallus is very highly calcified, although the apex continues its vegetative growth. The apex does not calcify until June (Fig. 9D), when the vegetative growth of the thallus stops and a cap of gametangial radial chambers is formed. The gametangial rays contain numerous cysts which when mature release pyriform biflagellate isogametes. The calcification process is not simultaneous; algae whose thalluses are calcified to different extents coexist in the same population at any one time, and different *A. acetabulum* populations also feature an unequal calcification period.

Elysia timida features a whole series of adaptations for optimum kleptoplast exploitation. *E. timida*'s photosynthetic production is one of the highest amongst the Mediterranean sacoglossans (Table IV), and in the laboratory it can survive for up to three months when starved (ROS &

RODRÍGUEZ, 1985; MARÍN & ROS, 1988, 1989, in press a and unpublished data). All these adaptations are related to the seasonal dynamics of the food used.

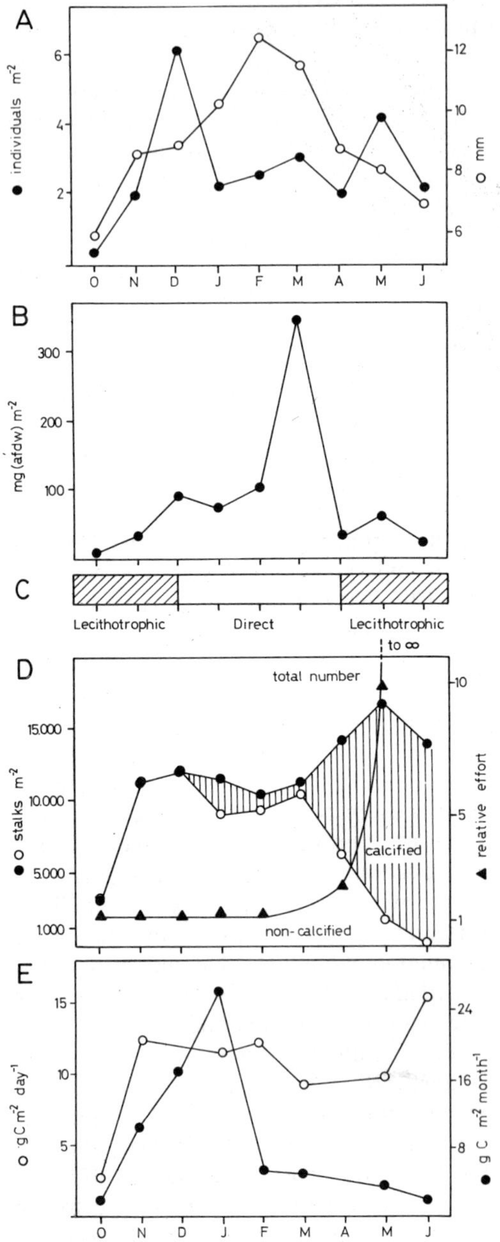

FIGURE 9 Seasonal variation of some population parameters of the *Acetabularia acetabulum* and *Elysia timida* populations from Mazarrón Bay (SE Spanish coast). A) Population density (individuals m^{-2}; full dots and scale at left) and individual length (mm; open dots and scale at right) of *E. timida*. B) *E. timida* biomass, in mg (afdw) m^{-2}. C) Type of development of *E. timida*. D) Population density (stalks m^{-2}; scale at left) of *A. acetabulum* (full dots, total algae; open dots, non-calcified algae) and relative effort by *E. timida* to obtain the same edible algal diet (triangles and scale at right, see Table VI). E) Net carbon fixation rate by *A. acetabulum* (g C m^{-2} day^{-1}; open dots and scale at left) and by *E. timida* (g C m^{-2} $month^{-1}$; full dots and scale at right). (Modified from MARÍN & ROS, in press a).

ii) THE PLAY

The young *E. timida* can be found on shallow rocks when the first *A. acetabulum* stalks appear in October. The population density of the sacoglossan species then increases to a maximum in December and a secondary maximum in May; the molluscs disappear in summer (Fig. 9A). The combination of number of individuals and individual growth produces a maximum sacoglossan biomass out of step with the population maximum (Fig. 9B). Calcification of the alga's cell wall first complicates browsing by *Elysia timida*, then impedes it altogether (Fig. 9D). Between January and June, the sacoglossan moves to the non-calcified apices of the algae (the only ones upon which it can feed), where it eats. From June, when all the thalli are totally calcified, the *E. timida* population falls drastically, until it disappears completely the following month.

Progressive calcification of the alga not only diminishes the food supply, but also makes it more difficult to find (Table VI). If we divide the total number of stalks of *A. acetabulum* found in a given month by the number of non-calcified stalks, we obtain a parameter measuring the energetic effort the mollusc must devote to obtain the same amount of food (F in Table VI). Even in March, virtually all attempts to find and eat a non-calcified stalk would have a positive result. In April, however, to obtain the same amount of food *E. timida* would have

TABLE VI. Some data on the *Acetabularia acetabulum* population in Mazarrón Bay (SE Spain coast) and on its consumption by *Elysia timida*, to show the effect of the calcification of the alga vegetative thallus on the efforts the mollusc must make to eat throughout its life cycle. A) Total number of *Acetabularia* stalks per m^2; B) Number of non-calcified stalks per m^2 ; C) Percentage of eatable *Acetabularia* stalks, (B/A) x 100; D) Number of stalks eaten per m^2 and month; E) Ratio of eatable *Acetabularia*, B/D; F) Relative effort by *Elysia* to obtain the same eatable diet. (Modified from MARÍN & ROS, in press a).

MONTH	A	B	C	D	E	F
October	2,764	2,764	100	89	31.0	1
November	11,333	11,333	100	510	22.2	1
December	12,000	12,000	100	368	32.6	1
January	11,555	9,111	78.8	285	31.9	1.3
February	10,444	9,555	91.4	521	18.3	1.1
March	11,333	10,666	94.1	1,809	5.9	1.1
April	14,222	6,222	43.7	143	43.3	2.3
May	16,950	1,753	10.3	312	5.6	9.7
June	13,998	0	0	103	0	∞

to double its efforts, and in May the effort necessary to obtain the same amount of food would be nearly ten times greater than in autumn and winter. The effort would become infinite in summer, when all *Acetabularia* stalks are calcified (Fig. 9D). This may explain the disappearance of the sacoglossans by mid-summer.

The monthly fixation of carbon by the chloroplasts retained by the sacoglossans amounts to 1% of the algae eaten during one month (Fig. 9E). This fraction is very low, but both the increasing difficulty in finding food as the year goes on and the relatively low abundance of the sacoglossans with respect to the algae must be borne in mind (Fig. 9A, D).

iii) THE PLOT (AS WE INTERPRET IT)

The reproductive pattern of *Elysia timida* is correlated to the annual growth cycle of its algal food. *E. timida* has a type 2 (of THOMPSON, 1976) development (lecithotrophic, with a short, non-feeding planktonic larval phase) in autumn and at the end of spring. In winter and early spring, however, the population changes to a direct, benthic development (type 3 of THOMPSON, 1976, without a planktonic phase; RAHAT, 1976; MARÍN & ROS, 1989, in press b; Fig. 9C). Direct development can be related to mollusc size and food availability: it occurs when the animal reaches maximum size (Fig. 9A) and *Acetabularia* is only partially calcified (Fig. 9D), thus allowing greater energy investment in yolk and accelerated development. Lecithotrophic development is restricted to less favourable periods as regards the energy investment pattern on the part of the mollusc (which is then of smaller average size) and/or the alga (which in summer is highly calcified).

An alternative interpretation is also possible, however: the planktonic lecithotrophic phase permits the mollusc to explore new areas when food is short (dispersion amongst the plankton ensures that a fraction of the larval population settles on stands of *Acetabularia acetabulum* which are not yet calcified). The direct, benthic development ensures that the population does not move far from the food-rich bottoms during the months in which the algal food is still available in large quantities (as it is either slightly calcified or not at all; Fig. 9D).

Recalling the theoretical situations mentioned above, it is easy to see that during its adult life, *Elysia timida* uses the subsidy supplied by the chloroplasts in three of the four ways possible. In its first stage (autumn), with a plentiful food supply, the energy subsidy is used as a complement; in fact, the juveniles of *E. timida* feed on *Cladophora* sp. in its first stage of development, but the chloroplasts of this green alga are not retained for long and the animal switches from the cladophoral to *Acetabularia* as soon (twelve days after its metamorphosis) as its radula enable it to perforate the cell wall of the dasycladal, which is more resistant than that of the cladophoral.

In winter and at the beginning of spring, the sacoglossan reaches its largest size (in the terms used above, Cc is invested in Pg). In synchrony with an efficient use of the chloroplasts (maximum photosynthesis; Fig. 9E), which leads us to believe that, at least in part, the photosynthates are now used as a supplement, it also changes to a metabolically more demanding type of development (Cc is invested in Pr). As spring goes on and summer draws near, the exploration effort of the sacoglossan requires greater energy expenditure, which may in part be subsidised by the kleptoplast supplement (Cc is invested in ΔRs and the mollusc reverts to lecithotrophy).

In summer, when more than 90% of *Acetabularia*'s stalks are calcified and the energy input contributed by the increasingly less successful forage trips dwindles beneath the subsistence threshold (Eo), *Elysia* can hardly survive using the usual herbivory process. Cc must then work as a partial substitute, in such a way that the typical laboratory experiment of keeping cryptogamic gastropods in starvation to determine their survival rate is to an extent reproduced in nature.

Research is needed to find those sacoglossan species (if they exist), which, even when using seasonal food, have an annual life cycle and do not disappear in the unfavourable period. If the possibility exists that Cc is used as total substitute it is in these species where such a situation would almost certainly occur. *Elysia timida* can endure up to three months in starved conditions in the laboratory, as has been stated. This is the nearest this species can go to filling the gap between June and October (Fig. 9A). Although we have not found a sole individual during this period (summer-early autumn) in Mazarrón Bay or in the remainder of the SE Spanish littoral, the possibility of this situation occurring (in this or in other sacoglossan species) cannot be entirely ruled out.

FOREIGN ORGANELLE RETENTION: CHROMOPLAST-SACOGLOSSAN "SYMBIOSIS", BUT ALGA-SACOGLOSSAN SYMBIOSIS

It therefore seems clear that the association between Sacoglossans and chromoplasts is an evolutionary solution directed toward avoiding the excessive dependency of the molluscs on food resources which are poor in energy, seasonally scarce or unobtainable for one reason or another. Other evolutionary solutions that have occurred in this order of opisthobranchs, characterised by notable polyphyletism and plasticity, have meant the appearance of oophagy in some species and families (JENSEN, 1980, 1981; see Table I), the absorption of dissolved organic matter (DeFREESE & CLARK, 1991) or the exploitation of lasting resources by others, that could have implied a secondary loss of chromoplast retention capacity (in *Limapontia capitata*, for instance; CLARK & BUSACCA, 1978).

The exploitation of the chloroplasts in their different degrees of adaptation and efficiency (structural and behavioural modifications of the sacoglossans, duration of the effective retention of the organelles, quantity of organic material synthesised by them and transferred to the mollusc, quality of photosynthates, etc.), seems to be closely correlated to the degree of difficulty in obtaining food (Table III). Moreover, the efficient exploitation of chromoplasts by sacoglossans that retain them requires various adaptations, some of which involve notable modifications of the basic structural, physiological or behavioural plan of a non-cryptogamic sacoglossan mollusc. (Incidentally, it is surprising that JENSEN,

1991, does not even mention the presence of chromoplasts in her comparative study of sacoglossan alimentary systems, as the digestive apparatus is one of the most modified by the chromoplast-sacoglossan association.)

The chromoplasts have also evolved during the sacoglossan evolution, in that they have increased their photosynthetic output (HINDE, 1983). We can clearly therefore speak of an alga-sacoglossan coevolution on a double level: specific (species-to-species) and organismic (chromoplast-to-individual mollusc).

If this is a clear case of coevolution, it is not so evident that the two consorts involved obtain mutual benefit. It has already been mentioned elsewhere that while the physiological consort of the mollusc is the "population" of chloroplasts, these do not have a genetic entity, a necessary premise to guarantee the passing on through inheritance of any evolutionary advantages produced by the mutual association (HINDE, 1983). We feel that it is necessary to pay attention to the mollusc's true consort, which is the food alga. Does the degree of alga-sacoglossan dependence shown, for example, in the high alimentary specificity or in the synchronisation of the biologial cycle of the mollusc with the phenological pattern of the alga, represent sufficient evidence of a mutualistic symbiosis? (If such an association has to be qualified as mutualistic or symbiotic it is not even clear to the specialists; see DOUGLAS & SMITH, 1989, for example. Here we adopt the traditional criteria of symbiosis: mutual benefit through the intimate physiological relationship between two species; CAULLERY, 1952; GEITLER, 1959; MARGALEF, 1974.)

Our hypothesis is that it is precisely the alga species involved in this coevolutionary "triangle" (sacoglossan-chromoplast-alga) which is the subject of the symbiosis. The alga, and not the chloroplast, is the consort which, at least in some cases, obtains benefits from the sacoglossan-chromoplast relation. Similarly, the whole algae are capable of evolving in response to the selection pressure exerted by the cryptogamic molluscs. However, while the molluscs and the cellular organelles have been studied in relation to a possible symbiosis, the same cannot be said of the chromoplasts' "donor" algae (TAYLOR, 1981). It would seem to be of greater interest to work out the plausible evolutionary changes of the algae in response to the aforementioned coevolution.

Two arguments which complement each other can be used to support the hypothesis of the algae as co-symbionts of the sacoglossans. First, an indirect reasoning, for which we shall use again the energy flux notations used above.

We could assume that the chloroplast-retaining sacoglossan has a metabolism higher than that of a sacoglossan species which does not retain them, because of the adaptations mentioned. The degree to which $Rs+\Delta Rs$ is higher than R may mean the partial or total use of the consumption increase brought about by the association. In the first case there would be no net benefit for the mollusc. A strategy in which

$$(Rs+\Delta Rs) - R > \Delta P$$

would be evolutionarily unviable. In other words, the selection of sacoglossans with captive and functional chromoplasts can only be understood if

$$(Rs+\Delta Rs) - R \leq \Delta P.$$

In this case, the symbiosis involves an increase in the evolutionary fitness of the molluscs, whatever they invest ΔP in.

Alternatively, a selection pressure towards the development of chromoplasts which are more resistant, longer lasting or more exploitable by the sacoglossans can only be explained if this involves an increase in the evolutionary fitness of the algae. Only the algae can be the object of selection for more efficient chromoplasts. The chromoplasts would be, then, simple thanatocretic instruments, such as the cnidocysts incorporated by the eolidaceans. The selection (in the sense of the eolidacean selecting actively the more urticant cnidocysts) produced by the eolidacean-hydrarian coevolution acts on these cnidaria and not on the cnidocysts themselves.

To sum up, if sacoglossans and algae are more efficient (the former because they obtain an input of energy additional to the input by typical herbivory, and the latter because they have the sacoglossans as an additional selection factor of more efficient chloroplasts), it is clear that this is a case of mutualistic symbiosis. This does not invalidate the possibility that the alga-sacoglossan coevolution has produced in the plant more efficient systems of defence (chemical, structural, etc.), or avoidance (calcification, seasonality, etc.) and a similarly more efficient herbivory in the animal.

Secondly, it has already been mentioned that, at least in one of the theoretical cases (Cc used as a food complement), the algae benefit in that they are freed from a fraction of the consumption by the herbivores. HINDE (1983) has already expounded this possibility of an alga-mollusc symbiosis. However, although she admits that "It is possible that selection pressure due to grazing by sacoglossans has produced chloroplasts capable of unusually extended survival in animal cells, thus decreasing grazing on the plants", she rapidly dismisses the possibility of such a relation thus: "If the plant from which the chloroplasts are obtained is considered to be the other member of the association, it is immediately clear that the relationship is simply a feeding one, not a mutualistic symbiosis. The animals feed on the algae, causing damage... that may... reduce the reproductive potential of the plants." However, she immediately admits that "There may have been some selective pressure on the food plants tending to produce chloroplasts that could survive longer in the Sacoglossa, thus reducing grazing on the algae." (HINDE, 1983: 104)

This decreased grazing is clearly of benefit to the algal population. The herbivory reduction may have been achieved, alternatively, by other means (those already mentioned of structural or chemical protection, for example). The final result is reduced grazing, achieved thanks to an alga-mollusc association (partial, given that only the chromoplasts are involved in it). A net benefit exists, therefore, which automatically qualifies the association as a mutualistic symbiosis.

As MARGALEF (1974) wrote, "The parallel evolution of the species united in a consortium.... is characterised, from the trophic point of view, by changes related to each other, within a general "progressive" direction. If a [subordinate] species finds itself at the limit of its turnover rate to subsist free, it can manage to subsist by increasing its energy flux.... or by reducing its mortality. Both modifications usually go hand-in-hand and frequently accompany a type of evolution in which a reduction in the turnover rate is permissible. Furthermore, due to the presence of the subordinate species, the principal species can reduce its turnover rate. The biochemical evolution has led to relations which are now indissoluble. It is necessary to consider not only the evolution of the isolated species, but also the evolution of groups of species bound in very intimate relations." (MARGALEF, 1974: 542).

Through the chromoplast "symbiosis", the alga experiences reduced mortality by herbivory, and its turnover rate also falls, as the pressure of exploitation that the mollusc exerts on it diminishes. The mollusc has a fraction of its metabolism subsidised, can invest in energetically expensive strategies and manages also to lengthen its life significantly, all of which also translates into a reduction of its turnover rate. The result is a true alga-sacoglossan symbiosis. The illustration which accompanies the above quotation (Fig. 15-16 in MARGALEF, 1974: 542), applies well to the situation described (MARÍN & ROS, 1989).

The chromoplast-retaining sacoglossans may be difficult experimental subjects when trying to establish reliable energy balances in the laboratory that universally confirm the assertions made here. In fact, this is one of the main difficulties for the precise delimitation of symbiosis situations (see the introductory quotation from MARGALEF, 1974; see also TAYLOR, 1981; HINDE, 1983; DYER, 1989; DOUGLAS & SMITH, 1989). The approach followed here, combining field data of population dynamics with theoretical arguments seems adequate, but leaves many paths open that demand further substantiation.

As the diet of the chromoplast-retaining sacoglossans is strictly herbivore and very specific, it is difficult to compare the metabolism of animals respectively with and without symbiotic chromoplasts (although the use of photosynthesis inhibitors may be a suitable method). On the other hand, the nudibranchs that retain zooxanthellae symbionts of the cnidaria that they consume (RIVA & VICENTE, 1978; RUDMAN, 1981, 1982; MARÍN & ROS, 1991), may be more suitable experimental subjects, as they can be fed with animal tissue either with or without algal symbionts. Whatever the immediate future of the research of the consortia in which sacoglossans and other opisthobranchs are involved, it will presumably provide as many surprises and as much satisfaction as in the last few years.

REFERENCES

AHMADJIAN, V. & PARACER, S. 1986. *Symbiosis*. University Press of New England.

1. BALLESTEROS, E. 1984. *Els vegetals i la zonació litoral: espècies, comunitats i factors que influeixen en la seva distribució*. Ph. D. Thesis. University of Barcelona.

BALLESTEROS, E. 1991. Structure and dynamics of north-western Mediterranean phytobenthic communities: A conceptual model. In: *Homage to Ramon Margalef; or, Why there is such pleasure in studying nature* (J.D. Ros & N. Prat, eds.). *Oecologia aquatica*, 10: 223-242.

2. BALLESTEROS, M. 1980. *Contribución al conocimiento de los Sacoglosos y Nudibranquios (Mollusca: Opisthobranchia)*. Ph. D. Thesis. University of Barcelona.

BLACKBOURN, D.J., TAYLOR, F. J. R. & BLACKBOURN, J. 1973. Foreign organelle retention by ciliates. *J. Protozoo.*, 20: 286-288.

BRACHET, J. & BONOTTO, S. (Eds.) 1970. *Biology of Acetabularia*. Academic Pres. New York.

BRANDT, K. 1985. Über die morphologische und physiologische Bedeutung des Chlorophylla bei Thieren. *Mitt. Zool. Sta. Neapel.*, 4: 191-302.

3. BURN, R. 1966. The opisthobranchs of a Caulerpan microfauna from Fiji. *Proc. malac. Soc. Lond.*, 37: 45-51.

4. CATALAN, J., MARÍ, J. & MILLET, X. 1980. *Estudio de la distribución de Acetabularia mediterranea en las costas baleares y parte norte de Cataluña*. Unpublished mimeo.

CAULLERY, M. 1952. *Parasitism and symbiosis*. Sidgwick & Jackson. London.

CIMINO, G. & SODANO, G. 1989. The chemical ecology of Mediterranean Opisthobranchs. *Chemica Scripta*, 29: 389-394.

CIMINO, G., SODANO, G. & VILLANI, G. 1990. Studio su basi chimiche dei comportamenti biologici dei Molluschi Opisthobranchi. *Atti Congr. Sorrento 29-31 magg. 1987*: 229-240. SIM. Napoli.

CIMINO, G., CRISPINO, A., DI MARZO, V., GAVAGNIN, M. & ROS, J. D. 1990. Oxytoxins, bioactive molecules produced by the marine opisthobranch mollusc *Oxynoe olivacea* from a diet-derived precursor. *Experientia*, 46: 767-770.

CLARK, K. B. & BUSACCA, M. 1978. Feeding specificity and chloroplast retention in four tropical ascoglossa, with a discussion of the extent of chloroplast symbiosis and the evolution of the order. *J. Moll. Stud.*, 44: 272-282.

5. CLARK, K. B. & DeFREESE, D. 1987. Population ecology of Caribbean Ascoglossa (Mollusca: Opisthobranchia): A study of specialized algal herbivores. *Amer. Malac. Bull.*, 5(2): 259-280.

CLARK, K. B. & GOETZFRIED, A. 1978. Zoogeographic influences on development patterns of North Atlantic Ascoglossa and Nudibranchia, with a discussion of factors affecting egg size and number. *J. Moll. Stud.*, 44: 283-294.

CLARK, K. B. & JENSEN, K. R. 1981. A comparison of egg size, capsule size and development patterns in the Order Ascoglossa (Sacoglossa) (Mollusca: Opisthobranchia). *Int. J. Invert. Repr.*, 3: 57-64.

CLARK, K. B., BUSACCA, M. & STIRTS, H. 1979. Nutritional aspects of development of the ascoglossan, *Elysia cauze*. In: *Reproductive ecology of marine invertebrates* (S. F. Stancyk, ed.): 11-14. Columbia University of S. Carolina Press.

CLARK, K.B., JENSEN, K.R. & STIRTS, H.M. 1990. Survey for functional kleptoplasty among West Atlantic Ascoglossa (= Sacoglossa) (Mollusca: Opisthobranchia). *Veliger*, 33: 339-345.

6. CLARK, K. B., JENSEN, K. R., STIRTS, H. M. & FERMIN, C. 1981. Chloroplast symbiosis in a non-Elysiid mollusc, *Costasiella lilianae* (Marcus) (Hermaeidae: Ascoglossa = Sacoglossa): Effects of temperature, light intensity and starvation on carbon fixation rate. *Biol. Bull.*, 160(1): 43-54.

7. COLGAN, N. 1911. Marine Mollusca. Clare Island Survey, part 22. *Proc. R. Ir. Acad.*, 31(2): 0-36.

CRISP, D. J. 1984. Energy flow measurements. In: *Methods for the study of marine benthos* (N. A. Holme & A. D. McIntyre, eds.): 284-372. Blackwell. Oxford.

DeFREESE, E. & CLARK, K.B. 1991. Transepidermal uptake of dissolved free amino acids from seawater by three ascoglossan opisthobranchs. *J. Moll. Stud. (T.E. Thompson Mem. Iss.)*, 57: 65-74.

DE NEGRI, A. & DE NEGRI, G. 1876. Farbstoff aus *Elysia viridis*. *Ber. Deut. Chem. Gesellsch.*, 9: 84.

DOUGLAS, A. E. & SMITH, D. C. 1989. Are endosymbiosis mutualistic? TREE, 4(1): 350-352.

DYER, B. D. 1989. Symbiosis and organismal boundaries. *Amer. Zoo.*, 29: 1085-1093.

EDMUNDS, M. 1966. Protective mechanisms in the Eolidacea (Mollusca: Nudibranchia). *J. Linn. Soc.*, (Zool.), 47: 27-71.

EDMUNDS, M. 1974. *Defence in animals. A survey of anti-predator defences*. Longman. Harlow.

FAULKNER, J. & GHISELIN, M. T. 1983. Chemical defense and evolutionary ecology of dorid nudibranchs and some other opisthobranch gastropods. *Mar. Ecol. prog. ser.*, 13: 295-301.

8. GASCOIGNE, T. 1956. Feeding and reproduction in the Limapontiidae. *Trans. Roy. Soc. Edinb.*, 63: 129-151.

GASCOIGNE, T. 1985. A provisional classification of families of the order Ascoglossa (Gastropoda: Nudibranchiata). *J. moll. Stud.*, 51: 8-22.

GEITLER, L. 1959. Syncyanösen. In: *Handbuch der Pflanzenphysiologie* (W. Ruhland, ed.): 530-536. Springer. Heidelberg.

GONOR, J. J. 1961. Observations on the biology of *Lobiger serradifalci*, a shelled sacoglossan opisthobranch from the Mediterranean. *Vie Milieu*, 12(3): 381-403.

9. GREENE, R. W. 1970a. Symbiosis in sacoglossan opisthobranchs: symbiosis with algal chloroplasts. *Malacologia*, 10(2): 357-368.

GREENE, R. W. 1970b. Symbiosis in sacoglossan opisthobranchs: Translocation of photosynthetic products from chloroplasts to host tissue. *Malacologia*, 10(2): 369-380.

GREENE, R. W. 1974. Sacoglossans and their chloroplast endosymbionts. In: *Symbiosis in the sea* (W. B. Vernberg, ed.): 21-27. University S. Carolina Press. Columbia, South Carolina.

10. GREENE, R. W. & MUSCATINE, L. 1972. Symbiosis in Sacoglossan Opisthobranchs: photosynthetic products of animal-chloroplast associations. *Mar. Biol.*, 14: 253-259.

GREENWOOD, P. G. & MARISCAL, R. N. 1984. The utilization of cnidarian nematocysts by aeolid nudibranchs: nematocyst maintenance and release in *Spurilla*. *Tissue & Cell.*, 16(5): 719-730.

HAEFELFINGER, H. R. 1969. El misterio de los nudibranquios. *Image Roche*, 29: 11-14.

HAWES, C. R. 1979. Ultrastructural aspects of the symbiosis between algal chloroplasts and *Elysia viridis*. *New Phytol.*, 83: 445-450.

HAWES, C. R. & COBB, A. H. 1980. The effects of starvation on the symbiotic chloroplasts in *Elysia viridis*: a fine structural study. *New. Phytol.*, 84: 375-379.

HÄMMERLING, J. 1931. Entwicklung und Formbildungsvermögen von *Acetabularia mediterranea*. *Biol. Zentralblatt.*, 51. 633-647.

HÄMMERLING, J. 1963. Nucleo-cytoplasmic interactions in *Acetabularia* and other cells. *Ann. Rev. Plant Physiol.*, 14: 65-92.

HINDE, R. 1978. The metabolism of photosinthetically fixed carbon by isolated chloroplasts from *Codium fragile* (Chlorophyta: Siphonales) and by *Elysia viridis* (Mollusca: Sacoglossa). *Biol. J. Linn. Soc.*, 10: 329-342.

HINDE, R. 1983. The retention of algal chloroplasts by molluscs. In: *Algal symbiosis. A continuum of interaction strategies* (L. J. Goff., ed.): 97-107. Cambridge University Press. Cambridge

11. HINDE, R. & SMITH, D. C. 1974. "Chloroplast symbiosis" and the extent to which it occurs in Sacoglossa (Gastropoda, Mollusca). *Biol. J. Linn. Soc.*, 6: 349-356.

12. HINDE, R. & SMITH, D. C. 1975. The role of photosynthesis in the nutrition of the mollusc *Elysia viridis*. *Biol. J. Linn. Soc.*, 7: 161-171.

13. JEFFREYS, J. G. 1863-69. *British Conchology or an account of the Mollusca which now inhabit the British Isles and the surrounding seas*, 5: 1-258.

JENSEN, K. R. 1980. A review of Sacoglossan diets, with comparative notes on radular and buccal anatomy. *Malacological Rev.*, 13: 55-77.

JENSEN, K. R. 1981. Observations on feeding methods in some Florida ascoglossans. *J. Moll. Stud.*, 47: 190-199.

JENSEN, K. R. 1983. Preliminary index of species of Ascoglossa. *Opisthobranch Newsletter*, 15(3): 9-16.

JENSEN, K. R. 1984. Defensive behavior and toxicity of ascoglossan opisthobranch *Mourgona germaineae* Marcus. *J. Chem. Ecol.*, 10(3): 475-486.

JENSEN, K. R. 1991. Comparison of alimentary systems in shelled and non-shelled Sacoglossa (Mollusca, Opisthobranchia). *Acta Zool.*, 72(3): 143-150.

KAWAGUTI, S. 1941. Study on invertebrates associating unicellular algae I. *Placobranchus ocellatus* von Hasselt, a nudibranch. *Japan Soc. Prom. Sci. Res. Tokyo*, 1941: 307-308.

KAWAGUTI, S. & YAMASU, T. 1965. Electron microscopy on the symbiosis between an elysioid gastropod and chloroplasts of a green alga. *Biol. J. Okayama Univ.*, 11(3-4): 57-65.

KAWAGUTI, S., YAMAMOTO, M & KAMISHIMA, Y. 1965. Electron microscopy on the symbiosis between blue-green algae and an opisthobranch,

Placobranchus. Proc. Japan Academy, 41(7): 614-617.
KEMPF, S. C. 1984. Symbiosis between the zooxhantella *Symbiodinium (Gymnodinium) microadriaticum* (Freudenthal) and four species of nudibranchs. *Biol. Bull.*, 166: 110-126.
KEMPF, S.C. 1991. A "primitive" symbiosis between the aeolid nudibranch *Berghia verrucicornis* (A. Costa, 1867) and a zooxanthella. *J. Moll. Stud. (T.E. Thompson Mem. Iss.)*, 57: 75-85.
LONG, S. J. 1983. Systematic list of the Opisthobranchia. *Opisthobranch Newsletter*, 154(6): 27-34.
MARCUS, E. 1982. Systematics of the genera of the order Ascoglossa (Gastropoda). *J. Moll. Stud.* suppl. 10: 1-31.
MARGALEF, R. 1974. *Ecología*. Omega. Barcelona.
MARÍN, A. 1988. *Moluscos gasterópodos del sudeste español. Faunística, ecología y estudio de la simbiosis con algas*. Ph. D. Thesis. University of Murcia.
MARÍN, A. 1991. La explotación de cloroplastos en moluscos marinos. *Mundo científico*, 112: 358-366.
14. MARÍN, A. & ROS, J. D. 1988. Los Sacoglosos (Mollusca: Opisthobranchia) del sudeste Ibérico. Catálogo de las especies y presencia de cloroplastos algales en las mismas. *Iberus*, 8(1): 25-49.
15. MARÍN, A. & ROS, J. D. 1989. The Chloroplast-animal association in four Iberian Sacoglossan Opisthobranchs: *Elysia timida, Elysia translucens, Thuridilla hopei* and *Bosellia mimetica*. In: *Topics in Marine Biology* (J. D. Ros, ed.) *Scient. Mar.*, 53(2-3): 429-440.
MARÍN, A. & ROS, J. D. 1991. Presence of intracellular zooxhantellae in Mediterranean Nudibranchs. *J. Moll. Stud. (T.E. Thompson Mem. Iss.)*, 57: 87-101.
MARÍN, A. & ROS, J. D. in press a. Dynamics of a peculiar plant-herbivore relationship: the chlorophycean *Acetabularia acetabulum* and the photosynthetic ascoglossan *Elysia timida. Mar. Biol.*
MARÍN, A. & ROS, J. D. in press b. Ecological and ultrastructural aspects of the development of chloroplasts retention in *Elysia timida. Mar. Ecol. Progr. Ser.*
16. MARÍN, A., ROS, J. D. & PÉREZ-RUZAFA, I. M. 1991. La alimentación de los gasterópodos bentónicos: aplicación de un método inmunológico y primeros resultados. *Actas V Simp. Ibér. Estud. Bentos Marino*, 1: 87-106.
MORTON, J. E. 1979. *Molluscs*. Hutchinson. London.
MURILLO, L., TEMPLADO, J. & TALAVERA, P. 1986. The ascoglossan opisthobranchs of a caulerpan fauna of the Mediterranean Sea. *Shell and Sea Life*, 17(11): 240-243.
NAVILLE, A. 1926. Notes sur les eolidiens. Un eolidien d'eau saumâtre. Origine des nématocystes. Zooxhantelles et homochromie. *Rev. Suisse Zool.*, 33: 251-289.
17. ORTEA, J. A. 1982. Moluscos Opistobranquios de las islas Canarias. Primera parte: Ascoglosos. *Bol. Inst. Esp. Oceanogr.*, 6: 180-199.
18. PAINE, R. T. & VADAS, R. L. 1969. Calorific values of benthic marine algae and their postulated relation to invertebrate food preference. *Mar. Biol.*, 4: 79-86.
PAUL, V. J. & FENICAL, W. 1986. Chemical defense in tropical green algae, order Caulerpales. *Mar. Ecol. Progr. Ser.*, 34: 157-169.
PETRUSEWICZ, K. & MACFADYEN, A. 1970. *Productivity of terrestrial animals: Principles and methods*. Blackwell. Oxford.
19. PORTMANN, A. 1958. *Bosellia mimetica* Trinchese. Opisthobranche retrouvé en Méditerranée. *Vie Milieu*, 9: 74-80.
20. RAHAT, M. 1976. Direct development and symbiotic chloroplasts in *Elysia timida* (Mollusca: Opistobranchia). *Israel J. Zool.*, 25: 186-193.
RAHAT, M. & MONSELISE, E. B. 1979. Photobiology of the chloroplast hosting mollusc *Elysia timida. J. exp. biol.*, 79: 225-233.
RIEDL, R. 1966. *Biologie der Meereshöhlen*. Paul Parey. Hamburg & Berlin.
RIVA, A. & VICENTE, N. 1978. Observation d'algues symbiotiques dans l'organisme d'*Aeolidia alderi, Spurilla neapolitana* et *Favorinus branchialis. Haliotis*, 7: 116-119.
ROS, J. D. 1976. Sistemas de defensa en los opistobranquios. *Oecologia aquatica*, 2: 41-77.
ROS, J. D. 1977. La defensa en los opistobranquios. *Investigación y Ciencia*, 12: 48-60.
ROS, J. D. 1978. La alimentación y el sustrato en los opistobranquios Ibéricos. *Oecologia aquatica*, 3: 153-166.
ROS, J. D. 1981. Desarrollo y estrategias bionómicas en los Opistobranquios. *Oecologia aquatica*, 5: 147-183.
ROS, J. D. 1982. Tipos biológicos en los Opistobranquios. In: *Actas Ier. Simp. Ibér. Est. Bentos Marino*, I (F. X. Niell & J. D. Ros, eds.): 413-440. Universidad de Bilbao.
21. ROS, J. D. & RODRÍGUEZ, J. 1985. La simbiosis algal en *Elysia timida* Risso 1818. Primeros resultados. *Anales de Biología*, 4 (B. Ambiental, 1): 37-47.
RUDMAN, W. B. 1981. The anatomy and biology of alcyonarian-feeding aeolid opisthobranch molluscs and thir development of symbiosis with zooxanthellae. *Zool. J. Linn. Soc.*, 72(3): 219-262.
RUDMAN, W. B. 1982. The taxonomy and biology of further aeolidacean and arminacean nudibranch molluscs with symbiotic zooxanthellae. *Zool. J. Linn. Soc.*, 74: 147-196.
22. SALVAT, F. 1968. *Hermaea paucicirra* Pruvot-Fol, 1953. *Bull. Mus. Nation. Hist. Nat.*, (2)40(2): 358-365.
23. SCHMEKEL, L. & PORTMANN, A. 1982. *Opisthobranchia des Mittelmeeres. Nudibranchia und Sacoglossa*. Springer. Berlin.
24. STIRTS, H. M. & CLARK, K. B. 1980. Effects of temperature on products of symbiotic chloroplasts in *Elysia tuca* Marcus (Opisthobranchia: Ascoglossa). *J. Exp. Mar. Biol. Ecol.*, 43: 39-47.
25. TAYLOR, D. L. 1967. The occurrence and significance of endosymbiotic chloroplasts in the digestive glands of herbivorous opisthobranchs. *J. Phycol.*, 3: 234-235.
26. TAYLOR, D. L. 1968. Chloroplasts as symbiotic organelles in the digestive gland of *Elysia viridis*

(Gastropoda: Opisthobranchia). *J. mar. biol. Ass. U.K.*, 48: 1-15.

27. TAYLOR, D. L. 1971. Symbiosis between the chloroplasts of *Griffithsia flosculosa* (Rhodophyta) and *Hermaea bifida* (Gastropoda: Opisthobranchia). *Pubbl. Staz. Zoo. Nap.*, 39: 116-120.

TAYLOR, D. L. 1973. The cellular interaction of algal-invertebrate symbiosis. *Adv. Mar. Biol.*, 11: 1-56.

TAYLOR, D. L. 1981. Evolutionary impact of intracellular symbiosis. *Ber. Deutsch. Bot. Ges.*, 94: 583-590.

28. TEMPLADO, J., LUQUE, A. A. & MORENO, D. 1988. Nuevas aportaciones al conocimiento de los opistobranquios (Gastropoda: Opisthobranchia) del Sudeste español. *Iberus*, 8(1): 15-23.

TEMPLADO, J., LUQUE, A. A. & ORTEA, J. 1990. A commented check-list of the amphiatlantic Ascoglossa and Nudibranchia (Mollusca: Opisthobranchia). *Atti Congr. Sorrento 29-31 magg. 1987*: 295-326. SIM. Napoli.

TERRADOS, J. 1986. *Pigmentos fotosintéticos y producción primaria de las comunidades macrofitobentónicas del Mar Menor (SE de España).* M. Sci. Thesis. University of Murcia.

TERRADOS, J. 1991. *Crecimiento y producción de las praderas de macrófitos del Mar Menor*. Ph. D. Thesis. University of Murcia.

TERRADOS, J. & ROS, J. D. in press. The influence of temperature on seasonal variation of *Caulerpa prolifera* (Forsskal) Lamouroux photosynthesis and respiration. *J. Exp. Mar. Biol. Ecol.*

THOMPSON, T. E. 1960. Defensive adaptation in opisthobranchs. *J. Mar. Biol. Ass. U. K.*, 39: 123-134.

THOMPSON, T. E. 1976. *Biology of Opisthobranch Molluscs. I.* Ray Society. London.

THORSON, T. E. 1950. Reproductive and larval ecology of marine bottom invertebrates. *Biol. Rev.*, 25(1): 1-45.

TODD, C. D. 1981. The ecology of Nudibranch molluscs. *Oceanogr. Mar. Biol. Ann. Rev.*, 19: 141-234.

TRENCH, R. K. 1975. Of "leaves that crawl": functional chloroplasts in animal cells. *Symp. Soc. Exper. Biol.*, 29: 229-265.

TRENCH, R. K. 1980. Uptake, retention and function of chloroplasts in animal cells. In: *Endocytobiology: Endosymbiosis and cell biology, a synthesis of recent research* (W. Schwemmler & H. E. A. Schenk, eds.): 703-727. De Gruyter. Berlin.

TRENCH, R. K. & SMITH, D. C. 1970. Synthesis of pigment in symbiotic chloroplasts. *Nature*, 227(5254): 196-197.

TRENCH, R. K., BOYLE, J. E. & SMITH, D. C. 1973a. The association between chloroplasts of *Codium fragile* and the mollusc *E. viridis*. I. Characteristics of isolated *Codium* chloroplasts. *Proc. R. Soc. Lond.*, series B, 184: 51-61.

TRENCH, R. K., BOYLE, J. E. & SMITH, D. C. 1973b. The association between chloroplasts of *Codium fragile* and the mollusc *E. viridis*. I. Chloroplast ultrastructure and photosynthetic carbon fixation in *E. viridis*. *Proc. R. Soc. Lond.*, series B, 184: 63-81.

TRENCH, M. E., TRENCH, R. K. & MUSCATINE, L. 1970. Utilization of photosynthetic products of symbiotic chloroplasts in mucus synthesis by *Placobranchus ianthobapsus* (Gould), Opisthobranchia, Sacoglossa. *Comp. Biochem. Physiol.*, 37: 113-117.

WAUGH, G. R. & CLARK, K. B. 1986. Seasonal and geographic variation in chlorophyll level of *Elysia tuca* (Ascoglossa: Opisthobranchia). *Mar. Biol.*, 92: 483-487.

WILLIAMS, M. L. 1986. *Metabolic studies of chloroplast symbiosis*. Ph. D. Thesis. Trent Polytechnic.

YONGE, C. M. 1957. Symbiosis. In: *Treatise on Marine Ecology and Paleoecology* (J. V. Hedgpeth, ed.). *Geol. Soc. America*, 167: 429-442.

YONGE, C. M. 1963. The biology of coral reefs. In: *Advances in marine biology*, 1 (F. S. Russell, ed.): 209-260. Academic Press. New York.

YONGE, C. M. & NICHOLAS, H. M. 1940. Structure and function of the gut and symbiosis with zooxanthellae in *Tridachia crispata* (Oerst.). *Bgh. Pap. Tortugas Lab.*, 32: 287-301.

Homage to Ramon Margalef; or, Why there is such pleasure in studying nature (J. D. Ros & N. Prat, eds.). 1991. *Oecologia aquatica*, 10: 299-316

PLANKTONIC GRADIENTS ALONG A MEDITERRANEAN SEA CAVE

MANUEL PALAU, CARME CORNET, TECLA RIERA & MIKEL ZABALA

Departament d'Ecologia. Facultat de Biologia. Universitat de Barcelona. Avgda. Diagonal, 645. 08028 Barcelona.

Received: November 1991

SUMMARY

The existence of gradients in the composition of planktonic communities through a 50 m long submarine cave of the Medes Islands (NW Mediterranean) has been verified. In spite of an active water exchange (previously evaluated turnover ranges between 12 and 24 hours), strong gradients, persisting throughout all year, were found in both phyto- and zooplanktonic fractions. Numbers of both individuals and taxa of diatoms, dinoflagellates, copepods, and meroplanktonic stages of benthic invertebrates decrease steeply from the entrance of the cave to the inner parts, thus producing an almost total absence of plankton at the end of the cave. On the other hand, benthic diatoms, fungi and spirorbida - all benthic groups accidentally sampled - neither decrease nor even increase to the inner stations, which suggests a tolerance to the "cave effect". These results are so clear-cut that by themselves they can explain the puzzling decay of suspension-feeders among the benthic fauna from the inner parts of submarine caves in two ways : i) scarcity of benthic larvae for settlement, and ii) scarcity of phyto- and zooplankton for food.

KEY WORDS: Plankton, caves, gradients, W Mediterranean.

INTRODUCTION

Studies dealing with benthic fauna from worldwide submarine caves (see HARMELIN *et al.*, 1985, for references) point to the same general feature: a strong gradient in both benthic species richness and benthic biomass along the outside-inside axis; the deeper inside the cave the poorer the fauna. In spite of the universality of this pattern, which seems to suggest powerful and conspicuous causes, our knowledge of the subject has progressed very slowly.

Three kinds of causes have been advanced, all of them supported by very poor experimental evidence : 1) strong physical gradients inhibiting benthic survival; 2) trophic stress by food depletion, and 3) water stagnation or reduced water flow which will be at the origin of causes 1 and/or 2 (CINELLI *et al.*, 1977; BALDUZZI *et al.*, 1980, HARMELIN *et al.*, 1985; GILI *et al.*, 1986; FICHEZ, 1989, 1990, 1991b, d; ZABALA *et al.*, 1989).

Physical gradients (i.e. salinity, temperature, oxygen content), if present, are unlikely to be so strong as to prevent, by themselves, the survival of benthic biota (GILI, *et al.*, 1986). Light, the most conspicuous of these factors, can explain the disappearance of the benthic flora but not - or not at all- the decay of the benthic fauna (RIEDL, 1966; CINELLI *et al.*, 1977; BALDUZZI *et al.*, 1989).

Suspension-feeders and sediment-feeders do not seem to be very dependent on light if suspended materials are available to them.

The generally accepted explanation assumes caves to be permanent strong oligotrophic systems (FICHEZ, 1989, 1991a, b, c, d) where food supply is reduced by both planktonic and POM depletion.

Reduced water turnover may be the origin of oligotrophy either by reducing food inputs or by enhancing sedimentation (FICHEZ, 1989; 1991b, d).

Unfortunately, studies dealing with the models of water circulation inside caves are scarce. Persistent thermic gradients have been invoked as evidence of long water stagnation (POULIQUEN, 1971; PASSELAIGUE & BOURDILLON, 1985; FICHEZ, 1989). The persistence of benthic fauna throughout tunnels, where water movement is assumed to be important, was proposed as further indirect evidence of the importance of this parameter (HARMELIN, 1969). Nevertheless, flow measurements by means of plaster balls showed patterns of moderate (BALDUZZI *et al.*, 1989) or negligible decay of water movement inside caves relative to outer waters or to tunnels (ZABALA *et al.*, 1989).

All the hypotheses summarized above rest on the assumption that benthic life inside the caves is limited by adult mortality. But alternative hypotheses can be formulated on the basis of stresses for the survival of larval stages. There is evidences that settlement is severely reduced inside submarine caves (HARMELIN, 1980; GARCÍA & ZABALA, in prep.) but at present it is not known whether this phenomenon is due to: 1) larval scarcity inside caves due to larval behaviour, mortality, predation or sedimentation; 2) settlement failure even if larvae are

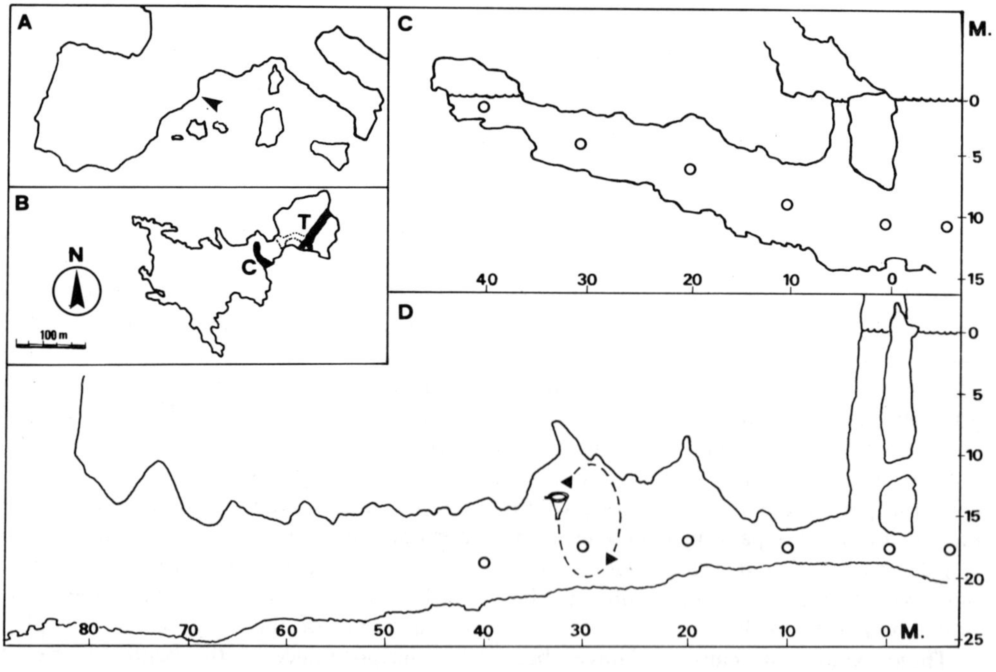

FIGURE 1. Sketch of study cave and tunnel. A) General location in the Western Mediterranean. B) Cave and tunnel situation in the Meda Petita island. C) Diagrammatic section of the cave showing sampling points (open circles). D) Diagrammatic section of the tunnel showing sampling points (open circles) and an example of the circular trajectories followed for zooplanktonic samples.

abundant, or 3) post-settlement mortality in the early stages of development.

To date, little attention has been paid to the study of planktonic communities living inside submarine caves. Only indirect estimates are available from values of chlorophyll and POM (FICHEZ, 1989, 1990; ZABALA *et al.*, 1989) or volume and size distribution of suspended particles as measured by Coulter counters (GILI *et al.*, 1986). But taxonomic details, based on direct microscopic examination, have rarely been reported. More precisely, the abundance of meroplanktonic larval stages of benthic fauna inside caves has not been evaluated.

The aim of this paper is to study the structure of planktonic communities along the axis of a cave, searching for the existence of a similar gradient to those previously reported for benthic communities (GILI *et al.*, 1986). As a part of the same project, an experiment on settlement on panels anchored outside and inside this cave was carried out simultaneously (GARCÍA & ZABALA, in prep.). The hypothesis invoked here assumes that a strong and persistent decay in phyto- and zooplanktonic communities inside the cave can explain, in itself, the equivalent decay in benthic fauna, in two ways: 1) food depletion, because phyto- and zooplanktonic organisms are the main trophic sources for the suspension-feeder dominant benthic biota, and 2) strongly reduced settlement of benthic organisms is due to the scarcity of meroplanktonic larvae.

MATERIAL AND METHODS

THE SITE

The study was carried out in a cave and a tunnel located in the Medes Islands (Catalonia, NW Mediterranean), where previous studies on benthic fauna, water flow and trophic-related parameters have been undertaken (GILI *et al.*, 1986; ZABALA *et al.*, 1989).

The cave is a karstic hole, roughly cylindrical in shape, less than 50 m long, 4 - 7 m in diameter, about 756 m^3 in volume (GILI *et al.*, 1986). It opens to the east side of the Meda Xica Island between 8 (upper) and 12 (lower) m depth. For 35 m after the entrance the "tube" runs straight E-W and rises slowly (the boulder-rocky bottom rises from a depth of 12 m to a depth of 6 m). The inner 10 m are characterized by a northward turn (45°) followed by a sudden increase in the bottom slope, which rises from a depth of 6 m to reach the sea level just at the end of the cave where an air-filled room develops (Fig. 1C). Light penetration in the cave displays a sharp gradient, 99 % of PAR disappearing within the first 10 m from the opening (GILI *et al.*, 1986).

One of the two mouths of the tunnel opens less than 50 m from the cave at similar orientation and depths (upper 16 m, lower 19 m; Fig. 1D). The tunnel is a roughly straight tube 83 m long, 8-12 m width, 4-9 m height, directed SW-NE and slowly descending to the northern mouth, which opens at 14 (upper) - 23 (lower) m depth (BIBILONI *et al.*, 1984; ZABALA *et al.*, 1989).

Macrobenthic fauna from the cave display the pattern of decay (GILI *et al.*, 1986) which characterizes submarine caves studied worldwide (see HARMELIN *et al.*, 1985 for references). Benthic assemblages of the tunnel do not show such a clear pattern of decay, at least from the point of view of biomass. Even if there is a conspicuous zonation with both species substitution and a decrease in species number, organic covering of the walls remains total (more than 100 %) throughout the tunnel (with the sole exception of the high vaults in the roof; BIBILONI *et al.*, 1984).

Water flow along the cave has been estimated as active enough to ensure the turnover of the whole mass of water enclosed in the cave every 12-24 h (RIERA *et al.*, 1985; ZABALA *et al.*, 1989). Moreover, when samples of water of equivalent depths from outside and inside the cave were compared for hydrographic parameters such as temperature, salinity or oxygen content, a surprising equivalence was found throughout the year (GILI *et al.*, 1986). In this case the common annual cycle described for the neritic waters of this region (PALAU, 1988; MARGALEF, 1964; JACQUES, 1969), with a stratified season during summer and a mixed season during winter, should be applicable to the cave.

Although water flow through tunnels is assumed to be higher than flow through caves, the only empirical evidence reported to date from this tunnel relative to the cave does not support with this assumption (ZABALA *et al.*, 1989).

Phytoplanktonic communities from waters surrounding the Medes Islands were studied by PALAU (1988). Changes in the composition of zooplankton over two 24-h cycles were also reported from a station just above the entrance to the cave (BARANGÉ, 1988; BARANGÉ & GILI, 1988). Previous data on chlorophyllic pigment concentration seem to support the expectation that phytoplankton would decrease inside the cave (ZABALA *et al.*, 1989). But data provided by Coulter counter on suspended particles offered a much more homogeneous pattern (GILI *et al.*, 1986).

SAMPLING

Five sampling points were established at 10 m intervals along the sagital axis of the cave. An additional outer point, 10 m from the entrance, was chosen as a control. For practical purposes these points will be referred to by their distance from the mouth of the cave: -10 (outer), 0 (mouth), 10, 20, 30 and 40 (at 10, 20, 30 and 40 m from the entrance; Fig. 1). All six points were sampled with SCUBA-diving around midday, 5 times at different seasons between years 1987 and 1989 (6-VI-87 with two replicates, 29-VI-88, 6-VII-88, 13-X-88, 24-I-89). The same procedure was repeated inside the tunnel once on 29-VI-88. Scale differences between phyto- and zooplanktonic fractions led us to choose two separate procedures of sampling which were both handled by SCUBA-divers. Phytoplankton was sampled by aspiration of 125 cm^3 from the central point of the cave section at each sampling point with a polypropylene syringe. The sample was fixed with lugol solution at a final concentration of 1%. Zooplankton was sampled by swimming around the sampling point for 3 minutes with a 74 μm mesh, 314 cm^2 in section (sample volume around 3.14 m^3) following a circular trajectory only 30 cm away from the walls and roof (Fig. 1). Samples were immediately fixed with formaldehyde at a final concentration of 5%.

STUDY

A few weeks after sampling, phytoplankton was studied through a WILD M40 inverted microscope (UTERMÖHL, 1958) after a 48-h sedimentation procedure. All cells present in a count chamber (100 cm^3) were counted (x 125) for big and rare species; but only a half chamber was surveyed for ciliates and rare and small cells. For each sample, an aliquot (0.3 to 1.6 cm^3 in volume) was counted at a higher magnification (x 600) for commonest cells. Ocasionally, when the characteristics of samples seemed to demand it, larger (6.4 cm^3 volume; x 150 magnification) or smaller (0.016 cm^3; x 600) aliquots were additionally censused. As total counts per sample always ranged between 200 and 700 cells, a noticeable accuracy with error around ±10 % was reached, and confidence level was estimated to be 95 % (cf. LUND *et al.*, 1958).

Zooplankton was totally censused with a M4 WILD stereoscopic microscope (x4 -

x40) and organisms were sorted into large groups of heterogeneous taxonomic level (from family to class). In particular, Copepods account for the whole population (including copepodite and nauplii stages).

STATISTICAL PROCEDURES

Because of taxonomic inaccuracies, diversity estimations (SHANNON-WEAVER, 1963) and evenness estimations (PIELOU, 1966 a,b) were carried out on a fraction of the planktonic community (only the cells wholly classified at the species level) comprising diatoms, armoured dinoflagellates, chlorophyceans, cryptophyceans, cyanophyceans, euglenophyceans and silicoflagellates.

Multivariate analyses were carried out over Q mode matrix (LEGENDRE & LEGENDRE, 1979). Principal Component Analyses (PCA) were made over the correlation matrix of transformed [x=log (x+1)] abundance numbers. Only the commonest species, which were present in a large percent of samples, were used in order to avoid the bias produced by double zeros.

The statistical significance of differences between the slope of curves generated when abundance numbers are plotted against distance from the entrance of the cave, were determined by a *F*-Snedecor test (SOKAL & ROHLF, 1969).

RESULTS

PHYTOPLANKTON

183 taxa were identified, mainly at the species level, among all samples. Some species (Table I) are new additions to the regional catalogue (PALAU, 1988).

Cell density in "outer" waters (station -10) was always between 519 and 2753 cell cm^{-3}, a rank which agrees well with data previously recorded in this area in an annual cycle (PALAU, 1988). Nanoplanktonic flagellates and monads usually accounted for the main fraction (Fig. 2A). Diatoms and large dinoflagellates were always lower in number; among the former, opportunistic taxa such as *Chaetoceros* spp., *Leptocylindrus danicus* and *Nitzschia delicatissima* were strikingly dominant during periods of large proliferation (VI-87, VI-88, X-88). Only at these moments were diatoms more abundant than dinoflagellates. The reverse was observed among the poorest samples (VII-88, I-89). These proliferations agree well both in time and structure, but densities were up to 2 times higher than those reported for the north Catalonian coastal area (MARGALEF, 1964; JACQUES, 1969; PALAU, 1988). Cryptophyceans were minor components which reached their highest values in summer and winter (Fig. 2A). Planktonic ciliates showed densities very close to the annual regional mean, and also similar temporal patterns (PALAU, 1988).

Along the cave a clear gradient of phytoplankton cell numbers arises. From outside to inside, the same pattern of decimation persists throughout all year, both when planktonic assemblages are considered as a whole (Fig. 3A) and as separate taxocoenoses of diatoms, dinoflagellates and ciliates (Fig. 4 A-C). In association with decay of nanoplanktonic cell density, species richness drops overwhelmingly toward the inner end of the cave (Fig. 5 A, B).

No "cave-dwelling" phytoplankton (with cave preference) was found. Occasionally, the diatoms *Nitzschia closterium* and *Scrippsiella trochoidea* were found in higher densities in the medium cave zone than in outer waters. A moderate (0.36 cell cm^{-3}) proliferation of *Heterodinium* cf. *milneri*, a typical representative of dark deep plankton (SOURNIA, 1982), was also detected at this area during October 1988. Moreover, the existence of fungi, which were usually provided with conidiophora (maximum number 1755 spores cm^{-3}) in

TABLE I. Phytoplankters censused in this study. Species marked with an asterisk were not previously recorded in the Medes Islands area.

BACILLARIOPHYCEAE

Achnanthes sp.
Amphora spp. *
Amphiprora alata Kütz. *
A. gigantea (O'Meara) Cleve.
Amphiprora sp. *
Anaulus spp. *
Asterionella bleakeleyi W. Sm. *
Asterionella glacialis Castracane; syn. *A. japonica* Cleve.
Bacteriastrum delicatulum Cleve.
B. hyalinum Lauder.
Campylodiscus sp. *
Cerataulina pelagica (Cleve) Hendey; syn. *C. bergonii* (Peragallo) Schütt.
Chaetoceros danicus Cleve. *
C. peruvianum Brightwell.
Chaetoceros spp.
Cocconeis placentula Ehrenberg. *
Corethron criophilum Castracane; syn. *C. hystrix* Hensen.
Cyclotella spp.
Cymbella sp.
Denticula seminae nov. nom.; syn. *D. marina* Semina.*
Diatoma vulgare Bory.
Diploneis cf. *interrupta* (Kütz) Cleve.
Diploneis spp. *
Ditylum brightwellii (West) Grunow.
Fragillaria sp.
Gramatophora sp.
Guinardia flaccida (Castracane) Peragallo.
Hemiaulus hauckii Grunow.
Hemidiscus cuneiformis Wallich.
Leptocylindrus danicus Cleve.
L. mediterraneus (Peragallo) Hasle; syn. *Dactyliosolen mediterraneus* Peragallo.
Licmophora spp.
Melosira granulata (Ehrenberg) Ralfs.
M. sulcata (Ehrenberg) Kützing; syn. *Paralia sulcata.*
M. varians Agardh.
Navicula distans (W. Smith) Ralfs.
Navicula spp.
Nitzschia closterium (Ehrenberg) W. Smith.
N. longissima (Brébisson in Kützing) Ralfs in Pritchard.
N. paradoxa (Gmelin) Grunow; syn. *Bacillaria paradoxa* Gmelin.
N. pseudodelicatissima Hasle.
N. cf. *seriata* Cleve.
Odontella mobiliensis (Bailey) Grunow; syn. *Biddulphia mobiliensis* Bailey
Pleurosigma spp.
Rhizosolenia alata f. *genuina* Brightwell.
R. alata f. *gracillina* (Cleve) Grunow.
R. calcaravis Schultze
R. fragilissima Bergon.
R. hebetata F. *semispina* (Hensen) Gran; syn. *R. semispina* Hensen.
R. imbricata v. *shrubsolei* (Cleve) Schröder; syn. *R. shrubsolei* Cleve.
R. robusta Norman; syn. *R. sigma* Schütt.
R. setigera Brightwell; syn. *R. hensenii* Schütt.
R. stolterfothii Peragallo.
Schöderella delicatula (H. Peragallo) Pavillard.*
Skeletonema costatum (Greville) Cleve.*
Striatella unipunctata (Lyngbye) Agardh.
Synedra undulata (Bailey) Gregory.
Synedra sp.
Tabellaria sp.*
Thalassionema nitzschioides (Grunow) Van Heurk; syn. *Thalassiothrix nitzschioides* (Grunow) Grunow in Van Heurk.
Thalassiosira eccentrica (Ehrenberg) Cleve; syn. *Coscinodiscus eccentricus* Ehrenberg.
Thalassiosira spp.
Thalassiothrix frauenfeldii (Grunow).
Triceratium sp.
Trigonium alternans (Bailey) Mann; syn. *Triceratium alternans* Bailey.
Tropidoneis sp.

DINOPHYCEAE

Amphidinium crassum Lohmann.
A. operculatum Claparède & Lachman.
Amphidinium sp.
Amphidoma caudata Halldal; syn. *Oxytoxum tonollii* Rampi, *O. margalefi* Rampi.
Blepharocysta splendor-maris (Ehrenberg) Stein.
Brachydinium sp.
Ceratium azoricum Cleve.
C. candelabrum (Ehrenberg) Stein.
C. declinatum (Karsten) Jörgensen.
C. furca (Ehrenberg) Claparède & Lachmann.
C. fusus (Ehrenberg) Dujardin.
C. *hirundinella* (O. F. Müller) Schrank.*
C. macroceros (Ehrenberg) Vanhöffen.
C. massiliense (Gourret) Jörgensen.
C. pentagonum Gourret.
C. trichoceros (Ehrenberg) Kofoid.
C. tripos (Müller) Nitzsch.
Ceratium sp.
Cladopyxis sp.
Cochlodinium spp.
Dinophysis caudata Saville-Kent.
D. operculata (Stein) Balech.*
D. parva Schiller.
D. rotundata Claparède & Lachmann; syn. *P. rotundatum* (Claparède & Lachmann) Kofoid & Michener.
D. scoederi Pavillard.
Diplopsalis lenticula Bergh.
Dossodium asymmetricum (Mangin) Loeblich; syn. *Diplopsalis asymmetrica* Drebes & Elbrachter.

TABLE I. Cont.

Erythropsis sp.*
Gonyaulax ligustica Rampi.*
G. polygramma Stein.
Gonyaulax spp.
Gymnodinium sp.
Gyrodinium fusiforme Kofoid & Swezy
G. spirale (Bergh) Kofoid & Swezy
Gyrodinium spp.
Heterocapsa triquetra (Ehrenberg) Stein.*
Heterodinium cf. *milneri* (Murray & Whitting) Kofoid.*
Kofoidinium velleloides Pavillard.
Mesoporus perforatus (Gran) Lillick.
Micracanthodinium sp.
Noctiluca scintillans (Macartney) Ehrenberg; syn. *N. miliaris* Suriray ex Lamarck.
Oxytoxum brunelli Rampi.
O. gracile Schiller.
O. longiceps Schiller.
O. mediterraneum Schiller.*
O. scolopax Stein.
O. variabilis Schiller.*
Oxytoxum sp.
Palaeophalocroma sp.*
Podolampas palmipes Stein.
Polykrikos schwartzii Bütschli.
Pronoctiluca spinifera (Lohmann) Schiller.
Pronoctiluca sp.
Prorocentrum balticum (Lohmann) Loeblich; syn. *Exuviaella baltica* Lohmann.
P. compressum (Bailey) Abe ex Dodge; syn. *E. compressa* (Bailey) Ostenfeld.
P. dactylus (Stein) Dodge.*
P. micans Ehrenberg.
P. minimum (Pavillard) Schiller.*
P. triestinum Schiller
P. vaginulum (Ehrenberg) Dodge; syn. *P. adriaticum* Schiller.*
Prorocentrum spp.
Protoceratium aerolatum Kofoid.
Protoperidinium bipes (Paulsen) Balech; syn. *Minuscula bipes* (Paulsen) Lebour.
P. breve (Paulsen) Balech.
P. brochi (Kofoid & Swezy) Balech.
P. cerasus (Paulsen) Balech.
P. conicum (Gran) Balech.
P. depressum (Bailey) Balech.
P. diabolum (Cleve) Balech.
P. elegans Cleve.
P. leonis (Pavillard) Balech.
P. mite (Pavillard) Balech.
P. ovatum Pouchet.
P. ovum Schiller.
P. pallidum (Ostenfeld) Balech.
P. pedunculatum Shütt.
P. pyriforme (Paulsen) Balech.
P. quarnerense Schröeder.
P. steinii (Jörgensen) Balech.
Protoperidinium spp.
Pseliodinium vaubanii Sournia.
Ptychodiscus noctiluca Stein; syn. *P. inflatus* Pavillard.
Pyrocystis lunula (Schütt) Schütt.
Pyrophacus horlogium Stein.
Scaphodinium mirabile Margalef; syn. *Leptospathium navicula* Cachon & Cachon-Enjumet.
Scrippsiella trochoidea (Stein) Loeblich III; syn. *Peridinium trochoideum* (Stein) Lemmermann.
Torodinium robustum Kofoid & Swezy.
T. teredo (Pouchet) Kofoid & Swezy.
Torodinium sp.
Triadinium polyedricum (Pouchet) Dodge; syn. *Goniodoma polyedricum* (Pouchet) Jörgensen.

CLOROPHYCEAE

Halosphaera viridis Schmitz.
Monorhaphidium spp.; syn. *Ankistrodesmus* spp.
Pediastrum boryanum (Turpin) Meneghini.
P. simplex Meyen; syn. *P. clatratum* (Schöeter) Lemmermann.
Scenedesmus spp.

CRYPTOPHYCEAE

Cryptomonas spp.

CYANOPHYCEAE

Lingbya sp.
Merismopedia spp.
Oscillatoria thiebautii (Gommont ex Gommont) Geitler.

DESMIDIACEAE

Staurastrum sp.*

EUGLENOPHYCEAE

Eutreptiella marina da Cunha.*
Euglena sp.

HAPTOPHYCEAE

Coccolithus huxleyi (Lohmann) Kamptner.
Phaeocystis poucheti Lagerh.
Solenicola setigera Pavillard.

SILICOFLAGELLATOPHYCEAE

Dyctyocha fibula Ehrenberg.
D. speculum Ehrenberg; syn. *Distephanus speculum* (Ehrenberg) Haeck.

the inner half of the cave was signalled. Slightly different is the case of a group of benthic diatoms such as *Amphora* spp., *Diploneis* spp., and *Navicula* spp. which were strikingly dense in the inner stations of the cave during June 1987, so making a shift from the pattern of decay above described (Fig. 4A). Synchronically (June 1987), these taxa were abundant over experimental panels from the inner end of the cave (GARCÍA & ZABALA, in prep.)

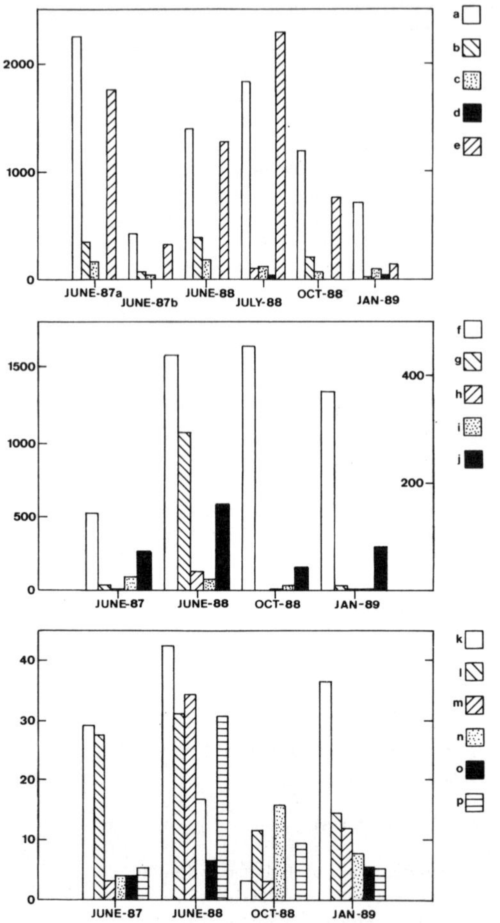

FIGURE 2. Density of planktonic groups in outer waters. A) Phytoplankton (cells cm^{-3}) and ciliates (cells dm^{-3}). B) Holoplankton (ind. m^{-3}) (scale at left, copepods; scale at right, other than copepods). C) meroplankton larvae (ind. m^{-3}). a) small flagellates; b) diatoms; c) dinoflagellates; d) cryptophyceans; e) ciliates; f) copepods; g) appendicularians; h) siphonophores; m) polychaetes; n) barnacles; o) echinoderms; p) others.

ZOOPLANKTON

If number of individuals is the descriptor of abundance chosen, zooplankton from outer waters was clearly dominated by copepods (densities from 527 to 1639 individuals m^{-3}), followed by the sum of the whole meroplankton and, occasionally (June 1988), by appendicularians. The remainder of the true holoplankton represented a minimal fraction of the whole (Fig. 2B). These proportions agree with the structure of zooplanktonic populations reported from coastal waters in this area (VIVES, 1986). No marked oscillations in copepod frequency during the year cycle were observed (Fig. 2B) except in spring (June 1988), when copepod frequency (75

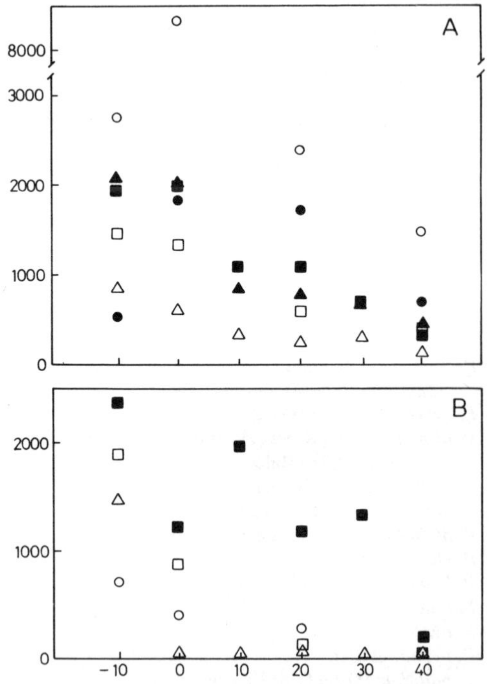

FIGURE 3. Densities of plankton along the cave for different seasons. A) Phytoplankton (cells cm^{-3}). B) Zooplankton (ind. m^{-3}) (open circles, June 87a; full circles, June 87b; full squares, June 88; full triangles, July 88; open squares, October 88; open triangles, January 89).

%) decreased to the benefit of appendicularians (14 %) and meroplankton (7.7 %).

11 high taxa were identified among the meroplanktonic fraction. Larvae of both bivalve and gastropod mollusks were the commonest organisms (Fig. 2C). Even if sampling was excessively reduced to assess the significance of seasonal differences, spring samples showed a peak in meroplanktonic density (162 ind. m^{-3}; 10 high taxa present). Bivalve and gastropod larvae, already provided with thin shells, and trocophora from polychaete worms were the dominant groups. Anthomedusae and barnacle larvae were also abundant but larvae of echinoderms were strikingly scarce; the rest of meroplankton was composed almost exclusively of ceriantid larvae. In contrast, the poorest samples were those from autumn (43 ind. m^{-3}; 9 taxa represented). In this season, all groups exhibit minimal densities with the only exception of barnacle larvae, which showed populations as dense as those from spring 1988; there are no echinoderm larvae. In

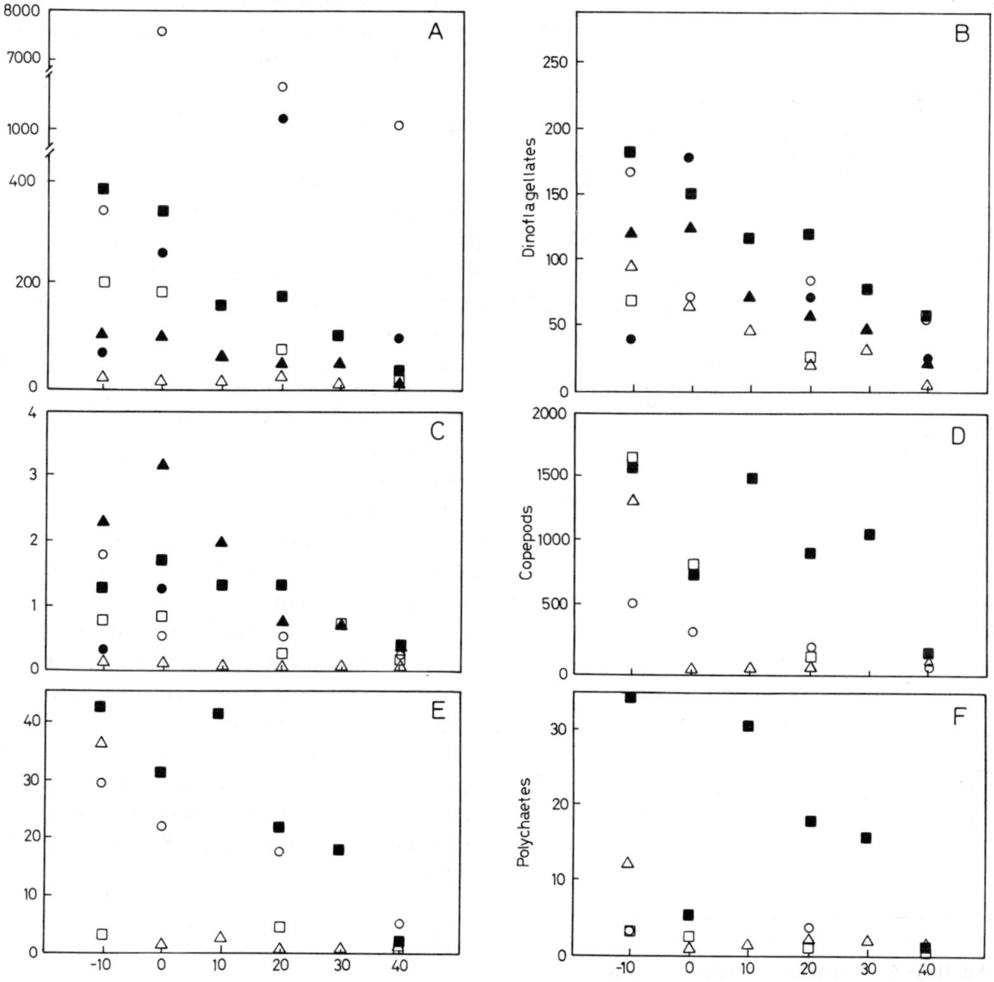

FIGURE 4. Densities of planktonic groups along the cave for different seasons. A) Diatoms (cells cm^{-3}). B) Dinoflagellates (cells cm^{-3}). C) Ciliates (cells cm^{-3}). D) copepods (ind. m^{-3}). E) Bivalves (ind. m^{-3}). F) Polychaetes (ind. m^{-3}). Symbols as in figure 3.

winter, meroplanktonic densities grew, partially due to the increase of polychaete worms, bivalve and echinoderm larvae. The latter reach densities similar to those from spring but gastropod larvae were as scarce as in autumn; barnacle larvae also showed their minimal annual density.

A strong gradient in zooplanktonic density was fairly clear along the cave throughout the year. Zooplanktonic density was higher in "outer" waters; it decreases progressively along the axis of the cave and reaches the minimal values at the inner end (Fig. 3B). Except in January 89, when there was a gentle increase in the medium part of the cave, richness (expressed as number of high taxa represented in each sample) also decays progressively from outer to inner waters (Fig. 5C). A similar pattern arises when each zooplanktonic group is represented separately (Fig. 4 D-F). Density of planktonic eggs of unidentified organisms showed the same general pattern of decay.

Larvae of clonal organisms such as sponges, bryozoans and ascidians were not found (inside or outside the cave) at any time during the annual cycle.

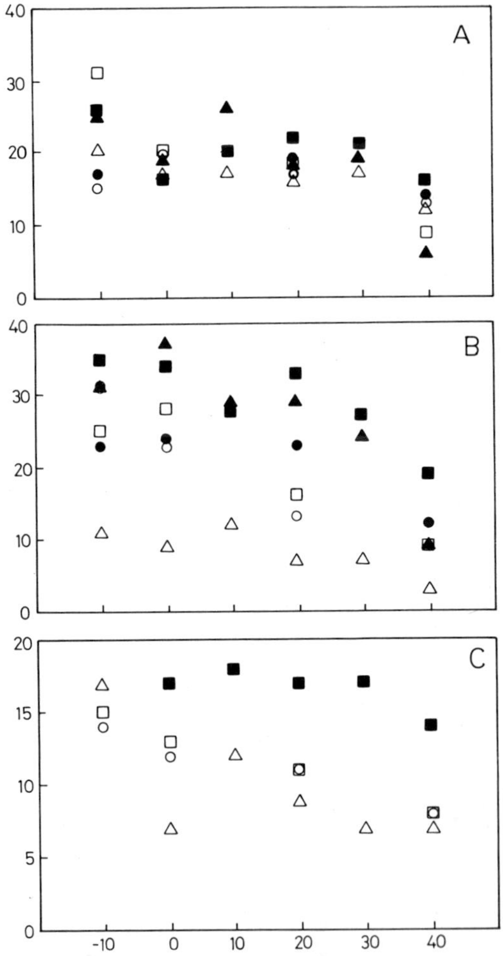

FIGURE 5. Taxonomic richness along the cave for different seasons. A) Diatoms (number of species per sample). B) Dinoflagellates (number of species per sample). C) Zooplankton (number of high taxa per sample). Symbols as in figure 3.

CAVE-TUNNEL COMPARISON

Samples from the tunnel also exhibit a pattern of decay in planktonic density, but the explanation of this pattern is unclear. When correlation between planktonic density *vs.* distance from the entrance is used as an estimator of the intensity of the planktonic gradient, samples coming from the cave always show good correlations between these two variables ($P<0.05$; Table II); in contrast, samples coming from the tunnel show far more erratic values. If diatoms, which are the only group to show a good correlation in both tunnel ($n=4$, $r=-0.96$, $P< 0.05$) and cave, are considered as reference, significant differences seem to arise between the slope of graphics representing the cave and the tunnel (Fig. 6; F-test, $P <0.01$). These results seem to suggest an agreement with the previous hypotheses about a higher water renewal in tunnels than in caves (HARMELIN, 1969), but contrasts with our previous results on dissolution of plaster balls (ZABALA *et al.*, 1989).

COMMUNITY STRUCTURE: FACTORS.

The structure of planktonic cave communities appears to be dependent on only a reduced number of factors as

TABLE II. Correlations between planktonic densities and the distances from the entrance of the cave and tunnel in June 1988. a) intercepts. b) Slopes (r, correlation coefficients; P, significance levels; N.S., non significative).

	Cave (n = 6)				Tunnel (n = 4)			
Planktonic taxa	a	b	r	P	a	b	r	P
Total phytoplankton	1713.40	-34.37	-0.960	0.01	1058.50	-29.83	0.898	N.S.
Diatoms	305.44	-7.12	-0.960	0.01	142.73	-3.41	-0.965	0.05
Dinoflagellates	152.65	-2.41	-0.980	0.001	95.24	-1.50	-0.888	N.S.
Ciliates	1.44	-0.02	-0.833	0.05	1.56	-0.02	-0.832	N.S.
Total zooplankton	1873.5	-32.48	-0.811	0.05	876.47	-9.52	-0.323	N.S.
Copepods	1290.16	-19.34	-0.690	N.S.	701.31	-7.24	-0.293	N.S.
Appendicularia	213.14	-5.51	-0.976	0.001	35.89	-0.47	-0.453	N.S.
Gastropods	28.94	-0.50	-0.855	0.05	7.74	0.04	0.136	N.S.

suggested by PCA applied to the group of the whole samples. PCA carried out on selected species of both phytoplanktonic (Table IIIa) and zooplanktonic fractions (Table IIIb) showed a very similar pattern. Phytoplankton analysis gives 3 eigenvalues higher than unity, these axes explaining 77.6 % of whole variance (47.6% for PC I; 18.5% for PC II; 11.5% for PC III). Zooplankton analysis gives only 2 eigenvalues higher than unity, both explaining 78.1% of whole variance (70.0% for PC I; 8.1% for PC II).

The highly significant correlation (n=20, r= 0.88, P< 0.001) between the component I of the two PCA also seems to suggest a similar interpretation for their significance (Fig. 7). Two facts facilitate this characterization:

1) With two highly significant exceptions (*Nitzschia closterium* and larvae of Spirorbidae, which are two "end cave" taxa), all species of both phyto- and zooplanktonic fractions load on positive values of the first component; then, a significant positive correlation between component I and planktonic density or biomass results.

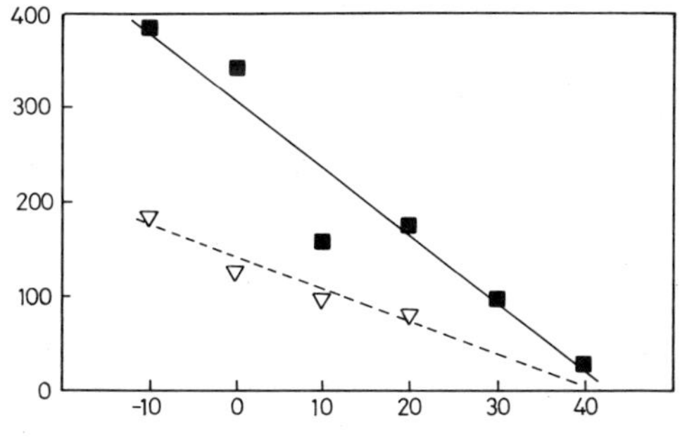

FIGURE 6. Comparison of densities of diatoms (cells cm^{-3}) from the cave (full squares) and the tunnel (inverted open triangles), in June 88, as a function of distance from the entrance. Differences between slopes are statistically significant (F-test; P< 0.01).

TABLE III. List of taxa used for Principal Component Analysis (PCA).

a) PHYTOPLANKTON

Chaetoceros spp.
Leptocylindrus danicus
Nitzschia closterium
Nitzschia pseudodelicatissima + *N. seriata*
Pleurosigma spp.
Rhizosolenia spp.
Thalassionema nitzschioides
Ceratium spp.
Diplopsalis lenticula + *Dissodium asymmetricum*
Gymnodinium spp.
Gyrodynium spp.
Prorocentrum spp.
Protoperidinium spp.
Scrippsiella trochoidea
Nanoplanktonic athecate dinoflagellates
Nanoplanktonic flagellates and monads
Planktonic cilliates

b) ZOOPLANKTON

Anthomedusae
Appendicularians
Bivalves
Cladocerans
Copepods
Echinoderms
Gastropods
Ostracods
Polychaetes
Siphonophorans
Spirorbidae
Unidentified eggs

2) There is a significant negative correlation between component I and the distance from the point where the sample was taken to the entrance of the cave. This is true for both phytoplankton (Fig. 8A; n=30, r=-0.57, P< 0.001) and zooplankton (Fig. 8B; n=20, r=-0.54, P< 0.001) analyses.

Thus we can confirm that the distance from the entrance of the cave is a prime factor in the structure of planktonic cave communities. This result reinforces the picture of a strong persistent planktonic gradient along the axis of the cave.

The second component of PCA on the phytoplanktonic fraction could be related to seasonality. This is fairly clear when factor scores of samples are plotted on the space defined by component I and component II (Fig. 9). Axis II separates samples from different seasons. Noticeably, the extreme high and low positions are occupied by samples from the outer station. Samples taken from the inner part of the cave tended to be closer together, pointing to the higher seasonal stability (or the lower seasonal variability) of planktonic communities in the cave. This trend is easily recognized when samples from each season are joined by straight segments following their relative position along the transect (Fig. 9). Irrespective of the position of the outer sample, they describe a convergent trajectory towards a restricted region of the space where all the inner samples overlap.

The higher seasonal heterogeneity of phytoplanktonic assemblages from the outer parts of the cave can be explained by the existence of largely dominant species which can exhibit strong seasonal fluctuations in density. With increasing distance from the entrance there is a reduction in the numbers of individuals of the commonest species, which is faster than the reduction of species numbers. This is suggested by comparing the evolution of both diversity and evenness indexes with increasing distance from the entrance (Fig. 10). Despite a net decrease in species richness, diversity (Fig. 10 A) rises because evenness (Fig. 10 B) rises quickly

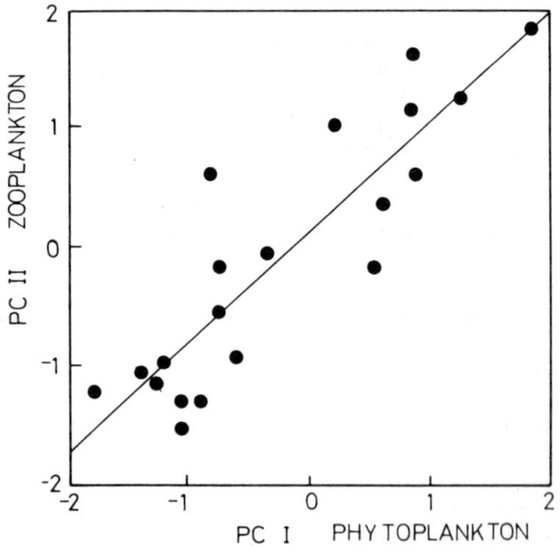

FIGURE 7. Relationship between the factor scores over the first component resulting from the PCAs carried out on the phytoplanktonic and the zooplanktonic samples (n=20; r=0.88; P< 0.01).

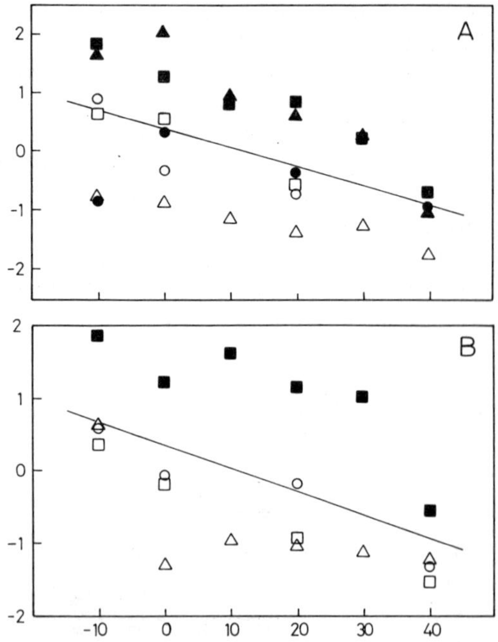

FIGURE 8. Relationship between the factor scores over the first components of PCAs and the distance from the entrance of the cave. A) Phytoplankton samples (n=20; r= -0.57; P< 0.001); B) Zooplankton samples (n=20; r=-0.54; P< 0.02). Symbols as in figure 3.

(only evenness shows a significant correlation with distance; n=30; r=0.40, P< 0.05).

DISCUSSION

STRONG PLANKTONIC GRADIENT

Although there is no clear taxonomic substitution within assemblages, both species number and density of planktonic communities show a strong gradient along the axis of the cave, the farther from the entrance the lower the values. This gradient is of special importance because it has rarely been reported (RIEDL, 1966).

The strength of this pattern relies on: i) its *magnitude*, ii) its *persistence* throughout an annual cycle, and iii) its *taxonomic catholicity*.

Magnitude. Decay values of the planktonic cells over a distance of only 50 m range between 60-100 % of the whole population. They range between 75-100 % for ciliates, copepodes, meroplanktonic larvae and diatoms (mean value 2 % ind. m^{-1}), or between 60-80 % for dinoflagellates and total phytoplanktonic cells. So high decay values are rarely reached on horizontal gradients from open waters. They have been reported from estuarine tidal fronts (MacALICE, 1970; INCZE & YENTSCH, 1981), or small-scale studies of neritic waters (WIEBE, 1970; FASHAM *et al.*, 1974; IBANEZ, 1976; ; PALAU, in prep.). Decay values are also several times higher than vertical gradients normally found in open waters as a result of productive, predatory, migratory and sedimentary processes (MARGALEF, 1969). Only gradients through summer thermoclines, estuarine pycnoclines (MacALICE, 1970) or deep chlorophyll maxima (ESTRADA, 1985; VENRICK, 1988) can equal such high values.

Persistence. Decay on planktonic communities along the axis of this cave is not a spureous event but a persistent

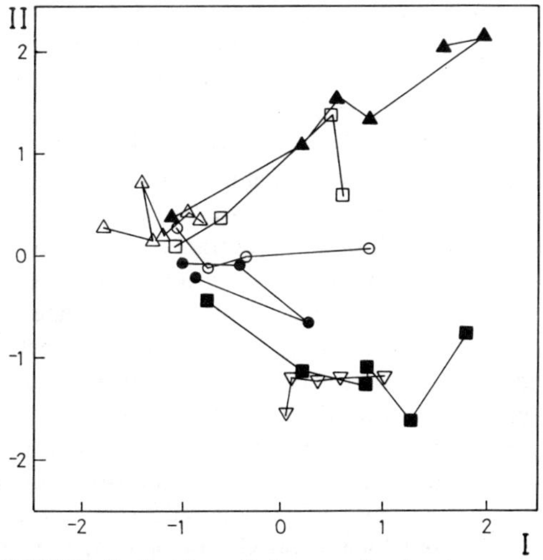

FIGURE 9. Position of the samples in the space defined by the first and second components resulting from the PCA carried out with the phytoplankton fraction. Trajectories join all contiguous samples from equal date with arrows pointing to inner samples. Symbols as in figures 3 and 6.

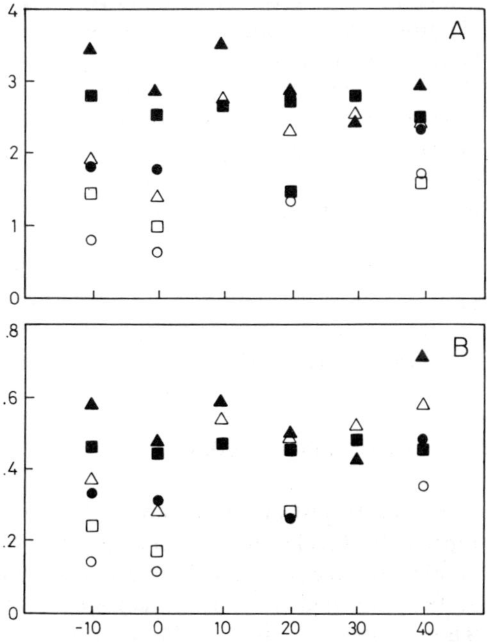

FIGURE 10. Relationship between the level of phytoplankton structure at different seasons and the distance from the entrance of the cave. A) H' diversity. B) Evenness. Symbols as in figure 3.

phenomenon which is reproduced accurately at each season and under very different environmental conditions. It also appears to be unrelated to absolute density of planktonic communities from surrounding outer waters which are representative of the neritic West Mediterranean planktonic communities. For example, results on the evolution of meroplanktonic organisms from outer samples during the annual cycle agree with those reported from neighbouring areas of Castellón (VIVES, 1966) and Banyuls-sur-Mer (THIRIOT- QUIÉVREUX, 1970).

The same pattern seems to be repeated in the tunnel, even if slightly smoother in its magnitude. As tunnels are assumed to be provided with higher water renewal (HARMELIN, 1969; HARMELIN *et al.*, 1985; but see ZABALA *et al.*, 1989) we are persuaded that this pattern will also be applicable to the submarine caves overall.

Taxonomic catholicity. Almost all taxonomic groups which have been identified show a similar pattern of decay. Only fungi, benthic diatoms and spirorbid larvae eventually display an inverted distribution, their density remaining stable or slightly increasing with increasing distance from the entrance. The taxonomic universality of this pattern strengthens our conviction that we are witness to a very general process which may equal the universality of the well documented pattern of decay in benthic assemblages from submarine caves (HARMELIN *et al.*, 1985).

THE "MOUTH BOUNDARY"

Station 0, just at the mouth of the cave, repeatedly breaks the general pattern of continuous decay in planktonic density by shifting to higher values than expected. This "anomaly" has been observed at different seasons, for the whole phytoplankton and also for the main taxa alone. The increase in phytoplanktonic density is coupled to a decrease in both diversity and evenness (Fig. 10). This suggests a proliferation of the commonest species more than the contribution of new taxa as the immediate cause of the increase in phytoplankton density. A similar "anomaly" emerges from an attentive examination of results on trophic-related parameters previously reported for the same cave (ZABALA *et al.*, 1989). This study reports an increase in chlorophyll concentrations, 665/665a ratios and oxygen concentration at this station; BOD_5 and ETS values also decrease, in agreement with expectations for a zone of phytoplanktonic proliferation. Nevertheless, the "anomaly" for the remaining parameters measured by these authors, e.g. POC, PON and bacterial activity ([^{3}H]-thymidine), consists in a conspicuous decrease at this station, which disagrees both with our data and with our expectations of it as a proliferative region.

But what are the causes of the persistence of this structure, if it is truly persistent? We suggest two untested clues which could overlap to produce a stronger effect. Dark caves seem to be zones where an active mineralization occurs, trapped organic matter being totally oxydized to inorganic dissolved compounds such as nitrate and phosphate nutrients, which are far more abundant inside than outside the cave (FICHEZ, 1989). When these nutrients, which cannot be used in photosynthesis because of the lack of light, reach the illuminated entrance to the cave, a minute process of enrichment should result. On the other hand, the entrance to the cave should be viewed as a kinetic boundary between two compartments provided with different degrees of freedom: an outer environment free in all three dimensions and an inner semi-restricted medium where advective processes should be limited by the proximity of walls. As a result, the existence of accumulative processes, which are characteristic of interphases, is not a remote possibility (MARGALEF, 1961; LEGENDRE, 1981).

A CAUSE FOR BENTHIC DECAY ?

The relevance of these results lies in the fact that the two gradients, the benthic and the planktonic, are probably not only parallel but also linked by a causal relationship. The planktonic impoverishment seems to be strong and persistent enough to account - by itself - for the decay of benthic macrofauna, which is dependent upon it, from the inner parts of the caves. There are two possible explanations :

1) **Low settlement.** Meroplanktonic larvae of the main benthic taxa were almost totally absent from samples of the inner parts of the cave even when larval density in outer waters was richer (in spring). The lack of larval stages of clonal organisms such as sponges, bryozoans and ascidians contrasts with the absolute dominance of these groups in benthic cave assemblages (GILI *et al.*, 1986). This paradox may be explained because these groups have mainly lecithotrophic larvae, which have a extremely short span of free life (JACKSON, 1986); but alternative explanations, such as those based on behavioural stresses (supposed), e.g. lack of light stimulus, cannot be rejected. This fact can easily explain the extremely low rates of benthic settlement on experimental panels reported by other authors (HARMELIN, 1980) and also detected by us (GARCÍA & ZABALA, in prep.). If not by itself, this low value of recruitment needs only the addition of a very weak factor of post-settlement or adult mortality to explain easily the entire paucity of benthic organisms in submarine caves.

2) **Trophic stress**. Planktonic paucity itself can be the cause of adult mortality of benthic organisms by enhancing trophic stresses. The relevance of meroplanktonic larvae - in addition to holoplanktonic biomass - in the food requirement of the dominant sessile benthic groups, which are mainly active filter-feeders, has recently been recognized (SEBENS & KOEHL, 1984).

Our present results strongly contrast with our previous results on the POM contents of water sampled in the same cave, which do not display a conspicuous gradient (GILI *et al.*, 1986; ZABALA *et al.*, 1989). The paradox has two possible explanations:

i) A high fraction of this POM could be tripton which has not been accounted for in this study. The decay of the planktonic fraction to the inner parts of the cave could be hidden by a reciprocal gradient of increasing triptonic fraction.

ii) Even if considerable amounts of POM reach the end of the cave, trophic quality is probably a constraint factor. At present, we believe that a large percentage of POM from the inner waters of the cave is due to the mineralization of faecal pellets egested by dense swarms of mysids inhabiting the cave (ZABALA *et al.*, in prep.) while planktonic POM seem to be genuinely lacking. Pellets should be inadequate as food for benthic filter-feeders settling over the walls and roofs of caves because of their high density, which also produces high sinking rates.

CAUSES OF PLANKTONIC DECAY

Thus the striking pattern of planktonic disappearance demands an explanation, but at present we are only able to forward and discuss several hypotheses. Two ways are possible :

1) Plankton enters the cave but it is unable to survive. Plankton death can arrive by: 1.1) predation; 1.2) sedimentation, or 1.3) physiological stress; or, alternatively,

2) The entrance of plankton to the cave is prevented, for example by behavioural constraints.

1.1. **Predation**. Predation by planktonic consumers is unlikely to be important because all planktonic groups, prey and predator, disappear at the same time. Predation by benthic suspension feeders from the entrance of the cave, where benthic biomass is very high, cannot be completely rejected. Even if values of benthic biomass are known (GILI *et al.*, 1986), data on consumption rates from suspension-feeders of these communities - mainly sponges, scleractiniaria, hydroidea and bryozoa - are lacking. Nevertheless, such a strict planktonic depletion is unlikely to be at work all the time or, at least, it is unlikely to be effective during planktonic blooms.

1.2. **Physiological stresses**, such as those produced on photosynthetic pigments by large periods of darkness, have been reported to kill plankton quickly (SMETACEK, 1985). However, diatoms are physiologically adapted to maintain viability in dark waters as resting cells occurring as seeding populations at considerable depths in the ocean (SMAYDA & MITCHELL-INNES, 1974; PLATT *et al.*, 1983; SMETACEK, 1985). Resting, degradated or unpigmented photosynthetic cells should still be identifiable in samples if a cave-water turnover ranging over about 12-24 h is accepted (GILI *et al.*, 1986; ZABALA *et al.*, 1989). But this is not the case. Further, if light could explain the phytoplanktonic disappearance no environmental stress is known to operate inside this cave to kill zooplanktonic organisms.

1.3. **Sedimentation** is expected to be higher inside caves than in outer bottoms. But is it strong enough to provide a complete explanation for such a severe gradient? This is at present the unsolved question. Sedimentation rates inside submarine caves have recently been evaluated from a northwestern Mediterranean locality near Marseilles (FICHEZ, 1989). From this author's data, a function between sedimentation rates and the distance from the entrance of the cave can be described [sedimentated weight (g m^{-2} d^{-1}) = -50.60 distance (m) + 4096.16; r^2 = 0.90]. A similar function can be found which relates concentration of sestonic matter and the distance from the entrance of the cave [sestonic weight (mg m^{-3}) = -7.46 distance (m) + 1783.45; r^2 = 0.99). Note that the slope of the last function suggests a strong gradient very near to those describing the gradients of planktonic disappearance found in our study (Table II). As suggested by these numbers, sedimentation seems to be able to deplete seston from water at rates which are greater than or equal to those deduced from our results. Unfortunately we lack data about sedimentation rates in our cave.

Therefore, by exclusion of the other hypotheses, we think that sedimentation must be the effectively important factor in the process of decimation found on planktonic communities of submarine caves. Consequently, we suggest that further studies concentrate on the role of sedimentary input to the cave bottoms.

ACKNOWLEDGMENTS

We would like to thank J. Flos, R. Margalef and J. D. Ros for helpful comments on the manuscript. We also thank A. M. Domingo for the preparation of drawings, and R. Coma for many shared hours spent underwater capturing plankton. This research was partially supported by a Caixa de Catalunya research grant.

REFERENCES

BALDUZZI, A., BOERO, F., CATTANEO, R, PANSINI, M., PESSANI, D., PRONZATO, R. & SARÀ, M. 1980. Ricerche sull'insediamento dello zoobentos in alcune grotte marine della Penisola Sorrentina. *Mem. Biol. Marina Oceanogr.* (Suppl.), 10: 121-127.

BALDUZZI, A., BIANCHI, C.N., BOERO, F., CATTANEO, R., PANSINI, M. & SARÀ, M. 1989. The suspension- feeder communities of a Mediterranean sea cave. In: *Topics in Marine Biology.* (J.D. Ros, ed.): 247-250. *Scient. Mar.*, 53(2-3).

BARANGÉ, M. 1988. Prey selection and capture strategies of the benthic hydroid *Eudendrium racemosum*. *Mar. Ecol. Prog. Ser.*, 47: 83-88.

BARANGÉ, M. & GILI, J.M. 1988. Feeding cycles and prey capture in *Eudendrium racemosum* (Cavolini, 1785). *J. exp. Mar. Biol. Ecol.*, 115: 281-293.

BIBILONI, M.A., GILI, J.M. & ROS, J.D. 1984. Les coves submarines de les illes Medes. In: *Els Sistemes Naturals de les Illes Medes*, (J.D. Ros, I. Olivella & J.M. Gili, eds.): 707-737. *Arxius de la Secció de Ciències*, 73.

CINELLI, F., FRESI, E., MAZZELLA, L., PANSINI, M., PRONZATO, R. & SVOBODA, A. 1977. Distribution of benthic phyto- and zoocenoses along a light gradient in a superficial marine cave. In: *Biology of Benthic Organisms* (B.F. Keegan & P. O'Ceidigh , eds.): 173-83. Pergamon Press. Oxford.

ESTRADA, M. 1985. Deep phytoplankton and chlorophyll maxima in the Western Mediterranean. In: *Mediterranean Marine Ecosystems* (M. Moraitou-Apostolopoulou & V. Kiortsis, eds.): 247-277. Plenum Press. New York.

FASHAM, M.J.R., ANGEL, M.V. & ROE, H.S.J. 1974. An investigation of spatial pattern of zooplankton using the Hardy-Longhurst plankton recorder. *J. Exp. Mar. Biol. Ecol.*, 16 :93-112.

FICHEZ, R. 1989. *Conditions d'oligotrophie en milieu aphotique, étude des grottes sous-marines, comparaison avec les milieux profonds et bilans énergétiques.* Ph. D. Thesis. University Aix-Marseille II.

FICHEZ, R. 1990. Les pigments chlorophylliens: indices d'oligotrophie dans les grottes sous-marines. *C. R. Acad. Sci. Paris*, 310 (3): 155-161.

FICHEZ, R. 1991a. Decrease in allochtonous organic inputs in dark submarine caves, connection with lowering in benthic community richness. *Hydrobiologia*, 207: 61-69 .

FICHEZ, R. 1991b. Absence of Redox Potential Discontinuity in Dark Submarine Cave Sediments as Evidence of Oligotrophic Conditions. *Estuar. Coast. Shelf Sci.*, 31: 875-881.

FICHEZ, R. 1991c. Benthic oxygen uptake and carbon cycling in aphotic, oligotrophic conditions. A submarine cave experiment. *Mar. Biol.*, 110: 137-143.

FICHEZ, R. 1991d Particulate organic matter in an aphotic and oligotrophic submarine cave. *Mar. Biol.* (submitted).

GARCÍA RUBIES, A. & ZABALA, M. in prep. The effect of the distance in benthic recruitment: the case of submarine caves.

GILI, J.M., RIERA, T. & ZABALA, M. 1986. Physical and biological gradients in a submarine cave on the Western Mediterranean coast (North East Spain). *Mar. Biol.*, 90: 291-297.

HARMELIN, J.G. 1969. Bryozoaires des grottes sous-marines obscures de la région marseillaise, faunistique et écologie. *Téthys*, 1: 793-806.

HARMELIN, J.G. 1980. Établissement des communautés de substrats durs en milieu obscur. Résultats préliminaires d'une expérience à long terme en Méditerranée. *Mem. Biol. Mar. Oceanogr.* (Suppl.), 10: 29-52.

HARMELIN, J.G., VACELET, J. & VASSEUR, P. 1985. Les grottes sous-marines obscures: un milieu extreme et un remarquable biotope refuge. *Téthys*, 11(3-4): 214-229.

IBANEZ, F. 1976. Contribution à l'analyse mathématique des événements en Écologie planctonique. *Bull. Inst. Oceanog. Monaco*, 72(1431): 1-96.

INCZE, L.S. & YENTSCH, C.M. 1981. Stable density fronts and dynoflagellate patches in a tidal estuary. *Estuar. Coast. Shelf Sci.*, 13: 547-556.

JACKSON, J.B.C. 1986. Modes of dispersal of clonal benthic invertebrates: consequences for species' distributions and genetic structure of local populations. *Bull. Mar. Science,* 39(2): 588-606.

JACQUES, G. 1969. Aspectes quantitatives du phytoplancton de Banyuls-sûr-Mer (golfe du Lion). III-Diatomées et dynoflagelées de juin 1965 à juin 1969. *Vie Milieu*, 20(1-B): 91-126.

JOHANES, R.E. & SATOMI, M. 1966. Composition and nutritive value of fecal pellets of a marine crustacean. *Limnol. Oceanogr.*, 2: 191-197.

LAMPITT, R.S., NOJI, T. & VON BODUNGEN, B. 1990. What happens to zooplankton faecal pellets? Implications for material flux. *Mar. Biol.*, 104: 15-23.

LEGENDRE, L. 1981. Hydrodinamic control of marine phytoplankton production: the paradox of stability. In: *Ecohydrodynamics* (J.C.J. Nihoul, ed.): 191-208. Elsevier. Amsterdam.

LEGENDRE, L. & LEGENDRE, P. 1979. *Écologie numèrique*. Masson. Paris.

LUND, J.W.G., KIPLING, C. & LE CREN, E.D. 1958. The inverted microscop method of estimating of algal numbers and the statistical basis of estimations by counting. *Hydrobiologia,* 11:1 43-170.

MacALICE, B.J. 1970. Observations on small-scale distribution of estuarine phytoplankton. *Mar. Biol.,* 7(2): 100-111.

MARGALEF, R. 1961. Communication of structure in planktonic populations. *Limnol. Oceanogr.*, 6(2): 124-128.

MARGALEF, R. 1964. Fitoplancton de las costas de Blanes (provincia de Gerona, Mediterráneo Occidental) de julio de 1955 a junio de 1963. *Inv. Pesq.*, 26: 131-164.

MARGALEF, R. 1969. Small scale distribution of phytoplankton in Western Mediterranean at the end of July. *Pubbl. Staz. zool. Napoli,* 37 (suppl.): 40-61.

OTT, J.A. & SVOBODA, A. 1976. Sea caves as model systems for energy flow studies in primary hard bottom communities. *Pubbl. Staz. Zool. Napoli*, 40: 478-485.

PALAU, M. 1988. *Evolución anual de la hidrografía, fitoplancton y ciliados planctónicos en las aguas de las Islas Medes.* Master Thesis. University of Barcelona.

PASSELAIGUE, F. & BOURDILLON, A. 1985. Thermographie de l'ecosystème de quelques grottes sous-marines obscures. *Rapp. Comm. int. Mer Medit.*, 29(5): 155-156.

PIELOU, E.C. 1966a. Shannon's formula as a measure of specific diversity: his use and measure. *Am. Nat.*, 100: 463-465.

PIELOU, E.C. 1966b. Species-diversity and

pattern-diversity in the study of ecological succession. *J. theor. Biol.*, 10: 370-383.

PLATT, T., SUBBA RAO, D.V., SMITH, J.C., LI, W.K., IRWIN, B., HORNE, E.P.W. & SAMEOTO, D.D. 1983. Photosynthetically-competent phytoplankton from the aphotic zone of the deep ocean. *Mar. Ecol. Prog. Ser.*, 10: 105-110.

POULIQUEN, L. 1971. Les spongiaires des grottes sous-marines de la région de Marseille: Écologie et systématique. *Téthys*, 3(4): 717-758.

REISWIG, H.M. 1981. Particulate organic carbon of bottom boundary and submarine cavern waters of tropical coral reefs. *Mar. Ecol. Prog. Ser.*, 5: 129-133.

RIEDL, R. 1966. *Biologie der Meereshöhlen.* Paul Parey. Hamburg und Berlin.

RIERA, T., GILI, J.M. & ZABALA, M. 1985. Respiration, teneur en oxygène et renouvellement de l'eau dans une grotte sous-marine. *Rapp. Comm. int. Mer Médit.*, 29(5): 163-166.

RIERA, T., ZABALA, M. & PEÑUELAS, J. 1991. Mysids from submarine cave emerge each night to feed. *Scient. Mar.*

SEBENS, K.P. & KOEHL, M.A.R. 1984. Predation on zooplankton by the benthic anthozoans *Alcyonium siderium* (Alcyonacea) and *Metridium senile* (Actiniaria) in the New England subtidal. *Mar. Biol.*, 81: 255-271.

SHANNON, C.E. & WEAVER, W. 1963. *The mathematical theory of communication.* University of Illinois Press. Urbana.

SMAYDA, T.J. & MITCHELL-INNES, B. 1974. Dark survival of autotrophic planktonic marine diatoms. *Mar. Biol.,* 25: 195-202.

SMETACEK, V. 1985. Role of sinking in diatom life-history cycles: ecological, evolutionary and geological significance. *Mar. Biol.*, 84: 239-251.

SOKAL, R.S. & ROHLF, F.J. 1969. *Biometry.* Freeman. New York.

SOURNIA, A. 1982. Is there a shade flora in the marine plankton? *J. Plankton. Res.*, 4(2): 391-399.

THIRIOT-QUIÉVREUX, C. 1970. Cycles annuelles des populations planctoniques des mollusques en 1968 dans la region de Banyuls-sur-Mer. Comparaison avec les années precedentes 1965-67. *Vie Milieu*, 21(2B): 311-336.

UTERMÖHL, H. 1958. Zur vervollkommung der Quantitative Phytoplankton Methodik. *Mitt. Int. Verein. Theor. Angew. Limnol.*, 9: 1-38

VENRICK, E.L. 1988. The vertical distributions of chlorophyll and phytoplankton species in the North Pacific central environment. *J. Plankton Res.*, 10(5): 987-998.

VIVES, F. 1966. Zooplancton nerítico de las aguas de Castellón (Mediterráneo occidental). *Inv. Pesq.*, 30: 49-166.

VIVES, F. 1986. Succession des différents groupes du zooplancton sur les côtes catalanes au cours d'un cycle annuel: juillet 1982-juillet 1983. *Rapp. Comm. int. Mer Médit.,* 30(2):200.

WIEBE, P.H. 1970. Small scale spatial distribution in oceanic zooplankton. *Limnol. Oceanogr.*, 15: 205-217.

ZABALA, M., RIERA, T., GILI, J.M., BARANGÉ, M., LOBO, A. & PEÑUELAS, J. 1989. Water flow, trophic depletion, and benthic macrofauna impoverishment in a submarine cave from the Western Mediterranean. *Mar. Ecol.*, 10(3): 271-287.

ZABALA, M., CAROLA, M., COMA, R. & RIERA, T. Horizontal transfer of matter by a cave-dwelling mysid. *Mar. Ecol.* (submitted).

Homage to Ramon Margalef; or, Why there is such pleasure in studying nature 317
(J. D. Ros & N. Prat, eds.). 1991. *Oecologia aquatica*, 10: 317-323

STEPHOS MARGALEFI SP. NOV. (COPEPODA: CALANOIDA) FROM A SUBMARINE CAVE OF MAJORCA ISLAND (WESTERN MEDITERRANEAN)

TECLA RIERA[1], FRANCISCO VIVES[2] & JOSEP-MARIA GILI[3]

[1]Departament d'Ecologia. Facultat de Biologia. Universitat de Barcelona. Av. Diagonal, 645. 08028 Barcelona. Spain.
[2] Institut d'Estudis Avançats de les Illes Balears (CSIC). Crtra. de Valldemossa, km 7.5. 07071 Palma de Mallorca. Spain.
[3] Institut de Ciències del Mar (CSIC). Pg. Nacional, s/n. 08039 Barcelona. Spain.

Received: February 1990

ABSTRACT

Stephos margalefi sp. nov. is described as a new species from a submarine cave of Majorca (Balearic Islands). The main diagnostic character is the structure of the fifth legs in both sexes, which distinguish it of the 19 earlier known species of the genus *Stephos*.

KEY WORDS: *Stephos margalefi*, Copepoda, Calanoida, submarine caves

INTRODUCTION

The genus *Stephos* is generally formed by species of small size (many of them hardly reach one millimeter) which live near the coast, in shallow waters, close to the bottom; some species colonize peculiar habitats, such as marine caves.

After the recent papers by STRÖMGREN (1969), FOSSHAGEN (1970), KOS (1972), ANDRONOV (1974), GREENWOOD (1978) and BOXSHALL *et al.* (1990) a total of 20 species are known, including one species, *S. seclusum*, recently described by Barr but not yet published.

The genus *Stephos* is distributed in different latitudes: from arctic to tropical zones and in the temperate zones, which is where most of their species have been described.

MATERIAL

The specimens were collected on 18 September 1987 in a submarine cave, La Catedral (NE of Majorca, Balearic Islands, Spain; see ROS & GILI, 1985, and BIBILONI *et al.*, 1989, for a detailed description) by means of plankton catches carried out at midday with a net trailed by a diver for three minutes. In this time, several forward and backward runs were carried out in the cave, covering a total distance of 200 m. The net was conical, with a length of 80 cm, a mesh size of 60 µm and a mouth of 20 cm diameter.

The holotype, one adult female, body length 0.77 mm, is in the Institut de Ciències del Mar, Barcelona, Spain. The paratype, one adult female, body length 0.78 mm, is in the Rijksmuseum van

Natuurlijke Historie, Leiden, the Netherlands.

HABITAT

The catches were made at different points and the copepods were always found in the same zone of the cave. This allows us to delimit the characteristic habitat where these animals live. The specimens are concentrated near the innermost part of the left tunnel of the cave (Fig. 1) and tend to be situated near the bottom. In this zone, the cave has a fine sand and mud substrate with continuous arrivals of detritus (like remains of sea-grasses and sea-weeds) from outside the cave. The entrance to the left tunnel of La Catedral cave is exposed to currents (due to the north winds dominating this zone), and thus the trophic inputs are assured.

The hydrodynamic effect of the currents determines the resuspension of the sediment, thus enhancing the feeding possibilities of the copepods. So their feeding habits are suspensivorous-detritivorous, and their localization in a well defined area of the cave would be a consequence of this.

The sample is composed of different zoological groups, especially of copepods: *Ridgewayia* spp., *Clausocalanus* spp., and one not yet identified harpacticoid genus.

ADULT FEMALE

The overall length of 4 individual females ranged between 0.77 and 0.80 mm, with an average of 0.78 mm. The length of the cephalosoma is 0.57 mm and of the abdomen is 0.21 mm with a ceph/abd ratio = 2.73. The maximum body width is 0.25 mm, which is found in the first thoracic segment.

The cephalosoma is hard and strong and in dorsal view reminds one of *Clausocalanus*. The head is attached to the first thoracic segment. So the anterior body consists of three thoracic segments plus the cephalosoma. The last segment, which is clearly delimited in the dorsal part, is formed by the union of the 4th and the 5th segments.

Each one of these segments has a pair of normal legs with endopods and exopods formed by three segments each, with the exception of the 5th pair, which has a single branch formed by an endopod divided into two segments.

CEPHALOTORAX OR ANTERIOR BODY

The first antennae (A_1) are made up of 24 segments. Each one of the four distal segments has a similar length and is a little more than twice the intermediary segments. The proximal segments are a little longer than the intermediary ones, but generally they are shorter than the distal segments (Fig. 2, b).

The strong thick rostrum is downturned and does not carry rostral filament.

The second antennae (A_2) have two basipods followed by a longer endopod and a shorter exopod; both are divided into two

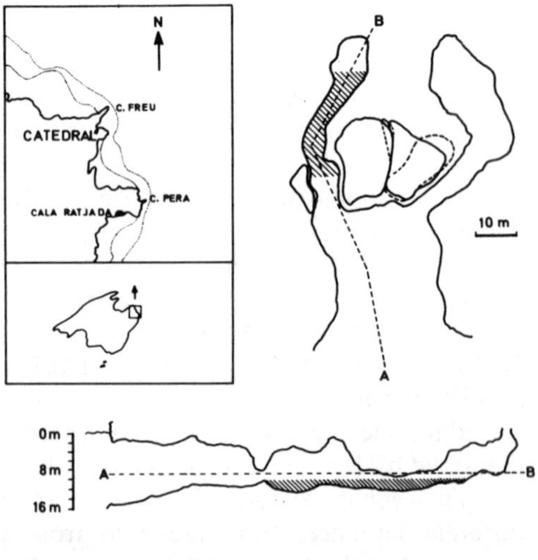

FIGURE 1. Geographic situation of the submarine cave La Catedral at the NE of Majorca (left). Axial projection (right) and section (bottom) of the cave. Sampling area: stripped.

segments. The second basipod holds one outer seta and the first segment of the exopod, two setae (one basal, very similar to that of the second basipod, and another shorter, situated distally, in the second third of the total length).

After that, we find a very short second segment which ends in two groups of five setae each. The endopod is formed by two similar segments, the distal one being shorter than the basal. This last segment holds a pair of setae situated near their union with the second segment.

This last segment shows a seta in the middle part of the outer margin and another one, very similar, on the inside margin, ending in five distal setae of equal length (Fig. 4, a).

The mandible (Md) has a strong cutting blade, with a cutting tooth, a longer, curved inside cutting tooth separated from the first by a semicircular cleft followed by 7 teeth of small and decreasing size, ending in sharp points.

One of the borders of the masticatorial plate is fine and uniform; the other one shows a group of fine hairs in its central part.

The palps carry three setae in the inner part, showing a smooth border without setae in the outer part.

The endopod is formed by two equal segments: the first one holds four setae in the inner border, while the second one has between 6-10 terminal setae. The exopod is formed by four segments of similar length but of decreasing width. Every segment has a basal seta and the distal segment ends in two setae of similar appearance and length to those of the appendix.

The maxillae (Mx) are formed by an exopod and an endopod, both inserted on a biarticulated basipod. The maxillae are constituted by several lobes with numerous setae (fig. 4, c).

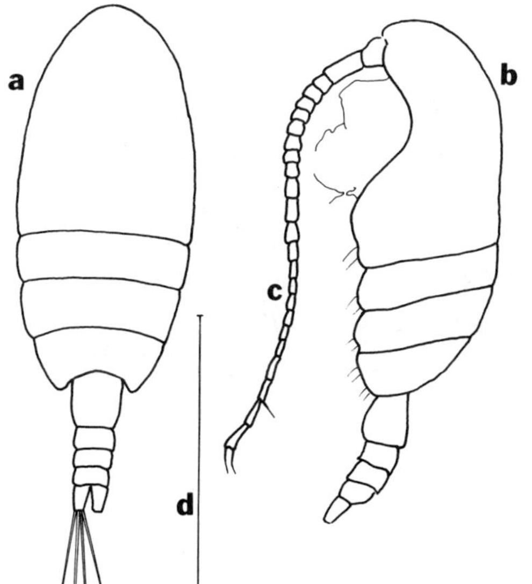

FIGURE 2. *Stephos margalefi* n. sp. (female). a: dorsal view; b: lateral view; c: antenna (A_1); d: 0.5 mm.

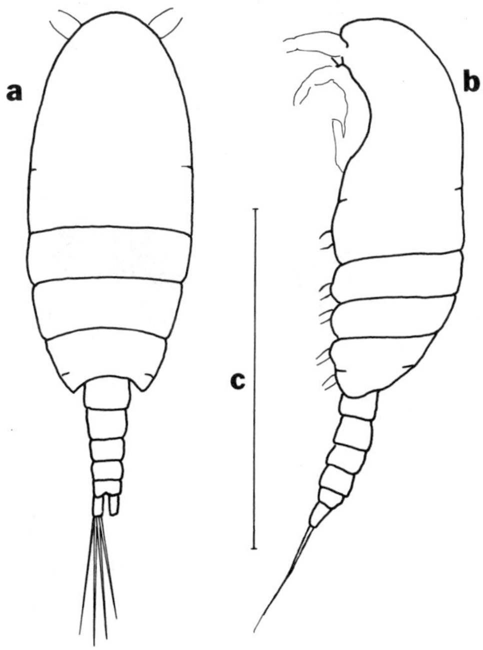

FIGURE 3. *Stephos margalefi* n. sp. (male). a: dorsal view; b: lateral view; c: 0.5 mm.

The first maxilliped (Mxp_1) shows a bisegmented basipod followed by the exopod. Both the B_1 and the B_2 hold several lobes (5) and all of them end in three plumous setae, but the first one carries four setae (Fig. 4, d).

The second maxilliped (Mxp_2) is formed by two basipods followed by the exopod with five segments. The distribution of the setae can be seen in figure 4, e.

The first pair of legs (L_1) is made up of an exopod formed in turn by three regular segments with no characteristic aspect. The endopod in one segment, shows a strong digitation in the outer upper part (Fig. 5, L_1). Both segments are linked to the second basipod.

The second pair of legs (L_2) is formed by three segmented exopods and two segmented endopods. The first segment of the exopod is longer than it is wide; the second one shows a length twice its width and the third one is three times longer. The endopod, on the other hand, has the first segment as long as it is wide, and the second one is about three times longer than the first.

Meanwhile, the exopod shows a strong terminal spine and one seta; the endopod, with no spine, ends in a pair of setae.

The third and fourth pair of legs (Fig. 5, L_3 and L_4) show normal three-segmented exopods and endopods. Both pairs end in a strong spine and one seta of similar length (in the 3rd pair) and with a spine, a little longer than the seta, in the 4th pair.

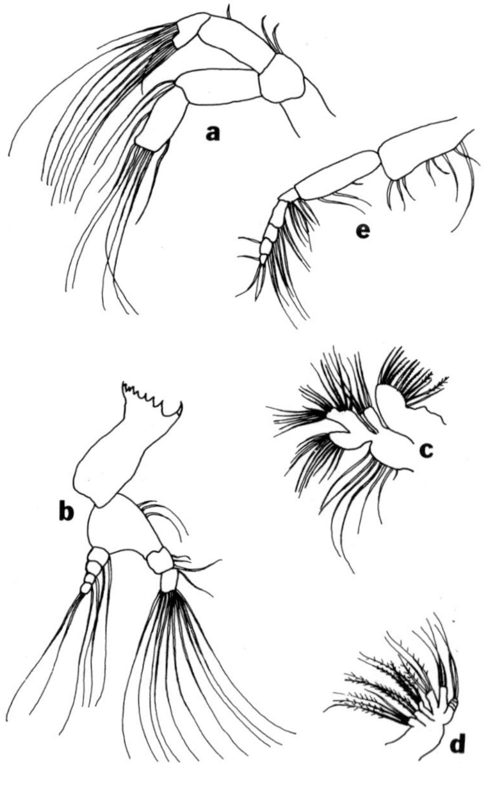

FIGURE 4. *Stephos margalefi* n. sp. (female). a: second antenna (A_2); b: Mandible (Md); c: maxillae (Mx); d: first maxilliped (Mxp_1). e: second maxilliped (Mxp_2).

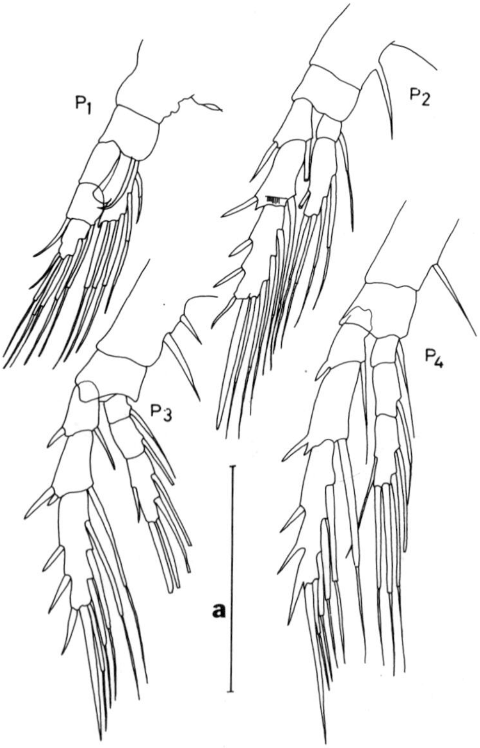

FIGURE 5. *Stephos margalefi* n. sp. (female). P_1-P_4: first, second, third and fourth pairs of legs. a = 100 mm.

TABLE I. Size (in μm) of male and female specimens of *Stephos margalefi* n. sp.

	Females				Mean	Males				Mean
Total length	780	772	804	778	783	747	739	731	731	731
Length of cephalosoma(c)	569	569	585	577	575	536	575	536	528	534
Length of abdomen (a)	211	199	219	211	210	211	203	195	203	203
Ratio c/a					2.73					2.63
Width (maximum)	244	247	260	252	251	236	228	228	229	230
Length of furcal rami (l)	32	32	32	32	32	32	37	36	36	35
Width of furcal rami (av)	24	24	24	24	24	23	23	21	23	22
Ratio l/av					1.3					1.5

Spines are similar in length to the distal segment. The first basipod of the third pair of legs shows setae of different length, whereas the first basipod of the fourth pair of legs has only one seta.

Except for the first segment of the first pair of legs, all the segments of the four legs show a strong spine at the distal part of the outer margin. In the inner margin of all segments of exopods of the four legs there are setae made up of two parts: the basal one, the ramus, of circular section, is well constituted; and the distal half part is plumous and much more feeble.

The 5th leg (L_5; Fig. 6) is formed by one branch only, composed of two segments of similar length and width. The distal segment ends in an inclined border showing several teeth in "saw" and in the outer margin has a strong spine of similar length to the width of the segment.

ABDOMEN OR LAST BODY

The abdomen is formed by four segments plus the furca. The genital segment, which shows a ventral and posterior protuberance, is twice as long as the 2nd and 3rd segments, which are very similar to it, and a little longer than the 4th (or anal) segment.

The furcal ramus is likewise longer than the anal segment and its length is about one and a half times its width. The furca shows four setae: the two central ones are longer than the lateral one, and not much longer than the total length of the abdomen.

ADULT MALE

The average length of the four individuals was 0.74 mm, with a range of 0.73-0.74.

The average length of cephalosoma was 0.53 mm, and the abdomen measured 0,20 mm with a ratio between them of 2.63. The sexes hardly show any difference in size (Table I).

The maximum body width was 0.23 mm.

The cephalosoma is very similar to that of the female (strong, robust and composed of the head and the first thoracic segment, although in some specimens a slight division can sometimes be distinguished). Like the female, the body is three-segmented, the last segment of which represents the fusion between the 4th and 5th segments, and, as in the head, a first segment. Here, in some specimens, we can also observe the union line between both segments.

Both the first antennae and the different mouth parts (mandible, maxillae and maxillipeds) show an identical constitution to the female. Likewise, the four first pairs of legs are also similar in both sexes. Only the 5th pair is different and very characteristic in the male: large and with an elaborate structure (Fig. 6, b).

The abdomen has 5 segments. The first and second are longer than the third and fourth. The anal segment is the shortests and its length is smaller than that of the furca. These, like the female, are provided with four setae of similar size.

REMARKS

The 20 known species of the genus *Stephos* are listed in Table II. The first nine species are distributed in temperate and boreal regions (except *S. morii*, collected in Australian waters), and all of them have a similar general size and morphology. The most similar species to *S. margalefi*, *S. gyrans*, was collected in the Mediterranean Sea. The 5th legs of the males form a complex grasping organ differing greatly from species to species. All temperate and boreal species have a swollen penultimate segment bearing "membranous lamellae" (J. Barr, pers. comm.). At the same time, the shape of the last prosomal segment, the 5th legs, and the genital segment of the females are important systematic characters (A. Fosshagen, pers. comm.) In this way, the structure of both male and female 5th legs shows a morphological form that distinguishes *S. margalefi* from other members of the genus described earlier. Specifically, the length of the segments in the right 5th leg and the terminal segment of the 5th right leg of males are very different from those of the other known species (Fig. 6).

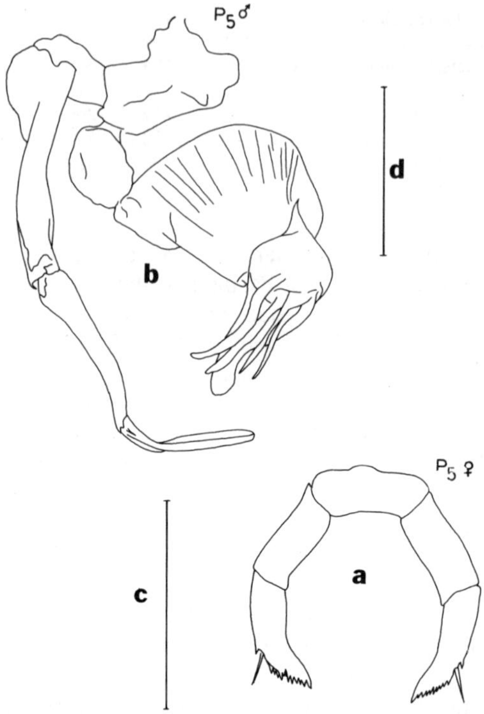

FIGURE 6. *Stephos margalefi* n. sp. a: fifth pair of legs (female); b: fifth pair of legs (male). c and d: 100 mm.

TABLE II. The 19 species of the genus *Stephos* known before the description of *S. margalefi*.

- *S. gyrans* (giesbrecht, 1892)
- *S. lamellatus* Sars, 1902
- *S. arcticus* Sars, 1902
- *S. morii* Greenwood, 1978
- *S. tsuvazakiensis* Tanaka, 1966
- *S. scotti Sars,* 1902
- *S. minor* Scott, 1892
- *S. kurilensis* Kos, 1972
- *S. fultoni* Scott, 1898
- *S. lucayensis* Fosshagen, 1970
- *S. deichmannae* Fleminger, 1957
- *S. rustadi* Strömgren, 1969
- *S. maculosus* Andronov, 1974
- *S. tropicus* Mori, 1942
- *S. exumensis* Fosshagen, 1970
- *S. antarcticum* Wolfenden, 1908
- *S. longipes* Giesbrecht, 1902
- *S. pentacanthos* Quin-Chao, 1974
- *S. canariensis* Boxshall, Stock & Sánchez, 1990
- *S. seclusum* Barr (not yet published)

DERIVATIO NOMINIS

The new species is named as an homage to Dr. Ramon Margalef, who at different times introduced the authors to the study of marine zooplankton.

ACKNOWLEDGEMENTS

Thanks are due to Dr. A. Fosshagen for valuable help and advice, and to Dr. D.J. Barr for his aid and helpful suggestions. Also thanks to Mr. J. Lenard, for his help during the plankton sampling.

REFERENCES

ANDRONOV, V.N. 1974. New species of Diaixidae and Stephidae (Copepoda) from the northern regions of the Indian Ocean. *Zool. Zh.*, 53: 460-464 (in Russian).

BIBILONI, M.A., URIZ, M.J. & GILI, J.M. 1989. Sponge communities in three submarine caves of the Balearic Islands (Western Mediterranean): Adaptations and faunistic composition. *P.S.Z.N.I.: Mar. Ecol.*, 10:317-334.

BOXSHALL, G. A., STOCK, J. M. & SÁNCHEZ, E. 1990. A new species of *Stephos* Scott, 1982 (Copepoda: Calanoida) from an anchialine lava pool on Lanzarote, Canary Islands. *Stygologia,* 5(1): 33-41.

FOSSHAGEN, A. 1970. Marine biological investigations in the Bahamas 12. Stephidae (Copepoda, Calanoida) from the Bahamas, with remarks on *Stephos sinuatus* and *S. arcticus* Sars, *Sarsia*, 41: 37-48.

GREENWOOD, J.G. 1978. A replacement name *Stephos morii* proposed for the preoccupied name *S. tropicus* (Copepoda, Calanoida). *Proc. R. Soc. Qd.*, 89: 131.

KOS, M.S. 1972. A new species of *Stephos* (Copepoda, Calanoida) from coastal waters of the Kuril Islands. *Crustaceana*, 23: 113-118.

ROS, J.D. & GILI, J.M. 1985. Opisthobranches des grottes sous-marines de l'Île de Majorque (Baléares). *Rapp. Com. Int. Mer. Médit.*, 29: 141-145.

STRØMGREN, T. 1969. A new species of *Stephos* (Copepoda, Calanoida) from the Norwegian West coast. *Sarsia, 37*: 1-8.

Homage to Ramon Margalef; or, Why there is such pleasure in studying nature (J. D. Ros & N. Prat, eds.). 1991. *Oecologia aquatica*, 10: 325-335

SURVIVAL, OPPORTUNISM AND CONVENIENCE IN THE PROCESSES OF CAVE COLONIZATION BY TERRESTRIAL FAUNAS

XAVIER BELLÉS

Centre d'Investigació i Desenvolupament (CSIC). Jordi Girona, 18. 08034 Barcelona. Spain.

Received: May 1990

SUMMARY

Modern discoveries of cave-restricted animals in many tropical limestone areas, in lava tubes and other non-calcareous caves, and in subterranean non-cave habitats, have led us to reconsider classic interpretations of cave colonization processes, which were mainly based on the direct influence of Pleistocene climatic changes. The relict concept (for cave species) and the refuge concept (for caves) have lost their value as paradigms, and cave colonization directly determined by Pleistocene glaciations should rather be considered as a localized model. Organisms can colonize caves for a variety of reasons, including survival (caves as immediate refuges), opportunism (caves as virgin spaces to be colonized) or convenience (caves as escape valves when new competitive pressures appear in surface habitats). All these styles of colonization can be well exemplified in temperate areas which suffered the direct impact of Pleistocene glaciations. However, all of them can also be postulated in other situations which have nothing to do with climatic revolutions.

KEY WORDS: Biospeleology, troglobites, cave colonization.

INTRODUCTION

The first cave-restricted animal to be studied scientifically was the amphibian *Proteus anguinus*, from Yugoslavian caves, which was described by Laurenti in 1768. However, the beginnings of a systematic research on cave biology date from the early twentieth century, when Racovitza and Jeannel sparked off the spectacular rise of modern biospeleology in Europe.

One of the first proposals made by this European school was a general division of cave animals into three categories: troglobites (cave-restricted or cave-obligated that are unable to survive in epigean habitats), troglophiles (cave facultative that can reproduce in caves but are also found in cryptic and humid epigean habitats), and trogloxenes (cave dwellers that cannot reproduce in caves, most of which usually return to epigean environment to feed). The most peculiar cave-adapted group is, of course, that of troglobites, which have been typified by a series of morphological features, including loss or reduction of eyes, wings and cuticular pigments, and an increase of sensorial devices, often accompanied by the development of longer appendages and a more slender body form: the so-called "troglobiomorphic" characters. From a physiological point of view, many troglobites tend to lose their circadian rhythms, show a lower metabolism, and display specialist reproductive and life cycle strategies: the so-called K-strategies.

In the 60's, data available on cave biology were summarized in two key monographs: VANDEL's *Biospéologie* (1964), with an encyclopaedic character although emphasizing European findings, and BARR's *Cave ecology and the evolution of troglobites* (1968), primarily focused on ecological and evolutionary studies, and emphasizing North American examples. Besides the intrinsic value of both monographs as landmarks in biospeleological literature, they had the additional value of generating a renewed interest in cave animals. Progressively, new data were added and old concepts became surpassed by new findings, especially in the last fifteen years, when there has been a virtual revolution in our understanding of cave biology. Therefore, fast progress and original research have been achieved in different domains, from organismic to evolutionary level, as shown by modern reviews (CULVER, 1982; HOWARTH, 1983; JUBERTHIE, 1984, 1989). The present essay will deal with recent discoveries related to cave colonization processes by terrestrial faunas, and how these findings have modified the classic concepts which dominated in the 60's.

CAVE FAUNAS FROM LIMESTONE TEMPERATE AREAS. THE CLASSICAL DATA

VANDEL (1964) provided a thorough overview of the whole discipline of biospeleology from a worldwide perspective. According to the information compiled in this monograph, the richest troglobitic faunas were those of calcareous regions in Southern Europe, Northern Africa and Southeastern United States. Outside these main regions, selected discoveries in lava tubes and in limestone caves in Japan, New Zealand and Mexico, suggested the occurrence of well developed troglobitic faunas in these areas as well.

Early work by Leleup in Afrotropical caves was also reviewed in VANDEL (1964), but the scarcity of terrestrial "troglobiomorphic" animals found in this region prompted the idea that the vast majority of troglobites were to be found in temperate limestone caves. This view was followed by other biospeleologists primarily working in North America, like BARR (1968), who also stressed the importance of limestone caves stating that "the presence of troglobites in lava tubes, gypsum caves, and sea caves has been established, but is exceptional" (BARR, 1968: 47).

These findings led to the proposal of *ad hoc* theories accounting for cave colonization processes, which were largely based on the direct influence of palaeoclimatic changes. Therefore, the general consensus was that the ancestors of terrestrial troglobites were stenotherm and hygrophilic species which lived in subepigean habitats, such as moss, leaf litter of deep soil, and in some way were "preadapted" to cave life.

Accordingly, it was argued that most cave animals presumably derive from subepigean thermohygrophilic ancestors which had colonized the caves before the Pleistocene to survive *in situ* during the transition from tropical to temperate climate which occurred between the Tertiary and the Quaternary (VANDEL, 1964; JEANNEL, 1965). In turn, Pleistocene glaciations should have prompted new processes of cave colonization protagonised by cryohygrophilic faunas as the glaciers retreated and the climate became progressively warmer and drier (VANDEL, 1964; BARR, 1968).

Many chorological data of selected cave species living in temperate areas fit well with these assumptions, since their distributions show a high coincidence with the limits of those areas which were ice-covered during the Pleistocene. Examples of such distributions have been shown both in Europe (VANDEL, 1964) and in North

America (BARR, 1968), and this correspondence associated the caves with the idea of refuge, which was largely exploited by the proponents of the orthogenetical theories, who postulated that troglobites were a sort of "live fossils" (JEANNEL, 1943) in a senescent stage of their phyletic evolution (VANDEL, 1964).

TROGLOBITES IN TROPICAL CAVES AND IN NON-LIMESTONE SUBTERRANEAN HABITATS

In the last fifteen years there has been an extraordinary increase in information about cave faunas in the tropics. HOWARTH's (1983) review of the ecology of cave arthropods provides a checklist of tropical areas where terrestrial troglobites from limestone caves have been discovered. These areas include Hawaii, New Guinea, Sarawak, Malaysia, Thailand, Malgasy, Jamaica, Puerto Rico, Cuba, Mexico and Central America, Venezuela and Brazil. New prospections allow us to extend this list to other tropical areas, especially in South America (Ecuador: BESSON *et al.*, 1982; Peru: MARTÍNEZ *et al.*, 1983) and the West Pacific (Australia, New Caledonia, Philippines, Fiji: STRINATI & AELLEN, 1983; PECK, 1983)

Concerning non-limestone caves, the pioneering work on lava tube faunas done in Japan (UENO, 1971) and in Hawaii (HOWARTH, 1972) was followed by new discoveries in volcanic caves in other areas, like the Galápagos Islands (see PECK &

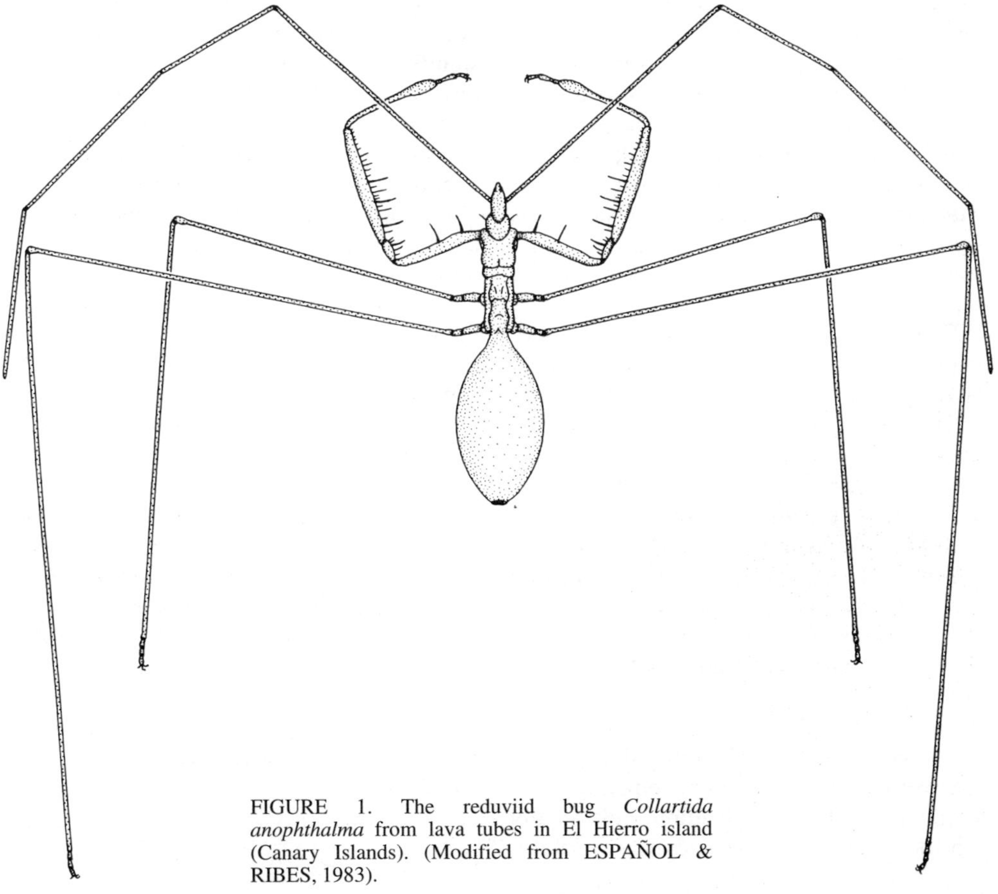

FIGURE 1. The reduviid bug *Collartida anophthalma* from lava tubes in El Hierro island (Canary Islands). (Modified from ESPAÑOL & RIBES, 1983).

KUKALOVA-PECK, 1986) or the Canary Islands (see HERNÁNDEZ *et al.*, 1986). Cave species have also been found in caves or mines excavated in different materials like sandstone, tuff, shale, chert, breccia, rhyolite, diorite and pegmatite (see HOWARTH, 1983).

In addition, the cave environment is now considered to be much larger than previously assumed, especially after the discovery of troglobites in the voids in fractured rock and talus slopes, both in calcareous and in non calcareous (volcanic and other) materials (UENO, 1977; HOWARTH, 1973; OROMÍ *et al.*, 1986), even at short distances from the surface (MSS: *milieu souterrain superficiel;* JUBERTHIE *et al.*, 1981).

It is worth noting that, although poorly modified species still dominate among tropical cave faunas, many of them show highly "troglobiomorphic" features. To give just a few examples, we can mention the reduviid bug *Collartida anophthalma* (Fig. 1) from lava tubes in the Canary Islands (ESPAÑOL & RIBES, 1983), or the paradoxosomatid millipede *Selminosoma chapmani* and the sundathelphusid crabs *Rouxana phreatica* and *Holthuisiana alba* from limestone caves in Papua-New Guinea (CHAPMAN, 1986).

From a physiological point of view, only a few studies have been made on tropical cave species. BOUSFIELD & HOWARD (1976) described some details of the reproductive biology of terrestrial talitrid amphipods from lava tubes in Hawaii, which show a tendency towards K-type strategies, just as in many troglobite species in temperate regions. Conversely, metabolic rates of cave-adapted crickets from the same islands did not show significant differences when compared with those of surface related species (AHEARN & HOWARTH, 1982).

In connexion with all these new discoveries, the realization emerged that morphophysiological features classically considered as characteristic of troglobites are neither exclusive nor extensive to all of them. Thus, morphophysiological criteria are not sufficient to assess the degree of dependence on cave environment, and even less to infer the time of cave colonization by a given species. In short, some poorly modified cave species, for example from tropical areas, could be more genuine troglobites, with a longer history in caves, than other "ultraevolved" (VANDEL, 1964) cavernicolous species living in temperate regions.

Another important piece of evidence afforded by biospeleological studies in tropical regions is that many of the troglobites discovered are not relicts, as closely related species can also be found in the surface habitats in the same area (UENO, 1977; HOWARTH, 1983; JUBERTHIE, 1984; CHAPMAN, 1986).

The discovery of cave species in the voids in fractured rock is especially significant, not only because it has an obvious ecological and chorological interest, but also because it casts new light on the reconstruction of cave invasion phenomena. Indeed, colonization of the more superficial fissures could, in most cases, be a previous stage towards cave invasion.

The fact that troglobites can be sampled in mines dug in limestone and other rock materials, as recognized by HUSSON (1936) (see also UENO, 1977), indicates that cave faunas may effectively migrate through the voids within fissured rock, and suggests that this mesocavernous system may be suitable for troglobites. In fact, quantitative studies on cave populations based on mark-recapture methods often indicate that most of these populations stay in inaccessible fissures (DELAY, 1978). These studies have also demonstrated effective migration between the cave and the voids of superficial rock (MSS) just beneath the soil level (Bellés & Gracia, unpublished).

CAVE COLONIZATION. THE PARADIGM OF CLIMATIC REVOLUTIONS REVISITED

These recent discoveries summarized above have led us to reconsider the role of climatic revolutions in the processes of cave colonization, and also the concept of "live fossils", which has been, and still is, generally applied to troglobites. It is now clear that not all the cave obligated species are relicts, but it is equally apparent that many of them can fit this definition. To give some selected examples from the Iberian Peninsula (see BELLÉS, 1987), the spider *Telema tenella,* the pseudoscorpion *Troglobisium racovitzai,* the isopod *Cantabroniscus primitivus,* the millipede *Cantabrodesmus lorioli*, the diplurans *Oncinocampa falcifer, Paratachycampa hispanica* and *P. peynoensis,* and the beetle *Ildobates neboti,* which show primitive features and are geographically very distant and isolated from their nearest relatives, can be considered as troglobite relict species.

Furthermore, recent studies combining biotaxonomical data with a good knowledge of Pleistocene glaciation timing in selected areas, and stressing the importance of the glacial-interglacial cycles, have cast new light on the paradigm of climatic revolutions. Studies by PECK (1981) based on the catopid beetles of the *Ptomaphagus hirtus* group from Grand Canyon caves in North America, and those of JUBERTHIE (1988) and coworkers carried out on the *Speonomus delarouzeei* group from Eastern Pyrenean caves, in the Iberian Peninsula, are especially representative.

According to these studies, it has been proposed that ancestors of these species invaded the cave habitat during glacial periods (when the forest extended down to the altitude of caves), and interglacial warming caused the extinction of surface populations of the species that have invaded the caves. Accordingly, species isolated in one interglacial period and supposed to be capable of overland dispersal in the next glacial, could have protagonised new processes of cave colonization, isolation and speciation, thus accounting for the taxonomic-chorological complexity which is usually found in these species groups (CULVER, 1982).

Although this scenario appears to be reasonable, it is worth underlining the fact that it serves for concrete and local cases which can be explained when conveniently contextualized, and not as a general rule. Let us examine some other situations which do not fit these explanations.

Firstly, cave species have been discovered in areas which were ice covered in Pleistocene times. In some cases, it can reasonably be argued that they might have survived extended periods of glaciation in deep subterranean habitats beneath the ice. The relict distribution of several freshwater amphipod species of *Stygobromus*, presently living in subterranean waters in Canada, could be explained by the above hypothesis (HOLSINGER & SHAW, 1986).

Secondly, and as already stated, the occurrence of cave species in tropical regions far from those which were glaciated

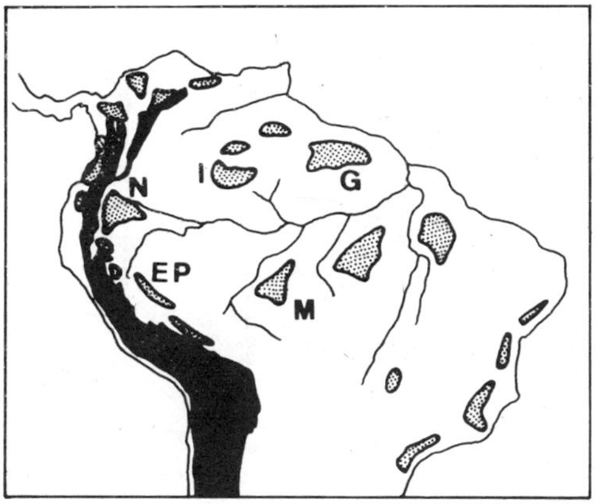

FIGURE 2. Map of tropical America showing the forest refugia during Pleistocene glacial maxima. The Andes (above 1000 m) are indicated as solid black. EP, East Perú; G, Guyana; I, Imeri; M, Madeira; N, Napo. (From LYNCH, 1988, slightly modified).

in Pleistocene times is largely demonstrated. This led us to examine the controversy between authors who claim that glaciations only influenced the glaciated areas and their boundaries (HOWARTH, 1982; JUBERTHIE, 1984), and those who maintain that glaciations influenced the whole planet (REDDELL, 1981).

Since Pleistocene refuge areas in tropical regions are well documented (LYNCH, 1988; Fig. 2), it thus seems reasonable to consider that glaciations influenced to a greater or lesser extent the entire globe. Therefore, it might be possible that, although indirectly, glaciations influenced cave colonization processes also in tropical areas.

In summary, the adventure of cave invasion could have been prompted by a variety of factors, including survival (caves as immediate refuges), opportunism (caves as virgin spaces to be colonized) or convenience (caves as escape valves when new competitive pressures appear in surface habitats). Indeed, all these styles of colonization can be well exemplified in temperate areas which suffered the impact of glaciations. However, and as we will see, all of them can also be reasonably postulated in other situations which have nothing to do with climatic revolutions.

SURVIVAL. THE CAVE AS A REFUGE

Certainly, caves could have served in some cases as immediate refuges to avoid surface abiotic stress. Many of the examples used to illustrate cave colonization by cryohygrophilous faunas prompted by the post-Pleistocene glaciers retreat seem to fit this hypothesis. Carabid beetles of the genus *Aphaenops* in the Pyrenees and of the genus *Pseudanophthalmus* in North America, are classic examples reviewed in detail by VANDEL (1963) and BARR (1968), respectively. An additional piece of evidence which tends to support the survival hypothesis for these carabid beetles is the occurrence of species living in the surface in high altitude forests (*Aphaenops ludovici* in the Northern Pyrenees and *Pseudanophthalmus sylvaticus* in West Virginia, for example) which are closely related to the cave obligated ones (CULVER, 1982; JUBERTHIE, 1984).

Habitat preferences as a function of temperature in several non-troglobite

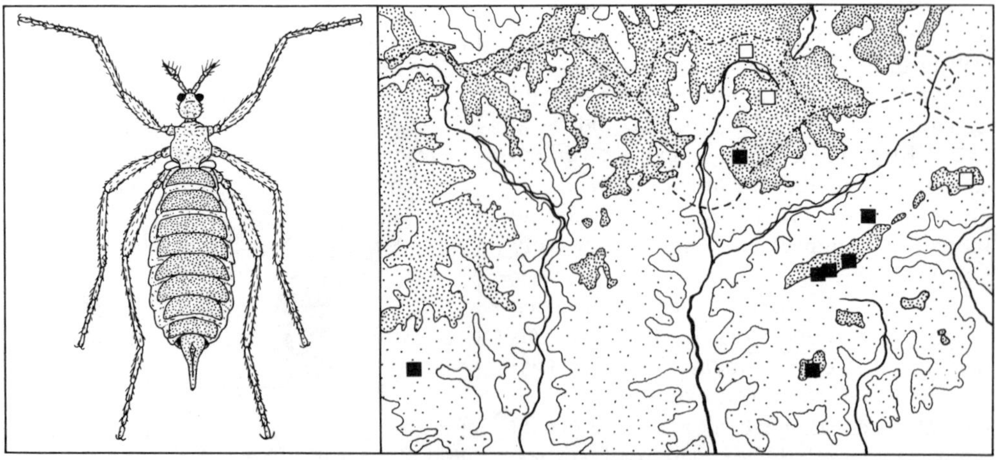

FIGURE 3. The tipulid dipteran *Niphadobata catalonica* and its distribution in the Eastern Pyrenees. The white squares indicate epigean localities and the black ones indicate caves. Areas above 2000 m are shown as densely dotted, and those between 1000 and 2000 m as sparsely dotted.

species also suggest the likelihood of such a mode of colonization being induced by abiotic stress in surface habitats. The Palaearctic carabid beetle *Porotachys bisulcatus* is an example. This species has clear preferences for forest habitats in northern localities, whereas in southern ones it shows a clear tendency to occupy caves. It finally becomes a typical troglophile in the Balearic Islands, Southern Iberian Peninsula and Northern Africa (BELLÉS, 1987). Also illustrative is the Eastern Pyrenean distribution of the dipteran *Niphadobata catalonica*, an apterous tipulid which, at altitudes higher than 2000 m, lives either in caves or in surface habitats, under stones or on the snow (for example in the Coll de la Devesa, at some 2500 m a.s.l.). However, between 1000 and 2000 m (for example in Sant Gervàs mountains), it is exclusively found in caves (Fig. 3) (ESCOLÀ & BELLÉS, 1977).

Further. In the National Park of Cutervo, established in 1961 near Cajamarca, in Peru, large areas of rain forest used to be destroyed in past decades to exploit the land for pasture or agricultural purposes. Both in the entrance and in the deeper zones of the caves located in the ancient forest areas live different species of carabid beetles (*Andinorites, Trechisibus, Incatrechus;* MATEU & BELLÉS, 1981) which are absent in epigean habitats of the same area (now occupied by pastures and agricultural land) and which show close relationships with forest species. Therefore, examples of cave colonization as a solution to survival under surface stress can also be found in present circumstances which have no relation to climatic revolutions.

OPPORTUNISM. THE CAVE AS AN EMPTY SPACE

A cave can also be envisaged as an empty space able to be colonized, also in absence of biotic or abiotic surface stress. Concepts from the island biogeography theory (SCHOENER, 1988) could be applied *a priori* to caves, at least to illustrate how the colonization of an "empty" cave could have occurred. The equilibrium theory in its strict sense (MAC ARTHUR & WILSON, 1967), which explains the number of species of an island as a balance between immigration and extinction rates, emphasizing the importance of the area effect, seems less aplicable to caves. On the one hand, because studies made on cave faunas in this respect (VILLEUMIER, 1973) failed to demonstrate any area effect; on the other, because species turnover in troglobite communities should presumably be very low, at least in terrestrial faunas. As Culver pointed out "If the time scale is evolutionary, the immigrations are the isolation of species in caves. By contrast, under a historical hypothesis, the number of species is not an equilibrium but is rather the unique result of the history of the area" (CULVER, 1982:155). However, the analogy between caves and islands applied to both evolutionary time and ecological time, is still controversial, which suggests that more data are badly needed (MAY, 1977; CULVER, 1982; JUBERTHIE, 1989).

Of course, the most favoured candidates to succeed in the adventure of colonization of a virgin cave should be those species which arrive first and which are already adapted to live in stenothermal habitats, with high humidity, and more or less cryptic as, for example, humicolous species.

Rapid colonization of mines, with the establishment of gradients and a spatial distribution of the newly arrived fauna which is similar to those found in natural caves is a clear example of a process of colonization of an empty subterranean space, although in this case most of the colonists are troglobites from the adjacent mesocavernous system.

Lava tubes offer another example of a subterranean space with a recent history of colonization. Lava tubes and their surroundings formed in each volcanic episode represent an empty space to be rapidly colonized. The existence of a mesocavernous volcanic system which harbours troglobites (OROMÍ *et al.*, 1986) suggests that dispersal between this system and lava tubes effectively occurs, just as in limestone areas. However, lava tubes usually flow near the surface and are relatively ephemeral due to erosion effects. Thus, the cycle of formation, colonization and destruction of a lava tube occurs in a relatively short time. This, and the fact that lava materials can be accurately dated, means that reconstructions of lava tube colonizations might be easier than limestone cave systems. Another difference with respect to cave faunas in limestone temperate regions is that lava tube troglobites usually have near relatives in the epigean habitats of the same area. It thus becomes clear that the concepts of refugia and relict do not have any sense in these circumstances.

In this context of invasion of virgin subterranean spaces, the influence of glaciations can also be argued, not only as promoter of cave colonization due to abiotic stress in the surface, but also as promoter of new subterranean habitats. As pointed out by JUBERTHIE (1984), mechanical erosion and gelifraction as consequences of glaciarism, generates new spaces suitable for colonization (new talus slopes and voids in fractured rock). The retreat of glaciers should be, indeed, accompanied by the availability of these new spaces susceptible to first colonizations.

CONVENIENCE. THE CAVE AS AN ESCAPE VALVE

Another way to analyze the biogeography of cavernicoles is to superimpose their distribution ranges on those areas which were forest refugia during Pleistocene. Often a remarkable concentration of troglobite species appears in these ancient refugia, whereas in zones which were tundra-like in the Pleistocene, cave obligated species are almost lacking. General distributions of terrestrial troglobites in the Iberian Peninsula offer good examples of such a correlation (BELLÉS, 1987).

Glacial-interglacial intermittencies during Pleistocene times provoked, among other things, deep modifications in the distribution of vegetation. With each glacial advance, the forest shrank into fragments and with the retreat of glaciers, these fragments expanded to cover the grassland areas again and extended considerably. Therefore, contraction (to form a sort of forest island in a sea of grassland) provoked geographical isolation and promoted allopatric speciation, whereas forest expansion led to new enlargement of species ranges and to eventual contacts with their newly differentiated relatives.

This contraction-expansion scheme paralleling the glacial-interglacial intermittencies has been exploited mainly to explain the occurrence of highly speciose species-groups in zones which were on the boundaries of Pleistocene glaciers, and to reconstruct the speciation processes. Representative of these concepts are the studies of PECK (1981) on *Ptomaphagus hirtus* complex from Grand Canyon caves, and those of JUBERTHIE (1989) and coworkers on *Speonomus delarouzeei* complex from Pyrenean caves.

In addition, it is worth noting that these ideas can also be applied in areas which were far from direct glacial influence. The speciation processes undergone by cockroaches belonging to the genus *Loboptera* in volcanic subterranean environments in the Canary Islands, also seem largely influenced by contraction-expansion cycles in the Laurel forest, which occurred from the Pleistocene to the Holocene (MARTÍN *et al.*, 1986).

However, formation of refuge areas

might also be of interest when searching for reasons to explain the cave colonization processes. As pointed out by JUBERTHIE (1984) and BELLÉS (1987), another consequence of the formation of refuges, apart from speciation phenomena, would have been the concentration of forest organisms in these privileged areas, which became crowded while new competitive pressures appeared. These circumstances could have prompted cave invasions to avoid these pressures and, again, those species already linked to cryptic and stenotherm habitats would have been the candidates most likely to succeed in such an adventure (Fig. 4). It therefore becomes apparent that abiotic stress on the surface has had a rather indirect influence on these cases of cave colonization.

Finally, over the last years the Pleistocene tropical lowland forest refugia hypothesis has gained extensive support, and different tropical areas corresponding to forest refuges during glacial maxima have been well documented (for example Napo, Imeri, Guyana, East Peru and Madeira refuge areas in the Neotropical region: LYNCH, 1988; Fig. 2). Unfortunately, though promising *a priori*, biospeleological research studies have not been focused on these special areas, and nothing is known about the relative diversity and richness of their cave animals.

SOME CONCLUSIONS

Recent discoveries which resulted in a major breakthrough in our former understanding of cave colonization processes can be summarized in the following points:

1. Not all troglobites are relicts.
2. Not all troglobites are "troglobiomorphic" or K-strategists.
3. There are troglobites with a recent history of cave colonization.
4. Troglobites inhabit areas which, during Pleistocene times, were either ice-covered or far from the glaciers.
6. Troglobites also live in non-calcareous caves (lava tubes, sandstone caves) and

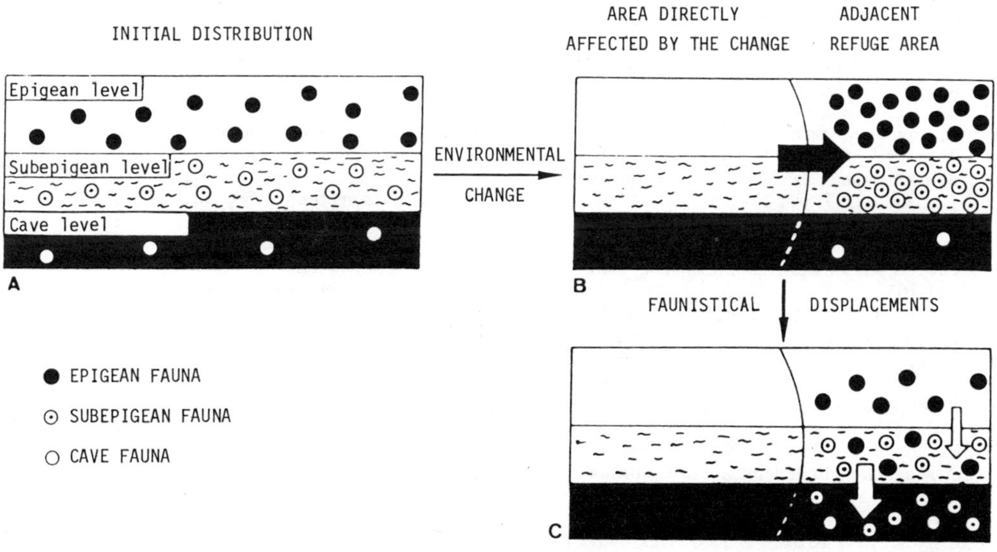

FIGURE 4. Hypothesis of competitive pressures in refuge areas. An environmental change (A) could lead to a faunal concentration in refuge areas (B), thus determining new competitive pressures and, in turn, faunistical displacements including cave colonizations (C). (From BELLÉS, 1987).

subterranean non-cave habitats, both in calcareous and in non-calcareous materials (voids in fractured rocks, talus slopes and so on).

The relict concept, for cave species, or the refuge concept, for caves, have lost their value as paradigms in cave biology; and cave colonization processes directly determined by the Pleistocene climatic revolution should rather be considered as a localized model.

It is clear that organisms colonize caves for a variety of reasons. Survival, opportunism and convenience have been pointed out here to show this variety of reasons, rather than to systematize and analyze them. Indeed, it seems probable that a cave colonization process should be more properly explained as a mixture of superimposed patterns rather than by a single pattern. Pleistocene glaciations prompted cave colonization for different reasons, including survival, opportunism and convenience. However, similar reasons can be found in other situations which have nothing to do with major climatic changes.

These new findings would, hopefully, provide a new point for more holistic reconstructions of cave colonization processes. We may have the impression that in the 60's biospeleologists were perhaps trying to run before they were able to walk. Now, although there is still much to be cleared up, we might be able to begin to walk without stumbling so often.

REFERENCES

AHEARN, G.A. & HOWARTH, F.G. 1982. Physiology of cave arthropods in Hawaii. *J. Exp. Zool.*, 222: 227-238.

BARR, T.C. 1968. Cave ecology and the evolution of troglobites. *Evol. Biol.*, 2: 35-102.

BELLÉS, X. 1987. *Fauna cavernícola i intersticial de la Península Ibèrica i les Illes Balears*. Moll-C.S.I.C. Palma de Mallorca, Madrid.

BESSON, J.P., LERA, D. & VALICOURT, E. 1982. *Ecuador 82. Expédition Spéléologique de la S.S.P.P.O.* Société de Spéléologie et de Préhistoire des Pyrénées Occidentales. Pau.

BOUSFIELD, E.L. & HOWARTH, F.G. 1976. The cavernicolous fauna of Hawaiian lava tubes, 8. Terrestrial Amphipoda (Talitridae), including a new genus and species with notes on its biology. *Pac. Insects*, 17: 144-154.

CHAPMAN, P. 1986. Non-relictual cavernicolous invertebrates in tropical Asian and Australasian caves. *Com. IX Congr. Int. Espeleol.*, Barcelona, 2: 161-163.

CULVER, C.D. 1982. *Cave life. Evolution and ecology*. Harvard University Press. Harvard.

DELAY, B. 1978. Milieu souterrain et écophysiologie de la reproduction et du developpement des Coléoptères Bathysciinae hypogés. *Mem. Biospéol.*, 5: 1-349.

ESCOLÀ, O. & BELLÉS, X. 1977. Sobre el gènere *Niphadobata* Enderlein a la Península Ibèrica. *Com. VI Symp. Espeleol., Biospeleol.*, Terrassa: 65-68.

ESPAÑOL, F. & RIBES, J. 1983. Una nueva especie troglobia de *Emesinae* (Heteroptera, Reduviidae) de las Islas Canarias. *Speleon*, 26-27: 57-60.

HERNÁNDEZ, J.J., MARTÍN, J.L. & MEDINA, A.L. 1986. La fauna de las cuevas volcánicas en Tenerife (Islas Canarias). *Com. IX Congr. Int. Espeleol.*, Barcelona, 2: 139-142.

HOLSINGER, J.R. & SHAW, D.P. 1986. A new stygobiont amphipod crustacean (Crangonyctidae, *Stygobromus*) from glaciated karst on Vancouver Island, Canada. *Com. IX Congr. Int. Espeleol.*, Barcelona, 2: 98-101.

HOWARTH, F.G. 1972. Cavernicoles in lava tubes on the island of Hawaii. *Science*, 175: 325-326.

HOWARTH, F.G. 1983. Ecology of cave arthropods. *Ann. Rev. Entomol.*, 28: 365-389.

HUSSON, R. 1936. Contribution à l'étude de la faune des cavités souterraines artificielles. *Ann. Sc. Nat. Zool.*, 19: 5-30.

JEANNEL, R. 1943. *Les fossiles vivants des cavernes*. Gallimard. Paris.

JEANNEL, R. 1965. La genèse du peuplement des milieux souterrains. *Rev. Ecol. Biol. Sol.*, 2 (1): 1-22.

JUBERTHIE, C. 1984. La colonisation du milieu souterrain; théories et modèles, relations avec la spéciation et l'évolution souterraine. *Mém. Biospéol.*, 11: 65-102.

JUBERTHIE, C, 1988. Paleoenvironment and speciation in the cave beetle complex *Speonomus delarouzeei* (Coleoptera, Bathysciinae). *Int. J. Speleol.*, 17 (1-4): 31-50.

JUBERTHIE, C. 1989. Insularité et spéciation souterraine. *Mém. Biospéol.*, 16: 3-14.

JUBERTHIE, C., DELAY, B. & BOUILLON, M. 1981. Extension du milieu souterrain en zone non calcaire: description d'un nouveau milieu et de son peuplement par les Coléoptères troglobies. *Mém. Biospéol.*, 7: 19-52.

LYNCH, J.D. 1988. Refugia. In: *Analytical biogeography* (A.A. Myers & P.S. Giller, eds.): 311-342. Chapman and Hall. London.

MAC ARTHUR, R.H. & WILSON, E.O. 1967. *The*

theory of island biogeography. Princeton Univ. Press. Princeton.

MARTÍN, J.L., IZQUIERDO, I. & OROMÍ, P. 1986. The genus *Loboptera* Brum. & W. (Blattaria, Blattellidae) in the Canary Islands and its distribution in the underground compartment. *Com. IX Congr. Int. Espeleol.*, Barcelona, 2: 142-145.

MARTÍNEZ, A., ROMERO, D., ROMERO, M. & RIBERA, C. 1983. El carst del nord del Perú. Expedicions Hirca-76 i Millpu-77. *Speleon*, 26-27: 147-180.

MATEU, J. & BELLÉS, X. 1981. Résultats faunistiques de diverses campagnes realisées en Amerique Latine. Espèces inédites de *Trechisibus* Motschulsky recoltées au Pérou. *Nouv. Rev. Ent.*, 11 (1): 39-45.

MAY, R.M. 1977. Caves as islands. *Nature*, 268: 589.

OROMÍ, P., MEDINA, A.L. & TEJEDOR, M.L. 1986. On the existence of a superficial underground compartment in the Canary Islands. *Com. IX Congr. Int. Espeleol.*, Barcelona, 2: 147-151.

PECK, S.B. 1981. Evolution of cave Cholevinae in North America (Coleoptera: Leiodidae). *Proc. VIII Int. Congr. Speleol.*, Bowling Green, 2: 503-505.

PECK, S.B. 1983. Recherches biospéologiques dans le Sud-Pacifique. *Mém. Speleo-Club Paris*, 9: 33-38.

PECK, S.B. & KUKALOVA-PECK, J. 1986. Preliminary summary of the subterranean fauna of the Galapagos Islands, Ecuador. Part II. The insects, evolution, and biogeography. *Com. IX Congr. Int. Espeleol.*, Barcelona, 2: 166-169.

REDDELL, J.R. 1981. A review of the cavernicole fauna of Mexico, Guatemala, and Belize. *Tex. Mem. Mus. Bull.*, 27: 1-327.

SCHOENER, A. 1988. Experimental island biogeography. In: *Analytical Biogeography* (Myers, A.A. & Giller, P.S., eds.): 483-512. Chapman and Hall. London.

STRINATI, P. & AELLEN, V. 1983. Voyage biospéologique autour du monde. *Mém. Speleo-club Paris*, 9: 7-32.

UENO, S.I. 1971. The fauna of the lava tubes around Mt. Fuji-san. I. Introductory and historical notes. *Bull. Natl. Sci. Mus.* Tokyo, 14: 201-218.

UENO, S.I. 1977. The biospeleological importance of non-calcareous caves. *Proc. VII Int. Congr. Speleol.*, Sheffield: 407-408.

VANDEL, A. 1964. *Biospéologie. La biologie des animaux cavernicoles*. Gauthier-Villars. Paris.

VUILLEUMIER, F. 1973. Insular biogeography in continental regions II. Cave faunas from Tessin, southern Switzerland. *Syst. Zool.*, 22: 64-76.

Homage to Ramon Margalef; or, Why there is such pleasure in studying nature (J. D. Ros & N. Prat, eds.). 1991. *Oecologia aquatica*, 10: 337-349

MEDITERRANEAN WOODY PLANT GROWTH-FORMS, BIOMASS AND PRODUCTION IN THE EASTERN PART OF THE IBERIAN PENINSULA

JAUME TERRADAS

Centre de Recerca Ecològica i Aplicacions Forestals (CREAF). Universitat Autònoma de Barcelona. 08193 Bellaterra. Barcelona. Spain.

Received: February 1990

SUMMARY

The major physiognomic features of Mediterranean vegetation in the eastern part of the Iberian Peninsula are described considering a coastal and a continental gradient and oro-Mediterranean formations. The main communities include broadleaved evergreen forest, maquis, pine woodlands, "garigue", matorral, thorn-scrub and dwarf-shrub steppes along mesic-xeric gradients, and pine-juniper woodlands and xeroacanthic cushion-like formations as altitudinal stages. We discuss some morphologico- functional traits of dominant plants from these different formations as related to environmental conditions, and their corresponding energy use strategies. Even though there are relatively few data on biomass and production, a broad synthetic approach is given, with some reference values.

KEY WORDS : Mediterranean vegetation, growth-forms, biomass, production, energy use strategy

INTRODUCTION

Two major mesic-xeric gradients are observed in the eastern part of the Iberian Peninsula (Fig. 1). One runs from the Pyrenees to the south, from areas with 700-1000 mm annual rainfall to areas with 190-350 mm. The second runs from the northeast coast to continental areas, and its annual rainfall ranges from 600-700 mm to 300-400. In what follows we refer to them as coastal gradient and continental gradient. Different temperature changes occur along both gradients. Along the coastal gradient, there is an increase in summer length, average temperature and winter minima. Along the continental gradient, both winter length and thermic amplitude increase, whereas winter minima decrease.

Different shifts could thus be expected in vegetation structure and plant-growth forms following each gradient. Other factors, such as wind, soil, etc., can also affect plant fitness and modify adaptation features. Moreover, there are local deviations from climatic general patterns due to local climatic circumstances that will not be considered here.

Mediterranean climax ecosystems have been discussed on phytosociological grounds by BOLÒS (1985) for most of the area (Catalan countries), RIVAS (1974) for the southern part, and RIVAS (1987) for all Spain. BOLÒS (1985) distinguishes between a boreo-Mediterranean province, with climax communities corresponding to *Quercion ilicis* alliance, an austro-Mediterranean province, with climax communities corresponding to *Rhamno-Quercion cocciferae* and *Oleo-Ceratonion* alliances, and a western oro-Mediterranean province, corresponding to high and moderately high Mediterranean mountains, with *Pino-Juniperion sabinae*, *Xeracantho-Erinaceion* and *Hypericion balearici* climax communities, the last one found only in Majorca.

In the southern end of the area, RIVAS (1974) identifies mature communities as included in *Asparago-Rhamnion oleoidis* and *Periplocion angustifoliae* alliances. Ebro river valley vegetation is puzzling, and a *Rhamno-Quercion cocciferae* climax has been classically assumed (BRAUN-BLANQUET & BOLÒS 1957), but this might be an incomplete view (TERRADAS, 1986).

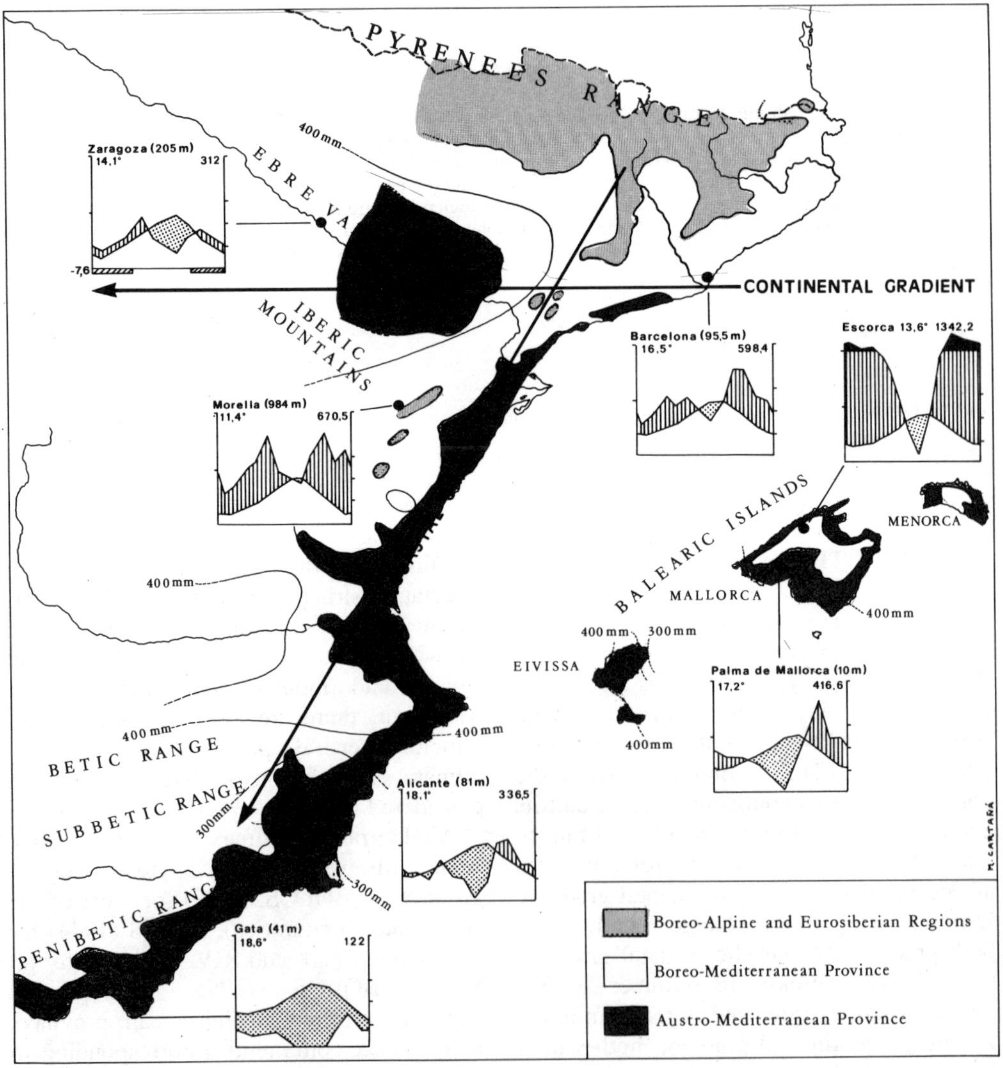

FIGURE 1. The eastern Iberian Peninsula and major features referred to in the text.

A PHYSIOGNOMIC DESCRIPTION OF VEGETATION

From a morphological point of view, mature vegetation along the coastal gradient changes from the broadleaved evergreen forest to tall thicket matorral and maquis (*Oleo-Ceratonion*), and thorn-scrub communities (*Oleo-Ceratonion* and *Asparago-Rhamnion*). Broadleaved evergreen forests are dominated by *Quercus ilex* ssp. *ilex* or, on sandy soils in the northern part of the area, by *Quercus suber*. These forests are often replaced by pine-oak mixed forests, by *Quercion-ilicis* maquis with *Arbutus unedo* and *Erica scoparia*, or by matorral: *Cistion mediomediterraneum* on siliceous soils and *Quercus coccifera* "garigues" or *Rosmarino-Ericion* scrub on limestone, all of them often with pines. Forests, maquis and "garigues" are communities with phanerophytes with plane leaves, often coriaceous or sclerophyllous and entire, as the dominant plants, whereas matorral communities are richer in nanophanerophytes or chamaephytes with soft (malacophyllous) leaves, sometimes plane (like in *Cistus* species) but in other cases leptophyllous (*Erica*) or linear.

The austro-Mediterranean *Oleo-Ceratonion* maquis is also dominated by phanerophytes with coriaceous or sclerophyllous leaves, mostly medium or small sized or divided (as in *Ceratonia siliqua* or *Pistacia lentiscus*). But this sclerophyllous vegetation has very often been replaced by *Cistion* or *Rosmarino-Ericion* communities. Malacophyllous plants become more frequent and diverse in this area.

In the southern part of the territory, maquis broadleaved evergreen phanerophytes like *Olea europaea*, *Quercus coccifera* and *Pistacia lentiscus* are progressively replaced in mature communities by a thorn-scrub. These communities are at first dominated by *Rhamnus lycioides*, a malacophyllous phanerophyte with long shoots and short shoots (dolichoblasts and brachiblasts) and marked seasonal heteromorphism, and the dwarf palm *Chamaerops humilis.* Floristically, they are similar to maquis and so they are included by phytosociologists in the *Rhamno-Quercion* alliance, but the physiognomy has shifted. We are near the tree border (which is found somewhere between 350 and 400 mm of annual rainfall). Also replacing communities after disturbance are clearly different: large surfaces are covered by chamaephytes, i. e. "tomillares" (*Thymo-Siderition leucanthae*), or halophytic, nitrophilous or gypsophylous communities, and by dry therophytic grasslands. The typical thorny shrub growth-form diversifies to the south, when climate becomes drier, with plants like *Maytenus europaeus*, *Periploca angustifolia*, *Ziziphus lotus*, *Asparagus albus* or *Lycium intricatum* added to *Rhamnus lycioides* (*Asparago-Rhamnion* and *Periplocion* alliances).

The continental gradient begins at the northeast edge with the broadleaved sclerophyllous forest or maquis near the coast, which is soon replaced by an impoverished *Quercus ilex* ssp. *rotundifolia* forest with *Quercus coccifera* "garigues", *Rosmarino-Ericion* matorral or pine woodlands as successional stages. At

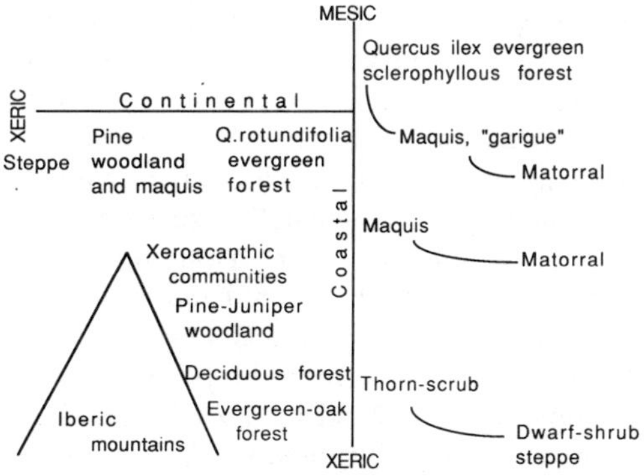

FIGURE 2. Major plant formations along the two mesic-xeric gradients and on Mediterranean mountains. For the coastal gradient, degradation series are shown.

the Ebro valley, rainfall decreases fast and broadleaved trees disappear, and are replaced by a pine woodland with a shrub understory also classified phytosociologically as *Rhamno-Quercion cocciferae.* Thermophyllous species of this community, including the pine itself, are gradually disappearing to the centre of the Ebro valley, where the winter climate becomes colder. There, on gypsic ground, slopes and shallow soils are mostly covered by a nitrophilous or gypsophilous chamaephyte semidesert or steppe (*Salsolo-Peganion, Gypsophylion*), whereas fine-textured loams are occupied by hemicryptophytic or rhizomatous geophytic perennial grasses (*Agropyro-Lygeion*). In this landscape, a high diversity of therophytes covers the free spaces between perennial grasses or shrubs during the short wet periods. Halophytic chamaephytes or nanophanerophytes appear on saline soils, near the bottom of undrained valleys.

Further up the mountains in the northern part of the territory, extra-Mediterranean conditions are soon found, which correspond to the Euro-Siberian or Boreo-Alpine regions. But in the Iberian and southern (sub-Betic, Betic and Penibetic) mountains, some features of Mediterranean climates reach higher altitudes. According to BOLÒS (1985), oro-Mediterranean vegetation covers a very limited area in some scattered isolated patches, occurring above *Quercus rotundifolia* and deciduous sub-Mediterranean forests. *Pino-Juniperion sabinae* communities of Iberian mountains are made up of large carpet-like patches of *Juniperus sabina* or *J. communis* ssp. *hemisphaerica*, with isolated individuals or clumps of *Pinus sylvestris* or *Juniperus thurifera* emerging from shrub cover and a therophytic grassland as a background. In more exposed environments this seems to be very dynamic vegetation, because pine growth overshadows *Juniperus* carpets and can suppress them, but pines germinate only under the carpet protection.

A dense forest is rarely obtained. Summits of Iberian mountains and southern "sierras" are occupied by spiny cushion-like *Xeracantho-Erinaceion* communities, in a poor form if compared with Sierra Nevada and Atlas equivalents.

Spiny cushion-like shrubs are also present above 1100 m in Majorca (*Hypericion balearici*), with a different floristic composition. Xeroacanthic shrubs can be considered as an interesting case of evolutionary convergence in oro-Mediterranean climates: different families are represented by species with similar morphology in these highly fluctuating windy environments. Similar forms are also found on the coasts of the Balearic islands, in windy environments too.

DOMINANT WOODY GROWTH FORM

We will now try to summarize the major woody plant growth forms that dominate Mediterranean landscapes in our territory. MARGARIS (1981) has suggested two main adaptation syndromes for Mediterranean environments: maquis plants, corresponding to typical sclerophyllous woody plants, and phrygana plants (equivalents are said to be Israeli "botha", Iberian "tomillares", Portuguese "barocales", Italian "gariga" and some other formations). Similar views have been discussed by ORSHAN (1983).

Maquis plants are called isomorphic because they maintain the same kind of leaves all through the year. The average life span of the leaf is usually between 1-2 years, but can be shorter in some species like *Pistacia lentiscus* (7.9-15.7 months according to DIAMANTOGLOU & MITRAKOS, 1981). Leaf fall peaks tend to be advanced in early summer. Leaves are rich in structural components (lignine, hemicelluloses) that make them rigid and have low nutrient contents. This last condition also means that nutrient requirements are low. In *Quercus ilex* we have found accumulation of nutrient reserves in branches (ESCARRÉ *et al,* 1987) and these reserves are used to build new leaves. Sclerophyllous plants may use wood reserves as a way to deaden the effects of environmental fluctuations. In any case, plant function is usually based on a conservative nutrient economy. Canopy structure is shaped by light availability acting as a key factor (GRACIA, 1983).

Phrygana strategy consists in a marked seasonal heteromorphism, with large leaves during the wet-season, appearing on dolicoblasts just with the first autumn rains, and showing high photosynthetic rates (almost twice the rates of the sclerophyllous leaves of maquis plants). During the spring, phrygana plants have their maximal growth rates. Then wet-season leaves are shed, and they are replaced by drought-season leaves. Meanwhile, dolicoblasts elongate and lignify. Before leaf fall, there is a retranslocation of most leaf nitrogen stocks to other plant organs (data on *Timbra capitata*, *Phlomis fruticosa*, *Cistus* spp. and *Sarcopoterium spinosum* are given by MARGARIS, 1981). The morphological traits of summer leaves differ from those of wet-season leaves (palissade parenchyma thickness, stomata distribution, etc.) and they are more xeromorphic. Dimorphism seems to be controlled by day length. Soft leaves can fold, which reduces their exposed area in transpiration. Light is not a major factor for canopy organization. A shallow root system enables these plants to respond faster than maquis plants to early autumn rains.

ORSHAN (1989) has pointed out that heteromorphic plants have much larger fluctuations of transpiratory rates during the year than isomorphic plants. In fact, isomorphic plants are regulated by deep water use at the end of summer (higher rates are thus possible in this period); moreover, the properties of sclerophyllous leaves impose some limitations on spring transpiration. These authors assume that isomorphic plants use soil water gradually by developing roots to deeper soil levels when surface horizons become dry. This process can help to control activity rates, thus allowing the plants to reach the end of summer with predawn water potentials that are not too negative : -5 MP has been measured in *Arbutus unedo* (TENHUNEN *et al.*, 1985), less stable than other sclerophylls like *Quercus coccifera*, whereas in the malacophyllous *Rosmarinus officinalis* MERINO (1988) measured -7 MP. Gas exchange can be controlled by a high number of stomata in isomorphic plants; stomata number in heteromorphic plants is usually lower. The ecophysiology of maquis plants has been studied by authors like LARCHER (1972) and BRECKLE (1966); recent important advances in that subject have been reported by Lange and co-workers in several papers on *Arbutus unedo, Ceratonia siliqua, Quercus suber, Phyllirea,* etc. (LANGE *et al.,* 1982; TENHUNEN *et al.*, 1987). Daily and seasonal cycles clearly show effective stomatal control of transpiratory and photosynthetic rates. Nevertheless, considerable differences are found between species: *Arbutus unedo* has larger decreases in both maximal photosynthetic rates and xylem potential than *Quercus coccifera*; this seems to be related to much shallower roots in *Arbutus* (TENHUNEN *et al.*, 1985). *Pistacia lentiscus*, which has low gas exchange control and short leaf span, can be considered as a maquis plant with some transition characters to the malacophyllous strategy, enabling it to live in rather xeric environments.

To these two morphologico-functional types, we must add some others. One is constituted by pines, with only a few species, but dominant over large areas. Pine growth-form is clearly not a specifically Mediterranean adaptation pattern. In fact, it is found in many other climates. Besides, it does not represent specific adaptations to any Mediterranean environment because it is found in most conditions. But we should not forget the success of the pine form in Mediterranean countries. Mediterranean pines have relatively shallow root systems. Nevertheless, *Pinus halepensis* can endure rather xeric conditions. In that case, morphological changes appear both in the canopy form and in leaf size. GINDEL (1973) found that leaf length was closely dependent on annual rainfall, both in *P. halepensis* and *P. pinea.* An interesting case to be studied is *Chamaerops humilis,* the only native European palm, which is a dominant plant in a large variety of maquis, matorral, "garigue" and thorn-scrub formations. It represents an atypical successful growth-form in Mediterranean environments, with very large sclerophyllous leaves. However, very scarce data exist on its ecophysiological characteristics, underground parts and other relevant features. Other morphologico-functional types are related to more local conditions. Let us discuss these conditions briefly.

With increasing water stress and unpredictability of water resources, woody plants can follow different strategies. One is that of nanophanerophytes and phanerophytes with both long and short shoots. These plants respond to rainfall by producing either leaves from axillary buds or from short shoots dispersed along the branches (SHMIDA & BURGESS, 1988). These leaves, generally small, are produced quickly after episodic rains because few resources are invested in short internodal growth, whereas during long wet periods long shoots elongate. So, leaf area can show pronounced changes (some plants with these characteristics grow in relatively dry habitats of temperate areas, such as *Berberis vulgaris*). In extreme cases, summer deciduous plants are found (for instance, *Asparagus albus*, *Withania frutescens*). Individual canopy structure is very open and leaves are arranged in a diffuse way, making convective exchange and cooling easy. Transpiration rates per leaf surface unit are relatively high; however, they are low on a soil surface basis. Also photosynthetic rates are high during periods of leaf growth. But during long periods, the carbon budget must be negative, because maintenance costs are high. Spines are probably an adaptation to reduce herbivory and maintain positive total carbon budgets. In a *Thymo-Siderition* community, we have observed *Rhamnus lycioides* roots passing through the petrocalcic layer, whereas chamaephyte dwarf-shrubs had only shallow roots branching within the top 15-30 cm of soil. The biomass of thorny shrubs can be greater underground than aboveground, as shown by MARTÍN (1984) in *Withania frutescens* and *Rhamnus lycioides.* If compared with other thorny shrubs, *Rhamnus lycioides* has some characteristics, such as high chlorophyll content, low nitrogen content and high specific leaf density, which are more similar to those of sclerophyllous plants. This might be related to the fact that its larger distribution range extends further to the north than most other thorn-shrub species.

In *Quercus ilex* forests evapotranspiration can amount to about 500 mm in the course of a year: by using water budgets in small catchments, 469 mm have been measured at La Castanya (Montseny) by ÀVILA (1989) and 495 mm at l'Avic (Prades) by ESCARRÉ *et al.* (1984). Under higher rainfall inputs, near Montpellier, RAMBAL (1986) calculates, from a water relationships model, a 529-665 mm yr^{-1} evapotranspiration for a *Quercus coccifera* "garigue". Values can certainly be reduced along mesic-xeric gradients, but closed woody sclerophyllic canopies (forest, maquis, garigues) do not seem to be able to develop below 350-400 mm of annual water supplies (from rain or lateral flow).

Below 300 mm of annual rainfall, or even above this value on slopes with shallow and eroded soils, nanophanerophytes are replaced by chamaephytes or hemicryptophytic and rhizomatous grasses. Differences between chamaephytes and grasses have recently been discussed by SHMIDA & BURGESS (1988). A high diversity of chamaephytic dwarf shrubs exists in our territory. Small microphyllous or nanophyllous leaves are largely dominant. Roots are often scarce and relatively long; so, they can exploit localized wet sites. Nevertheless, when a chalk crust exists they cannot usually pass through it and deep water resources are unavailable to these plants. An opportunistic strategy is the response to unpredictable and scarce water availability, either at the individual (leaf production and growth bursts) or community (fast population dynamics) level. High maintenance costs and short productive periods, as well as terminal shoot growth, would made chamaephytes vulnerable to strong herbivory. As a result, chemical defence and unpalatability are frequent.

Some woody plants can exploit more permanent, local water resources, such as rock crevices or soil under rock protection, by developing a system of long roots. This permits more conservative strategies of energy use, thus allowing plants of the maquis type, chamaephytes or thorny shrubs to survive under climates drier than expected.

Other stress causes besides drought are present in Mediterranean countries. MITRAKOS (1980) has noted the double stress, summer drought and winter cold, to which Mediterranean plants are subject. Winter cold depresses metabolism and can also provoke frost-drought injury, thus reducing production, as we have shown for *Quercus ilex* along an altitudinal gradient (SAVÉ *et al.*, 1986). Further inland or at higher altitudes, the Mediterranean flora becomes clearly impoverished in our territory, as indicated before. The vegetative period can be reduced either by increasing summer duration along the coastal gradient or by colder and longer winters as one proceeds into the continent or up onto the mountains. Along the continental gradient, the shift to colder and longer winters comes along with a change from mesic to xeric conditions, while the vegetational change shows analogies with the change along the coastal gradient. On the mountains, the effects of colder conditions could be expected to lead to different morphological types of plants. Nevertheless, carpet-like forms are also found in extra-Mediterranean climates and xeroacanthic communities are present in warmer Mediterranean climates on the Balearic coast. Pines and junipers dominate cold oro-Mediterranean climates and xeroacanthic communities seem to be more associated with strong wind exposure than with temperature values or fluctuations.

Obviously, there is also a variety of conditions linked to soil differences, and we can especially distinguish rocks, sands, marshes and saline soils. Sandy, waterlogged or saline environments sustain similar vegetation forms under very different climates, with only a few minor adaptations to local conditions. To some extent, the same is true for rocky environments, although they are often isolated refuges with a high level of endemism. Perhaps the only fact worth mentioning here is the increase in saline mainland habitats along the mesic-xeric gradients.

Both succulent leaves and stems are mostly restricted to rocks, saline soils and salt-sprayed coastal environments. Some succulent plants from xeric habitats have expanded after successful introductions (*Agave*, *Opuntia*). On gypsic soils we can find some species with succulent leaves (i.e. *Helianthemum squamatum*, *Ononis tridentata*). The thorny shrub *Lycium intricatum* also has succulent leaves. Only one species, the rhizomatous herb *Caralluma europaea*, can be considered a spontaneous stem succulent linked to xeric conditions, but it has a very restricted distribution in the southern end of the territory. No desert succulent formations exist directly related to increasing aridity.

The xeroacanthic cushion-like growth-form needs perhaps more comment. A general characteristic of this type of plants is the fact that they undergo not only the seasonal double stress, very marked in mountain plants, but also a continuous loss due to sprout mortality by wind mechanical effects or other associated physiological factors.

So, this form is a result of heavy exploitation by the physical environment, which adds to herbivorism, making it difficult to sustain a positive carbon budget. Spines, as well as very closed canopies, reduce losses due to herbivory. Individual canopies have an extreme monolayer leaf arrangement: leaves are densely packed in a 0.5-2 cm thick layer. Malacophyllous leaves are common, with some transition forms to sclerophylly, in true cushion-like plants. In coastal habitats under salt spray influence, succulent leaves sometimes occur. So, leaf overheating would be expected under high irradiance, due to low convective cooling, except under strong wind regimes that cause efficient eddy transfers. Another factor to be taken into account with these plants is their nutrient cycling.

Nutrient losses by both litter wind carrying and soil erosion are very high, but the closed structure of these plants tends to minimize them. For instance, under summit cushions in the Aytana Sierra we have measured litter stocks of 0.125 g cm^{-2}, a quantity similar to that found in developed forests with much larger aboveground biomass. An external but growth-form controlled nutrient reserve is thus obtained. We have measured underground biomass in some mountain and coastal cushion-like Balearic plants. *Astragalus balearica* has slightly more than 20 % of its total biomass under ground and this rate is 26 % in *Launaea cervicornis* individuals; these are not high values, if compared with many woody plants from arid zones.

Cushion forms are often found on mountains and coasts under other climates, but all the xeroacanthic Mediterranean cushions have a very closed and spiny canopy, which gives a characteristic physiognomy to their formations. Although cushions can be merely morphogenic structures, this is not always the case. Sometimes this morphology has a genetic basis (BRAUN-BLANQUET, 1962). We have tried to distinguish these two different origins in some cushion-like Balearic coastal plants (TERRADAS *et al.*, 1989). So, branching angles ranging from 10º to 15º in cushion shrubs are assumed to depend on genetic determinism, whereas branching angles ranging from 45º to 90º, or even more, are found in morphogenic cushions (M. Riba, pers. comm.).

GROWTH-FORMS AND ENERGY USE STRATEGIES

SHMIDA & BURGESS (1988) recognize different strategies of desert plants using a two axis space. The first axis is oriented from poor to rich resources, while the second axis goes from episodic to stable resources. This seems to us a useful approach. Along a mesic-xeric gradient, rich resource availability becomes more episodic, then unreliable, whereas reliable resources become lower in concentration or quality. Near the mesic end of our gradients, most dominant plants use relatively (for this area) stable and rich resources in a competitive strategy (GRIME, 1979). Maquis-type isomorphic plants in mature communities are *K* strategists, with slow energy use. Pioneer stages in disturbed sites can follow *r* strategies. When approaching the xeric end of the gradient, resource patchiness increases over time and space. As a result, *K* and *r* strategists (or competitive, ruderal and stress resistants if Grime's classification is preferred), find opportunities, and *r*-strategists are eventually more favoured. Resource availability fluctuations have effects similar to disturbance, promoting high energy strategies (in the sense of SHMIDA & BURGESS, 1988): short active periods with high production per time of activity. Transpiration and photosynthetic rates can be high, but leaf area is low and variable. For instance, *Rhamnus lycioides* leaves have on average much larger transpiration rates and water content changes than *Quercus coccifera*, *Pinus halepensis* or *Juniperus thurifera.* The leaves, however, have a diffuse arrangement and small coverage, and leaf area can decrease during drought periods (TERRADAS, 1973). So, *Rhamnus lycioides* has a higher energy strategy than *Quercus coccifera* and is adapted to fast responses to rich episodic resources, whereas *Q. coccifera* relies more on continuous use of poorer but stable resources.

All along the gradients, man's use of the landscape can promote pioneer and less mature communities. As a result, we can expect *r*-strategists to increase in importance. But other associated changes take place as well, thus reducing plant cover and modifying soil conditions. This can open spaces to species adapted to more xeric environments, and benefit the northwards advance of southern xerophytic *r*-strategists. A good example of this might be *Inula viscosa*. Original habitats of this Asteraceae deciduous shrub are the south-Mediterranean "ramblas" (dry bed watercourses), an environment under a high disturbance regime by episodic waterflows. A ruderal strategist, this plant has expanded its area to the north, colonizing old fields and waste lands. During the earlier years after field abandonment, it behaves like a chamaephyte, but as soil becomes compacted it shifts to a hemicryptophytic mode (SORIA, 1980).

True xeroacanthic cushion-like plants use stable and poor resources, with a high control of the external steps of nutrient cycles. But they cannot easily be classified as *K*-strategists, because they are under heavy exploitation by the physical environment. Photosynthetic rates have not been measured, but they are likely to be relatively high to balance losses and long inactive periods. Growth rate can be higher than expected from poor cover, after observations on *Centaurea balearica* colonizing disturbed sites on the coast of the island of Minorca. On the same, *Anthyllis hermanniae* and *Launaea cervicornis* are found on disturbed sites, being replaced by *Erica multiflora*, *Phillyrea media* var. *rodriguesii*, *Pistacia lentiscus* and *Pinus halepensis* scrubs and woodlands in the succession to more mature stages.

First growth is probably fast in cushions, but it must slow down and take place as short bursts after a certain size (ruled by local wind conditions) is reached. Studies in population dynamics could also be very useful in order to understand fully the energy strategy of these plants.

When considering energy use, different levels must be taken into account. Individual survival is important to phanerophytes and nanophanerophytes, because construction costs are high, but investment in structure assures either aerial or subterranean space control. But certain parts of the plants, such as leaves or fine roots, can be used in a way which is more or less economical in terms of energy. A certain degree of opportunism is always present. In an adult tree canopy, leaf morphology, leaf size, leaf behaviour and leaf span change along a vertical axis. We can consider woody plants as complex systems or populations of organs including hierarchically organized structures, some of them with high turnover rates protecting the others from external fluctuations or exploiting episodic resources. Under appropriate conditions, *Quercus* species can have more than one shoot growth period. The very high capacity of resprouting and increasing production rates after disturbance in *Q. coccifera* and other *K*-strategists (as a result of use of internalized resource stocks) allows them to conserve space control and dominance, succeeding each other as if there was a shift in energy use strategy within the same genetical individuals. Leaves of young resprouts show morphological and physiological differences from adult normal leaves (PERY *et al.*, 1989): higher Chl a/Chl b, transpiration rates, and osmotic potential, lower specific weight, cuticle and leaf thickness, proline and starch grain content. In fact, resprout leaves have many more available resources than adult normal leaves because they can use stocks. Besides, the rate between leaf area and fine root biomass is much higher in resprouts. So, in the same individual low energy use is dominant, but periods of high energy use occur seasonally or as a result of disturbances, when high quantities of resources are freely available. In phrygana and thorny shrub types, leaves conserve and always maintain the high energy pattern. Woody structure is also maintained, but leaf "populations" can fluctuate greatly, as can individual leaf area (by folding). Moreover, thorny shrubs have a similar leaf strategy but higher investment in woody parts: they use deeper soil resources and have longer individual life than phrygana-type plants or dwarf-shrubs. In very poor dry sites, as well as in heavily exploited and fluctuating environments, site conservation can become less important or not feasible, and individual turnover increase (steppe chamaephytes, therophytes), giving a classical *r*-strategy gradually adapted to the use of more episodic energy pulses.

To sum up, some plants tend to conserve space control by means of large investments in woody structures (which means a long individual life). We found a shift in these plants from slow gradual to fast pulsed changes both in leaf area and leaf activity as a response to increasingly episodic and unpredictable resource availability in the environment.

Other plants sacrifice space control and long individual life, invest relatively more in photosynthetic organs and show a higher turnover at population levels. But in all kinds of plants, when large resources by leaf area unit are either available in the environment or are mobilized as a result of disturbance, a pattern of high energy is developed. Relatively stable forest as well as maquis, no matter how rich they are in resources, usually show closed canopies and low resource availability by leaf unit; thus, they behave in a low energy manner. However, in transient conditions after canopy removal due to disturbance, large resource stocks are mobilized and root systems are always present. So, new leaves have much higher resource availability by leaf unit and they behave in a high energy manner. Other less stable communities limit their activity to episodic periods of high resource availability, and always have open

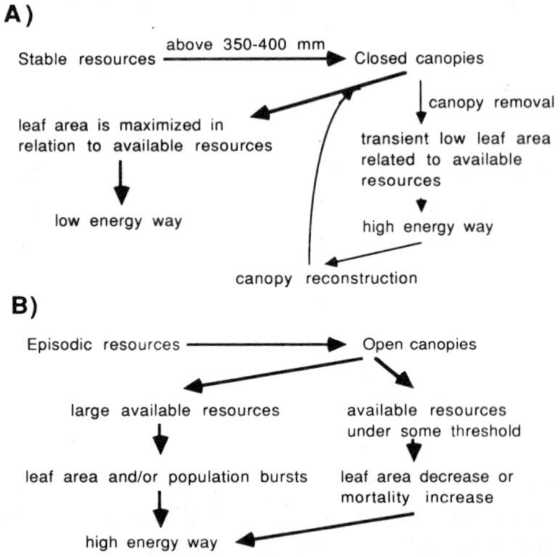

FIGURE 3. Two main ways of energy use are present along our mesic-xeric gradients. Sclerophyllous forest and maquis plants use a low energy strategy: they have low nutrient requirements, maintain long continuous periods of activity, and show some regulation due both to their internal stocks and their good stomatal control. Other plants, such as mallacophyllous phrygana and some thorn-scrub plants, have episodic pulses of high energy use due to their high dependence on external patched reserves; they show high nutrient requirements, discontinuous growth bursts and regulation of gas exchanges by leaf area changes.

canopies working in a pulsed high energy manner (Fig. 3).

BIOMASS AND PRODUCTION

There are few data on biomass and production for the area considered, except for forests. A number of evaluations have been made in the last few years, mostly unpublished, on different types of natural communities. They allow us a general overview.

As far as broadleaved forests are concerned, accurate studies on Montseny and Prades holm-oak forests have been carried out. Aboveground biomass of a high quality 60-90 years old stand from La Castanya (Montseny) is 165 t ha^{-1} (dry weight) for trees and 4.7 for understory, while root biomass is 65 t ha^{-1}. At Prades, a lower and denser holm-oak coppice has 100 t ha^{-1} aerial tree biomass and a 15 t ha^{-1} understory (ESCARRÉ *et al.*, 1984). Aboveground biomass values from 40 to 270 t ha^{-1} have been measured in different holm-oak forests in the Mediterranean region, and these values probably give us a range estimation for this type of forest in our territory. Values for cork-oak forests are probably a little lower, because these forests usually have more open canopies. *Arbutus unedo* and *Erica arborea* maquis or heaths lie somewhere between 10 and 60 t ha^{-1}, and *Quercus coccifera* "garigue" in the 10-50 t ha^{-1} range, probably with a lower average than maquis.

Oleo-Ceratonion mature maquis are hardly found as a continuous cover, and there are no good estimates of their biomass. A Balearic community with *Olea europaea* and *Pistacia lentiscus* in patches covering 40 % of the soil surface has a biomass of only 9 t ha^{-1}. Mediterranean pine forests and woodlands include very diverse communities, often rather open. A *Pinus halepensis* woodland at Sierra Espunya gave 61 t ha^{-1} of aboveground biomass. Average *P. halepensis* woodland biomass is almost certainly between 50 and 80 t ha^{-1} in most areas. RAPP (1978) has measured 151 t ha^{-1} near Montpellier for the same species. Under much wetter conditions, 146 t ha^{-1} was found in a *Pinus pinea* forest studied by CABANETTES & RAPP (1978) in southern France. Aboveground biomass in most *Rosmarino-Ericion* and *Cistion* matorral communities is less than 25 t ha^{-1}. *Thymo-Sideritition* dwarf shrub formations are clearly under 10 t ha^{-1} (we have measured 4.3 t ha^{-1} in such a community near Alicante; see also FISHER *et al.*, 1987), with underground biomass lower than aboveground biomass; similar low values are found in most *Salsolo-Peganion* communities. There are no evaluations of thorn-scrub community biomass in southern Spain, and in fact it is difficult to find well preserved representative formations. Isolated thorny shrub individuals usually appear in dwarf-shrubs communities.

Xeroacanthic formations usually have less than 30% plant cover and corresponding low biomass: we have measured 1 t ha^{-1} with 15-20 % cover in *Launaeetum cervicornis* from the Balearic coastal (much higher values can be expected in *Aro-Phillyreetum* communities, but these are constituted by morphogenic cushions of maquis species); in the *Hypericion balearici* from the Balearic mountains, with a 60% cover we have found 4.2-4.8 t ha^{-1} biomass; on the summit of Aytana sierra, cover, is 10-15 % and biomass is 3.3 t ha^{-1}.

Available data on primary production are very scarce. Moreover, spatial and temporal variations can be very large. Whereas in the high quality Montseny holm-oak stand production reaches a noticeable 9.7 t ha^{-1} y^{-1}, similar to the values found for most mid-European forests, many mature holm-oak forests do not exceed 4-5 t ha^{-1} yr^{-1} (3 of them being leaves). Pine forests can show higher figures: the Sierra Espunya forest, in the south of this area, reaches 8-9.5 t ha^{-1} yr^{-1}. Obviously, there is a marked north-south decreasing gradient of forest production associated with the climatic gradient. GRACIA (in TERRADAS *et al.*, 1989) gives a range of stem wood production between 12.2 and 0.64 t ha^{-1} yr^{-1} (provincial averages for any kind of forest, evergreen oaks excluded), following the same decreasing pattern.

In maquis the aboveground production is usually below 4 t ha^{-1} yr^{-1}, even though during the first year after disturbance (fire, for instance) it can be higher. In *Quercus coccifera* "garigue" we find 2-3 t ha^{-1} yr^{-1} one year after fire and 0.5-1.25 in most other cases. Higher productions can be found in old field scrub of *Inula viscosa* (SORIA, 1984), where 5-6 t ha^{-1} yr^{-1} has been measured in the early years after field culture abandonment, but these values decline rapidly. *Rosmarinus* matorral and dwarf shrub communities are probably below 1 t ha^{-1} yr^{-1}, and the same must be true for xeroacanthic communities, but we lack direct measurements.

The patterns of change in productivity along mesic-xeric gradients remain obscure because belowground productions are unknown. Leaf production decreases clearly from a maximum around 3 t ha^{-1} yr^{-1} in broadleaved, pine forests and tall shrub formations with complete soil cover to chamaephyte steppes, whereas the rate of photosynthetic organ production to total aboveground production increases. In general, increasing aridity implies more energy being invested in belowground and photosynthetic parts, as well as higher costs of maintenance.

Biomass and production are not simple responses to rainfall gradients. Man has exploited the Iberian Peninsula resources for thousands of years. As a result, we cannot know to what degree soil degradation and nutrient losses have been increased and water-holding capacity has decreased by overgrazing, fire and agriculture. A consequence of man's exploitation of the land could be a northwards shift of plant communities adapted to more xeric environments, especially along the coastal gradient.

AKNOWLEDGEMENTS

Most data for this study result from projects funded by Fundació Enciclopèdia Catalana and CICYT. Many colleagues and students of my biogeography courses have contributed to their collection. F. Rodà and A. Escarré have made helpful comments on the manuscript.

REFERENCES

ÀVILA, A. 1989. Balanç d'aigua i nutrients en una conca d'alzinar del Montseny. *Estudis i Monografies*, 13. Diputació de Barcelona.

BOLÒS, O. de. 1985. *Corologia de la flora dels Països Catalans*. ORCA, Notícies i Comentaris 1. Institut d'Estudis Catalans. Barcelona.

BRAUN-BLANQUET, J. 1979. *Fitosociología.* Blume. Madrid.

BRAUN-BLANQUET, J. & BOLÒS, O. de. 1957. Les groupements végétaux du bassin moyen de l'Ebre et leur dynamisme. *Anal. Est. Exp. Aula Dei*, 5: 1-266.

BRECKLE, S.W. 1966. *Ökologische Untersuchungen im Korkeichenwald Kataloniens.* Botanischen Institut der Landwirtschaftlichen Hochschule Hohenheim & Jardí Botànic Mar i Murtra de Blanes. Barcelona.

CABANETTES, A. & RAPP, M. 1978. Biomasse, minéralomasse et productivité d'un écosystème à Pins pignons *Pinus pinea* L. du littoral méditerranéen. I. Biomasse. *Oecol. Plant.*, 13 (3) : 271-286.

DIAMANTOGLOU, S. & K. MITRAKOS, K. 1981. Leaf longevity in Mediterranean evergreen sclerophylls. In: *Components of productivity of Mediterranean climate regions. Basic and applied aspects.* (N.S. Margaris & H.A. Mooney, eds.): 17-19. Junk. The Hague.

ESCARRÉ, A., FERRÉS, L, LÓPEZ, R., MARTÍN, J., RODÀ, F. & TERRADAS, J. 1987. Nutrient use strategy by evergreen-oak (*Quercus ilex* ssp. *ilex*) in NE Spain. In: *Plant response to stress.* (J.D. Tenhunen *et al.*, eds): 429-435. NATO ASI Series. Springer. Berlin.

ESCARRÉ, A., LLEDÓ, M.J., BELLOT, J., SÁNCHEZ, J.R., ESCLAPÉS, A,. CLEMENTE A. & ROVIRA, A. 1984. Compartimentos y flujos biogeoquímicos en un encinar. Distinto predominio de factores físicos y químicos en su control. *IV Conf. Int. Ciencias Naturales*, La Habana.

FERRÉS, L., RODÀ, F., VERDÚ A.M.C. & TERRADAS, J. 1984. Circulación de nutrientes en algunos ecosistemas forestales del Montseny (Barcelona). *Mediterranea.* Biol., 7: 139-166.

FISHER, G.C., ROMERO, A., LÓPEZ BERMÚDEZ, F., THORNES, J..B. & FRANCIS, C. 1987. La producción de biomasa y sus efectos en los procesos erosivos en un ecosistema semiárido del SE de España. *Anal. Biol.*, 12: 91-102.

GRACIA, C. 1983. *La clorofila en los encinares del Montseny*. Ph. D. Thesis. University of Barcelona.

GRIME, J.P. 1979. *Plant strategies and vegetation processes.* John Wiley. Chichester.

LANGE, O.L. TENHUNEN, J.D. & BRAUN, M. 1982. Midday stomatal closure in Mediterranean type sclerophylls under simulated habitat conditions in an environmental chamber. I. Comparison of the behaviour of various European Mediterranean species. *Flora*, 172: 563-579.

LARCHER, W. 1972. Der Wasserhaushalt immergrüner Holzpflanzen im Winter. *Ber. Dt. Bot. Ges.*, 85 : 315-327.

LLEDÓ, M.J. 1982. *Datos energéticos, estructurales y químicos de Pinus halepensis Mill. en el pinar de Sierra Espuña (provincia de Murcia).* Master Thesis. University of Alicante.

MARGARIS, N.S. 1981. Adaptative strategies in plants dominating Mediterranean-type ecosystems. In: *Mediterranean shrublands*: (F. di Castri, D.W. Goodall & R.L. Specht, eds.): 309-315. Elsevier. New York

MARTíN, J. 1984. *Variaciones estacionales en el contenido de nutrientes y pigmentos en cuatro especies del espinar costero de Alicante.* Ph. D. Thesis. University of Alicante

MITRAKOS, K. 1980. A theory for Mediterranean plant life. *Oecol. Plant.*, 1 (15) 3 : 245-252.

MERINO, O. 1988. Xylem water potential and phenological relationships in narrowleaf species growing on sand dunes in southern Spain. In: *Time scales and water stress.* (F. di Castri, C. Floret, S. Rambal & J. Roy, eds.): 227-232. Proc. 5th Int. Conf. on Mediterranean ecosystems, I.U.B.S., Paris.

ORSHAN, G. 1983. Approaches to the definition of Mediterranean growth-forms. In: *Mediterranean type ecosystems* (F.J. Kruger, D.T. Mitchell & J.U.M. Jarvis, eds.): 86-100. Ecological Studies, 43. Springer. Berlin

ORSHAN, G. (ed.) 1989. *Plant Pheno-morphological studies in Mediterranean type ecosystems.* Geobotany, 12. Kluwer. Dordrecht.

PERY, M., SAVE, R., ALEGRE L. & TERRADAS, J. 1989. Efectos del fuego sobre las relaciones hídricas y la morfología en brotes de *Arbutus unedo* L. *VIII Reunión Nac. S.E.F.V. and I Congr. Hispano-Luso de Fisiología Vegetal.* Barcelona.

RAPP, M. 1978. El ciclo biogeoquímico en un bosque de pino carrasco. In: *Ecología Forestal.* (P. Pesson, ed.): 97-117. Mundi-prensa. Madrid.

RIVAS-MARTÍNEZ, S. 1974. La vegetación de la clase *Quercetea ilicis* en España y Portugal. *Anal. Inst. Bot. Cavanilles*, 31 (2): 205-259

RIVAS-MARTÍNEZ, S. 1987. *Memoria del Mapa de Series de Vegetación de España*. ICONA. Madrid..

SAVÉ, R., RABELLA, R. & TERRADAS, J. 1988. Effects of low temperature on *Quercus ilex* L. ssp. *ilex* water relations. In: *Time scales and water stress.* (F. di Castri, C. Floret, S. Rambal & J. Roy, eds.): 103-105. Proc. 5th Int. Conf. on Mediterranean Ecosystems, I.U.B.S., Paris.

SORIA, L.L. 1980. Aspectos de la biología de poblaciones de *Inula viscosa* en un campo abandonado. *Mediterranea*, 4: 115-131.

SORIA, L.L. 1984. Biomasa y producción de poblaciones de *Inula viscosa* en campos abandonados. *Ecologia Mediterranea*, X (3-4): 75-83.

SHMIDA, A. & BURGESS, T. L. 1988. Plant growth-form strategies and vegetation types in arid environments. In: *Plant form and vegetation structure.* (P.J.M. van der Aart, H.J. During and J.T.A. Verhoeven, eds.): 211-241. Academic Publishing. The Hague.

TERRADAS, J. 1973. *Clima y economía hídrica en comunidades vegetales de los Monegros.* Ph.D. Thesis. University of Barcelona..

TERRADAS, J. 1986. El paisatge vegetal dels Monegros: assaig d'interpretació. *Orsis*, 2 : 71-95

TERRADAS, J., PRAT, N., ESCARRÉ, A. & MARGALEF, R. 1989. *Els sistemes naturals.* Història Natural dels Països Catalans, XIV. Fundació Enciclopèdia Catalana. Barcelona.

TENHUNEN, J., BEYCHSLAG, W., LANGE O.L. & HARLEY, P.C. 1987. Changes during summer drought in leaf CO_2 uptake rates of macchia shrubs growing in Portugal: limitations due to photosynthetic capacity, carboxylation efficiency and stomatal conductance. In: *Plant responses to stress* (J.D. Tenhunen *et al.*, eds.): 305-327. NATO ASI Series, 15. Springer. Berlin.

(J. D. Ros & N. Prat, eds.). 1991. *Oecologia aquatica*, 10: 351-366

THE SUBANTARCTIC *NOTHOFAGUS* FORESTS OF TIERRA DEL FUEGO: DISTRIBUTION, STRUCTURE AND PRODUCTION

EMILIA GUTIÉRREZ[1], V. RAMÓN VALLEJO[2], JOAN ROMAÑÀ[2] & JAUME FONS[2]

[1] Departament d'Ecologia. Facultat de Biologia. Universitat de Barcelona. Av. Diagonal, 645. 08028 Barcelona. Spain.
[2] Departament de Biologia Vegetal. Facultat de Biologia. Universitat de Barcelona. Av. Diagonal, 645. 08028 Barcelona. Spain.

Received: August 1990.

SUMMARY

Evergreen *Nothofagus betuloides* and deciduous *N. pumilio* form the main forest types in Tierra del Fuego. These forests were sampled along two altitudinal gradients to study their structure and dynamics and assess the causes of their distribution.

The distribution patterns of the two species of *Nothofagus* seem to respond to different climates and soils. The dominant soil processes are hydromorphy in the evergreen forest and podzolization in the deciduous one. *N. betuloides* is an evergreen resilient to short-term environmental fluctuations, due to its ability to retain nutrients. Leaves on the tree may last up to 7 years, with an average density of 17 mg cm^{-2}. In contrast, the leaves of *N. pumilio* are shed in autumn and reach only 8 mg cm^{-2}.

In both types of forests the following features can be outlined. Old-growth forest stands develop in the middle and lower slopes. The distribution of diameter sizes of the trees usually shows a pronounced bimodality. Recruitment is discontinuous as shown by the spatial pattern of tree sizes, and regeneration is vegetative in the upper slopes.

Leaf area indices range between 2.3 and 2.8 m^2 m^{-2} for the deciduous forest and from 2.5 to 4.5 m^2 m^{-2} for the evergreen forest. It appears that tree standing biomass ranges between 6.0 and 13.6 kg C m^{-2} in the deciduous forest, and between 10.5 and 15.6 Kg C m^{-2} in the evergreen forest. Production varies from 148.0 to 372.1 g C m^{-2} yr^{-1}, and from 204.6 to 346.4 g C m^{-2} yr^{-1}, respectively. In relation to boreal forests, biomass can be considered high, but production is medium to low. The relation of wood production to leaf biomass shows that *N. betuloides* behaves similar to coniferous forests and *N. pumilio* as hardwood forests.

There is much accumulation of organic matter on the forest floor, especially coarse wood debris, reflecting the stage of development of these forests and the unfavourable conditions for litter decomposition.

KEY WORDS: *Nothofagus*, production, forest structure, regeneration, forest soils, Tierra del Fuego.

INTRODUCTION

The forests of Tierra del Fuego extend along the southernmost foothills of the Andes, between 53° 30' and 56° South (Fig. 1). They are formed by three broad-leaved species of *Nothofagus*. *N. betuloides* (Mirb.) Oerst. is the main constituent of the evergreen Magellanic forest, whereas *N. pumilio* (Poepp. and Endl.) Krasser forms the deciduous forest. The third species, *N. antarctica* (Forstf.) Oerst., also deciduous, is restricted to the least favourable habitats. The forest ecosystems of Tierra del Fuego are poorly known, with the possible exception of their floral composition. Geographical location and climate have both contributed to this lack of studies, even though peculiar environmental conditions make this region a place of considerable scientific interest. It is the closest area to the South Pole not covered by ice. The influence of the Andes, Quaternary glaciations and the nature of the substrate formed mainly by metamorphic and volcanic rocks, have shaped a unique landscape.

Tierra del Fuego is a group of islands to the south of the Magellan Straits, and only 1,170 km from Antarctica. The disjunctive distribution of living *Nothofagus* attests to an ancient trans-Antarctic forest belt. Nowadays, the proximity to this continent is reflected in the climate, which is much colder than that of the Northern Hemisphere at the same latitude. The climate is cold-temperate without pronounced thermal oscillations during the year.

Measurements taken at Ushuaia, representative of the forest area, show an average temperature of 9.4 ºC in summer and 1.7 ºC in winter (Fig 2). Thus, the average annual variation is less than 8 ºC. Strong winds are the other climatic characteristic of the zone. Climatic subzones are caused by the mountains of the Andes, which in Tierra del Fuego change direction and run from west to east, bordering the western and southern coasts.

The Andes, as in the rest of the South American continent, are the great structural element of the landscape. The distribution and type of ecosystems are, to a great extent, determined by the presence, arrangement and elevation of the mountains, which were lifted in the Pliocene. The altitude of the Andes in Tierra del Fuego is not great, yet sufficient to slow down the clouds from the Pacific ocean.

Gradual precipitation has created a gradient ending in steppes. Consequently, the coastal strip occupied by evergreen forest has a cold-temperate climate with high humidity (Cfk'c, according to the Koppen classification; PISANO, 1981). The area occupied by *N. pumilio* runs between the steppe and the evergreen forests. The climate in this zone is cold-temperate with

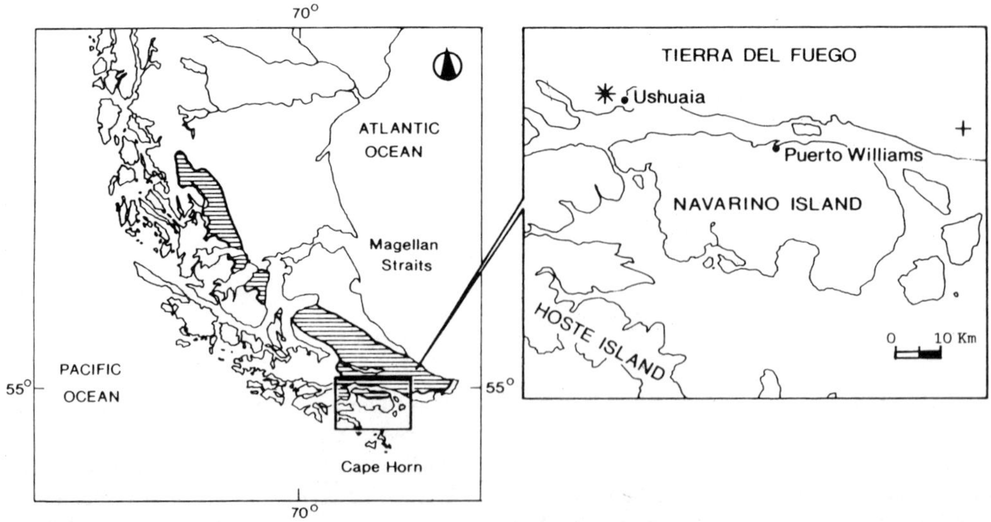

FIGURE 1. Location map. General distribution of *Nothofagus* forests on Tierra del Fuego. The striped area corresponds to the deciduous *N. pumilio* forest and the unstriped coast to the evergreen *N. betuloides* forest. The samples were taken in the Valle de Andorra (*) and the Costa de Moat (+).

steppe degeneration (Dfk'c; PISANO, 1981). Both types of forest are situated within the climatic limits of conifer forests (Fig. 3) (HAMMOND, 1972).

The flora and fauna of the different terrestrial ecosystems of Tierra del Fuego reflect the past and present effects of recent glaciations, climatic changes, and exogenous disturbances. This last factor is cited by PISANO (1975) as a lack of homeostasis.

On the forested mountain slopes numerous landslides bring about strong geomorphological and soil dynamism on different scales. Strong winds and shallow rooting cause many tree falls, creating gaps of different sizes and with the soil churned up and disturbed to varying degrees. Large quantities of wood debris are accumulated on the forest floor, in relation with the old-growth forest stands in some areas, and the unfavourable conditions for decomposition.

The severity of the climate marks an altitudinal gradient in the vegetation. The tree line rarely surpasses 700 m a.s.l. at these latitudes.

The morphology and degree of development of the trees are related to the altitudinal gradient. In the upper part of the forest and along a strip of more than 40 m above the timberline, trees take on a bushlike aspect, known as krummholz. At the tree line the height of the trees is less than 50 cm and *N. antarctica* usually appears as an accompanying species.

Studies on the forests of Tierra del Fuego have been mainly descriptive, especially biogeographic and taxonomic. Outstanding works are those of PISANO (1975, 1977, 1981) and DIMITRI (1982). Fluctuations in the extension of the forests based on pollen profiles as well as on other paleoecological features have been described by AUER (1960), MARKGRAF (1977) and VUILLEUMIER (1971). The climatic conditions, rainfall and temperature patterns have been analysed by ZAMORA & SANTANA (1979 a, b) and SANTANA (1984). Special reference should be paid to the work of SCHMIDT & URZÚA (1982) on the structure and production of *N. pumilio*, focused on forest management and timber production. The foliar anatomy of the species of *Nothofagus* has been studied by RAGONESE (1981) and the xylem anatomy by RIVERA (1987, 1988). We recently presented the first results on some aspects of soil-plant relationships for both types of forest (GUTIÉRREZ *et al.*, 1989).

There is, however, a lack of quantitive studies on most of the ecological aspects of these forests. The results of comparative studies carried out in the *N. pumilio* and *N. betuloides* forests are presented in this study. We have analysed the possible

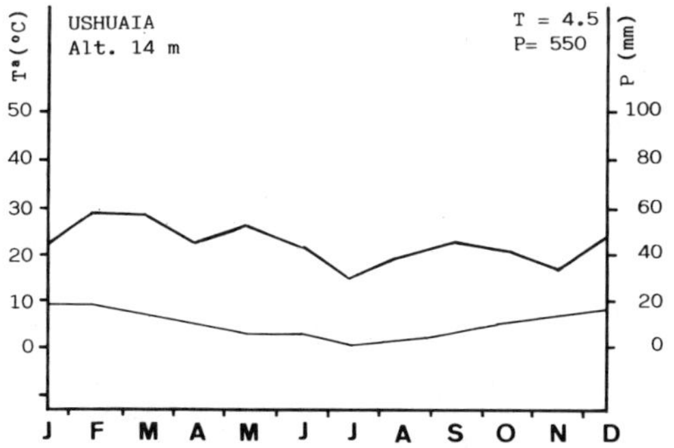

FIGURE 2. Climatic diagram according to the data from Ushuaia, Tierra del Fuego, 54° 48'S and 68° 19'W.

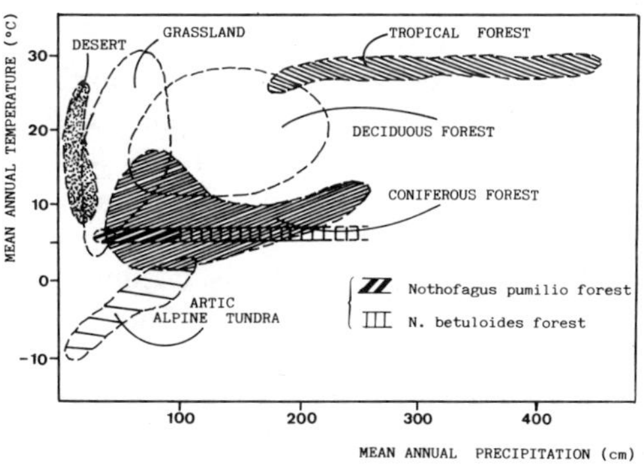

FIGURE 3. The ecoclimatic position of *Nothofagus* forests of Tierra del Fuego among the major terrestrial ecosystems in the ecoclimatic space defined by average temperature and precipitation (modified from HAMMOND, 1972).

causes of the distribution of the two species of *Nothofagus*, the former deciduous and the latter evergreen.

Climatic conditions are considered to be the primary cause of the distribution of these *Nothofagus* species, but other factors may also come into play. Soil characteristics may have a relevant role as well. On the other hand, apart from the research carried out by SCHMIDT & URZÚA (1982), no other studies have been undertaken concerning the structure and production of these forests and hence no data are available for the possibilities of exploitation and management techniques. The existence of great extensions of unexploited forest in Tierra del Fuego implies minimal human disturbance. Given the speed with which these ecosystems are being exploited for timber, crop and livestock all over South America, there is an urgent need for a database for a more rational management of the environment.

DESCRIPTION OF SAMPLING SITES

For the study of the two types of forest, two representative sites were selected on the Isla Grande de Tierra del Fuego (Republic of Argentina). The deciduous *N. pumilio* forest was sampled in the Valle de Andorra (54° 49'S & 68° 42'W) and the evergreen forest in the Costa de Moat (54° 49'S & 67° 21'W (Fig. 1).

The Valle de Andorra is 4 km north of Ushuaia. The annual average precipitation and temperature are 561 mm and 4.5 ºC (HOLDGATE, 1960). The tree stratum is represented fundamentally by *N. pumilio*. The bottom of the valley is occupied by peat bogs of *Sphagnum* sp., with the forests developing on the mountain slopes, from 240 to 500-600 m a.s.l. In the lower and middle parts of the forest there are large emerging trees and discontinuities in the canopy. On the floor there is a large accumulation of wood, and young shoots appear in the clearings. Trees are smaller in the higher slopes of the mountains.

The evergreen forest was sampled in the area of Costa de Moat where the Beagle Channel exits towards the Atlantic, at around 110 km east of Ushuaia. Although climatic data are not available for this site, it is wetter and warmer than the Valle de Andorra. The area receives between 700 and 1000 mm of rainfall annually (ZAMORA & SANTANA, 1979a). An altitudinal gradient is also evident in this forest.

A mixed forest is formed in the lower parts, with *N. betuloides* forming the canopy. *Drimys winteri* Forst. and *Maytenus magellanica* (Lam.) Hook. f. form the understory, both species being broad-leaved and evergreen. This type of formation disappears at an approximate altitude of 150 m a.s.l., giving way to monospecific stands of *N. betuloides*. Isolated populations of *N. pumilio* also appear in well drained zones. The topography is gentler and the mountains are lower than in the Valle de Andorra. The upper forest limit reaches 350-400 m a.s.l., and is often limited by peat bogs which, in this zone, cover the higher parts of the mountains and extend over the middle parts, which are interconnected by gentle slopes.

METHODS

The recognition of different phases of forest development is based on the analysis of population structure of dominant species. Plots of 26 m x 26 m were delimited wherever possible in both types of forest in the upper, middle and lower slopes. Samples were gathered in each plot for the study of both trees and soils.

Estimation of biomass and production was carried out using dimensional analysis (WHITTAKER & WOODWELL, 1968; WHITTAKER & MARKS, 1975) from field measurements of trees. Diameter at breast height (DBH), at 1.30 m above

TABLE I. Climatic characteristics of 6 representative sites of the *Nothofagus* forest area. Due to their location they can be considered to encompass the gradient over which the forests extend. The data from Isla Nueva are the most representative of the zone in which the evergreen forest was sampled. Med: average annual temperature; Mmed: average of the maxima; mmed: average of the minima; Mabs: absolute maximum registered; mabs: absolute minimum registered. All the temperatures in ºC. P, precipitation in mm yr^{-1}. Data from ZAMORA & SANTANA (1979a) except for those from Ushuaia.

SITE	LAT	LONG	Med	Mmed	mmed	Mabs	mabs	P (mm)
Punta Arenas	53º08'S	70º53'S	6.2	10.8	2.0	26.4	-11.8	438.9
San Isidro	53º47'S	70º59'W	5.9	9.0	3.1	22.0	-6.0	876.5
Ushuaia	54º48'S	68º19'W	5.4	9.6	1.2	29.5	-21.0	550.5
Puerto Navarino	54º57'S	68º20'W	6.4	8.9	3.3	23.2	-8.2	447.8
Puerto Williams	54º56'S	67º38'W	5.5	9.4	1.6	26.4	-10.0	553.4
Isla Nueva	55º10'S	66º36'W	5.6	9.3	2.8	23.8	-6.2	738.1

ground, tree height, and bark thickness were measured in all trees within the plots. Wood cores were extracted from the trunks at 1.30 m covering all the tree sizes in order to assess age structure, and to calculate increases in biomass and sapwood area. Production was estimated from the wood increments of the last ten annual rings. Branch and leaf biomass were calculated through allometric relationships from 6 trees of *N. pumilio* and 7 of *N. betuloides,* which were felled, measured and weighed.

All the trees in each plot were mapped in adjoining squares of 2 x 2 m, forming a lattice of 169 squares. Spatial structure was analysed with the hierarchical squares technique (GREIG-SMITH, 1964) and the distribution pattern was expressed as Morisita's index (WILLIAMSON, 1975; ELLIOT, 1983).

Total nitrogen and total carbon in leaves were measured with a Carlo Erba Autoanalyzer. Water-soluble phosphorus in the forest floor and mineral soil was determined using HUMPHREYS & PRITCHETT's (1972) procedure. A detailed explanation of the complete sampling methodology can be found in ROMAÑÀ *et al.* (1989).

THE DISTRIBUTION OF *NOTHOFAGUS* SPECIES

In a first approximation, the roles of continentality and of winter harshness in the dominance of deciduous over evergreen appear evident (Table I). This is also the explanation given from field observations. McQUEEN (1976), HUECK (1978) and PISANO (1977) attribute a strong resistance to low temperatures to *N. pumilio.* This explanation is supported by the presence of species with laurel-like leaves, such as *D. winteri* and *M. magellanica,* in the evergreen forest. These thermophile species are distributed along the coastline and at low altitude, in areas of relative climatic mildness where the evergreen forest is found. In the same way, evergreen flowering plants have a thermophile distribution compared with the deciduous flowering plants (WALTER, 1976), this character being interpreted as adaptation to extremely low temperatures.

The above explanation of the distribution of *Nothofagus* species is not entirely satisfactory. The existence of two distinct forest types is not related to a strong environmental gradient (Table I). Furthermore, both species are often found growing side by side. Yet, the transition from evergreen to deciduous forest takes place in just a few kilometers. The limits of each species could be both climatic and edaphic (PISANO, 1977). *N. betuloides* survives on soils with rather poor drainage, whereas *N. pumilio* is much less tolerant to waterlogging. The soils of the evergreen forest are predominantly hydromorphous, while the processes of podzolization are relevant in the deciduous forest (ROMAÑÀ *et al.*, 1989). The question thus arises as to

TABLE II. Specific leaf weight (mg cm^{-2}) for the two *Nothofagus* species. MO1, MO2 and MO3, sample sites from lowest to highest altitude in the *N. betuloides* forest on the Costa de Moat. VA1, VA2 and VA3, the same for *N. pumilio* in the Valle de Andorra. Date of sampling January 1989.

	species					
	N. betuloides			*N. pumilio*		
YEAR SITE	MO1	MO2	MO3	VA1	VA2	VA3
1 (1988)	14.35	10.89	12.55	8.28	8.61	9.14
2 (1987)	19.24	17.71	16.95			
3 (1986)	19.30	19.10	17.37			
4 (1985)	?16.10	19.33	18.15			
5 (1984)		18.54	18.59			
6 (1983)		18.91	19.20			
7 (1982)		20.04	19.61			

TABLE III. Percentage of carbon (C) and nitrogen (N) in leaves (one year leaves, * and two year leaves, +). MO1, MO2 and MO3, sample sites from lowest to highest altitude in the *N. betuloides* forest on the Costa de Moat. VA1, VA2 and VA3, the same for *N. pumilio* in the Valle de Andorra. LR, Lake Roca (near Ushuaia).

	% C	% N	C/N	Species
VA1	44.77	2.43	18.42	*N. pumilio*
VA2	45.22	2.10	21.53	*N. pumilio*
VA3	45.45	2.35	19.34	*N. pumilio*
MO1	49.50	1.67	29.64	*N. betuloides* (*)
	48.17	0.85	56.06	*D. winteri*
MO2	49.50	1.67	29.64	*N. betuloides* (*)
	49.58	1.18	42.02	*N. betuloides* (+)
MO3	49.80	1.32	37.73	*N. betuloides* (+)
LR	47.12	2.17	21.71	*N. antarctica*

the relative advantages of the deciduous or the evergreen habit.

The evergreen condition is advantageous for the absorption of nutrients when growth is limited by low soil nutrient concentrations. The evergreen character of *N. betuloides* gives it a certain independence from environmental fluctuations owing to a greater retention of nutrients in the tree. Furthermore, the longer lifespan of the leaves, up to seven years (Table II), allows greater photosynthetic production per nutrient unit in the leaves, increasing the efficiency of their use. This may represent an advantage both in habitats with low nutrient contents and those in which nutrient recycling is difficult. Sclerophyllous leaves of *N. betuloides* (RAGONESE, 1981) are adaptative in this kind of habitat. Average leaf density, an index of sclerophylly, is slightly higher than 8 mg cm^{-2} in *N. pumilio* and 17 mg cm^{-2} in *N. betuloides* (Table II). For comparison, *Fagus sylvatica* has values between 2.2 and 7 mg cm^{-2} in the Montseny mountains, NE Spain (TERRADAS, 1984), lower than those of the *Nothofagus* species. The index of sclerophylly calculated for *N. betuloides* (Table II) is similiar to that of the Mediterranean sclerophylles, whose values range between 10 and 25 mg cm^{-2} (SALA, 1986).

The differences in nutrient content between the two species are high. *N. pumilio* has large accumulations of nitrogen in its leaves (Table III). Thus, the C/N ratio reaches values around 20. In contrast, the nitrogen content in *N. betuloides* is low. Its C/N ratio is almost double, similar to the values found in *Quercus ilex*. In the soil, the C/N ratio is 40 in the Valle de Andorra, and 80 in the Costa de Moat, values much higher than those found in the leaves. The large accumulation of wood debris on the ground and the strong hydromorphism in the Costa de Moat, which retards the decomposition of organic matter, seem to cause these high concentrations of carbon on the ground.

In swamped boreal environments the sclerophyllous and evergreen habit of various species is associated with low nutrient contents in the plant tissues (LARSEN, 1982). This has been interpreted as an adaptation to restrictions imposed by a lack of O_2 in soil water, low temperatures in the radicular environment, absence of mycorrhizae, and the existence of free toxins. Mycorrhizae are indeed very abundant in the *N. pumilio* forest, and scarce under *N. betuloides*. The formation of mycorrhizae has been described for other *Nothofagus* species (MEYER, 1966). However, they have never been reported in waterlogged soils.

With respect to phosphorus, the differences between one type of soil and another are also pronounced (Table IV). The data for soluble phosphorus in the organic horizons of the deciduous forest are higher than average annual requirements of forest ecosystems in different environments (COLE & RAPP, 1981). Values for the evergreen forest range from average to moderately low. Sclerophylly is present in some Australian species in spite of high rainfall; the anomalous presence of this trait has been attributed to low soil fertility, because the sclerophylly index decreases after the addition of nitrogen and phosphorus (BEADLE, 1966). Likewise, in well aerated soils at mid-latitudes, sclerophylly has been related to low availabilty of nutrients, especially phosphorus (LOVELESS, 1961).

The decomposition of organic matter and the release of nutrients in waterlogged soils is slower than that expected under cold climates with low temperatures and minimum thermal fluctuation. Waterlogging of the soil appears to be related to the distribution of the two species in the Valle de Andorra and in the Costa de Moat. The characteristics of hydromorphic soil would promote physiological strategies for nutrient conservation in the trees, together with a greater efficiency in nutrient use. The maintenance of the leaves throughout the year, as well as their greater longevity and strong retranslocation of nutrients, can be interpreted as adaptative in this context. Such a strategy is possible under climatic conditions allowing the persistence of leaves in winter. Slight differences in the absolute minimum temperatures seem to be sufficient for the establishment of such conditions, as shown by the climatic data in Table I.

At the regional level, however, the hypothesis of hydromorphy is still questionable, as it is difficult to imagine that a parameter of this type could have such a pronounced geographical component. One element to consider is that the distribution area of *N. betuloides* roughly coincides with the rainier coastal areas. In addition, these southernmost areas seem to have been more affected by glaciation, producing gentle slopes which facilitate water accumulation.

TABLE IV. Water soluble phosphorus (Kg P ha^{-1}) in the organic layer, H, and in the first 15 cm of mineral soil. MO1, MO2 and MO3, sample sites from lowest to highest altitude in the *N. betuloides* forest on the Costa de Moat. VA1, VA2 and VA3, the same for *N. pumilio* in the Valle de Andorra.

	Available P (kg P ha^{-1})			
	H		0-15	
SITE	MEAN	STD	MEAN	STD
VA1	14.33	6.79	0.5	0.08
VA2	22.03	3.42	0.2	0.08
VA3	11.68	4.36	0.4	0.22
MO1	2.30	1.27	0.4	0.22
MO2	4.13	3.02	0.1	0.01
MO3	12.10	2.55		

STAND STRUCTURE

The distribution of size frequencies, either of DBH or of the biomass of individuals, is frequently used to characterize the structure of a forest stand, together with the ranges and values of the other variables. Although without common agreement regarding the size interval used by different authors, we have defined twelve size classes of DBH (FORD, 1975; MOHLER *et al.*, 1978; Figs. 4 and 5). These distributions reflect the differences of the forest stands according to altitude. The biggest trees grow in the lower and medium slopes, where better edaphic conditions and a milder climate prevail. Other parameters reflect these differences among stands. Yet the density of trees is quite high in the upper slopes. Thus, the basal area increases with elevation as a consequence of the greater packing together of individuals.

Dendrochronological values at 1.30 m tend to underestimate tree age. Hardwood of some cores was rotten so the age was estimated dividing the tree radius by the annual mean increment. To avoid over- or underestimation of tree ages, cores rotten to more than half of tree radius were rejected. In this manner, our results indicate that the biggest trees are not always the oldest (Fig. 6). Tree ages show a high variation with respect to size in both forests. This may be a consequence of the great heterogeneity of habitats. In the evergreen forest, a high variability is also found in the ages reached within a narrow category of DBH, especially above 45 cm. In contrast with *N. pumilio*, *N. betuloides* reaches older ages with smaller diameters. This may be related to the oligotrophic condition of the soil in which the latter species grows (Table IV). Nevertheless, differences exist among the stands according to the site characteristics. In the lower slopes of the evergreen forest, classes of DBH less than 25 cm do not exist and there is no clear relation between age and DBH. In the remaining stands the relationship between size and age is more evident and could be statistically described by allometric and/or exponential functions.

The structure of a forest is dynamic. It changes over time and depends on the ecological characteristics of both the component species and the whole habitat. The dependence on the characteristics of the species is manifest in the degree of tolerance to shade. This trait determines to a large extent the way in which regeneration is produced and is reflected in size and age structure of the stand. The more shade-tolerant species tend to form multiple-cohort stands and tree recruitment is continuous throughout time. On the other hand, when the trees vary little in size and have originated in a relatively short interval of time, the structure of the stand is even-aged, corresponding to single-cohort stands. This type of stand can be formed by shade-tolerant or shade-intolerant species, although even-aged stands of intolerant species are more common. According to the literature, *Nothofagus* species are shade-intolerant (HUECK, 1978). Thus, the expected distribution of DBH's would correspond to that of even-aged stand forests (cf. SCHMIDT & URZÚA, 1982). Most of the observed size distributions are bimodal (Figs. 4 and 5). The regenerating class, formed by trees with less than 7 cm of DBH, is scarce or lacking in many stands except the one in the lower part of the Valle de Andorra (Fig. 4) and in the upper slope of Moat (Fig. 5). But hardly any of these stands can be considered uneven-aged. The bimodal distributions correspond more to the final stage of

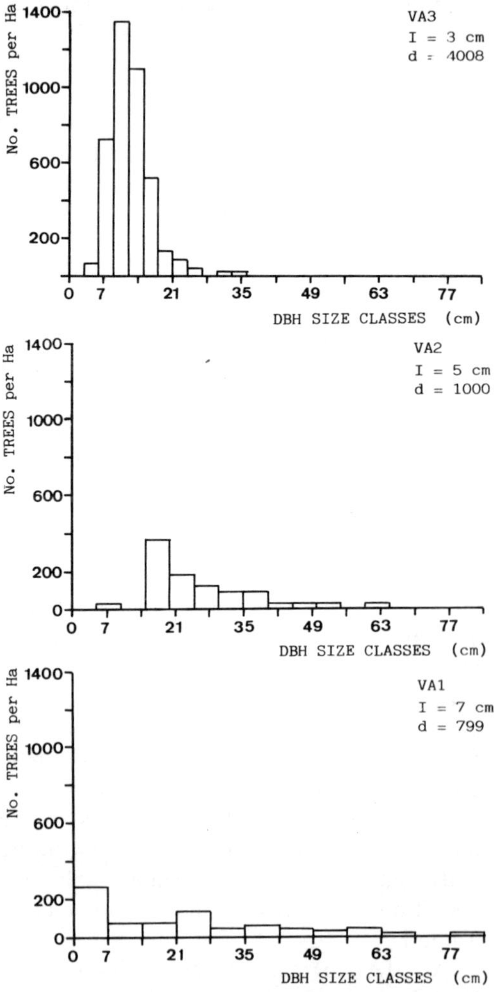

FIGURE 4. Distribution of trees according to the DBH sizes in the lower (VA1), medium (VA2) and upper (VA3) slopes of the deciduous *N. pumilio* forest in the Valle de Andorra. Notice that the interval (I) defining size classes and the density of the stand (d, number of trees per hectare) varies among sample plots.

self-thinning and even-aged stands of mature forest. In these stages of development, distributions are often indeed bimodal when twelve size categories are used (FORD, 1975; MOHLER *et al.*, 1978). Bimodality is a consequence of the aging of the stand in which the largest size categories in trunk diameter and tree height form a distinct group of trees. When age is considered (Fig. 6), most of the stands cannot be classified as even-aged, except the plot of the higher slope in the Valle de Andorra.

Analysis of the spatial distribution of all the trees in the plots gives clues to the regeneration pattern. Two examples of spatial distribution are shown in figure 7 for the lower part of the Valle de Andorra, VA1, and the higher part of the Costa de Moat, MO3. In the first case, regeneration is produced in areas free of large individuals. Therefore, the bimodality observed in DBH size distributions (Figs. 4 and 5) may be due to the recruitment in bursts of new individuals when the canopy has disappeared. Although it is not known up to what point the *Nothofagus* species are shade-intolerant, it is very significant that in the low part of the Costa de Moat where the *N. betuloides* forest has *D. winteri* and *M. magellanica* in the understory, there are

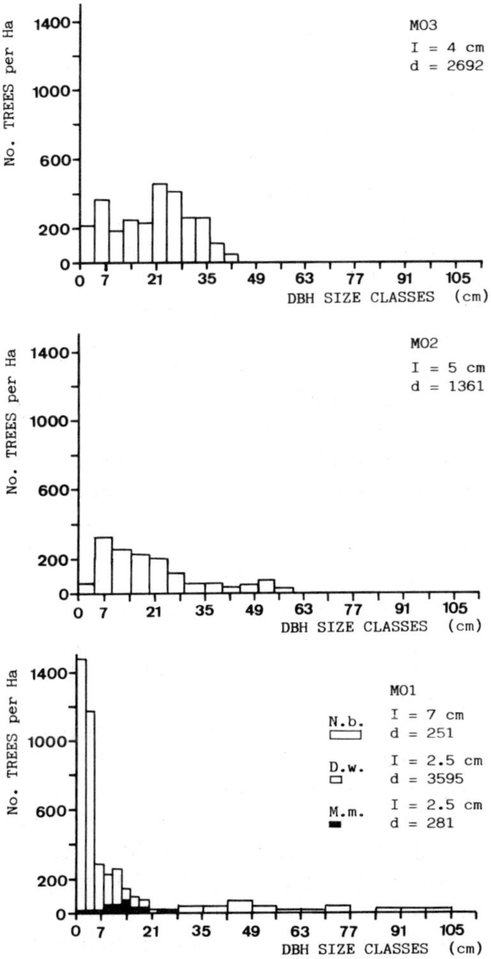

FIGURE 5. Distribution of trees according to the DBH sizes in the lower (MO1), medium (MO2) and upper (MO3) slopes of the evergreen *N. betuloides* forest on the Costa de Moat. Notice that the interval (I) defining size classes and the density of the stand (d, number of trees per hectare) varies among sample plots. Data are also given for *Drimys winteri* (D.w.) and *Maytenus magellanica* (M.m.).

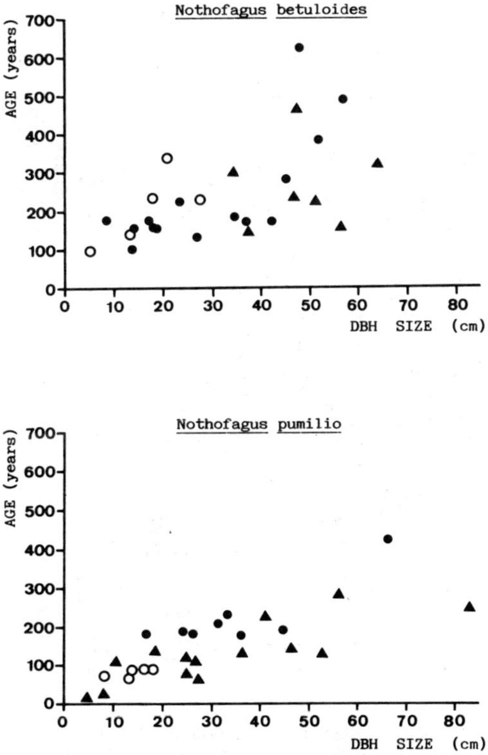

FIGURE 6. Relation between age and size of trees in 6 sampled plots. Data from the lower (triangles), middle (filled circles) and upper (open circles) slopes of each type of forest.

neither saplings nor poles (Fig. 5, MO1).

If the recruitment of new individuals is produced in bursts and these *Nothofagus* species do not regenerate beneath their canopy, the regenerating class should exhibit a spatial pattern in accordance with environmental minor disturbances. Spatial analysis was carried out twice, once for the large trees, with a DBH above 30 cm, and then including all the trees. The results are displayed in figure 8. The large trees feature a regular distribution in small areas (Morisita index values of less than one), which becomes random as the area increases (Morisita index values around one). However, when all the trees are included, the distribution is aggregate up to areas of 64 and 80 m^2 in the low and medium parts of both forests, and 16 and 36 m^2 in the higher parts. These cluster dimensions indicate that the recruitment of new individuals is produced in a discontinuous form and in a space defined by other physical attributes. The result is a mosaic of patches of sizes and ages, sometimes overlapping and sometimes distinct (cf. PEET & CHRISTENSEN, 1987). The development of these patches may be subpopulations or different groups of trees, but other new ones also appear as the distributions of diameters (Figs 4 and 5) and ages (Fig. 6) seem to indicate. On the other hand, the spatial distribution of trees is very different between the upper and lower slopes of both forests (Fig. 7). The distribution of large trees in the upper parts is aggregate, the clusters being of 16 and 32 m^2. When all the trees are included the size of the patches is 4 and 8 m^2, these being included in the former clusters. The smaller clusters present in the higher parts of the forest correspond to a different type of regeneration, i.e. vegetative. Regeneration is accomplished by sprouting at the base of the trunks of living trees. This type of regeneration renders a different pattern from those of the lower and middle parts of the forest.

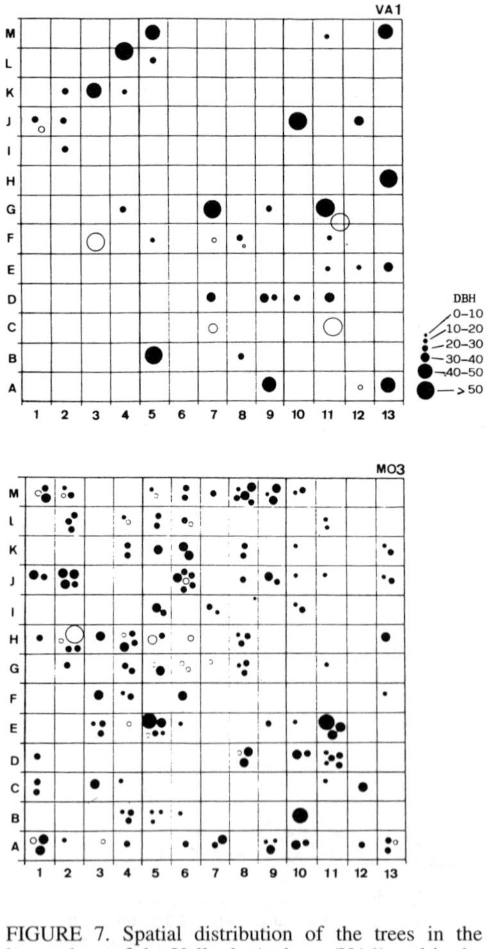

FIGURE 7. Spatial distribution of the trees in the lower slope of the Valle de Andorra (VA1) and in the upper slope of the Costa de Moat (MO3). The distribution at middle elevations is very similar to that of the lower slopes. DBH in cm, grid in m. White circles represent dead individuals.

Another very important structural parameter is the leaf area index, LAI. The leaf area index is fairly similar in all the sampled plots; values are between 2.3 and 4.5 m^2 m^{-2}, being higher in the evergreen forest (Fig. 9 and Table V). The range of values for the leaf area index are coherent with those expected according to latitude, climate, and the structure and dynamics of forest stands subjected to frequent treefalls. *Fagus sylvatica* features values of 3.2 in leaf area index at similar latitudes in the Northern Hemisphere owing to the harsher climate (De ANGELIS *et al.*, 1981).

The relation between leaf area and sapwood area is interesting due to their mutual functional implication. These *Nothofagus* species are diffuse-porous with only small-diameter vessels (RIVERA, 1987, 1988). In principle, this implies a smaller LAI per unit of sapwood with respect to ring-porous trees such as oaks. The values found for *N. pumilio* are 0.12 and 0.13 $m^2\ cm^{-2}$ in the stands situated in the medium and lower slopes and 0.07 $m^2\ cm^{-2}$ in the upper slopes. Results show that a greater sapwood surface is needed in the stand of the subalpine zone, i.e. the area of evaporation per unit of sapwood area is lower. These variations may reflect aspects related to the permeability of the wood, the increased viscosity of water, a possible adaptation to more dessicant conditions imposed by wind and a higher demand of carbohydrates. The values are similar to those found for *N. solandri* in montane and subalpine stands, respectively (BENECKE & NORDMAYER, 1982, in WARING & SCHLESINGER, 1985). The ratio of leaf to sapwood area calculated for the stands of *N. betuloides* forest, gives values of 0.32 $m^2\ cm^{-2}$ for the lower slope close to the sea and 0.11 and 0.09 $m^2\ cm^{-2}$ for the medium and upper slopes, respectively. The requirements of the sapwood area appear to be reduced in the milder climate, particularly in the lower coastal location. These and other previously mentioned differences among stands at different elevations may be due to either genetic differentiation (WARING & SCHLESINGER, 1985) or environmentally dependent different gene expressions.

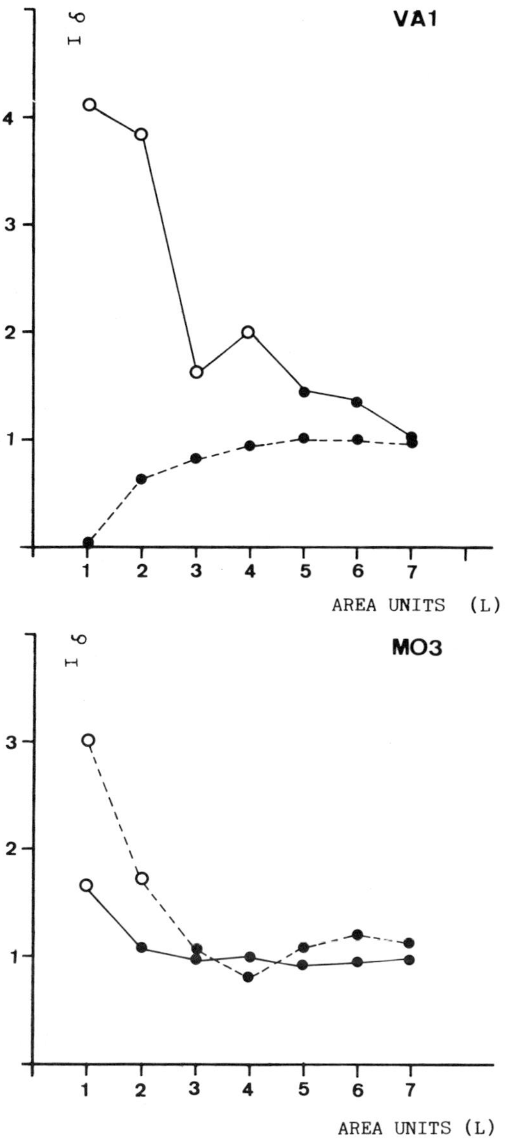

FIGURE 8. Representation of the values of the Morisita index (Id) according to area. Each area unit measures $(2 * L)^2\ m^2$, L being the unit of length. Open dots are data points significant to 95 % according to the F test. The values around 1 are those that take a random distribution, those over 1 appear in aggregates, and those below 1 are regular. The broken line corresponds to trees with DBH > 30 cm, the continuous line to all trees in the plot.

BIOMASS, PRODUCTION AND SOIL ORGANIC MATTER

The range of aboveground biomass and production values is shown in Table V. The evergreen forest features a greater biomass but a lower production than the deciduous forest. This difference translates into a lower biomass turnover rate for the former. The *Nothofagus* forests of Tierra del Fuego have climatic conditions corresponding to the zone of boreal conifer forests (Fig. 3), but show higher biomass values than the latter (Table V). This is not observed in the *N. pumilio* stands of the upper slope, with a production of 148.0 g C m^{-2} yr^{-1}, a figure that lies in the low range of observed values. High values of biomass are translated into a much lower productivity and a higher turnover rate in relation to the boreal forests. These data agree with the results and conclusions that refer to other structural parameters.

Relative to the incident radiation, the evergreen forest has a greater average production efficiency than the deciduous one. Values range from 0.32 to 0.54%, and

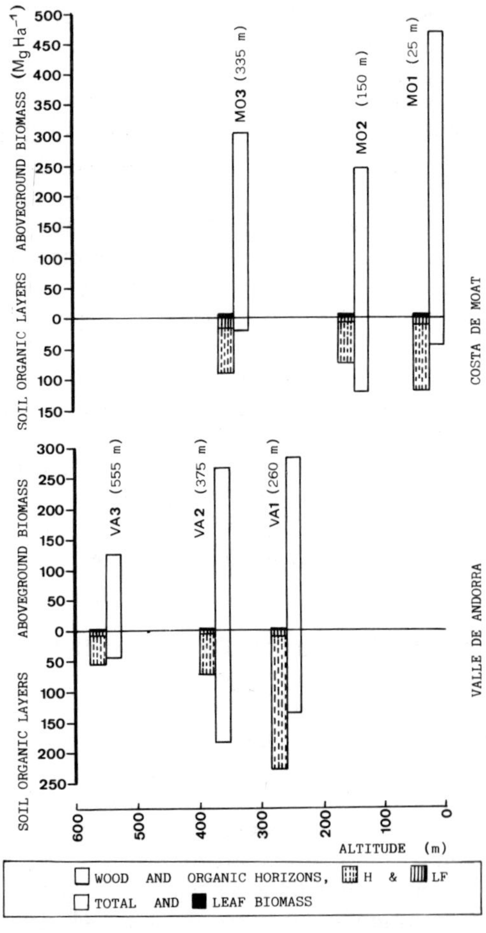

FIGURE 9. Distribution of biomass and soil organic matter, both in Mg ha^{-1}, in the 6 sampled plots.

TABLE V. Structural characteristics, including biomass and production, of the *Nothofagus* forests sampled in Tierra del Fuego.

	Deciduous forest	Evergreen forest
Age	91 - 247	339 - 628
Trees ha^{-1}	799 - 4008	251 - 2692
LAI (m^2 m^{-2})	2.3 - 2.8	2.5 - 4.5
Basal area (m^2 ha^{-1})	50.5 - 66.9	62.6 - 107.7
Biomass (kg C m^{-2})	6.0 - 13.6	10.5 - 15.6
Production (g C m^{-2} y^{-1})	148.0 - 372.1	204.6 - 346.4
Leaves (g C m^{-2} y^{-1})	85.3 - 100.6	90.2 - 143.5
P/B (yr-1)	0.022 - 0.031	0.018 - 0.027
B/P (yr)	46 - 32	56 - 37
	Boreal Forests	**Temperate Forests**
Biomass (kg C m^{-2})	2.5 - 16.6	2.5 - 25.7
Production (g C m^{-2} yr^{-1})	166.7 - 833.4	250.0 - 1041.8

from 0.17 to 0.44%, respectively. In the relevant calculations involved we considered the average incident radiation in each area (ZAMORA & SANTANA, 1979a). The average estimated incident radiation for the deciduous forest is 85 Kcal cm^{-2} yr^{-1}, and that of the evergreen forest is 65 Kcal cm^{-2} yr^{-1}. Although these data are not referred to the sampling stations, they provide an adequate quantification. The efficiency of these forests appears to be low, being more similar to that of hardwood than coniferous forests (WEBB *et al.*, 1983), in spite of having climatic similarities with the latter. But if we consider relations such as the production of wood relative to leaf biomass (Fig. 10), the *N. betuloides* forest has a lower wood production per unit of photosynthetic surface than *N. pumilio*. Furthermore, given that *N. betuloides* has a greater LAI its growth efficiency, expressed in g of wood per m^2 of leaves per year, is lower, being situated in an intermediate position between the hardwood and the coniferous forests. On the other hand, *N. pumilio* overlaps with hardwood forests (Fig. 10, O'NEILL & De ANGELIS, 1981). In general, the restrictions on accumulating and

maintaining high leaf area indices are greater in cold climates. Besides the effect of climate, in Tierra del Fuego the following factors should be considered: windfall disturbance, fungal disease, landslides and snow-related dynamism of soil, all of which prevent higher LAI and leaf biomass values. If indeed, forests that do not reach and maintain an LAI of 6 do not reach a steady state (WARING & SCHLESINGER, 1985), the forests of Tierra del Fuego may be an example of forests that do not reach a successional steady state.

In the general context of production in terrestrial ecosystems, the accumulation of biomass in the forests of Tierra del Fuego appears to be limited by low temperatures. Temperature is indeed one of the main factors limiting production although it is not adequately described by means of simple statistical relations (WARING & SCHLESINGER, 1985). This dependence on temperature is also conspicuous at a local level along the altitudinal gradient in the Valle de Andorra. The limitations on potential production in the upper slope of the deciduous forest can be related to a harsher climate. On the other hand, soil conditions may have a greater importance in the evergreen forest (Table IV, Fig. 9). Thus, the altitudinal gradient is blurred by the heterogeneity of soil hydromorphism.

In agreement with the structural and dynamic characteristics, the amount of organic material accumulated in the forest soil of Tierra del Fuego is high (Fig. 9). The climatic conditions, low temperatures and an average annual fluctuation that does not exceed 8 °C, favor a slow decomposition. The accumulation of organic matter on the soil results from the balance between litter inputs and the rate of decomposition, the former being dependent upon forest production. The rate of decomposition is determined by the climatic conditions, the nature of the litter, and the characteristics of the soil itself. The soil characteristics, e.g., waterlogging and chemical conditions, affect, in turn, biological activity. Thus, the rate of decomposition integrates the biological potential of the soil and determines both the turnover and availability of nutrients for plants, mainly N and P. The decomposition rate can be considered as a soil fertility index.

In the topographic sequence of the Valle de Andorra, the organic matter, and especially the wood debris in the organic horizons, decreases from the lower to the upper slopes, in accordance with biomass (Fig. 9). These results indicate that the differences in organic matter content are explained by the quantity of litter fall. Furthermore, the harsher climate in the upper slopes does not appear to delay the decomposition of organic matter. Since the estimated contribution from leaves is

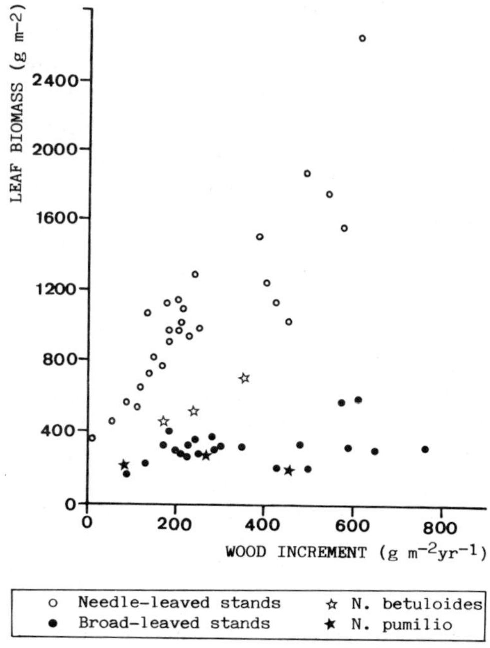

FIGURE 10. Wood production in relation to leaf biomass of the sampled *Nothofagus* forests in Tierra del Fuego, together with published data for other forests (O'NEILL & De ANGELIS, 1981). Notice the similarity of the evergreen *N. betuloides* with coniferous forests and of deciduous *N. pumilio* with hardwood forests.

similar in the three plots (Fig. 8, Table IV), the differences in litter accumulation must be accounted for by the wood fraction.

The accumulation of organic matter in the sampled stands of the evergreen forest, in contrast to the Valle de Andorra, increases from the lower to the upper parts. This difference is conditioned by the role of excess of water in the evergreen forest, where it causes a great spatial variability in the accumulation of organic matter. Decomposition takes place temporarily under anaerobic conditions, which are quite restrictive to mineralization. The H soil horizon in the stands on the lower slopes of both forests has a high mineral content, probably due to more intense biological activity. A lot of worms were observed during sampling.

High amounts of litter and wood debris may control the long-term productivity of forest communities (CHARLEY & RICHARDS, 1974). The accumulation of organic matter in the H horizon is very high in the forests studied: up to around 200 Mg ha^{-1} in Valle de Andorra and 135 Mg ha^{-1} in Costa de Moat (Fig. 9). These values are similar to those found in the boreal coniferous forests of Vancouver, 50° N of latitude (VALLEJO, 1987). In contrast, typical Mediterranean forests, such as those of evergreen *Quercus ilex* in Montseny, NE Spain, reach only 10 to 50 Mg ha^{-1} (HERETER, 1990).

The H horizon has plenty of tree roots in both *Nothofagus* forests. However, in Moat most of the roots usually do not reach the mineral horizons because of waterlogging. Also, the H horizon has the highest concentrations of nutrients in the soil (Table IV). Therefore, the H horizon plays a key role in the maintenance of soil fertility. Rational forest management practices should take this into account.

The rate of decomposition (k; OLSON, 1963), calculated for the LF horizon and taking into account the estimated contribution of fallen leaves from Table V, ranges between 0.20 and 0.38 in the stands sampled. These values indicate a relatively rapid incorporation of litter remains into the H horizon. The mean residence time of leaves in the LF horizon, that is 1/k, ranges between three and five years. However, the values of k, calculated for the leaf fraction, may be underestimates, since wood debris less than 2 cm in diameter were included. These woody remains were not included in the estimates of litter fall. The mean residence time for all litter fall including leaves and coarse wood debris ranges from 15 to 60 years in both *Nothofagus* forests.

CONCLUSIONS

The forest landscape of Tierra del Fuego has a mosaic pattern at different spatial scales. Within a given forest type, patches of younger trees are included in the biggest patches of old forest. This structure may be caused by strong disturbances like landslides and windthrown trees in the slopes. The gaps produced after such events are colonized by *Nothofagus* seedlings, while there is little tree regeneration under the old-growth forest canopy. The result is multicohort stands. There is also a clear altitudinal gradient in forest structure, the extreme conditions of the upper slopes preventing the development of old-growth forests.

Owing to the limititations imposed by climate, the *Nothofagus* forests of Tierra del Fuego have low productivity. These unexploited forests have a temperature-limited accumulation of biomass and production. Yet there is a huge accumulation of organic matter on the forest floor with an important contribution of decaying wood. The large amounts of accumulated litter reflect the slowness of the decomposition process and the stage of the forest.

In adverse soil conditions, such as waterlogging, the evergreen *N. betuloides* is the dominant forest species. However, it has a lower productivity than *N. pumilio*, even though it is distributed in the rainier and milder areas of Tierra del Fuego.

Broad-leaved evergreen trees are exceptional in extreme latitudes whereas they are common in very different habitats, such as under Mediterranean type conditions. This study shows that the evergreen and sclerophyllous characteristics of *N. betuloides* are related to shortage of nutrients in waterlogged soils. This is a physiological limitation similar to those attributed to typical Mediterranean ecosystems in much drier conditions.

ACKNOWLEDGEMENTS

We are grateful to the Servei d'Espectroscòpia de la Universitat de Barcelona, and the Centre de Recerca Ecològica i Aplicacions Forestals, CREAF, for the technical assistance provided. This study was funded by the Instituto de Cooperación Iberoamericana, ICI, Comisión Nacional V Centenario. Thanks are given to Robin Rycroft and Cristian Ruíz for their accurate translation and interesting comments.

REFERENCES

AUER, V. 1960. The quaternary history of Fuego-Patagonia. *Proc. Roy. Soc. (B)*, 152: 507-515.

BEADLE, N.C.W. 1966. Soil phosphate and its role in molding segments of the Australian flora and vegetation, with special reference to xeromorphy and sclerophylly. *Ecology*, 47: 992-1007.

CHARLEY, J.L. & RICHARDS, B.N. 1974. Energy and nutrient dynamics of forest floors. *Proc. 7th. Triennial Conf. IFA*. Calouncha, 1: 25-34.

COLE, D.W. & RAPP, M. 1981. Cycles élémentaires dans les ecosystèmes forestiers. In: *Dynamic properties of forests ecosystems*. (D.E. Reichle, ed.): 341-409. Cambridge Univ. Press. Cambridge.

De ANGELIS, D.L., GARDNER, R.H. & SHUGART Jr., H.H. 1981. Productivity of forests ecosystems studied during IBP: the woodlands data set. In: *Dynamic properties of forests ecosystems*. (D.E. Reichle, ed.): 567-672. Cambridge Univ. Press. Cambridge.

DIMITRI, M.J. 1982. *La región de los bosques andino-patagónicos*. INTA. Buenos Aires.

ELLIOTT, J.M. 1983. *Some methods for the statistical analysis of benthic brates*. Wilson. Kendal.

FORD, E.D. 1975. Competition and stand structure in some even-aged plant monocultures. *J. Ecol.*, 63: 311-333.

GREIG-SMITH, P. 1964. *Quantitative plant ecology*. 2nd ed. Butterworths. London.

GUTIÉRREZ, E., FONS, J., ROMAÑÀ, J. & VALLEJO, R.V. 1989. Soil-vegetation relationships in the *Nothofagus* forests of Tierra del Fuego. *Proc. XVI Pacific Inter. Congress*. Chile

HAMMOND, A.L. 1972. Ecosystem analysis: biome approach to environmental science. *Science*, 175: 46-48.

HERETER, A. 1990. *Els sòls forestals del Massís del Montseny*. Ph. D. Thesis, Universitat de Barcelona.

HOLDGATE, M.W. 1960. Vegetation and soils in the South Chilean Islands. *J. Ecol.*, 49: 559-580.

HUECK, K. 1978. *Los bosques de Sudamérica. Ecología, composición e importancia económica*. GTZ. Berlin.

HUMPHREYS, F.R. & PRITCHETT, W.L. 1972. A soil testing for predicting phosphorus status of *Pinus elliottii* plantations. *Plant and soil*, 37: 479-489.

LARSEN, J.A. 1982. *Ecology of the northern lowland bogs and conifer forests*. Academic Press. New York.

LOVELESS, A.R. 1961. A nutritional interpretation of sclerophylly based on differences in the chemical composition of sclerophyllous and mesophytic leaves. *Ann. Bot.*, 25: 179-184.

MARKGRAF, V. 1977. New data on the late- and postglacial vegetational history of "La Misión", Tierra del Fuego, Argentina. *Proc. IV Inter. Palyn. Conf.* Lucknow.

McQUEEN, A.R. 1976. The ecology of *Nothofagus* and associated vegetation in South America. *Tuatara*, 22: 36-38.

MEYER, F.H. 1966. Mycorrhiza and other plant symbiosis. In: *Symbiosis* (S.M. Henry, ed.): 171-255. Academic Press. London.

MOHLER, C.L., MARKS, P.L. & SPRUGEL, D.G. 1978. Stand structure and allometry of trees during self-thinning of pure stands. *J. Ecol.*, 66: 599-614.

OLSON, J.S. 1963. Energy storage and the balance of producers and decomposers in ecological systems. *Ecology*, 44: 322-331.

O'NEILL, R.V. & De ANGELIS, D.L. 1981. Comparative productivity and biomass relations of forest ecosystems. In: *Dynamics properties of forest ecosystems*. (D.E. Reichle, ed.): 411-450. Cambridge Univ. Press. Cambridge.

PEET, R.K. & CHRISTENSEN, N.L. 1987. Competition and tree death. *BioScience*, 37: 586-595.

PISANO, E.V. 1975. Características de la biota magallánica derivadas de factores especiales. *Ans. Inst. Pat.*, 6: 123-137.

PISANO, E.V. 1977. Fitogeografía de

Fuego-Patagonia chilena. I. Comunidades vegetales entre las latitudes 52° y 56° S. *Ans. Inst. Pat.*, 8: 121-250.

PISANO, E.V. 1981. Bosquejo fitogeográfico de Fuego-Patagonia. *Ans. Inst. Pat.*, 12: 159-171.

RAGONESE, A.M. 1981. Anatomía foliar de las especies sudamericanas de *Nothofagus*. *Darwiniana*, 23: 587-603.

RIVERA, S.M. 1987. Aspectos silológicos vinculados a la dendrocronología de *Nothofagus pumilio* (Poepp. et Endl.) Krasser. I. Area de vasos. *Parodiana*, 5: 9-16.

RIVERA, S.M. 1988. Revisión xilológica del género *Nothofagus* Bl. (Fagaceae) para la Argentina. Mon. *Acad. Nac. Cien. Exac. Fca. & Nat.*, 4: 73-84.

ROMAÑÀ, J., FONS, J., VALLEJO, R.V. & GUTIÉRREZ, E. 1989. *Condiciones edáficas en relación con la vegetación forestal en la Isla Grande de Tierra del Fuego*. Final Report. ICI, Madrid.

SALA, A. 1986. *Evolució de l'estructura vertical de la garriga desprès del foc: índex foliar i gradients morfològics*. Master Thesis. Universitat de Barcelona.

SANTANA, A. 1984. Variación de las precipitaciones de 97 años en Punta Arenas como índice de posibles cambios climáticos. *Ans. Inst. Pat.*, 15: 51-60.

SCHMIDT, H. & URZÚA, A. 1982. *Transformación y manejo de los bosques de lenga en Magallanes*. Ciencias Agrícolas, 11. Universidad de Chile.

TERRADAS, J. (ed.) 1984. *Introducció a l'ecologia del faig al Montseny*. Diputació de Barcelona, Barcelona.

VALLEJO, V.R. 1987. *Anàlisi i valoració de sòls forestals*. Report CIRIT. Barcelona.

VUILLEUMIER, B.S. 1971. Pleistocene changes in the fauna and flora of South America. *Science*, 173: 771-780.

WALTER, H. 1976. *Vegetació i climes del món*. Uniersitat de Barcelona. Barcelona.

WARING, R.H. & SCHLESINGER, W.H. 1985. *Forest ecosystems. Concepts and management*. Academic Press. Orlando.

WEBB, W.L., LAUENROTH, W.K., SZAREK, S.R. & KINERSON, R.S. 1983. Primary production and abiotic controls in forests, grasslands and desert ecosystems in the United States. *Ecology*, 64: 134-151.

WHITTAKER, R.M. & WOODWELL, G. M. 1968. Dimension and production relations of trees and shrubs in the Brookhaven forest, New York. *J. Ecol.*, 56: 1-25.

WHITTAKER, R.M. & MARKS, P.L. 1975. Methods of assessing terrestrial productivity. In: *Primary productivity of the biosphere*. (H. Leith & R.M. Whittaker, eds.): 55-118. Springer. Berlin.

WILLIAMSON, G.B. 1975. Pattern and seral composition in an old-growth beech-maple forest. *Ecology*, 56: 727-731.

ZAMORA, E. & SANTANA, A. 1979a. Características de la costa occidental de la Patagonia entre las latitudes 46° 40' y 56° 30' S. *Ans. Inst. Pat.*, 10: 109-144.

ZAMORA, E. & SANTANA, A. 1979b. Oscilaciones y tendencias térmicas en Punta Arenas entre 1888 y 1979. *Ans. Inst. Pat.*, 10: 145-154.

Homage to Ramon Margalef; or, Why there is such pleasure in studying nature (J. D. Ros & N. Prat, eds.). 1991. *Oecologia aquatica*, 10: 367-385

EFFECTS OF INCREASING LEVELS OF CO_2 AND O_2 DERIVED COMPOUNDS ON BIOGEOCHEMICAL CYCLES AND SHAPING OF THE BIOSPHERE

JOSEP PEÑUELAS

Institut de Recerca i Tecnologia Agroalimentàries. Centre de Cabrils. Crtra. de Cabrils, s/n. 08348 Barcelona. Spain.

Received: February 1990

SUMMARY

Plants reduce CO_2 to molecular species of increasing energy content. These compounds are then moved by heterotrophic organisms through decreasing energy levels as the organic matter is returned to inorganic form by successive oxidations. Thus, CO_2 and O_2 are the major substrates of biospheric reactions.

Oxygen is toxic for anaerobic and, in excess, for aerobic organisms. Its toxicity is overcome by aerobic organisms by using it as an electron terminal acceptor in respiration and by developing biochemical defenses. It still determines phenomena such as anaerobiosis, the depth limit of higher aquatic plants, or bioluminiscence, and perhaps others like the diurnal vertical migration of zooplankton.

Low photosynthetic efficiency in light use, shape and color of vegetation may result from the peculiar availability of light and CO_2. Therefore, slight changes in these factors would produce life shapes deeply alien to the world we now perceive.

Atmospheric CO_2 levels have increased by about 70 μmol mol^{-1} over the past 240 years, as has been shown by ice core studies and Mauna Loa data. Parallel to this increase, some plant changes have already been found: an overall decrease from 121% to the present-day 100% level in stomatal density, and from 144 % to present-day 100% leaf nitrogen content of 14 herbarium specimens of trees, shrubs and herbs collected over the last 240 years in Catalonia, a Mediterreanean area. Similar declines have also been reported in experimental studies on leaves of plants grown in enriched CO_2 environments. Consequently the C/N ratio has increased in leaves which together with the decrease in the specific leaf area (SLA) may have consequences for herbivores, decomposers and ecosystems.

Increases in oxygen-derived compounds, mainly ozone, have potentially large biospheric effects. Nowadays, they seem to play an important role in forest decline and some decreased crop productions, as has been shown in laboratory experiments. Increased photosynthetic rates are expected with increasing CO_2 levels in the decades to come, but regional ozone increase is expected to decrease plant production. Responses of ecological systems are however complex and difficult to predict.

Finally, a possible change in photosynthetic efficiency, plant shape and color is speculated to be expected because of increasing CO_2 levels, and the role of anthropospheric activities in the balance between reduction and oxidation in the biosphere is emphasized.

KEY WORDS: Carbon dioxide, oxygen, biogeochemical cycles, biosphere.

SOLAR RADIATION AND BIOGEOCHEMICAL CYCLES

Life is riding on the dissipation of solar radiation to the outer space, a dissipation that generates a gradient between reduced and oxidised substances. Photosynthesis takes advantage of solar radiation to generate molecules of more negative redox potential from which animals and bacteria then take the energy, transforming them into molecules of progressively more positive redox potential. The cycle of matter closes when these molecules are again used by plants thanks to the open energy flux from sun to space.

On a global scale, life charges the biospheric battery and afterwards, the differences of potential are not abruptly wasted in only one jump. Instead, there are a lot of little jumps among different components that have been multiplying and becoming more complex during evolution, generating viruses, bacteria, algae, trees, animals, and finally, beings as complex as humans. The reason for this complication of organisms, their communities and of their products must be sought in the action of selection forces that have been retarding the flux of solar energy in a non-uniform manner. In this way we can understand the existence of structures as complex as plants and animals in a biosphere that could otherwise be constituted only by procariota, thanks to their enormous metabolic versatility and capacity of adaptation to the use of available energy. The organization of living matter has been successful in increasing the persistence of complex structures, able to present stronger relations, domination of space and time and capacity of regulation. On one hand - in the surrounding space - more entropy has been generated, but on the other hand - the biosphere - more memory, information and organization have been stored.

With carbon, oxygen and hydrogen, three other elements - nitrogen, sulfur and phosphorus (with a chemical position quite different from the other two) - are of particular interest in the study of our planet's biosphere. They have been selected because of their unique atomic qualities. Through the active intervention of the biosphere, each of these elements follows a closed cycle, passing through molecular species of increasing energy content as the elements are incorporated into living tissue, and then moving through decreasing energy levels as the organic matter is returned to inorganic form. CO_2 reduction and organic matter oxidation are the major components of this global process. It is expected then that CO_2 and O_2 greatly influence the characteristics of the biosphere: CO_2 because of its scarcity and O_2 because of its toxicity for anaerobes and, in excess or through its derived compounds, for aerobes.

The approximation I am going to use here is to a certain extent taken from the field of economics, in order to provide insights into the selective pressures on characteristics such as leaf reflectivity, effective leaf size, stomatal conductance, size of photosynthetic enzyme pools, plant shape and color, nitrogen fixation, etc. This is a cost-benefit approach that has been criticized on a number of grounds (GOULD & LEWONTIN, 1979; HARPER, 1982). Principal among these criticisms are the non-falsability of the assumption of adaptation, the *post hoc* nature and untestability of optimality models, failure to consider nonselective factors shaping evolution, particularly the nature of available genetic variation and developmental constraints, the inappropriateness of an atomistic approach, and teleological and Panglossian assumptions. Analyses of different authors reviewed by GIVNISH (1986) have effectively rebutted many of these criticisms.

THE ROLE OF CO_2 AND LIGHT. PLANT PRODUCTION, PHOTOSYNTHETIC EFFICIENCY, SHAPE AND COLOR

From studying IBP (International Biological Program) data collected between 1964 and 1974, it can be seen that primary production in terrestrial environments follows a gradient from Poles to Equator that is not found in oceans. This fact shows that radiation, temperature, water and nutrients are the determinant factors of production in terrestrial ecosystems, and nutrients and radiation (and predators) are the determinants in oceanic ecosystems (BEGON *et al.*, 1986). Globally, however, a final limiting factor can be considered: the low CO_2 concentration in the Earth's atmosphere, specially compared with a primitive atmosphere.

Radiation, although used inefficiently, can be limiting during the day in the interior of the canopy, in C4 plants, under the shade of higher leaves, under the shade of the chloroplasts of the same leaf, in deep water, etc. Plants adapt to systematic changes in light environment (diurnal and seasonal) for example with senescence, falling or movement of leaves, or with special metabolic pathways and anatomies for shade and sun conditions. At the canopy level, they optimize LAI, leaf angle inclination, and density distribution.

Linked to radiation, temperature is another key factor. Apart from its effect on respiration, it determines transpiration, and therefore water availability.

Water is another of the most important limiting factors of primary production, as a carrier of necessary elements. It is also responsible for many of the anatomical and metabolic characteristics of plants.

Primary production also depends on a well-formed soil and its mineral nutrients, especially nitrates of mainly biological origin.

These previous considerations are well accepted by the scientific community, but besides those, many of the most familiar characteristics of plants - low photosynthetic efficiency, shape, disposition of branches and leaves and color - may also be consequences in the last instance of the present low levels of CO_2 in the Earth's atmosphere.

In optimal laboratory conditions, the highest light use efficiency that has been found is 20% under low light and 8% under intense light. In the field it is usually around 1%. These low field values can be explained by the loss of photons falling on soil and moribund materials, or consumed in water evaporation. But even under optimal laboratory conditions, efficiencies as low as 8% are obtained under intense light conditions. The explanation of this low efficiency seems to be the existing high amounts of photosynthetic pigments present in plants, with which they can take advantage of low light, the most usual condition plants must face: mornings, dawns, dusks, shady environments, shade from other plants, leaves or chloroplasts. Therefore, the excess of photons under intense light cannot find a sufficiently high number of production active centers. The pigment composition of the photosynthetic antennae is the result of an old decision. Photosystem I was originated under reducing conditions, with different proton suppliers. Later the capacity to split water was acquired by adding a Photosystem II. The high number of chlorophyll molecules per antenna (200 to 900) makes each reaction center act as a bottleneck, and thus a limit is set to the production of the biosphere. In ecology this important limitation is rarely considered (FALKOWSKY, 1983; MARGALEF, in press). Only some mutants of *Chlorella*, higher plants and the heterocysts of cyanobacteria - specialist cells in the avoidance of oxygen problems in their nitrogen fixation - show lower pigment ratios. They may represent the promise of a substantial increase in efficiency of primary producers.

But a more general explanation of the low efficiency comes from considering the very low concentration of CO_2 on present-day Earth. When photosynthesis appeared, it had to adapt to the availabilities of that time. Among them it should be pointed out that CO_2 was surely much more abundant than now, radiation was lower and the temperature was higher (WALKER, 1986). Photorespiration could be an added mechanism to get rid of oxygen, when there was much more CO_2 available. In relatively recent geological ages, atmospheric CO_2 concentration has been largely decreasing until this last century, in which man's activities are increasing it again. Now, with much less CO_2 availability, under high temperatures, photorespiration has been mainly reduced in C4 and CAM plants, mostly in tropical and dry environments. In temperate environments, lower temperatures and higher water availability make possible the persistence of C3 plants with high levels of photorespiration.

Plants have tended to take profit of CO_2 for the maximum time possible and as light is low most of time, they are adapted to do their best at these most frequent low light conditions. In fact, when the CO_2 supply increases, the photosynthesis rate rises too. More support for this theory comes from the existence of the mentioned metabolic and anatomical structures (C4 and CAM) specialized in the activation of CO_2 absorption under drought and heat conditions when the other important element, water, is more difficult to obtain. These mechanisms may also constitute an adaptation to increasing O_2 concentration. Aquatic plants have also acquired equivalent capacities by using bicarbonate as well as CO_2 as a source of carbon (PEÑUELAS, 1985; BOSTON *et al.*, 1989).

The fact that photosynthesis is not continuous but has a 24 hour oscillation has probably been important in building internal stores of usable CO_2, as an adaptation to anticipate change. CAM plants are to a certain extent like migratory animals that go in search of spring. CAM metabolism prepares them in the dark to use the rising sun immediately.

The shape of plants can also be understood as a result of the low efficiency of photosynthesis under high light and the low availability of CO_2 (COLINVAUX, 1986). Leaves and branches have arrangements in trees, as well as in herbs, that seem to be directed to diffuse the light in such a way that the maximum photosynthetic tissue receives light and CO_2 in optimal conditions. In light environments trees have small divided leaves, arranged in depth in a multilayer design, while those of shady environments tend to have bigger leaves arranged in a monolayer design to take advantage of all the low light already in the first interception. The same happens in herbs where erect leaves receive light in an inclined plane increasing the photosynthetic area that absorbs CO_2 under optimal radiation.

There are also two energetic tradeoffs that underlie many photosynthetic adaptations related to carbon source availability. The first arises from the inevitable association of carbon gain with water loss. CO_2 and water vapor diffuse in and out of the same pathway with parallel increases of photosynthesis and transpiration. The second results from the conflict between leaf photosynthetic capacity and the energetic costs of constructing and maintaining tissue capable of high photosynthetic rates (nitrogen, phosphorus, and so on; GIVNISH, 1984, GULMON & CHU, 1981). Together these tradeoffs link photosynthesis to the costs of transpiration and needed nutrients.

Plants also appear to have decreased leaf absorptance or increased leaf angle as a means of reducing water loss, thus to extend their period of activity into prolonged drought periods and as a means of reducing photon flux at irradiances higher than those necessary to saturate the photosynthesis process (EHLERINGER & WERK, 1986). For instance, the PAR incident on the vertical surfaces of unshaded cacti or on both sides of the leaves of unshaded agaves for various latitudes and seasons averages only about 20 mol m^{-2} d^{-1} on clear days (NOBEL, 1986).

The variation in orientation of the leaves of agaves, and many other plants from almost erect near the central stem to almost horizontal near the ground, helps to ensure a relatively even distribution of PAR throughout the canopy, which in turn tends to lead to a higher net productivity. Furthermore, these plants also have hairs to help.

At the other extreme, when light is low and limiting, velvety leaf surfaces, red undersurfaces, and blue iridiscence appear to enhance the absorption of light by leaves slightly and have thus been selected in many taxonomically unrelated groups ecologically restricted to extreme shade in tropical rain forests (LEE, 1986).

In the aquatic environment, although the most limiting factor does not seem to be CO_2 (PEÑUELAS, 1985) but light, the existence of C4 like, sediment CO_2 users, CAM and HCO_3^- users plants shows that CO_2 is also an importing limiting factor. In that medium, and relative to light availability, pigments that make some algae brown (Phaeophyceae) and red (Rhodophyceae) absorb wavelengths that terrestrial plants reflect. Green aquatic plants solve the problem by adopting a dense array of pigments (among them, also carotenoids that absorb partially in the green window).

The photosynthetic use of visible wavelengths in vegetation is the result of molecular compromise between sensitivity to destruction by energetic UV radiation, and the response of chemical bonds to low energy IR radiation. But, in fact, a possible partial explanation of green color in plants, with the luxury of not using that most abundant radiation, resides in the low availability of the other limiting factor of photosyntesis, CO_2, which means that light is in excess most of the time.

If shape and color of vegetation are then accepted to result from the peculiar availability of light and CO_2, then the consequent abundance of animals is also another result. Small changes in these factors would be expected to produce a selection of life shapes deeply alien to the world we now perceive (COLINVAUX, 1986).

THE ROLE OF OXYGEN AND ITS DERIVED COMPOUNDS. RESPIRATION AND BIOCHEMICAL DEFENSES

Oxygen was a kind of poison for primitive organisms. The first aerobes had to present or develop defenses and even more than this, to take advantage of it as a better respiratory electron terminal acceptor via cytochrome oxidase or via other alternative oxidases not sensitive to cyanide, sulfide or other reduced components of primitive Earth, when it first became an oxidated planet. Nowadays both respiratory paths remain in plants (LAMBERS, 1985; PEÑUELAS *et al.*, 1988). Several functions for the alternative oxidase have been proposed (AZCÓN-BIETO & PEÑUELAS, 1987) but a general physiological function is not accepted, and the ecological implications of having non-phosphorylating electron transport pathways are not well understood. The alternative oxidase of plant mitochondria could be an evolutionary relic derived from ancestral respiratory oxidases of primitive aerobes; these oxidases could have evolved as "safety valves" in the presence of compounds that occasionally inhibit cytochrome oxidase or the equivalent, or they might even be previous to it. In all cases, keeping electrons flowing within membranes while high energy conserving pathways are blocked increases the "flexibility" of electron transport (AZCÓN-BIETO & PEÑUELAS, 1989). Currently, there are still relevant environments where poisons of cytochrome c oxidase are present. This is the case of anoxic waters and sediments rich in sulfide. Aquatic plants that root under such conditions present an active alternative oxidase (PEÑUELAS *et al.*, 1988). However, the presence of poisons of cytochrome c oxidase in the environment is surely lower than in the early Earth, and alternative oxidase in plants could have developed secondary functions, such as thermogenesis in Aracea (DIAMOND, 1989) and others (LAMBERS, 1985), taking advantage of the flexibility of electron transport.

As far as respiration is concerned, another interesting biological phenomenon could have evolved: bioluminiscence, which provides a way for oxygen detoxification, through reactions that give off a great deal of energy in the form of light. It is difficult to find an alternative hypothesis to explain the origin of light-producing organisms although new uses have appeared in the course of the evolution (MARGALEF, in press).

Oxygen at concentrations higher than normal is still a poison for aerobic organisms as was already recognised by BERT (1878). In plants, WARBURG (1920) found that 100% O_2 inhibits the photosynthetic rate of *Chlorella* by 73%. Oxygen is a substrate of the enzyme Rubisco giving photorespiration waste, and besides it is toxic due to mechanisms that are still being discussed (SULLIVAN *et al.*, 1983). Because of spin restrictions, molecular oxygen preferentially undergoes univalent reductions, the products of which are the reactive superoxide radical, hydrogen peroxide, or the hydroxyl radical, an extremely aggressive compound with a high redox potential (1.8V) (ELSTNER, 1982). These are some of the most reactive substances in biological systems; they are linked to membrane lipid peroxidations, nucleic acid oxidation, hyaluronate depolymerization, and enzyme inactivation (CADET & TEOULE, 1978). The targets of active oxygen species are then membranes, pigments, enzymes and nucleic acids (CERUTTI, 1985). They also seem to be basic factors in organism senescence.

In order to eliminate active oxygen species, aerobic organisms have developed cellular defenses: enzymes like superoxidismutase, peroxidases and catalases and small molecules such as glutathion, ascorbic acid, hydroquinones, alpha-tocopherol, carotenoids, and unsaturated fatty acids. These defenses may be overriden by oxygen levels higher than atmospheric ones.

The oxygen biotoxicity determines biospheric characteristics in several ways, sometimes as clear as in the distinction between aerobic and anaerobic worlds, sometimes in a less expected way as, for example, in the mechanisms by which hydrostatic pressure acts on vascular plants in fresh water. Their inhibited growth below depths of 10-12 m is mainly related to the increase of oxygen partial pressure in their aerial spaces caused by the linear increase of hydrostatic pressure with depth (PEÑUELAS, 1987). The oxygen toxic effect, the carboxylation resistance and the CO_2 compensation concentration increase linearly with O_2 partial pressure.

Another unexpected example of its action which is still to be confirmed, is the diurnal migration of zooplankton up to the surface during night to eat phytoplankton and down to deep waters during the day, plausibly in order to avoid oxygen-derived active compounds formed as a consequence of chlorophyll receiving photons that cannot be processed photosynthetically (PEÑUELAS *et al.*, in press a).

Thus, changes in concentration of oxygen or its derived compounds, as well as those of CO_2, would make it very difficult for the shape of the biosphere to remain as it is now on Earth.

And, in fact, nowadays, concentrations of CO_2 and oxygen-derived compounds such as ozone are clearly increasing in the Earth's atmosphere.

ACTUAL CHANGES IN CO_2, OZONE AND OTHER ATMOSPHERIC GASES CONCENTRATION

CO_2 has changed in the course of the Earth's history but the present increase is fundamentally different from past changes because now there is an increase at rates and levels for which we have no historical or natural analogue back to the last two interglacial events (FRIEDLI *et al.*, 1986).

There is also increasing evidence that the abundance of an oxygen derived compound, tropospheric ozone, is increasing over large regions, and that the phenomenon of urban smog is no longer confined to cities and developed countries (PRINZ *et al.*, 1984). Nowadays it is also marked in Third World tropical countries because of frequent burning of savanna gramineae (GRAEDEL & CRUTZEN, 1989).

This has clear implications for productivity in impacted areas and may be expected to affect biogeochemical cycling significantly over extensive regions. The chemistry of tropospheric ozone assumes additional importance in that the abundance of the OH radical may be expected to change in response to exchanges in lower atmospheric ozone. The OH radical is the ultimate cleansing agent for a wide range of gases emitted to the atmosphere. It regulates oxidation of nitrogen and sulfur compounds, carbon monoxide, and hydrocarbons and is responsible for the removal of a wide variety of industrial halocarbons.

Together with the increase in concentration of these gases, the expected derived climatic changes are also relevant to terrestrial biogeochemistry, altering the balance between carbon fixation and release because photosynthesis responds differently to temperature than to respiration, and water availability strongly influences carbon storage and release. The atmospheric pollution changing the supply of carbon, nitrogen, phosphorus or sulfur can also change the storage and release of all these elements.

Some effects of all these present changes, especially of carbon dioxide and ozone increases, have already been noticed in vegetation.

PRESENT CHANGES IN VEGETATION BECAUSE OF CO_2 INCREASE, AND BIOSPHERIC CONSEQUENCES

Estimates of historical concentrations of atmospheric CO_2 suggest that plants have existed and evolved for at least 10 million years at CO_2 mole fractions in the range of 250 to 290 μmol mol^{-1} (GAMMON *et al.*, 1985). Thus, important physiological and anatomical changes are to be expected nowadays, at 348 μmol mol^{-1} CO_2. A decrease in stomatal density in herbarium specimens of tree leaves collected over the last 200 years (WOODWARD, 1987), and an increase in the growth rate of bristlecone pines (LaMARCHE *et al.*, 1984) have already been reported and have been related to the increase from 278 to 348 μmol mol^{-1} CO_2 in the last 240 years. Now, WOODWARD & BAZZAZ (1988) have also found in controlled experiments at different CO_2 levels that stomatal density and stomatal index increase markedly as the CO_2 mole fraction is reduced below 340 μmol mol^{-1}.

Both stomatal parameters have also been found to increase with altitude (KORNER *et al.*, 1979) and in simulated altitude conditions (WOODWARD & BAZZAZ, 1988), under lower CO_2 partial pressure conditions. So this is what we would expect to find in plants grown in previous centuries under lower CO_2 mole fractions.

Higher plant growth is also expected with increasing CO_2 so it would also impose an increased demand upon nutrients. Thus N, S or P could be among the first limiting factors (LEMON, 1983 and the references cited therein). The increased carbon gain associated with higher atmospheric CO_2 levels should increase the amount of carbon relative to other essential elements in plant tissue because it is questionable whether additional optimum amounts of other mineral nutrients will be obtained by all plants unless there is increased symbiotic nitrogen fixation (NORBY & SIGAL, 1989).

To check plausible changes in stomatal density, specific leaf area of plants and nitrogen and sulfur leaf content in natural environments along with the rising atmospheric CO_2 levels, all these parameters have been measured in the leaves of 14 species of trees, shrubs and herbs (*Pinus uncinata*, *P. pinea, Betula pendula, Juniperus communis, Ceratonia siliqua, Buxus sempervirens, Pistacia lentiscus, Helleborus foetidus, Rhododendron ferrugineum, Amaranthus caudatus, Papaver alpinum, Cynodon dactylon* and *Gentiana alpina*) collected over the last 240 years both from wet and dry habitats of mountains and plains in Catalonia and stored as dried herbarium specimens (PEÑUELAS & MATAMALA, 1990). Present vegetation was also studied for changes in response to the present day higher levels of CO_2 relative to the last centuries.

Although some of the herbarium specimens may well have been growing in widely varying conditions before collection, surely explaining part of the important intraspecific variability that was found, and although the relative changes along the years were not the same in all the studied species (PEÑUELAS & MATAMALA, 1990), the stomatal density and the specific leaf area were found to decrease in most species with a significant mean overall - all species together - reduction. The linear regression lines for overall relative changes of leaves showed a 121-100% reduction in the stomatal density, a 184-100% reduction in the specific leaf areas, and a reduction from 144 % to the present 100% in the N leaf content. Consequently, the C/N ratio has increased overall. The main changes took place under the first slower increases in concentration of CO_2. During the last years the change has been smaller. Sulfur content presented its highest values mostly in 1940s samples, when increased quantities of sulfur rich carbon were burnt.

The ice record data show that the CO_2 in the atmosphere has increased by about 70 $\mu mol\ mol^{-1}$ or 25 % (FRIEDLI *et al.*, 1986). This could well be one of the main reasons for these changes in stomatal density, SLA and C/N ratios of leaves.

The higher stomatal density can be understood because the lower CO_2 levels of the pre-industrial era made the uptake of CO_2 more expensive in terms of water loss than at present. The photosynthetic rate of plants from the last centuries may have been lower, indicating the likelihood of lower water use eficiencies than the present (WOODWARD, 1987).

The increase in CO_2 levels and the consequent decrease in leaf nitrogen concentration found in plants nowadays (PEÑUELAS & MATAMALA, 1990) have also been found experimentally to produce a decrease in SLA (LARIGAUDERIE *et al.*, 1988), as was found in those herbarium specimens. Although it might well be an artifact of studied material with more decomposition in elder specimens, it has been found that the growing leaf, which preferentially uses its own photosynthate (DICKSON & SHIVES, 1982), is a very efficient system for the use of extra assimilates from CO_2 enrichment (GAUDILLÈRE & MOUSSEAU, 1989). The higher rate of photosynthesis under present-day higher CO_2 levels would have increased the amount of carbohydrate available for plant growth. In some experiments, a decrease in SLA even nearly proportional to the increases in photosynthetic rate have also been found under artificially increased CO_2 levels (ACOCK *et al.*, 1982). The decrease in SLA could also be an indirect result of increased levels of CO_2 through higher temperatures and higher vapor pressure deficits.

Decreases in N leaf content have also been found experimentally. In effect, cotton (WONG, 1979) and soybean seedlings (WILLIAMS *et al.*, 1986) have been found to present much lower percent N values associated with starch accumulation under increased CO_2. Data relating N availability to plant growth and carbohydrate content indicate that under N-limiting conditions stored carbohydrates accumulate (WILLIAMS *et al.*, 1986). Several other authors have also reported decreases in leaf nitrogen concentration under CO_2 enrichment during vegetative growth (LARIGAUDERIE *et al.*, 1988 and references therein) in spite of increases in root/shoot ratio and in both, favorable and low level N availability (LARIGAUDERIE *et al.*, 1988). Because raising the CO_2 decreases the nitrogen requirements for maintaining a given level of leaf photosynthesis (PEARCY & BJORKMAN, 1983), the C:N ratio can increase. However, the reasons for this N decline are not yet clear. Several mechanisms such as the increase in starch and structural compound content or reduced N uptake as a result of stomatal closure at high CO_2 (MADSEN, 1975) have been proposed. An N availability limitation or N redistribution in higher leaf areas could also be added. The well demonstrated N deficiencies in many unfertilised plant communities show that nutrient uptake often cannot be increased in response to plant demand.

If these results were confirmed, as atmospheric CO_2 rises the areas of moderately infertile soils could become strongly nutrient-limited, and areas where moderate nutrient deficiencies occur would expand. The increase in CO_2 interacting with N, S and P would make many terrestrial communities progressively more oligotrophic and could also lead to nutrient deficiencies and acidification of susceptible soils and freshwater (SKIBA & CRESSER, 1988). And in agricultural lands, it would almost certainly be necessary to add more N fertilizer to crops or to develop new nitrogen fixing technologies in order to take advantage of the increased CO_2 (BAKER & LAMBERT, 1979). However, it is also true that nowadays there is already an excess of nitrogen compounds in man-ruled ecosystems. N and S content in precipitation has also increased, although not uniformly, and the natural plant nitrogen fixation is increased under higher CO_2 supplies (NORBY & SIGAL, 1989).

Changes in carbon supply can then simultaneously affect the costs of transpiration, nutrient uptake, mechanical support, and/or herbivory. Similarly, shifts in leaf nitrogen content would affect photosynthetic capacity, cost of nutrient absorption, and exposure to herbivory (MOONEY & GULMON, 1982).

At the biospheric level, such effects of increased CO_2 might then be important. The decrease in SLA together with the increase of C/N ratios (PEÑUELAS & MATAMALA, 1990) and possible increase in starch or structural and phenolic compounds (WULFF & STRAIN, 1982) could have substantial effects on feeding efficiency of herbivory (LINCOLN *et al.*, 1986; FAJER *et al.*, 1989). Polyphenols may reduce grazing because of toxicity or reduced digestibility. Increased C/N ratio and increased structural and phenolic compound production could also have important effects on decomposers and consequently profound effects on nutrient cycling and plant growth (OECHEL & STRAIN, 1985). It would also produce the accumulation of hard structures that make it easy for the plant to obtain new nitrogen and phosphorus resources. However, recent results (LARIGAUDERIE *et al.*, 1988) have shown that finally, at the end of the vegetative growth, leaf nitrogen contents do not differ as a consequence of higher CO_2 levels, so litter quality would probably be the same with respect to C/N ratio.

Together with previous studies (WOODWARD, 1987; LaMARCHE *et al.*, 1984; PEÑUELAS & MATAMALA, 1990) the results presented here emphasize that the overall rise in CO_2 is already having important effects on the biosphere. It can thus be added to the other important impacts of man on the biosphere.

PRESENT EFFECTS OF TROPOSPHERIC OZONE INCREASE ON VEGETATION

Ozone is considered to be the main air pollutant in the USA and its importance is increasing as much in Europe and even in tropical countries. It could have a persistent and long-lasting effect on plants. Besides the well known high concentrations in summer, monitoring of ozone has also shown potential phytotoxic concentrations under certain circumstances in autumn and even in winter (FRICKE, 1983; PRINZ *et al.*, 1984).

The widespread decline of conifers, hardwoods and field crops in Europe and USA is believed to be caused by a combination of stresses. The decline shows a clear gradient of increasing severity with higher elevation (JOHNSON & SICCAMA, 1983; PRINZ *et al.*, 1982; BLANK, 1985). This is consistent with a natural increase in environmental stress, altitudinal gradients in ozone concentration (McLAUGHLIN, 1985) and deposition rates of acidic cloud moisture (FRIEDLAND *et al.*, 1984). Ozone formation and distribution in the atmosphere correlate with this spatial and temporal development of forest decline (PRINZ *et al.*, 1982; ARNDT *et al.*, 1982). The levels may be high enough to cause acute injury to sensitive plant species (ROBERTS, 1984; SKARBY & SELLDEN, 1984; PEÑUELAS *et al.*, in press b).

Some decrease in the productivity of certain crops, especially summer ones, can also be understood as a consequence of elevated summer ozone concentrations. This may well be the case in crop fields of ozone rich areas such as the Ebro Delta or Maresme (Catalonia).

Research concerned with the effects of ozone on Norway spruce and Sitka spruce, of special interest to an understanding of European forest decline, including Pyrenean spruces, has been carried out in our laboratory and at Lancaster University Field Station (PEÑUELAS *et al.*, in press b). Three-year-old saplings of the spruces *Picea abies* (L.) Karst and *P. sitchensis* (Bong.) Carr. were exposed in solar domes to 70 ppb O_3 - a fairly usual concentration in summer - for all the summer, seven hours a day, under approximately natural conditions, and in cabinets at the same ozone concentrations under 14 h 250 µE m^{-2} s^{-1} at 20 °C day/ 15 °C night. No visible injury nor different growth rates were found in ozone fumigated trees. However, old needles presented a slight increase in the fluorescence stress index and an increased cyanide resistant respiration, and both young and old needles of fumigated trees presented more negative water potentials, higher dry weight/fresh weight ratios, lower chlorophyll concentrations, lower nitrogen and sulfur percentages, higher C/N and C/S ratios, surprisingly lower needle electrolyte leakage, and lower respiration rates.

Apparently, older leaves were more damaged by ozone, and this is consistent with symptoms of forest decline affecting high-altitude forests in acid soils, as is the case in the high Pyrenees mountains. Crown thinning and premature shedding of older needles are observed in damaged forests at high altitude (PRINZ *et al.*, 1982). The restriction of damage to previous year's needles may be due to the inability of older tissues to repair damaged membranes (LEVITT, 1980) or to the lower ability of older tissues to detoxify pollutants.

The changes in water potential, fresh weight loss, fresh weight / dry weight ratio and cuticle width strongly suggest that the leaves of spruces exposed to ozone are less efficient in their use of water. Ozone decreases photosynthetic pigment concentration, so it may be suggested that it contributes to chlorosis shown by declining forest trees and crop plants like watermelon in the Ebro Delta. A clear decrease was also found in the carbon, nitrogen and sulfur content of fumigated trees. In many investigations, decreases in total protein content due to exposure to pollution have also been reported (ROWLAND *et al.*, 1988). Such changes could either be the result of a decrease in the rate of protein synthesis and/or an increase in the rate of protein degradation. The question remains as to whether these changes indicate damage to the balanced metabolism or a plant response to mitigate any potential damage.

Currently it is believed that ozone causes deterioration of cell membranes and cuticular waxes, and other cellular membranes, such as the envelopes of the mitochondria or chloroplasts, which are similarly damaged but to a lesser extent as the effect of ozone and its toxic radicals are progressively diluted and absorbed from the outside inwards. This partially explains the disturbance of ultrastructure within both these organelles and inhibition of both photosynthesis and respiration caused by high doses of ozone (WELLBURN, 1988) that were found incipient under normal doses (higher values of fluorescence stress indexes - indicative of wasted light energy - and lower oxygen uptake) especially in old needles. These disturbed functions within spruces are also supported by the observed losses of chlorophyll in ozone fumigated spruces.

Little is known of interactions among environmental stresses involving ozone. Several authors have suggested that the decline, although involving air pollution, may be triggered by frost or drought or other stresses (REHFUESS, 1986). The circumstancial evidence in support of this view is compelling because of the temporal and spatial correlations with elevation, temperature inversions, low temperature minimum frequencies, drought years, frost episodes (reviewed in DAVISON *et al.*, 1987) but there is little experimental evidence. Spruces exposed to ozone in summer might show indirect effects many weeks later, for example reduced frost resistance (BROWN *et al.*, 1987).

In our laboratories, three-year-old saplings of the Sitka spruce *Picea sitchensis* were exposed in solar domes to 70 ppb O_3 for all the summer, seven hours a day, and afterwards they were hardened for 10 days at 20 ^oC, 10 hours light / 4 ^oC, 14 hours darkness and for further 10 days at 18 ^oC, 9 hours light / 2 ^oC, 15 hours darkness. Finally they were submitted to night temperatures of -2 ^oC after an accidental 32 ^oC maximum day temperature which lasted a few minutes. Both ozone fumigated and control trees were visibly damaged, but the damage was clearly greater in fumigated trees. They presented lower chlorophyll concentration, higher respiration, and lower N and S percentages, but no difference in electrolite leakage nor in dry weight / fresh weight ratio. These results indicated that present ozone levels may affect spruces, predisposing them to posterior stresses like drought, winter desiccation, freezing injury or parasites. Interaction among these factors may be an important element contributing to the decline of sub-alpine stands of spruces throughout Europe. New studies should be carried out in the direction of interaction between environmental stresses.

FORESEEABLE FUTURE CHANGES BECAUSE OF INCREASING LEVELS OF CO_2, OZONE AND OTHER GASES. FROM PLANTS TO ECOSYSTEMS

New gradual increases in CO_2 concentration are expected during the next decades. Current global circulation models differ in their projections of global mean temperature and precipitation with increasing CO_2. However, even with precise climate scenarios for the next century, accurate biological predictions still cannot be made because of the poor knowledge about the response of ecosystems to increasing CO_2.

There are powerful techniques to predict responses such as modelling (PASTOR &

POST, 1988) and historical reconstruction (DAVIS & ZABINSKI, in press, PEÑUELAS & MATAMALA, 1990), but they need to be supplemented by experimentation. Besides other ecological and climatic consequences, these increases are supposed to raise photosynthesis and plant growth in the case of a good supply of other limiting factors. Thus, changes in anatomical and physiological characteristics of vegetation (LEMON, 1983) like those cited up to now with consequent reduced water loss and changes in plant phenology (CARLSON & BAZZAZ, 1980) are also expected. In fact, most studies carried out with agricultural crops and species from natural communities indicate that productivity and yield increase with elevated CO_2 levels (STRAIN & CURE, 1985). There are also differences in response to elevated CO_2 among species and among genotypes within species, but the differences have not been fully investigated (BAZZAZ *et al.*, 1985, PEÑUELAS & MATAMALA, 1990).

What is currently known is that at the leaf level, net assimilation rates are increased instantaneously and, to a lesser degree, over time, that photosynthesis and respiration may be reduced or increased after prolonged exposure to elevated CO_2 (ACOCK & ALLEN, 1985; DRAKE *et al.*, 1988; OECHEL & STRAIN, 1985); that the effect of CO_2 enhancing photosynthesis increases with temperature, and is greater for C3 plants (PEARCY & BJORKMAN, 1983); that the conductance is decreased and water use efficiency increased, preferentially in C3 plants (PEARCY & BJORKMAN, 1983); that whole plant water use may, however, increase because of the increase in leaf area (ACOCK & ALLEN, 1985); that water stressed plants may have enhanced productivities (IDSO, 1988); that root-shoot ratio response, although variable, often increases (CURE, 1985; OECHEL & STRAIN, 1985); that leaf area increases although proportionally less than leaf specific weight (ACOCK & ALLEN, 1985); that C/N ratios of tissues increase (STRAIN & BAZZAZ, 1983); that nutrient deficiencies reduce CO_2 enhancement effects (CONROY *et al.*, 1988), even though nutrient use efficiencies increase; that the herbivore feeding rate on tissues may increase (LINCOLN *et al.*, 1986), competitive relations are altered in most instances (CURTIS *et al.*, 1989), and that phenological development may be accelerated, although there are examples of delayed senescence (OECHEL & STRAIN, 1985).

High CO_2 concentrations may change the pattern of carbon allocation to plant parts, but the available data are inadequate for describing general patterns (BAZZAZ *et al.*, 1985, 1987). There may also be shifts in the kinds of chemical defenses in plant tissues and changes in rates of plant tissue consumption by herbivores (LINCOLN *et al.*, 1986) if nitrogen leaf content decrease is confirmed (PEÑUELAS & MATAMALA, 1990). Decomposition rates and nutrient cycling may change in response to altered plant C/N and lignin/N ratios, and starch content, leading to negative feedback via decomposition restricting the positive effects of CO_2 enhancenment (MELILLO *et al.*, 1982).

The response to enhanced CO_2 concentrations at the ecosystem level has been the subject of only a few experiments (TISSUE & OECHEL, 1987; DAVIES, 1988) and is largely unknown.

Hardly anything is known either about the ecosystem response to long-term exposure (DRAKE *et al.*, 1988). Elevated CO_2 should have modest effects on light-limited systems because the light requirement of photosynthesis is only moderately sensitive to CO_2 (EHLERINGER & BJORKMAN, 1977). Elevated CO_2 should decrease the nutrient investment required for photosynthesis (though probably not the nutrient investment required for growth). Effects on water-limited systems should be quite dramatic, because increased CO_2 decreases the stomatal conductance - and hence water loss - required to sustain a given rate of photosynthesis. These few studies of whole communities, however, do not always show enhanced productivity. A community of short-stature annuals showed no response (WILLIAMS *et al.*, 1988) and short term measurements of some deciduous trees showed only a small enhancement in growth (WILLIAMS *et al.*, 1986). Whole ecosystems respond significantly, but in different ways - nutrient limited low temperature systems, such as the tundra, show little response (HILBERT *et al.*, 1987), whereas nutrient rich marsh systems show a large response (DRAKE *et al.*, 1988; CURTIS *et al.*, 1989). Experimental studies in which whole plant communities have been subjected to elevated CO_2 also show that competitive hierarchies change. The resulting community structure is influenced by other environmental factors such as moisture, light, nutrients and/or temperature (BAZZAZ *et al.*, 1985). It is important to note that the response at the community level may not be directly predictable from the response of individual species to elevated CO_2 or to other environmental factors because of changes in species composition and interactions with heterotrophic organisms.

Responses of ecological systems are complex because of the inherent intricacies of ecological systems and their interactions with the physical system, and because these processes depend on the influence of history and scale. Besides, they encompass a great temporal variability. Time-lags in the ecological response to global change result not only from differences in longevity or life cycles among organisms, but also from non-linear processes. For example, low availability of nitrogen, the most frequently limiting nutrient in terrestrial ecosystems (PASTOR *et al.*, 1984), may prevent ecosystems from changing in response to changing climate until after temperature thresholds are exceeded by dominant species. As previously dominant species are replaced by others, consequent changes in carbon and nitrogen cycles can happen rapidly, amplifying ecosystem response (PASTOR & POST, 1988). In general, rapid response within a year can be produced in biotic factors such as microbial communities, plankton density, reproductive success of plants and animals, transpiration, photosynthesis and respiration rates, pathogen outbreaks, animal behavior, and densities of annual plants and short-lived animals. A lag of several years or decades can be found in other processes, such as plant community structure. Moreover, changes in the genetic structure of populations depend on generation time. Whether redundancy within an ecosystem can delay responses of the system to environmental changes requires investigation, as this may be an important and perhaps unexpected influence of biodiversity on ecosystem processes (DAVIS, 1988). It is then a fundamental question in the context of global change whether changes in species composition and diversity will significantly affect ecosystem function.

The increase in CO_2 is also expected to change climate, as is well known. The enormous significance of the rate of temperature change can be appreciated when spatial displacement is considered. There would be lags of decades in the adjustment of ecological systems to rapidly changing climate.

Critical foci for experimental ecosystem studies on the effects of these changes are wet tundra, boreal forests, temperate forests receiving anthropogenic nitrogen and sulfur deposition, tropical forests and semiarid ecosystems. Rivers offer an excellent integration of biogeochemical processes operating in specific watersheds. Oceans must also be studied because they are by far the largest active reservoir of carbon. In order to obtain global data, remote sensing of vegetation and land form characteristics will be necessary (TUCKER *et al.*, 1986; PEÑUELAS *et al.*, in press c; GAMON *et al.*, in press).

Three research approaches should be used to increase our ability to predict biotic responses to global change and feedback to the global system: (1) laboratory and field experiments at the organism level, and compilation of existing data on population and community patterns; (2) experiments on intact ecosystems, using large-scale manipulations and taking advantage of natural experiments; and (3) ecosystem models assembled coupling population-community models with process -functional models for simulation of the response of ecosystems to large, rapid changes in environmental factors.

Present models assume that individuals are identical, but as more complex models are built, besides scaling up from single plants to landscape and incorporating spatial dimension, patterns of genetic variation will also have to be incorporated.

Our knowledge of past responses of natural systems to changed CO_2 concentrations is also limited. Data from previous centuries reviewed before are still scarce. The fossil record should also be inspected for changes at the end of the last glacial period that may be related to changing CO_2 concentration in the atmosphere.

The other anthropogenic changes in the concentration of other tropospheric gases like oxygen derived compounds also have potentially large, landscape-level effects on ecological systems, primarily through effects on ecosystem components, especially those with slow turnover rates such as soil organic matter and long lived organisms such as trees and fish. Synergistic effects of various oxygen derived pollutants (O_3, NO_x, SO_2.) are well known from simple laboratory experiments (MANSFIELD *et al.*, 1987; WELLBURN, 1988) and these effects will certainly be found in the study of ecosystems (MATHY, 1988).

Changes in stratospheric constituents, especially depletion of ozone concentration, would also importantly affect ecosystems as the intensity of ultraviolet radiation reaching the terrestrial and marine portion of the biosphere increases. Previous investigations in arctic and alpine regions indicate that the changes in ultraviolet irradiance would cause changes both in productivity and in the distribution of specific vegetation types (DAVIS, 1988). Intensified ultraviolet radiation would increase mutation rates, but it is not clear what effect this would have on population structure and viability (GRIBBIN, 1988). The subject of the ozone hole and the penetration of UV radiation has also to be considered in the context of the magnetosphere.

SOME SPECULATIONS ON FUTURE SHAPE AND COLOR OF LANDSCAPE IN AN ANTHROPOSPHERIC EARTH

Let me now speculate about what might happen in the long term as a consequence of the mentioned future changes in CO_2 and O_2 derived compounds. They might result in the development of new landscapes, new ecosystem types or new biomes.

If CO_2 determines the efficiency, shape and color of vegetation, changes in these characteristics might be expected because of higher availability of CO_2. For example, leaves may be bigger, not as much divided as nowadays, and tending more to the monolayer design because they are already going to produce more with the first light interception by disposing of more CO_2 substrate. More efficiency can be expected in the field because under higher light intensity, higher production will be obtained. Nowadays, when light is low and limiting, as previously mentioned, red undersurfaces and blue iridiscence appear to slightly enhance absorption of leaves restricted to extreme shade in tropical rain forests (LEE, 1986). Similarly, under future increased CO_2 availability a higher use of light might be necessary; under such conditions one could expect that luxurious plant green color might evolve to blues and reds: in order to take advantage of the higher availability of CO_2, plants might need to use the maximum possible radiation, and therefore the green wavelength range. In this way, they would not be able to reflect it; they would not be green. However, it must not be forgotten that life continues by preserving "information" and "structure", an activity in which it seems more "interested" than in growing "like mad".

All these atmospheric changes are consequences of man's activities, mainly burning fossil fuels. With his activity, man has also begun to alter the trend towards reduction that life was following until the beginning of the industrial era by accumulating reduced organic material and releasing oxygen to the atmosphere. Now, man is burning that fossil organic material and returning carbon to the atmosphere at such a rate that no natural reservoir can absorb all of it and therefore there is a hindrance to this trend of reduction. This anthropospheric action may impede the biosphere evolution towards a stronger redox gradient that could lead to an oxidated planet, with an explosive atmosphere, richer in oxygen, and with upper layers of the hydrosphere converted into a hydrogen peroxide soup (MARGALEF, in press). Mankind, with its combustion - and therefore oxidation of fossil fuels, now so criticized because of the generation of CO_2 increase in the atmosphere and the subsequent greenhouse effect - may in fact be avoiding this fatal oxidated destiny. Perhaps Gaia (LOVELOCK, 1979) has found in the anthroposphere (PEÑUELAS, 1988b) another of life's self-regulation mechanisms. Who knows?

It is because of anthropogenic causes of global change that possible continuation and/or even a surpassing of LOVELOCK's (1979) Gaia theory can be established. If the current trends of human behavior do not change abruptly - which seems improbable- soon it will not be possible to consider all the Earth acting as a kind of superorganism that regulates the physical, chemical, biological and geological conditions of the planet, in order to maintain itself homeostatically, as the Gaia theory postulates. The organization resulting from evolution and actuation of Gaia has given rise to mankind and its realizations, creating a special environment, a new planet layer, the anthroposphere (PEÑUELAS, 1988b, in press), that is now taking the main role in the regulation and organization of our planet.

ACKNOWLEDGEMENTS

I am grateful to R. Margalef, A. Caballero, J. Azcón-Bieto and J. Catalán for comments on the manuscript, and to CIRIT (Catalan Government) and Fundación Conde de Barcelona for financial support.

REFERENCES

ACOCK, B. & ALLEN, H. 1985. Crop responses to elevated carbon dioxide concentrations. In: *Direct effects of increasing carbon dioxide on vegetation.* (B.R. Strain & J.D. Cure, eds): 53-97. US Department of Energy ER-0238, Washington, DC

ACOCK B., BAKER, D.N., REDDY, V.R., McKINION, J.M., WHISLER, F.D., CASTILLO, D., & HODGES, H.F. 1982. Soybean responses to CO_2 In: *Response of Vegetation to Carbon Dioxide.* Dept. of Energy and Dept. Agriculture USA

ARNDT, U., SEUFERT, G. & NOBEL, W. 1982. Die Beteiligung von Ozon an der Komplexkrankheit der Tanne (*Abies alba* Mill) eine prufenswerte Hypothese. *Staub Reinhalt Luft*, 42: 243-247.

AZCÓN-BIETO, J. & PEÑUELAS, J. 1987. Cyanide resistant respiration in photosynthetic organs of freshwater aquatic plants. *Plant Physiology,* 84:

701-707.

BAKER, D.N. & LAMBERT, J.R. 1979. Report of the AAAS-DOE *Workshop on Environmental and Societal Consequences of a Possible CO_2-Induced Climate Change*. Annapolis, MD

BARNES, J.D. & DAVISON, A.W. 1986. Assessment of pollution-induced injury by chlorophyll fluorescence analysis. Direct effect of dry and wet deposition on forest ecosystems in particular canopy interactions. *Proc. COST- CE C Workshop.* Roskille. Goteborg, 1986

BARNES, J.D. & WILSON, J.M. 1984. Assessment of the frost sensitivity of *Trifolium* species by chlorophyll fluorescence analysis. *Ann. Appl. Biol.*, 105: 107-116

BAZZAZ, F.A., GARBUTT, K. & WILLIAMS, W.E. 1985. Effect of increased atmospheric carbon dioxide concentration on plant communities. In: *Direct Effects of increasing carbon dioxide on vegetation.* (B.R. Strain & J.D. Cure, eds.): 166-170. National Technical Information Service. Springfield Va.

BAZZAZ, F.A., CHIARELLO, N., COLEY, P.D. & PITELKA, L. 1987. The allocation of resources to reproduction and defense. *Bioscience*, 37: 58-67.

BEGON, M., HARPER, J.L. & TOWNSEND, C.R. 1986. *Ecology. Individuals, Populations and communities*. Blackwell. Oxford.

BERI, R.M. & LEMON, H.W. 1969. Chemical examination of the wax from needles of black spruce (*Picea marina*) and balsam fir (*Abies balsamea*). *Can. J. Bioch.*, 48: 67-69.

BERT, P. 1878. *La pression barométrique*. Masson. Paris (Reprinted 1979), CNRS, Paris.

BLANCK, L.W. 1985. A new type of forest decline in Germany. *Nature*, 314: 311-314.

BOSTON, H.L., ADAMS, M.S. & MADSEN, J.D. 1989. Photosynthetic strategies and productivity in aquatic ecosystems. *Aquatic Botany*, 34: 27-57.

BROWN, K.A., ROBERTS, T.M. & BLANK, L.W. 1987. Interactions between ozone and cold sensitivity in Norway Spruce: a factor contributing to the forest decline in central Europe? *New Phytologist*, 105: 149-155.

CADET, J. & TEOULE, R. 1978. Comparative study of oxidation of nucleic acid components by hydroxyl radicals, singlet oxygen and superoxide anion radicals. *Photochem. Photobiol.*, 28: 661-667.

CARLSON, R.W. & BAZZAZ, F.A. 1980. The effects of CO_2 concentrations on growth, photosynthesis, transpiration, and water use efficiency of plants. In: *Environmental and Climatic Impact of Coal Utilization* (J.J. Singh & A. Deepak, eds.): 609-623. Academic Press. New York.

CERUTTI, P.A. 1985. Prooxidant states and tumor promotion. *Science*, 227: 375-381.

CITERNE, A., BANVOY, J. & DIZENGREMEL, P. 1988. Pollution atmosphérique et métabolisme respiratoire de l'epicéa. In: *Air pollution and ecosystems* (P. Mathy, ed.): 678-683. Reidel. Dordrecht.

COLINVAUX, P. 1986. *Ecology*. John Wiley. New York

CONROY, J.P., KUPPERS, M., KUPPERS, B., VIRGONA, J. & BARLOW, E.W.R. 1988. The influence of CO_2 enrichment, phosphorus deficiency and water stress on the growth, conductance and water use of *Pinus radiata* D. Don. *Plant Cell & Environment,* 11: 91-98.

COYNE, P.I. & BINGHAM, G.E. 1982. Variation in photosynthesis and stomatal conductance in and O_3-stressed ponderosa pine stand. *For. Sci.*, 28:. 257-273

CROSSLEY, A. & FOWLER, D. 1986. The weathering of scots pine epicuticular wax in polluted and clean air. *New Phytologist*, 103: 207-218.

CURE J.D. 1985. Carbon dioxide doubling responses: a crop survey. In: *Direct effects of increasing carbon dioxide on vegetation.* (Strain, B.R. & Cure, J.D., eds): 99-116. US Department of Energy ER-0238, Washington DC.

CURTIS, P.S., DRAKE, B.G. & WHIGHAM, D.F. 1989. Nitrogen and carbon dynamics in C3 and C4 estuarine marsh plants grown under elevated CO_2 in situ. *Oecologia*, 78: 297-301.

DAVIS, M.B. 1988. Ecological systems and dynamics. In: *Towards an understanding of global change:* 69-106. National Academy Press. Washington D.C.

DAVIS, M.B. & ZABINSKI, C. in press. Changes in geographical range resulting from greenhouse warming effects on biodiversity in forests. In: *Consequences of the greenhouse effect for biological diversity*.

DAVISON, A.W. & BARNES, J.D. 1986. Effects of winter stress on pollutant responses. How are the effects of air pollutants on agricultural crops influenced by the interaction with other limiting factors? *Proc. COST-CE C Workshop*: 16-32. Roskilde.

DAVISON, A.W., BARNES, J.D. & RENNER, C.J. 1987. Interactions between air pollutants and cold stress. *Proc. 2nd Int. Symp. Air Pollutant and plant Metabolism*: 307-328.

DIAMOND, J.M. 1989. Hot sex in Aracea. *Nature*, 339: 258-259.

DICKSON, R.E., SHIVES, J.B.Jr. 1982 $^{14}CO_2$ fixation, translocation and carbon metabolism in rapid expanding leaves of *Populus deltoides. Ann. Bot.*, 50: 37-47.

DOMINY, P.J. & HEATH, R.L. 1985. Inhibition of the K^+ stimulated ATPase of the plasmalemma of pintobean leaves by ozone. *Plant Physiol.*, 77: 43.

DRAKE, B.G., LEADLEY, P.W., ARP, W., CURTIS, P. & WHIGHAM, D. 1988. The effect of elevated atmospheric CO_2 on C3 and C4 vegetation on Chesapeake Bay. In: *Proceedings of Symposium on the Physiological Ecology of Aquatic Plants.* Aarhus. Denmark.

EHLERINGER, J.R. & BJORKMAN, O. 1977. Quantum yields for CO_2 uptake in C3 and C4 plants: dependence on temperature, CO_2 and O_2 uptake. *Plant Physiology*, 59: 86-90.

EHLERINGER, J.R. & WERK, K.S. 1986. Modifications of solar-radiation absorption patterns and implications for carbon gain at the leaf level. In: *On the economy of plant form and function* (T. J. Givnish. ed.): 57-82. Cambridge University Press. Cambridge.

ELSTNER, E.F. 1982. Oxygen activation and oxygen toxicity. *Ann. Rev. Plant Physiology,* 33: 73-96.

FAJER, E.D., BOWERS, M.D. & BAZZAZ, F.A. 1989. The effects of enriched carbon dioxide atmospheres on plant-insect herbivore interactions. *Science,* 243: 1198-1200.

FALKOWSKI, P.G. 1983. Light-shade adaptation and vertical mixing of marine phytoplankton: A comparative field study. *J. Mar. Res.*, 41: 215-237.

FRICKE, W. 1983. Grossräumige Verteilung und Transport von Ozon und Varlaufern. VDI. Berichte 500 Säure, Niederschlage. Ursachen und Wirkungen VDI-Verlag.

FRIEDLAND, A.J., GREGORY, R.A., KARENLAMPI, L. & JOHNSON, A.H. 1984. Winter damage to foliage as a factor in red spruce decline. *Can. J. For. Res.*, 14: 963-965.

FRIEDLI, H., LOTSCHER, H., OESCHGER, H., SIEGENTHALER, U. & STAUFFER, B. 1986. Ice core record of the $^{13}C/^{12}C$ ratio of atmospheric CO_2 in the past two centuries. *Nature*, 324: 237-238.

GAMMON, R.H., SUNDQUIST, E.T. & FRASER, P.J. 1985. History of carbon dioxide in the atmosphere. In: *Atmospheric carbon dioxide and the overall carbon cycle* (J.R. Trabalka, ed.): 26-62. U.S. Department of Energy. Washington D C.

GAMON, J., PEÑUELAS, J., FIELD, C. & BJORKMAN, O. in press. Spectral shifts in vegetation reflectance under water and nutrient stress. *Plant, Cell and Environment.*

GAUDILLÈRE, J.P. & MOUSSEAU, M. 1989. Short term effect of CO_2 enrichment on leaf development and gas exchange of young poplars (*Populus euramericana* cv I 214). *Oecologia Plantarum* , 10(1): 95.

GIVNISH, T.J. 1984. Leaf and canopy adaptations in tropical forests. In: *Physiological Ecology of Plants of the Wet Tropics* (E. Medina, H.A. Mooney & C. Vásquez-Yáñez, eds.): 51-84. Junk. The Hague.

GIVNISH, T.J. (ed.). 1986. *On the economy of plant form and function.* Cambridge University Press. Cambridge.

GOULD, S.J. & LEWONTIN, R.C. 1979. The spandrels of San Marco and the Panglossian paradigm: a critique of the adaptationist programme. *Proc. R. Soc. Lond.*, Ser. B, 205: 581-598.

GRAEDEL, T.E. & CRUTZEN P.J. 1989. Una atmósfera cambiante. *Investigación y Ciencia*, 158: 22-31.

GRIBBIN, J. 1988. *The hole in the sky.* Corgi. London.

GULMON, S.L. & CHU, C.C. 1981. The effect of light and nitrogen on photosynthesis, leaf characteristics, and dry matter allocation in the chaparral shrub *Diplacus aurantiacus. Oecologia*, 49: 207-212.

HARPER, J.L. 1982. After description. In: *The plant community as a working mechanism.* (E. I. Newman, ed.): 11-25. Blackwell. Oxford.

HEATH, R.L. 1975. Ozone. In: *Response of plant to air pollution.* (J.B. Mudd, & T.T. Kozlowski, eds.): 23-28. Academic Press. New York.

HEATH, R.L. 1980. Initial events in injury to plants by air pollutants. *Ann. Rev. Physiol,* 31: 395-401.

HILBERT, D.W., PRUDHOMME, T.I. & OECHEL, W.C. 1987. Response of tussock tundra to elevated carbon dioxide regimes: analysis of ecosystem CO_2 flux through nonlinear modelling. *Oecologia,* 72: 466-472.

IDSO, S.B. 1988. Three phases of plant response to atmospheric CO_2 enrichment. *Plant Physiology*, 87: 5-7.

INSKEEP, W.P. & BLOOM P.R. 1985. Extinction coefficients of chlorophyll a and b in NN-dimethylformamide and 80% acetone. *Plant Physiol.*, 77(2): 483-492.

JOHNSON A.H. & SICCAMA, T.G. 1983. Acid deposition and forest decline. *Environmental Sci. Tech.*, 17: A244-A304.

KORNER, C., SCHEEL, J.A. & BAUER, H. 1979. Maximum leaf diffusive conductance in vascular plants. *Photosynthetica,* 13: 45-82

LaMARCHE, V.C.Jr, GRAYBILL, D.A., FRITTS, H.C. & ROSE, M.R. 1984. Increasing atmospheric carbon dioxide: Tree-ring evidence for growth enhancement in natural vegetation. *Science*, 225: 1019-1021.

LAMBERS, H. 1985. Respiration in intact plants and tissues: its regulation and dependence on environmental factors, metabolism and invaded organisms. In: *Encyclopedia of Plant Physiology,* 18 (R. Douce & D.A. Day, eds): 418-473. Springer. Heidelberg

LAMBERS, H., DAY, D.A. & AZCÓN-BIETO, J. 1983. Cyanide-resistant respiration in roots and leaves. Measurements with intact tissues and isolated mitochondria. *Physiol. Plant.*, 58: 148-154.

LARIGAUDERIE, A., HILBERT, D.A. & OECHEL, W. 1988. Effect of CO_2 enrichment and nitrogen availability on resource acquisition and resource allocation in a grass, *Bromus mollis. Oecologia*, 77: 544-549.

LAVOREL, J. & ETIENNE, A.L. 1977. In vivo chlorophyll fluorescence. In: *Primary processes of photosynthesis; topics in photosynthesis* (J. Barber, ed.): 203-268. Elsevier. Amsterdam.

LEACH, J.E., WOODHEAD, T. & DAY, W. 1982. Bias in pressure chamber measurements of leaf water potential. *Agric. Meteorol.*, 27: 257-263.

LEE, D.W. 1986. Unusual strategies of light absorption in rain-forest herbs. In: *On the economy of plant form and function* (T. J. Givnish, ed.): 105-132. Cambridge University Press. Cambridge.

LEE, T.T. 1967. Inhibition of oxidative phosphorylation and respiration by ozone in tobacco mitochondria. *Plant Physiol.*, 69: 1444.

LEE T.T. 1968. Effect of ozone on swelling on tobacco mitochondria. *Plant. Physiol.*, 43: 133.

LEMON, E.D. (ed.). 1983. *CO_2 and plants. The response of plants to rising levels of atmospheric carbon dioxide.* AAAS Selected Symposium 84. Westview Press. Boulder, Colorado

LEVITT, J. 1980. *Responses of plants to environmental stresses. 1: Chilling. Freezing and high temperature stress.* (T.T. Kozlowski, ed.). Academic Press. New York & London.

LINCOLN, D.E., COUVET, D. & SIONIT, N. 1986. Response of an insect herbivore to host plants grown in carbon dioxide enriched atmospheres. *Oecologia*, 69: 556-560.

LOVELOCK, J.E. 1979. *Gaia, a new look at life on*

Earth. Oxford University Press. Oxford.

LUCAS, P.W., COTTAM, D.A. & MANSFIELD, T.A. 1987. A large-scale fumigation system for investigating interactions between air pollution and cold stress on plants. *Environmental pollution*, 43: 15-28.

LUCAS, P.W., COTTAM, D.A., SHEPPARD, L.J. & FRANCIS, B.J. 1988. Growth responses and delayed winter hardening in Sitka spruce following summer exposure to ozone. *New Phytologist*, 108: 495-504.

MADSEN, E. 1975. Effect of CO_2 enrichment on growth, development, fruit production, and fruit quality of tomato from a physiological viewpoint. *Phytotronics*, III: 318-330.

MANSFIELD, T.A., WHITMORE, M.E., PANDE, P.C. & FREER-SMITH, P.H. 1987. Responses of herbaceous and woody plants to the dry deposition of SO_2 and NO_2. In: *Effects of Atmospheric Pollutants on Forests, Wetlands and Agricultural Ecosystems* (T.C. Hutchinson & K. Meema, eds.): 131-144. Springer. New York.

MARGALEF, R. 1974. *Ecología*. Omega. Barcelona.

MARGALEF, R. 1980. *La biosfera entre la termodinámica y el juego*. Omega. Barcelona.

MARGALEF, R. in press. *A view of the biosphere*. Chicago University Press.

MATHY, P. (ed.). 1988. *Air pollution and ecosystems*. D. Reidel. New York.

MCDOWALL, F.D.H. 1965. Stages of ozone damage to respiration of tobacco leaves. *Can. J. Bot*., 43: 419

McLAUGHLIN, S.B. 1985. Effects of air pollution on forests. A critical review. *J. Air Poll. Control Assoc*., 35: 512-521.

MELILLO, J.M., ABER, J.D. & MURATORE, J.F. 1982. Nitrogen and lignin control of hardwood leaf litter decomposition dynamics. *Ecology*, 63: 621-626.

MOONEY H.A. in press. Biological response to climate change: the scope of the problem and an agenda for research. *Oikos*.

MOONEY, H.A. & GULMON, S.L. 1982. Constraints on leaf structure in relation to herbivory. *BioScience*, 32: 198-206.

NOBEL, P.S. 1986. Form and orientation in relation to PAR interception by cacti and agaves. In: *On the economy of plant form and function* (T.J. Givnish, ed.): 83-104. Cambridge University Press. Cambridge.

NORBY, R. J. & SEGAL, L.L. 1989. Nitrogen fixation in the lichen *Lobaria pulmonaria* in elevated atmospheric carbon dioxide. *Oecologia*, 79: 566-568.

OECHEL, W. C. & STRAIN, B.R. 1985. Native species responses to increased carbon dioxide concentration. In: *Direct Effects of increasing carbon dioxide on vegetation* (G.R. Strain & J.D. Cure, eds.): 117-154. US Department of Energy. DOE/ER-0238.

PASTOR, J., ABER, J.D., McCLAUGHERTY, A. & MELILLO, J.M. 1984. Above ground production and N and P cycling along a nitrogen mineralization gradient on Blackhawk Island, Wisconsin. *Ecology*, 65: 256-268.

PASTOR, J. & POST, W.M. 1988. Responses of northern forests to CO_2-induced climate change. *Nature*, 334: 55-58.

PEARCY, R.W. & BJORKMAN, O. 1983. Physiological effects. In: *CO_2 and plants: the response of plants to rising levels of atmospheric carbon dioxide*. (E.R. Lemon, ed.): 65-106 Westview Press. Boulder, Co.

PEARCY, R.W., BJORKMAN, A., CALDWELL, M.M., KEELEY, J.E., MONSON, R.K. & STRAIN, B.R. 1987. Carbon Gain by Plants in Natural Environments. *Bioscience*, 37(1): 21-29.

PEÑUELAS, J. 1985. Bicarbonate as an inorganic carbon source for *Fontinalis antipyretica* and *Fissidens grandifrons*. *J. Exp. Bot.,* 36 (164): 441-448.

PEÑUELAS, J. 1987. High oxygen tension inhibits vascular aquatic plant growth in deep waters. *Photosynthetica,* 21(4): 494-502.

PEÑUELAS, J. 1988a. Effects of hydrostatic pressure and irradiance on two submerged aquatic plants. An experimental study. *Pol. Arch. Hydrobiol.*, 35(1): 23-32.

PEÑUELAS, J. 1988b. *De la biosfera a la antroposfera*. Barcanova. Barcelona.

PEÑUELAS, J. in press. Anthroposphere regulation of biosphere. *Environmental Conservation.*

PEÑUELAS, J., MURILLO, J. & AZCÓN-BIETO, J. 1988. Actual and potential dark respiration rates and different electron transport pathways in freshwater aquatic plants. *Aquatic Botany*, 30: 353-362.

PEÑUELAS, J., RIERA, T. & ZABALA, M. in press a. Mysids go back every night to cave to avoid light induced radicals formation. *Marine Biology*.

PEÑUELAS, J., AZCÓN-BIETO, J. & LUCAS, P. in press b. Effects of ozone normal summer concentration on Norway and Sitka spruces. *J. Environ. Exp. Bot.*

PEÑUELAS, J., GAMON, J. & FIELD, C. in pressc. Leaf reflectance related to remote sensing and biochemical and structural characteristics of sunflower crops. *Oecologia.*

PEÑUELAS, J. & MATAMALA, R. 1990. Changes in N and S leaf content, stomatal density and specific leaf area of 14 plant species along the last three centuries of CO_2 increase. *J. Exp. Bot.*. 41: 1119-1124

PRINZ, B., KRAUSE, G.H.M. & STRATMAN, H. 1982. *Waldschaden in der Bundesrepublik Deutschland*. LIS-Berichte, 28: 1-154. Landesanstalt für des Landes NRW, Essen, FRG.

PRINZ, B., JUNK, K.D. & KRAUS, G.H.M. 1984. Forest effects in West Germany. *Symposium on air pollution and the productivity of the forest.* Publ. Izaac Walton League of America.

REHFUESS, K.E. 1986. *Perceptions on frost diseases in Central Europe.* Foundation for Science and Technology, Royal Society of London, 15 January 1985.

ROBERTS, T. M. 1984. Effects of air pollutants on agriculture and forestry. *Atmospheric environment*, 18: 629-652.

ROWLAND, A.J., BORLAND, A.M. & LEA, P.J. 1988. Changes in aminoacids, amines and proteins in response to air pollutants. In: *Air Pollution and Plant Metabolism.* (S. Schulte-hosted, N.M.

Darrall, L.W. Blank & A.R. Wellburn, eds.): 189-215. Elsevier. London.
SCHONHERR, J. 1976. Water permeability of isolated cuticular membranes: the effect of cuticular waxes on diffusion of water. *Planta*, 131: 159-164.
SCHREIBER, U., VIDAVER, V., RUNECKLES, V. C. & ROSEN, P. 1978. Chlorophyll fluorescence assay for ozone injury in intact plants. *Pl.ant Physiol.*, 61: 80-84.
SKARBY, L. & SELLDEN, G. 1984. The effects of ozone on crops and forest. *Ambio,* 13: 68-72.
SKIBA, U. & CRESSER, M. 1988. The ecological significance of increasing atmospheric carbon dioxide. *Endeavour*, 12(3): 143-147
STRAIN, B.R. & BAZZAZ, F.A. 1983. Terrestrial plant communities. In: *CO_2 and plants. The response of plants to rising levels of atmospheric carbon dioxide.* (E.R. Lemon, ed.): 177-201. Westview Press. Boulder, Co.
STRAIN, B.R. & CURE, J.D. 1985. *Direct effects of increasing carbon dioxide on vegetation.* Duke University Press. Durham, N.C.
SULLIVAN, S.G., WINTERBOURN, C.C. & STERN, A. 1983. Hypothesis: oxygen, superoxide dismutase and catalase form a metabolic pathway that protects against oxidative damage in the red cell. In: *Oxygen Radicals and their Scavenger Systems.* (I. G. Cohen & R.A. Greenwald, eds.): 364-367. Elsevier. Amsterdam.
SWANSON, E.S., THOMSON, W.W. & MUDD J.B. 1973. The effect of ozone on leaves cell membranes. *Can. J. Bot.*, 51: 1213.
TISSUE, D. & OECHEL, W. 1987. Responses of *Eriophorum vaginatum* to elevated CO_2 and temperature in the Alaskan tussock tundra. *Ecology*, 68: 401-410.
TRANQUILLINI, W. 1979. *Physiological ecology of the alpine timberline*. Ecological studies 31. Springer. New York.
TUCKER, C.J. & SELLERS, P.J. 1986. Satellite remote sensing of primary production. *Intern. J. Remote Sensing,* 7(11): 1395-1416.
WALKER, J. C. G. 1986. *Earth History.* Jones & Bartlett. Boston.
WARBURG, O. 1920. Über die Geschwindigkeit der photochemischen Kohlensäurezersetzung in lebenden Zellen. II. *Biochem. Z.*, 103: 188-217.
WELLBURN, A. 1988. *Air pollution and acid rain. The biological impact.* Longman. New York.
WILLIAMS, W.E., GARBUTT, K., BAZZAZ, F.A. & VITOUSEK, P.M. 1986. The response of plants to elevated CO_2 IV. Two deciduous-forest tree communities. *Oecologia*, 69: 454-459
WILLIAMS, W. E., GARBUTT, K. & BAZZAZ, F.A. 1988. The response of plants to elevated CO_2. V. Perfomance of an assemblage of serpentine grassland herbs. *Env. Exp. Bot.*, 28: 123-130.
WONG, S.C. 1979. Elevated partial pressure of CO_2 and plant growth. *Oecologia*, 44: 68-74.
WOODWARD, F.I. 1986. Ecophysiological studies on the shrub *Vaccinium myrtillus* L. taken from a wide altitudinal range. *Oecologia*, 70:580-586.
WOODWARD, F.I. 1987. Stomatal numbers are sensitive to increases in CO_2 from pre-industrial levels. *Nature*, 327: 617-618.
WOODWARD, F.I. & BAZZAZ, F.A. 1988. The responses of stomatal density to CO_2 partial pressure. *J. Exp. Bot.*, 39: 1771-1781.
WULFF, R.A. & STRAIN, B.R. 1982. Effects of CO_2 enrichment on growth and photosynthesis in *Desmodium paniculatum*. Can. J. Bot., 60: 1084-1091.

Homage to Ramon Margalef; or, Why there is such pleasure in studying nature (J. D. Ros & N. Prat, eds.). 1991. *Oecologia aquatica*, 10: 387-391

FLOW-NETWORK ORGANIZATION IN ECOSYSTEMS AND THE MATHEMATICAL THEORY OF INFORMATION

JORGE WAGENSBERG[1], RICARD V. SOLÉ[2] & AMBROSIO GARCÍA[1]

[1] Departament de Física Fonamental. Facultat de Fisica. Universitat de Barcelona. Av. Diagonal, 647. 08028 Barcelona. Spain
[2] Departament de Física i Enginyeria Nuclear. Universitat Politècnica de Catalunya. Pau Gargallo, 5. 08026 Barcelona. Spain

Received: February 1990

ABSTRACT

Ecological flow networks are described by means of the Theory of Graphs and characterised by two quantities of the Mathematical Theory of Information, namely the Internal Information Transfer (a measure of the Specialization) and the Joint Entropy (a measure of Connectivity). Both indexes fall into a narrow interval of values. This tendency of real stable ecosystems is interpreted in the light of two variational adaptative principles.

KEY WORDS: Flow-network, information, ecosystem modelling.

INTRODUCTION

The use of Shannon's entropy,

$$S(p_1, ..., p_n) = -\sum_{i=1}^{n} p_i \log_2 p_i \qquad (1)$$

as a measure of ecological diversity was first introduced in the 50s by MacArthur and Margalef and extensively used later in the study of real ecosystems. Applications of this concept are today numerous and varied. Shannon's entropy provides, in physical sciences (WEHRL, 1978), a good approach to the equilibrium states in isolated systems. The final statistical structure in these situations can be obtained by means of the maximization of (1) under some physical constraints. For example, if p_i is the probablility of finding a given particle of a dilute gas (confined in a box) in the interval of energy $(E_j, E_j+\Delta E_j)$, we can use the constraints of normalization of $\{p_i\}$ -set,

$$\sum_{i=1}^{n} p_i = 1 \qquad (2)$$

and the energy-conservation condition,

$$\sum_{i=1}^{n} p_i E_i = \overline{E} \qquad (3)$$

(here $\overline{E}$ is the mean amount of energy) to find the equilibrium probability distribution, $\{p_{ieq}\}$. We then have,

$$\delta\left[S + \alpha\left(1 - \sum_{i=1}^{n} p_i\right) + \beta\left(\overline{E} - \sum_{i=1}^{n} p_i E_i\right)\right]=0 \qquad (4)$$

and performing the variation of (4), we

obtain the best known Gibbs canonical distribution

$$P_i = \frac{1}{z} e^{-Ei/kT} \quad (5)$$

where Z is the sum $\sum_{i=1}^{n} e^{-Ei/kT}$, called the partition function.

Entropy-like functions (1) can be defined in several systems where $\{p_i\}$ is known. Then, with appropriate constraints, it is possible to derive the equilibrium distributions in, for example, ecological systems when an extensive quantity is defined. The maximum entropy (MAXENT) principle, due to Jaynes, allows one to make these unbiased estimates of which otherwise only some averages (corresponding to macroscopic measures) are known. The MAXENT principle provides very elegant access to basic relations and concepts of thermodynamics and other fields (JAYNES, 1985) and has been applied in recent years in the study of biomass distribution and ecological adaptation. The use of correlation i.e. $F_{ij} = \langle q_i q_j \rangle$ as constraints allows us to extend this approach to systems far from equilibrium (HAKEN, 1985, 1987) as for example in laser physics, in a quite straightforward form.

The MAXENT algorithm has been used (LURIE & WAGENSBERG, 1983) in the study of the statistical structure of biomass diversity in fish populations. Defining probabilities as $p_i = P\,[m_i, m_i + \Delta m]$, and using (2) and constant value of mean biomass,

$$\sum_{i=1}^{n} p_i m_i = \overline{M} \quad (6)$$

as constraints, it can be shown, by using this variational procedure, that

$$p_i = \frac{1}{z} e^{-mi/\overline{m}} \quad (7)$$

where now Z is the equivalent partition function defined as

$$Z = \sum_{i=1}^{n} e^{-mi/\overline{m}}$$

and this predicted probability distribution was proved to conform closely to fishery data. Other applications of MAXENT theory also showed the possibility of recovering other real macroscopic results from this approach (WAGENSBERG & VALLS, 1988). We will now consider the problem of structure and organization of flow networks in ecosystems, using some concepts of information theory and physics.

FLOW-NETWORKS IN ECOLOGY

An ecosystem can be shown as a set of

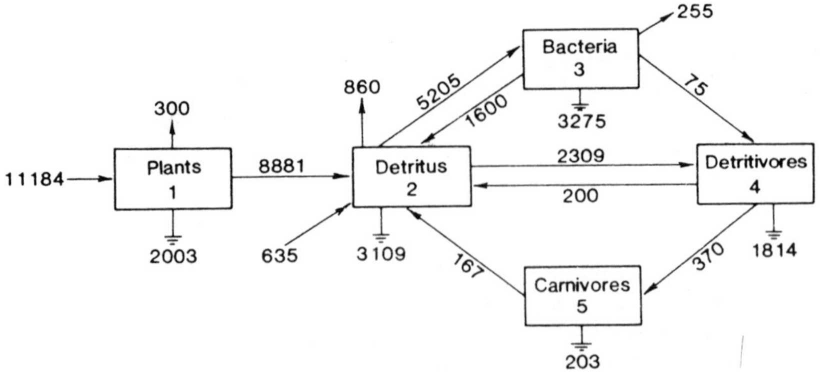

FIGURE 1. Flow energy in the Cone Spring ecosystem. Reprinted from ULANOWICZ (1986). Flows are given in kcal $m^{-2}\ y^{-1}$

components $\{C_i\}$ i = 1,2, ..., n between which an extensive quantity (energy, carbon, etc.) is exchanged. This picture can be represented by an oriented graph as is shown, for example, in figure 1. Here the arrows give the direction and magnitude of the flows. In our formal description, we will consider the boundaries (i.e. the origin and sink of resources) as another compartment and the system of figure 1 will now be represented by a graph as in figure 2. Physically, this consideration enables us to talk of a closed system. Two sets of probabilities of emission or reception of energy at nodes i = 1, 2, ..., n and a set of conditional probabilities, $\{p_{ij}\}$ that gives the amount of energy transfer between nodes i and j (i,j = 1, 2, ..., n) are necessary in our description.

Some studies of matter and energy flow in an ecosystem (ULANOWICZ, 1986) provide some regularities for this structure. But is it possible to define macroscopic quantities useful to describe a network-flow in an ecosystem? It has been found (WAGENSBERG *et al.*, 1989) that two central magnitudes of Mathematical Theory of Information such as the joint entropy H(X,Y):

$$H(X,Y)= -\sum_j p_j \log_z p_j - \sum_i \sum_j p_j p_{ij} \log_z p_{ij} \quad (8)$$

and the information transfer, I (X,Y),

$$I(X,Y)= -\sum_j p_j \log_2 p_j - \sum_i \sum_j p_j p_{ij} \log_2 p_{ij} \quad (9)$$

provide us a general and interesting approach to this problem.

Calculations of H(X,Y) and I(X,Y) from some real ecosystems showed important regularities (Table I). Real data provides us with the set of probabilities p_{ij} of the interaction matrix, say Σ. We can now generate all the possible states compatible with Σ, as is shown in figure 3. The steady state appears on this "space of states" indicated by a cross. Two essential facts should be emphasized after inspection of Table I: (1) H_{stat} always falls in a quite a narrow interval around the 3 bits and (2) I_{stat} tends to attain the maximum value that is compatible with the given H_{stat} except in the case of Wingra Lake. The case of this lake is quite important in our interpretation of these data. We believe that I (X,Y) values are some kind of measure of the global fitness of the ecosystem, as the result of a mechanism of selection. Wingra Lake is an eutrophic lake, and in this case a

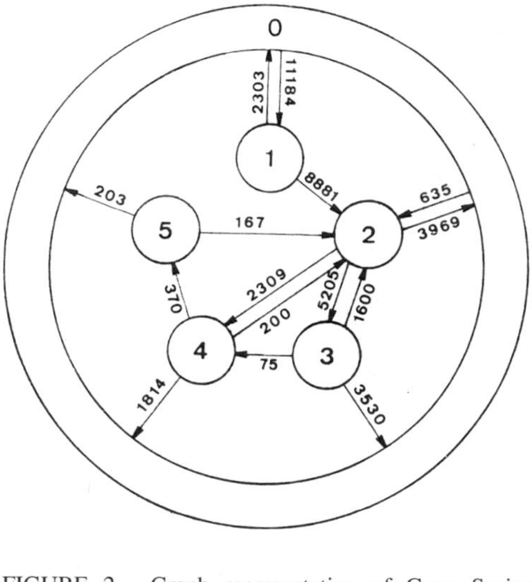

FIGURE 2. Graph representation of Cone Spring ecosystem including surroundings. 1, plants. 2, detritus. 3, bacteria. 4, detritivores. 5, carnivores.

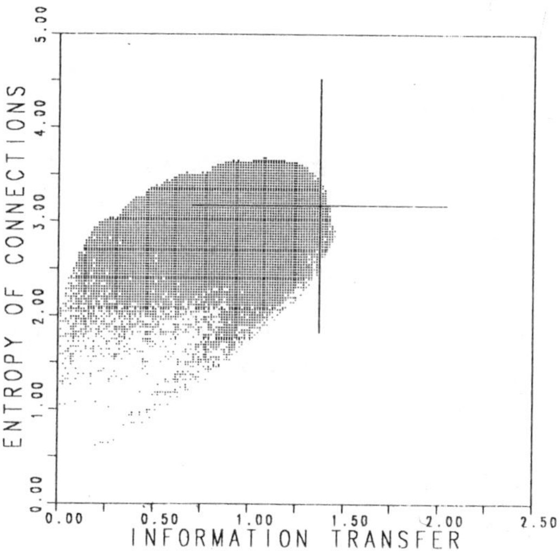

FIGURE 3. Domain of available states derived from the Cone-Spring matrix. The cross is the position of real stationary state.

TABLE I. H and I values at the stationary state (when the inflow equals the outflow at each node) for six different well known ecosystems (RICHNEY *et al.*, 1978; ULANOWICZ, 1986).

ECOSYSTEM	Nodes	Hstat	Hmax	Istat	Imax
Findlay Lake	4	3.08	3.28	1.16	1.25
Mirror Lake	5	3.17	3.57	1.25	1.38
Cone Spring	5	3.07	3.55	1.32	1.40
Wingra Lake	5	2.95	3.64	0.76	0.99
Arion Lake	5	3.01	3.63	0.97	1.13
Crystal River	7	2.93	4.08	1.25	1.35

strong hydrodynamic instability generates continuous resuspension of sediments.

These data enable us to talk about two "magic numbers" in ecological systems, in this sense of significant tendences in ecosystem organization. The first one, the joint entropy H(X,Y) (say entropy of connections) can be interpreted as the result of thermodynamical obliged dissipations at each node. If we look at the compartments as thermodynamic machines, in the light of the second law of the thermodynamics, some part of the inflow energy will always be lost in form of dissipation. The second one, the values of I(X,Y), are in agreement with our proposal of information transfer as a relevant quantity suggestive of an underlying variational principle.

MAXIMUM INFORMATION TRANSFER

We will now try to apply the MAXENT approach to our last results. Physical constraints are limiting factors on the entropy of connections H(X,Y). If we consider this fact as a constraint itself, we can find the maximum value of information transfer under the next set of conditions:

a) normalization of probabilities:

$$\{\sum_{i=1}^{n} p_{ij}\} = 1\} \quad j = 1, 2, ..., n \qquad (10.1)$$

and

$$\sum_{j=1}^{n} p_j = 1 \qquad (10.2)$$

and:

b) a fixed given value of H(X,Y), say H_0. If we perform the variation:

$$\delta\ [-\sum_i p_i \log_2 p_i + \sum_i \sum_j p_j p_{ij} \log_2 p_{ij} +$$

$$+\alpha\,(H_0 + \sum_j p_j \log_2 p_j + \sum_i \sum_j p_j p_{ij} \log_2 p_{ij}) +$$

$$+ \sum_j \beta_j (1 - \sum_i p_{ij}) + \gamma (1 - \sum_j p_j)] \qquad (11)$$

we obtain the following equations:

$$(1 + \log_2 p_i)\, p_j + p_j (1 + \log_2 p_{ij}) (1 + a) - - \beta_j = 0 \qquad (12.1)$$

and

$$- (1 + \log_2 p_{ij})\, p_{ij} - H(x/j) + \alpha (1 + \log_2 p_j) + \alpha\, H(x/j) - \gamma = 0 \qquad (12.2)$$

Equations (12) are difficult to solve, but a solution of this system with theoretical implications can be obtained by considering that: (a) In real ecosystems at stationary state, probabilities of inflows equal probabilities of outflows at each node; (b) H (X,Y) values suggest that essentially only three levels are significative in the energy transfer through the flow network, and (c) If H (X,Y) = Ho, maximization of I (X,Y) implies in fact maximization of H(X) + H(Y) + H_0, i.e. of H(X) + H(Y). It can be shown that, under the above-mentioned conditions, max {H(X) + H(Y)} holds for equal probabilities of inputs and outputs (uniform distributions) and Σ normalized by arrows and columns.

TABLE II. a) Flow matrix of the Cone Spring ecosystem. b) Interaction Matrix.

a)

Producers	Consumers	Environment	
0	0	11184	Producers
8881	2815.7	635	Consumers
2303	9516	0	Environment

b)

Producers	Consumers	Environment	
0.0000	0.0000	0.9463	Producers
0.7941	0.2283	0.0537	Consumers
0.2059	0.7717	0.0000	Environment

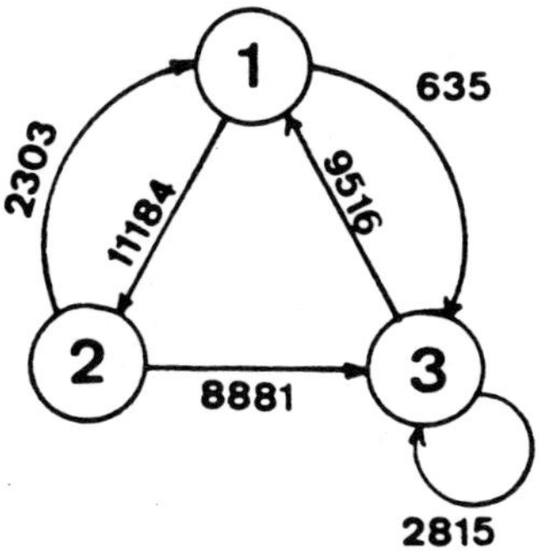

FIGURE 4. A typical three-compartment ecological graph corresponding to the Cone Spring ecosystem. 1, environment. 2, producers. 3, consumers.

If we describe the Cone Spring case using a three-level model, where only surroundings, primary producers and the others are considered (typically as in Fig. 4), approximately uniform distributions are obtained and the interaction matrix appears normalized by arrows and columns.

In Table II we can see the energy-flow matrix, the corresponding interaction matrix Σ; the probabilities of inflows (outflows) at each node in the stationary state are $p(1) = 0.3165$, $p(2) = 0.3490$ and $p(3) = 0.3345$.

DISCUSSION

Ecological flow-networks seem to be self-organized in a particular way, characterized by a high information Transfer value compatible with a thermodynamically limited Joint Entropy. This tendency is especially apparent when we reduce our description of an ecosystem to a graph with only three nodes: primary producers, consumers and surroundings. In the Cone Spring case, real data are close to theoretically predicted values. The preliminary analysis of some other real systems suggests that this might be a fairly general tendency in the organization of flow-networks in ecosystems. Our proposal is that some kind of global efficiency is related to $I(X,Y)$ and has been maximized in biological evolution. Interaction between sub-levels at the third node may have been selected to provide an adequate quantity of energy cycling. A compromise between persistence and global efficiency can explain then the special features of the Σ energy transfer matrix and the tendency towards the maximization of Information Transfer.

REFERENCES

JAYNES, E.T. 1985. Where do we go from here?. In: *Maximum Entropy and Bayesian Methods in Inverse Problems* (R. Smith & W.T. Grandy, eds.): 21- 58. Reidel. Dordrecht.

HAKEN, H. 1985. Application of the maximum information entropy principle to self-organizing systems. *Z. Phys. B.*, 61: 335-338.

HAKEN, H. 1988. *Information and self-organization.* Springer. Berlin, Heidelberg.

LURIE, D. & WAGENSBERG, J. 1983. On biomass diversity in ecology. *Bull. Math. Biol.*, 45: 287-294.

LURIE, D., VALLS, J. & WAGENSBERG, J. 1983. Thermodynamic approach to biomass distribution in ecological systems. *Bull. Math. Biol.*, 45: 869-873.

MACARTHUR, M. 1955. Fluctuations of populations, and a measure of community stability. *Ecology*, 36: 533-536.

MARGALEF, R. 1958. Information theory in ecology. *Gen. Syst*, 3: 36-71.

RICHEY, J.E., WISSMAR, R.C., DEVOL, A.H., LIKENS, G.E., EATON, J.S., WETZEL, R.G., ODUM, W.E., JOHNSON, N.M., LOUCKS, O.L., PRENTKY, R.T. & RICH, P.H. 1978. Carbon flow in four lake ecosystems: a structural approach. *Science,* 202: 1183-1186.

SHANNON, C.E. & WEAVER, W. 1962. *The mathematical theory of information.* University of Illinois Press. Urbana, Ill.

ULANOWICZ, R.E. 1986. *Growth and development: ecosystems phenomenology.* Springer. New York.

WEHRL, A. 1978. General properties of entropy. *Rev. Mod. Phys.*, 50: 221-246.

WAGENSBERG, J., VALLS, J. & BERMÚDEZ, J. 1988. Biological adaptation and the mathematical theory of information. *Bull. Math. Biol.*, 50: 445-464.

WAGENSBERG, J., GARCÍA, A. & SOLÉ, R.V. 1990. Connectivity and information transfer in flow-networkks: two magic numbers in nature? *Bull. Math. Biol.,* 52: 733-740

(J. D. Ros & N. Prat, eds.). 1991. *Oecologia aquatica*, 10: 393-409. 1990

ECOLOGY AT AN EXHIBITION: IMPACT AND INFORMAL LEARNING

CARME PRATS[1] & JORDI FLOS[2]

[1] Museo Nacional de Ciencias Naturales (CSIC). J. Gutiérrez Abascal, 2. 28006 Madrid. Spain.
[2] Department of Ecology. Faculty of Biology. University of Barcelona. Av. Diagonal, 645. 08028 Barcelona. Spain.

Received: April 1991

SUMMARY

During 1986 and 1987, an itinerant exhibition on "Ecology" written by Professor Margalef was followed in eight cities in Catalonia, (NE Spain). An extensive survey was carried out that included a survey on visitors' opinions and an experimental study on learning. This paper studies the relationships between opinions on the different environments of the exhibition (in terms of "Like", "Dislike" and "Surprise"), the physical and formal contents of the environments, and the time spent by visitors in front of each. We also analyse the results of an experiment on learning applied to students from four different age levels who used the exhibition as a complement to their formal studies. The experiment evaluated the most liked and the most disliked topics in the exhibition. The factors studied which could have some effect on marks obtained in an examination were: a) seeing the exhibition; b) having prepared something on ecology at school before visiting the exhibition; c) receiving a general introduction on how to look at an exhibition, just before the visit.

The frequency of positive opinions on a given environment is inversely related to its density of information and directly related to the time spent looking at it. Students using the exhibition as a complement to their formal education only benefit if they like the exhibition (studied environment by environment). Having prepared something on ecology at school before visiting the exhibition has striking effects on teacher trainees, which raises serious questions about the way formal teaching is often carried out.

KEY WORDS: Natural history, museum, exhibition, survey, learning, ecology.

INTRODUCTION

Science is an important part of culture, and as such has to be made available in a comprehensible form. This consideration makes it possible for two traditionally separate groups, the scientific community and the general public, to draw a little closer to each other.

The use of exhibitions as a vehicle for communication has grown in recent years. The value of exhibitions as an effective method of information and introducing new products or ideas is beyond doubt. The design of an exhibition allows the combination of a wide variety of market techniques, both in terms of presentation (graphic design, audiovisual systems and incorporation of three dimensional objects), and the possibility of provoking passive or participatory behaviour on the part of the visitor, in the use or otherwise of interactive programmes.

If this range of aesthetic and participatory possibilities make the exhibition an ideal means of communication, they also involve a number of instrumental restrictions, which have to be taken into account as they arise. An exhibition is like a theatre without actors, with the plot and performance of the piece lying in the scenic arrangement itself.

Traditional natural science museums have evolved to the point where they currently try to incorporate a didactic discourse (both precise and comprehensible) in their exhibitions. The objects on display (species and models) are not only offered as *objets d'art*, but form part of a more elaborate message; consequently the need arises to evaluate the exhibitions. It is necessary to know to what extent the intended message of the exhibition's initial concept has been communicated. Therefore, there are two clearly defined fields of research in natural science museums, which correspond to the two functional environments of the present day museum: 1) The museum's scientific research, which corresponds to its traditional functions (collecting, preserving and researching), which contribute to the documentary base and scientific updating which are indispensable to the museum; 2) Research into the "educational objective", which includes the whole body of public exhibitions and activities that stem from them, their collections, subject matter and the research activities of the museum departments. It corresponds to the function of the public projection of the museum (scientific communication) and contributes the empirical base in the understanding between the museum and its public.

This study links up with a line of museographical research (evaluation of exhibitions in Natural History, Science and Technology museums and zoos), mostly developed in the United States and Canada, but also in the United Kingdom, with other connotations (LUCAS, 1983; SCREEVEN, 1984).

Research studies in museums are normally structured around a series of questions, and use similar methods and evaluation resources that basically follow two patterns: direct observation and analysis of the verbal or written replies of the visitors.

Visitor surveys set out to discover the kind of public, and their motivations, interests and preferences (ELLIOTT & LOOMIS, 1975). When direct observation is used, information is obtained concerning the amount of time spent in the different sections of the exhibition, the behaviour and different itineraries followed by different kinds of visitors (CONRAN, 1977; SCREEVEN, 1976; FALK, 1982; McMANNUS, 1987; PEARL & KOOL, 1988). This information, while not giving direct indications related to learning, characterises the conditions in which this may occur.

The most complex objective is the evaluation of learning from a specific exhibition. This involves a stringent experiment design and the development of accurate measuring methods. Many studies, above all in American centres, have been carried out to measure the effect of different experimental conditions on the learning and motivation of schoolchildren (STONEBERG, 1981; GENNARO, 1981; TANCK, 1982; KORAN *et al.*, 1983; LEHMAN & LEHMAN, 1984; FLEXER & BORUN, 1984; PRATS & FLOS, 1984), as well as on behaviour and learning linked to interaction units (EASON & LINN, 1976; ARTH & CLAREMAN, 1977) and a number of experimental studies on learning on field excursions comparing different environments (FALK *et al.*, 1978; FALK, 1983; PRATS, 1985).

This paper contains results obtained from a research project (PRATS, 1990) based on the "Ecology" exhibition prepared by Professor Ramon Margalef for the Barcelona Country Council (MARGALEF, 1985). In creating and developing the exhibition, the author used the language of display as a didactic discourse to explain certain features of his own speciality, ecology.

There are three qualities which make the author particularly suited to the development of this exhibition: 1) His extensive knowledge of the subject; 2) An ability to select a line of argument which is not excessively simple and which is carefully defined in its general content and in each of its parts; 3) The imagination to translate concepts into examples from daily life which are rarely associated with ecology.

The characteristics of the author and the effort of the designers to provide the correct physical presentation for the theoretical discourse together produced the ideal study material from the point of view of scientific communication.

Here we present the opinions of the exhibition given by the visitors and their relationship with the physical characteristics of the exhibition and time spent in the visit. We also present the results of the study on learning about two of the eight topics that were developed in the exhibition and which received the best and worst opinions from the visitors.

It is hoped that information will be obtained from these studies which will lead to an improvement in exhibitions, making them more attractive while at the same time faithfully communicating the message that the author of the exhibition wishes to transmit. The next step will be to make potentially unattractive contents enjoyable for the public, rather than to increase the number of well-accepted elements in order to obtain easy success. Natural Science exhibitions should attract a wide public who will come to know and share an essential aspect of the culture of our time and acquire an overall, up-to-date vision of the world we live in.

MATERIAL AND METHODS

THE EXHIBITION

The Ecology exhibition consists of an introduction and 92 elements divided into eight highly differentiated parts, to which we shall refer as the eight environments of the exhibition: 1. The blue planet; 2. Energy and production; 3. Richness and variety of life; 4. The selection game; 5.The language of nature; 6. The rhythm of life; 7. An old setting: the forest; 8. An important actor: man. This ordering suggests an itinerary, although each environment can be read independently on its own merits.

The exhibition was planned to be itinerant. For this reason its basic design centred around simple, mobile metallic structures that allowed for a spatial rearrangement depending on the characteristics and dimensions of each place where it was put on show (Figs. 1 and 2). The visual impact is obtained essentially through large, high-quality photographs and full colour drawings. There are also elements that can be manipulated (games, maquettes) and three-dimensional models.

The contents are slightly removed from what is commonly understood as "ecology", although they set out to reflect this science in the events and phenomena of everyday life. Nevertheless, the exhibition contains a lot of information, often on a highly scientific level. In the general text the author selected fragments that were coloured on the exhibition panels in order to enhance and emphasize certain concrete ideas. There is complementary material in the form of a triptych and a book-catalogue (MARGALEF, 1985). A thorough description of each environment's contents is given in PRATS *et al.* (1989) and PRATS (1990).

In order to objectify the characteristics and conditions of each environment, we have taken into account the following descriptors (Table I): the number and dimensions of the panels that constitute each unit or environment; the number and dimensions of the photographs and drawings of each panel; the words and lines of text of each element, whether panel or illustration; the words and lines of text

FIGURE 1. A view of Environment 4, "The selection game", at the Museu de Zoologia (Barcelona) venue.

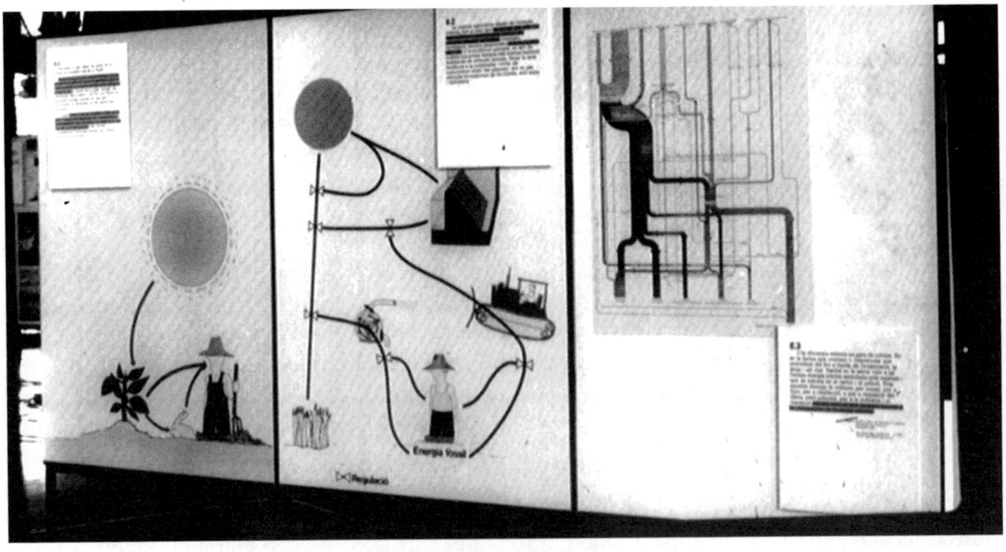

FIGURE 2. A view of Environment 8, "An important actor: man", at the same location.

TABLE I. Features of the exhibition environments. Panels, total number of panels forming each environment; Illus., number of photographs and drawings illustrating each part; Text, number of lines of text per total number of panels in each environment; Und., number of lines underlined in the total number of panels in each part; Mod., number of three-dimensional elements in the whole environment; Man., number of manipulable elements; Gam., number of games contained in each part of the exhibition; Vid., number of videos.

	Panels	Illus.	Text	Und.	Mod.	Man.	Gam.	Vid.
A1	10	37	58	16	-	-	-	1
A2	9	16	77	20	1	2	-	-
A3	7	12	69	13	-	1	-	-
A4	14	54	143	28	4	6	1	-
A5	7	15	70	15	2	1	-	-
A6	11	29	140	37	1	3	1	-
A7	9	41	105	15	3	2	1	-
A8	10	18	124	41	1	-	-	1

underlined (that is, a measure of the density of information) and the number of maquettes or models, manipulable elements, videos and games contained in each environment.

The exhibition travelled to eight different Catalan cities (Fig. 3). Although the specific characteristics of each place must be taken into account when some results are analysed (PRATS *et al.,* 1989), the results and conclusions presented in this paper are independent of the particular spatial layouts of the different cities.

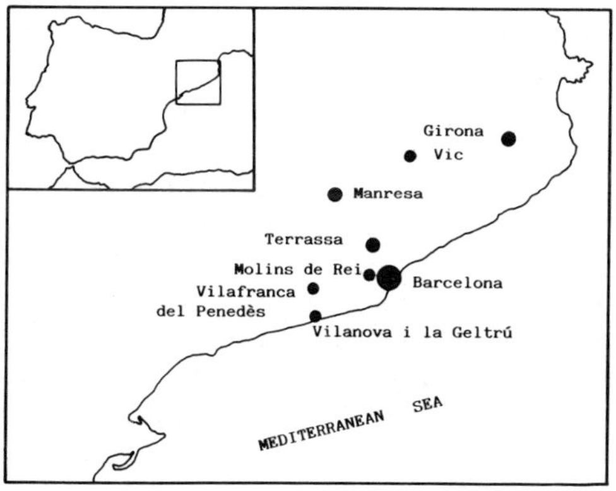

FIGURE 3. Situation of the eight Catalan cities to which the exhibition travelled. Number of inhabitants (1986 data) are given in brackets: Barcelona (1,701,812); Terrassa (156,458); Manresa (143,615); Vilanova i la Geltrú (45,039); Vic (28,399); Vilafranca del Penedès (26,433) and Molins de Rei (18,160).

OVERALL STUDY

QUESTIONNAIRES ON OPINION

Opinions were collected from a written questionnaire that visitors to the exhibition were invited to complete, and which consisted of both "open" and "closed" questions. Thus the results presented here correspond to the voluntary collaboration of a sector of the public that freely chose to register their opinions and suggestions. In the questionnaire, closed questions were used to obtain demographic information, while opinions and interests of visitors were obtained from the open questions, replies to which were subsequently analysed: What did you like/dislike/find surprising about the exhibition? Why?

Being conscious of the problems that can arise in the use of opinion study surveys (SCHUMAN & SCOTT, 1987), we decided not to use the direct interview method, thus avoiding lively and variable interference from the researcher, and prepared the questionnaire very carefully (PRATS *et al,* 1989; PRATS, 1990). The limited range of application of the questions, and the nuances of meaning obtained from the questions asking for reasons for the different opinions, facilitated the interpretation of replies. Furthermore, this opinion sounding project was complemented by an analysis of the itineraries followed by visitors to the exhibition, which was then related to the replies to the written questionnaire. Data concerning visitors' conduct and itineraries were obtained through direct observation of their movements and through timing them as they went from one environment to the next. These data were related to the demographic characteristics of the visitors observed. A very extensive experiment on learning was also conducted. We will consider the results of those items with relevance from the point of view of the opinions. The learning data were related to the opinions and behaviour data.

The written questionnaires have been published elsewhere (PRATS *et al.*, 1989). Here we use only some of the demographic data and the opinion replies to the question about like, dislike and surprise.

Having collected the questionnaire in all the cities to which the exhibition travelled, one of the most laborious tasks was to create a code that would allow the most objective possible transcription of the replies. This numerical transcription presented no problems in the case of the closed questions, in which the reply is chosen from a set of pre-established alternatives. On the other hand, the open questions, in which replies are developed freely, required a more elaborate process. The mathematical analysis of the codified replies was carried out by means of the SPSSx statistical package at the Calculation Centre of the University of Barcelona.

The code was very stringently applied. To start with, it was ratified and adjusted by four biologists who formed part of the study team, first individually and then together. Then, each questionnaire was corrected by two different people while a third (the same for all the questionnaires) revised and unified the definitive results of the whole survey. A total of 2,218 questionnaires were analysed.

OBSERVATION OF VISITORS' ROUTES

The routes followed by 296 individuals were observed and recorded by discreetly following them while they visited the exhibition. Their movements and their itineraries were recorded on a plan of the venue which featured a detailed layout of the Ecology exhibition.

Besides the time spent in each place, indicating what activity was being carried out, demographical information about the visitor was also given (sex and age group) as well as the social composition of the visit ("who they came with").

DESIGN OF LEARNING EXPERIMENT

The experiment was carried out with a total of 1,580 students in 161 groups from 28 different schools and colleges in Catalonia. Four different students levels were worked with, which were from the eighth year of EGB (13, 14 years old), the first year of BUP (14, 15), COU (18, 19) and teachers trainees (20 to 35 years old). None of the groups was specifically chosen for the experiment; once the teachers had requested a visit their collaboration was then asked for. The number of students in each level therefore represents the use made of the centre by each group.

The students participating in the exhibition were from particularly important levels within what is considered to be the most frequent route through formal education.

- The eight year of EGB is the last year of the second stage, in which all students study the same subjects. At the end of this year students have to make a choice between going on to study sciences or arts, or leaving to enrol in a course of professional training. The sample from this group included 220 students.

- The first year of BUP is the course which follows on from EGB, and represents, with age differences of between six months and a year, a group in their first year of specialisation in science. This level provided the most visitors to the exhibition and shows that the teachers regularly used it as a teaching aid. 945 students from this level took part in the experiment.

- Students of COU, from the science speciality, were the second largest group for whom the exhibition was appropriate to their study programme. Bearing in mind that this student population is much smaller and that there are fewer students in each centre, it is also a level which regularly used the exhibition as a complement to the natural science study programme. The sample included 238 students.

- The teacher trainees were also interesting to evaluate, as they had already decided to work in education, this field being one of the fundamental objectives of the exhibition. This group contained 177 students.

The main factor that we wanted to look at was the visit to the exhibition itself (a two-level factor), but we also wanted to take two other factors into account. The first was having prepared or studied something about ecology before attending the exhibition (two-level factor), and a second, which applied to the experimental groups who saw the exhibition, was receiving a brief introduction to museums before visiting the exhibition.

Samples were taken by making three random subgroups within each class group as they entered the museum. Each group was subjected to different experimental conditions (yellows, greens, reds, from Table II). The extent of their knowledge of ecology was measured by a questionnaire on concepts included in the exhibition. These questionnaires also contained the questions from the opinion survey addressed to voluntary visitors.

A pilot questionnaire was prepared and applied to six groups of between 35 and 40 students from four different schools at BUP level, and the answers used to make up the definitive questionnaire in three different formats. The pilot questionnaire enabled us to see whether the students had understood the questions, and from the content of the answers we could see if the questions were sufficient to reveal differences in the knowledge of students from the same level. The ease with which subjects understand the questions index (easibility index) is fundamental in evaluating knowledge (PRATS, 1985). Both those questions which nobody answers and those which everybody answers are useless in providing an evaluation of differences in individual knowledge.

TABLE II. Experiment on learning design scheme. See text.

Seeing the exhibition	Introduction to museums	Subject preparation NO	Subject preparation YES
NO		Yellows NP	Yellows P
YES	NO	Greens NP	Greens P
YES	YES	Reds NP	Reds P

By using the same questionnaire and the same marking scale for all levels of students, it was possible to contrast the answers within each level and observe the progression from one level to another (PRATS, 1989).

RESULTS AND DISCUSSION

IMPACT OF THE EXHIBITION

On the whole, the exhibition met with approval, since 89% said that they had enjoyed it and intended either to return or recommend it to friends. 40% of the visitors considered it suitable for everyone. The remainder of replies are divided up into a number of smaller percentages, with the exception of those who said that it was also ideal for students: 21%.

The data collected on opinion is presented in figures 4 and 5, in two sets: the whole voluntary population (eight cities taken together) and the students who participated in the learning experiment. This paper does not consider aspects outside the environments that the answers to the questionnaire also touched on (PRATS *et al.*, 1989). The most important tendencies of opinions of the eight items were similar for all the cities to which the exhibition travelled. There are two generalizable opinions, which referred to: "The selection game" (Environment 4) and "An important actor: man" (Environment 8).

"The selection game" (Environment 4) was widely liked, was found surprising to a lesser degree, and very few expressed dislike. Therefore, this exhibit was the most succesful, according to those who completed the questionnaire.

"An important actor: man" (Environment 8) was the opposite case, since no one mentioned having either liked or having been surprised by it; on the contrary, those that did mention it at all said that they had disliked it.

Considering other parts of the exhibition, "An old setting: the forest" (Environment 7) is the part of the exhibition most mentioned after Environment 4. Environments 2, 3, 5, and 6 were mentioned very little, while Environment 1 was mentioned with slight variations in reply, depending on the city.

REASONS FOR LIKE, SURPRISE AND DISLIKE

Each question concerning opinion was accompanied by another asking why the person had mentioned like, surprise or dislike regarding the exhibit. The scope of visitors was an interesting interpretation of the communication power of the exhibition as grasped by the public.

"The selection game" (Environment 4) can be considered, on the basis of the reasons given, an exhibit that was liked and which caused surprise because it is attractive and interesting, with a content that is easy to understand and that people did not expect to find. It also has an outstanding participative quality compared to the other aspects mentioned, which allowed the visitor to experiment. The participative element corresponds to the birdsong game, which is the part of the exhibition most mentioned in this environment.

"An old setting: the forest" (Environment 7) was also considered to be an attractive, well-presented part that, while not surprising, was liked by virtue of its didactic content. The most important reasons mentioned for this were that it is a subject people should know about and also that it is closely linked to nature and the present day. This way of showing nature was the aspect people found most surprising about this particular environment. On the other hand, no one mentioned its participative possibilities, even though objectively it was the section that contained most, with a game and an apparatus that could be manipulated to show slides of animals. It was this apparatus, however, that people mentioned as having liked most about this particular environment.

"An important actor: man" (Environment 8) was the only part of the exhibition which produced only unfavourable comments. Criticism was levelled above all at the presentation, which was found to be unattractive and boring, and at the subject itself or some of the aspects dealt with. It was specified that there was no mention of the deterioration of the ecosystem and there were even those who considered the subject as having nothing to do with ecology. Another important reason why it was disliked was that people found it difficult to understand and unclearly expressed.

We think that the author himself expressed a simpler message in Environments 4 and 7 than in Environment 8. "The selection game" and "An important actor: man" included some ideas that were new for many people, but the former Environment was better put together, with very adequate examples.

ENVIRONMENTS, IMPACT AND DESCRIPTORS

This section relates the formal characteristics of the exhibition's eight parts or environments (dimensions, texts, illustrations, etc.) to the opinions received. These have been transformed into probabilities: 1) PO: probability of giving an opinion on a given environment; 2) PL: probability of liking conditioned to giving

TABLE III. Descriptors of the environments that enter as independent variables in a step by step multiple regression (significance $p < 0.05$), where the dependent variable is that indicated on top: PO, probability of giving an opinion; PL, probablity of liking provided an opinion is given; PD, probability of disliking provided an opinion is given. The descriptors used are given in Table I. The sign + or - preceding the name of the descriptor means a positive or negative effect. A dashed line means that no variable was significant.

	Variables					
	PO		PL		PD	
total population	+SIZE	-UNDERL	+PHOTOS	+MANIP	-UNDERL	-MANIP
AGE 1	+ SIZE		+PHOTOS	+MANIP	-MANIP	-PHOTOS
AGE 2	+SIZE	-UNDERL	+GAMES		-MODELS	-MANIP
AGE 3	+PHOTOS		----------------	---------------	-MODELS	+TEXT
AGE 4	-UNDERL		----------------	---------------	----------------	-------------

an opinion; 3) PD: probability of disliking conditioned to giving an opinion.

To study the dependence between the probabilities of giving an opinion and the physical characteristics of the parts, a step by step multiple regression was carried out in which the independent variables were those of Table I, and the dependent variables the different probabilities.

Table III shows, by a sign followed by the name of the variable, those variables that enter in the regression, being introduced when they have a significance above 0.05 (standard method). It can be seen that the probability of giving an opinion is directly proportional to the size and inversely proportional to the density of information. The probability of giving a "like" opinion (having given an opinion), is proportional to the quantity of illustrations, manipulable elements and games for the two youngest groups, while for the other groups there is no significant dependence. For the probability of giving a negative opinion, and taking the respondents as a whole, the only two variables that affect the regression are the density of information and the manipulable elements, but if it is done according to age groups, the probability of giving a "dislike" opinion is inversely proportional to the density of illustrations, manipulable elements and models, and directly proportional to the quantity of text.

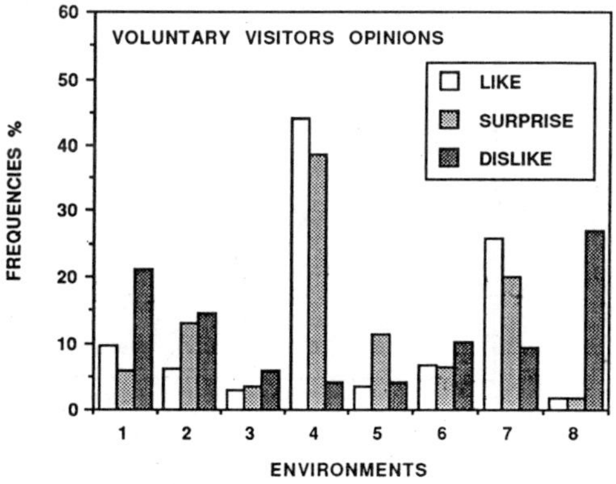

FIGURE 4. Frequencies of replies (like, surprise, dislike) of voluntary visitors for the eight environments of the "Ecology" exhibition. Percentages were calculated for each kind of reply, through the eight environments, for the whole population from the eight cities where the exhibition travelled.

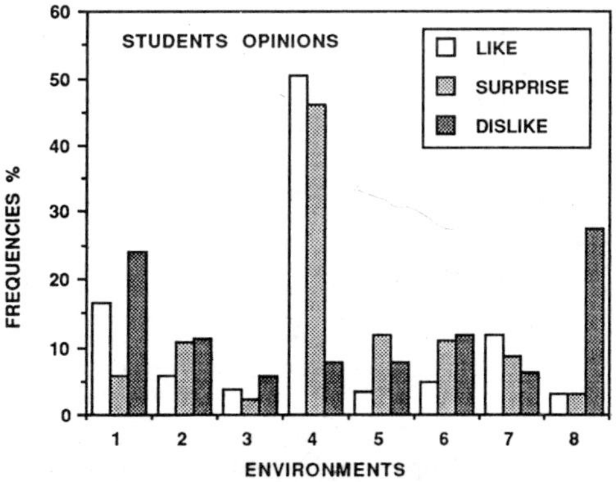

FIGURE 5. Frequencies of replies (like, surprise, dislike) of students participating in the learning experiment (the experiment was carried out in Barcelona).

We also calculated the partial correlations between the probabilities and the descriptive variables, controlled by the size of the environment and the number of photos; the importance of the density of information (underlined text), of manipulable elements and of the three dimensional models when qualifying an opinion is confirmed. It is interesting to note that the older visitors did not qualify their opinions in relation to the descriptors analysed and that the probability of giving an opinion is closely associated (inverse relationship) with the density of information. This is probably due to the older visitors being more accustomed to giving positive opinions, or rather, they tend to give an opinion only if they do not have to give the "dislike" opinion.

Another interesting observation is that the three dimensional models are more important for adults or older students (Classes 2 and 3) while the opinions of the youngest students were more closely tied to the presence of manipulable elements, games and illustrations.

There is, therefore, no such thing as an ideal exhibit, given that this prototype will depend on the content, will be different according to each exhibition and will probably also depend on the characteristics of each museum and even of each visitor. It is clear that the most attractive exhibits are those with a short and clear message, vividly displayed (ALT & SHAW, 1984).

TIME SPENT ON THE VISIT AND OPINION

A museum visit normally lasts no longer than two hours (FALK, 1982). No more than thirty minutes of this time is dedicated to the exhibition itself, with the rest of the time spent walking, in the shop, the cafeteria or the toilet. As regards observation of each exhibit, the maximum average time spent is thirty to forty seconds. The visitors that we observed in this study dedicated between 24 and 31.5 minutes to the exhibition, distributed in an inverse order to their respective ages. The average time spent on each particular exhibit was between 10 and 30 seconds.

One would expect that the probability of an opinion being expressed about an environment to be in direct proportion to the time spent looking at it. It is to be expected too that if visitors spend a long time in front of a panel or in an environment, it is because they like it. It can be considered that "like" and "surprise"

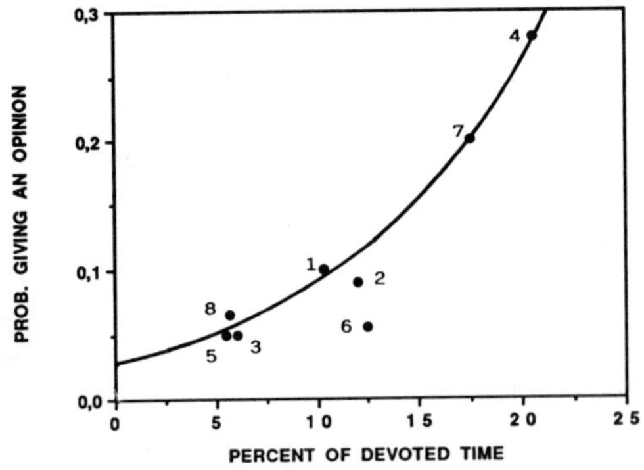

FIGURE 6. Dispersion diagram for the probability of giving an opinion versus the percent of time devoted to this environment. An exponential curve was fitted to the data (R^2=0.97) to show the general trend without including Environment 6 (see text for an explanation).

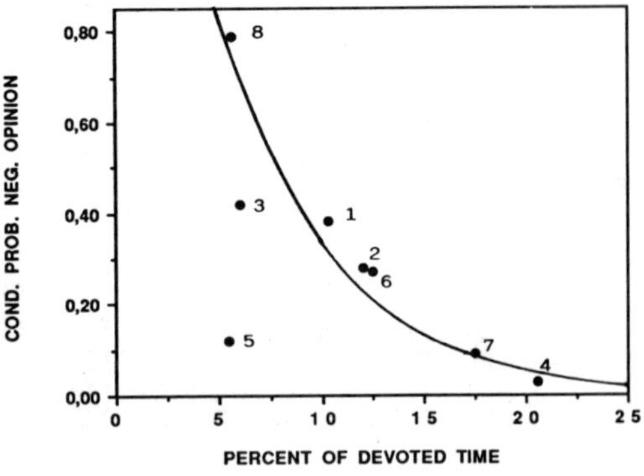

FIGURE 7. Dispersion diagram for the probability of expressing a negative opinion on an environment, conditionned to having expressed an opinion on that environment, versus the percent of time devoted to it. An exponential curve was fitted to the data (R^2=0.92) to show the general trend without including Environment 5 (see text for an explanation).

opinions are positive, while "dislike" is clearly negative. An analysis of the probability of negative opinions being expressed about an environment in relation to the time spent in front of it is the equivalent of an analysis of positive opinions (like and surprise) taken together. As regards the time spent in seeing the exhibition, we considered that to make comparable the different itineraries and behaviour patterns of the different classes of age and sex, as well as to understand that subjective time is what is in fact related to the kind of opinion, we have used the percentage of the total time of the itinerary devoted to each environment as our reference variable.

We have calculated for each of the environments (1 to 8) the probability of an opinion being expressed and the probability that such an opinion, if expressed, will be negative (figures 6 and 7). The data clearly show that there is a direct link between the percentage of time devoted and the expressing of an opinion. In figure 4 we have drawn an exponential curve that was fitted to the data after taking out Environment 6, as it was seen that it may be considered an exception that needs to be examined. This section was very attractive since it contained a game called the "tree-trunk game", a magnifying glass through which to see growth rings in the section of a trunk and also a number of metrometers associated with the rhythms of different animals. Although considerable time was devoted to this section, since otherwise it could not be properly exploited, people did not express opinions in the proportion to which one might expect on the basis of the time devoted to it. On the other hand, if the visitor expressed an opinion, this was negative in a proportion similar to what one would expect from the time spent (Fig. 7).

As regards negative opinions, the probability of an opinion expressed being negative falls as the time devoted to an environment increases, following the same exponential trend (but with a negative slope). The exception to the general tendency is Environment 5 (which was excluded from data used to fit the curve in figure 7). This section contained a water wheel made of Meccano, but it was a very small environment. People in general did not spend much time there, but it was unnecessary to do so in order to see it and the probability of opinions being expressed about this environment follows the overall tendency. The environment was liked, even though it did not attract much attention.

We can conclude that the relative time

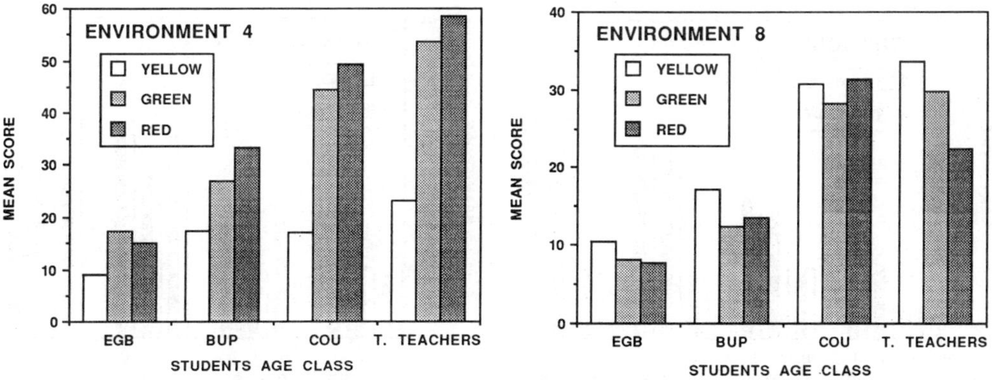

FIGURE 8. Distribution of the average scores of knowledge of Environment 4 (left) and Environment 8 (right), for each level of students and experimental subgroup (Yellows answered the questions before seeing the exhibition; Greens after seeing the exhibition and Reds received an introduction to museums, saw the exhibition and took the test).

Table IV. Number of individuals (n), mean score (m) and its standard deviation (st.dev.) obtained for the different levels of students (EGB, BUP, COU and Teachers trainees; see text) and experimental conditions, on Environments 4 and 8. Yellows answered the questions before seeing the exhibition; Greens after seeing the exhibition; Reds received an introduction to museums, saw the exhibition and took the test.

		ENVIRONMENT 4				ENVIRONMENT 8			
		EGB	BUP	COU	Teach. T.	EGB	BUP	COU	Teach. T.
	n	59	101	53	43	59	101	53	43
Yellows	m	9	17.4	17.2	23.3	10.4	17.2	30.8	33.6
	st dev.	13.5	13.8	15.6	18	13	14.5	14.9	18.4
	n	54	90	51	31	54	90	51	31
Greens	m	17.5	27.1	44.3	53.5	8.2	12.3	28.2	29.7
	st dev.	17.9	18.7	21	22.6	12.4	11.1	22	17.7
	n	49	99	56	42	49	99	56	42
Reds	m	15.1	33.3	49.3	58.6	7.8	13.6	31.3	22.4
	st dev.	16.1	20.5	25.3	24.8	11.3	16.4	24.8	18.2
Global Mean score		13.9	25.9	36.9	45.1	8.8	14.4	30.1	28.6

devoted to an environment is a clear indication that it was liked only when the proportion is higher than about 13% of the total time spent at the exhibition. This level of 13% is probably not independent of the size of the exhibition as we can note that a uniform distribution of time among the eight environments would give a 12.5 % of time devoted to each one of them.

Some subjective relationship must exist between the time devoted and the benefit gained (psychological, cultural, enjoy, etc.). This balance, which is difficult to set up *a priori,* is what we should attempt to achieve in exhibitions.

LEARNING OF EXPERIMENT GROUPS

We decided to study the acquisition of knowledge, or the amount learnt from the exhibition, using those two items for which the opinion and observation have produced opposite results: better for Environment 4 and worse for Environment 8, and with the same trends shown by all the visitors and students subject to the experiment.

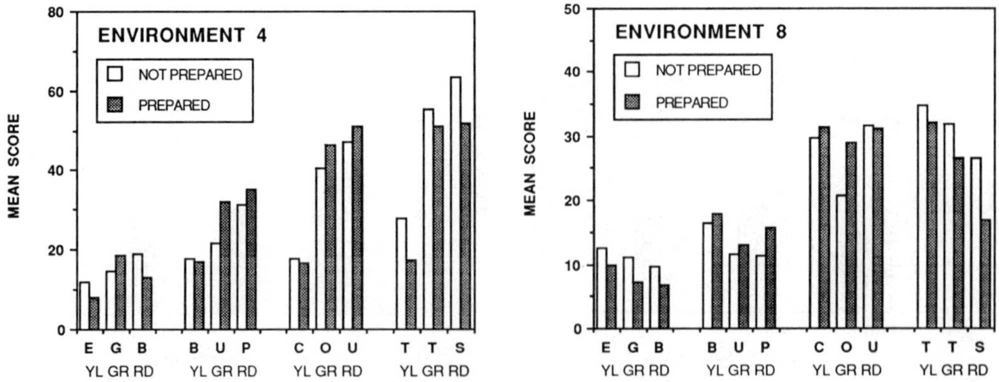

FIGURE 9. Distribution of the average scores of knowledge of Environments 4 (left) and 8 (right), for each level of students and experimental subgroup, for those who had prepared or not something on ecology at the school before the visit to the exhibition (Prepared and not prepared). At the bottom, YL, yellows; GR, Greens; RD, Reds (see also figure 6 and text).

TABLE V. Number of individuals (n), mean score (m) and its standard deviation (st.dev.) obtained for the different levels of students (EGB, BUP, COU and Teachers trainees; see text) and experimental conditions on Environments 4 and 8. See Table IV. Prep.= groups of students who had prepared something on ecology at school before the visit; Not Pr.= students who had not prepared anything on ecology before the visit.

n
mean
st.dev.

		ENVIRONMENT 4				ENVIRONMENT 8			
		EGB	BUP	COU	Teach.T .	EGB	BUP	COU	Teach.T.
		14	47	19	24	14	47	19	24
	Not Pr.	11.8	17.8	17.9	27.9	12.5	16.4	29.7	34.8
		13	14.1	18.7	20.2	12.8	11.3	15	17.7
Yellows									
		45	54	34	19	45	54	34	19
	Prep.	8.1	17	16.8	17.4	9.8	17.9	31.5	32.1
		13.7	13.6	14	12.9	13.2	16.8	15.1	19.5
		13	43	17	18	13	43	17	18
	Not Pr.	14.6	21.7	40.6	55.3	11.2	11.5	20.8	31.9
		15.6	16.2	19.9	22.5	16	10.7	23	13.4
Greens									
		41	47	34	13	41	47	34	13
	Prep.	18.4	32	46.2	51.2	7.3	13.1	29	26.5
		18.6	19.6	21.6	23.6	11.1	11.6	21.8	22.6
		16	47	23	24	16	47	23	24
	Not Pr.	18.8	31.3	47	63.5	9.7	11.3	31.7	26.5
		17.7	20.2	21.8	27.8	9.7	14.6	21.6	18.7
Reds									
		33	52	33	18	33	52	33	18
	Prep.	13.3	35.1	50.9	51.9	6.8	15.8	31.1	16.9
		15.2	20.8	27.7	19	12	17.7	27.2	16.3
Global Mean									
	Not Pr.	15.1	23.6	35.2	48.9	11.1	13.1	27.4	31.1
	Prep.	13.3	28	38	40.2	8	15.6	30.5	25.2

Figures 8 and 9 show the distribution of the average scores of knowledge of Environments 4 and 8 respectively, for each level of students and experimental subgroup (Yellows answered the questions before seeing the exhibition; Greens after seeing the exhibition and Reds received an introduction to museums, saw the exhibition and took a test). Tables IV and V show means and standard deviations of each group.

SIGNIFICANCE OF THE KNOWLEDGE SCORES

An analysis of the variance (ANOVA) was carried out for each of the knowledge measurements in order to assess the significance of the mean differences observed among the different experimental conditions. The tables of the ANOVAs are not shown, but the significant differences will be commented upon, along with those cases in which interactions amongst factors were also significant. Although we have taken a probability of 0.05 as a norm for the significance limit of the Statistic F, in most cases in which the differences are indicated as significant, the probability of the Statistic F are much less. In any case, there is no important conclusion based on differences that have dubious significance levels.

In the first ANOVA two factors were

considered: a) seeing the exhibition and b) preparing something on ecology at school. The second ANOVA performed was for those who had seen the exhibition and considers the effect of receiving an introduction to museums.

The trends for the scores for each item and level are shown in Table VI, with an asterisk indicating those which have a probability for F of less than 0.05. The signs + and - indicate that the score increases or decreases.

In EGB the only significant tendency is having seen Environment 4 of the exhibition. Those students who saw this area received higher marks than those who answered the questionnaire before seeing the exhibition.

BUP appeared to be the group which was most receptive to all the factors analysed. For Environment 4 it is significantly positive as those who have seen the exhibition, those who have done some preparation, and within the group that has seen the exhibition, those who received the introduction, receive higher knowledge scores. As regards Environment 8 they are also receptive to the exhibition, but in a negative sense. There are higher marks for before the exhibition than after. To interpret the diminishing knowledge scores, it is necessary to bear in mind that students of BUP respond less to the part of the questionnaire referring to Environment 8 after seeing the exhibition. It may be that the answers have not worsened with respect to those who have not seen the exhibition, but that they have been more selective in answering and have chosen questions that do not correspond to the subject matter of Environment 8.

As regards COU, only the improvement of scores for Environment 4 after seeing the exhibition is significant.

The teacher training students, as with all other levels evaluated, answered the item 4 questions significantly better after having seen the exhibition. Unlike the other levels, previous preparation has a negative effect for both Environments 4 and 8. Those students who have received some ecology preparation therefore receive lower knowledge scores than those who have done nothing on the subject. We also checked that it did not depend on the questions which were left unanswered. As regards seeing the exhibition, the results of knowledge of Environment 8 also had a negative effect and corresponded to a worsening of the answer.

TABLE VI. Asterisks indicate scores which have a probability for F of less than 0.05. The signs + and - indicate that the score increases or decreases.

Students level	EGB		BUP		COU		Teacher trainees	
Environment	A4	A8	A4	A8	A4	A8	A4	A8
Seeing the exhibition	*+	-	*+	*-	*+	-	*+	*-
Preparation	-	-	*+	+	+	+	*-	-
Introduction	-	-	*+	+	+	+	+	-

It seems that classroom preparation did not mesh with the concepts evaluated. Another factor to consider in these results is the teacher's awareness of being examined, which affected more those groups which had prepared than those which had not, with the latter group being more spontaneous and disposed to answer a subject from their own personal experiences. These results, which may seem surprising, were subsequently and meticulously contrasted using ANOVAs, for possible factors that were not considered that may have accidentally had a significant influence on the composition of the students of each experimental subgroup. It can be stated that these factors had no influence on the results provided by the students.

CONCLUSIONS

In overall terms, the exhibition on "Ecology" worked well, since 89% of the people who completed the questionnaire said they had liked it and intended either to come back or recommend it to others.

Environment 4, "The selection game", was the most successful part of the exhibition because people found it very attractive and interesting. It caused surprise because its content was easily understood, something that people had not expected. It also has a participative quality which allowed the visitor to experiment.

Environment 8, "An important actor: man" was not liked by visitors. Criticism was mainly directed at the presentation, which was found to be unattractive and boring. Its conceptual content did not match the expectations, and people found it difficult to understand, and poorly expressed.

The opinion of volunteers is the same as the opinion expressed by students for almost all the aspects studied.

It has been shown that the probability of an opinion being expressed about an environment in the exhibition is proportional to the time devoted to that environment. Similarly it has been observed that the opinion is more likely to be positive the more time is spent in the environment in question.

The formal characteristics of the presentation are related to the impact. For the whole population, the probability of giving an opinion on an Environment is directly related to its size and inversely related to the density of information (proportion of underlined text). The opinion of surprise is not related to any of the studied descriptors.

Public opinion seems to be shaped by series of factors and descriptors that act together, which in the case of the exhibition studied are the following: a) the complexity of the subject matter; b) previous knowledge and the interest of the public in the subject; c) the coherence between title and content from the point of view of public expectations; d) success in the quality of design and coherence amongst the parts and the overall composition; e) use of appropriate examples linked to everyday life.

There is a direct relationship between the opposite impact for Environments 4 and 8 and the knowledge remembered for each one of them. All groups of students (EGB, BUP, COU, Training teachers) obtained better marks for Environment 4 concepts after seeing the exhibition (the part most liked). None of the four groups of students improved their qualification for Environment 8 questions after visiting the exhibition. This was the part with the most negative impact. Furthermore, teacher trainees achieved worse results for knowledge of Environment 8, after seeing the exhibition.

Previous knowledge of "ecology" obtained at school affected (differently) the level of knowledge evaluated after seeing the exhibition. Differences are found among levels. Those students of BUP with a previous preparation on ecology achieved significantly better results than those who had not prepared it, while the teacher trainees who had not prepared ecology in class, obtained better marks after the exhibition than those who had been prepared.

A short explanation on how to look at an exhibition, given before the visit, had a positive and significant effect on the marks obtained by the students of BUP for Environment 4. Students of BUP seem those most receptive to all the factors considered.

Students using the exhibition as a complement to their formal education only benefited if they liked the exhibition (in this case studied environment by environment). The amount of conceptual information is critical. It seems that the proportion or density of written information must be the least possible and that the overall density of conceptual content of each environment or exhibit must be low. For practical purposes one sees that if a topic is complex and involves many new concepts, it has to be split into several environments or exhibits and has to be distributed in a large enough space in order to obtain a decreased density of information. This should be done with the subjects included in Environment 8 when possible in any future exhibition. Many of the ideas included in "An important actor: man", such as "external or exosomatic energy", did not reach the public, although we know the great importance the author gives to them (MARGALEF, 1985b).

Moreover, if the presentation and content do not match the expectations (the topic as it was learned formally at school) then people do not only dislike the exhibition, but they do not learn anything either. This result is in itself a major criticism on some widespread methods in formal learning (or teaching), where students are prepared against anything new or different from what is said in class or in textbooks. Students feel self-confident when they are aware of their ability to match the teacher's preferences or opinions (those of textbooks or school) but they are unable to express in an examination something that is not what they have been taught, even if it is not necessarily contradictory. The question is most alarming when we think that the students presenting this response to the experiment on learning about an exhibition are those who are being prepared to follow a career as teachers.

ACKNOWLEDGEMENTS

This study was conducted while one of the authors (Prats) was directing the Department of Education of the Museu de Zoologia in Barcelona. Thanks are due to all the people that participated in the acquisition and manipulation of rough data. We are very grateful to Arthur Lucas, Professor of the King's College (London), for his helpful assessment. The study received grants from the Generalitat de Catalunya and from the Fundació de la Caixa de Barcelona.

REFERENCES

ALT, M.B. & SHAW, K.M. 1984 Characteristics of ideal museum exhibits *British J. of Psychology*, 75:25-36.

ARTH, M. & CLAREMAN, L. 1977. The discovery Room *Curator*, 20(3): 169-180.

CONRAN, R. J. 1977. An investigation into the response of primary school children to the British Museum (Natural History). MSc (Education) 305 pp. University of Keele. Staffordshire.

EASON, P. & LINN, M.C.,1976. Evaluation of the effectiveness of participatory exhibits. *Curator*, 19(1): 45-62.

ELLIOTT, P. & LOOMIS, R.J. 1975. *Studies of visitor behavior in museums and exhibitions. An annotated bibliography of sources primarily in the English language*. Smithsonian Institution, Office of Museum Programs, Washington, D.C.

FALK, J. H. 1982. The use of time as a measure of visitors' behavior and exhibit effectiveness. Museum Roundtable Reports. *J. Museum Education*, 7(4): 22-28.

FALK, J. H. 1983. Field trips: A look at environmental effects on learning. *J. Biological Education*, 17(2): 137-142.

FALK, J.H., MARTIN, W.W., & BALLING, J.D. 1978. The novel field-trip phenomenon: Adjustment to novel settings interferes with task learning. *J. Research in Science Teaching*, 15(2) : 127-134.

FLEXER, B.K. & BORUN, M. 1984. The impact of a class visit to a participatory science museum exhibit and a classroom science lesson. *J. Research in Science Teaching*, 21(9): 863- 873.

GENNARO, E.D. 1981. The effectiveness of using pre-visit instructional materials on learning for a museum field trip experience. *J. Research in Science Teaching*, 18(4): 275- 279.

KORAN, J.J., LEHMAN, J. R., SHAFER, L.D. & KORAN, M.L. 1983. The relative effects of pre- and post-attention directing devices on learning from a "walk-through" museum exhibit. *J. Research in Science Teaching*, 20(4): 341-346.

LEHMAN, J. R. & LEHMAN, K.M. 1984. The relative effects of experimenter and subject generated questions on learning from museum case exhibits. *J. Research in Science Teaching*, 21(9) : 931-935.

LUCAS, A. M. 1983. Scientific Literacy and Informal Learning. *Studies in Science Education*, 10:1-36.

MCMANNUS, P.M. 1987. *Communication With and Between Visitors to a Science Museum*. PhD thesis. University of London.

MARGALEF, R. 1985a. *L'Ecologia*. Diputació de Barcelona. Barcelona

MARGALEF, R. 1985b Simple facts about life and the environment not to forget in preparing school books for our grandchildren. In: *Trends in ecological research for the 1980's* (Cooley, J.H. & F.B. Gooley, eds.): 299-320. Plenum Press. New York.

PEART, B. & KOOL, R. 1988. Analysis of a Natural History Exhibit: Are dioramas the answer?. *Int. J. Museum Management and Curatorship*, 7, 117-128.

PRATS, C. 1985. *L'ús d'exposicions d'un Museu de Zoologia com a complement escolar per l'ensenyament de les ciències biològiques: estudi experimental.* Master thesis. University of Barcelona.

PRATS, C. 1989. *Anàlisi de l'impacte d'una exposició d'ecologia i estudi experimental del seu efecte en l'adquisició de coneixements.* Ph. D. Thesis. University of Barcelona.

PRATS, C. & FLOS, J. 1984. La visita al Museu de Zoologia com a eina didàctica: estudi experimental. *Miscel.lània Zoològica,* 8: 273-285.

PRATS, C., FLOS, J., PIQUÉ, J., CARRAU, M.J. i DURO, A. 1989. *Analysis of an itinerant exhibition, Ecology: type of public, opinions and behaviour evaluated in eight towns and cities in Catalonia.* Estudis i Recerques, 3. Ajuntament de Barcelona.

SCHUMAN, H. & SCOTT, J. 1987 Problems in the use of survey questions to measure public opinion. *Science*, 236: 957-959.

SCREVEN, C. G. 1976. Exhibit Evaluation - A Goal-referenced Approach. *Curator* 19: 271-290.

SCREVEN, C. G. 1984. Educational Evaluation and Research in Museums and Public Exhibits: A Bibliography. *Curator*, 27/2: 147-165.

STONEBERG, S.A. 1981. *The effects of pre-visit, on site, and post-visit zoo activities upon the cognitive achievement and attitudes of sixth grade pupils.* Ph.D. Thesis. Univ. of Minnesota, Minneapolis.

TANCK, S. 1982. *The effect of pre-visit preparation on children's learning during an arboretum field trip.* Master Thesis. Univ. of Minnesota. Minneapolis.

RAMON MARGALEF,

LIMNOLOGIST, MARINE BIOLOGIST, ECOLOGIST, NATURALIST

Professor Ramon Margalef in 1990

Homage to Ramon Margalef; or, Why there is such pleasure in studying nature 413
(J. D. ROS & N. PRAT, eds.). 1991. *Oecologia aquatica*, 10: 413-423

RAMON MARGALEF, LIMNOLOGIST, MARINE BIOLOGIST, ECOLOGIST, NATURALIST

JOANDOMÈNEC ROS

Departament d'Ecologia. Facultat de Biologia. Universitat de Barcelona. Av. Diagonal, 645. 08028 Barcelona. Spain.

Received: December 1991

Professor Ramon Margalef has pioneered ecological research. Since the early forties, he has worked in nearly all aspects of aquatic ecology - limnology and oceanography - and has put together a very thorough, comprehensive theory of ecology, based mainly on his own work but universal enough to be widely accepted and praised as the main corpus of theory of a scientific discipline still in its early years.

His scientific ideas have had an influence on more than a generation of ecologists, through his university lectures and his papers and books, which number nearly 400 (see Appendix I). Of these, *Perspectives in ecological theory* (1968) and *La Biosfera, entre la termodinámica y el juego* (1980), represent what is generally considered to be one of the main efforts to establish ecological knowledge on a proper footing. His books *Ecología* (1974) and *Limnología* (1983) are considered by the specialists to be among the best text books ever written on these topics, and have had great influence worldwide, especially in Spanish-speaking countries.

Ramon Margalef was born in Barcelona in 1919, where he began his studies in business school, but soon developed an interest in natural history and biology. His education was interrupted by military service during the Spanish Civil War. At the end of his military duties (in 1943) he took a job with an insurance company, but continued his scientific interests at the Institut Botànic de Barcelona. He built his own microscope with spare parts bought in second-hand markets and began an intensive study of freshwater algae, while he read all the books he could get his hands on on biology, physics and other subjects to which, had he followed a stricter syllabus, he may not have had access.

At the same time as acquiring a wide self-education, Margalef published the results of his first research. His studies attracted the attention of university officials, who offered him a fellowship; he could thus follow a formal university curriculum. By 1951 he had received his doctorate (on "Temperature and morphology of living beings") and became a member of the Institut d'Investigacions Pesqueres (of the Spanish National Research Council, CSIC), which he would later direct and convert into an oceanographic centre of international repute.

At the beginning of the sixties, he formalized what would be the seeds of his text books on Ecology (*Comunidades naturales* dates from 1962) and began to give informal classes on Ecology. Later he would become the first Professor of Ecology in Spain (1967). Margalef has done all his scientific work in this country, where as a senior professor and a senior researcher he has had a hand in training a very numerous group of limnologists, oceanographers, ecologists and marine biologists (see in Appendix II the relation of the Ph. D. Theses he has directed). In turn, these people are now spreading their scientific message in ecology departments and research institutions in Spain and abroad. But Margalef's influence stretches to many countries around the world where he has worked (mainly in North and South America and Europe) and where he is currently invited to give lectures and courses and to participate in research programmes or scientific conventions. In turn, Margalef has entertained guest scientists from all over the world in his laboratory in Barcelona.

Professor Margalef contributed greatly to the international recognition of his Department of Ecology (University of Barcelona), and previously of the Institut d'Investigacions Pesqueres (now Institut de Ciències del Mar), which he directed in the sixties. He achieved this in part through the scientific journals of which he was editor: *Publicaciones del Instituto de Biología Aplicada* (now discontinued), *Investigación Pesquera* (now *Scientia marina*), and *Oecologia aquatica*, of which this is the tenth issue. For many years, under his direction, *Investigación Pesquera* was one of Spain's most internationally influential scientific journals.

Ramon Margalef is an oft-quoted authority in all fields of ecology and aquatic sciences, and the most quoted of all Spanish scientists (1978-82). *Perspectives in ecological theory* and the articles "On certain unifying principles in ecology" (*Amer. Natur.*, 1963) and "Life-forms of phytoplankton as survival alternatives in an unstable environment" (*Oceanol. Acta*, 1978) have been quotation classics. He has been awarded honorary doctorates by several universities and has received important scientific awards, among them the Prince Albert Medal (from the Oceanographic Institute of Paris), the Huntsman Award (considered as the equivalent of a Nobel Prize in Oceanography), the Naumann-Thienemann Medal (equivalent of a Nobel Prize in Limnology), the Ramón y Cajal Award (the highest honour for a Spanish scientist) and the Narcís Monturiol Award (the highest honour for a Catalan scientist).

This comes as no surprise to the scientific community, Margalef and the men in whose honour these prizes were established being (paraphrasing the words of G. E. Hutchinson, a distinguished ecologist and limnologist himself) "*facile principes* in their chosen fields". Professor Margalef reached retirement age in 1986, and was later appointed Emeritus Professor at the University of Barcelona.

The main fields of Professor Margalef's research and study can be listed as follows: Limnology, Biological Oceanography, General and Theoretical Ecology. He has also made contributions to Human Ecology and Theoretical Biology in general. In what follows the reader can find a summary of his scientific work. A list of his publications is given in Appendix I, and the doctoral theses he has directed are listed in Appendix II.

THE LIMNOLOGIST

Although Ramon Margalef owes his international reputation to his research on oceanography and to his development of a coherent theory of ecology, it is in the study of continental or fresh waters where we find his most extensive, most enduring and, at the same time, his most innovative work. Indeed, his first scientific papers, written nearly fifty years ago, dealt with the biology and ecology of both animal and plant aquatic organisms.

This is a field which Margalef truly pioneered in Spain. Beginning in the early forties, without academic or institutional support, he launched a one-man, multi-year study of virtually all the land waters of Spain. The extent of this was such that, out of all researchers on freshwater communities, it could be said that he is the one who has done the most work, covered the most fields, trained the most limnologists and, above all, in relation to ecology, the one who has formalised the immense range of his work into a coherent whole, creating the first corpus of limnology in his country and abroad. This work in limnology earned him the recognition of the Societatis Internationalis Limnologiae (SIL), which in 1989 awarded Margalef the Naumann-Thienemann Medal, *De limnologia optime merito*. The highest SIL award was conferred to Professor Margalef "for sharing his creative gifts of insight, intuition and synthesis of the ecological foundation of limnological phenomena and for his influence on limnology in the Spanish-speaking world". As proof of this international recognition he has been asked to organise the XXV SIL Congress, to be held in Barcelona in 1992.

River and lake phytoplankton algae constitute the first block of Margalef's limnological works. These studies are at the heart of his later dedication to marine phytoplankton (see below), and form the basis for his work on identifying ecological factors common to both types of aquatic communities. The task of identifying and drawing up an inventory of these organisms under the microscope involved enormous effort, kept up over many years; indeed, it still occupies Margalef, the limnologist, even today (especially the phytoplankton of some one hundred reservoirs all over Spain). This study forms part of a very ambitious piece of research, which will be further examined below.

The study of freshwater animals has been a similarly fertile area of research. The study of the biology and ecology of crustaceans, among other aquatic animals, enabled Margalef to produce a monograph as early as 1953 on *Los crustáceos de las aguas continentales ibéricas*, which continues to be used as a standard reference work in Spain.

But Margalef's greatest claim to fame is his global study of the environmental factors and the organisms of nearly all the freshwater and salty water bodies in the Iberian Peninsula and Balearic Islands, and certain places in North and South America, and Europe, although he concentrated on those of his own country. This integral study of lakes and water courses very soon enabled Margalef to write a monograph ("Limnosociología", 1947) which put Spanish limnology on a par with European limnology as far as basic knowledge is concerned. *Los organismos indicadores en la limnología* (1955) is a complement to his previously quoted treatise on Iberian crustaceans and, later on, various general works lay down the main lines of a regional form of limnology in this peninsula.

Within this field, special mention should be made of the ambitious Spanish reservoirs programme, which began twenty years ago through an agreement with the Spanish Ministry of Public Works. Since then, the Department of Ecology of the University of Barcelona has had a limnological team working on the physical, chemical, biological and ecological features of over a hundred reservoirs throughout the Iberian Peninsula (*Limnología de los embalses españoles*, 1976). Besides the numerous articles published on the results of this study (perhaps the most complete study of this type carried out anywhere in the world up to now), one should highlight the number of specialists (from Barcelona University and from other research centres in Spain) who have come together in a single limnological team that is both coherent and multidisciplinary. It is, above all, the first time that in-depth knowledge of freshwater ecology in Spain has been obtained. Besides directing and promoting this study, Margalef has published many papers, either alone or in collaboration with the above team. The study of Spanish reservoirs is currently being repeated, some fifteen years later, by another team in the Department of Ecology, and Professor Margalef is working on it once again.

The scope of Margalef's limnological work is remarkable; it includes improvements in methodology, taxonomy, ecology, palaeolimnology and biogeography. He has also examined pollution in continental waters. Mention should be made here of the achievements of his article "El concepto de polución y sus indicadores biológicos" (1969) and the already mentioned "*Los organismos indicadores en la limnología*", which contributed greatly to an increased awareness of the subject.

Margalef's encyclopaedic knowledge of limnology in general and of Iberian land waters in particular is embodied in his *Limnología* (1983), where the author has managed to put his enthusiastic stamp on a topic of study that has always been very dear to him.

THE MARINE BIOLOGIST

Margalef began to study marine organisms through the knowledge of continental water life-forms and after joining the Institut d'Investigacions Pesqueres (IIP). His concentrated efforts in the field of marine studies go far beyond the strict marine ecology or biological oceanography fields, and he has excelled in the study of marine unicellular algae. After publishing several works about the phytoplankton of the Catalan coast, he then collaborated with M. Massutí on a book entitled *Introducción al estudio del plancton marino* (1950), a monograph which is still widely used for identification purposes.

Professor Margalef's stays in various IIP laboratories around the Iberian Peninsula led to a series of studies on the hydrography, phytoplankton and primary production dynamics of the coasts of Blanes, Barcelona, Ebro delta, Castellón, Cádiz, Ría de Vigo, and of American (the Caribbean) and African (Sahara) coastal regions. Many publications and syntheses had their source in the quantitative study of an incredibly high number of plankton samples, whose inventories were the raw material on which a series of theoretical concepts about marine ecosystems, like species diversity and phytoplankton succession, for example, were based.

When thirty years ago the Institut d'Investigacions Pesqueres had no oceanographic ships available, invitations from other countries allowed Professor Margalef to take part in various international oceanographic cruises. At a time when automatic analysis in the field of oceanography was still very much in its infancy and computers were not available, his works reflected his pioneering attitude towards the development of automatic sampling, analysis and evaluation methods that allowed continuous readings of the marine environment and plankton characteristics to be taken. The use of multidimensional statistical analysis methods in his marine plankton studies were equally as pioneering.

When Margalef was researcher at the IIP (both before and after his short period as director of the Institut), he made considerable contributions to its consolidation as the Spanish Mediterranean's foremost laboratory and to its international prestige. He achieved this by surrounding himself with competent, enthusiastic collaborators and attracting internationally renowned researchers, involving them in high level research programmes and training courses, and raising the tone of his magazine, *Investigación Pesquera*. His many research stays in marine laboratories in Spain and abroad were always training courses for local students and young oceanographers.

One of Margalef's achievements was recognising the importance of phytoplankton organization - which had generally been considered as an unstructured suspension - in the study of marine communities. On this matter, he has often pointed out the faults in classical statistics which do not take space into account. Several of Margalef's most internationally renowned works have concentrated on the study of the small-scale spatial distribution of phytoplankton and on the role of auxiliary energy in phytoplankton production.

From the early seventies onwards, Margalef promoted the basic oceanographic study of the upwelling areas of the African coast, whose interest for fisheries had been highlighted by various IIP prospection cruises; he directed some of the oceanographic cruises in waters of the Saharan continental shelf and was personally responsible for the chemical and physical study of the upwelling, the analysis of phytoplankton and ciliate communities and primary production. In more recent years, his influence in the detection, study and interpretation of the Catalano-Balearic hydrographic front has also been significant.

Margalef also studied the population behaviour of phytoplankton algae cultivated in the laboratory in several types of recipient that allowed a certain local heterogeneity to be created in environmental conditions. His aim was to obtain conditions which were midway between the low volume classical shaken culture and natural sea ones in order to detect phytoplankton succession-related phenomena. His profound knowledge of plankton ecology - and phytoplankton ecology in particular - and various incursions into the study of production by seagrasses and corals, the biology of commercial fish species and other areas of oceanographic research enabled him to produce two multi-author books on marine ecology: *Ecología marina* (1967) and *Western Mediterranean* (1985).

THE ECOLOGIST

Perhaps Margalef's contributions to theoretical ecology are what have brought him international fame and prestige, and have made him one of the most frequently quoted contemporary ecologists. All his publications contain elements which reflect his concern to find general regularities, unifying principles (as he himself would call them in one of his best-known papers; see below), which explain how aquatic systems (both continental and marine) and terrestrial systems work or, to put it another way, how the entire biosphere works. He has been able to extrapolate from his immense work on aquatic environments to global problems of the first order concerning evolution, ecology and even human affairs.

In his introductory speech at the Reial Acadèmia de Ciències i Arts de Barcelona, Professor Margalef spoke about the use of information theory in Ecology ("La teoría de la información en ecología", 1957). He suggested applying certain information theory methods to the study of the variety or diversity of species of an ecosystem. The fact that this seminal paper was published in English just one year later ("Information theory in ecology", 1958) gives us some idea of the influence it had on ecologists at that time. Margalef had already set out his ideas in various articles about the regularities and the correlation between global or synthetic characters of ecosystems and their variations throughout the ecological succession.

"On certain unifying principles in ecology" (1963) was another paper that exerted a great deal of influence over contemporary ecology, as the numerous reprints and its inclusion in several *Readings in Ecology* demonstrate. As Margalef himself says, the article contained several points of view aimed at constructing a general frame of reference in which to order and unify ecological observations. This aspect of Margalef's work does not just affects basic ecological concepts; it also affects interesting problems related to nature management and the role man plays in the biosphere (see below).

Let us take a look at what Margalef himself has to say about his contribution to a more synthetic formalisation of the science of Ecology by referring to his seminal papers and books:

"At the beginning of the 1960s I was working in a research institute, with leisure and freedom, without administrative duties, and with just the supplementary stimulation provided by a small number of good students registered on a voluntary basis in a course on general ecology. A few years before I had shifted my principal interest from freshwater to the oceans, and had recently had the occasion to work in the Mediterranean and the Caribbean, as well as in the fjordlike rias of NW Spain. Weekly strolls over the countryside with my children were regular. At the time, I had the occasion to travel widely abroad, mostly in the US, and discuss my ideas in many personal contacts.

"I developed a feeling about the current trends in ecology and did not like them completely. Mathematical ecology was attractive, but I could not see any essential improvement since Volterra, and many current assumptions did not make much sense to me from the point of view of physics. I had been nurtured in the tradition of the old continental European approach to ecology. Thienemann, Braun-Blanquet, and, later, Hutchinson, influenced me strongly.

"My work leading to the speculations, or supporting the conclusions of this paper ["On certain unifying principles in ecology"], was published in Spanish and was almost unknown. So this first summary published in English conveyed ideas born and bred in relative isolation.

"From my discussions with colleagues and from the reaction to my seminars at the time, I feel that a part of the success of this paper was because it provided a relatively orthodox frame related to physics, that is, to thermodynamics, and to information theory to supplement a strictly individualistic approach to ecology. Most biologists accepted and accept that the success of an individual who has passed part of his genes to a new generation is everything that counts, and that no more questions should be asked. My point was that physical laws place a number of constraints that add to the proximity and interaction between organisms, in a way that, to some extent, we can anticipate who will overcome the selection test and what the ecosystem will look like. This was consistent with accepting that possibilities for life are almost boundless and that the approach called reductionistic is very effective in ecology. My approach could also be easily extended to humanized ecosystems.

"I have tried to present my recent views on the whole subject in *La Biosfera, entre la termodinámica y el juego*".

Margalef has also taken a very active part in the development of Ecology as a quantitative science, although he has always warned ecologists about the dangers of mathematical bewilderment that could make them lose sight of the rich biological reality. In this context, it should be mentioned that his sentence "We should try not to let statistics become the opium of the ecologist" has not always been correctly interpreted.

Any list of Margalef's works that involve important contributions to theoretical ecology would be very long. It should include his papers and books on species diversity and connectivity as a measure of ecosystem organization and complexity; the ecological succession as an evolutionary framework of ecosystem development; trophic level ecosystem models and network organization; the role of exosomatic or auxiliary energy in the production of organic matter in the biosphere; ecological equilibria, etc.

Margalef has periodically formalised his contribution to theoretical ecology in various books. The first of these was *Perspectives in ecological theory* (1968), which originated as a series of lectures given at the University of Chicago, and which meant a broadening of his above-mentioned corpus of theoretical work. "My purpose is only to express the conviction that some aspects of the solid ecology of yesterday and of today allow us to build a theoretical superstructure that, perhaps, is not irrelevant after all." (*Perspectives in ecological theory:* v). At that time, the book had an enormous impact on the world of ecological science (in Margalef's own words, "These works did provoke a certain amount of controversy and several points of view put forward in them have been accepted and have had some influence on present day Ecology") and it was translated into several languages, Japanese and Spanish to name but a few.

Ecología (1974), his monumental and authoritative text book, is permeated throughout his chapters by his conceptual framework and by creative and controversial thoughts about ecological science. Margalef himself explains this relative decrease in "eclecticism and objectivity" which are generally taken to be sensible text book features: "Personally, I have always been more interested in books that are highly personal and partial (as far as the selection of material or the points of view are concerned) than in those that claim to give a balanced, aseptic and neutral vision of science. Universities need creativity and a certain amount of controversial provocation. No science is static." (*Ecología*: vii-viii).

La Biosfera, entre la termodinamica y el juego (1980) is an update of *Perspectives* (with a "naturalist" bias and "somewhat passionate, in contrast to the most commonly accepted ecological theory"). The title itself already suggests two boundary levels in the happenings of the biosphere: the inexorability of physical laws and the randomness of fate. It is, in a sense, an ecological answer to the idea that is revealed in *Le hasard et la nécessité* (MONOD, 1970) about the evolution of living beings and the biosphere. Margalef understands that "ecology is more transdisciplinary than interdisciplinary, in as much that it suggests points of view that closely relate areas of knowledge which are traditionally kept separate", and thus bridges are built between traditional ecology and the above-mentioned sciences. At the same time, he removes some of the dead weights from the ecology he has helped to create and which contribute to its stagnation: "Several concepts used in ecology, and some of my favourites, such as niche, diversity, stability, succession... lose their value when they are critically examined, and I believe they are unnecessary" (*La Biosfera, entre la termodinámica y el juego*: vii).

Ecological theory and Margalef's own ideas on this matter also pervade *Limnología* (1983). Margalef's approach to limnology may be considered to be modelled on G. E. Hutchinson's, mainly by his multiple approach method to limnological and ecological problems and his willingness to master a great deal of detailed information.

As with *Perspectives*, *La Biosfera* stemmed from the organisation and formalisation of materials from postgraduate and specialist courses on theoretical ecology. *Teoría de los sistemas ecológicos* (1991) has the same origins, and it represents Margalef's third attempt (his own words) to offer a "personal view, and probably a heretical one, of what the biosphere is". Perhaps this "heretical" or heterodox feature is one of the reasons why this text has not appeared in English. Indeed, when he was asked to do an English version of *La Biosfera*, Margalef not only translated the original, but expanded and updated it, and renamed it *The Biosphere in the making*. As the American editors suggested modifications and changes and his colleagues made further suggestions, the manuscript became *A view of the Biosphere*, whose publication was delayed or postponed indefinitely on the basis of it being too speculative. However, it would not be at all hazardous to suggest that, as happened with *Perspectives* and *La Biosfera*, this most recent book is so far in advance of the thought of most ecologists that the just recognition of its merits will probably be delayed for some time.

In *Teoría de los sistemas ecológicos*, Professor Margalef develops "a theoretical conception in which a personal vision of what is known about the aspects of ecological systems and how they work can be framed, from individuals of different species to all of them as a whole within more or less large areas, and their connection with time examined. The biosphere is the living cover of the planet and it embraces all living organisms, their supports and the signs of their activity. The characteristics of the Earth as a planet, its materials and the availability of energy act as the framework for the biosphere. All the pieces of the puzzle fall into place in different ways. Their examination leads to a historical and organismic vision, which accepts the strong dependence of the physical environment, which, in turn, is strongly modulated by the existence of life...

"Perhaps there are very few 'laws of Nature' and, for the most part, they will be more like principles of impotence. The most important thing in nature - and life in particular - is its constructiveness: the tendency to complicate coherent or continuous structures... the tendency to increase information... strengthened by the constant selective pressure towards bigger structures having unified behaviour patterns, at the heart of which the value of the information is multiplied or becomes stronger. Recovery, in the form of information, of a part equivalent to the entropy produced in the system and its environment, explains most of the properties of life, organisms and ecosystems...

"Special emphasis is placed on trying to characterise organisms and ecosystems as physical systems, which forces the properties of the systems in general to be considered, especially as far as self-organisation is concerned. This took a big step... when systems formed by replicable subsystems were adopted. Ecosystems... combine inanimate, living subsystems and materials, whose interaction is clear in material cycles and in the history of the biosphere, in the role of external or exosomatic energy and in the apparent indetermination that enters any mechanism when it reaches a field of relatively low energy... The extraordinary capacity for making new things stems from the restrictions imposed by the few identifiable physical laws. Those related to thermodynamics are possibly the most constructive and the only ones that can be used to make any predictions. However, attempts at making predictions in ecology have not been very accurate..." (*Teoría de los sistemas ecológicos*: 18-19).

THE NATURALIST

In most of his books and articles, and in many other places too, Margalef defines himself as a naturalist, a profile which is surely dearer to him than any of the previous ones. This is so even though Margalef was "the" Spanish limnologist from the forties onwards, "the" marine biologist when he began to study marine plankton and, later, when nobody in Spain had even heard about ecology, "the" Spanish ecologist. Undoubtedly, the term "naturalist" is the one that fits best this universal Catalan who has become, in his own right, the most prominent naturalist in Spain - where nature scholars of whatever bent have always been thin on the ground.

Indeed, if we take a look at his publications (see Appendix I), especially those published during the first fifteen years, it is surprising to see that he has taken an interest in all animal and plant groups inhabiting aquatic environments, environments which he seems to have studied with panoptic vision. Professional profiles such as phycologist, carcinologist, phytoplanktologist and so on fit him perfectly. But a careful reading of his books on ecology, limnology and marine biology reveals that aquatic invertebrates and algae are not the only objects of his attention and study. Margalef is a keen entomologist and knows higher plants and terrestrial and aquatic vertebrates very well. His knowledge of physics, chemistry and geology is wide and this is apparent throughout the pages of his books. His contributions to biogeography (mainly of continental waters), micropaleontology and evolutionary biology (the causes of microevolution and large-scale patterns in the history of diversity through geological time) are by no means less substantial.

Other aspects of his character make him an out-of-the-ordinary naturalist, of he lineage Stephen Jay Gould calls Galilean, that is, the naturalist that "delights in nature's intellectual puzzles and our quest for explanation and understanding. Galileans do not deny the visceral beauty [of nature], but take greater delight in the joy of causal comprehension and its powerful theme of unification" (GOULD, 1991). The way Margalef the naturalist enjoys his research and his probing of the workings of the biosphere has inspired the subtitle of this volume of tribute.

At a time when the daily talk at universities is more about employment opportunities for the biologist rather than its vocation, Margalef is a constant reminder of what a scientist should be: someone who questions himself about the whys and wherefores of nature. He has also said and written on more than one occasion that nature should be looked at with the eyes of a child -inquisitive and searching - which do not take anything for granted and question everything. In this respect, it may be said that he is perhaps the archetypal naturalist - in the sense of a scholar and expert on nature in all its facets. We might call him a biologist of the biosphere, as others are scholars of a particular group of insects, plants or the Quaternary.

Margalef's contributions to a discipline that is still not clearly established linking ecology with human problems are also considerable. This is a discipline that goes beyond human ecology and social anthropology and could be defined as the application of ecological ideas to human problems, whether environmental or not. Consider the scope of such a definition (economics, demography, resource conservation and management, pollution, environmental degradation, etc.) and the importance of the basic postulate of Margalef's contribution to this field: man is to be viewed as a part of the biosphere and within a general ecological context. One of his most representative articles on this is "Ecological theory and prediction in the study of the interaction between man and the rest of the biosphere" (1973).

Margalef's clear vision of many human or anthropological problems has turned his lectures or articles into items which are highly sought after by institutions and magazines in the humanities field. Although he frequently despises his "sermons" (as he calls them, by analogy to the religious ones that many people hear but few take seriously), they are valuable and original pieces of thought directed to heterogeneous audiences who deeply appreciate the message delivered, be it ecological or otherwise. Margalef has also taken an active part in the activities of the Institut d'Humanitats de Barcelona, one of whose main objectives is to fill the gap that exists (especially in our country) between sciences and humanities.

This interest is patently clear in his popularizing work in the form of conferences, books or exhibitions for a very wide public. Although the books and articles for encyclopaedias and magazines of a divulgative nature have not been listed in the bibliography of Appendix I, they actually represent a considerable percentage of Margalef's output, characterised by a highly personal style that bears witness to the author's breadth and depth of knowledge and teaching qualities. *Ecología*

(1981) and *L'ecologia* (1985) are but two of Margalef's diverse range of popularizing books which are worthy of mention. His participation in the book *Natura, ús o abús? Llibre blanc de la gestió de la Natura als Països Catalans* (1976[1988]) and in the volume *Sistemes naturals* (1989) of the encyclopaedia *Història Natural dels Països Catalans* was outstanding.

Just in case it is wrongly thought that popularizing ecology is easy, Margalef himself has highlighted the limitations of the ecological discourse as far as leaving important basic matters to one side is concerned: the teaching of these matters is absolutely essential if future generations are to understand the way nature works, and manage its resources better than the present generation. In "Simple facts about life and environment not to forget in preparing schoolbooks for our grandchildren" (1984), to mention but one of his papers on that matter, he lists a series of concepts that must be given priority if ecology is to be made understandable. His advice: "It is necessary to go back to fundamentals - look at nature and state real and interesting problems in a simple and general form, if possible."

This ability to go back to the fundamental and state it in simple, understandable terms, combined with his ability to summarise his encyclopaedic knowledge, have made him a much sought after contributor to books related to his fields of interest, and have made him an expert disseminator of ecological information. Margalef is a charming writer, and his work as plain-science writer is second only to his serious scientific work.

In all his books, scientific or vulgarization articles and lectures, Margalef is always new and original and has the excellent teaching virtue of suggesting and upsetting: suggesting new relationships and general rules on the basis of data which are apparently unrelated; upsetting schemes and clichés which are taken for granted and which should be reformulated. Margalef's classes have never been famous for their clear and succinct construction; they are the classes that students most shy away from, because they can hardly ever take precise notes, maybe because Margalef jumps from one subject to another (or even one discipline to another) or because the listeners must take their time to digest the impact of the revelations, the questions and the paradoxes about nature that are provided at great speed. The great merit of these classes lies in the fact that they awaken the students' emotions and doubts, which unfailingly brings them to ask themselves about the ecological issue that is under consideration instead of taking it as a perfectly established solemn truth, as often happens in the "magisterial" lectures still commonly given in Spanish universities. Rather than attempting to persuade, Margalef attempts to provoke, which means that the student is obliged to think and not just memorise. His books also have this virtue.

Ramon Margalef is well-regarded everywhere as much for his kindness and modesty as for his science. It is difficult to define a boundary between his scientific interests, very broad as they are, and his personal commitment to human culture. The fact that his ideas cover both areas is an index of the breadth and depth of his scientific knowledge and of his own human character. In fact, he sees science as a part of human culture, and in this sense he can also be termed a humanist.

Margalef the man is a very warm and friendly human being; he has been always extremely generous with his time in teaching students and rearing young researchers from many countries, mainly Latin American. No doubt this willingness to help and his warm and continued interest in everyone he meets are the underlying reasons for his having stayed in Spain, although throughout his career he has received many tempting offers from some of the world's most prestigious teaching and research institutions. His attachment to his roots and the commitment to the idea that his teaching and research work should be undertaken in his own country are but a few more merits that should be emphasised, especially for a scientist of Margalef's calibre.

In summary, Ramon Margalef's work has been marked not only by scientific excellence and brilliant exposition, but also by far-reaching insight and a remarkable degree of personal candour. This Catalan naturalist, one of the finest world scientists of the 20th century, has served, and will continue to serve, as a model for scientific endeavour in the field of aquatic ecology.

REFERENCES not to be found in Appendix I

GOULD, S. J. 1991. *Bully for Brontosaurus. Reflections in Natural History*. W. W. Norton. New York.

MONOD, J. 1970. *Le hasard et la nécessité*. Seuil. Paris.

APPENDIX I. SCIENTIFIC PUBLICATIONS OF PROFESSOR RAMON MARGALEF

Papers in scientific reviews and divulgative books are not included.

1943

1. Los epibiontes en los animales de agua dulce. *Euclides*, 33, 3 pp.
2. Sobre la ecología de las larvas de algunos culícidos. *Graellsia*, 1: 7-12.
3. Contribución al conocimiento de los diaptómidos (Crustáceos copépodos) de España. *Bol. Real Soc. Esp. Hist. Nat.*, 41: 379-383.

1944

4. Datos para la flora algológica de nuestras aguas dulces. *Pub. Inst. Bot. Barcelona*, 4 (1): 130 pp, 22 cuadros y 7 láminas.
5. Notas sobre quironómidos. *Graellsia*, 2: 3-13, 65-76, 165-181; 3: 13-22.
6. Contribución al conocimiento de los crustáceos anfípodos que viven en las aguas dulces y salobres de España. *Bol. Real. Soc. Esp. Hist. Nat.,* 42: 199-209.

1945

7. Fitoplancton nerítico de la Costa Brava catalana (Sector de Blanes). Inst. Esp. Estudios Mediterráneo. *Publ. Biol. Mediterránea,* I, 48 pp.
8. Infusorios de las aguas continentales de la provincia de Barcelona. *Bol. Real Soc. Esp. Hist. Nat.*, 43 369- 381.
9. Un curioso tipo de biocenosis dulciacuícola: la fitotélmica. *Euclides*, 5: 398-399.

1946

10. Observaciones sobre el régimen alimenticio de varios pequeños animales de agua dulce. *R. Esp. Fisiol.*, 1: 245- 250.
11. Materiales para el estudio de la biología del lago de Bañolas (Gerona). *P. Inst. Biol. Apl.*, 1: 27-78.
12. Sobre dos supuestas protófitas. *Collectanea Botanica*, 1: 49-50.
13. Contribución al conocimiento del género *Platymonas* (Volvocales). *Collectanea Botanica*, 1: 95- 105.
14. Contribución al conocimiento hidrobiológico del país vasco-navarro. Estación de estudios Pirenaicos: *Aportación al estudio de la fauna y flora vasco-navarras (Sierra de Aralar):* 7- 44.
15. Algunos ostrácodos de las aguas continentales del NE de España. *P. Inst. Biol. Apl.*, 2: 89-95.
16. Fitoplancton nerítico estival de Cadaqués (Mediterráneo catalán). *P. Inst. Biol. Apl.*, 2: 89-95.
17. Sobre la morfología, afinidades y ecología de una interesante pleurocapsácea (cianofícea). *P. Inst. Biol. Apl.*, 2: 89- 95.
18. Miscelánea de zoología dulciacuícola. *P. Inst. Biol. Apl., 2: 117-120.*
19. (with M. BASSEDAS) Algunos branquiópodos del NE de España y consideraciones sobre la fauna ibérica de cladóceros. *P. Inst. Biol. Apl.*, 2: 127-148.

1947

20. Una oscilatoria notable por la estructura de sus vainas. *An. Jardín Bot. Madrid*, 6: 187-192.
21. Limnosociología. *Monografías de Ciencia Moderna*, 10. 93 pp.
22. Observaciones micropaleontológicas sobre los sedimentos lacustres miocénicos de Libros (Teruel). *Estudios Geológicos*, 5: 171-177.
23. Materiales para una flora de las algas del NE de España. I, Zignemales. *Collectanea Botanica*, 1: 107-121.
24. Observaciones sobre el desarrollo de la vida en pequeños volúmenes de agua dulce y sobre la ecología de las larvas de *Aedes aegypti*. *P. Inst. Biol. Apl.*, 3: 79-112.
25. Datos zoogeográficas sobre ostracódos de agua dulce de Cataluña y descripción de la nueva especie *Potamocypris pyrenaica*. *P. Inst. Biol. Apl.*, 3: 163-171.

1948

26. Primera nota sobre la biología de las aguas estancadas del bajo Urgel. *Ilerda*, 5, 65 pp.
27. Los métodos para la investigación de las comunidades acuáticas adnadas y especialmente las formadas por organismos microscópicos (perifiton, pecton). *Collectanea Botanica*, 1: 247-259.
28. A new limnological method for the investigation of thin layered epilithic communities. *Trans. Amer. Micr. Soc.*, 67: 153-154.
29. Le phytoplancton estival de la Costa Brava catalane. *Hydrobiologia*, 1: 15-21.
30. Flora, fauna y comunidades bióticas de las aguas

dulces del Pirineo de la Cerdaña. *Monogr. Est. Estud. Pirenaicos*, 11, 226 pp.
31. Estudios sobre la vida en las aguas continentales de la región endorreica manchega. *P. Inst. Biol. Apl.*, 45: 5- 51.
32. Algas de agua dulce de la laguna de Ariguanabo (Isla de Cuba). *P. Inst. Biol. Apl.*, 4: 79-89.
33. Notas sobre algunos rotíferos. *P. Inst. Biol. Apl.*, 4: 136-148.
34. Materiales para una flora de las algas del NE de España, II. *Collectanea Botanica*, 2: 99-130.

1949

35. Miscelánea de zoología dulciacuícola. *P. Inst. Biol. Apl.,* 5: 69-76.
36. Fitoplancton nerítico de la Costa Brava en 1947-48. *P. Inst. Biol. Apl.*, 5: 41-51.
37. Sobre el régimen alimenticio de los animales de agua dulce. *R. esp. Fisiol.*, 4: 307-213.
38. Las asociaciones de algas en las aguas dulces de pequeño volumen del Noreste de España. *Vegetatio*, 1: 258-284.
39. Algunos ciclópidos (Crustáceos copépodos) de Cataluña. *Real. Soc. esp. Hist. Nat.*, tomo 75 aniv.: 177-188.
40. Materiales para una flora de las algas del NE de España, III. *Collectanea Botanica*, 2: 233-250.
41. Datos para la hidrobiología de la Sierra de Guadarrama. *P. Inst. Biol. Apl.*, 6: 5-21.
42. Importancia de la neotenia en la evolución de los crustáceos de agua dulce. *P. Inst. Biol. Apl.*, 6: 41-51.
43. Una aplicación de las series logarítmicas a la fitosociología. *P. Inst. Biol. Apl.*, 6: 63-102.
44. Sobre la ecología de las larvas del mosquito *Aedes mariae. P. Inst. Biol. Apl.*, 6: 63-102.

1950

45. Caractères biogéographiques intéressants des biotes des eaux continentales de l'Espagne. *Verh. Int. Ver. theor. ang. Limnol.,* 10: 280-283.
46. Las papilas anales de las larvas de los culícidos y la absorción de tóxicos. *Rev. Esp. Fisiol.*, 5: 251-257.
47. (with M. MASSUTI) *Introducción al estudio del plancton marino*. Patronato Juan de la Cierva. 182 pp.
48. Datos para la hidrobiología de la cordilera cantábrica, especialmente del macizo de los Picos de Europa. *P. Inst. Biol. Apl.,* 7: 37-76.
49. Algunos crustáceos interesantes de las aguas dulces y salobres de España. *P. Inst. Biol. Apl.*, 7: 131-152.
50. Materiales para una flora de las algas del NE de España, IIIb. *Collectanea Botanica*, 2: 273-293.
51. Datos para la hidrobiología del estanque de Montcortés (Provincia de Lérida). *Bol. Real. Soc. Esp. Hist. Nat.*, 48: 209-218.
52. Segunda nota sobre la biología de las aguas estancadas del bajo Urgel. *Ilerda*, 13: 329-375.
53. (with F. PONZ) Actividad fosfatásica de los cladóceros, en particular sobre substancias disueltas en el agua ambiente. *Rev. esp. Fisiol., 6: 227-237.*

1951

54. Materiales para la hidrobiología de la Isla de Ibiza. *P. Inst. Biol. Apl.*, 8: 9-70.
55. Un sincárido del género *Parabathynella* en las Baleares. *P. Inst. Biol. Apl.*, 8: 151-153.
56. La base actual de la biogeografía. *Arbor*, 66/67, 17 pp..
57. Rôle des Entomostracés dans la régéneration des phosphates. *Verh. Inter. Ver. Theor. ang. Limnol.*, 11: 246-247.
58. Ueber die biogeographische Stellung der Binnengewasserlebewelt Spaniens. *Arch. Hydrobiol.*, 45: 304-313.
59. Diversidad de especies en las comunidades naturales. *P. Inst. Biol Apl.*, 9: 83-118.
60. Plancton de las costas de Castellón durante el año 1950. *P. Inst. Biol Apl.*, 9: 49-62.
61. Ciclo anual del fitoplancton marino en la costa NE de la Península Ibérica. *P. Inst. Biol. Apl.*, 9: 83-118.
62. (with J.R. RODA, J.J. GOMEZ, B. ANDREU) Datos para una crítica del significado de la medida vertebral de las poblaciones de la sardina. *P. Inst. Biol. Apl.*, 9: 185-192.
63. Observaciones sobre *Triops* (*Apus*) *cancriformis* de una localidad catalana. *P. Inst. Biol. Apl.,* 9: 247-254.
64. Sobre *Gammarus* interesantes de España. *P. Inst. Biol. Apl.*, 9: 255-269.

1952

65. (with F. SAIZ, J. RODRIGUEZ-RODA, C. TOLL, B. VALLES). Fitoplancton de las costas de Castellón durante el año 1951. *P. Inst. Biol. Apl., 10: 133-143.*
66. Materiales para la hidrobiología de la isla de Menorca. *P. Inst. Biol Apl.*, 11: 5-12.
67. Materiales para una flora de las algas del NE de España, IV. *Collectanea Botanica*, 3: 209-239.
68. Quelques remarques biogéographiques au sujet des crustacés d'eau douce des Baléares. *Vie et Milieu*, suppl. 2: 248-252.
69. Regiones limnológicas de Cataluña y ensayo de sistematización de las asociaciones de algas. *Collectanea Botanica*, 3: 43-67.
70. La vida en las aguas dulces de los alrededores del Santuario de Nuestra Señora de Aránzazu (Guipúzcoa). *Munibe*, 4: 73-108.
71. Une *Jaëra* dans les eaux douces des Baléares, *Jaëra balearica* nov. sp. (Isopoda Asellota). *Hydrobiologia*, 4: 289-213.
72. La vida en las aguas dulces de Andorra. *Monogr. Inst. Est. Pirenaicos,* 69: 1-107.

1953

73. Algunos orgasnismos interesantes de las aguas dulces de los Pirineos. *Pirineos*, 9: 407-420.
74. Algas de agua dulce y salobre de la Romagna

(Colección Zangheri). *Archivie Botanice*, 29, 3ª s., 13(4): 1- 19.
75. Estudios experimentales sobre las modificaciones inducidas por diferentes temperatures en células de clorofíceas. *Publ. Inst. Biol. Apl.,* 12: 5-78.
76. Caracteres ligados a las magnitudes de los organismos y su significado sistemático y evolutivo. *P. Inst. Biol Apl.*, 12: 111-121.
77. (with M. DURAN) Microplancton de Vigo, de octubre de 1951 a septiembre de 1952. *P. Inst. Biol Apl.*, 13: 5-78.
78. Materiales para la hidrobiología de la isla de Mallorca. *P. Inst. Biol. Apl.*, 15: 5-111.
79. Estudio sumario del fitoplancton de la Ría de Vigo (1948-1950). *Bol. Inst. Esp. Oceanogr.*, 47: 1-15.
80. Los crustáceos de las aguas continentales ibéricas. *Instituto Forestal Invest. Exper.*, 243 pp.
81. Modifications induced by different temperatures on the cells of *Scenedesmus obliquus* (Chlorophyceae). *Hybrobiologia*, 83-94.
82. Observaciones paleoecológicas y geocronológicas sobre los sedimentos lacustres miocénicos de Hellín (Albacete). *Mem. Com. Inst. Geol. Prov. Barcelona*, 10: 53-72.
83. Materiales para una flora de las algas del NE de España, IVb. *Collectanea Botanica*, 3: 231-260.

1954

84. Materiales para una flora de las algas del NE de España, V. *Collectanea Botanica*, 4: 53-79, 103-201.
85. Consideraciones sobre la determinación cuantitativa del fitoplancton por la valoración de pigmentos solubles y los factores que afectan a la relación entre cantidad de pigmento y peso seco. *P. Inst. Biol. Apl.*, 16: 71-84.
86. Un aparato para el cultivo de algas en condiciones regulables. *P. Inst. Biol. Apl.*, 17: 65-69.
87. (with J. HERRERA, J. RODRIGUEZ-RODA & J.J. GOMEZ). Fitoplancton de las costas de Castellón durante el año 1952. *P. Inst. Biol. Apl.,* 17: 131-134.
88. Una técnica de filtración para el estudio cualitativo y cuantitativo del fitoplancton. *P. Inst. Biol. Apl.*, 17: 65-69.
89. Diferencias interanuales en la producción de fitoplancton, 1: 10.
90. (with M. DURAN). Métodos de trabajo en el mar, práctica del laboratorio de Vigo. *Reunión sobre productividad y pesquerías*, I, 32.
91. Algunos crustáceos de agua dulce y salobre de la Romagna. *Boll. Soc. Entom. Italiana*, 84: 146-150.

1955

92. Temperatura, dimensiones y evolución. *Publ. Inst. Biol. Apl.*, 19: 13-90.
93. Los organismos indicadores en la limnología. *Instituto Forestal Inv. Exper.*, 308 pp.
94. (with J. HERRERA & F. MUÑOZ). Fitoplancton de las costas de Castellón durante el año 1953. *Inv. Pesq.*, 1: 17-29.
95. (with M. DURAN & F. SAIZ). El fitoplancton de la Ría de Vigo, de enero de 1953 a marzo de 1954. *Inv. Pesq.*, 2: 85- 129.
96. Temperature and morphology in freshwater organisms. *Verhandl. intern Ver. theor. agew. Limnol.*, 12: 507-514.
97. Comunidades bióticas de las aguas dulces del noroeste de España. *P. Inst. Biol. Apl.*, 21: 5-85.
98. Contribución al estudio de la fauna de las aguas dulces del Noroeste de España. *P. Inst. Biol. Apl.*, 21: 137- 171.
99. Datos para el estudio de la distribución de los crustáceos en las aguas continentales españolas. *P. Inst. Biol. Apl.*, 21: 173-177.
100. Métodos de trabajo en el mar, laboratorio de Vigo. 15 pp.
101. (with P. ARTE). Experiencias efectuadas con una sonda "Sadir- Carpentier" tipo S. 24 pp.
102. Dinámica de las poblaciones de fitoplancton. Variaciones interanuales en el fitoplancton de Castellón. *II Reunión Productividad y Pesquerías*, 48 pp.

1956

103. Materiales para una flora de las algas del NE de España, VI. *Collectanea Botanica*, 4: 319-329.
104. La diversidad de especies en las poblaciones mixtas naturales y en el estudio del dinamismo de las mismas. *Pub. Univ. Barcelona*, T. Hom. F. Pardillo: 229-243.
105. Cultivos experimentales de algas unicelylares. *Inv. Pesq.*, 3: 3-19.
106. (with F. MUÑOZ & J. HERRERA). Fitoplancton de las costas de Castellón durante el año 1954. *Inv. Pesq.*, 3: 75-90.
107. Información y diversidad específica en las comunidades de organismos. *Inv. Pesq.*, 3: 99-106.
108. (with M. DURAN, F. SAIZ & J.J. LOPEZ). El fitoplancton de la Ría de Vigo de abril de 1954 a junio de 1955. *Inv. Pesq.*, 4: 67-95.
109. Rotíferos marinos del plancton de la Ría de Vigo. *Inv. Pesq.*, 4: 133-1 .
110. Algas de agua dulce del noroeste de España. *P. Inst. Biol. Apl.*, 32: 43-152.
111. Paleoecología postglacial de la Ría de Vigo. *Inv. Pesq.*, 5: 89-112.
112. Estructura y dinámica de la "purga de mar" en la Ría de Vigo. *Inv. Pesq.*, 5: 113-134.
113. Estudios hidrobiológicos en los valles de Bohí. (Prov. de Lèrida). *Actes 2ème Congrès Intern. Et. Pyrennés: 87-108.*
114. Dos nuevos *Gammarus* de las aguas dulces españolas. *P. Inst. Biol. Apl.*, 23: 31-36.

1957

115. Comentarios biogeográficos sobre la distribución de la fauna de agua dulce en las montañas del norte de España. *Volumen homenaje a D. Joaquín Mendizbal (Aranzadi)*. 3 pp.
116. Materiales para una flora de las algas de España, VII suppl. *Collectanea Botanica*, 5: 87-107.
117. La vida en las aguas de elevado residuo salino de la provincia de Zamora. *P. Inst. Biol. Apl.*, 24:

123-137.
118. (with J. HERRERA). Hidrografía y fitoplancton de las costas de Castellón, de julio de 1956 a junio de 1957. *Inv. Pesq.*, 18: 17-44.
119. (with F. MUÑOZ & J. HERRERA). Fitoplancton de las costas de Castellón de enero de 1955 a junio de 1956. *Inv. Pesq.*, 7: 3-31.
120. Los microfósiles del lago miocénico de la Cerdaña como indicadores ecológicos. *Cursillos y Conf. Inst. Lucas Mallada*, 4: 13-17.
121. Paleoecología del lago de la Cerdaña. *P. Inst. Biol. Apl.*, 25: 131-137.
122. La teoría de la información en Ecología. *Mem. Real. Acad. Ciencias Artes Barcelona*, 32: 373-449. (Translated into english and published in *General Systems*, 3).
123. Nuevos aspectos del problema de la suspensión en los organismos planctónicos. *Inv. Pesq.*, 7: 105-116.
124. Un modelo para el estudio de la distribución de organismos de gran dispersabilidad y área de reproducción muy localizada. *Inv. Pesq.*, 7: 117-125.
125. Fitoplancton de las costas de Blanes (Gerona) de agosto de 1952 a junio de 1956. *Inv. Pesq.*, 7: 105-116.
126. Variación local e interanual en la secuencia de poblaciones de fitoplancton de red en las aguas superficiales de la costa mediterránea española. *Inv. Pesq.*, 9: 65-95.
127. Fitoplancton de las costas de Puerto Rico. *Inv. Pesq.*, 6: 39-52.

1958

128. Temporal succession and spatial heterogeneity in natural phytoplankton. *Perspectives in marine Biology*. Univ. California Press.
129. Trophic typology versus biotic typology, as exemplified in the regional limnology of Northern Spain. *Verh. Internat. Ver. Limnol.*, 13: 339-249.
130. Algunos crustáceos de las agua continentales de España y norte de Africa. *Miscelánea zoológica,* 1: 3-12.
131. Distribución de los crustáceos en las aguas continentales españolas. Grado de asociación entre las especies en relación con factores ecológicos e históricos. *P. Inst. Biol. Apl.*, 27: 17-31.
132. (with B. ANDREU). Componente vertical de los movimientos del agua en la Ría de Vigo y su posible relación con la entrada de la sardina. *Inv. Pesq.*, 11: 105-126.
133. La sedimentación orgánica y la vida en los fondos fangosos de la Ría de Vigo. *Inv. Pesq.*, 11: 67-100.
134. Productividad y sucesión. Determinación continua a través de cable de características fisicoquímicas del agua. *III Reunión sobre Productividad y Pesquerías* (1957), 5: 98.
135. Materiales para una flora de las algas del NE de España. VIIb, Supl. *Collectanea Botanica,* 5: 535-365.
136. Materiales para el estudio de las comunidades bióticas de las aguas dulces y salobres, principalmente del NE de España. *P. Inst. Biol. Apl.*, 28: 5-47.
137. (with J. HERRERA & F. MUÑOZ). La production du phytoplancton dans les eaux cotières du Levant espagnol. *Rapp. Proc. Verb. R. CIESMM*, 14: 269-272.

1959

138. Mode of evolution of species in relation to their places in ecological succession. *XV Intern. Congress. Zool.*, sect. X, 17 pp.
139. Nuevos datos para el estudio de la historia de la Ría de Vigo, desde la última glaciación. *Mem. Real Acad. Ciencias Artes Barcelona*, 33: 23-35.
140. (with J. HERRERA & J. ARIAS) Hidrografía y fitoplancton de las costas de Castellón, de julio de 1957 a junio de 1958. *Inv. Pesq.*, 15: 3-38.
141. Ecología, biogeografía y evolución. *Revista Univ. Madrid*, 8 (29/31): 221-273.
142. Pigmentos asimiladores extraídos de las colonias de celentéreos de los arrecifes de coral y su significado ecológico. *Inv. Pesq.*, 15: 81-101.

1960

143. Valeur indicatrice de la composition des pigments du phytoplancton sur la productivité, composition taxonomiques et proprietés dynamiques des populations. *Rapp. Proc. Verb. CIESMM,* 15: 227-281.
144. Ideas for a synthetic approach to the ecology of running waters. *Int. Rev. ges. Hydrobiol.*, 45: 133-153.
145. Silicoflagellatae populations in the plankton of Cape Cod area, past and present. *Biol. Bull.*, 119: 326.
146. The size of the marine diatom *Melosira sulcata* (Ehrenb.) Kuetz. in the Cape Cod area. *Biol. Bull.*, 119: 326.
147. (with J.H. RYTHER). Pigment composition and productivity as related to succession in experimental populations of phytoplankton. *Biol. Bull.*, 119: 326-327.
148. Avances y problemas técnicos en el estudio del plancton y de hidrografía. *IV Reunión sobre Productividad y Pesquerías*. (1960): 4-7.
149. Mecanismo de fertilización de las aguas costeras de Castellón. *IV Reunión sobre Productividad y Pesquerías* (1960): 23-25.
150. Los pigmentos vegetales en el estudio de la sucesión. *IV Reunión sobre Productividad y Pesquerías* (1960): 40-42.

1961

151. Communication of structure in planktonic populations. *Limnol. and Oceanogr.*, 6: 124-128.
152. La vida en los charcos de agua dulce de Nueva Esparta (Venezuela). *Mem. Soc. C.N. La Salle*, 21: 75-110.
153. Recientes progresos en el estudio de las comunidades vegetales por medio de la extracción de pigmentos. *Bol. Real Soc. Esp. Hist. Nat.*, 68:

291-300.

154. (with A. MORALES). Fitoplancton de las costas de Blanes, de julio de 1956 a junio de 1959. *Inv. Pesq.*, 16: 3-31.

155. (with F. CERVIGON & J. YEPEZ). Exploración preliminar de las características hidrogeográficas y de la distribución del fitoplancton en el área de la Isla Margarita (Venezuela). *Mem. Soc. C. N. La Salle*, 20: 211-221.

156. Méthode d'extraction des pigments dans l'étude de la végétation benthique. *Ann. Stat. Centr. Hydrob. apal.*, 8: 99- 104.

157. Algunas aplicaciones de la teoría de la información en el campo de la Biología y concretamente a la Ecología y al estudio de la evolución. *Scientia*, 55: 1-7.

158. Velocidad de sedimentación de organismos pasivos del fitoplancton. *Inv. Pesq*, 18: 3-6.

159. Hidrografía y fitoplancton de un área marítima de la costa meridional de Puerto Rico. *Inv. Pesq.*, 18: 33-96.

160. Fluctuations in abundance and availability caused by biotic factors. *Proceedings of the World Scient. meeting on the biol. of Sardines and related species*, 3: 1265-1285. F.A.O.

161. Correlations entre certains caractres synthétiques des populations de phytoplancton. *Hydrobiologia*, 18: 155-164.

162. Distribució i dinàmica de les poblacions de plàncton vegetal en relació amb les superfícies de transició i discontinuitat. *Miscel.lània Fontseré*: 245-257.

163. (with J. HERRERA). El nivel del mar en Castellón. *Inv. Pesq.*, 19: 55-63.

164. Distribución ecológica y geográfica de las especies del fitoplancton marino. *Inv. Pesq.*, 19: 81-101.

165. Variaciones intraespecíficas de los pigmentos asimiladores en clorofíceas y fanerógamas acuáticas. *Inv. Pesq.*, 19: 111-118.

166. (with J. HERRERA). Hidrografía y fitoplancton de las costas de Castellón, de julio de 1958 a junio de 1959. *Inv. Pesq.*, 20: 17-63.

167. Fitoplancton atlántico de las costas de Mauritania y Senegal. *Inv. Pesq.*, 20: 131-143.

168. (with J. HERRERA). Hidrografía y fitoplancton de las costas de Castellón, de julio de 1959 a junio de 1960. *Inv. Pesq.*, 22: 49-109.

169. Modalités de l'evolution en rapport avec la simplification des biocéneses insulaires. *Le peuplement des îles Méditerranéennes et le problème de l'insularité, C.N.R.S.*: 313-320.

170. Distribution du phytoplancton dans une échelle moyenne de dimensions et signification de ses pigments assimilateurs dans l'interprétation de la dynamique des configurations. *Rapp. Proc. verb. CIESMM,* 16: 139-140.

171. Caractéristiques et signification des zooxanthelles du "phytoplancton prisonnier" des acanthaires. *Rapp. Proc. verb. CIESMM*, 16: 141-142.

172. Nouveaux développements dans la technique de l'extraction des pigments du phytoplancton. *Rapp. Proc. verb. CIESMM., 16 (2): 223-224.*

1962

173. Modelos físicos simplificados de poblaciones de organismos. *Mem. Acad. Ciencias Artes Barcelona*, 34: 83-146.

174. Registro fósil de fluctuaciones climáticas de corto período en el área pirenaica y en época miocénica. *Actas Tercer Congr. int. Estudios Pirenaicos*: 217-226.

175. Diversità dello zooplancton del lago Maggiore. *Mem. Ist. Ital. Idrobiol*, 15: 137-151.

176. Organisation spatiale et temporelle des populations de phytoplancton dans un secteur du littoral méditerranéen espagnol. *Pubbl. staz. zool. Napoli*, 32 suppl.: 336-348.

177. Succession in marine populations. *Advancing frontiers of plant sciences*, 2: 137-186.

178. Adaptación, ecología y evolución: nuevas formas de plantear antiguos problemas. *Real. Soc. Esp. Hist. Nat.*, 60: 231-246.

179. *Comunidades naturales*. Instituto de Biología Marina de la Universidad de Puerto Rico. Mayagüez. 469 pp.

1963

180. Contribuciones al texto de Marine Biology I, First International Interdisciplinary Conference on Marine Biology, Princeton, 1961. *Amer. Inst. Biol. Sci.*, Wahsington, 286 pp.

181. Note technique sur les prélévement d'échantillons pour l'étude du phytoplancton. *Rapp. Proc. verb. CIESMM*, 17(2): 463.

182. Rôle des ciliés dans le cycle de la vie pélagique en Méditerranée. *Rapp. Proc. verb. CIESMM*, 17(2): 511- 512.

183. Ecologie marine: nouvelles vues sur de vieux problémes. *Année Biologique*, 1963: 3-16.

184. Desarrollo experimental de picnoclinas en pequeños volúmenes de agua. *Inv. Pesq.*, 23: 3-10.

185. Modelos simplificados del ambiente marino para el estudio de la sucesión y distribución del fitoplancton y del valor indicador de sus pigmentos. *Inv. Pesq.*, 23: 11-52.

186. Algunas regularidades en la distribución a escala pequeña y media de las poblaciones marinas de fitoplancton y en sus características dencionales. *Inv. Pesq.*, 23: 169-230.

187. *Scaphodinium mirabile*, nov. gen. nov. sp., un nuevo dinoflagelado aberrante del plancton marino. *Miscelánea zoológica*, 1(5): 3-4.

188. Un *Supraniphargus i*nteresante de Vizcaya (Amphipoda Gammaridae). *Miscelánea zoológica*, 1(5): 33-34.

189. On certain unifying principles in ecology. *American Naturalist*, 97: 357-374.

190. El ecosistema pelágico de un área costera del Mediterráneo occidental. *Mem. Real Acad. Ciencias Artes Barcelona*, 35: 3-48.

191. (with J. HERRERA). Hidrografía y fitoplancton de la costa comprendida entre Castellón y la desembocadura del Ebro, de julio de 1960 a junio de 1961. *Inv. Pesq.*, 24: 33-112.

1964

192. (with R. ESTABLIER). Fitoplancton e hidrografía de las costas de Cádiz (Barbate) de junio de 1961 a agosto de 1962. *Inv. Pesq.*, 25: 5-31.
193. (with J. HERRERA). Hidrografía y fitoplancton de la costa comprendida entre Castellón y la desembocadura del Ebro, de julio de 1961 a julio de 1962. *Inv. Pesq.*, 26: 49-90.
194. Fitoplancton de las costas de Blanes (Provincia de Gerona, Mediterráneo occidental), de julio de 1959 a junio de 1963. *Inv. Pesq.*, 26: 131-164.
195. Modelos experimentales de poblaciones de fitoplancton: nuevas observaciones sobre fitoplancton y sobre fijación de carbono inorgánico. *Inv. Pesq.*, 26: 195-203.
196. Sobre una *Spirogyra* de agua salobre. *Collectanea Botanica*, 6: 455-456.
197. Correspondence between the classic types of lakes and the structure and dynamic properties of their populations. *Verh. Internat. Verein. Limnol.*, 15: 169-175.
198. *The scope and objectives of Ecology*. Organization for economic cooperation and development. Directorate for scientific affairs. DAS/BS/64.19 (restricted), 33 pp.

1965

199. Evaluación contínua y automática de las poblaciones planctónicas. *V Reunión sobre Productividad y Pesquerías* (1965): 19-21.
200. Análisis de las comunidades de organismos y su significado como indicadores ecológicos. *V Reunión sobre Productividad y Pesquerías* (1965): 28-29.
201. Diversidad de las muestras de poblaciones de peces en función de la madurez del ecosistema, de la intensidad de explotación y de la selectividad de las artes. *V Reunión sobre Productividad y Pesquerías* (1965): 113-115.
202. Distribución ecológica de las especies del fitoplancton marino en un área del Mediterráneo occidental. *Inv.. Pesq.*, 28: 117-231.
203. Ecological correlations and the relationship between primary productivity and community structure. *Mem. Ist. Ital. Idrobiol.*, 18 suppl.: 355-364.
204. Distribution des espèces du phytoplancton méditerranéen par rapport aux différentes combinaisons des facteurs du milieu. *Rapp. Proc. verb. CIESMM*, 18(2): 349-352.
205. Krassilnikoviae et fausses krassilnikoviae. *Rapp. Proc. verb. CIESMM*, 18(2): 339-340.
206. Limnología. *Gran enciclopedia del Mundo*. Durvan. Bilbao, 11: 1048-1059.

1966

207. Variabilidad del número de radios en diatomeas y otras algas unicelulares. *Mem. Real. Acad. Ciencias Artes Barcelona*, 37: 207-220.
208. (with J. HERRERA). Estimación de la actividad total añadida y de la autoabsorción en las determinaciones de producción del fitoplancton con ^{14}C. *Inv. Pesq.*, 30: 37-44.
209. (with J. HERRERA). La producción primaria en las aguas del puerto de Barcelona. *Inv. Pesq.*, 30: 45-48.
210. Análisis y valor indicador de las comunidades de fitoplancton mediterráneo. *Inv. Pesq.*, 30: 429-482.
211. (with J. HERRERA & J. STEYAERT). Distribución et caractéristiques des communautés phytoplantoniques dans le bassin Tyrrénien de la Méditerranée en fonction des facteurs ambiants et la fin de la stratification estivale de l'année 1963. *Inst Royal Sci. Nat. Belgique Bull.*, 42(5): 1-56.
212. Aplicació de la cibernètica a l'estudi de sistemes biològics al nivell de l'ecosistema. *Treb. Soc. Catal. Biol.*, 20: 75-81.
213. Fluctuations of many years period in West Mediterranean phytoplankton. *Second Intern. Congress Oceanogr., Moskow*. Abstracts of papers: 243-244.

1967

214. (with A. BALLESTER). Fitoplancton y producción primaria de la costa catalana, de junio de 1965 a junio de 1966. *Inv. Pesq.*, 31: 165-182.
215. Correlaciones entre parámetros oceanográficos del Caribe. *Inv. Pesq*, 31: 1-15.
216. Significado de las diferencias verticales de potencial eléctrico en el mar. *Inv. Pesq.*, 15: 259-263.
217. Laboratory analogues of estuarine plankton systems. *Estuaries*, Amer. Ass. Advanc. Sci.
218. The food-web in pelagic environment. *Helgoland. Wiss. Meeresunt*, 15: 548-559.
219. Some concepts relative to the organization of plankton. *Oceanogr. Mar. Biol. Ann. Rev.*, 5: 257-289.
220. Luz y Temperatura. Las algas inferiores. Biogeografía histórica. el ecosistema. ritmos, fluctuaciones y sucesión. La vida suspendida en las aguas. In: *Ecología Marina*,. Fundación La Salle. Caracas: 100-129, 230-272, 356-376, 377-43, 454-492, 493-532.
221. Las formas inferiores de vida. In: *La Evolución*, Biblioteca Autores Cristianos. Madrid: 174-202.
222. Nous aspectes de l'estudi estadístic de les comunitats d'organismes. *Treb. Soc. Cat. Biologia*, 21: 93-100.
223. *La Ciencia del Mar*. Consejo Superior de Investigaciones Científicas. Barcelona, 24 pp.
224. Un ejemplo de diversidad química. *Inv. Pesq.*, 31: 489-490.
225. (with J. CASTELLVI). Fitoplancton y producción primaria de la costa catalana, de julio de 1966 a julio de 1967. *Inv. Pesq.*, 31: 491-502.
226. Experiencias con contadores de seston en el crucero del "Mechelen", "Mediterranean outflow 1965", y formas de elaborar los resultados. *Inv. Pesq.*, 31: 503-538.

1968

227. Fluctuaciones de varios años de período medio en la producción de fitoplancton en el Mediterráneo occidental. *Collectanea Botanica*, 7: 727-747.
228. Estudios sobre el ecosistema pelágico del NE de Venezuela. Composición y distribución del fitoplancton. *Mem. Soc. Cienc. Nat. La Salle*, 70-72: 139-205.
229. (with A. BALLESTER). Estudios sobre el ecosistema pelágico del NE de Venezuela. Producción primaria. *Mem. Soc. Cienc. Nat. La Salle*, 70-72: 207-221.
230. *Perspectives in ecological theory*. University of Chicago Press. 111 pp.

1969

231. Diversidad del fitoplancton de red en dos áreas del Atlántico. *Inv. Pesq.*, 33: 275-286.
232. (with F. GÓNZALEZ BERNÁLDEZ). Grupos de especies asociadas en el fitoplancton del Mar Caribe, NE de Venezuela. *Inv. Pesq.*, 33: 287-312.
233. Composición específica del fitoplancton de la costa catalana-levantina (Mediterráneo Occidental), en 1962-67. *Inv. Pesq.*, 33: 345-380.
234. (with J.M. SAN FELIU). Expeditions Antarctiques Belgo-Neerlandaises en 1966-67. Resultat d'analyses du seston à l'aide du compteur dimentionnel de particules, *I. Bull. Inst. R. Sci. Nat. Belg.*, 44(31): 1-5, with 4 folders.
235. Nouvelles observations sur la distribution des Ciliés oligotriches dans le plancton de la Meditérranée occidentale. *Rapp. Comm. int. Mer Médit.*, 19(3): 565-566.
236. Présence de *Chatonella subsalsa* Biecheler dans le port de Barcelone. *Rapp. Proc. verb. Comm int. Expl. Mer. Médit., 19(3): 581-582.*
237. El concepto de polución y sus indicadores biológicos. *Documentos de Investigación Hidrológica. Supl. Agua*, 7: 105-133.

1970

238. Seminario de ecología Matemática. *Inv. Pesq.*, 34: 1-148: 14, 18, 26, 29, 32, 33, 46, 54-56, 60, 63, 66, 69, 76, 79, 81, 82, 111, 113, 116, 124, 127, 128, 135, 140).
239. El ecosistema pelágico del mar Caribe. *Mem. Soc. C. Nat. La Salle*, 29(82): 5-36.
240. Estudios sobre la distribución a pequeña escala del fitoplancton marino. *Mem. Acad. Cien. Art. Barcelona,* 40(1): 3-22.
241. Size of centric diatoms as an ecological indicator. *Mitt. Intern. Verein Limnol.*, 17: 202-210.
242. Diversity and stability: A practical proposal and a model of interdependence. *Brookhaven Symp. Biology*, 22: 25-37.
243. Anfípodos de las aguas subterráneas del país vasco. *Munibe.*, 22: 169-174
244. Anfípodos recolectados en aguas subterráneas ibéricas. *Speleon*, 17: 63-65.
245. Régimen futuro de los océanos. *Arbor*, 297- 298: 49-61.
246. Exploitation and management in ecology. *Intern. Symp. on the ecological bases for environmental management*, Set. 1970. Abstracts: 34-36.
247. Comunidades planctónicas en lagunas litorales. In: *Lagunas costeras, un simposio* (A. Ayala & F.B. Phleger, eds.): 545-562. México, D.F.
248. Diversidad y productividad del fitoplancton en el Mediterráneo occidental. *Inv. Pesq.*, 34: 565-573.
249. (with M.D. BLASCO). Influencia del puerto de Barcelona sobre el fitoplancton de las reas vecinas: Una mancha del plancton de gran densidad con dominancia de *Thalassiosira*, observada en agosto de 1969. *Inv. Pesq.*, 34: 575-580.

1971

250. Distribución del fitoplancton entre Córcega y Barcelona, en relación con la mezcla vertical del agua, en marzo de 1970. *Ibnv. Pesq.*, 35(2): 687-698.
251. L'Home part de la biosfera i objecte d'estudi de l'Ecologia. *Treb. Soc. Cat. Biologia*, 30: 15-25.
252. Explotación y gestión en Ecología. *Pirineos*, 98: 103-120.
253. Una campaña oceanográfica del "Cornide de Saavedra" en la región de afloramiento de NO Africano. *Inv. Pesq.*, (Supl.): 1-39.
254. (with M. ESTRADA). Simple approaches to a pattern analysis of phytoplankton. *Inv. Pesq.*, 35(1): 297-369.

1972

255. Fitoplancton de la región de afloramiento del noroeste de Africa. I. Pigmentos y producción (Campaña "Sahara II" del Cornide de Saavedra). *Res. Exp. Cient. B/O Cornide*, 1: 23-51.
256. Regularidades en la distribución de la diversidad del fitoplancton en un área del mar Caribe. *Inv. Pesq.*, 36(2): 241-264.
257. Interpretaciones no estrictamente estadísticas de la representación de entidades biológicas en un espacio multifactorial. *Inv. Pesq.*, 36(1): 183-190.
258. Composition et analyse par groupes du phytoplankton au large des côtes méditerranéennes espagnoles, en 1965-67. *Rapp. Comm. int. Mer. Médit., 30(3): 307-310.*
259. Ecosistemi marini. *Enciclopedia della Sciencia e della Tecnica Mondadori*: 103-110.
260. Some remarks from the point of view of a general ecologist. *Pacem in Maribus*, V., *The Ocean environments*: 13-16. The Royal University of Malta Press.
261. Spatial manifestation on a small scale of the relations between nutrient supply and phytoplankton composition. *Proc. Joint Ocean Assembly.* (Tokyo, 1970): 24.
262. Homage to Evelyn Hutchinson, or why is there an upper limit to diversity. In: *Growth by Intussusception. Ecological Essays in honor of G. E. Hutchinson.* E. S. Deevey, ed. Trans. Connecticut Acad. Arts & Sci., 44: 1-443 (:211-235).

263. Regularidades en la distribución de la diversidad en un área del Mar Caribe. *Inv. Pesq.*, 36: 241-264.

264. Ecologia i problemes humans. *Perspectiva social*, 4: 35-46.

1973

265. Ecological theory and prediction in the study of the interaction between man and the rest of the biosphere. In: *Ökologie und Lebensschutz in internationaler Sicht*. H. Sioli, ed. Rombach. Freiburg: 307-353.

266. Some critical remarks on the usual approaches to ecological modelling. *Inv. Pesq.*, 37: 621-640.

267. Fitoplancton marino de la región de afloramiento del NW de Africa. II. Composición y distribución del fitoplancton (Campaña "Sahara II" del "Cornide de Saavedra"). *Res. Exp. Cient. B/O Cornide*, 2: 65-94.

268. Distribución de los ciliados planctónicos en la región de afloramiento del noroeste de Africa (Campaña "Sahara II" del "Cornide de Saavedra"). *Res. exp. Cient. B/O Cornide,* 2: 125-132.

269. Estima acústica de la densidad relativa de animales pelágicos en la región de afloramiento del noroeste de Africa (Campaña "Sahara II del "Cornide de Saavedra"). *Res. exp. Cient. B/O Cornide*: 2: 125-132.

270. (with M. MIR). Indicadors de canvis de salinitat en els sediments de l'Albufera de València. *Treb. Soc. Cat. Biol.*, 32: 109-117.

271. Composició del plàncton vegetal en diversos llacs artificials de Catalunya. *Treb. Soc. Cat. Biol.,* 32: 171-179.

1974

272. *Ecología.* Omega. Barcelona. 951 pages.

273. Asociación o exclusión en la distribución de especies del mismo género en algas unicelulares. *Mem. Real. Acad. Cien. Artes Barcelona*, XLII(8): 353-372.

274. Human impact on transportation and diversity in ecosystems. How far is extrapolation valid. In: *Proc. First Intern. Congress of Ecology*, The Hague. 7 pp.

1975

275. Assessment of the effects on plankton. In: *Marine Pollution & Marine Waste Disposal*. Pearson & Frangipane, eds. Pergamon. Oxford & New York: 301-306.

276. Typology of reservoirs. *Verh. Internat. Verein. Limol.*, 19: 1841-1848.

277. (with M. ESTRADA, J. ARMENGOL, A. GUISET, M.D. PLANAS, J. TOJA, F. VALLESPINOS). Observaciones limnológicas en las lagunas de La Mancha. *Bol. Est. central Ecología*, 8: 11-27.

278. (with L. CAMPAS, M.R. MIRACLE, J.M. VILASECA) Introdución al estudio de los lagos pirenaicos. *Naturalia Hispanica*, 4: 1- 47.

279. Diversity, stability and maturity in natural ecosystems. In: *Unifying Concepts in Ecology*. Van Dobben & Lowe-McConnell, eds. Junk. The Hague: 161-170.

280. External factors and ecosystem stability. *Schw. Z. Hydrol.*, 37: 102-117.

281. Composición y distribución del fitoplancton marino en la región de afloramiento del NW de Africa, en marzo de 1973. *Res. Exp. Cient. B/O Cornide*, 4: 145-170.

282. Ciliados planctónicos de la región de afloramiento del NW de Africa, según datos de la campaña "Atlor II". *Res. Exp. Cient. B/O Cornide*, 4: 171-173.

283. Fitoplancton invernal de la laguna costera de Alvarado (México). *Anal. Inst. Bot. A.J. Cavanilles*, 32(2): 381-387.

284. Distribution du seston dans la région d'affleurement du nord-ouest de l'Afrique en mars 1973. *Téthys*, 6: 77-88.

1976

285. (with J. ARMENGOL, M. ESTRADA, A. GUISET, M.D. PLANAS, N. PRAT, J. TOJA, A. VIDAL). *Limnología de los embalses españoles.* Ministerio de Obras Públicas. Madrid.

286. Bases ecològiques per a una gestió de la Natura. In: *Natura, ús o abús? Llibre Blanc de la Gestió de la Natura als Pasos Catalans*. R. Folch, Ed. Barcino. Barcelona: 25-64.

287. Biología de los Embalses. *Investigación y Ciencia*, 1: 50-62.

288. Algas de agua dulce de Doñana. *Oecol. Aquatica*, 2: 79-91.

289. Ecosystem diversity differences: Poles and Tropics. In: *Polar Oceans Conference*, Montreal, 1974.

290. Distribución horizontal del plancton marino a escala media (1 a 10 km) ilustrada con un ejemplo de la zona de afloramiento del NO de Africa. *Mem. Real. Acad. Cienc. Artes Barcelona*, 43: 131-148.

1977

291. Comunicación y engaño. Aspectos e implicaciones de la cripsis, advertencia y mimetismo. *Graellsia*, 31: 341-356.

292. Las fronteras de la ecología. *Bol. Inf. F. Juan March*, 58: 3-24.

293. *Ecología.* Univ. Nacional Ed. a distancia. Madrid.

294. Some elements for the construction of models of plankton systems. *Reunion ICES Cm 1975/2/L; 6. Plankton Commitee*, 6 pp.

295. L'inquinamento in un contesto ecologico generale. In: *Aspetti scientifici dell'inquinamento dei mari italiani.* Roma, gennaio 1976. Accademia Nazionale dei Lincei. pp: 11-22.

1978

296. Life-forms of phytoplankton as survival alternatives in an unstable environment.

Oceanologica Acta, 1(4): 493-509.
297. What is an upwelling ecosystem? In: *Upwelling ecosystems.* (R. Boje & M. Tomczak, eds.): 12-14 Springer. Berlin.
298. Phytoplankton communitties in upwelling areas. The example of NW Africa. *Oecologia Aquatica*, 3: 97-132.
299. General concepts of population dynamics and food links. In: *Marine Ecology*. O. Kinne, ed. Wiley. London, 4: 617- 704.
300. Models matemàtics i models físics de poblacions i d'ecosistemes. *Col.loq. Soc. Cat. biol., XI. Lògica i conflicte*; 169-177.

1979

301. Alternative approaches in the modelling of populations. *Inv. Pesq.*, 43(1) 337-350.
302. The organization of space. *Oikos*, 33: 152- 159.
303. Basic ecological parameters, monitoring and biological monitors in the aquatic environment. Pollutants and their ecotoxicological significance for European regions. *VI European Summerschool.*
304. Las rías gallegas. In: *Estudio y explotación del mar en Galicia.* Universidad de Santiago. Santiago: 101-121.
305. Los productores primarios de materia viva. In: *Estudio y explotación del mar en Galicia.* Universidad de Santiago. Santiago: 123-168.
306. El precio de la supervivencia. Consideraciones ecológicas sobre las poblaciones humanas. *Ethnica*, 15: 103- 115.
307. Caràcter de les diferències entre el plàncton vegetal marí i el d'aigua dolça. *Col.loq. Soc. Cat. Biol., XII. Biologia Molecular i ecologia*: 17-20.
308. Expressió senzilla de la regulació en sistemes dinàmics. *Col.loq. Soc. Cat. Biol., XIII. Matemàtica i Biologia*: 53-55.
309. Les ecosystmes aquatiques. In: *Proc. Second Intern. Congress of Ecology.* Venezia: 71-105.
310. (with N. PRAT). La Limnologia. *Quad. Ecol. Apl.*, 4: 9-23.

1980

311. Composición y fenología de la vegetación algal de un arroyo del Montseny (Barcelona). *Oecologia Aquatica*, 4: 111-112.
312. *La Biosfera: entre la termodinámica y el juego.* Omega. Barcelona.
313. (with J. ARMENGOL). La teoria de la biogeografía insular de MacArthur i Wilson. El poblament de llacs artificials com a prova d'una situació reflexa. *Butll. Soc. Cat. Biol.*, III-IV: 117-123.
314. La contaminació en un context ecològic general. *XI Congrès de metges i biòlegs de llengua catalana*: 5-15.
315. La teoria de la informació en ecologia a vint anys de distància. *Butll. Soc. Cat. Biol.*, III-IV: 159-186.
316. (with M. ESTRADA). Las áreas oceánicas más productivas. *Investigación y Ciencia*, 49: 8-20.
317. Comparabilidad de las distribuciones ecológicas en el tiempo y según una dirección del espacio, y uso de pruebas sencillas para definir propiedades demográficas de las respectivas poblaciones. *Studia Oecologica*, 1: 7-11.
318. Sampling design: Some examples. Interpreting observations: Diversity. In: *Phytoplankton Manual.* A Sournia, ed. UNESCO. París: 17-32.
319. *Ecología.* In: *Enciclopedia temática Planeta.* Planeta. Barcelona (Published as a book in 1981).
320. (with M. ESTRADA & D. BLASCO). Functional morphology of organisms involved in red tides, as adapted to decaying turbulence. In: *Toxic Dinoflagellate Blooms*. Elsevier. North Holland.
321. La capacidad de producción de los océanos. *Kobie,* 10: 695-699.
322. Sobre diversidad y conectividad en ecosistemas y en artefactos que funcionan. *Munibe*, 32: 397-300.
323. Cap a una teoria de la biosfera. *Ciència*, 1: 28-40. Barcelona.

1981

324. (with M. MIR). Phytoplankton of Spain reservoirs as dependent from environmental factors and as potential indicators of water properties. *Atti Congresso sui Baccini Lacustri Artificiali.* Sassari: 191-205
325. Bentos y plancton. explotación y evolución. *II Simposio Ibérico Estudios Bentos Marino.* Barcelona Abstracts: 5.
326. (with M. ESTRADA). On upwelling, eutrophic lakes, the primitive biosphere, and biological membranes. *Coastal and Estuarine Sciences*, 1: 522-529.
327. Stress in Ecosystems: a future approach. In: *Stress effects on natural ecosystems.* G.W. Barret & R. Rosenberg (eds.). Wiley. New York.
328. Asimetrías inducidas por la operación de energía externa en secuencias de sedimentos y de poblaciones. *Acta Geologica Hispanica,* 16: 35-38.
329. *El proyecto Paraná Medio y su incidencia sobre la ecología regional. La reserva de áreas naturales.* Agua y Energía, Sociedad del Estado, Gerencia proyecto Paraná Medio. Informes 10, 11, 19 + 4pp.
330. Meditació sobre la recerca a la Universitat. In: *Sobre les formes de l'activitat universitària.* Acte inaugural del curs 1981-1982. Universitat de Barcelona: 41-63.

1982

331. (with M. MIR & M. ESTRADA). Phytoplankton composition and distribution as an expression of proprieties of reservoirs. *Canad. Water Resourc. J.*, 7: 26-50.
332. Consecuencias previsibles de la regulación de los rios sobre las condiciones de producción de las áreas marinas costeras. *The Siren, UNEP.* Octubre 1982, pp: 7-14.
333. Instabilities in Ecology. In: *Stability of thermodinamic Systems.* (J. CASAS VAZQUEZ and G. LEBON, eds.) Springer, Lecture notes in Physics, 164: 12 pp.
334. Some thoughts on the dynamics of populations of

ciliates. *Ann. Inst. Ocanogr., N.S.,* 58 (suppl): 15-18.

335. Energía y medio ambiente. *Simposium Asamblea Anual del Comité Nacional Español de la Conferencia Mundial de la Energía.* 10 pp.

1983

336. (with E. GUTIERREZ). How to introduce connectance in the frame of an expression for diversity. *The American Naturalist,* 121(5): 601-607.
337. Algas de una laguna salobre mediterránea (Lago di Patria, Nápoles). *Collectanea Botanica,* 14: 305-315.
338. *Limnología.* Omega. Barcelona. 1010 pages.
339. La ciencia ecológica y los problemas ambientales técnicos, sociales y humanos. *Diez años después de Estocolmo.* CIFCA, Madrid. pp: 177-200. (Reeditado en *El desafio ecológico: Ecología y Humanismos.* Universidad Pontificia de Salamanca, 1985, pp: 21-87.

1984

340. El ecosistema pelágico del Mediterráneo. *Mundo Científico,* 4: 146-157.
341. Visión del paisaje desde la ecología. *Monogr. de l'Equip,* 1: 101-117.
342. Simple facts about life and the environment not to forget in preparing schoolbooks for our grandchildren. In: *Trends in ecological research for the 1980's* (J.H. COOLEY and F.B. GOLLEY, eds.) NATO Conf. Series. Plenum Press. New York and London: 299- 320.
343. Le plancton de la Méditerranée. *La Recherche,* 15: 1082-1094 (n. 158, sept. 1984).
344. Summary of the main arguments from a theoretical perspective. Workshop Agroecology. *Options Méditerranéennes.* Zaragoza. Série études 1984: 161-166.
345. Avances en Ecología. *Avances del Saber,* 5: 113-137. Labor, Barcelona

1985

346. From hydrodynamic processes to structure (information) and from information to process. *Can. Bull. Fish. Aquat. Sci.,* 213: 200-220.
347. La limnologia a Catalunya. *Mem. R. Acad. Cien. Artes Barcelona,* XLVI(8): 203-224.
348. Environmental control of the mesoscale distribution of primary producers and its bearing to primary production in the Western Mediterranean. *Med. Mar. Ecosyst., NATO Conf. Ser.,* I: 213-229.
349. Primary production in upwelling areas. Energy, global ecology and resources. *Simp. Internat. sobre áreas de afloramiento más importantes del oeste africano,* I: 225-232. CSIC. Barcelona.
350. Basic ecological parameters, monitoring and biological monitors in the aquatic environment. In: *Pollutants and their ecotoxicological significance,* H.N. NURNBERG, ed. John Wiley. New York: 149-175.
351. Ecosystems diversity and connectivity as measurable components of their complication. In: *The Science and praxis of Complexity.* The United Nations University. Tokyo: 228-244.
352. *L'Ecologia.* Diputació de Barcelona, Servei del Medi Ambient. 124 pp.
353. Introduction to the Mediterranean. In: *Western Mediterranean,* R. MARGALEF (ed.). Pergamon Press, Oxford. pp: 1-16.

1986

354. Limnología del lago Enriquillo (República Dominicana). *Oec. Aq.,* 8: 1-10.
355. Sucesión y evolución: su proyección biogeográfica. *Paleontologia i Evolució,* 20: 7-26.
356. Variaciones sobre el tema de la selección natural. Exploración, selección y decisión en sistemas complejos de baja energia. In: *Proceso al azar.* J.WAGENSBERG (Ed.). Tusquets. Barcelona: 121-140. (Also in: *Saber,* 7: 7-13.)

1987

357. Algunas contribuciones de la limnología al desarrollo de la ecología teórica. *Actas IV Congreso Español de Limnología:* 1-12. Sevilla.
358. Teoría y modelado de los sistema fluctuantes. *Seminario sobre bases científicas para la protección de los humedales de España.* R. Acad. Cien. Exactas, Físicas y Nat. Madrid: 31-41.
359. (with M. ESTRADA). Synoptic distribution of summer microplankton (Algae and Protozoa) across the principal front in the Western Mediterranean. *Inv. Pesq.,* 51(1): 121-140.
360. Història Natural del canvi en sistemes organitzats. *Saber,*14: 4-12.
361. Divagacions sobre el concepte de conservació. *Arrels,* 19: 6-12.

1988

362. Ways to differentiation and diversity in ecosystems. *Proceed. of the Conf. on Synergetics, Order and Chaos.* M.G. VELARDE (ed.). World Scientific, Singapore. London: 441-447.
363. La relación entre producción reciclada y la producción nueva en sistemass planctónicos. *Perspectiva en Oceanografía.* CAICYT, Univ. Menéndez Pelayo, julio 1987: 49-54.
364. (with M. ESTRADA) Supply of nutrientss to the Mediterranean photic zone along a persistent front. *Oceanologica Acta, Ocean. Pel. med:* 133-142.
365. La ecología como marco conceptual de reflexión sobre el hombre. *Ecología y Culturas,* Asoc. Interdisciplinar José de Acosta, Madrid: 15-39.
366. Manipulació a l'escala de l'ecosistema. Gestió de recursos naturals. *Tretze congrès de metges i biòlegs de llengua catalana.* Llibre de ponències, Andorra, nov. 1988: 53-64.

1989

367. Reset successions and suspected chaos in models of marine populations. *Int. Symp. long-term changes marine fish popul.* (Vigo, 1986): 321-343.
368. Evolución de los macrófitos y su coevolución con los herbívoros. *Homenaje a Pedro Montserrat:* 637-642. Instituto de Estudios Altoaragoneses. Instituto Pirenaico de Ecología. Jaca.
369. D'escarabats i d'altres coses. Una conversa amb Francesc Español. *Homenatge a Francesc Español.*
370. On diversity and connectivity, as historical expressions of ecosystems. *Coenoses*, 4: 121-126.
371. Condiciones de aparición de la purga de mar y presiones de selección de sus componentes. *Cuadernos de Áreas de Ciencias Mariñas. Seminario de Estudios Galegos*, 4: 13-20.
372. Le plancton: Survivre et s'organiser dans un fluide mobile. In: *Océanologie, actualité et prospective.* (M. Denis, ed.).: 169-185. Centre d'Oceanologie de Marseille.
373. Introducció al coneixement de la biosfera. Els ecosistemes pelàgics. In: *Història Natural dels Països Catalans*, 14 (J. Terradas, N. Prat, A. Escarré & R. Margalef, eds.): 15-25, 73-118. Enciclòpedia Catalana. Barcelona.

1990

374. Is there a "balance of nature"? In: *Environmental Ethics.* (P. Bourdeau, P.M. Fasella & A. Teller, eds.): 225-233. Comm. European Communities, ECSC-EEC-EAEC. Brussels. Luxembourg.
375. La Biosfera sota l'imperi de l'home: Canvis en fluxes energètics i cicles biogeoquímics. *Mem. R. Acad. Cien. Art. Barcelona*, 49: 295-332
376. Viure a la Terra: els límits i de les regles del joc. *In: Poblacions, societats i entorn*: 23-55. Els llibres de l'Institut d'Humanitats, 1. Barcanova. Barcelona.
377. Les xarxes en Ecologia. In: *Poblacions, societats i entorn:* 151-180. Els llibres de l'Institut d'Humanitats, 1. Barcanova. Barcelona.
378. Ecosistemas fluviales, transporte horizontal y dinámica sucesional evolutiva. *Interciencia*, 15 (6): 334-336.
379. Historia natural del cambio en sistemas organizados. In: *Sobre la imaginación científica* (J. Wagensberg, ed.): 140-169. Tusquets. Barcelona.

1991

380. Networks in Ecology. In: *Theoretical studies of ecosystems. The Network perspective* (M. Higashi & T.P. Burns, eds.): 41-57. Cambridge Univ.Press. Cambridge. New York.
381. Limnology: reconsidering ways and goals. In: *Scientific perspectives in theoretical and applied limnology* (R. de Bernardi, G. Giussani & L. Barbanti, eds.), 47: 57-76. Pallanza.
382. (with C. GOLDMAN). Concluding remarks. In: *Scientific perspectives in theoretical and applied Limnology.* (R. de Bernardi, G. Giussani & L. Barbanti, eds.), 47: 373-378. Pallanza.
383. Energía exosomática y crecimiento cero. *O ambiente na Peninsula Ibérica, perspectivas a montante.* Universidade de Trás-os-montes e Alto Douro. Universidade Internacional, 1991: 37-46.
384. L'homme dans la Biosphere. Les artifacts culturels en tant que commutateurs et amplificateurs sur les voies de l'energie exosomatique. *Actes du Forum Science, technologie et Societé: Le Defi de la Grande Europe*, 8-10 novembre 1990, Strasburg, pp. 145-155.
385. La predicción ecológica: incertidumbre y riesgo. In: *Forum Deusto. Tercer ciclo: El presente discontinuo.* Universidad de Deusto, Bilbao.
386. *Teoría de los sistemas ecológicos*. Publicacions de la Universitat de Barcelona, 290 pages.

APPENDIX II. PH. D. THESES DIRECTED BY PROFESSOR RAMON MARGALEF

1971. Dolors Blasco. Acumulación de nitritos en determinados niveles marinos por acción del fitoplancton.

1972. M. Dolors Planas. Composición, ciclo y productividad del fitoplancton del lago de Banyoles.

1973. Joandomènec Ros. Opistobranquios (*Gastropoda: Euthyneura*) del litoral Ibérico: Estudio faunístico y ecológico.

1974. M. Rosa Miracle. Estructura y dinámica de las poblaciones de la comunidad zooplanctónica del lago de Banyoles.

1976. Antonio Cruzado. Afloramiento costero en el Atlántico nororiental. Análisis del ecosistema en sus aspectos de hidrografía y producción primaria.

1976. Marta Estrada. Estudios sobre las poblaciones de organismos acuáticos en medio no uniforme.

1976. Felipe Fernández. Influencia de la luz, temperatura y materia orgánica particulada en la actividad metabólica y en la alimentación de los copépodos planctónicos.

1976. F. Xavier Niell. Estudios sobre la estructura, dinámica y producción del fitobentos intermareal (facies rocosa) de la Ría de Vigo.

1976. Julia Toja. Estudio limnológico comparado de dos embalses con distinto grado de eutrofia (Aracena y La Minilla).

1976. Ferran Vallespinós. Estudio comparado del metabolismo del nitrógeno en bacterias y cianofíceas: relaciones con el ciclo del carbono e importancia ecológica.

1977. Miquel Alcaraz. Ecología, competencia y segregación en especies de copépodos (*Acartia*).

1977. Joan Armengol. Los crustáceos planctónicos en los embalses españoles.

1978. Narcís Prat. Ecología y sistemática de los Quironómidos (*Insecta, Diptera*) de los embalses españoles.

1979. Francesc Amat. Diferenciación y distribución de las poblaciones de *Artemia (Crustacea, Branchiopoda)* de España.

1980. Jordi Flos. Material en suspensió oceànic i la seva distribució en el Mediterrani Occidental.

1981. Francisco A. Comín. Limnología de las lagunas costeras del delta del Ebro: características físico-químicas y fitoplancton de la Encañizada.

1981. Tecla Riera. Variabilitat morfològica de *Tropocyclops* i *Temora*: Aproximació crítica a l'ús generalitzat de la biometria en la taxonomia dels copèpodes.

1983. Carlos A. Gracia. La clorofila en los encinares del Montseny: Interpretación como una optimización del aprovechamiento de la luz.

1983. M. Pilar López. Aguas salinas epicontinentales próximas a la costa mediterránea española: estudio del medio.

1984. Enric Ballesteros. Els vegetals i la zonació litoral: espècies i factors que influeixen en la seva distribució.

1985. Miguel Alonso. Las lagunas de la España peninsular: taxonomía, ecología y distribución de los cladóceros.

1985. Francisco Gómez. Ecología del fitoplancton de la Ría de Pontevedra.

1985. Josep Peñuelas. Briòfits i fanerògames com a invasors de les aigües dolces: distribució, pigments, fonts de carboni i l'obstacle dels espais aeris.

1985. Javier Romero. Estudio ecológico de las fanerógamas marinas de la costa catalana: producción primaria de *Posidonia oceanica* (L.) Delile en las Islas Medes.

1985. Luís Zúñiga. Ecología de los entomostráceos planctónicos de los lagos nord-patagónicos.

1986. Cèlia Marrasé. Experimentos multifactoriales con plancton marino en microcosmos.

1987. Jordi Catalan. Limnologia de l'estany Redó (Pirineu Central). El sistema pelàgic d'un llac profund d'alta muntanya.

1987. Ricardo Génova. Análisis y significado de los anillos de crecimiento en dos especies forestales: *Pinus uncinata* y *Pinus silvestris* en la Península Ibérica.

1987. Emilia Gutiérrez. Dendrocronología de *Fagus silvatica*, *Pinus uncinata* y *P. silvestris* en Catalunya.

1987. Sergi Sabater. Estudi de les poblacions d'algues del riu Ter.

1988. Xavier Tomàs. Diatomeas de las aguas epicontinentales saladas del litoral mediterráneo de la Península Ibérica.

1988. Ricardo Iglesias. Diversidad taxonómica y ataxonómica en poblaciones de insectos: un ejemplo del ecosistema restinga.

1988. Antoni Palau. El embalse de Baserca (Pirineos centrales, Lérida). Estudio limnológico de un embalse nuevo de alta montaña con bombeo.

1989. Olga Delgado. Sistema carbónico-carbonatos en el Mediterráneo noroccidental y relación con las principales comunidades fitobentónicas productoras de carbonatos.

1990. Agustín Lobo. Estudios sobre la organización espacial del bentos marino. Una aproximación a través del análisis de imagen digital y de la modelización.

1990. M. José Gil. Estudios sobre ácaros de aguas continentales españolas.

CONTRIBUTORS

Oecologia aquatica publishes original research papers on different aspects of the ecology of marine and epicontinental waters. Their pages contain the production of the members of the Department of Ecology of the University of Barcelona, but other contributions are not excluded.

The authors wishing to publish in *Oecologia aquatica* must send their original papers to the Journal Editor. The Editorial Advisory Board will decide on the convenience of publication of the papers received, and will inform the authors. These must adapt to the following

INSTRUCTIONS FOR PUBLICATION

The original papers should be sent on magnetic support, floppy disk of 5 1/4 or 3 1/2 inches. The text should be contained in one or more files. The text should be written, in English or Spanish, using any processor which codifies with ASCII, preferably Wordstar or Wordperfect. The list of references cited in the text must be included at the end of the text, as well as the figure captions and table headings in both languages, English and Spanish. The tables must be in a different file. It should be better if they are typped using tabs. If not, the author will leave only one space between the columns. The maximum size allowed for the tables is 60 lines and 85 columns, interlines and intercolumns spaces included.

No extra spaces will appear between paragraphs neither indents, outdents or any other typographic controls. Boldface or underline are not allowed in the file. The authors can codify italic type for the latin names of species, genus, etc. Super and subscripts can be indicated as well.

The paper will contain always the following parts: Title, author, address, title of the summary (in Spanish if the paper is written in English and viceversa), summary (the same as the title), keywords (in Spanish and English), Introduction, Material and Methods, Results and Discussion. Subparts may be included inside each one of these.

Three print copies should be sent with the floppy disk, where the author must indicate the categories of the subparts or any other considerations.

The Editorial Advisory Board will not be in charge to check the correctness of the language. If necessary the paper will be returned back to the author.

The references will follow these exemples:

ELDERFIELD, H., LUEDTKE, N. & McCAFFREY, R.J. 1981. Benthic fluxes studies in Narragansett Bay. *Am. J. Sci.*, 281: 768-787. (for papers).

GOLTERMAN, H. L. 1977. Sediment as a source of phosphate for algal growth. In: *Interaction between sediments and freshwater* (Golterman, H.L., ed.): 286-293. (for papers in books).

MARGALEF, R. 1983. *Limnología.* Omega. Barcelona. (for books).

If the author is not able to print or/and codify italic type, he should indicate it with underline in the print copy.

The figures will be correctly laid, caring that legends can be easily read when reduced (60%). The authors are adviced to glance through the last issues of the journal to observe legends, drawings and formats. Photographs (b/w only) will be admited only under special conditions. Each author will receive galley-proofs of its paper, and when published, 50 reprints free of charge.

Under special circumstances, the Redaction will accept typed papers without magnetic support.